ELEMENTS OF ORGANIC PHOTOCHEMISTRY

ELEMENTS OF ORGANIC PHOTOCHEMISTRY

Dwaine O. Cowan
The Johns Hopkins University
Baltimore, Maryland

and

Ronald L. Drisko
Essex Community College
Baltimore, Maryland

PLENUM PRESS · NEW YORK AND LONDON

Library of Congress Cataloging in Publication Data

Cowan, Dwaine O
Elements of organic photochemistry.

Includes bibliographical references and index.
1. Photochemistry. 2. Chemistry, Physical organic. I. Drisko, Ronald L., 1942- joint author. II. Title.
QD708.2.C67 547'.1'35 75-28173
ISBN 0-306-30821-5

A Division of Plenum Publishing Corporation
227 West 17th Street, New York, N.Y. 10011

Printed in the United States of America

Preface

In the past fifteen years organic photochemistry has undergone a greater change and has stimulated more interest than probably any other area of organic chemistry. What has resulted is a population explosion, that is, an ever-increasing number of organic chemists are publishing important and exciting research papers in this area. Professor Bryce-Smith in the introduction to a recent volume of the Specialist Periodical Report (*Photochemistry*, Volume 6), which reviews the photochemical literature in yearly intervals, states that "the flood of photochemical literature is showing some signs of abatement from the high levels of two or three years ago . . . ." However, Volume 6 of that periodical contains 764 pages of excellent but very concise reviews.

We expect the development of the mechanistic aspects of organic photochemistry to continue at the present pace as new methods are developed to probe in increasing detail and shorter time scales the photochemical dynamics of both old and new photoreactions. Since photochemistry is no longer the sole domain of the specialist, it is relatively safe to predict a dramatic increase in the near future of the synthetic and industrial uses of organic photochemistry.

In part, this book has grown out of photochemistry courses given at Hopkins during the past 13 years. In order to maintain a reasonable length, we have found it necessary to select our topics carefully, and consequently many interesting photoreactions have had to be omitted. Our goal was not to produce an encyclopedic volume but rather a textbook designed so that a student, having a background in organic and physical chemistry, can start this text and proceed to the point where he or she can read and comprehend the significance of current photochemical literature and initiate research in this area. To accomplish our goal we have interwoven, in a gradual manner, photochemical and photophysical concepts along with the chemical and physical techniques necessary to understand a variety of photoreactions. Many of the specific examples presented in detail reflect the interests of the authors. However, the extensive documentation provided in the text will allow the student to seek out other examples of interest.

We wish to thank Ms. Tami Isaacs for her help in proofreading the text and to acknowledge our great debt to former teachers, our colleagues, and the photochemical pioneers of each generation.

Dwaine O. Cowan
Ronald L. Drisko

Baltimore, Maryland

Contents

1

Basic Photophysical and Photochemical Concepts

1.1. INTRODUCTION

Although man was not present to observe the event, the first photochemical reaction probably occurred billions of years ago. High-energy solar radiation ($\lambda < 2000$ Å) undoubtedly was an important factor in the development of large molecules, polymers, and eventually polypeptides from the primitive earth's reducing atmosphere (methane, ammonia, and hydrogen). However, from man's point of view, the greatest photochemical "breakthrough" occurred when the first few quanta of light were absorbed by a rudimentary photosynthetic unit, resulting in the release of molecular oxygen to the atmosphere, thus paving the way for all higher life:

$$nH_2O + nCO_2 + \text{light absorber} \xrightarrow{h\nu} (CH_2O)_n + nO_2 \qquad (1.1)$$

Considering the importance of photochemistry in man's history, it is surprising that it is only within the past ten to fifteen years that photochemical reactions have been extensively investigated. However, during this period the field has developed at such an explosive rate that the amount of literature reporting new photochemical discoveries increases daily. This is not to say that photochemistry was totally neglected prior to the 1950's. In fact, as far back as 1912, one of the earliest photochemists, Giacomo Ciamician, realizing the enormous potential of photochemistry, predicted: "Solar energy is not evenly distributed over the surface of the earth; there are privileged regions, and others that are less favored by the climate . . . On the arid lands there will spring up industrial colonies without smoke and without smokestacks; forests of glass tubes will extend over the plains and glass buildings will rise everywhere."* Although the realization of Ciamician's prediction would be attractive in view of today's pollution-troubled industry, much of the recent rapid progress in photochemistry has arisen from the technological development of convenient and efficient high-intensity light sources, making dependence on the whims of the sun unnecessary. Consequently a reaction that fifty years ago might have taken weeks of irradiation with sunlight to obtain a yield of product large enough to characterize can now be done in the laboratory in a few hours on a rainy day. Of equal importance in the burgeoning of this field have been the development and utilization of spectroscopic techniques and the development of molecular orbital theory. Using information obtained from these sources, the organic photochemist may in the future be able to both predict and control the products of his photochemical reaction.

Future developments are expected not only to yield a greater understanding of the mechanisms of many photochemical reactions, but also to provide a means for the adaptation of these reactions to large-scale industrial syntheses. A glimpse of the latter is seen in the production of ε-caprolactam (Nylon 6 monomer) by the Toyo Rayon Company using the photonitrosation of cyclohexane. In this process nitrosyl chloride is cleaved by light and the following sequence of reactions takes place:

$$NOCl + h\nu \longrightarrow NO + Cl$$

$$\text{(S)}\text{—H} + Cl \longrightarrow \text{(S)}\cdot + HCl \qquad (1.2)$$

$$\text{(S)}\cdot + NO\cdot \longrightarrow \text{(S)}\text{—NO}$$

$$\text{(S)}\text{—NO} \longrightarrow \text{C}_6\text{H}_{10}\text{=N—OH} \xrightarrow[\text{Beckman rearrangement}]{H_2SO_4} \text{ε-Caprolactam}$$

ε-Caprolactam

One plant in the Nagoya (Japan) region alone is capable of producing 120 tons of caprolactam *per day*. In addition, the Nylon 12 monomer, lauryl lactam, is being prepared by a similar route from cyclododecane.

In this chapter we introduce many of the basic concepts and the vocabulary necessary for the study of photochemistry.

* Quoted in: R. C. Cookson, *Quart. Rev.* **22**, 423 (1968).

1.2. ENERGY DISTRIBUTION IN THE EXCITED MOLECULE

The various intramolecular processes initiated by light absorption are illustrated schematically in Figure 1.1. Such a schematic representation of the energy levels and photophysical processes which can occur in the excited

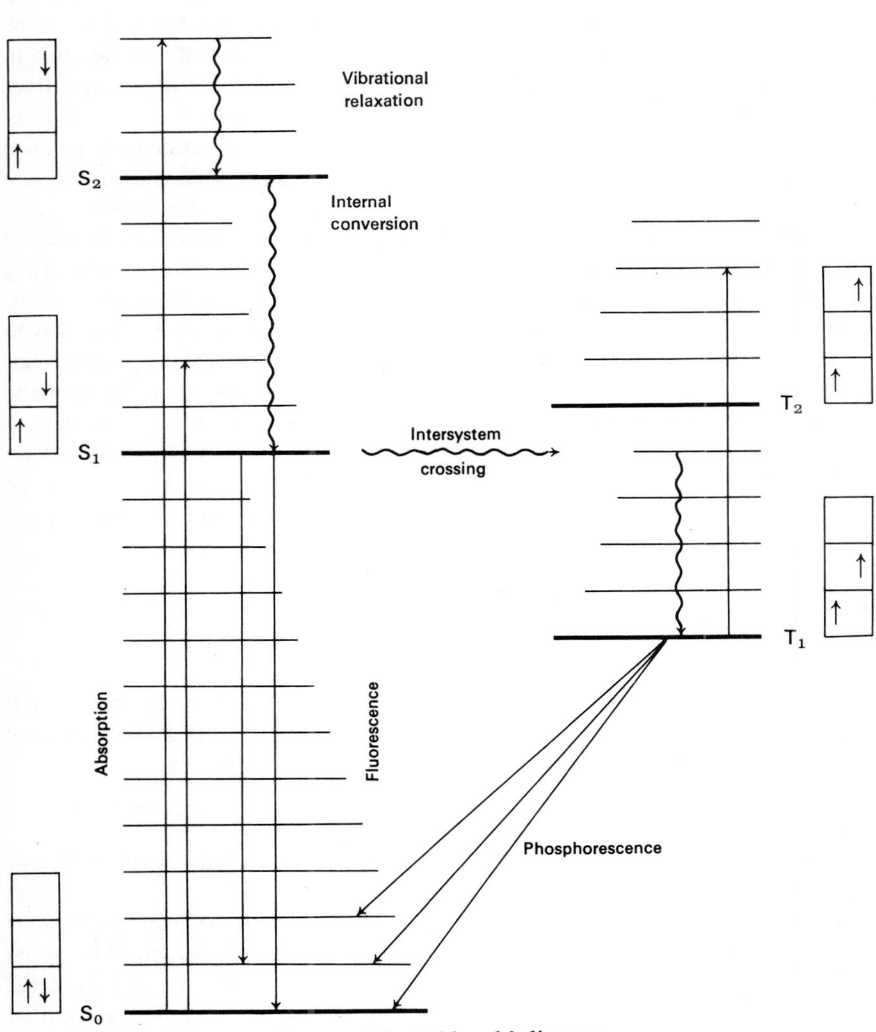

FIGURE 1.1. Jablonski diagram.

molecule is commonly called a *Jablonski diagram*. In this diagram the vertical direction corresponds to increasing energy; the horizontal direction has no physical significance. The electronic states are represented by the heavy horizontal lines; the symbols S_0, S_1, and S_2 represent the ground state and first and second excited singlet states (electron spins antiparallel or paired), respectively, and T_1 and T_2 represent the triplet states (electron spins parallel or unpaired). These sets of energy levels are frequently referred to as the singlet and triplet manifolds. Excitation is represented by the promotion of one electron from the ground state (S_0) to a higher electronic level. The position and spin of the excited electron relative to those of the electron in the ground state are noted in the boxes. Finally, the light horizontal lines correspond to the vibrational levels of the electronic states.

The lowest triplet level (T_1) is placed below the lowest excited singlet level (S_1), in accordance with Hund's rule, which states that the state of maximum multiplicity (highest number of unpaired electrons) lies at the lowest energy. Since two electrons having the same spin quantum number (spins unpaired) cannot be at the same place at the same time (Pauli exclusion principle), they will tend to avoid each other. This results in a lowering of the energy of the state since there is less repulsion between the negatively charged electrons in the triplet relative to the corresponding singlet state in which the electrons can theoretically be in the same region in space simultaneously. Each of the radiative-type processes (light absorption, fluorescence, and phosphorescence), designated by straight lines, and nonradiative processes (internal conversion and intersystem crossing), designated by wavy lines in Figure 1.1, will be briefly described in this section and discussed in greater detail in later chapters.

1.2a. *Light Absorption*

Molecular excitation by light absorption takes place during the period of one vibration of the exciting light wave. For light with a wavelength λ equal to 300 nanometers (nm), this corresponds to 10^{-15} sec:

$$\text{frequency } \nu = c/\lambda = (3 \times 10^{10} \text{ cm/sec})/(3 \times 10^{-5} \text{ cm}) = 10^{15}/\text{sec}$$

The time for excitation equals the time for one cycle, which equals $1/\nu$, or 10^{-15} sec.

This time period is too short for a change in geometry to occur (molecular vibrations are much slower). Hence the initially formed excited state must have the same geometry as the ground state. This is illustrated in Figure 1.2 for a simple diatomic molecule. The curves shown in this figure are called Morse curves and represent the relative energy of the diatomic system as a

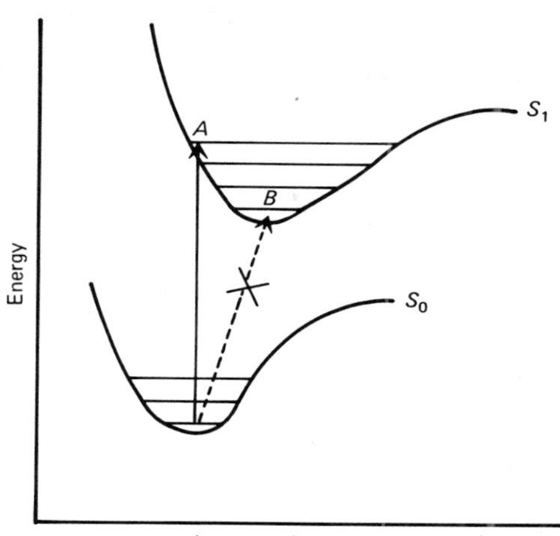

FIGURE 1.2. Diatomic potential energy curves and Franck–Condon transition.

function of the distance between the nuclei of the atoms. It can be seen that the energy is a minimum at a particular internuclear distance that corresponds to the equilibrium interatomic distance for each state. The higher energy state (S_1) has been drawn with a large equilibrium internuclear distance. The reason for this will be discussed later in this chapter.

For all points along the curve the nuclei of the molecule are motionless, that is, the energy is completely potential. The horizontal lines in the wells of these potential curves represent the vibrational levels of the states. At points along these horizontal lines intermediate to the intercepts the nuclei are in motion (vibration) and the total energy (ordinate) becomes the sum of the molecule's potential and kinetic energies. At 25°C most molecules of interest to us will be in their lowest vibrational level; hence excitation will occur from this level. Since the geometry cannot change during excitation, the molecule must find itself with the same internuclear distance after light absorption as it had before (Franck–Condon principle). This corresponds to the vertical arrow in Figure 1.2 and is termed a *vertical* or *Franck–Condon transition.*

The energy associated with the absorbed light quanta is proportional to the frequency ν and is given by the following equation:

$$\boxed{E = h\nu = hc/\lambda = hc\bar{\nu}} \tag{1.3}$$

For example,

$$E_{\lambda=800\,\text{nm}} = 35.75 \text{ kcal/mole}$$

$$E_{\lambda=200\,\text{nm}} = 143 \text{ kcal/mole}$$

A number of useful conversion factors are collected in Table 1.1.

TABLE 1.1. *Conversion Factors*

Quantity	Abbreviation or function	Conversion factor
WAVELENGTH, angstroms	λ, Å	1 Å = 10^{-8} cm = 10^{-4} μ = 10^{-1} nm
ENERGY, calories	*E*, cal	1 cal = 4.2 × 10^7 ergs
For a single quantum (λ in nm)	$h\nu = hc/\lambda = hc\bar{\nu}$	$(1.986 \times 10^{-9}/\lambda)$ ergs = $(1240/\lambda)$ eV
For a mole of quanta (i.e., per einstein; *N* is Avogadro's number; λ in nm, $\bar{\nu}$ in cm^{-1})	$Nhc/\lambda = Nhc\bar{\nu}$	$(12.01 \times 10^{14}/\lambda)$ ergs/einstein = $(2.86 \times 10^7/\lambda)$ cal/einstein = $2.86\bar{\nu}$ cal/einstein
POWER, watts	*P*, W	1 W = 1 J/sec = 10^7 ergs/sec
INTENSITY	Power/area	1 W/cm^2 = 10^7 ergs/cm^2 sec = $8.326 \times 10^{-9}\lambda$ einstein/cm^2 sec (λ in nm)

In order to do quantitative photochemistry, one must know how much of the light incident upon a sample is absorbed. For most systems this can be conveniently determined using Beer's law,

$$\mathrm{OD} = \varepsilon c d$$

where the optical density OD is defined by

$$\mathrm{OD} = \log(I_0/I_t)$$

ε is the molar absorptivity (also called the molar extinction coefficient), a constant characteristic of the molecule of interest, c is the molar concentration,* d is the path length, I_0 is the intensity of the incident light, and I_t is the intensity of the light transmitted through the sample. If the incident intensity is known, the absorbed intensity I_a can be calculated since

$$I_a = I_0 - I_t$$

and thus from Beer's law

$$\boxed{I_a = I_0(1 - e^{-2.3\varepsilon c d})} \tag{1.4}$$

Two conditions make the use of this equation easier: (1) If all the incident light is absorbed (OD > 2), then $I_a = I_0$; (2) if very little of the light is absorbed (OD ≥ 0.05), then I_a is approximated by $2.3\varepsilon c d I_0$.

* Note that c is used both for molar concentration and for the speed of light. The meaning of the symbol each time it is used should be clear from the context.

1.2b. *Internal Conversion and Intersystem Crossing*

After excitation has occurred, there are several processes which are important in the deactivation of the excited states. Those discussed in this section will be nonradiative, that is, they do not involve the emission of light.

As shown in Figure 1.2, a molecule undergoing a vertical transition upon excitation can arrive in the excited state with an internuclear distance (point A) considerably different from that corresponding to the minimum energy for the state (point B). In moving back to the equilibrium nuclear distance, the molecule finds itself three vibrational levels above its minimum energy. In other words, the molecule is vibrationally hot. The excess vibrational energy can be dissipated via bimolecular collisions with solvent molecules. This process is called *vibrational relaxation* and is usually faster by several orders of magnitude than intramolecular processes involving transitions between electronic states. This means that most processes involving a change in electronic state take place from low vibrational levels. For example, fluorescence is indicated in Figure 1.1 as a radiative transition that originates from the lowest vibrational level of the first excited singlet state.

Internal conversion is a nonradiative transition between states of like multiplicity (e.g., singlet to singlet, triplet to triplet, but not singlet to triplet):

$$S_2 \rightsquigarrow S_1$$

$$S_1 \rightsquigarrow S_0 \qquad (k_{\text{ic}})$$

The conversion from S_i to S_j is an isoenergetic process that is followed by vibrational relaxation of the new vibrationally hot state.

A nonradiative transition between states of different multiplicity is called *intersystem crossing*:

$$S_1 \rightsquigarrow T_1 \qquad (k_{\text{isc}})$$

$$T_1 \rightsquigarrow S_0 \qquad (k_d)$$

Figure 1.3 illustrates intersystem crossing as an isoenergetic process for a simple diatomic molecule. The crossing from the singlet to the triplet state has a maximum probability at point A in Figure 1.3, where the energy and geometry of the two states are equal.

Two important factors that influence the rate of these nonradiative processes are:

(a) Energy separation: the larger the energy gap (difference between the lowest vibrational levels of the two states), the slower the rate:

$$k(S_2 \rightsquigarrow S_1) > k(S_1 \rightsquigarrow S_0)$$

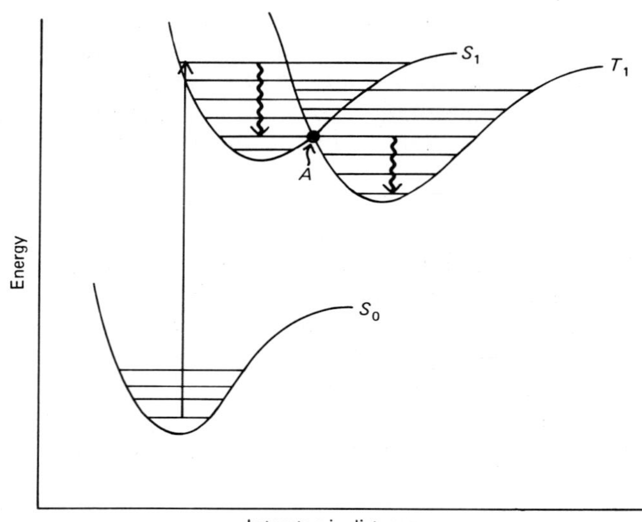

FIGURE 1.3. **Diatomic potential energy curves and intersystem crossing ($S_1 \rightsquigarrow T_1$) at point A.**

(b) Conservation of spin: transitions between states of different multiplicity are forbidden. However, as shall be seen later, these transitions can often compete very favorably with other (allowed) processes:

$$k(S_2 \rightsquigarrow S_1) > k(S_1 \rightsquigarrow T_1)$$

The ranges of rate constants for these various processes for typical organic molecules are

$$\boxed{\begin{aligned} k(S_2 &\rightsquigarrow S_1) = 10^{11}\text{–}10^{14}\ \text{sec}^{-1} \\ k(S_1 &\rightsquigarrow S_0) = 10^{5}\text{–}10^{8}\ \text{sec}^{-1} \\ k(S_1 &\rightsquigarrow T_1) = 10^{7}\text{–}10^{8}\ \text{sec}^{-1} \\ k(T_1 &\rightsquigarrow S_0) = 10^{-1}\text{–}10^{4}\ \text{sec}^{-1} \end{aligned}} \tag{1.5}$$

The reciprocal of the rate constant ($1/k$) for a particular process is called the eth life. This corresponds to the time necessary for the population of a state to decay to $1/e$ times the original population via that particular process.

1.2c. *Fluorescence and Phosphorescence*

In contrast to internal conversion and intersystem crossing, these processes are radiative (result in the emission of light). Kasha's rule states that the light-emitting level of a given multiplicity is the lowest excited level of that multiplicity (e.g., S_1 or T_1). This is a reflection of the relatively large rate constants for the nonradiative processes $S_2 \rightsquigarrow S_1$ and $S_1 \rightsquigarrow T_1$ in comparison with those for $S_1 \rightsquigarrow S_0$ and $T_1 \rightsquigarrow S_0$. There are exceptions to this rule but they are very few in number. Typical rate constants for radiative transitions between states of like multiplicity (*fluorescence*) and unlike multiplicity (*phosphorescence*) are in the following ranges:

$$\boxed{k_f = 10^7\text{–}10^9 \text{ sec}^{-1} \qquad k_p = 10^{-1}\text{–}10^4 \text{ sec}^{-1}} \tag{1.6}$$

The efficiency of these radiative processes often increase at low temperatures or in solvents of high viscosity. Consequently emission spectra are generally run in a low-temperature matrix (glass) or in a rigid polymer at room temperature. The variation in efficiency of these processes as a function of temperature and viscosity of the medium indicates that collisional processes compete with radiative and unimolecular nonradiative processes for deactivation of the lowest singlet and triplet states.

Kasha's rule and the rate constants discussed are applicable only for molecules in solution. In the gas phase, where collisions are few, transitions from vibrationally hot states and from electronic states other than the lowest state in each manifold are common.

1.3. PHOTOCHEMICAL KINETICS: CONCENTRATIONS, RATES, YIELDS, AND QUANTUM YIELDS

For a molecule A undergoing light absorption and reaction in its lowest excited singlet state to form a product P, we can write the following hypothetical mechanism, where A^s and A^t are the lowest excited singlet and triplet states, respectively:

$$\begin{array}{lll} & & \text{rate} \\ A + h\nu & \longrightarrow A^s & I_a \\ A^s & \longrightarrow A + h\nu & [A^s]k_f \\ A^s & \rightsquigarrow A & [A^s]k_{ic} \\ A^s & \rightsquigarrow A^t & [A^s]k_{isc} \\ A^t & \longrightarrow A + h\nu & [A^t]k_p \\ A^t & \rightsquigarrow A & [A^t]k_d \\ A^s & \longrightarrow P & [A^s]k_r \end{array}$$

For photochemical reactions and photophysical processes the efficiency is determined by the *quantum yield* Φ, which is defined as the number of molecules undergoing a particular process divided by the number of quanta of light absorbed:

$$\Phi_r = \frac{\text{number of molecules produced}}{\text{number of quanta absorbed}} = \frac{\text{rate}}{I_a}$$

If the rates of all the processes in the above mechanism are known, the quantum yield for the formation of P can be calculated:

$$\Phi_r = \text{rate}/I_a$$
$$d[\text{P}]/dt = k_r[\text{A}^s]$$

Using the steady-state approximation for $[\text{A}^s]$, we obtain

$$d[\text{A}^s]/dt = 0 = I_a - k_f[\text{A}^s] - k_{ic}[\text{A}^s] - k_{isc}[\text{A}^s] - k_r[\text{A}^s]$$
$$[\text{A}^s] = I_a/(k_f + k_{ic} + k_{isc} + k_r)$$
$$\Phi_r = (d[\text{P}]/dt)/I_a = k_r/(k_f + k_{isc} + k_{ic} + k_r)$$

Thus the efficiency of the reaction depends only upon the rate of reaction relative to the rates of all processes leading to deactivation of the excited state responsible for the reaction. The same is true for all other processes. Therefore an expression for the quantum yield for any particular process can be written down without going through the above kinetics:

$$\Phi_f = k_f/(k_r + k_f + k_{isc} + k_{ic})$$
$$\Phi_{isc} = k_{isc}/(k_r + k_{ic} + k_{isc} + k_f)$$

For a triplet state reaction the quantum yield is not only dependent upon the relative rates of the process and other processes leading to deactivation of the triplet state, but also to the efficiency of population of the triplet state (Φ_{isc}):

$$\Phi_p = \Phi_{isc}k_p/(k_p + k_d)$$

It is important to point out at this point that the rate constant k and the quantum yield for a photochemical reaction are not fundamentally related. Since the quantum yield depends upon relative rates, the reactivity may be very high (large k_r), but if other processes are competing with larger rates, the quantum yield efficiency of the reaction will be very small. That there is no direct correlation between the quantum yield and the rate is clearly seen from the data in Table 1.2 for the photoreduction of some substituted aromatic ketones in isopropanol:

$$2\text{RC(=O)R} + \text{CH}_3\text{CHOHCH}_3 \xrightarrow{h\nu} \text{R}_2\text{C(OH)}\text{—C(OH)—R}_2 + \text{CH}_3\text{C(=O)CH}_3$$

TABLE 1.2

Compound	Φ_{acetone}	k_r, sec^{-1} mole^{-1} liter
4,4′-Dimethylbenzophenone	0.71	0.7×10^6
Benzophenone	0.72	1.3×10^6
4-Trifluoromethylacetophenone	0.72	2.8×10^6

Thus the rates of production of acetone via the photoreduction of 4-trifluoromethylacetophenone and 4,4′-dimethylbenzophenone differ by a factor of four, while the quantum yields are essentially identical.

For reactions yielding more than one product the *percent yield* of a particular product can be calculated from the quantum yields of the various reactions:

$$\%\ \text{yield} = 100[\Phi_X/(\Phi_X + \sum_i \Phi_i)]$$

where $\sum_i \Phi_i$ represents the summation of the quantum yields of all other competing reactions.

The percent yield calculated in this way corresponds to the amount of product X relative to all products and is not the same as the percent of the starting material converted to X (this latter expression is given by $\Phi_X I_a/C_0$, where C_0 is the number of moles of reactant). It should be noted that similar to the relationship between rate and efficiency, a low quantum yield for a photochemical reaction does not necessarily mean that it will be formed in a low yield (the converse is true, however). If, for example, $\Phi_X = 0.005$ and $\sum_i \Phi_i = 0.00005$, then

$$\%\ \text{yield} = 100[0.005/(0.005 + 0.00005)] \approx 100\%$$

If the reaction were allowed to proceed long enough, all the starting material would be converted to product although the quantum yield is very low. On the other hand, if $\Phi_X = 0.4$ and $\sum_i \Phi_i = 0.6$, then

$$\%\ \text{yield} = 100[0.4/(0.4 + 0.6)] = 40\%$$

The yield of product X could never exceed 40% regardless of how long the reaction was run.

The time for a reaction to be completed is determined by the *rate*. For the formation of X above

$$\begin{aligned} \text{rate} &= k_r[\text{A}^s] \\ &= k_r I_a/(k_t + k_{isc} + k_{ic} + k_r) \\ &= \Phi_X I_a \end{aligned}$$

It is the specific rate constant k that is a direct reflection of the reactivity of a molecule toward a particular process, not the product yield or quantum yield.

The rate of reaction is dependent upon both the rate constant and the concentration of reactant molecules. Photochemical reactions occur through transformations of molecules which have a new distribution of electron density due to light excitation. The steady-state concentration of these excited molecules is given by

$$\boxed{N = I_a\ (\text{photons/sec}) \cdot \tau\ (\text{sec})}$$

where τ is the *lifetime of the excited state*;

$$\tau^s = 1/(k_{ic} + k_{isc} + k_f + k_r)$$
$$\tau^t = 1/(k_p + k_d)$$

Singlet lifetimes are of the order of 10^{-9} sec, whereas triplets are considerably longer lived, having τ's as large as 10 sec. With a very intense light source I_a may be as large as 10^{19}. With this intensity the number of molecules in the singlet and triplet excited states at any time during irradiation are

$$N^s = 10^{19} \times 10^{-9} = 10^{10} \text{ molecules } (\sim 10^{-13} \text{ mole})$$
$$N^t = 10^{19} \times 10 = 10^{20} \text{ molecules } (\sim 10^{-3} \text{ mole})$$

For a triplet state with a lifetime of 10 sec or longer the concentration of triplet molecules ($\sim 10^{-3}$ mole) is sufficiently high under continuous illumination that one can obtain a UV spectrum of the excited molecules.

1.4. CLASSIFICATION OF MOLECULAR ELECTRONIC TRANSITIONS AND EXCITED STATES

The physical and chemical properties of a molecule are determined primarily by the electron distribution in that molecule. Reactions that take place photochemically (excited state) but not thermally (ground state) result from the new electron distribution in the molecule resulting from excitation. The electron distributions and energies of a molecule in both the ground state and excited states are in principle available by solution of the Schrödinger equation. The total energy is the result of electrostatic attractive interactions between nuclei and electrons, repulsive interactions between electrons and electrons and between nuclei and nuclei, vibration of the nuclear masses, molecular rotation, and magnetic interactions resulting from the spinning electrons and nuclei. In practice a number of approximations are introduced to obtain solutions to the many electron, many nucleus Schrödinger equation. If the electronic orbital, spin, and nuclear motions can be treated separately, solution of the separate electronic part of the wave equation can yield electronic energy levels (orbitals) and approximate electron distributions. It is customary, although not rigorously correct, to place electrons into the molecular orbitals and to consider light absorption as a one-electron transition between the appropriate orbitals:

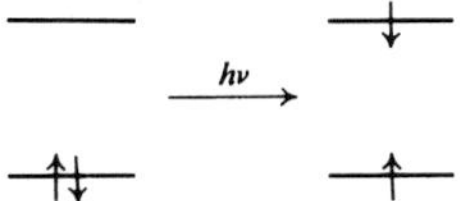

Some of these orbitals are highly localized and may resemble atomic orbitals. Others may be delocalized over a large number of atoms.

Kasha's classification of electronic transitions in terms of σ (bonding), σ^* (antibonding), π (bonding), π^* (antibonding), and lone pair (nonbonding, designated by n) orbitals is almost universally used by organic photochemists. Simple examples to help illustrate these transitions are given in Figures 1.4 and 1.5 for ethylene and formaldehyde. In these figures the structures on the left represent the atomic orbitals assuming there is no interaction. The structures in the center represent the result of interaction of the atomic orbitals in the molecule. The transitions occurring upon excitation are shown schematically on the right.

1.4a. $\pi \rightarrow \pi^*$ *Transitions*

In order for a transition to occur, the transition moment integral must be an even function. This integral is the product of the wave function corresponding to the ground state Ψ_g, an operator R, and the wave function of the excited state Ψ_e. Since R is antisymmetric, the product of the wave

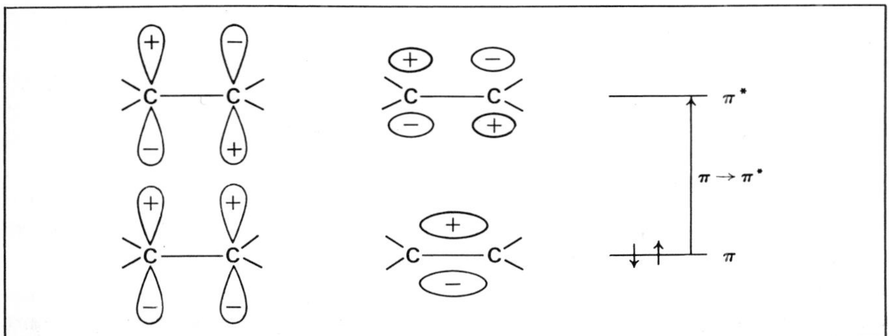

FIGURE 1.4. The π molecular orbitals of ethylene formed from the atomic p orbitals and the $\pi \rightarrow \pi^*$ electronic transition.

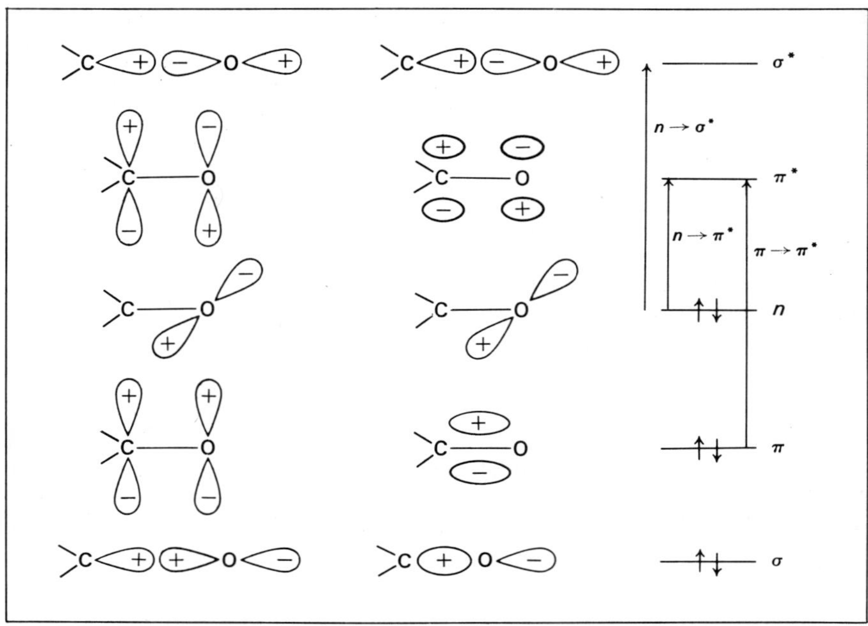

FIGURE 1.5. Localized molecular orbitals formed from the atomic basis orbitals and electronic transitions for the carbonyl group.

functions must also be antisymmetric. This means that there must be a change in symmetry of the wave function in proceeding from the ground to the excited state. $\pi \rightarrow \pi^*$ transitions are often symmetry allowed and consequently are usually intense:

$$\varepsilon_{\max}^{\pi\rightarrow\pi^*} \approx 10^4\text{–}10^5 \text{ liters/mole-cm}$$

A term frequently used in place of the molar absorptivity ε to describe the probability of a transition is the oscillator strength f, which can be related to ε in the following way:

$$\boxed{f = 4.3 \times 10^{-5} \int \varepsilon \, d\bar{\nu} \approx 10^{-5}\varepsilon_{\max}} \tag{1.8}$$

where $\int \varepsilon \, d\bar{\nu}$ is the integrated value of ε over the entire absorption band and is usually about one-fourth of the value of $\varepsilon_{\max}$. It can be shown that the more highly probable an absorption is, the more highly probable emission

TABLE 1.3. *Classification of Molecular Transitions*

Property	$\pi \rightarrow \pi^*$	$l \rightarrow a_\pi$	CT	$n \rightarrow \pi^*$
INTENSITY, log ε_{max}, liters/mole-cm	3–5	3–4	3–4	2–3
OSCILLATOR STRENGTH f	10^{-2}–10^{0}	10^{-2}–10^{-1}	10^{-2}–10^{-1}	10^{-3}–10^{-2}
$\tau_F{}^0$, sec	10^{-7}–10^{-9}	10^{-7}–10^{-8}	10^{-7}–10^{-8}	10^{-6}–10^{-7}
S–T SPLIT, kcal/mole	22.9	14.3	8.7	8.7
SOLVENT SHIFT,[a] cm^{-1}	−600	+600	−2500[b]	+800

[a] Nonpolar to polar.
[b] For aromatic carbonyl compounds.

from the upper state will be. The *mean radiative lifetime* $\tau_r{}^0$ is approximated by the following equation:

$$\boxed{1/\tau_r{}^0 = k_r \approx 10^4 \varepsilon_{max}} \tag{1.9}$$

where k_r corresponds to the rate constant for fluorescence or phosphorescence. Lifetimes calculated in this way are the radiative lifetimes or the lifetimes that would be observed in the absence of all other unimolecular and bimolecular deactivation processes and differ from the true lifetimes as follows:

$$\tau_f{}^0 = 1/k_f$$
$$\tau_f = 1/(k_f + k_{ic} + k_{isc} + \sum_i k_i[\mathrm{A}])$$

The lifetimes, molar absorptivities, and oscillator strengths for various transitions are summarized in Table 1.3.

The corresponding $\pi \rightarrow \pi^*$ triplet state energy levels can often be determined from the wavelength of phosphorescence emission at low temperature. The energy separation between the $(\pi \rightarrow \pi^*)^s$ and $(\pi \rightarrow \pi^*)^t$ levels is usually large (8000 cm^{-1}, ~ 23 kcal/mole). Another aspect of the $\pi \rightarrow \pi^*$ transition that can be helpful in its classification is the shift of its λ_{max} upon changing from nonpolar (hydrocarbon) to polar (alcohol) solvents. Since the excited state has more charge separation (is more polar) than the ground state, a polar solvent lowers the energy of the excited state relative to the ground state with a resulting shift to longer wavelength (red shift) (Figure 1.6).

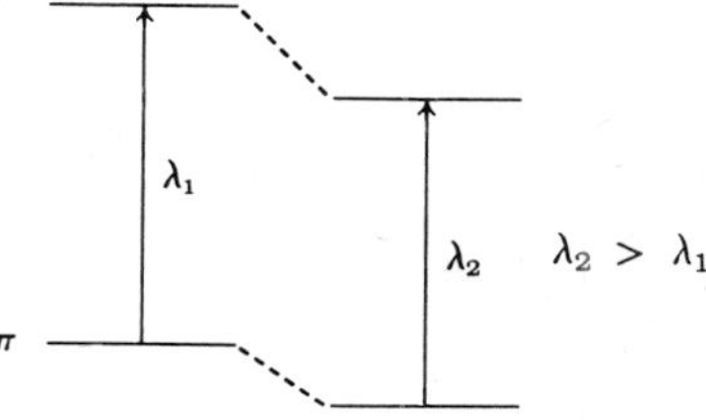

FIGURE 1.6. Shift in the $\pi \rightarrow \pi^*$ energy levels upon changing from a nonpolar to a polar solvent.

Since upon excitation an electron is promoted from a bonding orbital to an antibonding orbital, the bond length is expected to increase. Hence, for $\pi \rightarrow \pi^*$ transitions, in diagrams such as Figure 1.2, the excited state is drawn with a larger internuclear distance than the ground state.

1.4b. $n \rightarrow \pi^*$ *and* $l \rightarrow a_\pi$ *Transitions*

As Figure 1.5 suggests, the lowest energy transition for molecules with essentially nonbonding electron pairs is of $n \rightarrow \pi^*$ character. These transitions are common for compounds containing oxygen, nitrogen, and sulfur.

$n \approx p$ orbital $\qquad$ $n \approx sp^2$ hybrid orbital

The $n \rightarrow \pi^*$ transitions are weak ($\varepsilon \approx 100$ liter/mole-cm) since the lone pair does not overlap in space with the π system. In other words, the transition is spatial-overlap forbidden. Singlet–triplet splitting for $n \rightarrow \pi^*$ excited states is small in comparison to that for $\pi \rightarrow \pi^*$ states and an increase in solvent polarity produces a shift to shorter wavelength (blue shift) since in this case the excited state is less polar than the ground state (transfer of an electron formally localized on the negatively polarized oxygen atom to the region between the carbon and the oxygen in formaldehyde). Since the energy separation between the singlet and triplet states is small, the rate constant for intersystem crossing is usually large.

For molecules in which the lone pair is not totally nonbonding, transitions intermediate to $n \rightarrow \pi^*$ and $\pi \rightarrow \pi^*$ transitions result. These have been designated as $l \rightarrow a_\pi$ transitions. An excellent example to consider is aniline

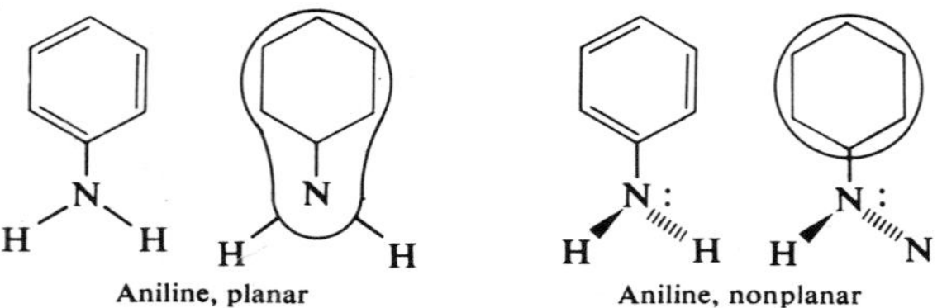

Aniline, planar $\qquad$ Aniline, nonplanar

in planar and nonplanar configurations. In the planar form there is extended conjugation between the ring and the nitrogen lone pair and $\pi \rightarrow \pi^*$ transitions occur. When the NH_2 group is rotated 90°, however, conjugation of the lone pair with the ring is impossible but an $n \rightarrow \sigma^*$ transition (very high energy) can occur. At intermediate angles of twist, the lone pair is aligned such that $n \rightarrow \pi^*$ transitions can occur. Consequently the absorption observed contains contributions from all of these configurations. Inasmuch as the orbitals of π origin cannot be classified strictly as π^*, Kasha has used the notation a_π.

1.4c. *Intramolecular Charge-Transfer Transitions* (*CT*)

When both electron donor (D) and acceptor (A) groups are attached to a π-electron system it is not possible to consider the transition in terms of the excitation of one electron since it is a composite of several different one-electron excitation types. An example of a molecule which has electronic transitions of this type is Michler's ketone:

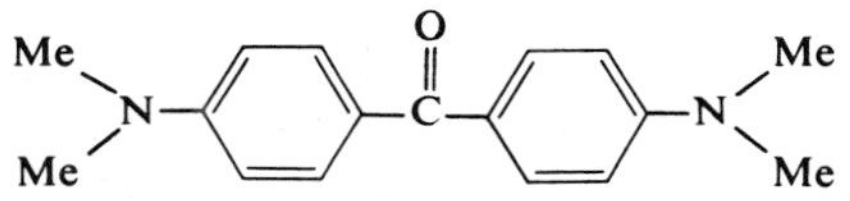

Porter has termed these transitions *charge-transfer excitations* (CT). Possible one-electron contributions to the excitation of a molecule represented as DRA (D, donor; *R*, chromophore; A, acceptor) are

$$D^+R^-A \qquad D^+RA^- \qquad DR^+A^- \qquad DR^*A$$

Where the $+-$ terms refer to $l \rightarrow a_\pi$ type excitations and the * to a $\pi \rightarrow \pi^*$ type transition. These absorptions occur at longer wavelengths than the related model compounds (benzene and dimethylamine for Michler's ketone), have a high intensity, $\varepsilon_{max} \approx 10^4$ liter/mole-cm, a small singlet–triplet splitting, and undergo a red shift of the absorption on going to a more polar solvent.

PROBLEMS

1. Calculate the energy (kcal/mole) corresponding to light of wavelength (a) 365 nm; (b) 313 nm; (c) 254 nm.
2. (a) Calculate the shift in λ_{max} (cm^{-1}) for a $\pi \rightarrow \pi^*$ transition if λ_{max} in cyclohexane is 300 nm and λ_{max} in ethanol is 305 nm. What change in energy does this correspond to?
   (b) Why is it not possible to convert a wavelength shift reported in cm^{-1} (e.g., $+800\ cm^{-1}$) directly into wavelength (say in nm)?

3. For the photochemical reaction

$$A \xrightarrow{h\nu} B$$

0.05 mole of B are produced from a 0.1 N solution of A in 1 hr of irradiation at 254 nm. If the intensity of the light (I_0) at 254 nm is 2×10^{19} quanta/sec and $\varepsilon_{254}^{A} = 10^4$ liters/mole-cm, calculate the quantum yield for the reaction. The reaction cell has a 5-cm light path and a volume of 1 liter.

4. Given the data for a unimolecular photochemical reaction, $k_r = 5 \times 10^5$ mole/sec, $\Phi_p = 0.75$, and $I_a = 3 \times 10^{19}$ quanta/sec, calculate the time required for the reaction to produce 1 mole of product.

5. If the quantum yield for the reaction

$$\text{S-ring–H} + NOCl \xrightarrow{h\nu} \text{S-ring–NO} + HCl$$

is 0.72, calculate the amount of product that can be produced in one day using a 1000-W high-pressure mercury source. Assume that all the light is absorbed by the reactants at the effective wavelength of 366 nm and that the lamp emits only 6% of its input at this wavelength.

6. A particular substituted benzene molecule is *o–p* directing in the ground state and is predicted by MO theory to be *m* directing in the first excited singlet state. If the most intense light source you have is one which produces 10^{19} photons/sec, the lifetime of the singlet state (τ_s) is 10^{-9} sec, and the specific rate constants for the bromination of the ground and excited state are approximately equal, how much *m*-bromination can you expect? How much *m*-bromination would you expect if the reaction is from the triplet state with a lifetime (τ_t) of 10 sec?

7. Derive a quantum yield expression (Φ_{RBr}) for the following reaction sequence:

$$\begin{array}{rcll} Br_2 + h\nu & \longrightarrow & 2Br\cdot & I_a \\ Br\cdot + RH & \longrightarrow & HBr + R\cdot & k_1 \\ R\cdot + Br_2 & \longrightarrow & RBr + Br\cdot & k_2 \\ M + Br + Br & \longrightarrow & M + Br_2 & k_3 \end{array}$$

8. A He–Ne continuous wave laser provides an 8-mW source of radiation at 6328 Å. Calculate the intensity of the light in photons/sec. The light beam has a 3 mm diameter. How many hours would be required to isomerize 1 millimole of compound X with $\Phi_r = 0.5$, $\varepsilon_{6328\,\text{Å}} = 10^3$? What assumptions must be made to solve this problem?

2

Photochemical Techniques and the Photodimerization of Anthracene and Related Compounds

In this chapter we will endeavor to answer the following questions: What does the photochemist do in preparing to investigate the photochemical behavior of a molecule? What equipment does he use to carry out his experiments? Once he has determined the results of the reaction, how can he develop a mechanism to account for these results? In answering these questions we will be concerned mainly with the photochemistry of anthracene and related compounds.

2.1. ABSORPTION AND EMISSION SPECTRA

Generally the first thing to be done in preparation for the photochemical study of a compound is to determine the visible and ultraviolet absorption spectrum of the compound. Besides furnishing information concerning the nature of the excited state potentially involved in the photochemistry (see Section 1.4), the absorption spectrum furnishes information of a more applied nature as to the wavelength range in which the material absorbs and its molar absorptivity ε. From this information it is possible to decide what type of light source to use for the irradiation, what solvents can be used to avoid solvent absorption at wavelengths absorbed by the reactant, and what minimum concentration of reactant to use to ensure complete light absorption.

2.1a. *Transition Probability*

We have seen in Chapter 1 that absorption and emission spectra are controlled at least in part by the Franck–Condon principle. However, this is only one of three major factors that must be considered.

A light wave can be described in terms of electric and magnetic fields with electric and magnetic vectors corresponding to these fields mutually perpendicular to each other and to the direction of propagation of the wave. Light interacts with the electrons of an absorbing molecule. The spin of these electrons (and the resulting magnetic properties) is relatively unimportant compared with the properties resulting from the charge on the electrons. The electric field component of the light wave interacts with an electron of a molecule by moving it, that is, by changing it from the initial state ψ_i to a final state ψ_j. Since this constitutes a relocation of charge, there must be a change in dipole moment during the transition. The transition is then said to be allowed. The probability for the transition is proportional to the square of the integral of the product of the wave function for the ground state ψ_i multiplied by an operator M that changes the position of the charge, multiplied by the wave function for the final state ψ_j:

$$\boxed{\mathrm{Pr} \propto R_{ij}^2 = \left(\int \psi_i M \psi_j \, d\tau\right)^2 = \langle\psi_i|M|\psi_j\rangle^2} \tag{2.1}$$

Pr is the probability for the transition and R_{ij} is the ***transition dipole moment integral.***

The operator M (dipole moment operator) is composed of x, y, and z components:

$$M = M_x + M_y + M_z \tag{2.2}$$

If we consider only the x direction, we see that a position operator equal to x multiplied by the charge e is required to move the electron. The operator then has the form

$$M_x = \sum_i eX_i \tag{2.3}$$

where e is the electronic charge and X_i is the vector distance of the ith electron from the origin of a fixed coordinate system for the molecule.

To aid our understanding of absorption and emission processes, Eq. (2.1) can be expanded in terms of electronic, vibronic (vibrational components of an electronic transition), and spin wave functions:

$$\psi_i = \psi_{e_i}\psi_{v_i}\psi_{s_i} \tag{2.4}$$

The transition moment integral now has the following form:

$$\boxed{R_{ij}^2 = \langle\psi_{e_i}|M|\psi_{e_j}\rangle^2\langle\psi_{v_i}|\psi_{v_j}\rangle^2\langle\psi_{s_i}|\psi_{s_j}\rangle^2} \tag{2.5}$$

If any one of these integrals (expectation value equations) is zero, the transition is said to be forbidden. For the electronic and spin wave functions, it is not necessary to evaluate the integral but only to note that an odd function integrated from minus infinity to infinity is zero, while an even function integrated within these limits results in a nonzero value. For example (Figure 2.1),

$$\int_{-\infty}^{\infty} \sin x \, dx = \int_{-\infty}^{\infty} \text{odd function} = 0$$

Electronic Integral

Since M (or er, where r is the distance the electron is moved during excitation) is an odd function, then

$$R_e = \langle\psi_{e_1}|M|\psi_{e_2}\rangle^2 \tag{2.6}$$

to be nonzero, ψ_{e_1} and ψ_{e_2} must have different symmetries:

$$\langle\text{odd}|\text{odd}|\text{odd}\rangle^2 = \left(\int \text{odd}\right)^2 = 0$$

$$\langle\text{even}|\text{odd}|\text{odd}\rangle^2 = \left(\int \text{even}\right)^2 \neq 0$$

The determination of symmetries can be done using group theory. However, we will apply a somewhat simpler method. For this we need only

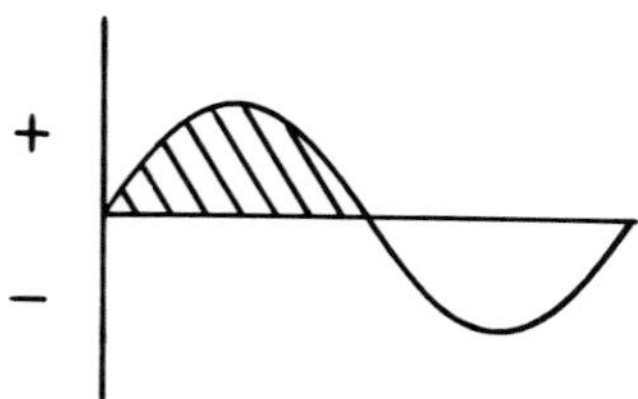

FIGURE 2.1. Plot of sin x.

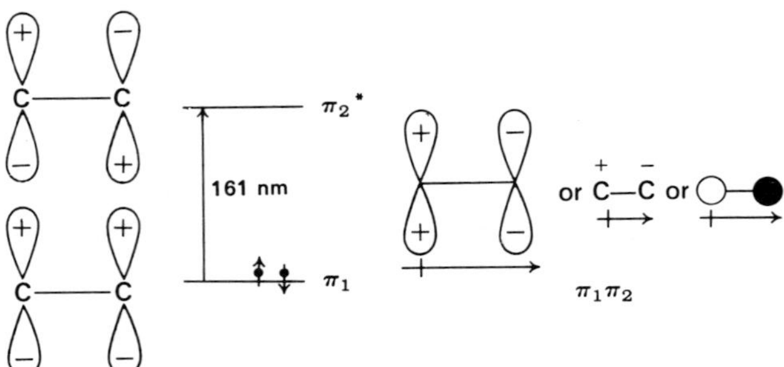

FIGURE 2.2

consider the highest occupied one-electron molecular orbitals (HOMO) and the lowest unoccupied molecular orbitals (LUMO) inasmuch as these will control the symmetry of the wave functions. Filled molecular orbitals must be even functions (odd × odd = even).

As examples let us consider ethylene, butadiene, and anthracene. The energy levels and wave functions for ethylene based on the LCAO MO approximation are shown in Figure 2.2. To determine if the transition is allowed (dipole moment), we multiply the wave functions. That is, we multiply the signs of the lobes for the two carbons in both states respectively. The result is given on the right in Figure 2.2 (open circle is positive, filled circle is negative). We can see that the transition is allowed in the direction of the molecular axis. Indeed a strong transition is observed in ethylene at 161 nm that corresponds to this one-electron transition.

A similar diagram for butadiene is shown in Figure 2.3. For butadiene the $\pi_2 \rightarrow \pi_4$ transition is forbidden for the *trans* form (no resultant dipole

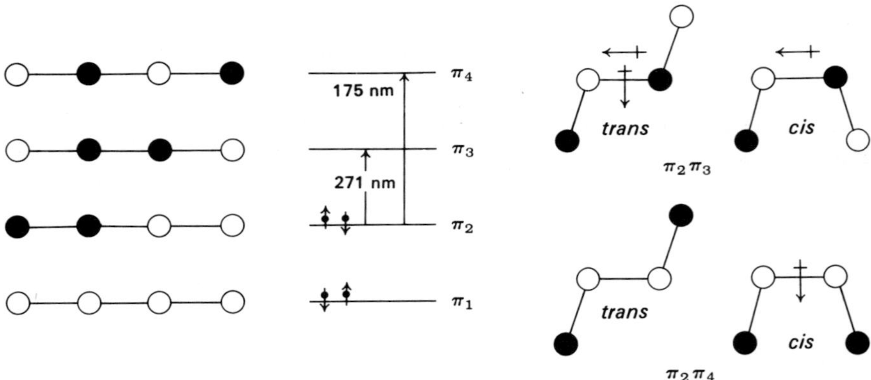

FIGURE 2.3

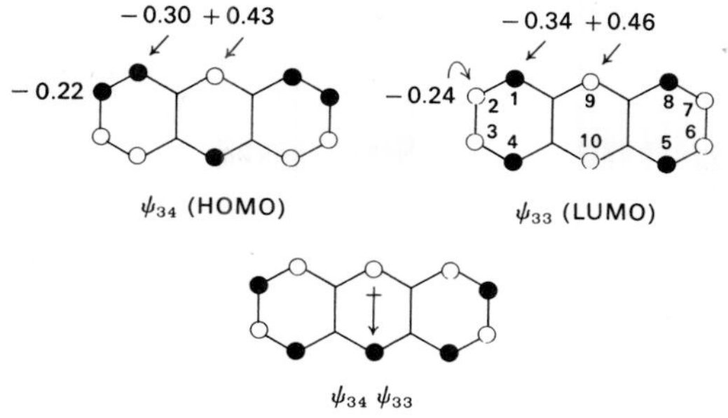

FIGURE 2.4

moment) but allowed for the *cis* form. In agreement this transition is observed for *cis*-butadiene in the 175 nm region but no corresponding absorption for the *trans* isomer is observed. The $\pi_2 \rightarrow \pi_3$ transition is observed for both the *cis* and the *trans* dienes (~270 nm) but is more intense for the *trans* form.

The HOMO and LUMO for anthracene are shown in Figure 2.4.[1] The $\pi_{34} \rightarrow \pi_{33}$ anthracene transition is allowed in the direction of the short axis of the molecule. The square of the coefficients provides a measure of the probability of finding an electron at a particular carbon atom. Note that for both ψ_{34} and ψ_{33} the greatest electron density is at the 9, 10 carbon atoms. This suggests that any chemical reaction involving these electrons will likely occur at these positions. This is indeed the case, as we shall see later when we consider the photochemistry of anthracene.

The lowest energy absorption for anthracene has been shown to be due to a transition dipole moment along the short axis of the molecule by polarization studies described in Section 2.1b.[2]

Vibronic Integral (Franck–Condon Factors)

The intensity of a vibronic transition depends upon the square of the overlap integral of the vibrational wave functions,

$$\boxed{I \propto \langle \psi_{v_1} | \psi_{v_2} \rangle^2} \tag{2.7}$$

To help visualize this process, let us consider a diatomic molecule with the energy curves shown in Figure 2.5(a). In this example the ground and excited states have the same equilibrium internuclear distance r_a. Since in solution at room temperature almost all the molecules will be in the lowest vibrational level of the ground state $v_1{}^0$ (subscripts refer to the electronic

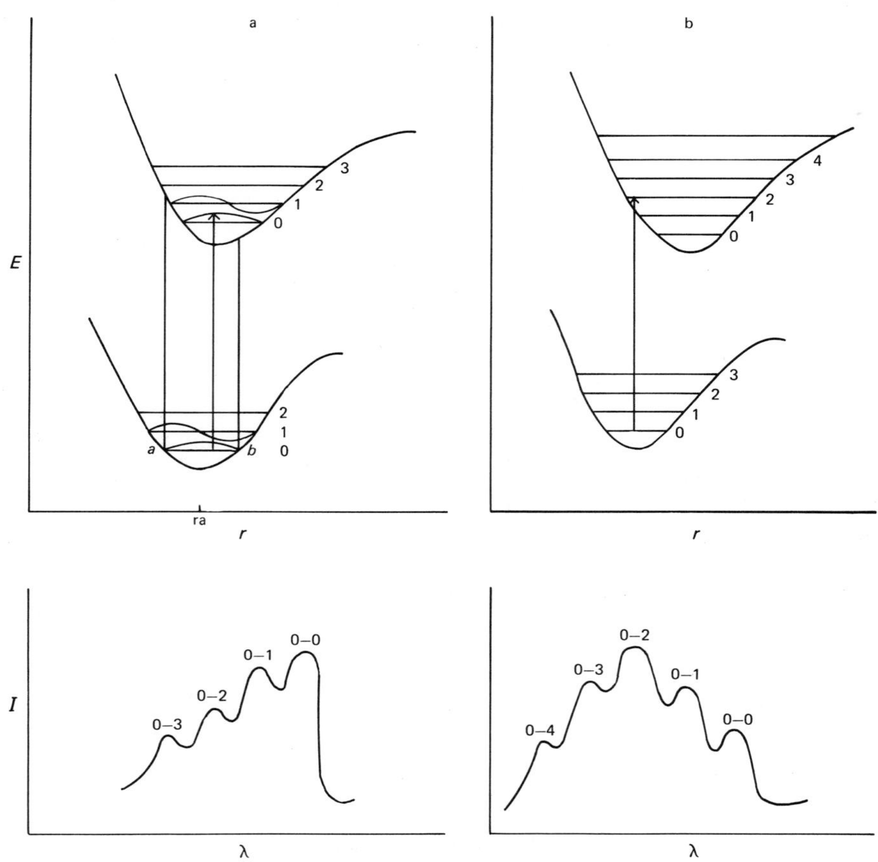

FIGURE 2.5

level, superscripts to the vibrational level of the particular electronic state), the molecule will have a range of bond distances from a to b, with the most probable value being r_a.

The overlap integral $\int \psi^0_{v_1}\psi^j_{v_2}\, dr$ will be a maximum for $j = 0$ for the case described in Figure 2.5(a) since the positive and negative contributions to $\psi^1_{v_2}$, $\psi^2_{v_2}$, and all higher levels tend to cancel each other. When there is little change in the geometry of the excited state relative to the ground state (similar r_a's), we expect that the $0 \rightarrow 0$ transition will always be the most intense. However, when there is a large change in the equilibrium geometry of the two electronic states as indicated in Figure 2.5(b), the most intense transition may not be the $0 \rightarrow 0$ transition. In Figure 2.5(b) the most intense absorption will result from the $0 \rightarrow 2$ or $0 \rightarrow 3$ transition. These transitions are drawn to the edge of the curve and are not extended to higher vibrational levels because the maximum probability for the geometry of the molecule in all vibrational levels (except the $i = 0$ level) corresponds to the turning point of the vibration (as approximated by a classical harmonic oscillator). The square of the vibrational wave function $(\psi_v)^2$, shown in Figure 2.6, is proportional to the probability that the molecule will have a given geometry.

When the ground and excited states have similar geometries the absorption and fluorescence spectra have mirror image symmetry. This is illustrated schematically in Figure 2.7. This type of spectrum is observed for a number of anthracene derivatives. An example is 9-anthramide shown in Figure 2.8. If, however, there is a large change in the equilibrium internuclear distance for the ground and excited states, the mirror image symmetry is lost. This is then a useful means of deducing some structural information regarding the relative geometries of the ground and excited states. A geometry change in going from the ground to the first excited state is

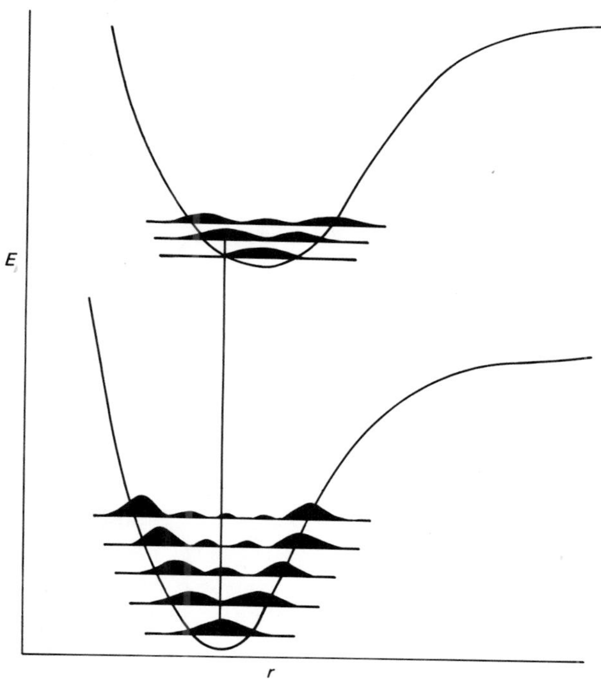

FIGURE 2.6

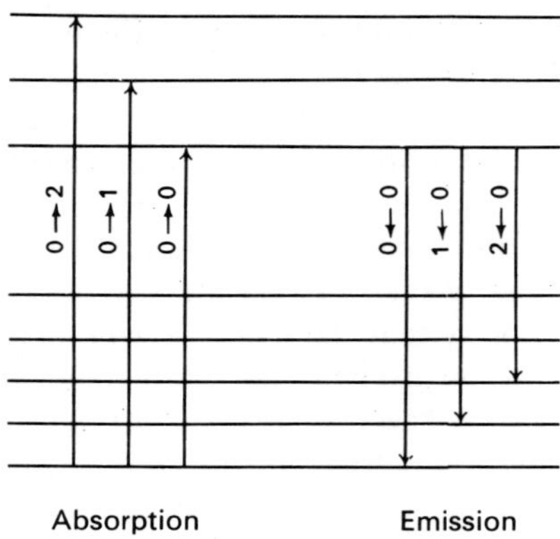

FIGURE 2.7

proposed for 9-anthroate esters. This conclusion results from spectral measurements like the one shown in Figure 2.9 for cyclohexyl-9-anthroate.

Spin Restrictions

For $S \rightarrow S$ and $T \rightarrow T$ transitions, the spin integral is unity, while for $S \rightarrow T$ and $T \rightarrow S$ transitions the value of the integral is zero; writing Pr

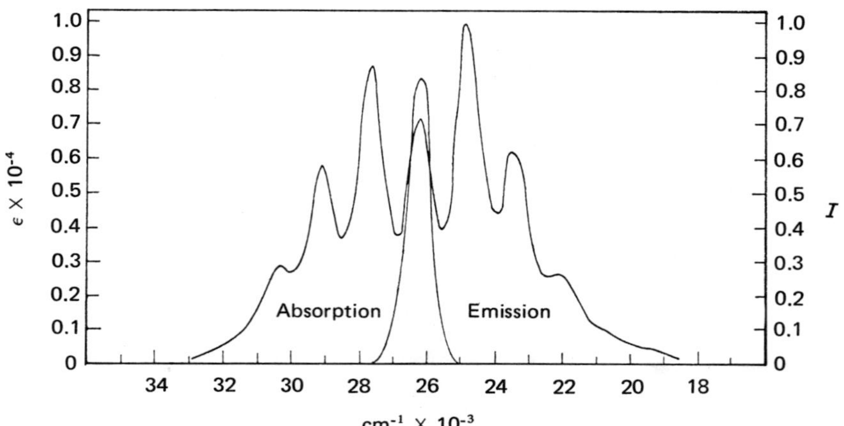

FIGURE 2.8. Absorption and fluorescence spectra of 9-anthramide. Note the symmetry of the absorption and emission. Reproduced with permission from Ref. 8.

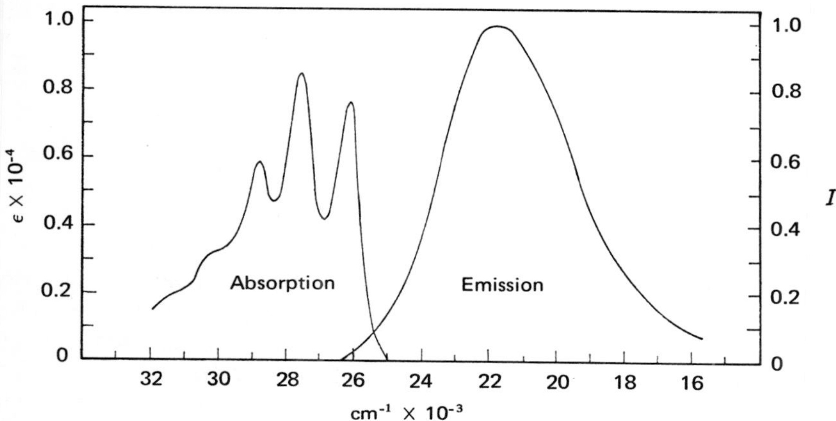

FIGURE 2.9. Absorption and fluorescence spectra of cyclohexyl-9-anthroate. Reproduced with permission from Ref. 8.

for that portion of the transition probability determined by the spin wave functions [see Eq. (2.5)], we have

$$\boxed{\mathrm{Pr} \propto \langle \psi_{s_1} | \psi_{s_2} \rangle^2} \tag{2.8}$$

How these functions can be perturbed so as to make $S \rightarrow T$ transitions possible will be discussed in another chapter.

2.1b. *Polarization Spectra*

We have seen that the transition dipole moment occurring upon excitation of a molecule has a distinct orientation with regard to the molecular axis. This orientation can be determined by measuring the absorption of polarized light (oscillating in only one plane) by oriented single crystals,

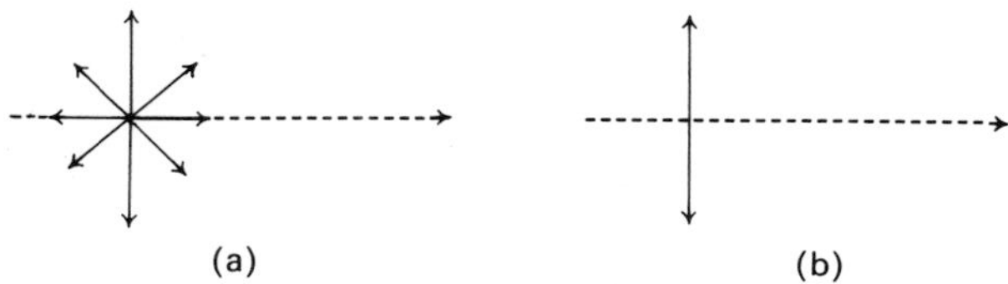

FIGURE 2.10. Direction of electric vector in (a) unpolarized and (b) polarized light; the direction of propagation of the light wave is along the horizontal from left to right.

no absorption absorption

FIGURE 2.11

where the orientation of the molecules in the crystal is known by x-ray studies, or by "doping" a guest crystal with the molecules of interest. Alternatively, two recent techniques replace the guest crystal with a liquid crystal or a stretched polymer[(3)] film to orient the molecules.

The absorption of polarized light is illustrated in Figures 2.10 and 2.11. Polarized emission from randomly oriented molecules held in a rigid matrix can provide information on the relative orientation of the absorption and emission dipole moments. This is shown in Figure 2.12. If the exciting light is entering from the y direction and is polarized in the xy plane, only molecules oriented such that their transition dipole moments are parallel to (or have a component parallel to) the xy plane will absorb the light. Molecules whose transition moment is parallel to the electric vector of the light are more likely to be excited than those that are inclined at an angle θ. If, following absorption, the molecule emits, some polarization of the emission will be observed. Observation of the intensity of the emitted light is made in the yz plane from the z direction and this intensity is measured in terms of its components parallel ($I_{\|}$) and perpendicular ($I_{\perp}$) to the plane. The degree of polarization P is defined as follows, where the second expression holds

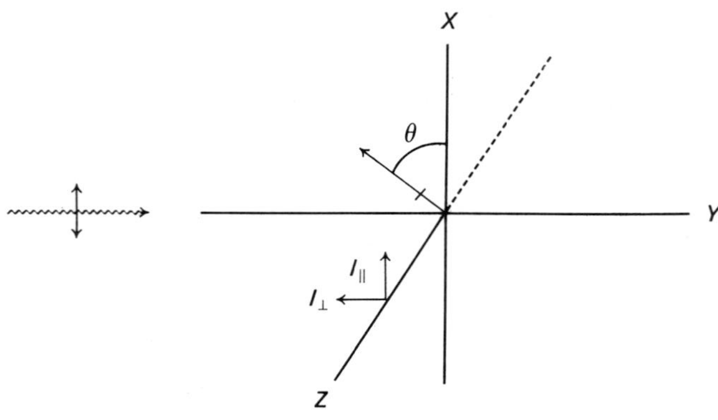

FIGURE 2.12

if there is an angle β between the absorption and emission dipole moments:

$$P = \frac{I_{\parallel} - I_{\perp}}{I_{\parallel} + I_{\perp}} = \frac{3\cos^2\beta - 1}{\cos^2\beta + 3} \tag{2.9}$$

The values of P range from $+\frac{1}{2}$ to $-\frac{1}{3}$ ($\beta = 0$ or $90°$). This equation would be considerably simpler if *only* those molecules with their transition moments parallel to the electric vector were capable of absorption or if the molecules were perfectly aligned, that is, $\theta = 0$. Then the angle between the two transition moments could be directly determined from the observed degree of polarization P.

By measuring the degree of polarization of the emission of a randomly oriented sample as a function of the excitation wavelength, one can often determine the existence and position of a weak transition which is buried beneath the envelope of a stronger transition of similar energy or the relative positions of overlapping transitions. Ideally, in this case the principal polarization spectrum should consist of regions of constant polarization joined by regions of rapidly changing polarization where electronic bands overlap. For instance, phenol, tyrosine, and cresol show polarization spectra consisting of two regions, one positive at the longer wavelengths and one negative at the shorter wavelengths.[4] The transition with a maximum at 275 nm in phenol is attributed to an $n \rightarrow \pi^*$ transition. It is expected that if in this type of transition an electron from the oxygen lone pair is promoted to an excited orbital localized primarily in the ring, the direction of the transition moment will be perpendicular to the plane of the ring and therefore at right angles to the transition moment arising from the ring-localized $\pi \rightarrow \pi^*$ absorption. This is borne out by the polarization spectrum of phenol.

2.1c. *The Measurement of Fluorescence Spectra and Fluorescence Quantum Yields*

A schematic diagram of a typical fluorescence spectrometer is shown in Figure 2.13.

Emission Spectra

To determine the fluorescence spectrum of a compound, one selects an exciting wavelength known to be within the absorption band of the compound with the excitation monochromator (generally the longest wavelength λ_{max} is chosen) and scans the emitted light at right angles with the emission mono-

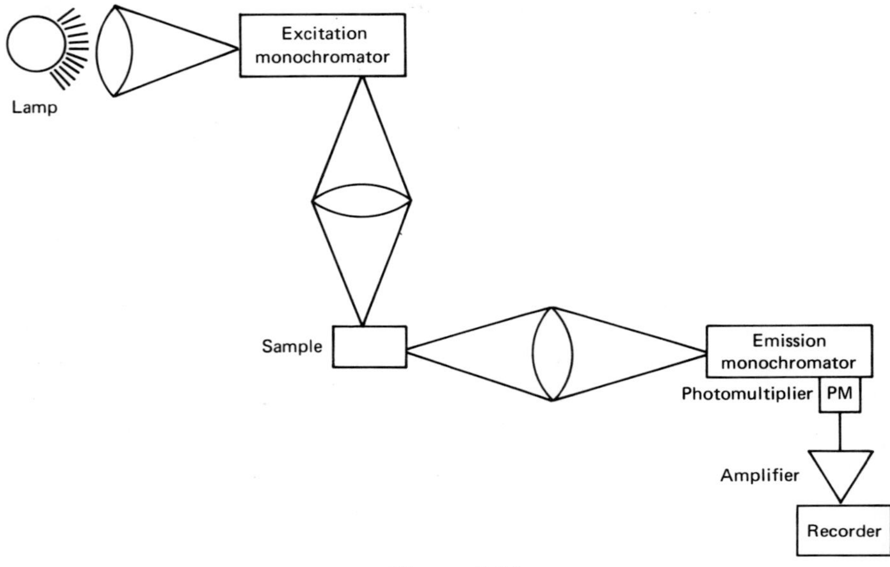

FIGURE 2.13

chromator. The signal from the photomultiplier is fed into an amplifier and displayed on a recorder.

Generally in solution at room temperature only fluorescence emission is observed. To obtain phosphorescence spectra, low temperatures are used and a rotating can with a narrow slit in the side is placed around the sample. Since the lifetime of the fluorescent state is generally much shorter than that of the phosphorescent state, the time required for the slit in the can to rotate 90° ensures that only the longer lived phosphorescence will be observed.

Since the fluorescence spectrometer is a very sensitive instrument designed to detect very small intensities of emitted light, one must be careful to determine that the results are meaningful and correctly interpreted. It is of primary importance that the sample be pure and free of foreign substances that may be fluorescent or act as fluorescence quenchers. Similarly, the solvent used must be pure. It is essential that the fluorescence of a blank using the same solvent as the initial determination be run as a routine part of a fluorescence study. Other factors which must be considered involve light scattering. Raman scattering from the solvent is detected as a small band that shifts as the wavelength is changed. Similarly, some of the exciting light finds its way into the emission monochromator through Rayleigh–Tyndall scattering. Obviously it is important to ensure that solutions are free of dust and other insoluble foreign matter, such as filter paper lint.

Excitation Spectra

To determine the fluorescence excitation spectrum, one selects a wavelength on the emission monochromator (generally the fluorescence λ_{max}) and scans the exciting wavelength with the excitation monochromator. Normally the fluorescence excitation spectrum will have the same shape and wavelength distribution as the absorption spectrum. However, in cases where the absorption band for a fluorescent state is hidden beneath that of a nonfluorescent state or when insufficient sample is available to run a routine absorption spectrum, the fluorescence excitation spectrum can provide very valuable information. Similarly, for phosphorescence where direct absorption to the triplet level does not occur one can learn a great deal from the phosphorescence excitation spectrum. This will be discussed in more detail later.

Emission Quantum Yields

Fluorescence quantum yields, although not easy to obtain, can provide valuable information. We showed in Chapter 1 that Φ_f can be expressed as

$$\Phi_f = k_f \Big/ \left(\sum_i k_i + \sum_j k_j[\mathrm{A}] \right) \tag{2.10}$$

When only unimolecular (first order) processes serve to deactivate the fluorescent state (e.g., at very low concentrations) the quantum yield becomes

$$\Phi_f{}^0 = k_f \Big/ \sum_i k_i \tag{2.11}$$

where $k_i = k_{ic} + k_{isc} + k_f$. The fluorescence lifetime is then given by

$$\tau_f = 1 \Big/ \sum_i k_i \tag{2.12}$$

Therefore

$$\Phi_f{}^0/\tau_f = k_f \tag{2.13}$$

From these two experimentally measurable quantities, $\Phi_f{}^0$ and τ_f, one can obtain the fluorescence rate constant. For many compounds $k_{ic} = 0$; then with the knowledge of k_f, k_{isc} can be determined.

In practice it is much simpler to determine the relative quantum yield of fluorescence than the absolute quantum yield (see Table 2.1).* This is done by comparing the fluorescence intensity of a given sample to that of a compound whose fluorescence quantum yield is known. For this one must

* For a discussion of fluorescence techniques see Ref. 5.

TABLE 2.1. *Absolute Fluorescence Quantum Yields*[a]

Compound	Solvent	Excitation wavelength, Å	Φ_f
Quinine	1.0 N H_2SO_4	3655	0.54 ± 0.02
Quinine	0.1 N H_2SO_4	3655	0.50 ± 0.02
Perylene	Ethanol	2537–4358[b]	0.94
Perylene	Benzene	3655	0.99 ± 0.03
Acridone	Ethanol	3655	0.72 ± 0.02
Anthracene[c,d]	Ethanol	3655	0.27 ± 0.01
Anthracene	Benzene	3655	0.27 ± 0.01
Fluorene	Ethanol	2537	0.68 ± 0.04
Naphthalene	Ethanol	2537	0.205 ± 0.014
Phenanthrene	Ethanol	2537–4358[b]	0.125
Aminoacridine	Water	2537–4358[b]	0.81 ± 0.02
Fluorescein[c]	0.1 N NaOH	2537–4358[b]	0.87
Triphenylene	Ethanol	2537	0.065 ± 0.006
Chrysene	Ethanol	2537–4358[b]	0.17
Pyrene	Ethanol	3131	0.53 ± 0.02
Pyrene	Benzene	3131	0.60 ± 0.03
Pyrene	Cyclohexane	3131	0.58 ± 0.01
Benzene	Hexane	2537	0.053 ± 0.008
9,10-Dichloro-anthracene	Benzene	3655	0.71 ± 0.04

[a] All values from the work of W. R. Dawson and M. W. Windsor, *J. Phys. Chem.* **72**, 3251 (1968).
[b] Φ_f values are averaged for the various excitation wavelengths inasmuch as the values of Φ_f were independent of excitation wavelength.
[c] Recommended as best fluorescence standards.
[d] A small decrease in Φ_f (from 0.27 to 0.24) was observed upon deuteration of anthracene.

know the relative optical densities OD of the two compounds (subscripts 1 and 2) in solution at the excitation wavelength employed:

$$\frac{\Phi_2 \text{OD}_2}{\Phi_1 \text{OD}_1} = \frac{\Phi_2 \varepsilon_2 c_2 \, d I_0}{\Phi_1 \varepsilon_1 c_1 \, d I_0} = \frac{\text{area under fluor. peak 2}}{\text{area under fluor. peak 1}} \tag{2.14}$$

Here use is made of the approximation $\varepsilon c d I_0 \approx I_a$ for OD ≤ 0.05. Low concentrations of both compounds must be used to avoid absorption of the emitted light by an inner filter effect.

We see then that the relative fluorescence quantum yield can be determined by measuring the areas under the fluorescence bands of the sample and the fluorescent standard. However, these spectra must be corrected before their true areas can be determined. Several factors are responsible for this. The most important of these are the phototube and monochromator responses. For most phototubes the maximum response occurs within a limited wavelength range, falling off rather sharply in some cases at the short- and long-wavelength ends. This is illustrated in Figure 2.14. Similarly,

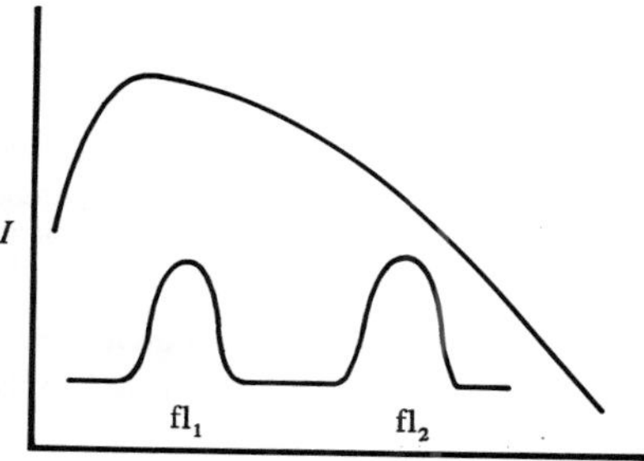

FIGURE 2.14. Phototube response.

monochromators do not pass all wavelengths with equal efficiency. Correction factors to compensate for these effects can be determined by using a standard (NBS) lamp.

If the refractive indices of the solvents used for the sample and the fluorescence standard are not the same, a further correction must be made. For example, quinine sulfate in 0.1 N H_2SO_4 ($\Phi_f = 0.5$) is commonly used as a fluorescence standard. If the fluorescence of the sample whose relative quantum yield is desired is determined in benzene, a correction factor of 27% must be applied in determining the relative areas under the fluorescence bands. If ethanol is used, this correction is only 5.5%.

Since the area under the curve is directly proportional to the intensity of the exciting light, one must correct by using relative intensities if two different exciting wavelengths are employed.

Finally, if the monochromator drive is linear in wavelength rather than in energy, a further correction is necessary. Since the number of corrections required for each point on a fluorescence spectrum is large, most workers, after determining the appropriate correction factors for their instrument, write a simple computer program to do the time-consuming spectrum correction. After the data are fed into the computer, it performs the required corrections and not only gives the quantum yield and band maximum of the fluorescence, but also the corrected spectrum, graphed in terms of cm^{-1}.

2.1d. *The Measurement of Fluorescence Lifetimes*

Once the fluorescence quantum yield has been determined, all that is required to calculate the fluorescence rate constant k_f is the fluorescence lifetime τ_f. Direct measurement of this quantity, like the measurement of the fluorescence quantum yield, is difficult, in this case because of the short lifetime of the fluorescent state (shorter than the normal flash from a flash lamp!). There are, however, several methods which have been developed to determine fluorescence lifetimes and these will be the subject of this section.

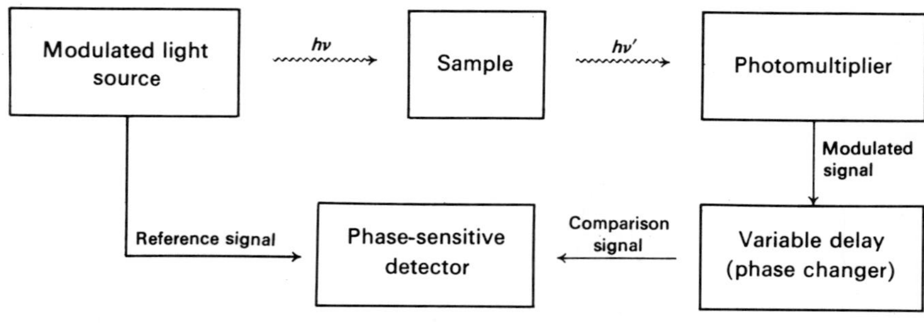

FIGURE 2.15

Phase Method

In this method the fluorescence is excited by an intensity-modulated light beam. Consequently, the fluorescence emission will also be modulated. Since there is a delay between excitation and emission corresponding to the fluorescence lifetime, the modulation of the emitted light will be out of phase with that of the exciting light. The phase of the fluorescence modulation relative to the exciting light is given by

$$\boxed{\omega\tau = \tan\phi} \tag{2.15}$$

where τ is the fluorescence lifetime, ω is the angular frequency ($2\pi\nu$ in cycles per second) of the exciting radiation, and ϕ is the phase angle between the exciting and emitted light.

A schematic diagram of the instrumentation[(6)] used to obtain fluorescence lifetimes by the phase method is given in Figure 2.15. There are a number of ways a system of this type can be set up. Modulation of the light source was originally achieved by using a Kerr cell, that is, an electrooptical shutter that could be pulsed to turn the excitation on and off rapidly. A later method used the reflection of light from ultrasonic standing waves. The best method and the one currently used is to modulate the light directly through the power supply to the lamp.

The variable delay can be as simple as an *RC* network. Often the variable delay line is calibrated directly in terms of lifetime units (nanoseconds). When the reference and comparison signals are in phase the fluorescence lifetimes can simply be read off the calibrated variable delay.

There are, however, problems associated with this method of determining fluorescence lifetimes. First, the phase method is not generally applicable for nonexponential signals and, as we shall see later, there are many cases where the observed fluorescence decay is indeed nonexponential. Second, the method is very sensitive to scattered light, which reduces the apparent fluorescence lifetime.

Single-Photon Counting Technique

This is the best method for determining fluorescence lifetimes. Developed only five or so years ago, this method has achieved wide acceptance.[(6)] A schematic diagram of the instrumentation used in this method is shown in Figure 2.16.

A pulsed light source is used in combination with two photomultipliers. The first photomultiplier observes the exciting sources directly. Its pulse is used to establish the zero time origin (the "on" cue) and to open the gate to the output from the second photomultiplier. Emission from the sample is received at the second photomultiplier as single photons (having passed through a tiny aperture in the light attenuator) and are converted into single-photon pulses. These pulses are fed into a pulse discriminator where pulses corresponding to spurious signals (noise) are discarded. The correct pulses are then fed through the gate (20 nsec window) to a time-to-height converter (supplying the "off" cue) and then to a multichannel analyzer where the pulses are counted as a function of time and are displayed.

The principal advantages of this technique are its very good time resolution, allowing the determination of lifetimes ranging from 10^{-6} to 10^{-10} sec, and the fact that single photons are counted. Thus good results can be obtained even with very weakly fluorescent materials.

TABLE 2.2

Compound	$\tau_f \times 10^9$, sec	Φ_f	$k_f \times 10^{-7}$, sec^{-1}	Solvent
9-Methylanthracene	5.2	0.29	5.45	EtOH
9,10-Dimethylanthracene	11.0	0.63	5.71	EtOH
9-Phenylanthracene	5.0	0.45	8.83	EtOH
9,10-Diphenylanthracene	6.8	0.84	12.40	EtOH
9-Chloroanthracene	2.8	0.11	3.94	EtOH
9-Bromoanthracene	1.1	0.017	1.54	EtOH
9,10-Dibromoanthracene	1.9	0.095	5.00	EtOH
9-Aminoanthracene	10.0	0.29	2.80	EtOH
9-Methoxyanthracene	3.9	0.17	4.35	EtOH
9-Anthroic acid	4.55	0.218	4.79	EtOH, H^+
9-Anthroic acid	12.3	0.646	5.24	Φ-CN
Methyl-9-anthroate	4.1	0.165	4.02	EtOH
9-Anthramide	2.5	0.100	4.00	EtOH
Sodium 9-anthroate	1.5	0.0689	4.59	H_2O
Sodium 9-anthroate	1.35	0.0554	4.17	D_2O

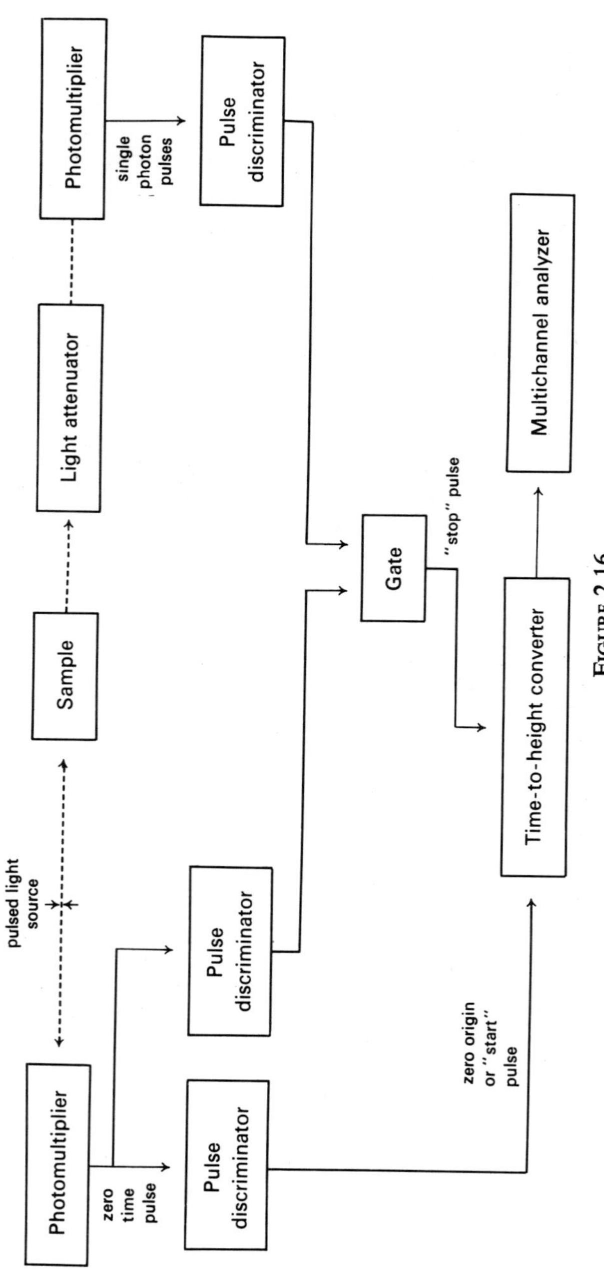

Photomultiplier
Light attenuator
Sample
pulsed light source
Photomultiplier
single photon pulses
Pulse discriminator
zero time pulse
Pulse discriminator
Pulse discriminator
Gate
"stop" pulse
zero origin or "start" pulse
Time-to-height converter
Multichannel analyzer

FIGURE 2.16

We have now seen how fluorescence quantum yields and lifetimes are experimentally determined, along with some of the strengths and weaknesses of the methods used. For anthracene these constants have been determined to be

$$\Phi_f = 0.22$$

$$\tau_f = 4.5 \times 10^{-9} \text{ sec}$$

$$k_f = \Phi_f/\tau_f = 5 \times 10^{7} \text{ sec}^{-1}$$

Data for various anthracene derivatives are given in Table 2.2.[7,8]

2.2. THE PHOTODIMERIZATION OF ANTHRACENE AND RELATED COMPOUNDS

The photodimerization of anthracene, having been first studied by Fritzsche in 1867 (two years after Kekulé proposed his revolutionary structure for benzene), was one of the first photochemical systems to be extensively investigated. Fritzsche found that upon exposure to sunlight, benzene solutions of anthracene yielded an insoluble substance which he called "Paraphoten." Observing that the photoproduct yielded anthracene upon melting, he concluded that he had obtained a polymer of anthracene.[9]

Further investigation of this interesting reaction carried out much later by other workers indicated that the photoproduct was not an anthracene polymer but was dimeric in nature. This conclusion was supported by the following data:

(a) Analyses for carbon and hydrogen yielded the empirical formula $C_{14}H_{10}$, the same as for anthracene.

(b) The molecular weight corresponded to twice that of anthracene.[10a,10b]

(c) The melting point of the product was greater than that of anthracene, but anthracene was produced upon melting.[11]

The structure of the dimer, correctly proposed by Linebarger in 1892,[12] is

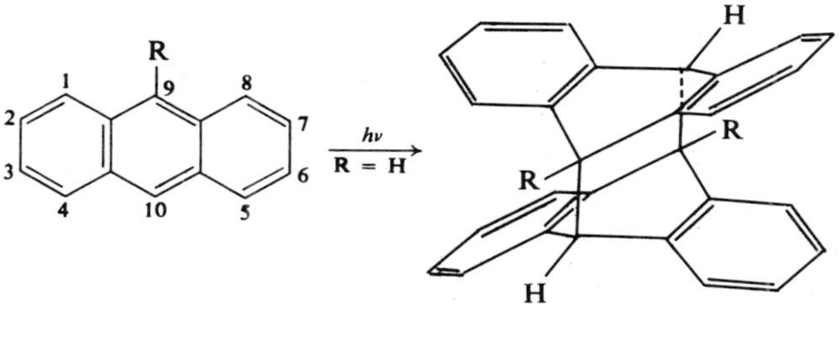

R, 1, 2, 3, 4, 9, 10, 5, 6, 7, 8 $\xrightarrow[R = H]{h\nu}$ H, R, R, H (2.16)

That dimerization occurs across the 9, 10 (*meso*) positions follows from the following experimental results:

(a) The ultraviolet spectrum of the photoproduct resembles that of 1,2-dimethylbenzene. The low-wavelength bands characteristic of anthracene are absent from the spectrum.(13)

(b) The nmr spectrum of the dimer shows a singlet at $\delta \simeq 5$ (bridgehead protons) and an aromatic multiplet at $\delta \simeq 7.3$ with integral ratios of 4:16 (1:4), respectively.

(c) Three-dimensional x-ray data indicate the photoproduct to be a dimer joined across the 9, 10 carbon atoms. The 9–9′, 10–10′ bond length was determined to be slightly longer than normal (1.62 versus 1.54 Å for a normal C—C bond) and the rings to be inclined at an angle of 23° in the plane normal to the molecular axis.(14)

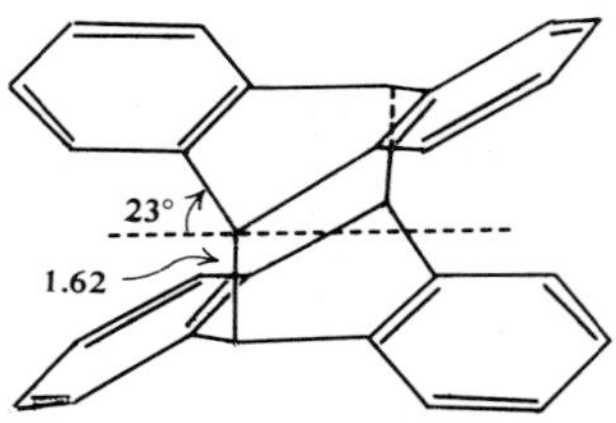

Anthracene photodimer

2.2a. *Structural Aspects: The Effect of Substituents on the Photodimerization*

Derivatives of anthracene bearing substituents on the 1 or 2 position can be photodimerized with efficiencies comparable to that for the unsubstituted molecule. However, with substituents at the 9 (*meso*) or 9, 10 (*dimeso*) positions a very interesting photochemical problem results. Since dimerization occurs across the 9, 10 positions, substituents at these positions exert a first-order effect on the photochemical reaction. The *meso*-substituted anthracenes examined include the following:(15–19)

Meso substituents yielding dimers:

—Me, —Et, —*n*Pr, —*i*Pr, —*n*Bu, —cyclohexyl, $—CH_2\phi$, —OMe, —OH,

$—\overset{O}{\overset{\|}{C}}H$, $—\overset{O}{\overset{\|}{C}}CH_3$, $—\overset{O}{\overset{\|}{C}}OH$, $—O\overset{O}{\overset{\|}{C}}CH_3$, —CN, $—NO_2$, —F, —Cl, —Br, —I

Meso substituents not yielding dimers:

$—\phi$, $—\overset{O}{\overset{\|}{C}}\phi$, $—O^-$

It can be seen that a large variety of 9-substituted anthracenes dimerize upon irradiation. There are a few, however, from which no dimers have yet been isolated. Although at this point it is difficult to say what determines whether a particular derivative will dimerize or not, it would appear from the above lists that the controlling factor cannot be steric in nature since the relatively crowded 9-cyclohexyl anthracene dimerizes whereas 9-phenylanthracene does not. This result would tend to indicate that electronic effects of the substituents may influence the dimerization.

In all of the cases in which dimerization has been observed, the yield of dimer formed, ranging from 40 to 70% under similar conditions, is less than that observed for anthracene. It should be recalled, however, from Chapter 1 that the yield is not a direct measure of the reactivity since it is dependent upon all other competing processes.

For *meso*-substituted anthracenes there are two possible stereöisomeric products, the head-to-head dimer and the head-to-tail dimer. All known

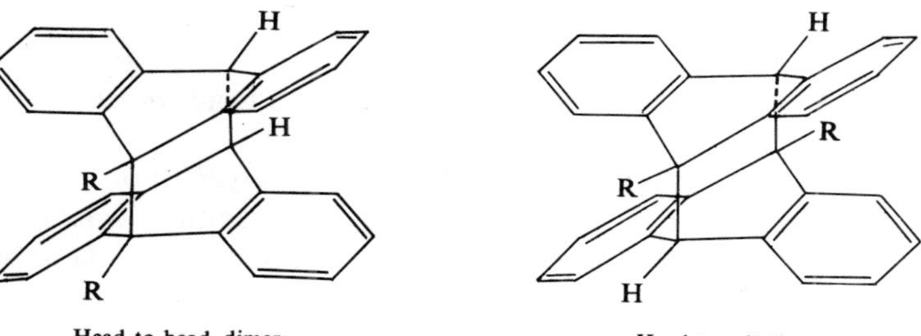

Head-to-head dimer Head-to-tail dimer

dimers formed intermolecularly, however, have been found to have the head-to-tail structure shown above.

The structure of the dimers from *meso*-substituted derivatives was initially determined by comparison of the observed and calculated dipole moments. For a head-to-tail dimer the dipole moments resulting from the 9 and 9′ substituents should cancel each other and the resultant dipole moment should be essentially zero. For a head-to-head arrangement the dipoles would be in the same direction and the resultant should be considerably greater than zero. The dimers produced upon irradiation of 9-chloro and 9-bromoanthracene solutions were observed to be 0.36 and 0.60 D, respectively. Since these values are much less than expected for a head-to-head arrangement for these derivatives (3.8 D), it was concluded that both of these dimers were formed in a head-to-tail configuration.[20]

Direct chemical evidence for the head-to-tail structure was furnished in 1959 by Applequist *et al.* using a stereospecific elimination,[21]

Br

$\xrightarrow[R\ =\ Br]{h\nu}$

H R R H

(2.17)

Zn, $LiAlH_4$, RLi
(transannular elimination)

H H

Earlier chemical evidence used to suggest that a head-to-head dimer was formed is incorrect.[20]

Further evidence was deduced by an NMR study.[21] For 9-substituted derivatives dimerization in a head-to-head configuration would yield a product with protons on adjacent bridgehead carbons. Since these protons would be in identical environments, this would correspond to an A_2 system with coupling equal to J_{AA} (expected to be 6–10 Hz in these eclipsed bridgehead positions). The J_{AA} for the head-to-tail isomer would be equal to zero since the bridgehead protons are now on opposite sides of the molecule. Since J_{AA} cannot be measured from the NMR of an A_2 system, a ^{13}C—H NMR spectrum was run using the natural abundance of ^{13}C ($\sim 1\%$). The head-to-head isomer now becomes an A_2X pattern, with X corresponding to ^{13}C. Thus J_{AA} can be determined from this experiment; it was observed that $J_{AA} = 0$ with $J_{^{13}C-H} = 136$ Hz. The head-to-tail structure of the dimer was therefore again confirmed.

Disubstituted Anthracenes[22]

In general, symmetrically substituted di*meso*-anthracenes produce only very low yields of dimers upon irradiation. On the other hand, asymmetrically substituted derivatives dimerize, although these products are less stable than those formed from the mono-substituted anthracenes.

Of the derivatives in Table 2.3, the dimer arising from the 9-methyl-10-methoxyanthracene has been found to be the most stable. Dimers produced from the others are quite thermally labile.

TABLE 2.3

Di*meso* derivatives yielding dimers		Di*meso* derivatives not yielding dimers	
9-R	10-R	9-R	10-R
—CN	—OC(=O)CH$_3$	—ϕ	—ϕ
—Me	—OMe	—OMe	—OMe
—Me	—OC(=O)CH$_3$	—CN	—CN
—Et	—CN	—	—

The stereochemistry of the dimers produced from the di*meso*-anthracenes is assumed to be *trans*, analogous to the head-to-tail dimers observed for the *meso* derivatives:

OMe / Me anthracene $\xrightarrow[R\,=\,Me;\ R'\,=\,OMe]{h\nu}$ dimer (R, R′) (2.18)

Again it appears that electronic factors (e.g., a polar transition state) are important in determining the dimerization.

Mixed Dimers

Photoproducts from mixtures of various substituted anthracenes in solution have been isolated and characterized. Some of these are given in Table 2.4.[21,24–26]

When photolyzing a mixture of compounds it is in principle possible to obtain photoproducts corresponding to dimerization of each of the

TABLE 2.4

9-R	10-R	9′-R	10′-R
—Me	—	—	—
—Br	—	—	—
—Me	—	—Et	—
—Me	—OMe	—	—
—Me	—Me	—CN	—
—Me	—Me	—OMe	—OMe

components with itself and also to addition of unlike derivatives (cross-dimerization). For example, if one irradiates a solution containing 9,10-dimethylanthracene and anthracene-9-carbonitrile, one might expect to obtain a mixture of dimers containing a tetramethyl derivative, corresponding to dimerization of the dimethylanthracene, a dicyanoderivative, corresponding to dimerization of 9-cyanoanthracene, and a dimethyl-cyano derivative, corresponding to the cross-dimerization of the two molecules. Only the mixed dimer is obtained[27]:

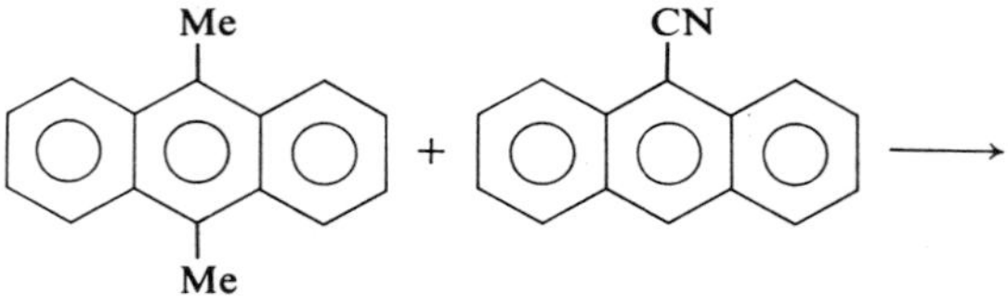

A + B ⟶ AB (99.5%) + B_2 (0.5%) (2.19)

Anthracene Photoisomers

The photodimerization of anthracene has been utilized to produce a number of interesting synthetic derivatives which are essentially photoisomers of the starting materials,[29–35]

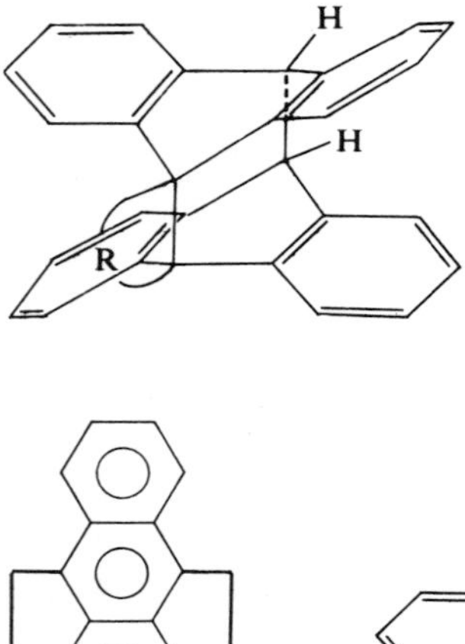

R = —CO_2CO—, —CH_2CH_2—, —CHOH—, —CONH—, —OCO_2—, —N=N—, —OC(=O)O— (2.20)

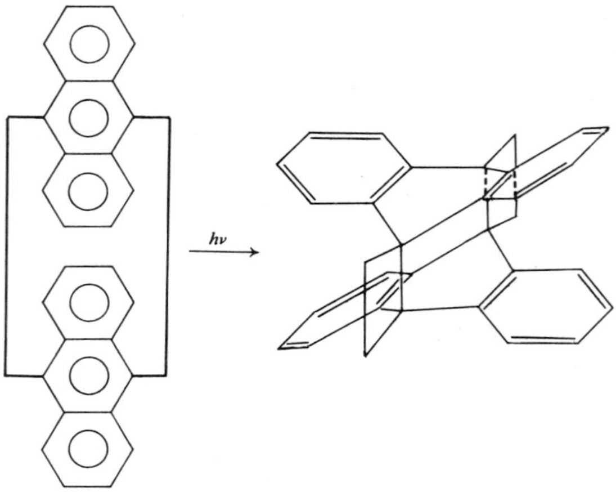

$\xrightarrow{h\nu}$ (2.21)

The intermediate formation of a dianthracene derivative was utilized in the synthesis of the highly strained olefin 9,9′-dehydrodianthracene,[37]

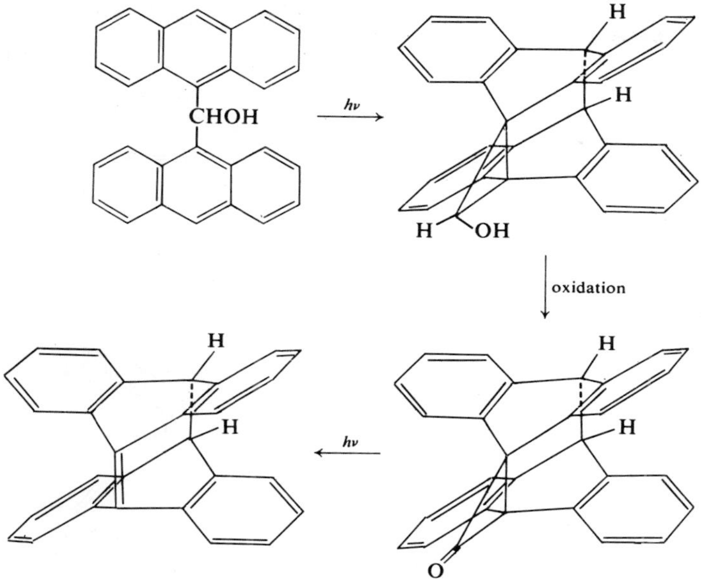

(2.22)

The Photoaddition and Photodimerization of Other Compounds

A number of other aromatic compounds have been observed to undergo photoaddition to anthracene to yield products similar in structure to the anthracene dimer:

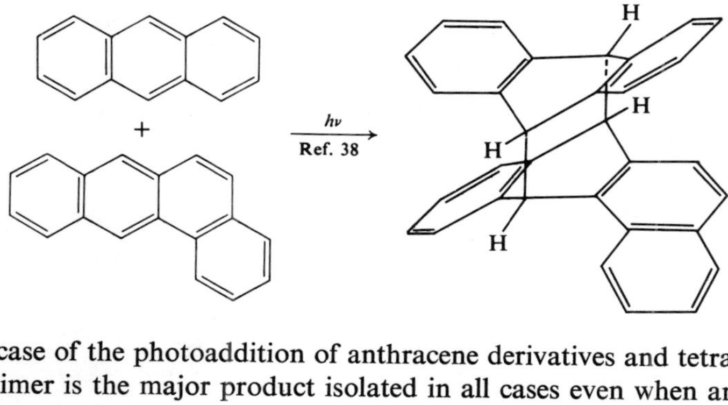

(2.23)

In the case of the photoaddition of anthracene derivatives and tetracene, the cross-dimer is the major product isolated in all cases even when anthracene itself is used.

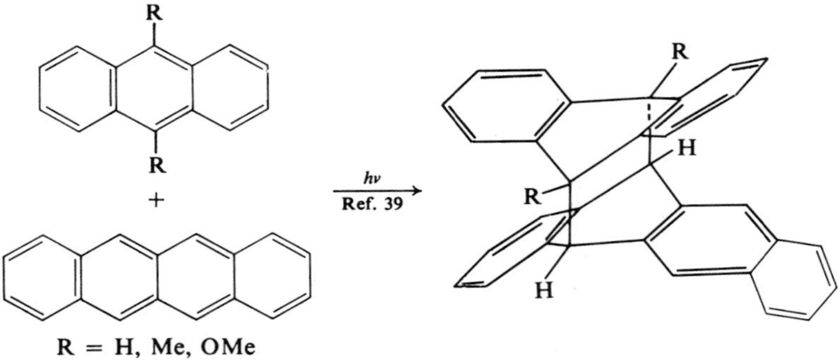

R
R
+
R = H, Me, OMe

$\xrightarrow[\text{Ref. 39}]{h\nu}$

R
H
R
H

(2.24)

Each of the following aromatic compounds is known to undergo an anthracene-like dimerization:

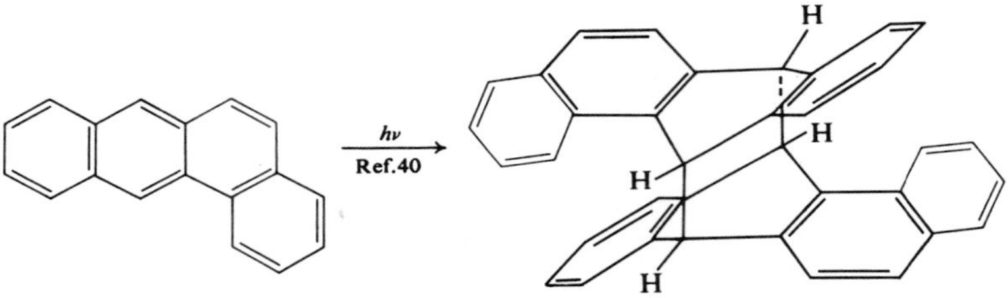

$\xrightarrow[\text{Ref.40}]{h\nu}$

H
H
H
H

(2.25)

$\xrightarrow[\text{Ref. 41}]{h\nu}$

H
H
H
H

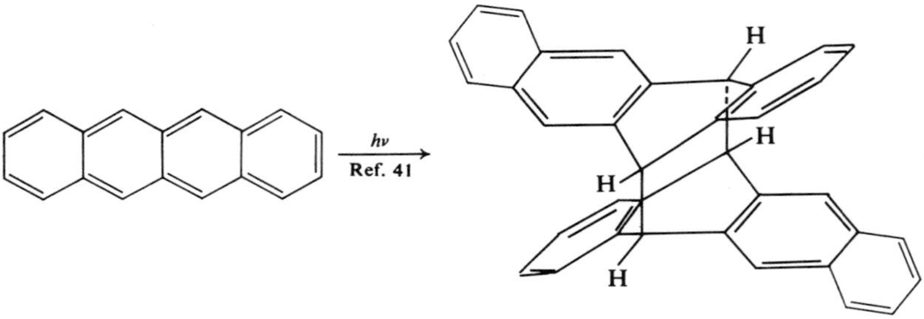

(2.26)

In fact, one does not need an anthracene-like nucleus to observe similar photochemical behavior[42–45]:

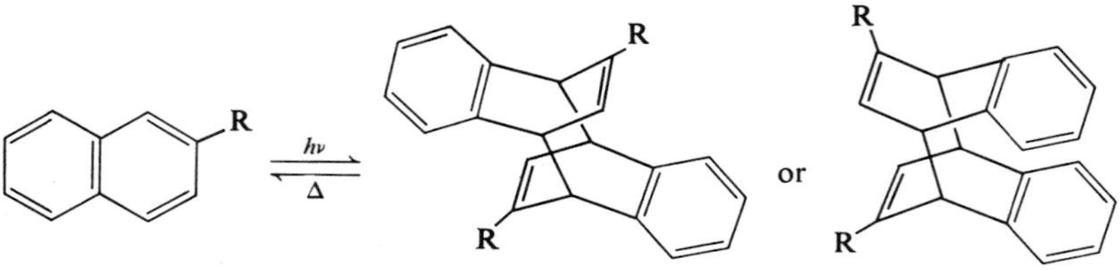

R

$\underset{\Delta}{\overset{h\nu}{\rightleftharpoons}}$

R
R
or
R
R

R = OCH_3, OCH_2CH_3, CN

(2.27)

Even some single-ring heterocycles undergo the dimerization:

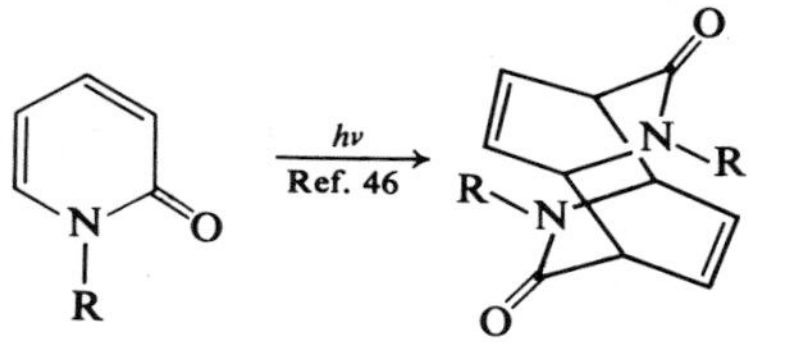

$h\nu$ Ref. 46 (2.28)

$h\nu$ Ref. 47 (2.29)

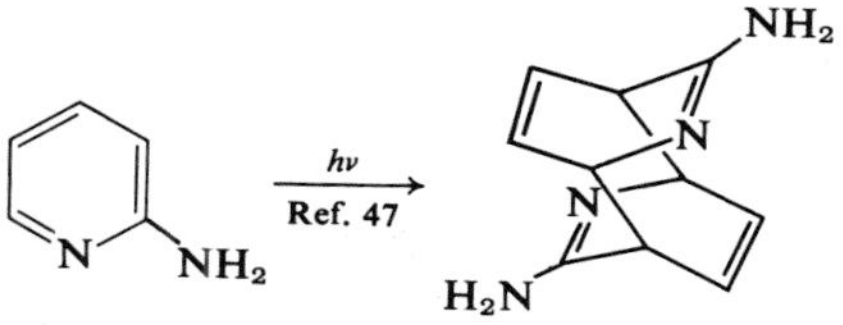

$h\nu$ Ref. 48 (2.30)

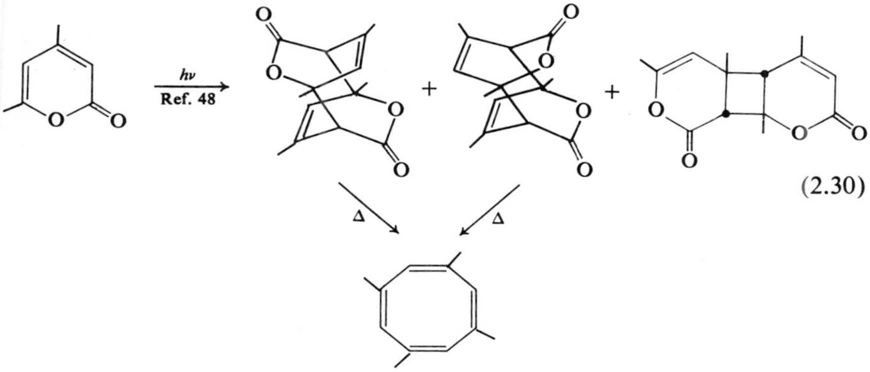

Recently the photoaddition of anthracene and simple dienes was discovered[49]:

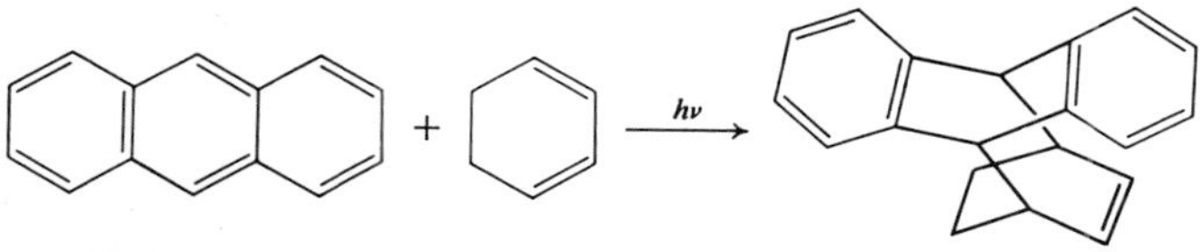

$h\nu$ (2.31)

The stereospecific addition of a *trans–trans* diene below indicates that the addition is concerted[50]:

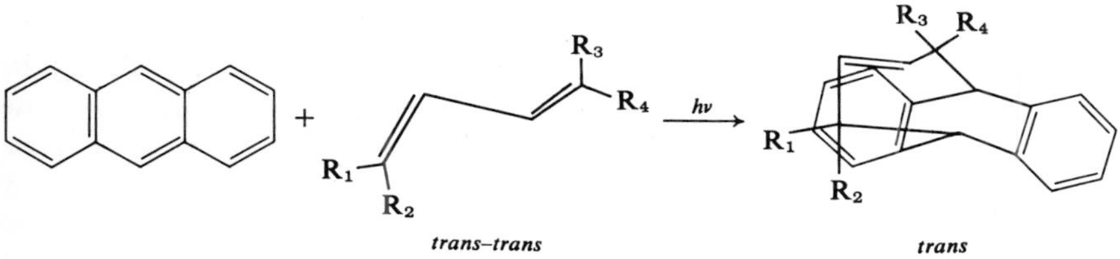

trans–trans *trans*

(2.32)

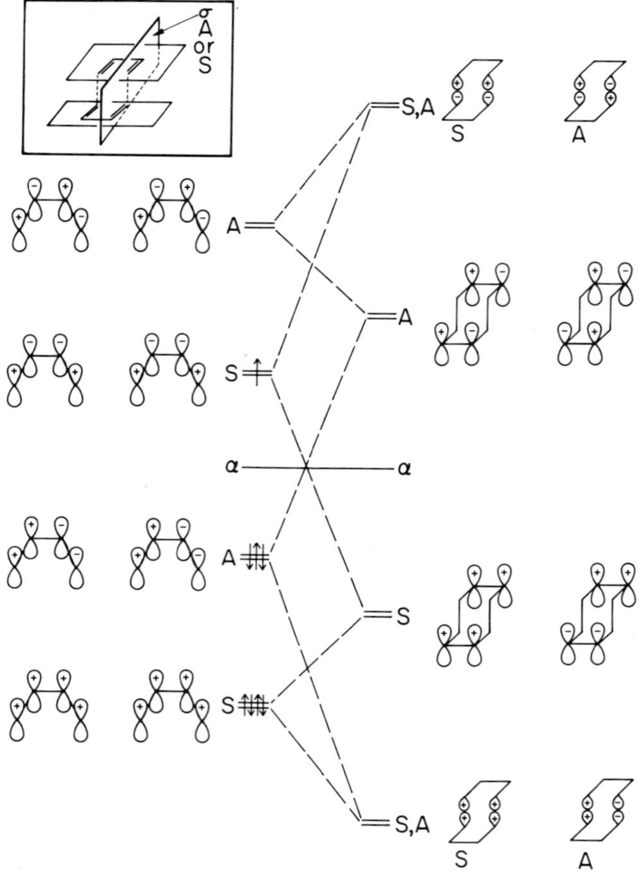

FIGURE 2.17. Reproduced with permission from Ref. 65.

While Woodward–Hoffman diagrams will be discussed in detail in a later chapter, the thermally forbidden, photochemically allowed $4\pi + 4\pi$ electrocyclic addition is shown in Figure 2.17. All orbitals undergoing change are classified using the only symmetry element preserved in the transition state, the mirror plane σ. The orbitals of S symmetry on the left side of the figure are correlated with the orbitals of S symmetry on the right side such that there is no crossing of orbitals with the S symmetry. The same type of correlation is carried out for A symmetry orbitals and eight electrons added such that either the ground state is obtained or the lowest excited state. If the electrons are added to form the ground state (left side of the figure) and we ask what will take place upon electrocyclic addition, we see that an excited state must be formed $[S^2A^2S^2(S^0)A^2]$ and this would require a large amount of energy. However, from the lowest excited state, the electrocyclic addition can occur to form an excited product $[S^2A^2S^2S^1A^1]$ with little cost in energy.

Summary

1. All of the photodimerizations and photoadditions mentioned up to this point have one common feature. That is, they result from the addition of two 4π-electron systems:

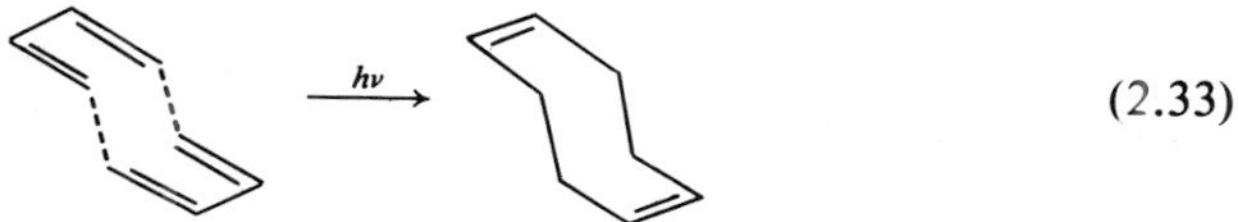

(2.33)

However, as we have seen, all derivatives containing 4π-electron systems do not behave similarly—some yield dimers and some do not, even though all should be equally allowed according to the Woodward–Hoffmann rules.

Many of the reactions have not been studied in detail. Therefore the following points apply only to the anthracene system, for which more is known.

2. *Meso* substitution generally lowers the yield of dimer produced, although dimerization still occurs. The three derivatives currently thought to be exceptions are 9–ϕ, $9\text{-}\overset{\overset{\text{O}}{\|}}{\text{C}}\text{-}\phi$, and $9\text{-}O^{\ominus}$. These are compounds where no dimer has been isolated from the photoreaction. The reasons for the low reactivity of these derivatives cannot be entirely steric but must be electronic in nature. However, a word of caution is required. The instability of a product may be responsible for failure to isolate dimer from these reactions inasmuch as other derivatives formerly thought to be photochemically inactive have been found to yield dimer when care has been taken to ensure product stability during isolation.

3. All of the *meso*-substituted photodimers have been found to have the head-to-tail structure.

4. Unsymmetrical di*meso* derivatives dimerize while symmetrical di*meso* derivatives give very low yields of dimer.

5. In the photolysis of mixtures of derivatives, the cross-dimer is produced as the major product.

Points 2–5 tend to indicate that electronic effects are important in the dimerization. More will be said about this after the kinetic data for the dimerization are added to the picture. First, however, let us see how preparative photochemical reactions are carried out.

2.2b. *Preparative Photochemical Techniques*

While there are perhaps as many possible ways of performing photochemical syntheses as there are photochemical reactions, only two basic designs for photochemical reactions are in wide use today. One design utilizes an external light source, while the other has the light source immersed in the solution being irradiated. For syntheses involving excitation with light of wavelengths in the visible region of the spectrum, an external source is commonly employed. Convenient sources of visible radiation are tungsten–iodide lamps and the high-intensity incandescent projection lamps used in most modern slide projectors. This latter source is especially convenient since the lamp can be cooled by the projector fan and the light can be focused by the projection lens. For wavelengths in the ultraviolet region the light sources most commonly employed are low-, medium-, or high-pressure mercury vapor lamps. The low-pressure mercury lamp emits radiation primarily as narrow bands centered at 1847, 2482, and 2537 Å. As the pressure is increased, the low-pressure resonance lines are reversed (dark bands) due to reabsorption of the emission by mercury and a number of secondary, pressure-broadened lines appear (3130, 3650, 4045 Å). As the pressure is increased even more, the emission appears as a continuum with the pressure-broadened mercury lines superimposed. These lamps can be used either as internal or external light sources.

The energy distribution for the commonly used 450-W Hanovia medium-pressure mercury lamp is given in Table 2.5.

Light Source External to the Reaction Vessel

The simplest and of course the most inexpensive external light source is the sun. This source, however, is not always as dependable and reproducible as the photochemist would like. Alternatively, a sun lamp, whose

TABLE 2.5. *Energy Distribution for a Hanovia 450-W Lamp*

Wavelength, Å	Energy, W	Wavelength, Å	Energy, W
Ultraviolet		Visible	
2482	2.3	4045	11.0 violet
2537 (reversed)	5.8	4358	20.2 blue
2652	4.0	5461	24.5 green
2804	2.4	5780	20.0 yellow
2894	1.6		
2967	4.3	Infrared	
3025	7.2	10140	10.5
3130	13.2	11287	3.3
3341	2.4	13673	2.6
3660	25.6		

FIGURE 2.18. Reproduced with permission from The Southern New England Ultraviolet Co.

emission closely resembles that of the sun ($\lambda > 3000$ Å), may be employed. Usually the photochemist sacrifices the economy of these sources in favor of the greater reproducibility of the more expensive sources currently available. A typical external source involves a series of parallel low-pressure mercury vapor lamps that ring the reaction vessel as shown in Figure 2.18. The Rayonet Photochemical Reactor (PPR-100) shown in Figure 2.18 uses 16 low-pressure mercury lamps to obtain high intensities of radiation consisting almost entirely of 2537-Å light. In addition, the manufacturer offers lamps coated with phosphors that absorb the 2537-Å light emitted by mercury and reemit the energy in the 3000–3600-Å range.

One disadvantage to using external light sources is that for irradiations using 2537-Å light, a Pyrex reaction vessel cannot be used since Pyrex absorbs strongly at wavelengths less than 3000 Å. Therefore more expensive quartz vessels must be used.

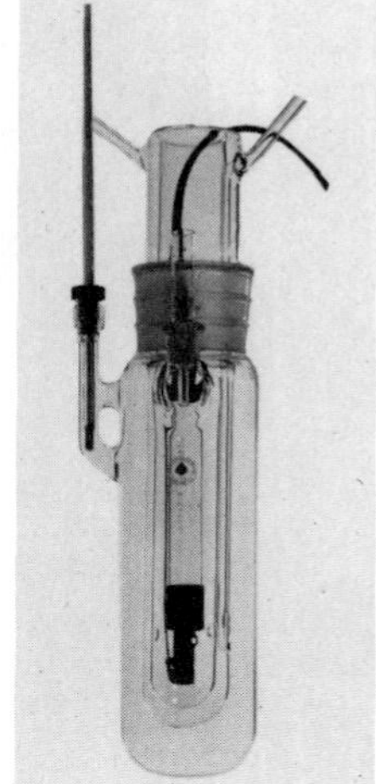

FIGURE 2.19 FIGURE 2.20

Reproduced with permission from Ace Glass.

Light Source Immersed in the Reaction Vessel

Probably the most widely used apparatus for preparative photolyses is the Hanovia immersion well, water-cooled jacket, filter sleeve, and 450-W medium-pressure mercury lamp combination shown in Figures 2.19 and 2.20. This unit has the advantages that (a) light does not have to pass through the walls of the reaction vessel, so Pyrex containers of various sizes and configurations can be used, (b) the solution and lamp can be kept relatively cool by the circulating water, (c) the light can be filtered to eliminate undesired high-energy wavelengths by using various filter sleeves, and (d) the intensity of the radiation emitted by the source is fairly constant over a long period of time.

Points to Consider Before Carrying Out a Photochemical Reaction

In order to avoid loss of time and possible erroneous conclusions regarding the results of a photolysis, the following points should be considered before a photochemical reaction is attempted.

1. Does the compound absorb at the wavelengths emitted by the light source to be used?

TABLE 2.6. *Filter Sleeves for Hanovia Immersion Apparatus*

Type	High-energy cutoff, nm
Vycor (7910)	220
Corex (9700)[a]	270 (~220)[a]
Pyrex (7740)	290
Uranium glass	330

[a] Maleski and Morrison(91) report a change in the properties of Corex.

2. Is the solvent transparent at wavelengths absorbed by the solute?

3. Does the product, if known, also absorb at these wavelengths? This point is important to consider since as a product is formed it will compete with the starting material for light (filter effect) and perhaps complicate the reaction by itself undergoing a photochemical reaction. Even if the product is itself photochemically inactive, the time to achieve complete reaction will be greatly prolonged since the effective light intensity is reduced by the product absorption. If possible, it is best to filter the light so that wavelengths absorbed by product are removed. The cutoff points for several commonly used filter sleeves are given in Table 2.6.

4. Is the product thermally labile? Most reactions using an immersion source are run at room temperature. Since fairly large amounts of infrared radiation are emitted from the light source, cooling is necessary.

5. Should oxygen be excluded by flushing with nitrogen or argon or by degassing with several freeze–thaw cycles? The presence of oxygen in the solution is an important factor. As will be seen later, oxygen often is an efficient quencher of the excited state leading to product. If the effect of oxygen on the reaction is not known, it is best to exclude the possibility of its playing a part by removing the dissolved gas by purging with an inert gas or by degassing.

6. Will the solvent react with the excited state to yield undesirable side-products? Often there is a real possibility that the solvent will enter into the picture through reaction with the excited solute. A common example of this is the abstraction of hydrogen atoms from solvents by excited ketones. Several solvents often used for a preliminary examination due to their relative inertness are benzene, *t*-butanol, carbon disulfide, carbon tetrachloride, and cyclohexane.

7. How long should the irradiation be allowed to proceed? Over-irradiation may produce secondary photolysis products as a result of primary product absorption. It is best in cases where the products are likely to absorb to monitor the reaction as a function of time.

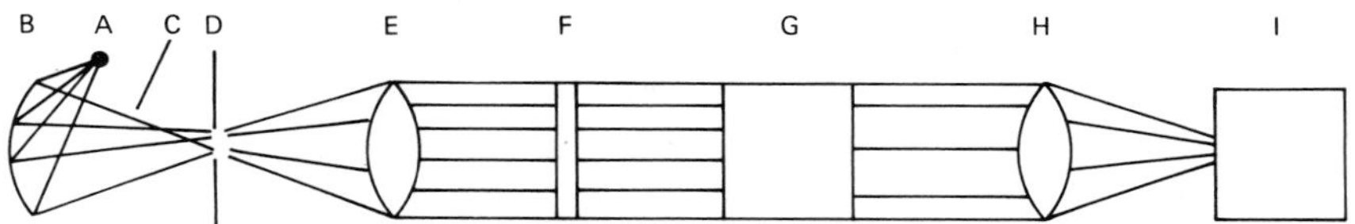

FIGURE 2.21. A, High-intensity point source lamp; B, parabolic mirror; C, light baffle; D, narrow slit; E, collimating lens; F, Corning filters; G, reaction cell or series of cells; H, focusing lens; I, photomultiplier.

The Determination of Quantum Yields

As we saw in Chapter 1, the quantum yield for a photochemical reaction is simply the amount of product produced per amount of light absorbed:

$$\Phi = \text{moles product/amount of light absorbed in einsteins}$$

There are two general types of experimental setup commonly used for the determination of photochemical quantum yields. The more elaborate of the two is the optical bench. A diagram of an optical bench with a good geometry is shown in Figure 2.21.

A simpler but less versatile apparatus that is becoming increasingly popular is the "merry-go-round" diagrammed in Figure 2.22.

With the optical bench, a monochromator and additional focusing and collimating lenses are often used in place of the filters to isolate a nearly monochromatic beam of light. The "merry-go-round" apparatus offers the advantage that several different solutions can be irradiated simultaneously with the assurance that the average incident intensities are equal. Generally with an optical bench only one or two samples can be irradiated at one time. The characteristics of a typical Corning Glass filter combination for the isolation of 365-nm radiation are shown in Figure 2.23.

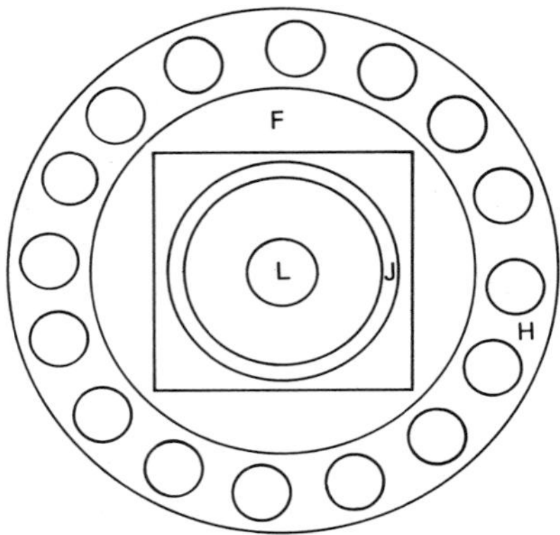

FIGURE 2.22. L, lamp; J, cooling jacket; F, filter holder; H, rotating cell holder.

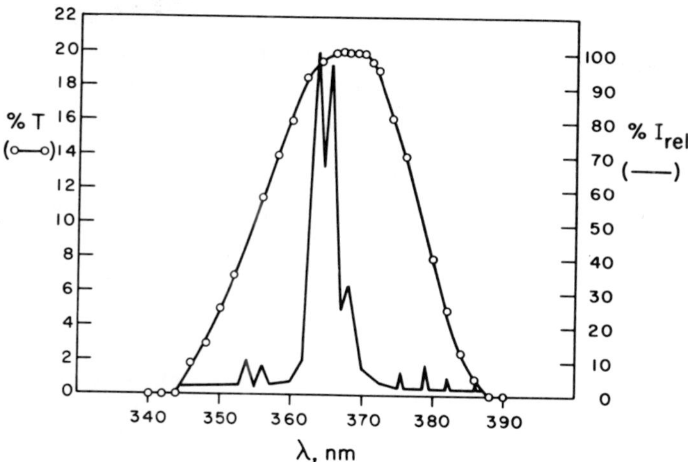

FIGURE 2.23. Percent transmission of sum of CS-7-37 and CS-O-52 filters and relative spectral intensity of Hg–Xe source 977-B-1 Hanovia lamp.

Measurement of the light intensity under conditions identical to those used in the photolysis of the compound of interest is essential for the determination of a quantum yield. Although a number of instrumental methods for measuring light intensities are available, unless these are carefully calibrated, the most accurate means is to use a chemical actinometer. This can be any photochemical reaction for which the quantum yield at the wavelength of interest is accurately known. The following photochemical systems are most commonly used for solution actinometry.

Uranyl Oxalate(51–53)

The uranyl ion (UO_2^{2+}) absorbs light in both the visible and ultraviolet spectral regions. In the presence of oxalic acid in excess of the uranyl ion concentration, the excited ion transfers its energy to the oxalic acid, which decomposes to form water, carbon dioxide, and carbon monoxide:

$$\begin{aligned} UO_2^{2+} + h\nu &\longrightarrow UO_2^{2+*} \\ UO_2^{2+*} + H_2C_2O_4 &\longrightarrow UO_2^{2+} + H_2C_2O_4^* \\ H_2C_2O_4^* &\longrightarrow H_2O + CO_2 + CO \end{aligned} \tag{2.34}$$

The amount of oxalic acid thus decomposed under conditions where all the light is absorbed by the uranyl ion is determined by titrating a sample of the solution with potassium permanganate before and after irradiation. Since the results depend upon the difference between two measurements, in order to obtain accurate values, fairly large conversions of oxalic acid must be made. The quantum yield for this actinometer solution is

λ, nm	470	436	365	260
Φ	0.40	0.573	0.492	0.58

With this actinometer reasonable exposure periods and solution volumes result with total absorbed intensities on the order of 10^{17} photons.

Potassium Ferrioxalate[54]

The most accurate solution actinometer currently available is the potassium ferrioxalate actinometer. Potassium ferrioxalate solutions absorb light in the range 250–509 nm. This broad range is both an advantage and a disadvantage since the solutions are sensitive to room light and must be carefully shielded from light until the intensity determination is made:

$$\begin{aligned}
[Fe^{III}(C_2O_4)_3]^{3-} &\xrightarrow{h\nu} [Fe^{II}(C_2O_4)_2]^{2-} + C_2O_4^{\dot{-}} \\
C_2O_4^{\dot{-}} + [Fe^{III}(C_2O_4)_3]^{3-} &\longrightarrow (C_2O_4)^{2-} + [Fe^{III}(C_2O_4)_3]^{2-} \\
[Fe^{III}(C_2O_4)_3]^{2-} &\longrightarrow [Fe^{II}(C_2O_4)_2]^{2-} + 2CO_2 \\
2[Fe(C_2O_4)_3]^{3-} &\xrightarrow{h\nu} 2[Fe(C_2O_4)_2]^{2-} + (C_2O_4)^{2-} + 2CO_2
\end{aligned} \quad (2.35)$$

overall reaction

The amount of ferrous ion produced is determined spectroscopically (at 510 nm) by formation of a highly colored complex with 1,10-phenanthroline.

The products of the photodecomposition of the ferrioxalate absorb only weakly at wavelengths absorbed by ferrioxalate and therefore provide no problem with regard to a product filter effect. This can be seen from Table 2.7.

If the products of an actinometer solution do interfere by competitive light absorption, one is limited to low conversion such that the concentration of product is much less than that of the initial reactant.

The quantum yield for the potassium ferrioxalate actinometer as a function of wavelength is shown in Table 2.8.

Lee and Seliger[54c] have estimated the error involved in the determination of the quantum yield for the ferrioxalate actinometer (at 365 nm) to be $\pm 2.5\%$. This then constitutes the minimum limit of error involved in the

TABLE 2.7

Wavelength, nm	400	350	300	250	230	200
$\varepsilon_{Fe(C_2O_4)_2^{2-}}/\varepsilon_{Fe(C_2O_4)_3^{3-}}$	0.02	0.01	0.01	0.04	0.05	0.38

TABLE 2.8. *Ferrioxalate Actinometer*

	Φ			
λ[a]	HP[b]	BB[c]	LS[d]	WA[e]
436	1.11	1.04	—	—
405	1.14	—	—	—
392	—	—	—	1.13
358	—	—	—	1.25
365	1.21	1.20	1.26	—
334	1.23	—	—	—
313	1.24	—	—	—
303	1.24	—	—	—
297	1.24	—	—	—
254	1.25	1.22	—	—

[a] $[Fe(C_2O_4)_3]^{-3}$ = 0.006 *M*.
[b] Data of Hatchard and Parker.[(54a)]
[c] Data of Baxendate and Bridge.[(54b)]
[d] Data of Lee and Seliger.[(54c)]
[e] Data of Wegner and Adamson[(54d)]

determination of the quantum yield of a photochemical reaction using ferrioxalate actinometry.

The quantum yield of an actinometer may be affected by temperature. For potassium ferrioxalate this temperature effect is very small, as indicated in Table 2.9.

Monochloroacetic Acid[(55)]

A chemical actinometer suitable for use at 254 nm is found in aqueous solutions of monochloroacetic acid,

$$ClCH_2COOH + H_2O \xrightarrow{h\nu} HOCH_2COOH + HCl \tag{2.36}$$

The amount of hydrochloric acid produced is determined by the addition of Ag^+. This reaction has been found to be quite temperature sensitive. For example, for λ = 254 nm

T, °C	25	56	69
Φ	0.31	0.61	0.69

Reinecke's Salt[(54d)]

The photoaquation of a transition metal coordination compound (Reinecke's Salt) has recently been developed as an actinometer for the

TABLE 2.9

λ	$\Phi(32°C)/\Phi(22°C)$
405	1.05
360	1.00
334	0.998
313	1.009
254	1.009

visible region of the spectrum. It is useful in the range 316–600 nm without a correction for transmitted light and out to 750 nm with corrections:

$$Cr(NH_3)_2(NCS)_4^- \xrightarrow[H_2O]{h\nu} Cr(NH_3)_2(NCS)_3(H_2O) + NCS^- \tag{2.37}$$

The free thiocyanate ions produced are determined spectroscopically by addition of Fe^{3+} and measurement of the absorbance at 450 nm. The quantum yield for this actinometer as a function of wavelength is given in Table 2.10.

Many other actinometer systems have been developed, although those described here cover a sufficiently wide region of the spectrum to meet most needs.

2.2c. *Kinetic and Mechanistic Aspects of the Anthracene Photodimerization*

We now address ourselves to the problem of accounting for the observations noted in Section 2.2a by building a mechanism for the dimerization of anthracene and related compounds.

Fluorescence as a Function of Concentration

We saw in Section 2.1c that fluorescence spectra are routinely determined by monitoring the emission at a right angle to the exciting light.

TABLE 2.10

λ	$[KCr(NH_3)_2(NCS)_4]$	Φ_{NCS^-}, 23°
360	0.003	0.388
416	0.008	0.310
452	0.010	0.311
504	0.005	0.299
570	0.004	0.286
585	0.010	0.270
600	0.025	0.276

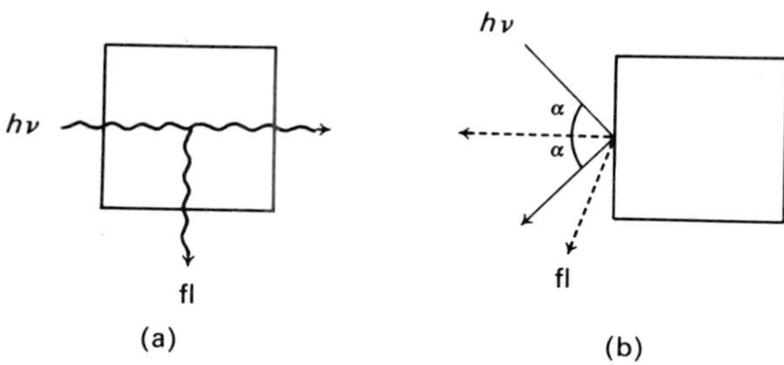

FIGURE 2.24. (a) Right-angle method; (*b*) front-surface method.

This method is perfectly suitable for low concentrations of fluorescent materials. However, in order to study factors which affect the fluorescence quantum yield, such as molecular association or photochemical reactions, much higher concentrations than can be used in the right-angle fluorescence method are required. This follows from the fact that the $0 \rightarrow 0$ vibrational bands in the absorption and emission spectra often overlap. Therefore at relatively high concentrations light emitted at these overlapping wavelengths will be reabsorbed.

A method has been devised by which fluorescence can be studied as a function of concentration. This is shown in Figure 2.24. The exciting light beam is introduced as an angle α to a line normal to the cell surface. Part of the beam is reflected back at angle α. Fluorescence is observed at an angle chosen to minimize the amount of reflected light. If the concentration of fluorescent molecules is high, essentially all of the light passed into the cell is absorbed in the first millimeter or so of solution. By making small changes in concentration one can thus determine the changes in observed fluorescence intensity. A plot of the reciprocal of the fluorescence intensity versus concentration is shown in Figure 2.25 for 9-anthroic acid.[63,64] At low concentrations the light beam travels too far into the cell to allow accurate determination of the fluorescence intensity. However, one can extrapolate to zero concentration to obtain the fluorescence intensity at infinite dilution. This intercept is by definition equal to $1/\Phi_f{}^0$ (where $\Phi_f{}^0$ corresponds to the fluorescence quantum yield determined by the right-angle method).[7] The significant feature to note in Figure 2.25 is that as the concentration of 9-anthroic acid is *increased*, the fluorescence intensity *decreases*. This indicates that some concentration-dependent process is competing with fluorescence to deactivate the 9-anthroic acid singlet state. Since we know that anthroic acid dimers are produced upon irradiation, this suggests (but does not prove) that the dimerization (concentration dependent) may occur by a singlet mechanism.

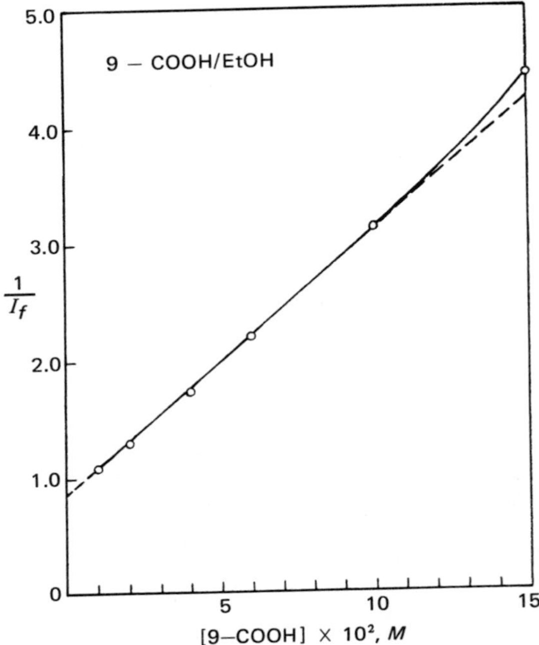

FIGURE 2.25. Reproduced with permission from Ref. 65.

The Dimerization Quantum Yield as a Function of Concentration

The dimerization of anthracene or its various derivatives can be conveniently followed spectroscopically by monitoring the reduction of the long-wavelength uv band as a function of time (the product absorbs only at much shorter wavelengths). If the intensity of the exciting light is known and is constant, one can obtain the quantum yield for dimerization. From the product study it is known that the quantum yield for the disappearance of anthracene is equal to twice that for the appearance of product (Φ_D). It has been found that a plot of $1/\Phi_D$ versus the reciprocal of the concentration for low conversions to dimer is a linear function. This type of graph is shown in Figure 2.26.[63] In this case the quantum yield increases with concentration.

Assuming for the moment that dimerization occurs from the excited singlet state, one can write the following simple mechanism[52–65,67]:

	rate
$A + h\nu \longrightarrow A^s$	I_a
$A^s \longrightarrow A + h\nu'$	$k_f[A^s]$
$A^s \longrightarrow A$	$(k_{isc} + k_{ic})[A^s]$
$A^s + A \longrightarrow \text{dimer}$	$k_D[A][A^s]$

From this mechanism we obtain

$$\Phi_D = k_D[A]/(k_f + k_{isc} + k_{ic} + k_D[A]) \tag{2.38}$$

$$1/\Phi_D = 1 + (k_f + k_{isc} + k_{ic})/k_D[A] \tag{2.39}$$

It can be seen that this mechanism allows a linear plot of $1/\Phi_D$ versus $1/[A]$ similar to Figure 2.26 with a slope of

$$\text{slope} = (k_f + k_{isc} + k_{ic})/k_D \tag{2.40}$$

and an intercept of unity.

However, in most cases the observed intercept is greater than unity. This indicates that if our mechanism is correct thus far, we must add another process by which singlet energy is lost. If, without going into any conjectures as to the nature of the process, we include a concentration quenching step,

$$A^s + A \longrightarrow 2A, \qquad \text{rate} = k_{cq}[A^s][A]$$

the kinetic expression becomes

$$\Phi_D = k_D[A]/(k_f + k_{isc} + k_{ic} + k_D[A] + k_{cq}[A]) \tag{2.41}$$

or

$$\frac{1}{\Phi_D} = \frac{k_D + k_{cq}}{k_D} + \frac{(k_f + k_{isc} + k_{ic})}{k_D[A]} \tag{2.42}$$

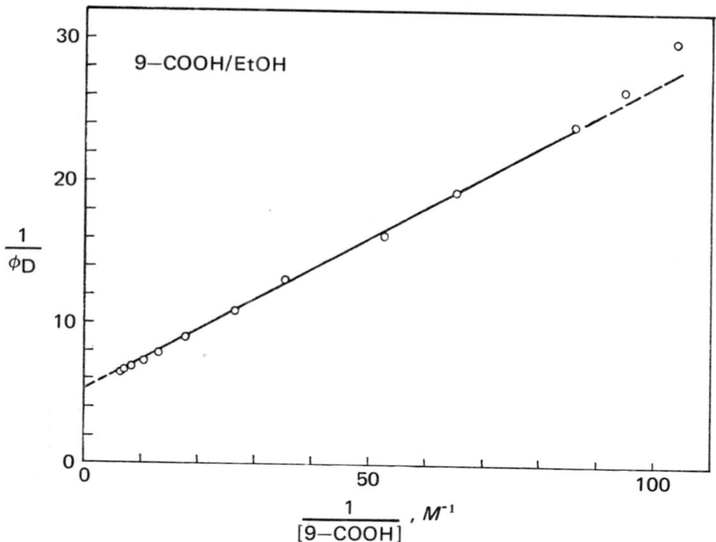

FIGURE 2.26. Reproduced with permission from Ref. 65.

We have now obtained an expression of the same form but where the intercept is dependent upon the magnitude of the rate constant k_{cq}. Furthermore, our slope is the same as before,

$$\text{slope} = (k_f + k_{isc} + k_{ic})/k_D \tag{2.43}$$

Since, as we have seen

$$\tau_f = 1/(k_f + k_{isc} + k_{ic}) \tag{2.44}$$

and

$$\text{slope} \times \tau_f = 1/k_D \tag{2.45}$$

the rate constant for dimerization can be determined. In addition, k_{cq} can be determined from the intercept:

$$\text{intercept} = (k_D + k_{cq})/k_D \tag{2.46}$$

It appears therefore that our proposed mechanism is sufficient to explain the observed experimental results. However, if we had assumed dimerization from a *triplet* state precursor,

$$\begin{array}{rcll} A + h\nu & \longrightarrow & A^s & I_a \\ A^s & \longrightarrow & A + h\nu' & k_f[A^s] \\ A^s & \longrightarrow & A^t & k_{isc}[A] \\ A^t + A & \longrightarrow & \text{dimer} & k_D[A][A^t] \\ A^t & \longrightarrow & A & k'_{isc}[A^t] \end{array}$$

we would have obtained the expression

$$\Phi_D = \Phi_{isc} \frac{k_D[A]}{k'_{isc} + k_D[A]} \tag{2.47}$$

where the first term on the right represents the efficiency of triplet formation and the second term represents the way the triplet energy is partitioned. Thus

$$\frac{1}{\Phi_D} = \frac{k_D[A] + k'_{isc}}{\Phi_{isc}k_D[A]} = \frac{1}{\Phi_{isc}} + \frac{k'_{isc}}{\Phi_{isc}k_D[A]} \tag{2.48}$$

It is seen that this expression has the same form as that obtained from the singlet dimerization mechanism although the slope and intercept in the kinetic expressions for these two mechanisms have different meanings:

$$\text{intercept} = 1/\Phi_{isc}, \qquad \Phi_{isc} = k_{isc}/(k_{isc} + k_f + k_{ic}) \tag{2.49}$$

$$\text{slope} = k'_{isc}/\Phi_{isc}k_D \tag{2.50}$$

This illustrates the point that although a mechanism can be devised to fit the experimentally observed details, *this does not prove that the reaction proceeds through that particular mechanism.*

If, however, we compare the expressions obtained for the fluorescence and dimerization data for the singlet mechanism,

dimerization
$$\frac{1}{\Phi_D} = \frac{k_D + k_{cq}}{k_D} + \frac{k_{isc} + k_{ic} + k_f}{k_D[A]} \tag{2.51}$$

fluorescence
$$\frac{1}{\Phi_f} = \frac{k_f + k_{isc} + k_{ic}}{k_f} + \frac{(k_D + k_{cq})[A]}{k_f} \tag{2.52}$$

$$\frac{1}{\Phi_f} = \frac{1}{\Phi_f^0} + \frac{(k_D + k_{cq})[A]}{k_f} \tag{2.53}$$

we see that with knowledge of Φ_f^0 and τ_f (and therefore k_f and k_D) we can obtain the sum $k_D + k_{cq}$ independently from the *intercept* of the dimerization plot and also from the *slope* of the fluorescence plot. If the sums obtained via the experimental data from these two methods are the same, the *reaction* must proceed from the singlet state! For the dimerization of 9-anthroic acid k_D calculated from the slope of the fluorescence plot equals $108 \times 10^7\ M^{-1}$ sec^{-1}, while k_D calculated from the intercept of the dimerization plot equals $102 \times 10^7\ M^{-1}\ \text{sec}^{-1}$.[(63–65)] Thus it appears that the dimerization of anthracene and its derivatives arises from the excited singlet state.

The Concentration Quenching Process

We should now examine the nature of the concentration quenching process that we proposed in our singlet dimerization mechanism. There are a number of possibilities, as follows.

(a) *Decomposition of a Vibrationally Excited Ground State Dimer.* It is important to note that the singlet excited anthracene molecule must react with a ground state anthracene molecule to produce a ground state dimer,

$$A^s + A \longrightarrow \text{dimer}$$

The alternative

$$A^s + A \longrightarrow \text{dimer}^s$$
$$\text{dimer}^s \longrightarrow \text{dimer}$$

is not possible since the singlet energy of the dimer, being essentially like that of benzene, is considerably greater than that available from the anthracene singlet. If, however, the energy available from the excited singlet is more than that necessary to form the bonds in the dimer, the dimer will be produced in a vibrationally excited ground state,

$$A^s + A \longrightarrow \text{dimer(vibrationally excited)}$$

Since anthracene dimers split to form anthracene monomers upon melting, it may be possible that if the excess vibrational energy in the "hot" dimer is not dissipated by solvent collisions[66] rapidly enough the bonds may rupture and monomers may be produced,

$$\text{dimer(vibrationally excited)} \longrightarrow 2\text{A}$$

(b) *Diradical Formation.* If the dimerization process proceeds through an intermediate diradical, there are two possibilities for *meso*-substituted anthracenes:

$$\text{A}^s + \text{A} \longrightarrow [\text{diradical I}] \text{ or } [\text{diradical II}] \tag{2.54}$$

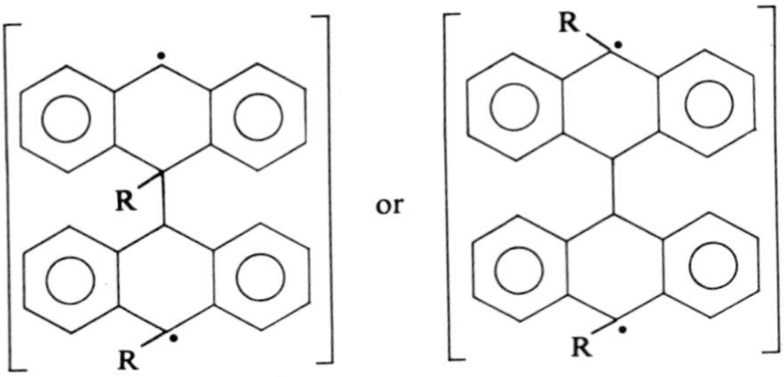

If the first diradical were to close to form dimer, the head-to-tail product would be obtained. This indeed is the product that is experimentally observed. If, on the other hand, the second diradical were to close, the head-to-head dimer would be produced. Since no head-to-head product is observed, it is logical to postulate that *if* this diradical were formed, it must decompose to ground state monomers,[63]

$$\text{diradical} \longrightarrow 2\text{A} \tag{2.55}$$

(c) *Excimer Mechanism.* An excimer is defined as two like molecules which have a common bound excited state but unbound ground states.[69,70] This is illustrated in Figure 2.27. Why excimers are formed can be seen in Figure 2.28. For a bound ground state, the energy levels of the two molecules would be split into two levels ΔE above and below the ground state energies of the isolated molecules. Since there would be two electrons in each of these states, the net gain in stabilization energy would be zero. For the excited state, however, two electrons are placed in the lower level and only one in the higher. The excited singlet, furthermore, is split and one electron is lowered in energy by the amount $\Delta E'$. Thus there is a net gain in stabilization energy for the bound excited state or excimer, $\Delta E + \Delta E'$.

Excimers are often characterized by a broad emission band containing no vibrational structure, occurring at longer wavelengths than emission corresponding to the monomeric singlet state.[41,67–69,71–73]

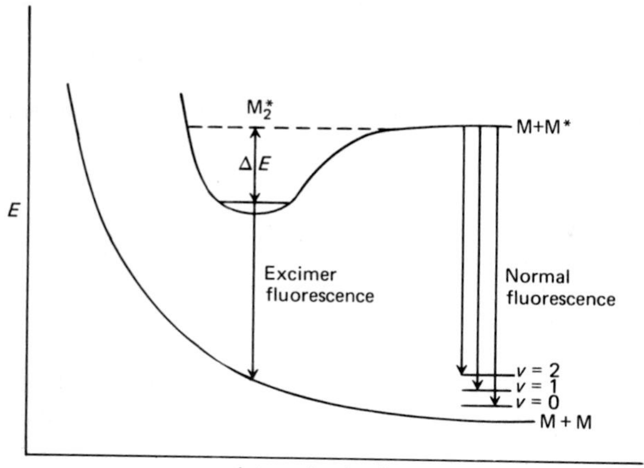

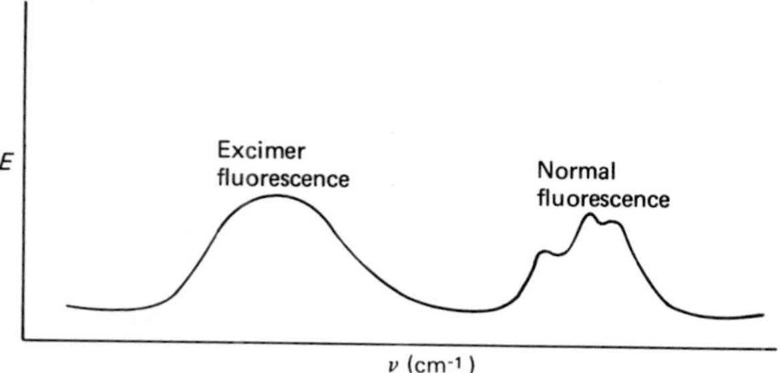

FIGURE 2.27

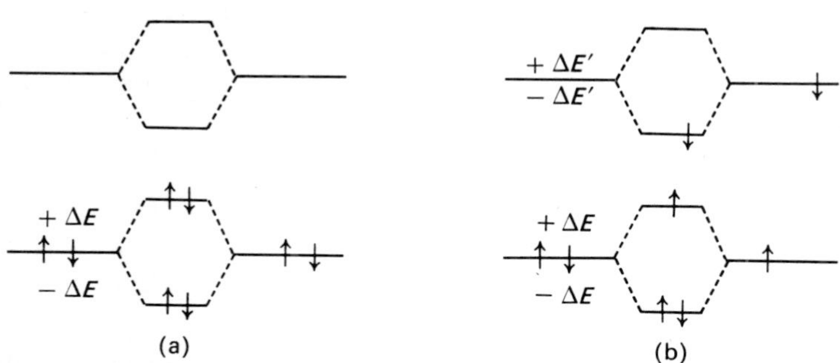

FIGURE 2.28. (a) Unbound ground state; $\Delta E_T = 2(\Delta E) - 2(\Delta E) = 0$. (b) Bound excited state; $\Delta E_T = -2(\Delta E) + \Delta E - \Delta E' = -(\Delta E + \Delta E')$.

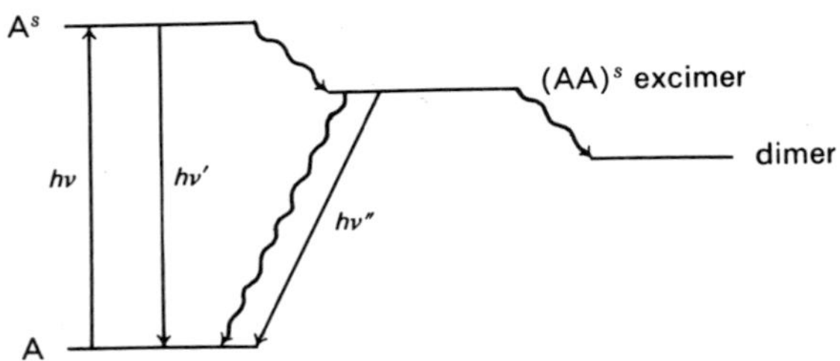

FIGURE 2.29

For anthracene the concentration quenching process could be the decomposition of the excimer to two ground state anthracenes, as shown in Figure 2.29[73,74,76]:

			rate
$A^s + A$	$\longrightarrow$	excimer	$k_E[A^s][A]$
excimer	$\longrightarrow$	dimer	$k_D[E]$
excimer	$\longrightarrow$	2A	$k_{cq}[E]$

Alternatively, one could write

			rate
$A^s + A$	$\longrightarrow$	dimer	$k_D[A^s][A]$
$A^s + A$	$\longrightarrow$	excimer	$k_E[A^s][A]$
excimer	$\longrightarrow$	2A	$k_{cq}[E]$

In the first mechanism the excimer is represented as a common intermediate for the formation of dimer and the deactivation of the excited anthracene. In the second, excimer formation is totally nonproductive with regard to dimer formation. Again as in paragraph (b) one can think of the excimer in the second mechanism as having a structure that, if dimerization proceeded, would yield the unobserved head-to-head product:

$$A^s + A \longrightarrow \text{excimer(h–h)} \\ \downarrow \\ 2A$$

If a common state (excimer) is formed which leads to both product dimer and deactivation, how would this change the meaning of the slope and intercept in our plots of $1/\Phi_D$ vs. $1/[A]$? If we write a simple excimer mechanism

		rate
$A + h\nu \longrightarrow A^s$		I_a
$A^s \longrightarrow A + h\nu'$		$k_f[A^s]$
$A^s \longrightarrow A$		$(k_{ic} + k_{isc})[A^s]$
$A^s + A \longrightarrow$ excimer		$k_E[A^s][A]$
excimer $\longrightarrow$ dimer		$k_D[E]$
excimer $\longrightarrow$ 2A		$k_{cq}[E]$

the following expression is obtained:

$$\frac{1}{\Phi_D} = \frac{k_D + k_{cq}}{k_D} + \frac{(k_D + k_{cq})(k_{isc} + k_{ic} + k_f)}{k_E k_D[A]} \tag{2.56}$$

We see then that a plot of $1/\Phi_D$ vs. $1/[A]$ from this mechanism would have the same intercept (with k_D and k_{cq} redefined, however) but apparently a different slope:

$$\text{slope} = \frac{(k_D + k_{cq})(k_{isc} + k_{ic} + k_f)}{k_E k_D} \tag{2.57}$$

However

$$\text{slope} = b/\tau_f k_E, \qquad b = \text{intercept} = (k_D + k_{cq})/k_D \tag{2.58}$$

k_D can therefore be determined from τ_f and the slope and intercept for the fluorescence,

$$\frac{1}{\Phi_f} = \frac{k_f + k_{isc} + k_{ic}}{k_f} + \frac{k_E[A]}{k_f} \tag{2.59}$$

Since k_f can be determined as seen previously, k_E can be obtained from the fluorescence plot from the slope and the value of k_f:

$$k_E(\text{from dimerization}) = \text{intercept}/(\text{slope} \times \tau_f) \tag{2.60}$$

$$k_E(\text{from fluorescence}) = \Phi_f \times \text{slope}/\tau_f \tag{2.61}$$

However, we have seen from our simple singlet mechanism that

$$k_D + k_{cq} = \text{intercept}/(\text{slope} \times \tau_f) \qquad \text{from dimerization} \tag{2.62}$$

$$k_D + k_{cq} = \text{slope} \times \Phi_f/\tau_f \qquad \text{from fluorescence} \tag{2.63}$$

Thus for both sets of experimentally derived data

$$\overset{\text{excimer}}{k_E} = \overset{\text{simple}}{k_D + k_{cq}} \tag{2.64}$$

It is impossible, therefore, to determine which mechanism is operative from the dimerization and fluorescence kinetic data.

By examining any correlation between excimer formation (as evidenced by characteristic excimer fluorescence) and dimerization quantum yield, one could perhaps determine whether dimerization is dependent upon prior excimer formation. Excimer fluorescence from anthracene solutions at room temperature is negligible although it has been observed in the solid state at low temperature.[(75)] Unfortunately, the data for substituted anthracenes allow no firm conclusions to be drawn. Some derivatives dimerize but do not exhibit excimer fluorescence. Others both dimerize and show excimer fluorescence. Still others show excimer fluorescence but do not dimerize; and finally, some neither dimerize nor show excimer fluorescence. Hopefully, further work will determine what role excimer formation plays in this photodimerization.

Kinetic Variations as a Function of Structure[(63–65)]

The rate constants for a number of anthracene derivatives in various solvents are given in Table 2.11. A number of common points can be seen.

(a) For all the compounds listed (with the exception of 9-anthroic acid) the sum of $k_{ic} + k_{isc}$ is greater than k_D. Since k_{ic} is expected to be small, most of this sum is probably due to k_{isc}. Therefore it appears that a considerable number of triplet states are formed in the irradiation of anthracene and its derivatives.

(b) For all the compounds, k_{cq} is larger than k_D. This indicates that the probability of molecular collisions leading to ground state molecules is higher than that for collisions leading to stable dimer.

(c) The rate constant for dimerization decreases in the order $k_D^{9\text{-Me}} > k_D^{9\text{-H}} > k_D^{9\text{-COOH}} > k_D^{9\text{-COONa}}$. The electron-withdrawing ability of these groups

TABLE 2.11[a]

System	k_f	$k_{ic} + k_{isc}$	k_D[b]	k_D[c]	k_{cq}	$k_D + k_{cq}$	k_{diffn}
9-COOH/EtOH,H^+	4.79	17.2	102	108	428	530	613
9-COOH/ϕCN	5.24	2.88	35.9	37.9	298	334	532
9-COONa/ϕCN	1.84	31.9	11.3	13.5	1440	1451	532
9-COONa/H_2O	4.59	62.1	73.0	112	657	730	739
9-COONa/D_2O	4.17	69.8	76.9	204	307	384	598
A/Ph-H	7.4	24	100	—	400	500	1100
9-Me/EtOH	5.56	8.8	220	—	850	1070	613
9-Et	5.85	9.6	—	—	—	708	613
9-*n*Pr/EtOH	5.68	9.2	—	—	—	670	613

[a] All data × 10^{-7}.
[b] From dimerization.
[c] From fluorescence.

TABLE 2.12

Solvent	k_{diffn}(20°C), $M^{-1}\,\text{sec}^{-1}$
Hexane	2.0×10^{10}
Chloroform	1.2×10^{10}
Benzene	1.0×10^{10}
Cyclohexane	6.9×10^{9}
Water	6.5×10^{9}
Ethanol	5.4×10^{9}
Glycerol[a]	6.0×10^{6}

[a] For viscous solvents the rate calculated using Eq. (2.65) is often too small. See P. J. Wagner and I. Kocherar, *J. Amer. Chem. Soc.* **90**, 2232 (1968).

increases in this same order. It appears, then, that electron-withdrawing groups in the 9 position decrease the rate of dimerization by removing electron density from the reaction center.

(d) The sum $k_{\text{D}} + k_{\text{cq}}$ is approximately equal to the diffusion rate constant k_{diffn}. This latter value corresponds to the fastest rate for a collisional bimolecular process. Its value depends upon how fast the molecules can diffuse together and can be estimated from the equation

$$k_{\text{diffn}} = \frac{8RT}{3000\eta}\frac{1}{4}\left(2 + \frac{r_a}{r_b} + \frac{r_b}{r_a}\right)$$

where η is the viscosity of the solvent and r_a and r_b are the collision radii of molecules a and b. For like molecules $r_a = r_b$ and this equation becomes

$$\boxed{k_{\text{diffn}} = 8RT/3000\eta} \qquad (2.65)$$

The values of k_{diffn} for several commonly used solvents are given in Table 2.12.

The data for sodium 9-anthroate in benzonitrile do not fit the pattern of the other derivatives since in this case $k_{\text{D}} + k_{\text{cq}} > k_{\text{diffn}}$. This effect cannot be due to k_{D} since this value is less than those of the other derivatives. Therefore k_{cq} must be greatly increased for the salt. This effect is thought to arise from both dynamic quenching and static quenching due to ion pairs.

Another interesting feature arises from the fact that

$$(k_{\text{isc}} + k_{\text{ic}})_{\text{acid}} < (k_{\text{isc}} + k_{\text{ic}})_{\text{salt}}$$

This is thought to reflect the fact that $k_{\text{isc}}^{\text{9-COOH}} < k_{\text{isc}}^{\text{9-COONa}}$. An explanation of this effect follows from an examination of the fluorescence spectra of these

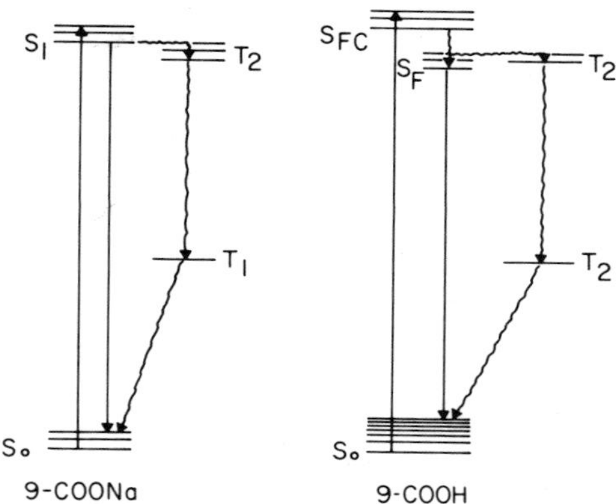

FIGURE 2.30. S_{FC}, Franck–Condon state; S_F, fluorescence state. Reproduced with permission from Ref. 65.

two derivatives. The sodium salt has a normal fluorescence having a small Stokes shift or essentially a mirror image relationship to the absorption spectrum. The fluorescence spectrum for the free acid, on the other hand, is shifted to longer wavelengths and is broad and structureless. Thus, at least the *fluorescent* singlet state for the acid is at longer wavelength (lower energy) than that for the salt. Since the second triplet state for anthracene is known to be essentially isoenergetic with the fluorescent singlet, intersystem crossing from the singlet to the triplet state should be efficient for "normal" anthracene derivatives such as the acid salt (Figure 2.30).

For the free acid the fluorescent singlet is lower in energy than that of the salt and hence probably lower than the second triplet state. Since the efficiency of a nonradiative process depends upon the energy separation between the two states, intersystem crossing from the fluorescent state in the acid to the lowest triplet state (T_1) should be less efficient than the corresponding process to T_2 in the salt. The lowering of the fluorescent singlet state of the acid is thought to arise from rotation of the carboxyl group in the excited state (spectroscopic singlet) permitting conjugation with the ring.[(73)]

2.3 THE ANTHRACENE TRIPLET STATE

Up to this point we have concerned ourselves almost exclusively with the anthracene singlet state and its resulting photodimerization. However, we

Table 2.13. *Some Physical Characteristics of the Anthracene Molecule*

Quantity	Value	Reference
E_t, kcal/mole	42.0	78
ΔE_{st}, kcal/mole	32	79
Φ_{isc}	0.75	
Φ_f	0.31	66*d*
τ_p, sec	0.014[a]	80
	0.1[b]	80
	0.1[c]	
Φ_p	~0.00002	79

[a] In isopentane at 20°C by oxygen perturbation method.[80]
[b] In isopentane at 77°K by oxygen perturbation method.[80]
[c] From phosphorescence decay.

have seen that the kinetics arising from the photodimerization and fluorescence studies suggest that fairly high triplet populations must be produced upon photolysis of anthracene. Some properties of the anthracene triplet are listed in Table 2.13. These data were obtained by techniques which will be discussed in detail in Chapter 3.

The Photochemistry of the Anthracene Triplet State

It can be seen from Table 2.13 that the anthracene triplet state is much longer lived than its singlet state (0.014 and 4.5×10^{-9} sec at room temperature, respectively). Consequently one might expect a great deal more photochemistry to arise from the triplet than from the singlet just from the simple fact that it is around longer. Surprisingly, however, if one is careful to remove oxygen from the solutions by several freeze–thaw cycles under vacuum, no photochemical products that can be ascribed to the triplet state reaction are observed. In air-saturated or oxygen-flushed solutions, on the other hand, one observes the formation of an oxygenated product (in addition to the singlet-derived dimer):

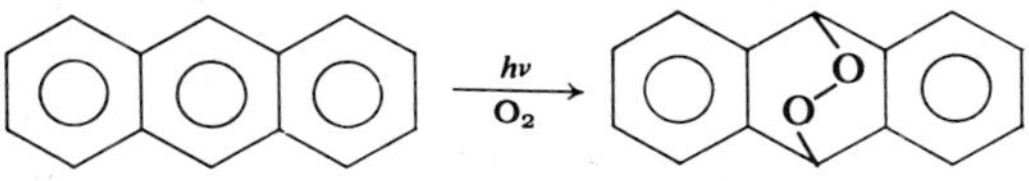

$$\xrightarrow[O_2]{h\nu} \qquad (2.66)$$

The amount of this endoperoxide produced depends upon the solvent and the anthracene concentration.

Two mechanisms have been proposed to account for the formation of this product[60,67,81–84] (for our purposes in this section we will neglect those steps leading to the dimer):

Mechanism A

	rate
$A \xrightarrow{h\nu} A^s$	I_a
$A^s \longrightarrow A$	$k_{ic}[A^s]$
$A^s \longrightarrow A + h\nu'$	$k_f[A^s]$
$A^s + O_2 \longrightarrow \cdot AOO \cdot$	$k_1[A^s][O_2]$
$A^s \longrightarrow A^t$	$k_{isc}[A^s]$
$A^t + O_2 \longrightarrow \cdot AOO \cdot$	$k_2[A^t][O_2]$
$\cdot AOO \cdot + A \longrightarrow AO_2 + A$	$k_3[\cdot AOO \cdot][A]$
$A^t \longrightarrow A$	$k_D[A^t]$

Mechanism B

$A \xrightarrow{h\nu} A^s$	I_a
$A^s \longrightarrow A + h\nu$	$k_f[A^s]$
$A^s \longrightarrow A$	$k_{ic}[A^s]$
$A^s \longrightarrow A^t$	$k_{isc}[A^s]$
$A^t + O_2{}^t \longrightarrow A + O_2{}^s$	$k_{et}[A^t][O_2{}^t]$
$O_2{}^s + A \longrightarrow AO_2$	$k_r[O_2{}^s][A]$
$A^t \longrightarrow A$	$k_D[A^t]$

In mechanism A the endoperoxide results from an intermediate diradical adduct of oxygen with either a singlet or triplet anthracene. The importance of the fourth step relative to the sixth is postulated to depend largly upon the solvent, being more important for those solvents favoring the fluorescence of anthracene (benzene) and less so in solvents in which the fluorescence is quenched and large triplet populations result (carbon disulfide). In the former case a strong rate dependence on $[O_2]$ is observed. In the latter case the rate is nearly independent of $[O_2]$. The formation of the stable endoperoxide from the intermediate diradical adduct (mole-oxide) depends upon collision with another ground state anthracene. This step is required to account for the anthracene dependence in the rate of product formation.

Mechanism B[85–88] involves a transfer of energy from a triplet excited anthracene to a ground state oxygen molecule (triplet) to produce a ground state anthracene and a singlet excited oxygen molecule. This corresponds to a spin-allowed process. There are two possible states for this singlet oxygen species, the $^1\Delta_g$ ($E = 22.5$ kcal/mole) and the $^1\Sigma_g{}^+$ ($E = 37.5$ kcal/mole).[89]

The energy available from the anthracene triplet (42 kcal/mole) is sufficient to produce either of these states. The singlet excited molecule subsequently attacks a ground state anthracene to produce the observed endoperoxide. The $^1\Delta_g$ state is believed to be responsible for the addition to anthracene to form the endoperoxide since it closely resembles a diradical species, while the $^1\Sigma_g{}^+$ state more closely resembles a dipolar ion.

Proponents of each of these mechanisms attack the other with numerous criticisms. Those favoring the mole-oxide intermediate (mechanism A) claim that (1) the singlet oxygen mechanism does not adequately describe the observed kinetics[(67)] and (2), it is eliminated since a chlorophyll derivative with a triplet energy too low to excite oxygen to its singlet level is still effective in producing product by the steps[(90)]

$$C^t + O_2 \longrightarrow \cdot COO\cdot$$
$$\cdot COO\cdot + A \longrightarrow C + AO_2$$

Those favoring the singlet oxygen mechanism (mechanism B) counter by asking why, if a mole-oxide is formed, must the oxygen be transferred to another anthracene to obtain a stable product?[(85)] It would appear that if the intermediate has the structure shown below, that a single closure would produce the endoperoxide directly:

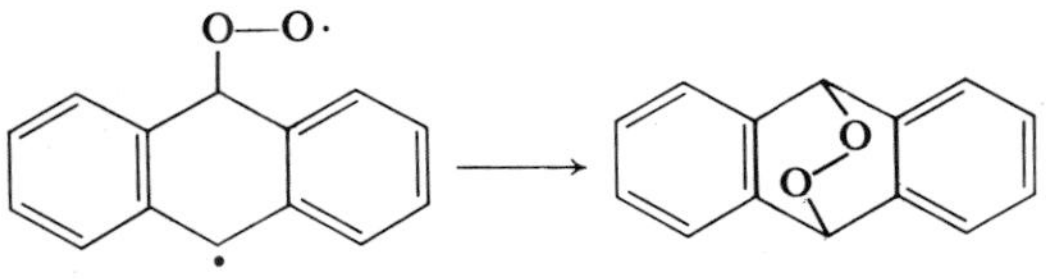

O—O·

(2.67)

Strong evidence in favor of mechanism B was obtained when it was discovered that singlet oxygen produced chemically by the reaction of hydrogen peroxide and sodium hydrochlorite or from gaseous oxygen excited by an electrodeless discharge yields the same products as the direct photolysis.[(85–88)]

Clearly all the answers to the questions of the mechanisms of both the photodimerization and photooxidation of anthracene are not yet known. Hopefully, what we have seen in this chapter will serve to convince the reader that photochemistry is still an exciting and challenging field of study in which there is ample room for further research.

PROBLEM

1. Derive an expression for $1/\Phi_D$ *vs.* $1/[A]$ assuming first that k_{-E} is not important and then that k_{-E} is an important mode of decay for the excimer:

$$
\begin{array}{rl}
A \xrightarrow{h\nu} A^s & I_a \\
A^s \rightarrow A & k_d \\
A^s \rightarrow A^t & k_{isc} \\
A^s + A \rightarrow E & k_E \\
E \rightarrow A^s + A & k_{-E} \\
E \rightarrow 2A & k_{cq} \\
E \rightarrow D & k_D
\end{array}
$$

REFERENCES

1. C. A. Coulson and A. Streitweiser, Jr., *Dictionary of π-Electron Calculations*, Freeman, San Francisco (1965), pp. 29–30.
2. H. H. Jaffe and M. Orchin, *Theory and Application of Ultraviolet Spectroscopy*, Wiley, New York (1962), p. 216.
3. E. W. Thulstrup, J. Michl, and J. H. Eggers, *J. Phys. Chem.* **74**, 3868 (1970).
4. G. Weber, *Biochem. J.* **75**, 335 (1960).
5. C. A. Parker, *Photoluminescence of Solutions*, Elsevier, New York (1968).
6. W. R. Ware, in *Creation and Detection of the Excited State*, A. Lamola, ed., Marcel Dekker, New York (1971).
7. A. S. Cherkasov, V. A. Molchanov, T. M. Vember, and K. G. Voldaikina, *Sov. Phys.—Doklady* **1**, 427 (1956).
8. R. Shon, D. O. Cowan, and W. Schmiegel, *J. Phys. Chem.* **79**, 2087 (1975).
9. J. Fritzsche, *J. Prakt. Chem.* **101**, 333 (1867); **106**, 274 (1869).
10. (a) K. Ellis, *J. Prakt. Chem.* **44**, 467 (1891); (b) W. Orndorff and F. Cameron, *Amer. Chem. J.* **17**, 658 (1895).
11. J. Houben, *Das Anthracen und die Anthrachinone*, Verlag, Leipzig (1929), p. 135.
12. C. Linebarger, *Amer. Chem. J.* **14**, 597 (1892).
13. C. A. Coulson, L. E. Orgel, W. Taylor, and J. Weiss, *J. Chem. Soc.*, 2961 (1955).
14. M. Ehrenberg, *Acta Cryst.* **20**, 177 (1966).
15. R. Calas and R. Lalande, *Bull. Soc. Chim. Fr.*, 763 (1959).
16. R. Lalande and R. Calas, *Bull. Soc. Chim. Fr.*, 766 (1959).
17. R. Lalande and R. Calas, *Bull. Soc. Chim. Fr.*, 144 (1960).
18. R. Calas, R. Lalande, and P. Mauret, *Bull. Soc. Chim. Fr.*, 148 (1960).
19. R. Calas, R. Lalande, J.-G. Faugere, and F. Moulines, *Bull. Soc. Chim. Fr.*, **119**, 121 (1965).
20. D. E. Applequist, E. C. Friedrich, and M. T. Rogers, *J. Amer. Chem. Soc.* **81**, 457 (1959).
21. D. E. Applequist, R. L. Little, E. C. Friedrich, and R. E. Wall, *J. Amer. Chem. Soc.* **81**, 452 (1959).

22. O. C. Chapman and K. Lee, *J. Org. Chem.* **34**, 4166 (1969).
23. R. Lalande and R. Calas, *Bull. Soc. Chim. Fr.*, 770 (1959).
24. T. M. Vember and A. S. Cherkasov, *Opt. Spectrosc.* **6**, 148 (1959).
25. T. M. Vember, *Opt. Spectrosc.* **20**, 188 (1966).
26. R. Lapouyade, A. Castellan, and H. Bouas-Laurent, *C. R. Acad. Sci. (Paris) C* **268**, 217 (1969).
27. H. Bouas-Laurent and R. Lapouyade, *Chem. Comm.*, 817 (1969).
28. F. D. Greene, S. C. Misrock, and J. R. Wolfe, *J. Amer. Chem. Soc.* **77**, 3852 (1955).
29. D. E. Applequist, T. L. Brown, J. P. Kleiman, and S. T. Young, *Chem. Ind. (London)*, 850 (1959).
30. F. D. Greene, *Bull. Soc. Chim. Fr.*, 1365 (1960).
31. I. Roit and W. Waters, *J. Chem. Soc.*, 2695 (1952).
32. W. Henderson, Doctoral Thesis, University of Minnesota (1962).
33. R. Livingston and K. S. Wei, *J. Amer. Chem. Soc.* **89**, 3098 (1967).
34. L. Kaminski, Doctoral Thesis, Massachusetts Institute of Technology (1959).
35. D. E. Applequist, M. A. Lintner, and R. Searle, *J. Org. Chem.* **33**, 254 (1968).
36. J. H. Golden, *J. Chem. Soc.*, 3741 (1961).
37. N. S. Weinshenker and F. D. Greene, *J. Amer. Chem. Soc.* **90**, 506 (1968).
38. R. Lapouyade, A. Castellan, and H. Bouas-Laurent, *Tetrahedron Lett.*, 3537 (1969).
39. H. Bouas-Laurent and A. Castellan, *Chem. Comm.*, 1648 (1970).
40. A. Schonberg, A. Mustafa, M. Z. Barakat, N. Latif, R. Moubasher, and A. Mustafa, *J. Chem. Soc.*, 2126 (1948).
41. J. B. Birks, J. H. Appleyard, and R. Pope, *Photochem. Photobiol.* **2**, 493 (1963).
42. J. S. Bradshaw and G. S. Hammond, *J. Amer. Chem. Soc.* **85**, 3953 (1963).
43. P. Wilanat and B. K. Selinger, *Austral. J. Chem.* **21**, 733 (1968).
44. J. S. Bradshaw, N. B. Nielsen, and D. P. Rees, *J. Org. Chem.* **33**, 259 (1968).
45. T. W. Mattingly, Jr., J. E. Lancaster, and A. Zweig, *Chem. Comm.*, 595 (1971).
46. (a) E. C. Taylor and W. W. Paudler, *Tetrahedron Lett.*, **25**, 1 (1960); (b) W. A. Ayer, R. Hoyatsu, P. DeMayo, S. T. Reid, and J. B. Stothers, *Tetrahedron Lett.*, 648 (1961); (c) G. Slomp, F. A. Mackiller, and L. A. Paquette, *J. Amer. Chem. Soc.* **83**, 4472 (1961).
47. E. C. Taylor and R. O. Kan, *J. Amer. Chem. Soc.* **85**, 776 (1963).
48. P. DeMayo and R. W. Yip, *Proc. Chem. Soc.* **84** (1964).
49. N. C. Yang and J. Libman, *J. Amer. Chem. Soc.* **94**, 1405 (1972).
50. N. C. Yang, J. Libman, L. Barrett, Jr., M. H. Hui, and R. L. Loeschin, *J. Amer. Chem. Soc.* **94**, 1406 (1972).
51. (a) W. G. Leighton and G. S. Forbes, *J. Amer. Chem. Soc.* **52**, 3139 (1930); (b) G. S. Forbes, G. B. Kistiakowsky, and L. J. Heidt, *J. Amer. Chem. Soc.* **54**, 3246 (1932); (c) F. P. Brackett, Jr., and G. S. Forbes, *J. Amer. Chem. Soc.* **55**, 4459 (1933); (d) B. M. Norton, *J. Amer. Chem. Soc.* **56**, 2294 (1934); (e) C. A. Discher, P. E. Smith, I. Lippman, and R. Turse, *J. Phys. Chem.* **67**, 2501 (1963).
52. J. N. Pitts, Jr., J. D. Margerum, R. P. Taylor, and W. Brim, *J. Amer. Chem. Soc.* **77**, 5499 (1955).
53. D. A. Volman and J. R. Seed, *J. Amer. Chem. Soc.* **86**, 1879 (1940).
54. (a) C. G. Hatchard and C. A. Parker, *Proc. Roy. Soc. (London) A* **235**, 58 (1956); (b) Baxendate and Bridge, *J. Phys. Chem.* **59**, 783 (1955); (c) J. Lee and H. H. Seliger, *J. Chem. Phys.* **40**, 519 (1964); (d) E. E. Wegner and A. W. Adamson, *J. Amer. Chem. Soc.* **88**, 394 (1966).
55. (a) R. N. Smith, P. A. Leighton, and W. G. Leighton, *J. Amer. Chem. Soc.* **61**, 2299 (1939); (b) L. B. Thomas, *J. Amer. Chem. Soc.* **62**, 1879 (1940).

56. E. J. *Bowen, Disc. Faraday Soc.* **14**, 143 (1953).
57. E. J. Bowen and N. Norton, *Trans. Faraday Soc.* **35**, 44 (1939).
58. E. J. Bowen, *Trans. Faraday Soc.* **42**, 133 (1946).
59. E. J. Bowen, *Trans. Faraday Soc.* **50**, 97 (1954).
60. E. J. Bowen and D. W. Tarner, *Trans. Faraday Soc.* **51**, 475 (1955).
61. A. S. Cherkasov and T. M. Vember, *Opt. Spektrosk.* **4**, 203 (1958).
62. A. S. Cherkasov and T. M. Vember, *Opt. Spectrosc.* **6**, 319 (1959).
63. W. W. Schmiegel, Doctoral Thesis, The Johns Hopkins University (1970).
64. D. O. Cowan and W. W. Schmiegel, *Angew. Chem.* (*Intern. Ed.*) **10**, 517 (1971).
65. D. O. Cowan and W. W. Schmiegel, *J. Amer. Chem. Soc.* **94**, 6779 (1972).
66. (a) G. W. Robinson and R. P. Frosch, *J. Chem. Phys.* **37**, 1962 (1962); (b) G. W. Robinson and R. P. Frosch, *J. Chem. Phys.* **38**, 1187 (1963); (c) M. A. El-Sayed, *J. Chem. Phys.* **38**, 2834 (1963); (d) T. Medinger and F. Wilkinson, *Trans. Faraday Soc.* **61**, 620 (1965).
67. R. S. Livingston, in *Photochemistry in the Liquid and Solid State*, F. Daniels, ed., Wiley, New York (1960), pp. 76–86.
68. T. Forster, in *Proceedings of the International Conference on Luminescence*, Vol. 1, G. Szigett, ed., Akademiai Kiado, Budapest (1968), pp. 160–165.
69. B. Stevens and M. I. Ban, *Trans. Faraday Soc.* **66**, 1515 (1964).
70. T. Forster and K. Kasper, *Z. Physik. Chem.* (*Frankfurt*) **1**, 275 (1964).
71. J. B. Birks and J. B. Aladekome, *Photochem. Photobiol.* **2**, 415 (1963).
72. T. Azumi, A. T. Armstrong, and S. P. McGlynn, *J. Chem. Phys.* **41**, 3839 (1964).
73. J. B. Birks and L. G. Christophorou, *Proc. Roy. Soc.* (*London*) *A* **274**, 552 (1963).
74. J. B. Birks, D. S. Dyson, and T. A. King, *Proc. Roy. Soc.* (*London*) *A* **277**, 270 (1965).
75. B. Stevens, *Spectrochim. Acta* **18**, 439 (1962).
76. E. Chandross and J. Ferguson, *J. Chem. Phys.* **45**, 3560 (1966).
77. T. C. Werner and D. M. Hercules, *J. Phys. Chem.* **73**, 2005 (1969); **74**, 1030 (1970).
78. (a) D. Evans, *J. Chem. Soc.* 1351, (1957); (b) G. N. Lewis and M. Kasha, *J. Amer. Chem. Soc.* **66**, 2100 (1944).
79. D. S. McClure, *J. Chem. Phys.* **17**, 905 (1949).
80. J. W. Hilpern, G. Porter, and C. J. Stief, *Proc. Roy. Soc. A* **277**, 437 (1964).
81. A. S. Cherkasov and T. M. Vember, *Opt. Spectrosc.* **7**, 207 (1959).
82. R. Livingston and V. S. Rao, *J. Phys. Chem.* **63**, 794 (1959).
83. E. J. Bowen, in *Advances in Photochemistry*, Vol. 1, W. A. Noyes, Jr., G. S. Hammond, and J. N. Pitts, Jr., eds., Interscience, New York (1963), p. 23.
84. K. Gollnich and G. O. Schenck, in *1,4-Cycloaddition Reactions*, J. Hamer, ed., Academic Press, New York (1967).
85. C. S. Foote and S. Wexler, *J. Amer. Chem. Soc.* **86**, 3879, 3880 (1964).
86. C. S. Foote, S. Wexler, W. Ando, and R. Higgins, *J. Amer. Chem. Soc.* **90**, 975 (1968).
87. S. N. Foner and R. L. Hudson, *J. Chem. Phys.* **23**, 1974 (1955); **25**, 601 (1956).
88. E. J. Corey and W. C. Taylor, *J. Amer. Chem. Soc.*, 3881 (1964).
89. (a) H. Kautsky and H. deBruijn, *Naturwiss.* **19**, 1043 (1931); (b) H. Kautsky, H. deBruijn, R. Neuwirth, and W. Baumeister, *Ber. Deut. Chem. Ges.* **66**, 1588 (1933); (c) H. Kautsky, *Biochem.* **2**, 271, 291 (1937).
90. H. Gaffron, *Biochem. Z.* **287**, 130 (1936).
91. R. Maleski and H. Morrison, *Mol. Photochem.* **4**, 507 (1972).

3

Photochemical Techniques and the Photochemistry of Ketones

In Chapter 2 we discussed a number of techniques used to study the various photophysical and photochemical processes occurring in anthracene and similar molecules. In that discussion we were primarily interested in the singlet state. In this chapter we will discuss some of the techniques available for studying the photophysical and photochemical properties of the triplet state. Most of our discussion will be directed to the photochemistry of simple ketones.

In general ketones undergo three main types of photochemical reactions. These are:

(a) Cleavage of a bond α to the carbonyl (α or Type I cleavage):

$$(\mathrm{R\overset{O}{\overset{\|}{C}}R})^* \longrightarrow \mathrm{R\overset{O}{\overset{\|}{C}}\cdot} + \mathrm{R\cdot}$$

$$\text{cyclobutanone} \xrightarrow{h\nu} [\text{acyl–alkyl biradical}] \begin{cases} \xrightarrow{-\mathrm{CO}} \text{propene} + \text{cyclopropane} + \mathrm{CO} \\ \longrightarrow \mathrm{CH_2{=}CH_2} + \mathrm{CH_2{=}C{=}O} \\ \longrightarrow [\text{oxacarbene}] \xrightarrow{\mathrm{ROH}} \text{2-alkoxytetrahydrofuran (H, OR)} \end{cases} \qquad (3.1)$$

(b) Hydrogen abstraction (intermolecular and intramolecular):

$$(R{-}\overset{\overset{O}{\|}}{C}R)^* + RH \longrightarrow R\overset{\overset{OH}{|}}{\underset{\cdot}{C}}R + R\cdot \tag{3.2}$$

$$\phi{-}\overset{\overset{O}{\|}}{C}{-}\phi + \phi{-}\overset{\overset{OH}{|}}{\underset{\underset{H}{|}}{C}}{-}\phi \xrightarrow{h\nu} 2\phi{-}\overset{\overset{OH}{|}}{\underset{\cdot}{C}}{-}\phi \longrightarrow \phi{-}\overset{\overset{\phi}{|}}{\underset{\underset{OH}{|}}{C}}{-}\overset{\overset{\phi}{|}}{\underset{\underset{OH}{|}}{C}}{-}\phi \tag{3.3}$$

$$\text{Me}C(=O\cdots H{-}CH_2CH_2CH_2) \xrightarrow{h\nu} R{-}\overset{\overset{OH}{|}}{\underset{\cdot}{C}}{-}CH_2{-}CH_2{-}\dot{C}H_2 \longrightarrow \text{Me}{-}\overset{\overset{OH}{|}}{C}{-}CH_2 \text{ (cyclobutane ring: } C{-}CH_2{-}CH_2{-}CH_2) \tag{3.4}$$

$$\downarrow$$

$$\text{Me}\overset{\overset{OH}{|}}{C}{=}CH_2 + CH_2{=}CH_2$$

$$\updownarrow$$

$$\text{Me}\overset{\overset{O}{\|}}{C}{-}CH_3$$

(c) Addition of an excited-state carbonyl compound to an olefin to form an oxetane:

$$(R{-}\overset{\overset{O}{\|}}{C}{-}R)^* + CH_2{=}CH\phi \longrightarrow \text{oxetane: } O{-}CH_2{-}\underset{\underset{H}{|}}{C}({-}\phi){-}CR_2{-}O \tag{3.5}$$

3.1 THE PHOTOREDUCTION OF ARYL KETONES: NATURE OF THE EXCITED STATE

We have already discussed one of the earliest photoreactions to be studied, that is, the $(4\pi + 4\pi)$ photodimerization of anthracene. That the singlet state was involved in this reaction was conclusively shown in the period 1955–1957. The first reaction in which the triplet state of the molecule was shown to be involved was the photoreduction of benzophenone by Hammond and co-workers(1) and Bäckstrom and co-workers(2) 1959–1961. This was the first in a series of many papers from Hammond's laboratory

(Hammond and his former students have published several hundred mechanistic photochemical papers since 1960). The reaction studied was given as an example in equation (3.3):

$$\phi\text{—}\overset{\overset{\displaystyle O}{\|}}{C}\text{—}\phi + \phi\text{—}\overset{\overset{\displaystyle OH}{|}}{\underset{\underset{\displaystyle H}{|}}{C}}\text{—}\phi \xrightarrow{h\nu} \phi\text{—}\overset{\overset{\displaystyle \phi}{|}}{\underset{\underset{\displaystyle OH}{|}}{C}}\text{——}\overset{\overset{\displaystyle \phi}{|}}{\underset{\underset{\displaystyle OH}{|}}{C}}\text{—}\phi \tag{3.6}$$

Assuming for the moment without proof that the reaction arises from the excited triplet benzophenone (B), one can write the following simplified mechanism:

		rate
$B + h\nu \longrightarrow B^s$		I_a
$B^s \longrightarrow B^t$		Φ_{isc}
$B^t \longrightarrow B$		$k_d[B^t]$
$B^t + BH_2 \longrightarrow 2\phi\text{—}\overset{\overset{\displaystyle OH}{|}}{\dot{C}}\text{—}\phi$		$k_r[B^t][BH_2]$
$2\phi\text{—}\overset{\overset{\displaystyle OH}{|}}{\dot{C}}\text{—}\phi \longrightarrow \phi\text{—}\overset{\overset{\displaystyle \phi}{|}}{\underset{\underset{\displaystyle OH}{|}}{C}}\text{—}\overset{\overset{\displaystyle \phi}{|}}{\underset{\underset{\displaystyle OH}{|}}{C}}\text{—}\phi$		$k_c\left[\phi\text{—}\overset{\overset{\displaystyle OH}{|}}{\dot{C}}\text{—}\phi\right]^2$

If the last step, the coupling of the ketyl radicals to form the product benzpinacol, is fast, the rate of product formation will be determined by the rate of hydrogen abstraction from benzhydrol to produce the ketyl radicals. Therefore

$$\Phi_r = \frac{\Phi_{isc}k_r[BH_2]}{k_d + k_r[BH_2]} \tag{3.7}$$

or

$$\frac{1}{\Phi_r} = \frac{1}{\Phi_{isc}} + \frac{k_d}{\Phi_{isc}k_r[BH_2]} \tag{3.8}$$

This mechanism would predict that a plot of $1/\Phi_r$ vs. $1/[BH_2]$ should be linear with an intercept equal to $1/\Phi_{isc}$. The experimental results[(1)] have been plotted in this way in Figure 3.1. The straight lines in Figure 3.1 are consistent with but do not prove the suggested mechanism. The slope for the lower line (BH_2) is equal to 0.05, while that for the upper is 0.133. This difference in rate upon replacement of a hydrogen atom with deuterium indicates that the reaction is subject to an isotope effect. Therefore the abstraction

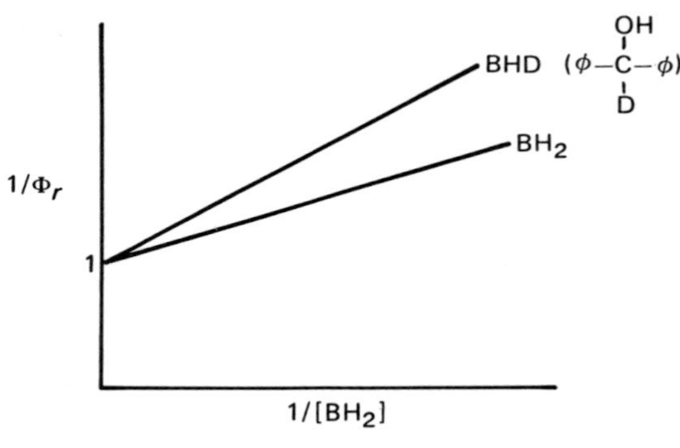

FIGURE 3.1

cannot be a diffusion-controlled reaction. The intercepts of both lines are the same and are equal to unity. If our mechanism is correct, the intercept is equal to $1/\Phi_{\text{isc}}$. Thus it would appear that $\Phi_{\text{isc}} = 1$ for benzophenone.

Hammond and co-workers[1] used the following reasoning in ruling out the possibility of singlet reaction and in implicating the triplet.

(a) Since hydrogen abstraction is a bimolecular process, the maximum rate constant possible is the diffusion rate constant k_{diffn}. We saw in Chapter 2 that this is given by

$$k_{\text{diffn}} = 8RT/3000\eta \tag{3.9}$$

For benzene solution, k_{diffn} calculated from this equation is 1×10^{10} liters/mole-sec. Actually Hammond used a formulation[3,4] that gave $k_{\text{diffn}} = 2 \times 10^9$ liters/mole-sec. This is probably a more realistic value than that calculated from the Stokes–Einstein equation and will be used for this discussion.

(b) Since an isotope effect on the rate of reaction was observed, the reaction cannot be diffusion controlled and $k_r \neq 2 \times 10^9$. The magnitude of the isotope effect indicates furthermore that k_r must be at least a factor of ten less than k_{diffn}:

$$k_r(\text{max}) \approx 2 \times 10^8 \text{ liters/mole-sec} \tag{3.10}$$

(c) Since the slope $= k_d/\Phi_{\text{isc}}k_r$ and $\Phi_{\text{isc}} = 1$,

$$k_d/k_r = 0.05 \quad \text{for BH}_2 \tag{3.11}$$

Therefore

$$k_d(\text{max}) = 0.05k_r = 0.05(2 \times 10^8) = 1 \times 10^7 \text{ sec}^{-1} \tag{3.12}$$

(d) If k_d corresponds to deactivation of a *singlet* state, it is possible to estimate the maximum expected value. Several formulas have been devised to relate the singlet radiative lifetime to the extinction coefficient of absorption,[(5)] e.g.,

$$1/\tau_r = 3 \times 10^{-9} n^2 \bar{\nu}_0^{\,2} (g_l/g_u) \int \varepsilon \, d\nu \tag{3.13}$$

where g_l and g_u are the multiplicities of the lower and upper states, respectively, n is the refractive index of the solvent, $\bar{\nu}_0$ is the band center expressed in wave numbers, and ε is the molar absorptivity. This equation, developed by Bowen, was refined by Birks and Dyson to compensate for the lack of mirror image symmetry in the absorption and emission spectra of some molecules:

$$\frac{1}{\tau_r} = 2.88 \times 10^{-9} n^2 \frac{g_l}{g_u} \frac{\int F(\bar{\nu})\, d\bar{\nu}}{\int \bar{\nu}^{-3} F(\bar{\nu})\, d\bar{\nu}} \int \frac{\varepsilon \, d\bar{\nu}}{\bar{\nu}} \tag{3.14}$$

where F is the intensity of the fluorescence at wave number $\bar{\nu}$.

From these equations one can approximate k_f for benzophenone to be $\sim 5 \times 10^6 \text{ sec}^{-1}$. This, however, is the expected rate constant for fluorescence, which should be in competition with radiationless deactivation of the excited state k_d. In actuality *no* fluorescence is observed for benzophenone although the fluorescence techniques are sensitive enough to detect fluorescence occurring with a quantum yield as low as $\Phi_f = 0.001$. Therefore k_d must be at least 1000 times greater than k_f! We have

$$k_d \geq 1000 k_f \geq 5 \times 10^9 \text{ sec}^{-1} \tag{3.15}$$

We have already seen, however, that k_d as determined from the experimental data has a maximum value of 1×10^7. Therefore the reactive species in the photoreduction of benzophenone *cannot be the singlet state.*

Evidence that the reactive state is a triplet follows from the use of triplet quenchers (oxygen and paramagnetic salts). For this we must add a further step to the mechanism:

$$B^t + Q \longrightarrow B + Q \qquad \text{rate} = k_q[B^t][Q] \tag{3.16}$$

The expression for the quantum yield now becomes

$$\Phi_r = \frac{\Phi_{isc} k_r [BH_2]}{k_d + k_r[BH_2] + k_q[Q]} \tag{3.17}$$

or

$$\frac{1}{\Phi_r} = \frac{1}{\Phi_{isc}} + \frac{k_d + k_q[Q]}{\Phi_{isc} k_r [BH_2]} \tag{3.18}$$

Since $\Phi_{isc} = 1$,

$$\frac{1}{\Phi_r} = 1 + \frac{k_d}{k_r[BH_2]} + \frac{k_q[Q]}{k_r[BH_2]} \tag{3.19}$$

TABLE 3.1. *Approximate Rate Constants for the Photoreduction of Benzophenone*[1]

Rate constant	Quencher	Value
k_q	$Fe[DPM]_3$	2×10^9 liters/mole-sec
k_q	O_2	1×10^9 liters/mole-sec
$k_r(H)$		5×10^6 liters/mole-sec
$k_r(D)$		1.8×10^6 liters/mole-sec
k_d		2.6×10^5 sec^{-1}

It is possible to obtain k_q/k_r from this expression either by keeping $[BH_2]$ constant and varying [Q] or by measuring the slope of $1/\Phi$ vs. $1/[BH_2]$ with and without quencher. The former is the preferred procedure. In this way one finds, using oxygen as triplet quencher,

$$k_q/k_r = 230 \text{ moles/liter} \qquad (Q = O_2) \tag{3.20}$$

and using ferric dipivaloylmethide ($Fe[DPM]_3$) as triplet quencher,

$$k_q/k_r = 400 \text{ moles/liter} \qquad (Q = Fe[DPM]_3) \tag{3.21}$$

Since these values are of the same order of magnitude, Hammond assumed that the quenching was diffusion controlled with $k_q = 2 \times 10^9$ liters/mole-sec in (3.21). With this value for k_q, the rates of the other processes can be determined and are shown in Table 3.1. As we shall see, these values agree well with those obtained by Linschitz using the technique of flash photolysis.

3.2. FLASH PHOTOLYSIS

The technique of flash photolysis was originally developed by Norrish and Porter as a method for studying reactive species such as triplets and radicals with relatively short lifetimes ($\tau \geq 1 \times 10^{-6}$ sec).[6] The beauty of this technique is that it involves the direct observation of the species of interest. The principal problem, however, is to determine the identity of the species causing the new electronic absorption. For their efforts in the development of this technique Norrish and Porter, along with Eigen, received the Nobel Prize in chemistry in 1961.

The method utilizes a high-intensity (10^{20} photons) pulse of light to populate the triplet state via intersystem crossing from the first excited singlet state. Before the molecules have an opportunity to decay back to the ground state a second light source is used to detect the triplets by promoting them to higher triplet levels. If the second flash source is used in conjunction with a spectrograph, an entire absorption spectrum can be recorded

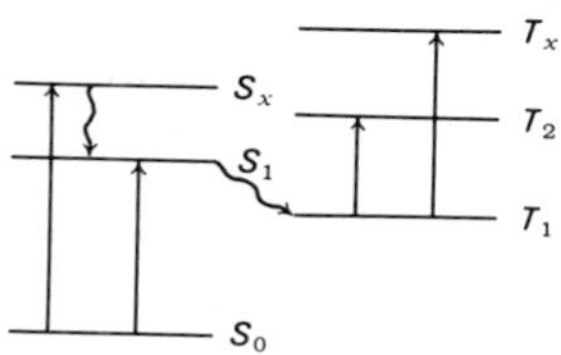

FIGURE 3.2

($S^0 \rightarrow S_x$, $T_1 \rightarrow T_x$). Figures 3.2 and 3.3 illustrate the principle of flash spectroscopy.[(6)] If the second light source is continuous, the change in optical density due to the transient species can be monitored as a function of time at a particular wavelength selected on a monochromator. This type of system is illustrated in Figure 3.4.

The absorption bands measured by the flash spectrographic method are often assigned by (a) comparison with known singlet–singlet absorption spectra, (b) comparison of the lifetime of the species responsible for the absorption with the phosphorescence lifetime, (c) comparison with calculated energies and intensities of the various possible absorptions by semi-empirical molecular orbital methods, and (d) comparison with published triplet absorption spectra and decay kinetics of model compounds.

In order to obtain a better understanding of what actually is involved in these flash photolysis techniques, let us elaborate on Figures 3.3 and 3.4 to show the instrumentation which is used. An expanded diagram of the flash spectrographic apparatus is presented in Figure 3.5[(7)] In the center of Figure 3.5 is a 0–20-kV power supply which charges either the 1-μF or the 2-μF capacitor. The 2-μF capacitor supplies the power to the initial flash lamp which surrounds the sample cell, and the 1-μF capacitor is connected to the delayed spectroflash lamp. The most elaborate part of the apparatus is concerned with triggering the light flashes. Since one is charging a 1- or

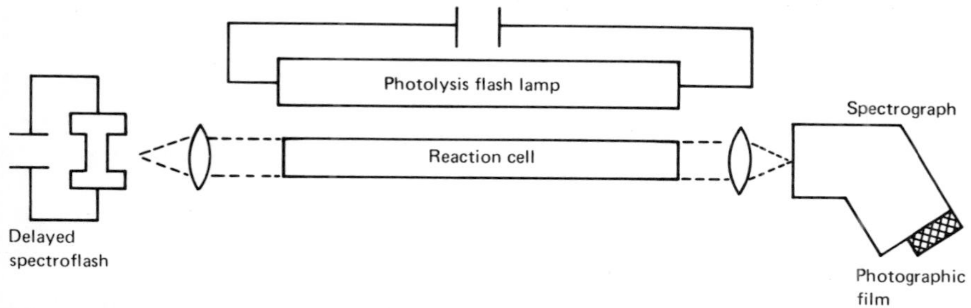

FIGURE 3.3. Flash spectrographic measurement of triplet spectra. (Adapted from Porter.[(6)])

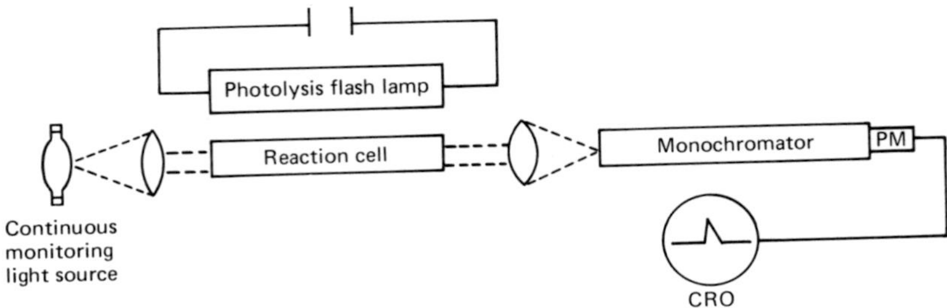

FIGURE 3.4. Flash kinetic spectrophotometry.

2-μF capacitor to about 20 kV and the resistance of the lamp is no greater than 5 Ω, it is clear that the number of kiloamps flowing through the apparatus is great indeed. Thus, high-energy electronic switching devices capable of withstanding high voltages without current conduction must be used. The triggering device consists of two ignitrons, one for each capacitor. These are actually large, mercury-filled tubes which act as switches. The ignitrons are in turn fired by the hydrogen thyratrons which receive pulses from the phantastron. The phantastron can be accurately delayed from 10 μsec to

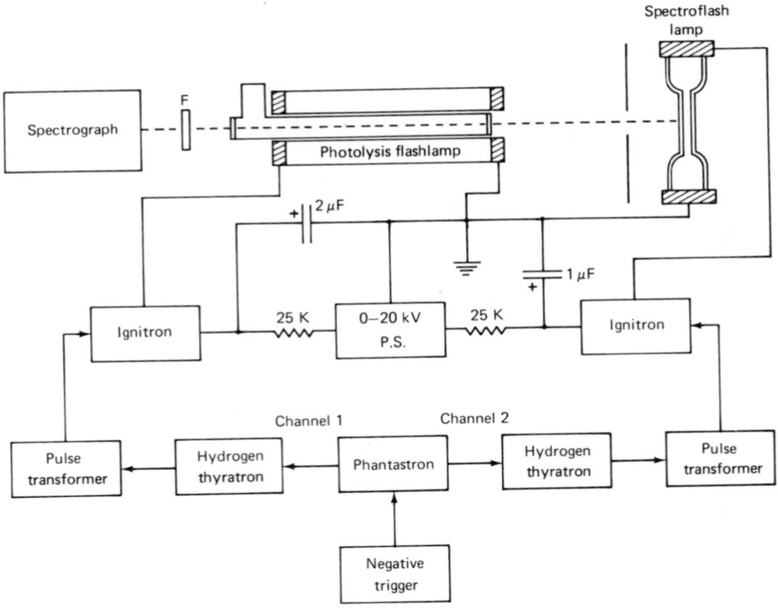

FIGURE 3.5. Block diagram of flash spectroscopic apparatus. Reproduced with permission from Ref. 7.

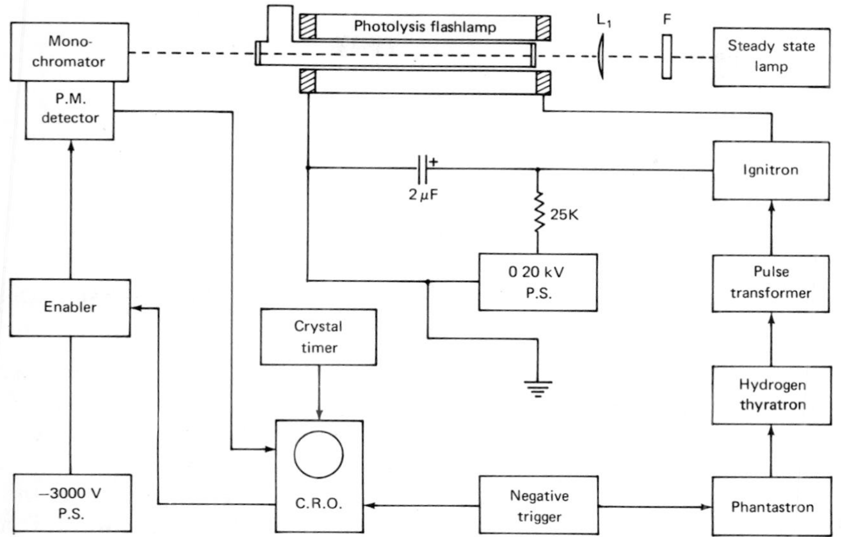

FIGURE 3.6. Block diagram of flash kinetic apparatus. Reproduced with permission from Ref. 7.

about 10 msec before triggering the hydrogen thyratron, which then fires the ignitron.

A similar diagram of a flash kinetic apparatus is given in Figure 3.6[7] In this method the second flash lamp has been replaced by a steady-state lamp. Consequently only one capacitor (2 μF) is necessary. The triggering device is the same as that used in Figure 3.5.[7] However, the negative trigger, which supplies a trigger pulse to the phantastron, also is connected to the oscilloscope, which after a variable delay of from 0.05 μsec to 50 sec switches on the photomultiplier through the enabler circuit. This is to avoid exposing the photomultiplier at high voltage to scattered light from the initial flash. This allows the photomultiplier to be operated at high voltage at some predetermined time after the flash in order to measure the rate of decay of any transient species formed during the flash.

What one is actually dealing with in these cases may be fairly well approximated by a simple *RCL* circuit (Figure 3.7). In Figure 3.7 *R* is the

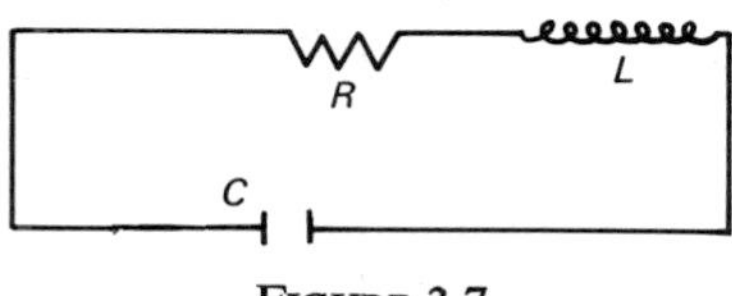

FIGURE 3.7

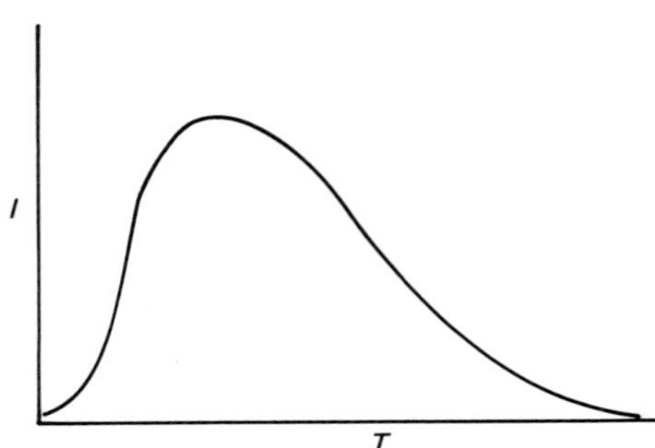

FIGURE 3.8. Damped flash mode.

resistance due to the lamp after discharge has started. Although this resistance depends upon the length of the lamp, generally R will be on the order of 5 Ω or less. L is the inductance due to the capacitor and the lamp leads and C is the capacitance, which is variable but generally ranges from 1 to 20 μF.

An attempt is made to keep the inductance as low as possible since this affects the mode of the discharge. There are two possible modes which can be obtained after flash generation. One, called the damped mode, has a current versus time plot similar to that shown in Figure 3.8.

The other possible mode is called the damped oscillatory mode and has a current versus time plot similar to Figure 3.9. This results in an irregular intensity versus time plot and the consequent loss of a considerable amount of light energy. A profile similar to Figure 3.8 results if R is high in comparison to L. Since R itself is only about 5 Ω or less, the inductance must therefore be kept to a minimum.

Under these conditions the time constant (or time for the flash intensity to decay to $1/e$ of the initial value) is given by

$$\tau_{1/e} = RC \tag{3.22}$$

For a 5-Ω resistance and capacitor charged to 2 μF, the time constant would

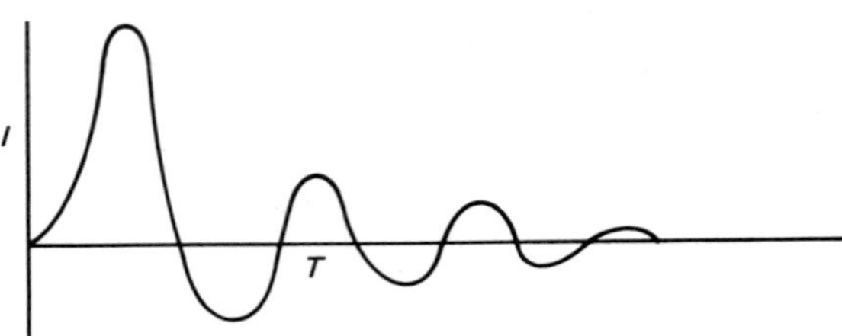

FIGURE 3.9. Damped oscillatory mode.

correspond to 10 μsec. Consequently the shortest lived species that can be conveniently studied under these conditions must have a lifetime somewhat in excess of 10 μsec. The best time constant that one is able to obtain by this apparatus is about 1 μsec.

The energy obtained from the pulse is given by

$$E = CV^2/2 \tag{3.23}$$

where V is the voltage applied across the capacitor and E is obtained in joules. For a 2-μF capacitor charged to 10 kV, $E = 100$ J. For the same capacitor charged to 20 kV, $E = 400$ J. Typically the energies used range from 100 to 500 J.

Analysis of the flash kinetic data is based on the assumption that the photomultiplier response is linear with respect to the incident light intensity. A typical trace is shown in Figure 3.10.

The optical density for the triplet absorption is

$$D_t = \log[I_0/I(t)] = \varepsilon d[\mathrm{A}^3] \tag{3.24}$$

When the decay of the triplet species is first order with respect to the triplet concentration, the following equations are applicable:

$$-d[\mathrm{A}^3]/dt = k_{\text{first}}[\mathrm{A}^3] \tag{3.25}$$

$$\ln([\mathrm{A}_1{}^3]/[\mathrm{A}_2{}^3]) = k_{\text{first}}(t_2 - t_1) \tag{3.26}$$

$$\ln[D_t(1)/D_t(2)] = k_{\text{first}}(t_2 - t_1) \tag{3.27}$$

where k_{first} is the sum of all first order processes leading to deactivation of the triplet state ($\sum k_i$).

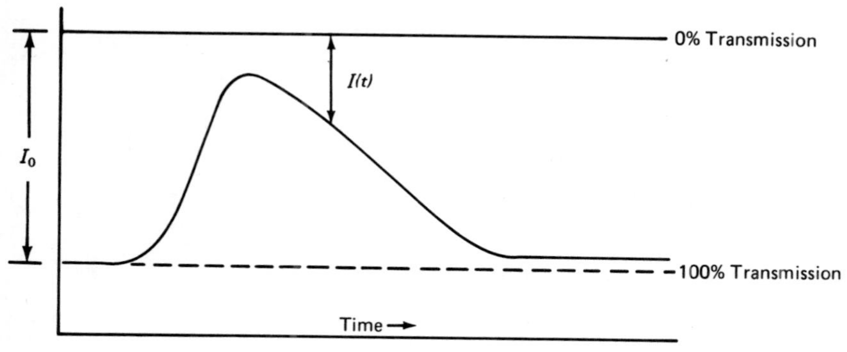

FIGURE 3.10

When decay is second order in the triplet state concentration the following kinetics is obeyed:

$$-d[A^3]/dt = k_{\text{second}}[A^3]^2 \tag{3.28}$$

$$(1/[A_1{}^3]) - (1/[A_2{}^3]) = k_{\text{second}}(t_2 - t_1) \tag{3.29}$$

$$[1/D_t(1)] - [1/D_t(2)] = (k_{\text{second}}/\varepsilon d)(t_2 - t_1) \tag{3.30}$$

In the general case where triplet decay occurs by both first and second order processes we obtain

$$-d[A^3]/dt = k_{\text{first}}[A^3] + k_{\text{second}}[A^3]^2 \tag{3.31}$$

$$[A^3] = D_t/\varepsilon d \tag{3.32}$$

$$-dD_t/dt = k_{\text{first}}D_t + (k_{\text{second}}/\varepsilon d)D_t^2 \tag{3.33}$$

In this case an initial estimate of k_{first} and k_{second} can be obtained from a plot of $(-1/D_t)\,dD_t/dt$ versus (A^3), where the intercept is k_{first} and the slope is $k_{\text{second}}/\varepsilon d$. However, an accurate fit to the integrated equation requires a somewhat complicated statistical analysis based on the following equation:

$$D_t(t) = \{[1/D_t(0) + \beta/\alpha]e^{\alpha t} - (\beta/\alpha)\}^{-1} \tag{3.34}$$

where $\beta = k_{\text{second}}/\varepsilon d$ and $\alpha = k_{\text{first}}$.

Bell and Linschitz[8] subjected the benzophenone–benzhydrol system to a flash spectroscopic study and found two transient species, a long-lived one that is believed to be the benzhydrol radical and a shorter-lived one believed to be the benzophenone triplet. Kinetic analysis of the system revealed the rate data shown in Table 3.2.

The agreement between the indirect measurements of Hammond and the direct measurements of Bell and Linschitz is surprisingly good and the fact that a triplet species is involved in the photoreduction is nicely confirmed by these complimentary techniques. There remains, however, one rather

TABLE 3.2. *Comparison of Rate Constants for the Photoreduction of Benzophenone*

Rate constant[a]	Hammond[b]	Bell and Linschitz[c]
k_q,[d] liters/mole-sec	2×10^9	0.9×10^9
k_r, liters/mole-sec	5×10^6	2×10^6
k_d, sec^{-1}	2.6×10^5	1×10^5

[a] Rate constants are for the photoreaction of benzophenone and benzhydrol in benzene at 25°C.
[b] Indirect determination based on an assumed value for k_q.
[c] Direct determination by flash photolysis.
[d] With $Fe(DPM)_3$ as quencher.

puzzling factor related to this system. This arises from the observation that the lifetime of the benzophenone triplet in benzene (without of course benzyhydrol) is much shorter than that measured in solvents such as carbon tetrachloride or perfluoroalkanes even though there is no appreciable destruction of the benzophenone.[8,9–13] One would expect the triplet lifetime to be about the same in all inert solvents in which there can be no hydrogen abstraction and consequently no triplet chemistry. We can solve this mystery after a short discussion of a fairly recent development by Porter and Topp which rather dramatically extends the utility of the flash photolysis technique.[14]

We have seen that the practical lower limit for the time constant $\tau_{1/e}$ for the normal flash kinetic apparatus is about 1 μsec. In order to study singlet state processes as well as other very reactive intermediates one needs time constants in the nanosecond range. This can now be achieved by using a *Q*-switched ruby laser. A diagram of the apparatus developed by Porter and Topp[14] is shown in Figure 3.11. In this method a flash lamp is used to pump a ruby rod into its excited state, achieving a population inversion. The normal ruby-stimulated emission has a lifetime approximately the same as that of a normal flash lamp, i.e., on the order of microseconds. However, using the laser in combination with a *Q*-switch, which is a cell of a bleachable dye (vanadyl phthalocyanine in this case) placed between the ruby rod and a reflecting mirror, prevents a large stimulation of emission and permits pumping of the laser excited states to population levels much greater than the normal lasing threshold. Upon bleaching of the dye, light is reflected from the mirror and stimulates the laser emission in a giant pulse which may last less than 20 nsec. This light pulse is then passed through certain types of crystal (such as ammonium dihydrogen phosphate, ADP) to obtain frequency doubling. About 20% of the 694-nm radiation that enters the frequency doubler will emerge at a wavelength equal to 347 nm, which is in a more generally useful range. The light then travels through a series of filters and a lens to a beam splitter. Part of the light pulse is directed to the sample cell; the remainder travels through the beam splitter to a mirror and from there back. This light is then delayed relative to that which was split off to the sample cell and the extent of the delay corresponds to twice the distance of the mirror from the beam splitter. After returning again to the beam splitter, a portion of the light is reflected to a solution of a highly fluorescent compound (e.g., 1,1′,4,4′-tetraphenylbuta-1,3-diene) which emits in a continuum at longer wavelengths (400–600 nm) than the exciting pulse. This fluorescence pulse then passes through the beam splitter, into the reaction vessel, and finally into the spectrograph. The delay of this monitoring pulse relative to the exciting pulse can be up to 100 nsec.

Using this apparatus, Porter and Topp were able to detect the following

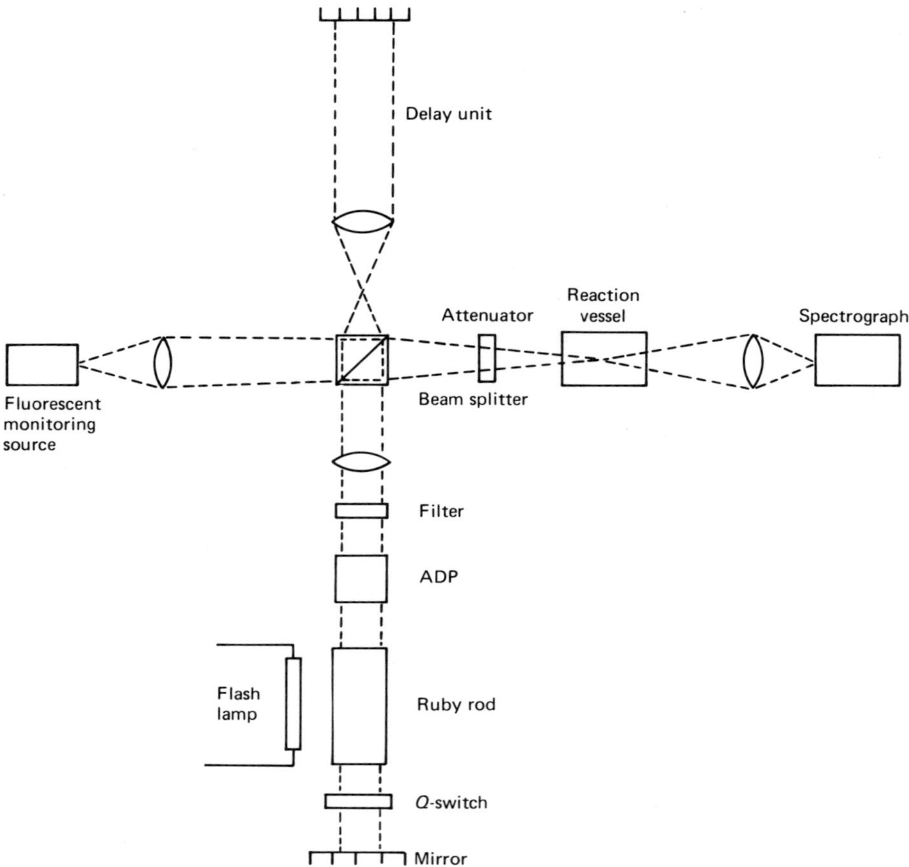

FIGURE 3.11. Nanosecond flash spectrographic apparatus. (Adapted from Porter and Topp.[14a])

processes occurring in 3,4-benzpyrene in cyclohexane: (a) an absorption band due to $T_1 \rightarrow T_x$, (b) an emission band corresponding to normal fluorescence, $S_1 \rightarrow S_0 + h\nu$, and (c) an absorption band corresponding to $S_1 \rightarrow S_x$. That this latter absorption band involved the first excited singlet and not the ground state followed from the fact that the lifetime of the observed transition was equal to the fluorescence lifetime under the same conditions. In addition, the rate of decay of the intermediate species was found to be equal to the rate of increase in the triplet population. Data for these transitions and those for other compounds studied are given in Table 3.3.[14b] The shortest lifetime successfully determined for a transient species so far is that of the 3,4-benzpyrene singlet, having a decay lifetime of about 39 nsec.

TABLE 3.3. *High Singlet and Triplet Transitions by Flash Photolysis*[a]

Compound	Wavelength, nm		τ_s, nsec	τ_f, nsec
	$S_1 \rightarrow S_x$	$T_1 \rightarrow T_x$		
Phenanthrene	545, 515	520, 510, 481	65	67
Triphenylene	500, 465, 433	428	45	44
3,4-Benzpyrene	590, 535, 510	480	39	41
1,2,3,4-Dibenzanthracene	540, 500	445	50	52.5
Coronene	600, 570, 530, 495, 465	525, 480	390	380
Pyrene	515, 480, 470	520, 483, 416	326	319

[a] Determined in polymethylmethacrylate.

The latest system Porter has developed for flash kinetic photometry is diagrammed in Figure 3.12.[15] This apparatus is similar to that of Figure 3.11 in that a ruby laser pulse (half-peak width of 13 nsec) is used as the flash source and the resulting emission is frequency-doubled with ADP. However, in this case part of the beam is deflected by the beam splitter to the trigger photomultiplier, which activates the oscilloscope and applies full voltage to the monitoring photomultiplier. In addition, a continuous arc lamp is used to provide the monitoring source. A pulsing unit consisting of 0.02-F capacitors is placed in circuit with the steady-state lamp to provide short pulses of up to 100 times higher light intensity. This maximizes the signal-to-noise ratio, which is proportional to the square of the light intensity. Although the monitoring source is not actually continuous in intensity, the

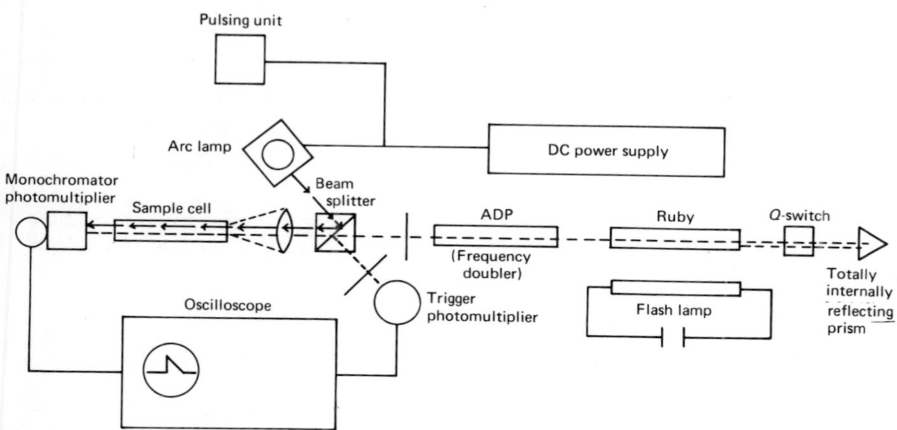

FIGURE 3.12. Nanosecond flash kinetic spectrophotometric apparatus. (Adapted from Formosinho *et al.*[15])

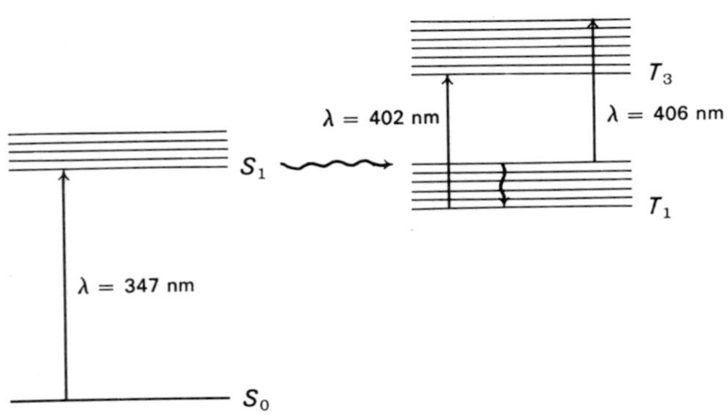

FIGURE 3.13

pulses being superimposed on the normal emitted intensity, on a nanosecond time scale one sees only the higher intensity occurring during the pulse.

By placing an oven which could be heated to 400°C around the sample cell, Porter has been able to study the electronic events occurring in gases. This research was stimulated by the observation obtained from a microsecond flash study that as the pressure of a gas increases, its fluorescence quantum yield decreases and its intersystem crossing quantum yield increases. With the nanosecond technique it was possible to observe the growth and decay of the triplet population by monitoring the intensity of the triplet to triplet absorption as a function of time after the initial flash. For anthracene vapors at 200°C at total gas pressures of 100 Torr or greater, the triplet population had already reached its maximum value by the time monitoring could be performed (about 20 nsec), as evidenced by the decay of the $T_1 \rightarrow T_3$ absorption ($\lambda = 402$ nm). At lower pressures there occurred an instantaneous rise in triplet population (<20 nsec) followed by a slower rise in the 402-nm absorption and a decrease in absorption above 408 nm. The duration of this second rise increased as the pressure was decreased. This has been attributed to pressure-dependent vibrational relaxation of the vibrationally excited T_1 state as shown in Figure 3.13.

The rise in Φ_{isc} with pressure (and the resulting decrease in Φ_f) can now be interpreted as arising from faster vibrational relaxation of the triplet from the point of crossover from the singlet with increasing numbers of molecular collisions. At lower pressures, where vibrationally deactivating collisions are few, the triplet has the possibility of crossing back over to the singlet level before deactivation has occurred.

Let us now return to the question of the benzophenone triplet state. The problem was to explain why the benzophenone triplet lifetime is shorter in benzene (10^{-5}–10^{-6} sec) than in other inert solvents (10^{-3} sec). A number of possibilities could explain this phenomenon. The first and most obvious is that the benzophenone is undergoing photoreduction by hydrogen abstraction from benzene.[8,12] As we have seen, however, this cannot be the case since the maximum quantum yield for reaction of benzophenone in benzene has been found to be 0.005 ± 0.001. Alternatively, one could propose that a hydrogen radical is abstracted from benzene but the reaction is reversed before the resulting radical pair can escape from the solvent cage. Another possibility is that of a charge transfer interaction with the benzophenone as acceptor and benzene as donor.[8,12,17] Still another is the reversible formation of an unstable diradical adduct such as[10,13,18–20]

$$\left(\phi_2C{=}O\right)^t + C_6H_6 \longrightarrow \left[\phi_2\dot{C}{-}O{-}C_6H_6{\cdot}{-}H\right] \qquad (3.35)$$

Finally, one has to consider that an unknown impurity present in the benzene is responsible for quenching the triplet state.[8,12,16] We then have four viable possibilities to explain this effect.

On flash photolysis of benzophenone in benzene two transients are observed.[9–11] One of these, the shorter lived, is assumed to be the excited triplet of benzophenone, and the longer lived transient would appear to be the ketyl radical by comparison of its spectrum with that of authentic ketyl radical formed in solvents that are good hydrogen donors. This, however, may be an entirely new species which fortuitously absorbs in the same region as the ketyl radical.

Some further information on this system has been obtained by nanosecond flash photolysis. These data are shown in Table 3.4.[14b,21] By using

TABLE 3.4

Solvent	k_d,[a] sec^{-1}	D_0/D_∞[b]
C_6H_6	$(1.0 \pm 0.1) \times 10^5$	90 ± 0.5
C_6D_6	$(0.8 \pm 0.14) \times 10^5$	12.5 ± 1.0
C_6F_6	$(1.4 \pm 0.01) \times 10^6$	15.0 ± 2.0
IPA	2.2×10^7	1.7 ± 0.2
Ethanol	9.7×10^6	1.5 ± 0.2
Dioxane	5.0×10^6	1.5 ± 0.2
Cyclohexane	3.3×10^6	2.0 ± 0.3

[a] Rate constant for decay of the benzophenone triplet.
[b] Ratio of the optical density of the benzophenone triplet to that of the ketyl radical.

highly purified benzene, the possibility that impurity quenching of the benzophenone triplet state is responsible for the short triplet lifetime was eliminated. The data in Table 3.4 indicate that the rate of decay of the triplet in benzene and that in deuteriobenzene are the same.[(21)] This is in marked contrast to what occurs when deuteriobenzhydrol is used in place of benzhydrol as a hydrogen donor in the photoreduction, a case that exhibits a strong isotope effect. Hence the lack of an isotope effect with deuteriobenzene effectively eliminates the possibility that a reversible hydrogen abstraction is responsible for the shortened lifetime. Additional evidence that hydrogen abstraction is not involved comes from a comparison of the D_0/D_∞ values for benzene and for the good hydrogen donors in which the ketyl radical is undoubtedly formed.[(14b)] Clearly these values are quite different.

That charge-transfer effects are not involved follows from the fact that the rate of triplet decay in perfluorobenzene is larger than that in benzene. If the benzophenone triplet were to act as acceptor and the benzene derivative as donor in a charge-transfer complex, the substitution of perfluorobenzene for benzene should render this type of process *much less* probable due to the strongly electron-withdrawing character of the fluorine atoms.

The only conclusion that can be drawn is that the lifetime of the benzophenone triplet in benzene is shorter than in other inert solvents due to the formation of a diradical species with the solvent which happens to absorb in the same region as the benzophenone triplet itself and the ketyl radical. Thus the advent of nanosecond flash spectrometry has allowed the solution of an interesting and difficult problem.

3.3. THE PHOTOREDUCTION OF ARYL KETONES: STRUCTURAL ASPECTS

So far in this chapter we have discussed only the photoreduction of benzophenone. We will now concern ourselves with the effect of structure on the photoreduction of aromatic ketones in general.

The following three carbonyl compounds are photoreduced in isopropyl alchohol as solvent or in the presence of benzhydrol[(1,22–28)]:

On the other hand, the following compounds do not undergo photoreduction under these conditions[27,29–32]:

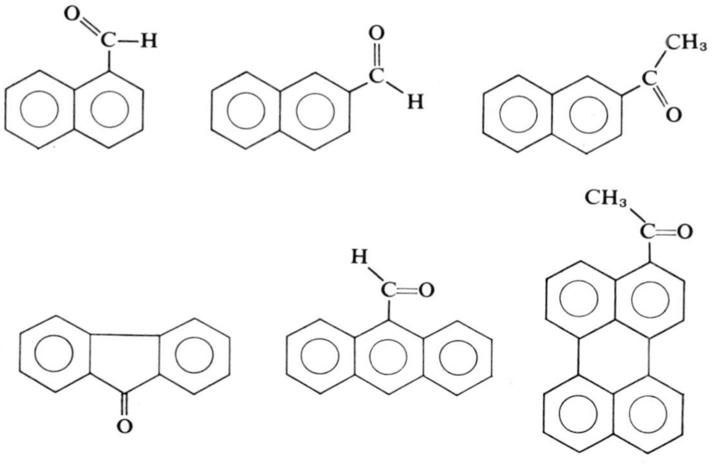

The question is, what difference (or differences) is responsible for the fact that photoreduction occurs in the first group of compounds but not in the second? There are four possible reasons for this behavior. We shall deal with each of these in turn and attempt to decide which factor is responsible.

(a) *Energetics.* As stated earlier, the rate-limiting step in the photoreduction of benzophenone is the abstraction of a hydrogen atom by the excited triplet ketone to produce a ketyl radical,

$$R_1-\overset{\overset{\displaystyle O}{\|}}{C}-R_2' + \phi-\underset{\underset{\displaystyle H}{|}}{\overset{\overset{\displaystyle \phi}{|}}{C}}-OH \longrightarrow R_1-\overset{\overset{\displaystyle OH}{|}}{\dot{C}}-R_2 + \phi-\overset{\overset{\displaystyle \phi}{|}}{\dot{C}}-OH \tag{3.36}$$

Possibly for those compounds that do not photoreduce the triplet does not have enough energy to abstract a proton to form the ketyl radical. To explore this possibility, one must look at the overall energetics of the abstraction reaction.

The triplet energies of some compounds of interest are[29,34–38]:

benzophenone	$E_t = 68.5$ kcal/mole
acetophenone	$E_t = 73.6$ kcal/mole
fluorenone	$E_t = 53.3$ kcal/mole

Some C—H bond energies are given in Table 3.5.

TABLE 3.5. *C—H Bond Energies*[38,40]

Compound	C—H, kcal/mole
CH_3CH_2—H	98
$(CH_3)_2CH$—H	94.5
$(CH_3)_3C$—H	91
ϕ—H	103
ϕ—CH_2—H	85
ϕ_3C—H	75
$HOCH_2$—H	93

Let us now look at the energetics of the abstraction reaction in detail for benzophenone[40]:

$$(\phi_2C{=}O)^t \longrightarrow \phi_2C{=}O \qquad \Delta H = -69 \text{ kcal/mole}$$
$$\phi_2C{=}O + H_2 \longrightarrow \phi_2CHOH \qquad \Delta H = -9 \text{ kcal/mole}$$
$$2H\cdot \longrightarrow H_2 \qquad \Delta H = -104 \text{ kcal/mole}$$
$$\phi_2CHOH \longrightarrow H\cdot + \phi_2\dot{C}{-}OH \qquad \Delta H \approx +78 \text{ kcal/mole}$$

$$(\phi_2C{=}O)^t + H\cdot \longrightarrow \phi_2\dot{C}{-}OH \qquad \Delta H = -104 \text{ kcal/mole}$$
$$RH \longrightarrow R\cdot + H\cdot \qquad \Delta H = +78 \text{ kcal/mole}$$

$$(\phi_2C{=}O)^t + RH \longrightarrow \phi_2\dot{C}{-}OH + R\cdot \qquad \Delta H = -26 \text{ kcal/mole}$$

Thus for the photoreduction of benzophenone in the presence of benzhydrol (RH = ϕ_2CHOH, C—H $\approx$ 78 kcal/mole) the overall reaction is *exothermic* by 26 kcal/mole. For the photoreduction of acetophenone the value calculated in this way is -31 kcal/mole, while for fluorenone the abstraction is still exothermic by 10 kcal/mole. Clearly, the energetics of the abstraction reaction is not the reason why benzophenone and acetophenone photoreduce but fluorenone does not, since the values for the overall reaction would indicate that all three compounds should react. However, this conclusion is not necessarily valid if large activation energies are involved.*

(b) *Reversible Reaction.* Another possibility to be examined is that the

* It has been found that fluorenone, 1-naphthaldehyde, and 2-acetonaphthone can be smoothly photoreduced in the presence of tri-*n*-butylstannane, an extremely good hydrogen donor.[29,33] Apparently hydrogen abstraction for fluorenone and the other compounds in this group does require a relatively high activation energy. Compounds in the first group do not seem to possess a high activation energy (benzophenone is photoreduced in cyclohexane).

abstraction reaction is reversible with the stability of the ketyl radical being strongly dependent on its substituents:

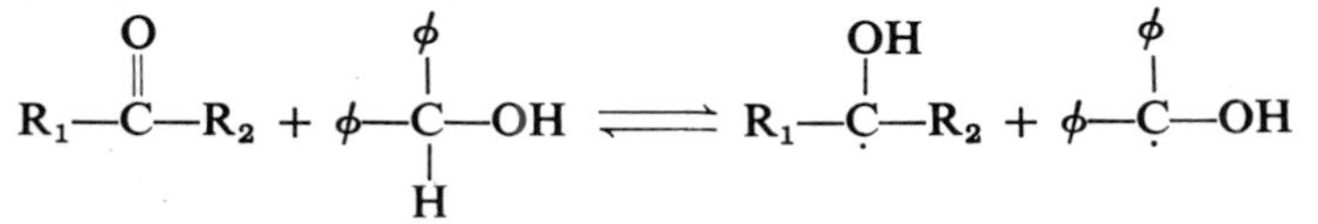

$$R_1-\overset{\displaystyle O}{\overset{\|}{C}}-R_2 + \phi-\overset{\displaystyle \phi}{\underset{\displaystyle H}{C}}-OH \rightleftharpoons R_1-\overset{\displaystyle OH}{\dot{C}}-R_2 + \phi-\overset{\displaystyle \phi}{\dot{C}}-OH \qquad (3.37)$$

To test this hypothesis, α-naphthaldehyde, which is apparently inert toward photoreduction, was irradiated in the presence of optically active 2-octanol. If a reversible hydrogen abstraction were to occur, a loss of optical activity in the 2-octanol should result. The results showed no loss in optical activity; thus the question of a reversible reaction has been answered.[33]

(c) *Competitive Reactions*. Although it is difficult to give a definite answer to the question of the possibility of competitive reactions being responsible for the failure to observe photoreductions of those compounds mentioned previously, this cannot be a general reason for the lack of reactivity toward hydrogen abstraction since some of these compounds can be isolated totally unchanged after photolysis. In some cases, however, other photochemical products are produced and these competing reactions could prevent the photoreductions if they are more efficient. An example of where a competing process does indeed prevent photoreduction is in the case of *o*-methylbenzophenone. Since the methyl substituent should not appreciably affect the energetics of the photoreduction of the benzophenone nucleus, the most logical explanation for the lack of reactivity of this compound involves hydrogen abstraction from the *o*-methyl group[41]:

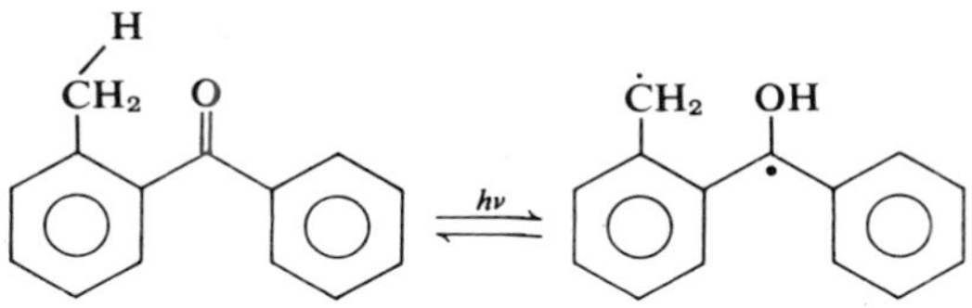

H
CH₂ O ⇌(hν) ĊH₂ OH (3.38)

(d) *The Nature of the Excited State*. In the case of aromatic ketones we are actually dealing with two distinct chromophores: that due to the aromatic nucleus ($\pi \rightarrow \pi^*$ transitions) and that due to the carbonyl group ($n \rightarrow \pi^*$ transitions). For a simple aromatic carbonyl compound such as benzaldehyde or benzophenone the lowest triplet state is $n \rightarrow \pi^*$ in character. However, as the conjugation in the aromatic nucleus becomes greater (e.g., a naphthyl group), the $\pi \rightarrow \pi^*$ triplet is lowered in energy and in many cases is actually the lowest triplet state. Since the chemistry we are observing is localized at the carbonyl group, one might expect that a transition localized generally within this same area ($n \rightarrow \pi^*$) would be responsible for the photoreduction, and one localized primarily in the aromatic nucleus would be inactive

TABLE 3.6. *Quantum Yields for the Photoreduction of Benzophenones*[a]

Ketone	Φ_r in IPA	T_1	EPA, IPA E_T, cm^{-1}	EPA $\tau^{77°}$, msec	Φ_r in C_6H_{12}	T_1	MC E_T, cm^{-1}	MC $\tau^{77°}$, msec
Benzophenone	1.0	$n \to \pi^*$	24,400 (413 nm)	5.4	0.5	$n \to \pi^*$	24,000 (417 nm)	5.1
4-F—	1.0	$n \to \pi^*$	24,400	5.6	—	$n \to \pi^*$	24,100	13
4-Cl—	1.0	$n \to \pi^*$			—	$n \to \pi^*$		
4-Br—	1.0	$n \to \pi^*$	24,400	5.4	—	$n \to \pi^*$	24,100	21
4-ϕ— (and 3-ϕ—)	0.1	$\pi \to \pi^*$	22,000	200	0.05	$\pi \to \pi^*$	21,500	150
4,4′-MeO—	1.0	$n \to \pi^*$	24,400	32	—	$n \to \pi^*$	24,100	25
4-HO—	0.02	CT	24,400	36	0.9	$n \to \pi^*$	24,100	34
3-H_2N—	0.03	CT	20,000	145	0.03	CT, $\pi \to \pi^*$	22,300	100
4-H_2N—	0.00	CT	22,000	200	0.20	?	23,500	33
4,4′-Me_2N—	0.00	CT	22,000	106	0.60	$n \to \pi^*$	23,000	41

[a] IPA, isopropyl alcohol; MC, methyl cyclohexane; EPA, 5:5:2, ether: isopentane: alcohol; $\tau^{77°}$ is the lifetime in rigid glasses at 77°K (see Section 5.1 and Table 5.1).

toward photoreduction. This might then explain why some compounds photoreduce and some do not.

To obtain more information on this point, let us examine the data given in Table 3.6[42–47] for some substituted benzophenones. The data in Table 3.6 indicate that benzophenone derivatives having lowest triplet states of $n \to \pi^*$ character undergo very efficient photoreduction in isopropyl alcohol. Those derivatives having a lowest $\pi \to \pi^*$ triplet, on the other hand, are only poorly photoreduced, while those having lowest triplets of the charge-transfer type are the least reactive toward photoreduction. In additon, in some cases photoreduction is more efficient in the nonpolar solvent cyclohexane than in isopropanol. This arises from the solvent effect on the transition energies for $n \to \pi^*$, $\pi \to \pi^*$, and CT transitions discussed in Chapter 1 (see also Table 3.7).

Since CT and $\pi \to \pi^*$ transitions are red-shifted (i.e., toward lower energy) in polar solvents such as isopropyl alcohol and $n \to \pi^*$ transitions

TABLE 3.7. *Solvent Shifts for Various Transitions in p-Substituted Benzophenones*[44],[a]

State	Intensity (log ε)	Solvent shift, cm^{-1}	Energy range, cm^{-1}	S–T splitting, cm^{-1}
$n \to \pi^*$	2	—800	30,000	$<3,000$
$\pi \to \pi^*$	4	+600	40,000	10,000
CT	3–4	+2500	30,000	$>3,000$

[a] In proceeding from a polar to a nonpolar solvent.

are blue-shifted, replacement of the polar solvent with cyclohexane lowers the energy of the $n \rightarrow \pi^*$ triplet state so that, in the case of 4,4′-dimethylaminobenzophenone, the lowest triplet is of the $n \rightarrow \pi^*$ type and the quantum yield for photoreduction rises from zero in isopropanol to 0.6 in cyclohexane.[(44)]

Coincident with the decrease in the quantum yield of photoreduction with lowest $\pi \rightarrow \pi^*$ and CT triplet states, there is an increase in the triplet lifetime. This behavior is also observed in the photoreduction of substituted acetophenones, the data for which are presented in Table 3.8.[(49.50)] In this case the quantum yield for photoreduction in isopropanol drops by a factor of 1000 from that of acetophenone itself to that of *m*-methoxyacetophenone although there is only a minor difference in their triplet energies. These facts again indicate that a factor related to the electronic structure of the lowest triplet state must be responsible for the change in reactivity with substitution. We note in Table 3.8 that the lowest triplet of acetophenone is an $n \rightarrow \pi^*$ state while that of *m*-methoxyacetophenone is $\pi \rightarrow \pi^*$.[(49)] A qualitative idea of the differences in electronic structure in these three types of triplet states can be obtained from the following exaggerated structures:

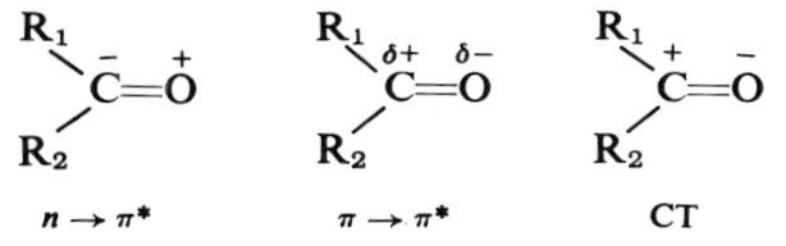

$$\underset{n \rightarrow \pi^*}{(R_1)(R_2)\overset{-}{C}=\overset{+}{O}} \qquad \underset{\pi \rightarrow \pi^*}{(R_1)(R_2)\overset{\delta+}{C}=\overset{\delta-}{O}} \qquad \underset{\text{CT}}{(R_1)(R_2)\overset{+}{C}=\overset{-}{O}}$$

In an $n \rightarrow \pi^*$ transition an electron originally located on the oxygen atom is removed to a region between the carbon and the oxygen, leaving the latter with a partial formal positive charge. In a $\pi \rightarrow \pi^*$ transition of an aromatic ketone the carbonyl group does not participate to any great extent; hence the normal polarization of the carbonyl is preserved. In a CT transition where the carbonyl is the electron acceptor a formal negative charge appears on the oxygen atom. Thus it would appear that in order for photoreduction to be efficient the carbonyl oxygen must bear a positive charge (or at least a greater amount of positive charge than in the ground state configuration) such as would result from an $n \rightarrow \pi^*$ transition.

This raises the question as to why one should observe any photoreduction at all with those compounds in which the lowest triplet clearly is of $\pi \rightarrow \pi^*$ character. The answer to this question lies in the fact that the electronic states are not pure in character. That is, an $n \rightarrow \pi^*$ state has some $\pi \rightarrow \pi^*$ character obtained from a mixing of other electronic levels. This mixing results from vibronic and spin–orbit coupling.[(51,52)] Likewise a $\pi \rightarrow \pi^*$ state has a degree of $n \rightarrow \pi^*$ character obtained by mixing. It is this degree of $n \rightarrow \pi^*$ character which is responsible for the observation of photoreduction of a compound with a lowest $\pi \rightarrow \pi^*$ triplet. One can write for the triplet state

$$T = a(n \rightarrow \pi^*) + b(\pi \rightarrow \pi^*) \tag{3.39}$$

where a and b are mixing coefficients, the magnitudes of which determine the relative contribution of each type of transition to the final state T. In Table 3.8 one finds that although the quantum yields for acetophenone and *p*-methylacetopheneone photoreduction in 2 *M* isopropanol in benzene are essentially the same, the compounds differ in the nature of their lowest triplet states, as indicated also by the differences in their phosphorescence (triplet) lifetimes and photoreduction rate constants. This must result from differing amounts of $n \rightarrow \pi^*$ and $\pi \rightarrow \pi^*$ character in the lowest triplets of these molecules.[50]

From this discussion we can draw the following conclusions:

1. Differences in reactivity toward photoreduction result from differences in the character of the lowest excited triplet state.

2. Compounds with a lowest $n \rightarrow \pi^*$ triplet will be the most reactive toward photoreduction.

3. Factors that reduce the energy gap between upper $\pi \rightarrow \pi^*$ triplet states and the lowest $n \rightarrow \pi^*$ triplet will lower the photoreactivity.

4. For compounds having lowest $\pi \rightarrow \pi^*$ triplet states, factors that reduce the energy gap between the upper $n \rightarrow \pi^*$ state and the lowest triplet will increase the reactivity toward photoreduction.

5. For compounds having lowest $\pi \rightarrow \pi^*$ triplet states, factors that increase vibronic coupling with higher $n \rightarrow \pi^*$ triplets and spin–orbit coupling with $n \rightarrow \pi^*$ singlets will increase reactivity through mixing of the states. Since vibronic and spin–orbit coupling depend inversely upon the energy separation between the states, closely lying $n \rightarrow \pi^*$ singlet and triplet states will couple more strongly with the lowest $\pi \rightarrow \pi^*$ triplet than higher lying states.

TABLE 3.8. *Quantum Yields for the Photoreduction of Acetophenones*

Compound	2-Propanol 313 nm Φ ketone	E_T, cm^{-1} ($\times 10^{-4}$)	E_T, kcal	τ_p	Φ_p
Acetophenone	0.68	2.59	74.1	0.004	0.74
p-Me—	0.66	2.55	72.8	0.084	0.61
m-Me—	0.46	2.54	72.5	0.074	0.64
3,4-diMe—	0.12	2.50	71.5	0.17	0.56
p-MeO—	0.04	2.50	71.5	0.26	0.68
3,5-diMe—	0.018	2.49	71.3	0.11	0.51
3,4,5-triMe—	0.018	2.46	70.8	0.20	0.46
m-MeO—	0.006	2.53	72.4	0.25	0.35

At this point let us briefly consider the relationship between the carbonyl triplet state and another system capable of hydrogen atom abstraction: alkoxy radicals. A comparison of the differences and/or similarities between the reactivity of the carbonyl triplet and that of an alkoxy radical should indicate whether the triplet state behaves as a normal ground state radical or if electronic excitation imparts unique properties leading to reactions not characteristic of ground state radicals.

As seen earlier in this section, the energy released upon formation of the benzophenone ketyl radical is about 104 kcal/mole. This value is almost identical to that released upon formation of *t*-butyl alcohol from the *t*-butoxy radical,

$$CH_3-\underset{CH_3}{\overset{CH_3}{\underset{|}{\overset{|}{C}}}}-O\cdot + \cdot H \longrightarrow CH_3-\underset{CH_3}{\overset{CH_3}{\underset{|}{\overset{|}{C}}}}-OH \qquad \Delta H = -104 \text{ kcal/mole} \tag{3.40}$$

The *t*-butoxy radical, therefore, should serve as an excellent model to compare with the benzophenone triplet since the energetics are essentially the same.

The *t*-butoxy radical can be produced from the radical decomposition of *t*-butylhypochlorite using azobisisobutyronitrile (AIBN) as catalyst[33]:

$$(CH_3)_2\overset{CN}{\overset{|}{C}}N{=}N\overset{CN}{\overset{|}{C}}(CH_3)_2 \xrightarrow{\Delta} 2(CH_3)_2\overset{CN}{\overset{|}{C}}\cdot + N_2 \tag{3.41}$$

$$(CH_3)_2\overset{CN}{\overset{|}{C}}\cdot + t\text{-BuOCl} \longrightarrow (CH_3)_2\overset{CN}{\overset{|}{C}}Cl + t\text{-BuO}\cdot \tag{3.42}$$

The *t*-butoxy radicals thus formed then abstract hydrogen from hydrocarbons to generate hydrocarbon radicals and *t*-butyl alcohol. The hydrocarbon radicals in turn generate more *t*-butoxy radicals by attacking the *t*-butylhypochlorite:

$$t\text{-BuO}\cdot + RH \longrightarrow t\text{-BuOH} + R\cdot \tag{3.43}$$

$$R\cdot + t\text{-BuOCl} \longrightarrow RCl + t\text{-BuO}\cdot \tag{3.44}$$

For example, the reaction of toluene with *t*-butoxy radicals yields the following products:

$$t\text{-BuO}\cdot + C_6H_5CH_3 \longrightarrow \underset{84\%}{C_6H_5CH_2Cl} + \underset{97\%}{t\text{-BuOH}} + \underset{4.7\%}{C_6H_5CHCl_2} \tag{3.45}$$

The percentages of products are based upon the total amount of *t*-butylhypochlorite decomposed.

TABLE 3.9. *Relative Reactivities of Substrates*[a] *toward Benzophenone Triplet and t-Butoxy Radicals*[40]

Substrate	$(\phi_2C{=}O)^t$	t-BuO·[b]
Benzhydrol	750	—
Isopropanol	29.1	—
Cumene	9.9	6.8
Ethyl benzene	4.6	3.2
2,3-Dimethylbutane (1°)	<0.01	0.10
(3°)	3.0	4.2
Cyclohexane (2°)	0.50	1.5
p-Methyl toluene	2.15	1.53
m-Methyl toluene	1.45	1.17
Toluene	1.00	1.00
p-Chlorotoluene	0.97	0.71

[a] Per reactive hydrogen relative to toluene as standard.
[b] At 40°C.

A comparison of the reactivities of the benzophenone triplet state and the *t*-butoxy radical toward various hydrocarbons is made in Table 3.9. The data for the two systems are indeed very similar although the benzophenone triplet exhibits somewhat higher selectivity, preferring to abstract the tertiary hydrogen from 2,3-dimethylbutane over the primary hydrogen by a factor of more than 300 (the *t*-butoxy radical prefers this hydrogen only by factor of about 40). The greater selectivity of the benzophenone triplet also appears in the case of cyclohexane, the tertiary hydrogen of 2,3-dimethylbutane being six times more reactive than this secondary hydrogen. In general, the agreement between these two systems is very good and one can conclude that the benzophenone triplet state behaves like a ground state radical, at least toward hydrogen abstraction. Thus it appears that upon electronic excitation the resultant triplet parallels a ground state radical in reactions involving similar energetics and *no unique chemical properties result from excitation under these conditions.*[40]

3.4. THE PHOTOREDUCTION OF ARYL KETONES: SECONDARY REACTIONS

As seen earlier in this chapter, the primary chemical process in the photoreduction of aromatic ketones is the abstraction of a hydrogen atom from the solvent, as in the case of the photoreduction of benzophenone in isopropyl alcohol, or from some other molecule present in the reaction media, by the triplet excited ketone to form a ketyl radical. The ultimate product of the photoreduction in the case of benzophenone is benzpinacol, for which the quantum yield of formation approaches one in many cases. In this section we will consider the secondary reactions occurring after the initial formation of the ketyl radical which lead to products derived both from the ketone and from the original hydrogen source. We will start with some general observations concerning these secondary reactions[(57)] and then attempt to account for these observations by describing the experimental approaches which have led to the complete elucidation of the mechanism of photoreduction.

We saw earlier that when benzophenone is photoreduced in the presence of optically active 2-butanol, the alcohol recovered from the reaction loses no optical activity.[(54)] This was presented as evidence that there could be no appreciable reversibility of the initial hydrogen abstraction since this should lead to racemization of the unreacted alcohol. However, if one uses ^{14}C-labeled benzhydrol and examines the initially produced benzpinacol for the presence of the label, one finds that the product pinacol contains no ^{14}C. This would indicate that there must be some type of rapid transfer of the hydrogen radical from the ketyl radical produced upon abstraction from benzhydrol,

$$\phi\text{—}\overset{\text{OH}}{\underset{\text{H}}{\text{C}_*}}\text{—}\phi \xrightarrow{\left(\phi\text{—}\overset{\text{O}}{\overset{\|}{\text{C}}}\text{—}\phi\right)^t} \phi\text{—}\overset{\text{OH}}{\underset{\cdot}{\text{C}^*}}\text{—}\phi + \phi\text{—}\overset{\text{OH}}{\underset{\cdot}{\text{C}}}\text{—}\phi \quad \not\downarrow \quad \phi\text{—}\overset{\text{OH}}{\underset{\phi}{\text{C}_*}}\text{—}\overset{\text{OH}}{\underset{\phi}{\text{C}}}\text{—}\phi \tag{3.46}$$

A possibility is fast exchange with ground state benzophenone:

$$\phi\text{—}\overset{\text{OH}}{\underset{\cdot}{\text{C}^*}}\text{—}\phi + \phi\text{—}\overset{\text{O}}{\overset{\|}{\text{C}}}\text{—}\phi \longrightarrow \phi\text{—}\overset{\text{O}}{\underset{*}{\overset{\|}{\text{C}}}}\text{—}\phi + \phi\text{—}\overset{\text{OH}}{\underset{\cdot}{\text{C}}}\text{—}\phi \tag{3.47}$$

A similar result has been obtained by Schenck *et al.*[(56)] When perdeuterobenzophenone is photoreduced with benzhydrol, only perdeuterated benzpinacol is isolated,

$$2(C_6D_5)_2CO + (C_6H_5)_2CHOH \xrightarrow{h\nu} (C_6D_5)_2\overset{\text{OH}}{\text{C}}\text{——}\overset{\text{OH}}{\text{C}}(C_6D_5)_2 + (C_6H_5)_2CO \tag{3.48}$$

When benzophenone is photoreduced in isopropanol as solvent, benzpinacol and acetone are produced:

$$2\phi-\overset{\overset{\displaystyle O}{\|}}{C}-\phi + CH_3-\underset{\underset{\displaystyle H}{|}}{\overset{\overset{\displaystyle OH}{|}}{C}}-CH_3 \xrightarrow{h\nu} \phi-\underset{\underset{\displaystyle \phi}{|}}{\overset{\overset{\displaystyle OH}{|}}{C}}-\underset{\underset{\displaystyle \phi}{|}}{\overset{\overset{\displaystyle OH}{|}}{C}}-\phi + CH_3-\overset{\overset{\displaystyle O}{\|}}{C}-CH_3 \quad (3.49)$$

$$\tfrac{1}{2}\Phi\left(\phi-\overset{\overset{\displaystyle O}{\|}}{C}-\phi\right) = \Phi\left(CH_3\overset{\overset{\displaystyle O}{\|}}{C}CH_3\right) = \Phi(\phi_2COHCOH\phi_2) \quad (3.50)$$

The quantum yield for the production of benzpinacol is equal to that for the production of acetone and also equal to one-half the quantum yield for the disappearance of benzophenone (this latter value is about two).[58]

If the primary chemical process is still abstraction of a hydrogen atom to produce the benzophenone ketyl radical, one would expect to observe some mixed pinacol product arising from coupling of the benzophenone ketyl and acetone ketyl radicals,

$$\left(\phi-\overset{\overset{\displaystyle O}{\|}}{C}-\phi\right)^t + CH_3-\underset{\underset{\displaystyle H}{|}}{\overset{\overset{\displaystyle OH}{|}}{C}}-CH_3 \longrightarrow \phi-\overset{\overset{\displaystyle OH}{|}}{\dot{C}}-\phi + CH_3-\overset{\overset{\displaystyle OH}{|}}{\dot{C}}-CH_3 \;\not\downarrow\; \phi-\underset{\underset{\displaystyle \phi}{|}}{\overset{\overset{\displaystyle OH}{|}}{C}}-\underset{\underset{\displaystyle CH_3}{|}}{\overset{\overset{\displaystyle OH}{|}}{C}}-CH_3 \quad (3.51)$$

It has been stated (incorrectly, as we shall see) that one observes only acetone and benzpinacol.

In Section 3.1 it was shown that the photoreduction of benzophenone can be quenched by addition of small amounts of triplet quenchers such as oxygen or ferric dipivaloylmethide.[60] In fact this was presented as evidence that the benzophenone triplet was involved in the photoreduction. This reaction can also be quenched by naphthalene. In the presence of naphthalene, light is still absorbed by benzophenone and thus benzophenone triplets are produced. However, photoreduction products are decreased. On examining this reaction with flash photolysis, triplet–triplet absorptions were observed but these absorptions corresponded to those of the naphthalene triplet. Thus the triplet excitation energy originally present in the benzophenone triplet must have been transferred to naphthalene and since little of the photoreduction product was observed, this transfer must have been fast in relation to hydrogen abstraction.[58,61–63] A similar direct quenching of the excited triplet is expected for oxygen and FeDPM.

The photoreduction is also quenched by the addition of small amounts of mercaptans or disulfides. However, this type of quenching is thought to be different from that observed in the presence of naphthalene. This follows from the fact that in the absence of a mercaptan or a disulfide, photoreduction of benzophenone in optically active 2-octanol occurs normally and no racemization of the unreacted alcohol results. In the presence of only small amounts of mercaptan, however, the quantum yield of photoreduction is drastically reduced and the unreacted alcohol is racemized.[64] This indicates that the quenching effect of the sulfur compounds occurs on the secondary reactions rather than on the primary process as in the case of naphthalene or oxygen. Additional evidence that quenching is of a different type in these two cases is obtained by examining the effect of using sodium isopropoxide in place of isopropyl alcohol with and without the mercaptan quencher. When benzophenone is irradiated in the presence of sodium isopropoxide, photoreduction occurs although the product formed is not the pinacol but rather benzhydrol. Using naphthalene as quencher, one is able to quench the formation of benzhydrol as effectively as the formation of benzpinacol when isopropanol is used, indicating that a common precursor (the ketone triplet) is being quenched. However, with mercaptan as quencher, the products are not quenched with equal effectiveness, indicating that two different secondary intermediates are involved. These intermediates are proposed to be (1) for photoreduction in isopropanol and (2) for photoreduction with sodium isopropoxide.[65]

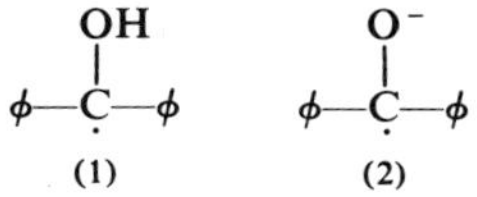

$$\underset{(1)}{\phi-\dot{C}(OH)-\phi} \qquad \underset{(2)}{\phi-\dot{C}(O^-)-\phi}$$

Let us now discuss some of the characteristics of this quenching with mercaptans and disulfides. Interestingly, both sulfur derivatives are equally effective in inhibiting the photoreduction and are in fact interconverted during the reaction. The same equilibrium mixture of mercaptan and disulfide is obtained regardless of which was initially added to the reaction mixture. Furthermore, there appears to be no appreciable consumption of the sulfur compounds.[64] When benzophenone is irradiated in the presence of isopropanol (OD) and mercaptan, isopropanol containing two deuterium atoms is isolated,

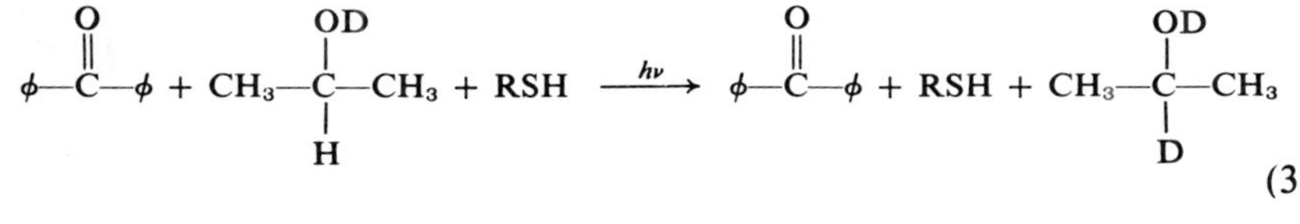

$$\phi-\overset{O}{\overset{\|}{C}}-\phi + CH_3-\underset{H}{\underset{|}{\overset{OD}{\overset{|}{C}}}}-CH_3 + RSH \xrightarrow{h\nu} \phi-\overset{O}{\overset{\|}{C}}-\phi + RSH + CH_3-\underset{D}{\underset{|}{\overset{OD}{\overset{|}{C}}}}-CH_3 \qquad (3.52)$$

In the absence of mercaptan or in the presence of naphthalene as quencher no product of this type is observed.[66] Since the presence of the mercaptan does not prevent the formation of ketyl radicals, the incorporation of the second deuterium must result from transfer of the following type:

$$\left(\phi\text{—}\overset{\displaystyle O}{\overset{\|}{C}}\text{—}\phi\right)^t + CH_3\text{—}\underset{\displaystyle H}{\underset{|}{\overset{\displaystyle OD}{\overset{|}{C}}}}\text{—}CH_3 \longrightarrow \phi\text{—}\overset{\displaystyle OH}{\overset{|}{\dot{C}}}\text{—}\phi + CH_3\text{—}\overset{\displaystyle OD}{\overset{|}{\dot{C}}}\text{—}CH_3 \qquad (3.53)$$

$$CH_3\text{—}\overset{\displaystyle OD}{\overset{|}{\dot{C}}}\text{—}CH_3 + RSSR \longrightarrow CH_3\text{—}\overset{\displaystyle O}{\overset{\|}{C}}\text{—}CH_3 + RS\cdot + RSD \qquad (3.54)$$

$$RSD + CH_3\text{—}\overset{\displaystyle OD}{\overset{|}{\dot{C}}}\text{—}CH_3 \longrightarrow RS\cdot + CH_3\text{—}\underset{\displaystyle D}{\underset{|}{\overset{\displaystyle OD}{\overset{|}{C}}}}\text{—}CH_3 \qquad (3.55)$$

$$2RS\cdot \longrightarrow RSSR \qquad (3.56)$$

The inhibitory function of the sulfur compounds then would appear to be to prevent the subsequent reaction of the initially formed ketyl radicals by catalyzing hydrogen transfer reactions. With this in mind, one can write the following mechanism for the inhibition of the photoreduction of benzophenone by sulfur compounds:

$$\phi\text{—}\overset{\displaystyle O}{\overset{\|}{C}}\text{—}\phi + CH_3\text{—}\underset{\displaystyle H}{\underset{|}{\overset{\displaystyle OH}{\overset{|}{C}}}}\text{—}CH_3 \xrightarrow{h\nu} \phi\text{—}\overset{\displaystyle OH}{\overset{|}{\dot{C}}}\text{—}\phi + CH_3\text{—}\overset{\displaystyle OH}{\overset{|}{\dot{C}}}\text{—}CH_3 \qquad (3.57)$$

$$CH_3\text{—}\overset{\displaystyle OH}{\overset{|}{\dot{C}}}\text{—}CH_3 + \phi\text{—}\overset{\displaystyle O}{\overset{\|}{C}}\text{—}\phi \longrightarrow \phi\text{—}\overset{\displaystyle OH}{\overset{|}{\dot{C}}}\text{—}\phi + CH_3\text{—}\overset{\displaystyle O}{\overset{\|}{C}}\text{—}CH_3 \qquad (3.58)$$

$$CH_3\text{—}\overset{\displaystyle OH}{\overset{|}{\dot{C}}}\text{—}CH_3 + RSH \longrightarrow CH_3\text{—}\underset{\displaystyle H}{\underset{|}{\overset{\displaystyle OH}{\overset{|}{C}}}}\text{—}CH_3 + RS\cdot \qquad (3.59)$$

$$RS\cdot + \phi\text{—}\overset{\displaystyle OH}{\overset{|}{\dot{C}}}\text{—}\phi \longrightarrow RSH + \phi\text{—}\overset{\displaystyle O}{\overset{\|}{C}}\text{—}\phi \qquad (3.60)$$

$$2\phi\text{—}\overset{\displaystyle OH}{\overset{|}{\dot{C}}}\text{—}\phi \longrightarrow \phi\text{—}\underset{\displaystyle \phi}{\underset{|}{\overset{\displaystyle OH}{\overset{|}{C}}}}\text{—}\underset{\displaystyle \phi}{\underset{|}{\overset{\displaystyle OH}{\overset{|}{C}}}}\text{—}\phi \qquad (3.61)$$

$$2RS\cdot \longrightarrow RSSR \qquad (3.62)$$

$$\phi\text{—}\dot{\text{C}}(\text{OH})\text{—}\phi + \text{RSSR} \longrightarrow \text{RSH} + \text{RS}\cdot + \phi\text{—C(=O)—}\phi \tag{3.63}$$

The net result of these reactions (besides a greatly reduced yield of reduction products) is

$$\phi\text{—}\dot{\text{C}}(\text{OH})\text{—}\phi + \text{CH}_3\text{—}\dot{\text{C}}(\text{OH})\text{—CH}_3 \longrightarrow \phi\text{—C(=O)—}\phi + \text{CH}_3\text{—C(OH)(H)—CH}_3 \tag{3.64}$$

This reaction is the reverse of the initial ketyl radical formation by the benzophenone triplet and is thermodynamically favorable. The experiments using optically active alcohols as source of hydrogen atoms show, however, that under normal conditions this reaction is unimportant. This is probably due to other, more efficient pathways for reaction of the ketyl radicals or perhaps to diffusion rates which separate the radicals before reverse transfer can occur. That this reaction can be important in some cases even without the presence of sulfur compounds was shown by studying the photoreduction of benzophenone in optically active ethers.[68] Although the reaction of benzophenone in methyl 2-octyl ether is only 0.17 times as fast as that in isopropanol, ethers can be used as sources of hydrogen atoms for photoreduction:

$$\left(\phi\text{—C(=O)—}\phi\right)^t + \text{CH}_3\text{OCH(CH}_3\text{)—(CH}_2\text{)}_5\text{CH}_3 \xrightarrow{h\nu} \phi\text{—}\dot{\text{C}}(\text{OH})\text{—}\phi + \text{CH}_3\text{O}\dot{\text{C}}(\text{CH}_3)\text{—(CH}_2\text{)}_5\text{CH}_3 \tag{3.65}$$

Using optically active methyl 2-octyl ether, an appreciable racemization of the unreacted ether isolated was observed, in contrast to the result using an alcohol, indicating that about half of the initially produced radicals underwent reverse transfer. The presence of mercaptan or disulfide greatly increased the amount of racemization:

$$\phi\text{—}\dot{\text{C}}(\text{OH})\text{—}\phi + \text{CH}_3\text{O}\dot{\text{C}}(\text{CH}_3)\text{—(CH}_2\text{)}_5\text{CH}_3 \longrightarrow \phi\text{—C(=O)—}\phi + \text{CH}_3\text{OC(CH}_3\text{)(H)—(CH}_2\text{)}_5\text{CH}_3 \tag{3.66}$$

Since transfer of a second hydrogen atom from the ether radical is unreasonable, a pathway available to the acetone ketyl radical in the photoreduction in isopropanol is removed in this system and reverse transfer can occur:

$$\text{CH}_3\text{O}\dot{\text{C}}(\text{CH}_3)\text{—(CH}_2\text{)}_5\text{CH}_3 + \phi\text{—C(=O)—}\phi \not\longrightarrow \phi\text{—}\dot{\text{C}}(\text{OH})\text{—}\phi + \text{ether diradical} \tag{3.67}$$

$$\text{CH}_3\text{—}\dot{\text{C}}(\text{OH})\text{—CH}_3 + \phi\text{—C(=O)—}\phi \longrightarrow \text{CH}_3\text{C(=O)CH}_3 + \phi\text{—}\dot{\text{C}}(\text{OH})\text{—}\phi \tag{3.68}$$

The rate of this step then must be responsible for the lack of racemization of optically active alcohols during photoreduction.

Benzophenone has also been found to be photoreduced in the presence of amines as hydrogen donors, although less efficiently than in the presence of benzhydrol or isopropyl alcohol. The photoreduction of ketones in aromatic amines is thought not to go by the same mechanism as the photoreduction in alcohols, for the following reasons:

(a) The photoreduction with amines shows less selectivity toward the nature of the hydrogen atom than observed with triplet abstraction of hydrogen radicals from alkanes.[68]

(b) The photoreduction of benzophenone by dimethylphenylamine and methyldiphenylamine can be quenched by triplet quenchers such as naphthalene and ferric dibenzoylmethanate but the quenching is much less efficient than for photoreduction with benzpinacol.[69]

(c) Fluorenone can be photoreduced by amines although it is inert in the presence of alcohols.[70,72]

(d) The photoreduction of benzophenone with benzhydrol is inhibited by the presence of tertiary amines such as triphenylamine or tri-*p*-tolylamine. Tri-*p*-tolylamine ($k_q/k_r = 525$) is a more effective quencher than triphenylamine ($k_q/k_r = 44$).[69]

(e) The quenching of the photoreduction by amines is greater in polar solvents than in nonpolar solvents.[69]

These last two points are consistent with electron transfer from the tertiary amines to the triplet ketone in competition with hydrogen abstraction from benzhydrol:

$$\left(\phi\text{—}\overset{\overset{\displaystyle O}{\|}}{C}\text{—}\phi\right)^{t} + \phi_3N\colon \longrightarrow \left[\phi_3\overset{\cdot}{N}{}^{+}\,\bar{O}\text{—}\dot{C}(\phi)_2\right]$$

$$\downarrow\ \phi\text{—}C(OH)(H)\text{—}\phi \qquad\qquad \downarrow \tag{3.69}$$

$$2\phi\text{—}\overset{\overset{\displaystyle OH}{|}}{\dot{C}}\text{—}\phi \qquad\qquad \phi_3N + (\phi)_2C{=}O$$

It is logical that tri-*p*-tolylamine should be a better quencher than triphenylamine since it is better able to stabilize the resulting radical cation. Likewise, a more polar solvent would tend to stabilize the ion pair.[71]

If similar electron transfer occurs from dimethylphenylamine which photoreduces benzophenone, the C—H bonds adjacent to the nitrogen atom may become very labile and proton transfer may occur,

$$\left(\phi\!-\!\overset{\overset{\displaystyle O}{\|}}{C}\!-\!\phi\right)^{t} + (CH_3)_2N\phi \longrightarrow \left[\begin{matrix}\phi\\ \phi\end{matrix}\!\!>\!\dot{C}\!-\!\bar{O}\quad \overset{+}{\dot{N}}\!\begin{matrix}\phi\\ -CH_3\\ CH_3\end{matrix}\right]$$

$$\Big\downarrow \text{proton transfer}$$

$$\begin{matrix}\phi\\ \phi\end{matrix}\!\!>\!\dot{C}\!-\!OH + CH_2\!=\!N\!\!<\!\begin{matrix}\phi\\ CH_3\end{matrix}$$

$$\Big\downarrow$$

$$\phi\!-\!\underset{\underset{\displaystyle OH}{|}}{\overset{\overset{\displaystyle \phi}{|}}{C}}\!-\!\!-\!\underset{\underset{\displaystyle OH}{|}}{\overset{\overset{\displaystyle \phi}{|}}{C}}\!-\!\phi \qquad (3.70)$$

If electron transfer is very rapid, it may be competitive with energy transfer from the ketone triplet to the naphthalene quencher. Hence smaller quenching efficiencies for naphthalene and other quenchers would be observed. Similarly if electron transfer to fluorenone occurs and the amine hydrogens are sufficiently activated, photoreduction of this compound could be observed. The photoreduction of fluorenone by other compounds containing weak bonds to hydrogen, such as tri-*n*-butylstannane, has indeed been reported.[29] What is observed in the case of amines, however, is probably a proton transfer rather than a hydrogen radical transfer and hence, by initial electron transfer followed by proton loss, even relatively strong C—H bonds are efficient in the photoreduction of fluorenone. Similar results are observed using sulfides instead of amines.

We saw earlier that when benzophenone is photoreduced in isopropyl alcohol only benzpinacol and acetone are produced although one would also expect to observe the formation of a mixed pinacol[73]:

$$\phi\!-\!\overset{\overset{\displaystyle O}{\|}}{C}\!-\!\phi + CH_3\!-\!\overset{\overset{\displaystyle OH}{|}}{C}HCH_3 \xrightarrow{h\nu} \phi\!-\!\overset{\overset{\displaystyle OH}{|}}{\dot{C}}\!-\!\phi + CH_3\!-\!\overset{\overset{\displaystyle OH}{|}}{\dot{C}}\!-\!CH_3 \qquad (3.71)$$

$$CH_3\!-\!\overset{\overset{\displaystyle OH}{|}}{\dot{C}}\!-\!CH_3 + \phi\!-\!\overset{\overset{\displaystyle O}{\|}}{C}\!-\!\phi \longrightarrow \phi\!-\!\overset{\overset{\displaystyle OH}{|}}{\dot{C}}\!-\!\phi + CH_3\!-\!\overset{\overset{\displaystyle O}{\|}}{C}\!-\!CH_3 \qquad (3.72)$$

$$2\phi\!-\!\overset{\overset{\displaystyle OH}{|}}{\dot{C}}\!-\!\phi \longrightarrow \phi\!-\!\underset{\underset{\displaystyle \phi}{|}}{\overset{\overset{\displaystyle OH}{|}}{C}}\!-\!\!-\!\underset{\underset{\displaystyle \phi}{|}}{\overset{\overset{\displaystyle OH}{|}}{C}}\!-\!\phi \qquad (3.73)$$

$$\phi-\underset{\cdot}{\overset{OH}{C}}-\phi + CH_3-\underset{\cdot}{\overset{OH}{C}}-CH_3 \not\longrightarrow \phi-\underset{\phi}{\overset{OH}{C}}-\underset{CH_3}{\overset{OH}{C}}-CH_3 \quad (3.74)$$

A number of different mechanisms have been proposed to account for the fact that this product is not observed. Recently, however, a report appeared that described the formation of the mixed pinacol from the photoreduction of benzophenone with isopropyl alcohol and the photoreduction of acetone with benzhydrol. The data from this study are presented in Table 3.10.[73]

When camphorquinone, which is known to scavenge all free ketyl radicals by hydrogen transfer,[74,75] was added to the photolysis mixture only the mixed pinacol and benzhydrol were produced, in the ratio of 1 to 0.8, respectively:

$$\text{camphorquinone} + \phi-\underset{\cdot}{\overset{OH}{C}}-\phi \ (\text{or } CH_3-\underset{\cdot}{\overset{OH}{C}}-CH_3) \longrightarrow \text{(semiquinone radical, O}\cdots\text{H}\cdots\text{O)} + \phi-\overset{O}{\overset{\|}{C}}-\phi \quad (3.75)$$

Since the mixed pinacol is formed in the presence of the ketyl radical scavenger camphorquinone, its formation must occur under conditions where the ketyl radicals are not available for scavenging, that is, in a solvent cage. Since benzpinacol is not observed under these conditions, its formation must occur outside of the initial solvent cage. The following mechanism is proposed:

$$\left(\phi-\overset{O}{\overset{\|}{C}}-\phi\right)^t + CH_3-\underset{H}{\overset{CH_3}{C}}-OH \longrightarrow \overline{[\phi_2\dot{C}OH + (CH_3)_2\dot{C}OH]}_{\text{solvent cage}} \quad (3.76)$$

$$\overline{[\phi_2\dot{C}OH + (CH_3)_2\dot{C}OH]}_{\text{solvent cage}} \xrightarrow{k_1} \phi_2-\overset{OH}{C}-\overset{OH}{C}-(CH_3)_2 \quad (3.77)$$

$$\xrightarrow{k_2} \phi_2CHOH + (CH_3)_2CO \quad (3.78)$$

$$\xrightarrow{k_3} \phi_2\dot{C}-OH + (CH_3)_2\dot{C}OH \ (\text{noncage}) \quad (3.79)$$

The fraction of products arising from the cage reaction then is given by

$$\alpha = (k_1 + k_2)/(k_1 + k_2 + k_3) \quad (3.80)$$

From the relative amounts of mixed pinacol and of benzpinacol in the absence of the scavenger we obtain

$$\frac{k_1}{k_3} = \frac{\text{mixed pinacol}}{\text{benzpinacol}} = a = 0.07 \quad (3.81)$$

TABLE 3.10. *Ratio of "Mixed" to "Normal" Pinacol in the Photoreduction of Benzophenone with Isopropyl Alcohol and Acetone with Benzhydrol*

Time, min	System: $\phi_2C{=}O$, IPA	$(CH_3)_2C{=}O$, Benzhydrol
40	0.066 ± 0.007	—
122	—	0.33 ± 0.02
922	0.070 ± 0.007	0.34 ± 0.02

and the ratio of benzhydrol to mixed pinacol is

$$\frac{k_2}{k_1} = \frac{\text{benzhydrol}}{\text{mixed pinacol}} = b = 0.8 \tag{3.82}$$

Then

$$\alpha = \frac{k_1 + k_2}{k_1 + k_2 + k_3} = \frac{k_1 + bk_1}{k_1 + bk_1 + (k_1/a)} = \frac{1 + b}{1 + b + (1/a)} = 0.11 \tag{3.83}$$

Therefore about 11% of the reaction proceeds within a solvent cage.

This brings up another important question. Since the triplet state of benzophenone has been shown to be responsible for photoreduction, the caged radical pair must initially have unpaired electron spins. Since the products of the cage reaction must be formed in their ground states, one electron must flip its spin. From the data for the photoreduction of acetone with benzhydrol it is possible to estimate a value for α for thermally generated free radicals (that is, singlet radical pairs). A comparison of this value ($\alpha = 0.05$) with that from the photochemical reduction of benzophenone in acetone ($\alpha = 0.11$) indicates that spin flipping is fast compared with diffusion from the cage. Thus the cage radical pair behaves identically whether generated photochemically (triplet state) or thermally (singlet state).[73]

3.5. THE PHOTOREDUCTION OF ARYL KETONES: SYNTHETIC APPLICATIONS

To conclude our study of the photoreduction of aryl ketones, let us now briefly consider some of the ways that this photoreduction has been applied to synthesis.

As part of their study of enzyme models capable of "remote oxidation," Breslow and co-workers have used a benzophenone derivative to functionalize a remote methylene group,[76]

$$C_6H_5-\overset{O}{\overset{\|}{C}}-C_6H_4-\overset{O}{\overset{\|}{C}}-O-CH_2(CH_2)_nCH_3 \xrightarrow{h\nu} CH_3(CH_2)_x-\overset{H}{\overset{|}{C}}-(CH_2)_y-CH_2-O-C(=O)-C_6H_4-\underset{OH}{\underset{|}{C}}(C_6H_5)-$$

1. dehydration
2. ozonolysis
3. hydrolysis

$$CH_3-(CH_2)_x-\overset{O}{\overset{\|}{C}}-(CH_2)_y-CH_2O\overset{O}{\overset{\|}{C}}-C_6H_4-\overset{O}{\overset{\|}{C}}-C_6H_5 \qquad (3.84)$$

As might be expected, a complex mixture of lactones corresponding to abstraction of a hydrogen atom from various sites along the methylene chain was obtained from the photolysis. The mixture of lactones was converted by dehydration, ozonolysis, and hydrolysis to a mixture of ketones. It was found that no functionalization occurs with ester side chains of less than nine carbon atoms. This is probably due to the inability of the carbonyl to approach any methylene closely enough to abstract a hydrogen. The data for side chains of nine carbons or greater is presented in Table 3.11.

The data in Table 3.11 indicate a rather surprising degree of selectivity of abstraction at carbon-12 (especially in a C-14 side chain) and carbon-14 in a C-16 side chain.

TABLE 3.11. *Relative Amount of Functionalization at Various Carbon Atoms as a Function of Chain Length*[76]

	Number of carbon atoms in side chain			
Oxidation site	14	16	18	20
C-9	1.4	1.1	0.1	2.0
C-10	3.0	7.8	8.0	5.0
C-11	11.0	12.0	17.0	15.0
C-12	49	13	21	20
C-13	22	10	18	19
C-14		56	12	19
C-15		7	5	13
C-16			13	8
C-17			6	0.7

To determine whether a more rigid conformation would increase the degree of selectivity of this remote oxidation, Breslow and Baldwin studied a similar system using the 3-α-cholestanol moiety,[77]

O O O —(CH$_2$)$_n$— O C hν OH O

(3.85)

When $n = 0$ or 1, the system appeared to be too rigid to allow the radical pair created upon hydrogen abstraction to form a carbon–carbon bond. Hence a considerable amount of chlorine appears in the product from radical abstraction from the solvent, carbon tetrachloride. When $n = 2$ the radicals are able to form a carbon–carbon bond. After a five-step workup of the crude irradiation product including reduction with LiAlH$_4$, acetylation, dehydration, oxidation with ruthenium tetroxide, and hydrolysis a 16% yield of previously unreported 12-keto-3α-chlorestanol was obtained. However,

O C—CH$_2$CH$_2$— O C O hν R O O=C—CH$_2$CH$_2$— C OH

(3.86)

R HO + R HO

35%

even when $n = 2$ about 67% of the product contained chlorine derived from attack on the solvent. To eliminate this possibility, acetonitrile was used as solvent. This led to an increased yield of carbon–carbon bond formation but workup failed to yield any ketone products. Apparently changing the solvent also changed the place of attack by the excited carbonyl. One obtained, after direct lead tetraacetate cleavage and hydrolysis, instead of ketone products the two steroidal olefins shown in equation (3.86), probably derived from an intermediate lactone.

A further analog ($n = 4$) was studied by Baldwin, Bhatnagar, and Harper.[78] In addition to the two olefins isolated above (44% combined yield), these workers isolated the Δ-16 olefin shown below, probably from attack on the 17 position:

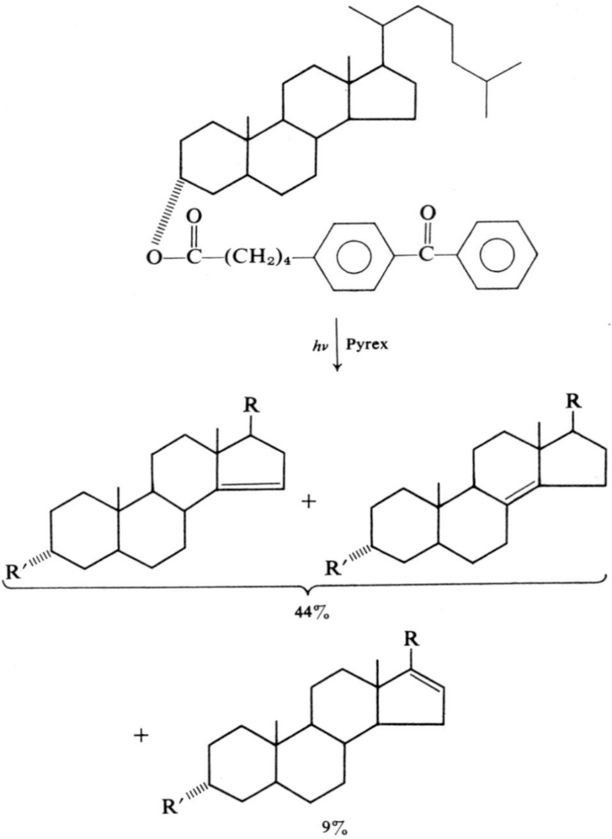

O—C(=O)—$(CH_2)_4$—C(=O)—

hν Pyrex

R R

R′ R′

44%

(3.87)

R

R′

9%

In addition to these olefinic products, a combined yield of 25% of the following lactones was isolated from the reaction mixture:

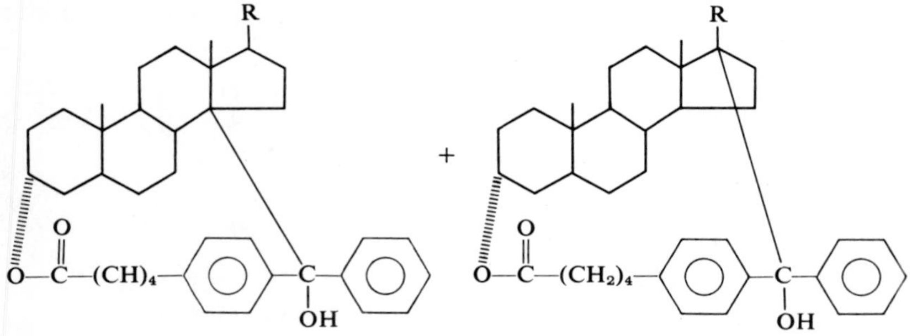

Two more examples of the use of photoreduction as a step in a synthetic preparation are[79,80]

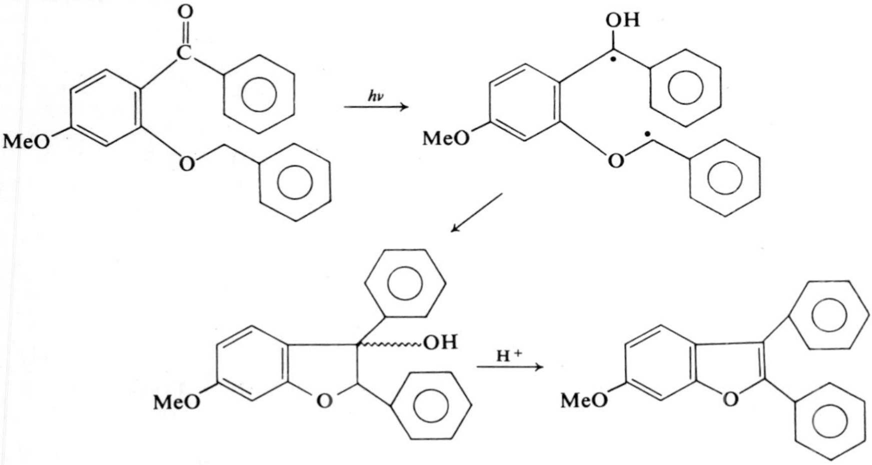

(3.88)

and[80]

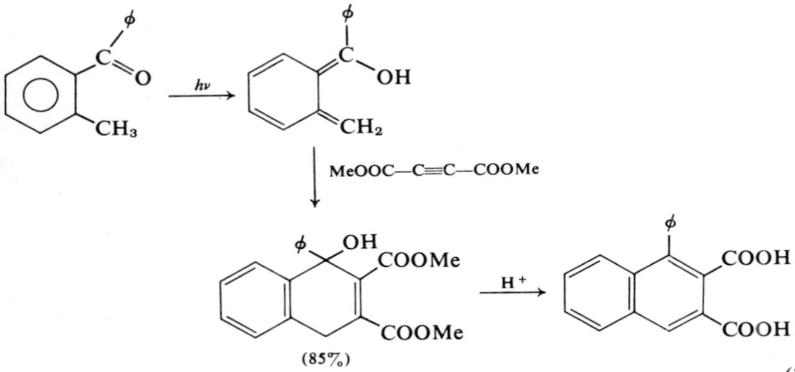

(3.89)

Hindered phenols react with triplet-excited benzophenones in acetone solution to yield fuchsone derivatives,

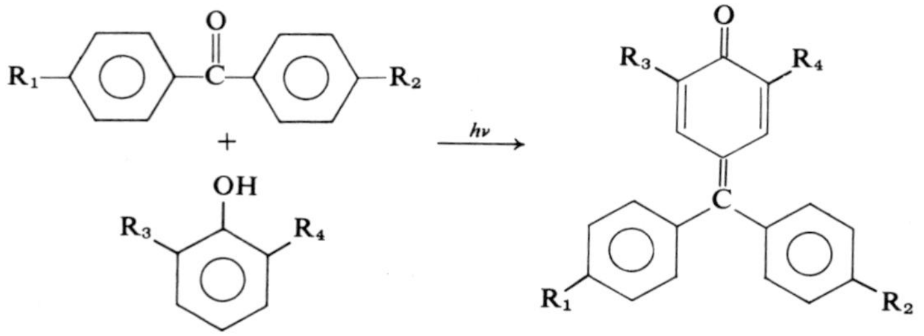

(3.90)

In the presence of catalytic amounts of mineral acid in methanol this reaction yields tetraphenylmethanes,

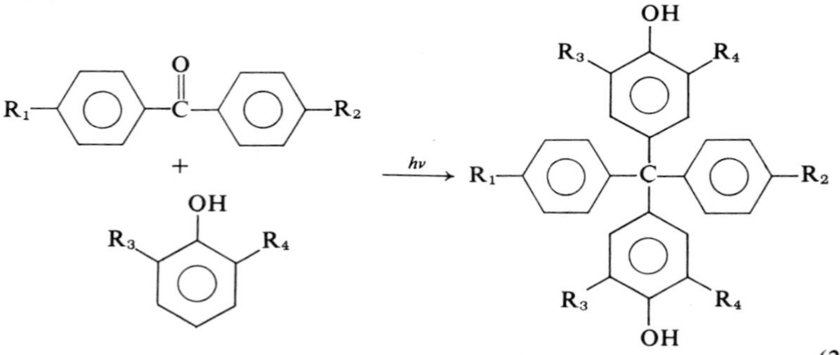

(3.91)

The yields of fuchsone produced as a function of substitution on the ketone and phenol are shown in Table 3.12.

TABLE 3.12. *Yields of Fuchsone as a Function of Substituents*[(81b)]

R_1	R_2	R_3	R_4	% Yield
H	H	*t*-Butyl	*t*-Butyl	47
H	H	Isopropyl	Isopropyl	38
H	H	Cyclohexyl	Cyclohexyl	34
H	CH_3	*t*-Butyl	*t*-Butyl	57
H	CH_3	Isopropyl	Isopropyl	40
H	CH_3	Cyclohexyl	Cyclohexyl	52
H	Cl	*t*-Butyl	*t*-Butyl	52
H	Br	*t*-Butyl	*t*-Butyl	65
H	OH	*t*-Butyl	*t*-Butyl	51
H	$COOCH_3$	*t*-Butyl	*t*-Butyl	54
CH_3	CH_3	*t*-Butyl	*t*-Butyl	41
Cl	Cl	*t*-Butyl	*t*-Butyl	50
CH_3COO	CH_3COO	*t*-Butyl	*t*-Butyl	63

[a] Irradiation in acetone solution followed by acidification.

These very interesting reactions were postulated to occur through the following mechanism:

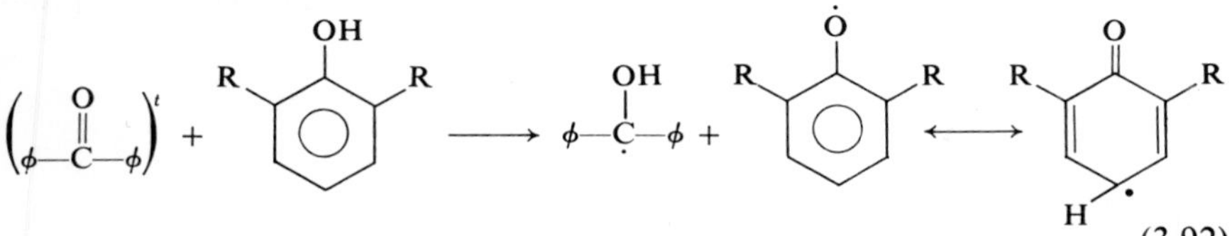

(3.92)

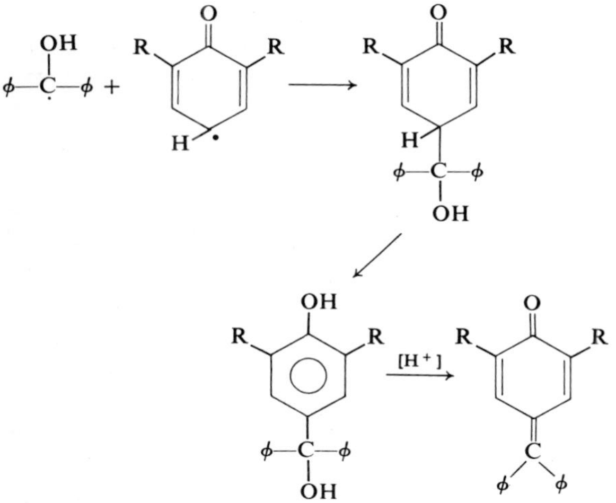

(3.93)

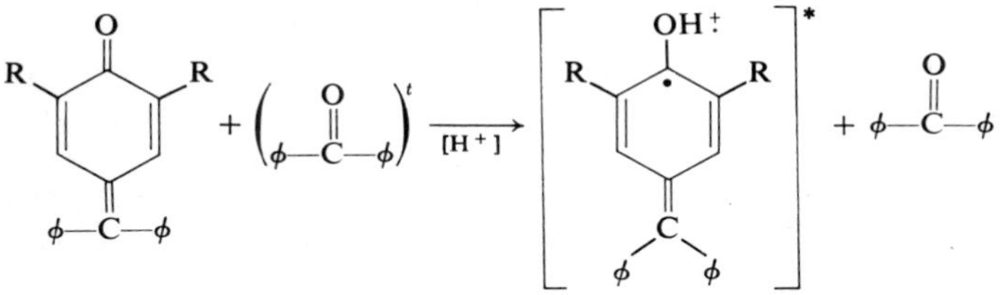

(3.94)

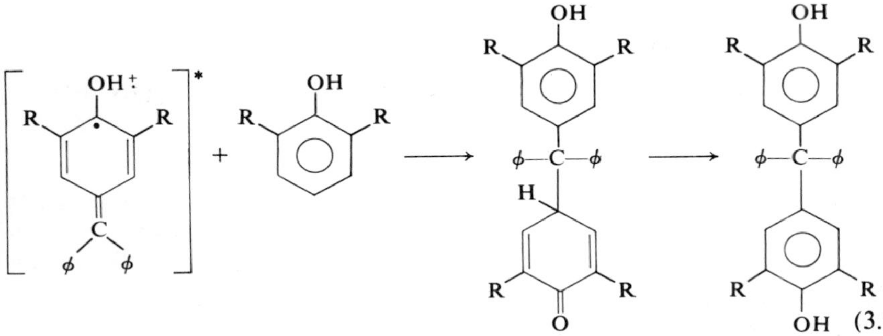

(3.95)

The second light-requiring step ultimately leading to the tetraphenyl methane was shown to occur by energy transfer from triplet benzophenone to the fuchsone, resulting in triplet fuchsone. The mechanisms of energy transfer will be discussed in a later chapter.

3.6. THE PHOTOREDUCTION OF ALKANONES

Although the photoreduction of aryl ketones has been extensively investigated, as we have seen in the preceding sections, few detailed reports have been made concerning the photoreduction of alkanones.[(82e)]

Wagner[(82a)] has studied the photoreduction of acetone to isopropanol by *n*-hexane as solvent (Φ for the formation of isopropanol is 0.53*) and by tri-*n*-butyl tin hydride (Φ for isopropanol is 1.00*). Both of these photoreductions were shown to occur exclusively from the acetone triplet state by total quenching of the reaction with 2,5-dimethyl-2,4-hexadiene and by 1,3-pentadiene, respectively. Similarly, 2-hexanone can be photoreduced by tri-*n*-butyl tin hydride in hexane solution (Φ for alcohol is 0.29).[(82b)] By comparing the quantum yields of singlet-derived type II cleavage (see Section 3.7) in the presence and absence of quencher (isoprene), Wagner was able to show that isoprene does not significantly quench 2-hexanone singlet states under these conditions and to conclude that photoreduction proceeds totally from the alkanone triplet state.

The photoreduction of cyclobutanone, cyclopentanone, and cyclohexanone by tri-*n*-butyl tin hydride was reported by Turro and McDaniel.[(82c)] Quantum yields for the formation of the corresponding alcohols were 0.01, 0.31, and 0.82, respectively. Although the results for cyclopentanone and cyclohexanone quenching were not clear-cut (deviations from linearity of the Stern–Volmer plots were noted at quencher concentrations $>0.6\ M$), all three ketone photoreductions were quenched by 1,3-pentadiene, again indicating that triplets are involved in the photoreduction.

3.7. INTRAMOLECULAR HYDROGEN ABSTRACTION BY KETONES (TYPE II CLEAVAGE)

With compounds containing C—H bonds γ to the carbonyl moiety, one often observes intramolecular hydrogen abstraction, in contrast to the intermolecular hydrogen abstraction by aryl ketones discussed earlier in this

* Quantum yield based on the value of $\Phi = 0.33$ for the 2-hexanone actinometer used in this study.[(84)]

chapter. This reaction is commonly called type II cleavage or the Norrish type II photoelimination:

$$\text{R—C(=O)—C(}\alpha\text{)—C(}\beta\text{)—C(}\gamma\text{)—H} \longrightarrow \text{R—C(OH)=CH}_2 + \text{>C=C<} + \text{(R, OH)cyclobutane} \qquad (3.96)$$

$$\text{R—C(OH)=CH}_2 \downarrow \text{R—C(=O)—CH}_3$$

A lower molecular weight methyl ketone and an olefin are isolated as products of this reaction. That the enol is formed as a primary product which rearranges to the ketone follows from its detection in the IR spectrum of gaseous 2-pentanone upon photolysis.[83] In addition to the ketone and olefinic products, one usually obtains varying amounts of cyclobutanols.

3.7a. *The Multiplicity of the Excited State*

We saw in Section 3.1 that the use of triplet quenchers can provide valuable information regarding the multiplicity of the excited state responsible for reaction. The fact that the photoreduction of benzophenone is strongly quenched by the presence of oxygen or soluble ferric salts was cited as evidence that this reaction proceeded from the triplet state of the ketone. A similar study applied to the type II cleavage of aromatic ketones indicated that this, too, results from the ketone triplet state.[88] With aliphatic ketones, however, only a portion of the total reaction could be eliminated with even the most efficient triplet quenchers.[84–87] Since these quenchers would not be expected to quench singlet states, it appears that in aliphatic ketones product is produced from both the excited singlet and triplet states. The amount of unquenchable (singlet) reaction appears to be dependent upon the bond strength of the γ C—H bond as indicated in Table 3.13.[87]

TABLE 3.13. *Percent Singlet Reaction as a Function of the γ C—H Bond Strength*[87]

Compound	γD_{C-H}, kcal/mole	Percent singlet reaction
2-Pentanone (1°)	98	6.5
2-Hexanone (2°)	94.5	33.3
5-Methyl-2-hexanone (3°)	91	52.6

In principle, the triplet state of ketones could react in either of two modes to produce the product ketone and olefin. These are via a concerted pathway or by a distinct diradical species:

$$\text{concerted}\quad \phi\text{—}\overset{\overset{\displaystyle O\cdots H}{\|}}{C}\text{(}CH_2CH_2CH_2\text{, cyclic)} \xrightarrow{h\nu} \phi\text{—}\overset{\overset{\displaystyle OH}{|}}{C}{=}CH_2 + CH_2{=}CH_2 \longrightarrow \phi\text{—}\overset{\overset{\displaystyle O}{\|}}{C}\text{—}CH_3 \tag{3.97}$$

$$\text{biradical}\quad \phi\text{—}\overset{\overset{\displaystyle O}{\|}}{C}\text{—}CH_2CH_2CH_3 \xrightarrow{h\nu} \phi\text{—}\overset{\overset{\displaystyle OH}{|}}{\dot{C}}\text{—}CH_2CH_2\dot{C}H_2 \longrightarrow \phi\text{—}\overset{\overset{\displaystyle OH}{|}}{C}{=}CH_2 + CH_2{=}CH_2 \tag{3.98}$$

Evidence that the triplet reaction of aromatic ketones proceeds through a distinct biradical intermediate was obtained by studying the photolysis of an optically active ketone, *S*-(+)-4-methyl-1-phenyl-1-hexanone, having its active center on the γ-carbon. If reaction were concerted, no racemization of unreacted ketone should be observed. On the other hand, if a biradical is formed and is able to disproportionate back to the starting ketone, racemization should result. The data in Table 3.14 indicate that racemization is indeed very high for the ketone in benzene solution.[89] Additional evidence for a biradical intermediate follows from the data for the photolysis in the presence of alcohol. In *t*-butanol as solvent, cleavage is essentially of unit efficiency and no racemization occurs. This dramatic decrease in disproportionation leading to racemization is thought to be due to hydrogen bonding between the hydroxy biradical and the alcohol molecules.[90]

TABLE 3.14. *Photoreaction of S-(+)-4-Methyl-1-phenyl-1-hexanone*[89],[a]

Solvent	Φ_{II}	Φ	Φ_{rac}
Benzene	0.23	0.03	0.78
Benzene + 2% *t*-butanol	0.33	0.03	0.57
t-Butanol	0.94	0.06	0.03

[a] Φ_{II}, quantum yield for type II cleavage; Φ, quantum yield for cyclobutanol formation; Φ_{rac}, quantum yield for racemization.

One might expect that in cases where the product olefin triplet lies lower in energy than that of the parent ketone, concerted cleavage to yield the triplet olefin should occur. For instance,

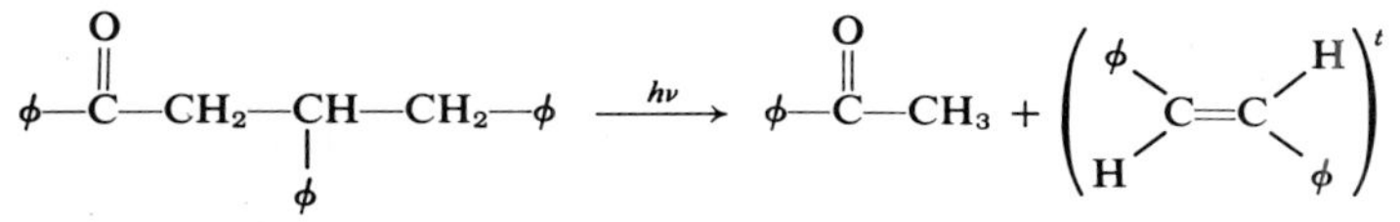

$$\phi\text{—}\overset{\overset{\displaystyle O}{\|}}{C}\text{—}CH_2\text{—}\underset{\underset{\displaystyle \phi}{|}}{CH}\text{—}CH_2\text{—}\phi \xrightarrow{h\nu} \phi\text{—}\overset{\overset{\displaystyle O}{\|}}{C}\text{—}CH_3 + \left(\begin{matrix}\phi & & H\\ & C{=}C & \\ H & & \phi\end{matrix}\right)^{t} \qquad (3.99)$$

Stilbene is known to undergo *cis–trans* isomerization in its triplet state to yield a photostationary state which is 60% *cis* and 40% *trans*,

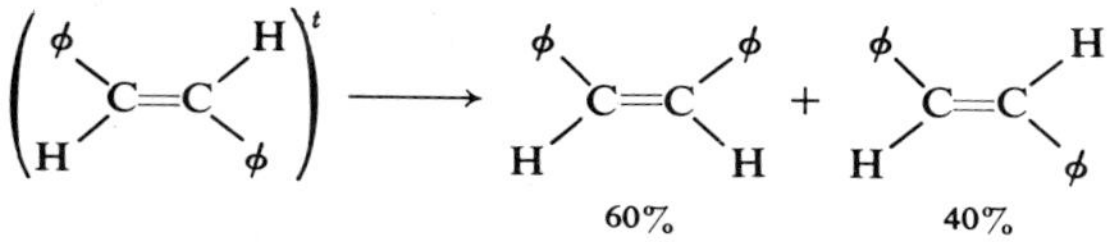

$$\left(\begin{matrix}\phi & & H\\ & C{=}C & \\ H & & \phi\end{matrix}\right)^{t} \longrightarrow \underset{60\%}{\begin{matrix}\phi & & \phi\\ & C{=}C & \\ H & & H\end{matrix}} + \underset{40\%}{\begin{matrix}\phi & & H\\ & C{=}C & \\ H & & \phi\end{matrix}} \qquad (3.100)$$

If triplet stilbene were formed in this way, analysis of the product ratio of steroisomers should indicate a 60:40 ratio. In fact, however, the stilbene produced was found to be about 99% *trans* although the reaction is about 50 kcal exothermic, enough energy to produce the triplet stilbene concertedly. Thus it appears that the triplet cleavage prefers to go through a biradical intermediate even when a concerted process is energetically possible.[(91)]

The preceding discussion applied to aromatic ketone triplet reactions. With aliphatic ketones the situation is quite different. As stated previously, aliphatic ketones undergo type II cleavage from both the excited singlet and the triplet state. By studying the reaction with and without added quencher, one can determine the characteristics of the reaction for each state, that is, the singlet reaction can be studied in the presence of a strong triplet quencher while the triplet reaction characteristics can be obtained by the difference between the reaction without quencher and that when quencher is added. For example, for the reaction

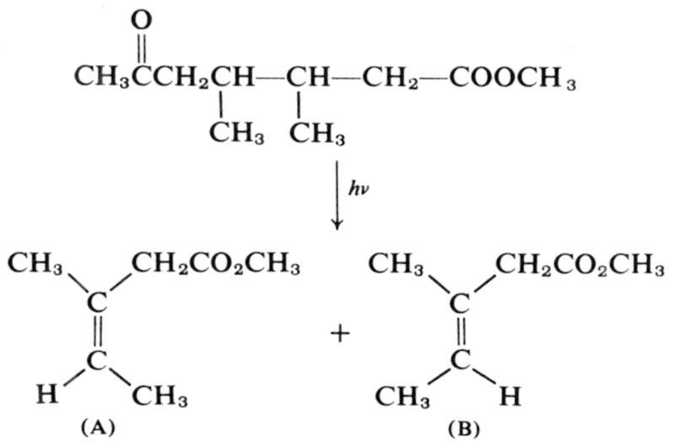

$$CH_3\overset{\overset{\displaystyle O}{\|}}{C}CH_2\underset{\underset{\displaystyle CH_3}{|}}{CH}\text{—}\underset{\underset{\displaystyle CH_3}{|}}{CH}\text{—}CH_2\text{—}COOCH_3 \xrightarrow{h\nu} \underset{(A)}{\begin{matrix}CH_3 & & CH_2CO_2CH_3\\ & C & \\ & \| & \\ & C & \\ H & & CH_3\end{matrix}} + \underset{(B)}{\begin{matrix}CH_3 & & CH_2CO_2CH_3\\ & C & \\ & \| & \\ & C & \\ CH_3 & & H\end{matrix}} \qquad (3.101)$$

the two diastereomers react to produce the following product distribution[92]:

erythro (triplet state) ⟶ 51% A + 49% B

threo (triplet state) ⟶ 35% A + 65% B

erythro (singlet state) ⟶ 98.9% A + 1.1% B

threo (singlet state) ⟶ 1.1% A + 98.9% B

Thus the triplet states of the two diastereomers react to yield different product distributions although this effect is far less marked for the triplet than for the singlet reaction, which is essentially stereospecific. The singlet reaction could be either concerted or due to an extremely shortlived biradical. Since the product distributions of the triple reaction of these two diastereomers are different, it is clear that cleavage must occur before complete equilibration. Thus the lifetime of the aliphatic ketone derived biradical must be considerably shorter than the corresponding biradical derived from an aryl ketone.

3.7b. *Stereoelectronic Effects*

Assuming that both singlet and triplet reactions proceed through discrete biradical intermediates, the efficiency of type II cleavage should be dependent upon the degree to which the *p*-orbitals of the radicals are parallel to the β bond; the more nearly parallel these are, the more overlap of the *p* orbitals with the developing π orbitals of the double bonds and the more efficient the cleavage. Two conformations best fulfilling this requirement are as follows[93]:

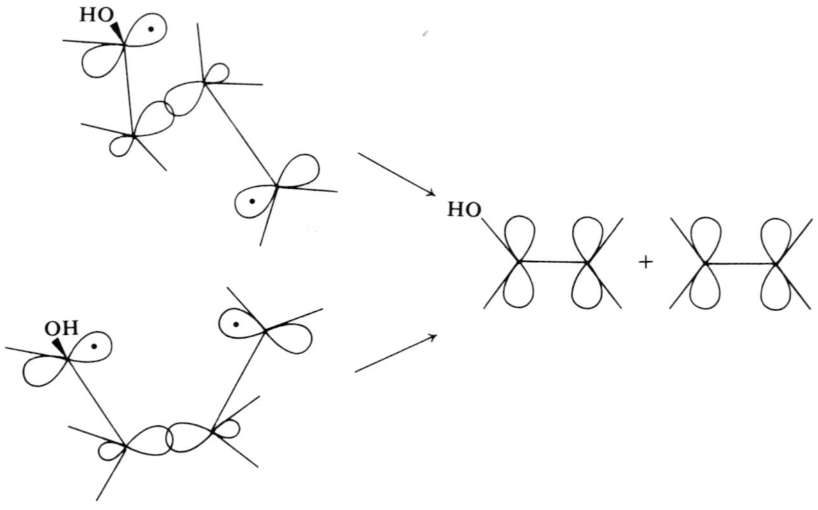

The importance of the stereoelectronic conformation in determining the path of reaction may be seen in the example of phenyl cyclobutyl ketone. Photolysis of this compound yields 60% of the highly strained bicyclopentane shown below and only 40% of open-chain olefin[94]:

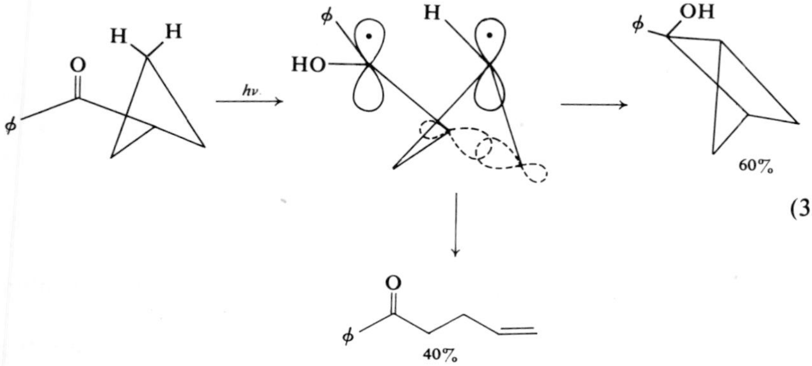

(3.102)

The reason for this behavior can be seen in the structure of the intermediate biradical. The rigidity of the cyclobutyl ring prevents a parallel alignment of the *p* orbitals with the β bond, which is held practically perpendicular. In order for type II cleavage to occur, an initially severely strained olefin must be formed. Hence radical recombination to yield the bicyclopentane system predominates.

A further example of the importance of stereoelectronic effects results from the behavior of *cis*- and *trans*-2-*n*-propyl-4-*t*-butylcyclohexanone,[95]

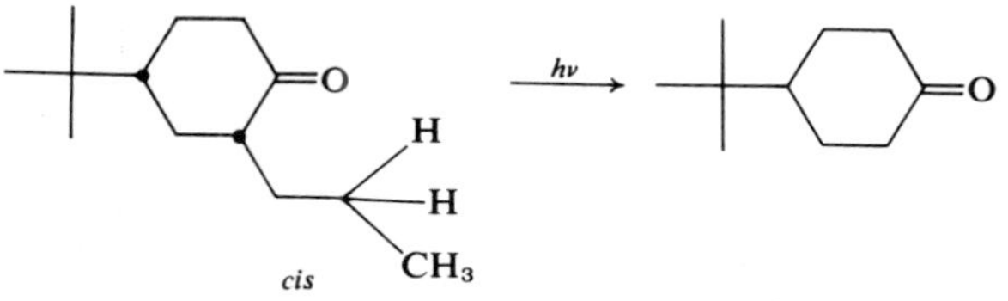

(3.103)

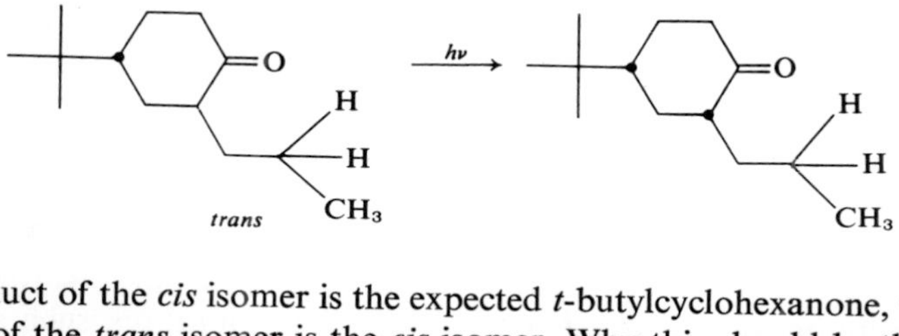

(3.104)

The product of the *cis* isomer is the expected *t*-butylcyclohexanone, while the product of the *trans* isomer is the *cis* isomer. Why this should be the case is seen by examining the probable stereochemistry of these isomers in their $n \rightarrow \pi^*$ excited states:

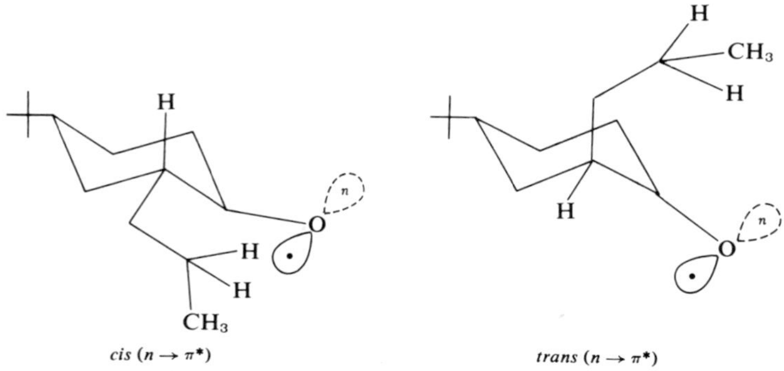

cis ($n \rightarrow \pi^*$) *trans* ($n \rightarrow \pi^*$)

It can be seen that in the *cis* isomer the geometry is favorable for the formation of the six-membered ring transition state for transfer of the hydrogen atom to the carbonyl. Hence type II cleavage is observed for this isomer. For the *trans* isomer, on the other hand, the hydrogen atom lies in the nodal plane of the carbonyl group and abstraction should be much less efficient. The *trans* isomer therefore isomerizes to the *cis*, probably by α-cleavage and epimerization. Turro applied this behavior in an interesting way by performing the selective removal of one side chain from 2,6-di-*n*-propyl-4-*t*-butylcyclohexanone,[96]

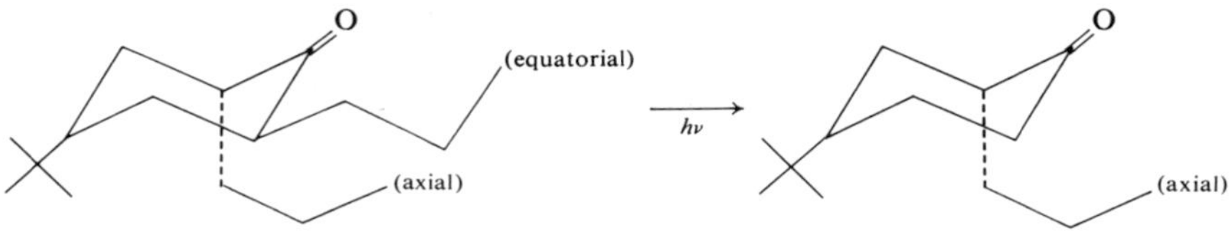

(3.105)

This reaction was run in the presence of the triplet quencher 1,3-pentadiene to prevent epimerization due to α-cleavage (a triplet reaction).

3.7c. *Substituent Effects*

The effects of substitutents on the γ-carbon on the efficiency of the type II cleavage are presented in Table 3.15.[89] These data indicate that the rate constant of cleavage increases as the strength of the γ C—H bond decreases, that is, from a primary to a secondary to a tertiary hydrogen atom. The substitution of groups capable of radical stabilization, such as —ϕ or

TABLE 3.15. *Effects of γ-Carbon Substitutents on Type II Cleavage*[89]

$\phi-\overset{O}{\overset{\|}{C}}-CH_2CH_2CHR_1R_2$

R_1	R_2	Φ_{II} [a]	k_r, $sec^{-1} \times 10^{-8}$
H	H	0.36	0.08
CH_3	H	0.33	1.3
CH_3	CH_3	0.25	4.8
$C(CH_3)_3$	H	0.24	2.0
ϕ	H	0.50	4.2
$CH_2{=}CH$	H	0.26	5.0
OH	H	0.31	3.9
OCH_3	H	0.23	6.2
$N(CH_3)_2$	H	0.03	4.0
Cl	H	0.09	0.3
$COOCH_3$	H	0.50	0.1
CN	H	0.30	0.05

[a] In benzene solution with excitation at 3130 Å.

—$CH_2{=}CH_2$, is also seen to increase the reactivity. Electron-donating groups, such as OH, OCH_3, and $N(CH_3)_2$, have a marked increasing effect on the rate constant, whereas the electron-withdrawing groups —$COOCH_3$ and —CN have a retarding effect. Thus it would appear that radical stabilizing effects and inductive effects of —CN and —$COOCH_3$ influence the efficiency in opposite directions.

The effect of substituents on the δ carbon is seen in Table 3.16.[89] The data indicate that the reaction is subject to the inductive effects of substituents even when substituted on the δ-carbon atom. The correlation

TABLE 3.16. *Effect of Substituents on the δ-Carbon*[89]

$\phi-\overset{O}{\overset{\|}{C}}-CH_2CH_2CH_2CH_2R$

R	Φ_{II}	k_r, $sec^{-1} \times 10^{-8}$	σ_I
H	0.33	1.3	0
Me, Et	0.30	1.5	−0.05
—COϕ	0.34	0.5	0.28
CO_2CH_3	0.63	0.4	0.30
Cl	0.58	0.2	0.47
CN	0.46	0.1	0.56
CH_2Cl	0.44	0.6	~0.20

TABLE 3.17. *The Effect of Ring Substituents on Type II Cleavage*

$R—\phi—\overset{O}{\overset{\|}{C}}—CH_2CH_2CH_3$ [89]

Ring substituent	Φ_{max} [a]	k_r, $sec^{-1} \times 10^{-7}$
p-CH_3O	0.26	0.06
p-CH_3	1.0	1.6
p-Cl	0.8	3.0
H	1.0	14.0
o-CF_3	1.0	13.0
p-CF_3	1.0	29.0

[a] In alcohol solution.

of the rate of cleavage with σ_i indicates that the reaction has a Hammett ρ of -2 for δ substituents. The triplet carbonyl of the benzoyl group must therefore be a rather electrophilic species.

Finally, we turn to the effect of ring substituents on the efficiency of the type II cleavage. This is shown in Table 3.17.[89] It can be seen from Table 3.17 that the rate constant is severely reduced when the aromatic ring bears an electron-donating substituent.[97] In section 3.3d we saw that the effect of substituents on photoreduction of aryl ketones was due to a raising or lowering of the $n \rightarrow \pi^*$ triplet state relative to the $\pi \rightarrow \pi^*$ triplet state. Derivatives whose $\pi \rightarrow \pi^*$ triplet became the lowest energy triplet were much less reactive than those with lowest $n \rightarrow \pi^*$ triplets. Since, as we have seen, type II cleavage in aromatic ketones occurs from the triplet state, a similar behavior might be expected here. Evidence that this is indeed the case follows from the fact that unsubstituted derivatives and those substituted with electron-withdrawing groups have triplet lifetimes less than 10 msec (indicative of $n \rightarrow \pi^*$ triplets) while those bearing electron-donating substituents have triplet lifetimes longer than 50 msec (indicative of $\pi \rightarrow \pi^*$ triplets). It appears therefore that, like photoreduction, type II cleavage is most efficient when the lowest triplet is of the $n \rightarrow \pi^*$ type.[98]

3.7d. *Synthetic Applications*

Intramolecular hydrogen abstraction has been successfully utilized in a number of interesting synthetic preparations. In general the recombination of the biradical resulting from intramolecular hydrogen abstraction to form a cyclobutanol has received more attention from the point of view of photochemical synthesis than type II cleavages. Some synthetic applications of these reactions follow in this section.

The photolysis of methylisopulegon in cyclohexane by Cookson *et al.* resulted in the interesting methylene cyclobutanol shown below in a 70% yield[(99)]:

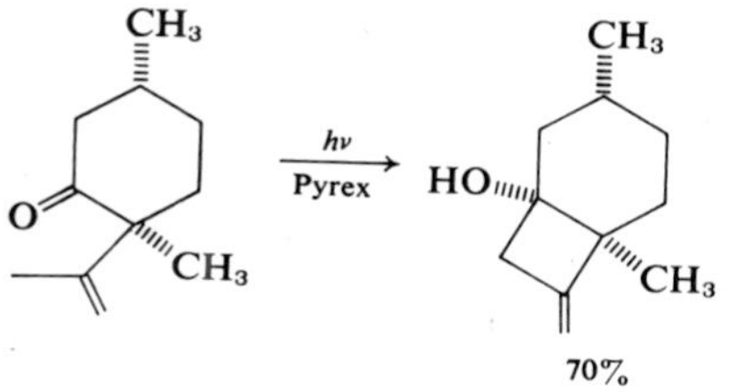

$h\nu$, Pyrex — 70% (3.106)

Padwa and Eisenberg used this reaction to obtain the following highly strained tricyclo compound[(100)]:

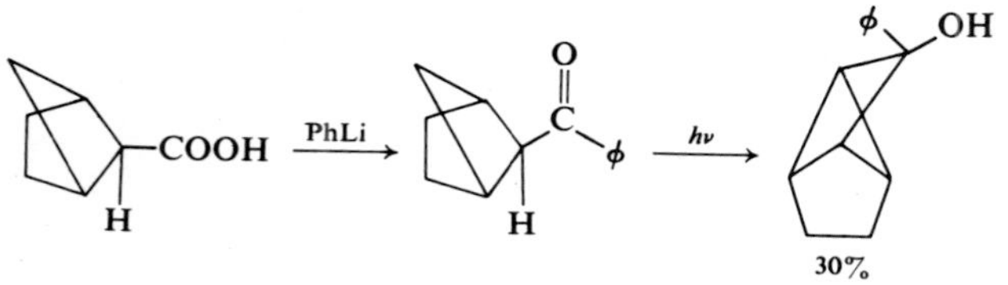

COOH —PhLi→ —$h\nu$→ 30% (3.107)

The photolysis of the following steroid system resulted in two products corresponding to the Norrish type II reaction and one product due to α-cleavage (Norrish type I cleavage)[(101)]:

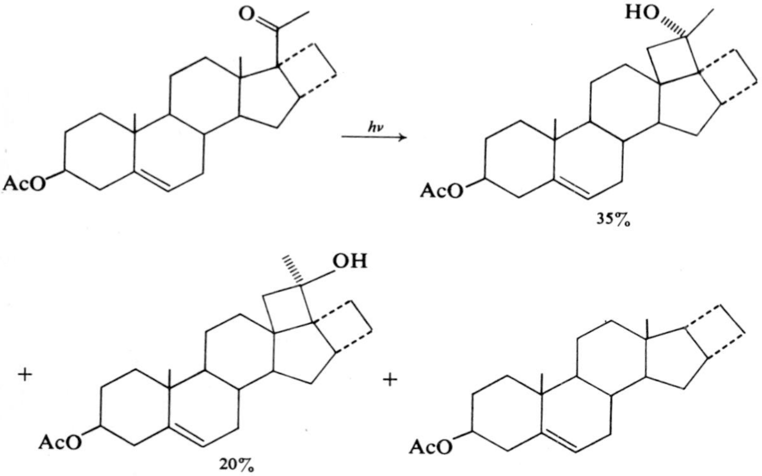

AcO —$h\nu$→ AcO 35% + AcO 20% + AcO (3.108)

Intramolecular hydrogen transfer is also important in the photolysis of large-ring cycloalkanones such as shown below. The singlet state is thought to be the reactive species in these reactions[102]:

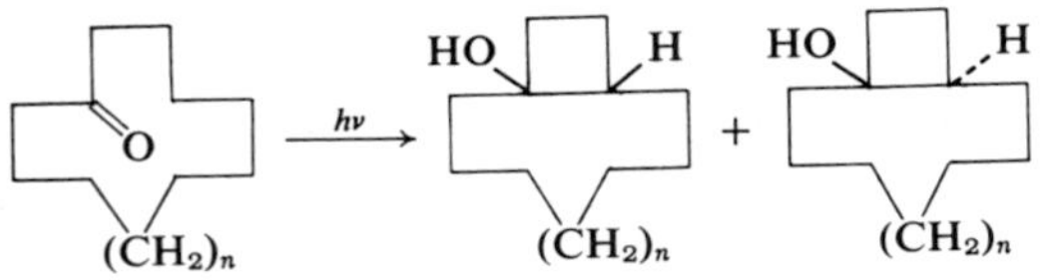

(3.109)

Ring opening due to intramolecular hydrogen abstraction has been demonstrated for cyclopropyl ketone derivatives[103]:

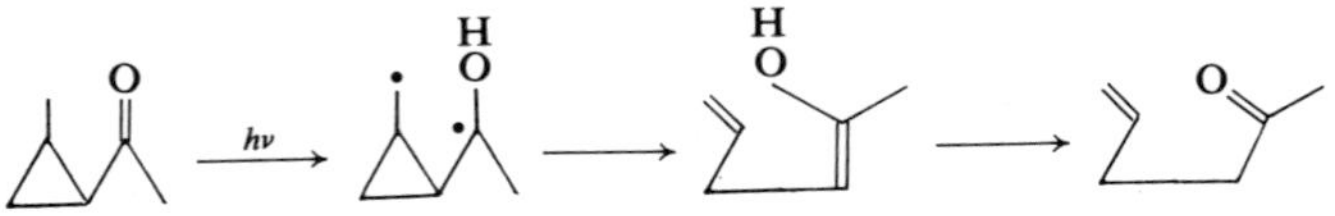

(3.110)

Similarly, photolysis of (+)-*cis*-caran-5-one yielded the following products[104]:

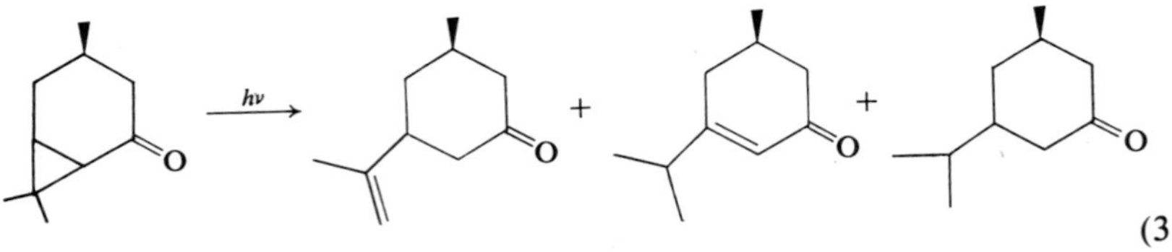

(3.111)

Intramolecular hydrogen abstraction has been utilized to remove a methoxy group from a pyranosidulose,[105]

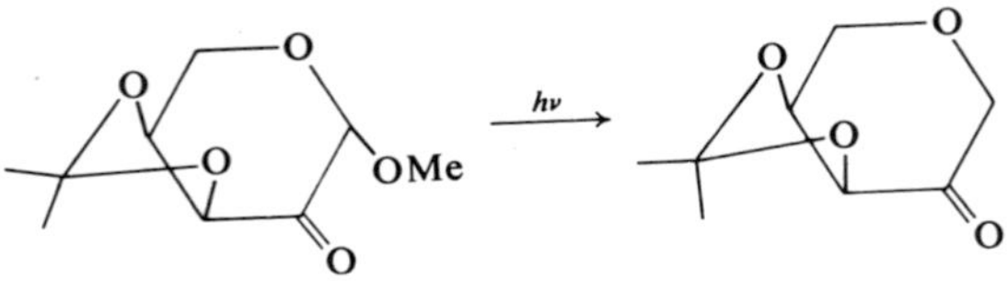

(3.112)

Rando and von E. Doering have investigated the synthetic utility of double bond positional isomerization in the photolysis of α,β-unsaturated esters,[106]

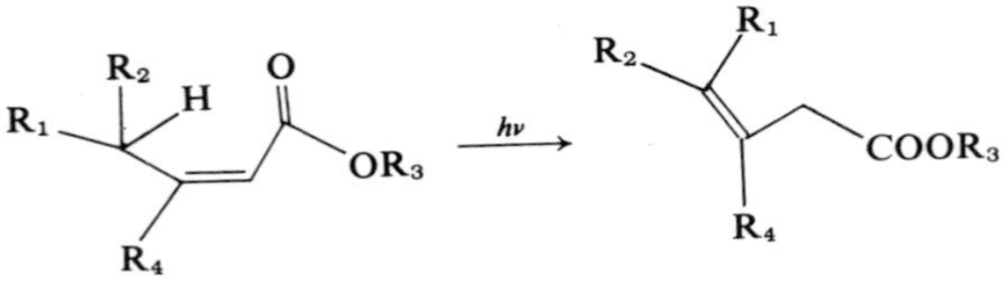

(3.113)

TABLE 3.18

R_1	R_2	R_3	R_4	*Cis/trans*	% Yield
Me	Me	Et	H	Not determined	85
Me	H	Me	H	Not determined	85
H	H	Me	H	Not determined	20
n-C_6H_{13}	H	H	H	0.5	94
n-$C_{12}H_{25}$	H	H	H	0.5	95

Table 3.18 gives the ratio of formation (where determined) of the *cis* and *trans* isomers and the percentage yields for various possible R_1, R_2, R_3, and R_4.

The photolysis of an azetidine has been found to yield pyrole derivatives by the Norrish type II reaction,[107]

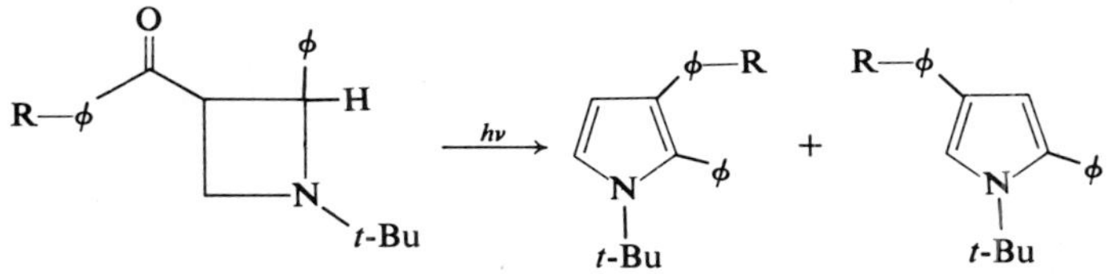

(3.114)

These products are thought to arise by electron transfer to the nitrogen atom followed by proton transfer and electron reorganization,[108]

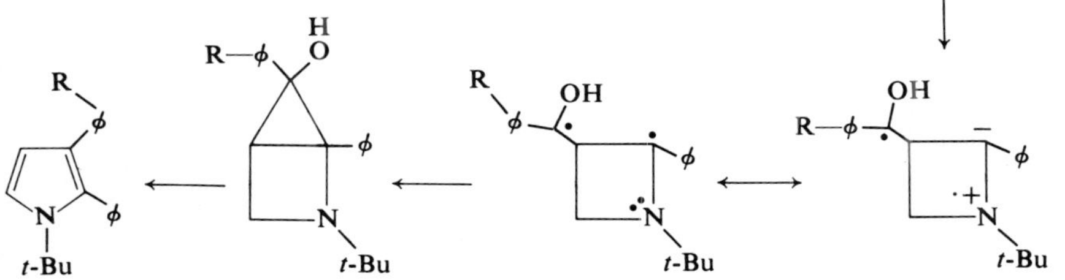

(3.115)

The other photochemical reactions of simple carbonyls mentioned earlier in this chapter—type I cleavage (α-cleavage) and oxetane formation—will be discussed in Chapter 4.

3.8. HYDROGEN ABSTRACTION BY GROUPS OTHER THAN THE CARBONYL

A number of other function groups are capable of photochemical hydrogen abstraction similar to that observed for the carbonyl. These reactions will be briefly described in this section.

(a) *Hydrogen Abstractions by Nitro Groups.* As is the case with simple carbonyls, nitro groups have lowest lying $n \rightarrow \pi^*$ singlet and triplet states. Studies of these excited states by flash photolysis have yielded transient spectra indicative of aci-nitro derivatives,[109,110]

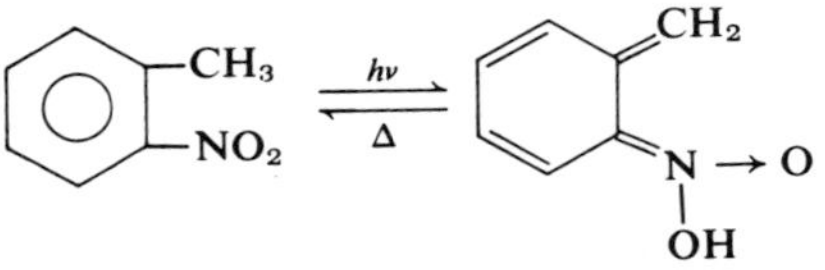

(3.116)

The presence of this species can be explained by hydrogen abstraction by the excited nitro group to yield the unstable derivative shown above. Although the product cannot be isolated and identified, due to its short lifetime, evidence for its existence (other than the flash spectrographic evidence) was obtained by running the photolysis in D_2O and observing incorporation of deuterium atoms into the methyl group.[111] No deuterium exchange occurred under the same conditions in the dark.

Compounds such as *o*-nitrotoluene which undergo photochemical reactions that are rapidly thermally reversed are called *photochromic* or *phototropic*. Nitro compounds comprise a very important group of compounds which exhibit this behavior. Further examples are as follows[112,113]:

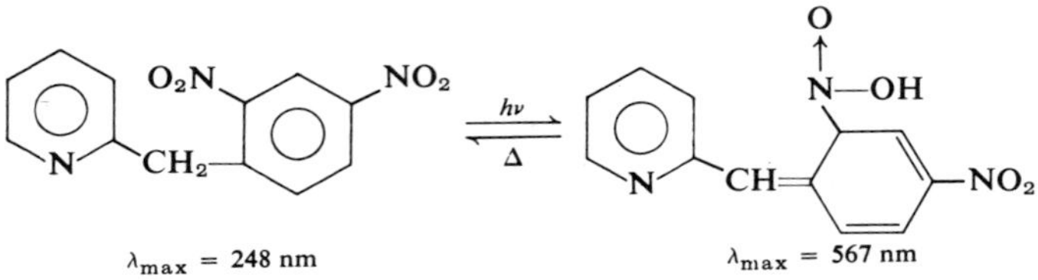

(3.117)

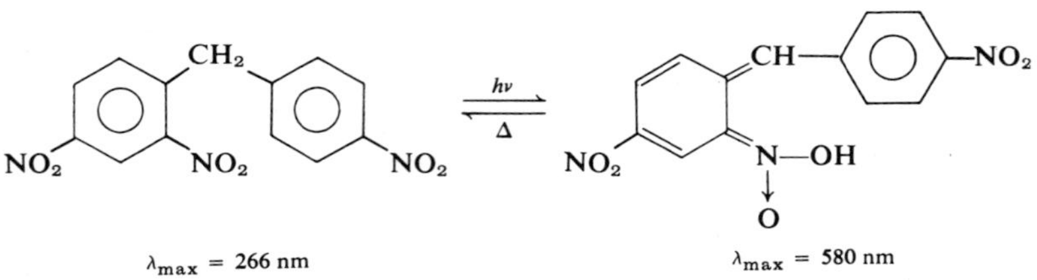

(3.118)

Highly conjugated nitro derivatives are often intensely colored, making them potentially useful for information storage.

(b) *Hydrogen Abstraction by Anils.* Anils behave photochemically in a manner very similar to nitro compounds in that intramolecular hydrogen abstraction occurs to yield derivatives which are rapidly thermally reversed,[114,115]

OH CH=N OCH$_3$ $\underset{\Delta}{\overset{h\nu}{\rightleftharpoons}}$ O CH—NH OCH$_3$

(3.119)

(c) *Hydrogen Abstraction by Olefins.* Certain dienes have been found to undergo double bond migration and cyclobutene formation upon photolysis, probably through intramolecular hydrogen abstraction,[116,117]

$$CH_2{=}C(R){-}CH{=}CH{-}CH_3 \xrightarrow{h\nu} R(CH_3)C{=}CHCH{=}CH_2 + \text{(R, CH}_3\text{-cyclobutene)} \qquad (3.120)$$

Excitation to produce a diradical-like intermediate (excited state) can result in either hydrogen abstraction or rearrangement and closure to form the cyclobutene:

$$CH_2{=}C(R){-}CH{=}CH{-}CH_3$$

(3.121)

PROBLEMS

1. Write a mechanism for the following transformations:

   (a) [J. R. Scheffer, J. Trotter, R. A. Wostradowski, C. S. Gibbons, and K. S. Bhandari, *J. Amer. Chem. Soc.* **93**, 3813 (1971); and *Tetrahedron Lett.*, 677 (1972).]

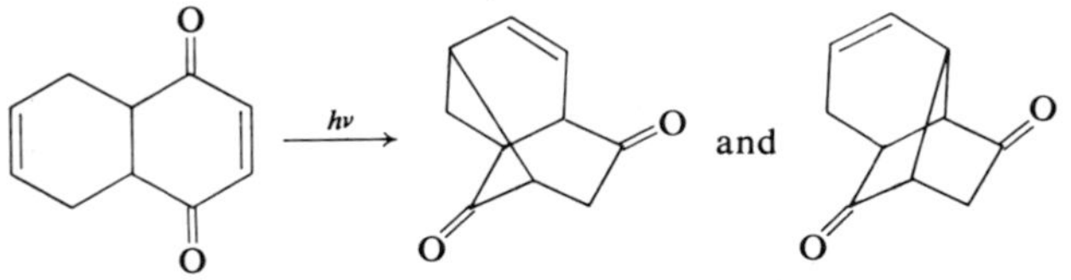

   $h\nu$ → and

   (b) [A. Padwa, F. Albrecht, P. Singh, and E. Vega, *J. Amer. Chem. Soc.* **93**, 2928 (1971).]

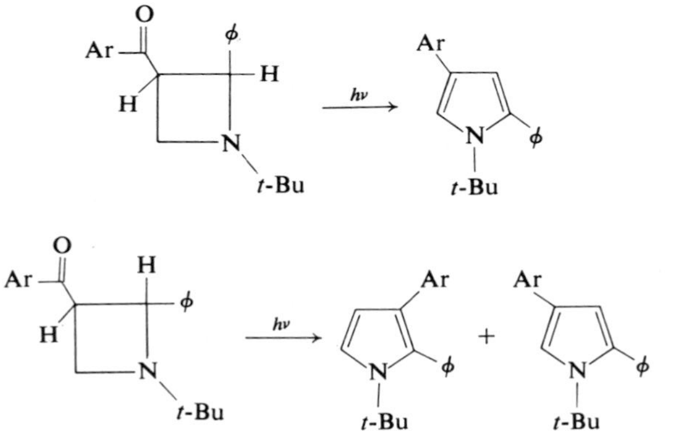

   Ar, O, ϕ, H, H, N, t-Bu $\xrightarrow{h\nu}$ Ar, N, ϕ, t-Bu

   Ar, O, H, ϕ, H, N, t-Bu $\xrightarrow{h\nu}$ Ar, N, ϕ, t-Bu + Ar, N, ϕ, t-Bu

REFERENCES

1. W. M. Moore, G. S. Hammond, and R. P. Foss, *J. Amer. Chem. Soc.* **83**, 2789 (1961).
2. H. L. J. Bäckstrom and K. Sandros, *Acta Chem. Scand.* **14**, 48 (1960).
3. G. V. Schultz, *Z. Physik. Chem.* **92**, 284 (1956).
4. M. V. Smoluchowski, *Z. Physik. Chem.* **8**, 129 (1918).
5. Th. Forster, *Fluoreszent Organischer Verbindungen*, Vendenhoeck and Ruprecht, Göttingen, 1951; J. B. Birks and D. J. Dyson, *Proc. Roy. Soc. A* **275**, 135 (1963); E. J. Bowen, *Luminescence in Chemistry*, van Nostrand, London (1968), pp. 13, 121.
6. G. Porter, in *Techniques of Organic Chemistry*, Vol. 8, Wiley, New York (1963), p. 1055.
7. R. W. Glass, Ph.D. Dissertation, Vanderbilt University (June 1968).
8. J. A. Bell and H. Linschitz, *J. Amer. Chem. Soc.* **85**, 528 (1963).
9. G. Porter and F. Wilkinson, *Trans. Faraday Soc.* **57**, 1686 (1961).
10. A. V. Buettner and J. Dedinas, *J. Phys. Chem.* **75**, 187 (1971).

11. W. D. K. Clark, A. D. Litt, and C. Steel, *J. Amer. Chem. Soc.* **91**, 5413 (1969).
12. C. A. Parker and T. A. Joyce, *Chem. Comm.*, 749 (1968); *Trans. Faraday Soc.* **65**, 2823 (1969).
13. J. Saltiel, H. C. Curtis, L. Metts, J. W. Miley, J. Winterle, and M. Wrighton, *J. Amer. Chem. Soc.* **92**, 410 (1970).
14. G. Porter and M. R. Topp, (a) *Nature* **220**, 1228 (1968); (b) *Proc. Roy. Soc. Lond. A* **315**, 163 (1970).
15. S. J. Formosinho, G. Porter, and M. A. West, *Chem. Phys. Lett.* **6**, 7 (1970).
16. P. J. Wagner, *Mol. Photochem.* **1**, 71 (1969).
17. P. J. Wagner and R. A. Levitt, *J. Amer. Chem. Soc.* **92**, 5806 (1970).
18. E. J. Baum and R. O. C. Norman, *J. Chem. Soc.* (B), 749 (1968).
19. D. I. Schuster and D. F. Brizzolara, *J. Amer. Chem. Soc.* **92**, 4357 (1970).
20. J. Saltiel, H. C. Curtis, and B. Jones, *Mol. Photochem.* **2**, 331 (1970).
21. D. I. Schuster, T. M. Weil and M. R. Topp, *Chem. Comm.*, 1212 (1971).
22. C. Weizmann, E. Bergmann, and Y. Hirshberg, *J. Amer. Chem. Soc.* **60**, 1530 (1938).
23. G. S. Hammond, W. P. Baker, and W. M. Moore, *J. Amer. Chem. Soc.* **83**, 2795 (1961).
24. S. G. Cohen, D. A. Laufer, and W. V. Sherman, *J. Amer. Chem. Soc.* **86**, 3060 (1964).
25. H.-D. Becker, *J. Org. Chem.* **32**, 2140 (1967).
26. N. C. Yang, M. Nussim, M. J. Jorgenson, and S. Murov, *Tetrahedron Lett.*, 3657 (1964).
27. N. C. Yang, *Pure Appl. Chem.* **9**, 591 (1964).
28. J. S. Bradshaw, *J. Org. Chem.* **31**, 237 (1966).
29. G. A. Davis, P. A. Carapellucci, K. Szoc, and J. D. Gresser, *J. Amer. Chem. Soc.* **91**, 2264 (1969).
30. F. Bergmann and Y. Hirshberg, *J. Amer. Chem. Soc.* **65**, 1429 (1943).
31. N. C. Yang, R. Loeschen, and D. Michel, *J. Amer. Chem. Soc.* **89**, 5465 (1967).
32. N. C. Yang and R. L. Loeschen, *Tetrahedron Lett.*, 2571 (1968).
33. G. S. Hammond and P. A. Leermakers, *J. Amer. Chem. Soc.* **84**, 207 (1962).
34. D. R. Kearns and W. A. Case, *J. Amer. Chem. Soc.* **88**, 5087 (1966).
35. K. Yoshihara and D. R. Kearns, *J. Chem. Phys.* **45**, 1991 (1966).
36. A. Kuboyama, *Bull. Chem. Soc. Japan* **37**, 1540 (1964).
37. W. G. Herkstroeter (unpublished).
38. W. G. Herkstroeter, A. A. Lamola, and G. S. Hammond, *J. Amer. Chem. Soc.* **86**, 4537 (1964).
39. C. Walling, *Free Radicals in Solution*, Wiley, New York (1957).
40. C. Walling and M. J. Gibian, *J. Amer. Chem. Soc.* **87**, 3361 (1965).
41. K. R. Hoffman, M. Loy, and E. F. Ullman, *J. Amer. Chem. Soc.* **87**, 5417 (1965).
42. A. Beckett and G. Porter, *Trans. Faraday Soc.* **59**, 2051 (1963).
43. N. C. Yang and C. Rivas, *J. Amer. Chem. Soc.* **83**, 2213 (1961).
44. G. Porter and P. Suppan, *Trans. Faraday. Soc.* **61**, 1664 (1965).
45. G. Porter and P. Suppan, *Pure Appl. Chem.* **9**, 499 (1964).
46. S. G. Cohen and M. N. Saddiqui, *J. Amer. Chem. Soc.* **86**, 5047 (1964).
47. S. G. Cohen and J. I. Cohen, *J. Phys. Chem.* **72**, 3782 (1968).
48. J. Petrushka, *J. Chem. Phys.* **34**, 1120 (1961).
49. N. C. Yang, D. S. McClure, S. L. Murov, J. J. Houser, and R. Dusenberg, *J. Amer. Chem. Soc.* **89**, 5466 (1967).
50. N. C. Yang and R. L. Dusenberg, *J. Amer. Chem. Soc.* **90**, 5899 (1968).
51. M. A. El-Sayed, *J. Chem. Phys.* **36**, 573 (1962); **38**, 2834, 3032 (1963).
52. A. A. Lamola, *J. Chem. Phys.* **47**, 4810 (1967).

53. C. Walling and B. B. Jacknow, *J. Amer. Chem. Soc.* **82**, 6108 (1960).
54. J. N. Pitts, Jr., R. L. Letsinger, R. P. Taylor, J. M. Patterson, G. Recktenwald, and R. B. Martin, *J. Amer. Chem. Soc.* **81**, 1068 (1959).
55. V. Franzen, *Liebigs Ann. Chem.* **633,** 1 (1960).
56. G. O. Schenck, G. Koltzenburg, and E. Roselius, *Z. Naturforsch.* **24**(b), 222 (1969); G. O. Schenck, G. Matthias, M. Papi, M. Cziesla, G. van Bunau, E. Roselius, and G. Koltzenburg, *Liebigs Ann. Chem.* **719**, 80 (1968).
57. D. C. Neckers, *Mechanistic Organic Photochemistry*, Reinhold, New York (1967), Chapter 7.
58. J. N. Pitts, H. W. Johnson, and T. Kuwana, *J. Phys. Chem.* **66**, 2456 (1962).
59. W. M. Moore and M. Ketchum, *J. Amer. Chem. Soc.* **84**, 1368 (1962).
60. G. S. Hammond and P. A. Leermakers, *J. Phys. Chem.* **66**, 1148 (1962).
61. A. Terenin and V. Ermolaev, *Trans. Faraday Soc.* **52**, 1042 (1956).
62. J. B. Farmer, C. L. Gardner, and C. A. McDowell, *J. Chem. Phys.* **34**, 1058 (1961).
63. G. Porter, *Proc. Chem. Soc.*, 291 (1959).
64. S. G. Cohen, S. Orman, and D. A. Laufer, *J. Amer. Chem. Soc.* **84**, 3905 (1962).
65. S. G. Cohen and W. V. Sherman, *J. Amer. Chem. Soc.* **85**, 1642 (1963).
66. S. G. Cohen, D. A. Laufer, and W. V. Sherman, *J. Amer. Chem. Soc.* **86**, 3060 (1964).
67. S. G. Cohen and S. Aktipis, *J. Amer. Chem. Soc.* **88**, 3587 (1966).
68. S. G. Cohen and H. M. Chao, *J. Amer. Chem. Soc.* **90**, 165 (1968).
69. R. S. Davidson and P. F. Lambeth, *Chem. Comm.*, 511 (1968).
70. S. G. Cohen and J. B. Gutlenplan, *Tetrahedron Lett.*, 5353 (1968), 2125 (1969).
71. H. Leonhardt and A. Weller, *Z. Phys. Chem.* (*Frankfurt*) **29**, 277 (1961).
72. R. S. Davidson and P. F. Lambeth, *Chem. Comm.*, 1265 (1967).
73. S. A. Weiner, *J. Amer. Chem. Soc.* **93**, 425 (1971).
74. B. M. Monroe, S. A. Weiner, and G. S. Hammond, *J. Amer. Chem. Soc.* **90**, 1913 (1968).
75. B. M. Monroe and S. A. Weiner, *J. Amer. Chem. Soc.* **91**, 450 (1969).
76. R. Breslow and M. Winnik, *J. Amer. Chem. Soc.* **91**, 3083 (1969).
77. R. Breslow and S. W. Baldwin, *J. Amer. Chem. Soc.* **92**, 732 (1970).
78. J. E. Baldwin, A. K. Bhatnagar, and R. W. Harper, *Chem. Comm.*, 659 (1970).
79. G. R. Lappin and J. S. Zannucci, *Chem. Comm.*, 1113 (1969); S. P. Pappas and J. E. Blackwell, *Tetrahedron Lett.*, 1171 (1966).
80. N. C. Yang and C. Rivas, *J. Amer. Chem. Soc.* **83**, 2213 (1961).
81. (a) H.-D. Becker, *J. Org. Chem.* **32**, 2115 (1967); (b) H. D. Becker, *J. Org. Chem.* **32**, 2124 (1967).
82. (a) P. J. Wagner, *J. Amer. Chem. Soc.* **88**, 5672 (1966); (b) P. J. Wagner, *J. Amer. Chem. Soc.* **89**, 2503 (1967); (c) N. J. Turro and D. M. McDaniel, *Mol. Photochem.* **2**, 39 (1970); (d) R. Simonaitis, G. W. Cowell, and J. N. Pitts, Jr., *Tetrahedron Lett.*, 3751 (1967); (e) J. C. Dalton and N. J. Turro, *Ann. Rev. Phys. Chem.* **21**, 499 (1970).
83. G. R. McMillan, J. G. Calvert, and J. N. Pitts, Jr., *J. Amer. Chem. Soc.* **86**, 3602 (1964).
84. D. R. Coulson and N. C. Yang, *J. Amer. Chem. Soc.* **88**, 4511 (1966).
85. P. J. Wagner and G. S. Hammond, *J. Amer. Chem. Soc.* **87**, 4009 (1965).
86. T. J. Dougherty, *J. Amer. Chem. Soc.* **87**, 4111 (1965).
87. N. C. Yang, S. P. Elliott, and B. Kim, *J. Amer. Chem. Soc.* **91**, 7551 (1969).
88. P. J. Wagner and G. S. Hammond, *J. Amer. Chem. Soc.* **88**, 1245 (1966).
89. P. J. Wagner, *Acct. Chem. Res.* **4**, 169 (1971).

90. P. J. Wagner, *J. Amer. Chem. Soc.* **89**, 5898 (1967).
91. P. J. Wagner and P. A. Kelso, *Tetrahedron Lett.*, 4151 (1969); R. A. Caldwell and P. Fink, *Tetrahedron Lett.*, 2987 (1969).
92. L. M. Stephenson, P. R. Cavigli, and J. L. Parlett, *J. Amer. Chem. Soc.* **93**, 1984 (1971).
93. N. J. Turro, J. C. Dalton, K. Dawes, G. Farrington, R. Hautala, D. Morton, M. Niemczyk, and N. Schore, *Acct. Chem. Res.* **5**, 92 (1972).
94. A. Padwa, E. Alexander, and M. Niemcyzk, *J. Amer. Chem. Soc.* **91**, 456 (1969).
95. N. J. Turro and D. S. Weiss, *J. Amer. Chem. Soc.* **90**, 2185 (1968).
96. K. Dawes, J. C. Dalton, and N. J. Turro, *Mol. Photochem.* **3**, 71 (1971).
97. P. J. Wagner and G. Capen, *Mol. Photochem.* **1**, 173 (1969).
98. P. J. Wagner and A. E. Kemppainen, *J. Amer. Chem. Soc.* **90**, 5898 (1968).
99. R. C. Cookson, J. Judec, A. Szabo, and G. E. Usher, *Tetrahedron* **24**, 4353 (1968).
100. A. Padwa and W. Eisenberg, *J. Amer. Chem. Soç.* **92**, 2590 (1970).
101. P. Sander-Plassman, P. H. Nelson, P. H. Boyle, A. Cruz, J. Iriarte, P. Crabbe, J. A. Zderic, J. A. Edwards, and J. H. Fried, *J. Org. Chem.* **34**, 3779 (1969).
102. T. Mori, K. Matsui, and H. Nozaki, *Tetrahedron Lett.*, 1175 (1970).
103. W. G. Dauben, L. Schutte, and R. E. Wolf, *J. Org. Chem.* **34**, 1849 (1969).
104. M. S. Carson, W. Cocker, S. M. Evans, and P. V. R. Shannon, *Chem. Comm.* 726 (1969).
105. P. M. Collins and P. Gupta, *Chem. Comm.*, 90 (1969).
106. R. R. Rando and W. von E. Doering, *J. Org. Chem.* **33**, 1671 (1968).
107. A. Padwa and R. Gruber, *J. Amer. Chem. Soc.* **90**, 4456 (1968).
108. A. Padwa and R. Gruber, *J. Amer. Chem. Soc.* **92**, 107 (1970).
109. G. Wettermark, *J. Amer. Chem. Soc.* **84**, 3658 (1962); *J. Phys. Chem.* **66**, 2560 (1962); *Nature* **194**, 677 (1962).
110. G. Wettermark, E. Black, and L. Dogliotti, *Photochem. Photobiol.* **4**, 229 (1965).
111. H. Morrison and B. H. Migdalof, *J. Org. Chem.* **30**, 3996 (1965).
112. J. D. Margerum, L. J. Miller, E. Saito, M. S. Brown, H. S. Mosher, and R. Hardwick, *J. Phys. Chem.* **66**, 2434 (1962).
113. J. D. Margerum, *J. Amer. Chem. Soc.* **87**, 3772 (1965).
114. M. D. Cohen, G. M. J. Schmidt, and S. Flavian, *J. Chem. Soc.*, 2041 (1964).
115. M. D. Cohen, D. Y. Hirshberg, and G. M. J. Schmidt, *J. Chem. Soc.*, 2051, 2060 (1964).
116. K. J. Crowley, *Proc, Chem. Soc.*, 334 (1962).
117. J. Srinivasan, *J. Amer. Chem. Soc.* **84**, 4141 (1962).

4

The Photochemistry of Simple Carbonyl Compounds: Type I Cleavage and Oxetane Formation

In Chapter 3 we discussed two photochemical reactions characteristic of simple carbonyl compounds, namely type II cleavage and photoreduction. We saw that photoreduction appears to arise only from carbonyl triplet states, whereas type II cleavage often arises from both the excited singlet and triplet states. Each process was found to occur from discrete biradical intermediates. In this chapter we will discuss two other reactions observed in the photochemistry of carbonyls, type I cleavage and oxetane formation.

4.1. TYPE I CLEAVAGE

Type I cleavage (also often called the Norrish type I process) is a term used to denote cleavage of a carbon–carbon bond α to the carbonyl group to yield two radicals. *α-Cleavage then is the process common to all reactions involving type I cleavage.* After cleavage has occurred the radicals can undergo several different dark reactions to produce a variety of products.

The various routes available to these radicals are shown in the following scheme:

Each of these reactions will be discussed in greater detail after we investigate the nature of the excited state involved in type I cleavage.

4.1a. *The Nature of the Excited State: Part I*

Cyclohexanones undergo type I cleavage to produce a mixture of ketenes and aldehydes by hydrogen transfer,[1–3]

(4.1)

A mechanistic study of this system was carried out by Wagner and Spoerke.[4] Under the conditions of analysis used in this study (glpc) only the quantum yields of alkenal formation could be determined. It was found that with most of the compounds studied, the products of type I cleavage could be essentially eliminated by the presence of the triplet quencher 1,3-pentadiene.[5] Thus the reactive state must be a triplet state.[6] The following mechanism was proposed:

	rate
$K \xrightarrow{h\nu} K^s$	I_a
$K^s \longrightarrow K^t$	Φ_{isc}
$K^t \longrightarrow D$	$k_r[K^t]$
$K^t + Q \longrightarrow K$	$k_q[K^t][Q]$
$D \longrightarrow K$	$k_c[D]$
$D \longrightarrow KT$	$k_K[D]$
$D \longrightarrow A$	$k_A[D]$

This mechanism yields the following expression for Φ_A:

$$\Phi_A = \Phi_{isc}\left(\frac{k_r}{k_r + k_q[Q]}\right)\left(\frac{k_A}{k_A + k_c + k_K}\right) \tag{4.2}$$

or

$$\frac{1}{\Phi_A} = \frac{(k_r + k_q[Q])(k_A + k_c + k_K)}{\Phi_{isc}k_A k_r} \tag{4.3}$$

In the absence of quencher the quantum yield of alkenal $\Phi_A{}^0$ becomes

$$\Phi_A{}^0 = \Phi_{isc}k_A/(k_A + k_c + k_K) \tag{4.4}$$

Multiplying this and the previous expression together, we obtain

$$\Phi_A{}^0/\Phi_A = 1 + (k_q/k_r)[Q] \tag{4.5}$$

Using this equation, Wagner and Spoerke assumed that $k_r = k_A + k_c + k_K = 1/\tau$, that is, the triplet lifetime is determined solely by the triplet reaction. In other words, deactivation of the triplet ketone by internal conversion back to the ground state was assumed to be unimportant. The expression then becomes

$$\Phi_A{}^0/\Phi_A = 1 + k_q\tau[Q] \tag{4.6}$$

It should be noted that this expression is a general one that can be used for any photochemical reaction that can be quenched. It is commonly called the Stern–Volmer equation. This equation predicts that if the proposed mechanism is correct, the data, when plotted as $\Phi_A{}^0/\Phi_A$ vs. [Q], should be linear with an intercept equal to unity and a slope equal to $k_q\tau$. Linear plots were indeed observed out to large Φ^0/Φ values. Assuming a value of $5 \times 10^9\ M^{-1}\ \text{sec}^{-1}$ for the quenching rate constant,[7] the data presented in Table 4.1 were obtained.

TABLE 4.1. *Rate Data for the Formation of Alkenals from Cyclic Ketones by Type I Cleavage*

Ketone[a]	Φ_{-K}[b]	Φ_A	$k_q\tau$, M^{-1}	$(1/\tau) \times 10^{-8}$,[c] sec^{-1}
Cyclopentanone	0.28	0.24	47	1.1
Cyclohexanone (CH)	0.24	0.09	152	0.33
3-Methyl-CH	0.083	0.033	209	0.25
3,5-Dimethyl-CH	0.033	0.005	206	0.24
3,3,5-Trimethyl-CH	0.024	0.002	200	0.25
2-Methyl-CH	0.50	0.42	10.6	4.7
2-Phenyl-CH	0.51	0.04	15.2	3.3
2,6-Dimethyl-CH	0.55	0.40	5.4	9.3
2,2-Dimethyl-CH	0.52	0.41	2.8	18.0

[a] Photolysis in benzene solution at 313 nm.
[b] Quantum yield for the disappearance of ketone.
[c] Calculated assuming $k_q = 5 \times 10^9\ M^{-1}\ sec^{-1}$ in benzene solution.

These data reveal several interesting trends. First of all, the quantum yields are considerably higher for α-substituted derivatives than for unsubstituted or β-substituted derivatives. This undoubtedly arises from the ability of the α-substituents to stabilize the alkyl portion of the biradical and thus to enhance the rate of cleavage (and therefore to increase $1/\tau$). Second, β-substituted derivatives have quantum yields considerably smaller than the cyclohexanone itself, although their triplet lifetimes are very similar and are also similar to that of cyclohexanone. In order for the lifetime to remain essentially the same with k_A and k_K decreasing relative to cyclohexanone, k_c must increase. Thus it appears that the effect of β-substituents is to increase the amount of radical recoupling relative to product formation. Part of this behavior may be due to the decreased availability of β-hydrogen atoms (to be abstracted) upon methyl substitution in these positions. However, the decreases in quantum yields are much too drastic for this to be the sole answer. A further complication arises from the fact that the product formed from 3-methylcyclohexanone corresponds to a selective cleavage at one bond, although either of two bonds could be broken:

O t 50% O CHO 50% O CHO (4.7)

It is possible that the effect of β-substituents arises from small rotational barriers which impede hydrogen transfer in these short-lived biradicals.

The magnitude of the values of $1/\tau$ indicates that the original assumption that radiationless decay of the triplet back to the ground state is unimportant in these reactions is correct since the triplet states of aliphatic ketones commonly have rate constants for radiationless decay from the triplet state on the order of $10^5\ sec^{-1}$ ($\tau = 10^{-5}$ sec).[(8)]

Finally, the fact that the more highly strained cyclopentanone has a larger value of $1/\tau$ ($=k_r$) than cyclohexanone indicates that relief of strain may be an important factor in the rate of type I cleavage. Similar results were reported by Dalton *et al.*,[(8)] who found the rates of α-cleavage of a series of 2-alkylcyclopentanones to be an order of magnitude higher in all cases than those of the corresponding cyclohexanones. These workers also studied the effect of γ-substituents on the quantum yield of alkenal formation from cyclohexanones. These results are shown in Table 4.2. As can be seen, the presence of γ-substituents reduces the quantum yield of alkenal formation, similar to the presence of β-substituents. Derivatives containing the bulky *t*-butyl group in the 4 position gave little or no alkenal. In methanol, however, a second product of type I cleavage results from addition of the solvent to a ketene to produce an ester,

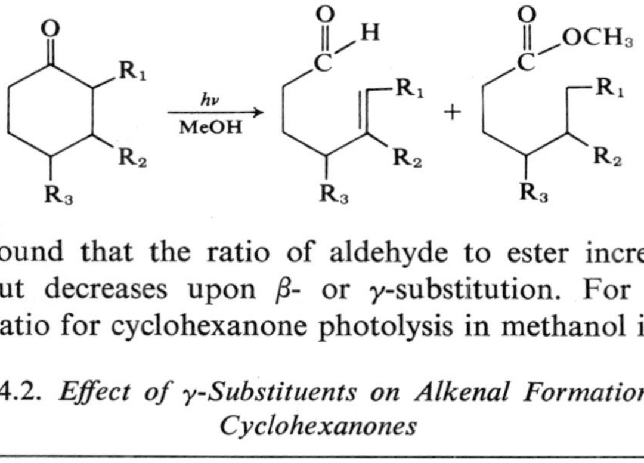

(4.8)

Coyle[(8)] has found that the ratio of aldehyde to ester increases upon α-substitution but decreases upon β- or γ-substitution. For example, the alkenal:ester ratio for cyclohexanone photolysis in methanol is 1.6, that for

TABLE 4.2. *Effect of γ-Substituents on Alkenal Formation from Cyclohexanones*

Ketone	Φ_A	$k_q\tau$[c]	$(1/\tau) \times 10^{-8}$, sec^{-1}
Cyclohexanone (CH)	0.09[a]	450	0.11[d]
4-Methyl-CH	0.07[b]	>100	<0.66[e]
4-*t*-Butyl-CH	0.0[b]	>100	<0.66[e]

[a] In benzene solution.
[b] In cyclohexane.
[c] With 1,3-pentadiene as quencher.
[d] Assuming $k_q = 5 \times 10^9\ M^{-1}\ sec^{-1}$ for benzene.
[e] Assuming $k_q = 6.6 \times 10^9\ M^{-1}\ sec^{-1}$ for cyclohexane.

2-alkylcyclohexanone is 2.5, while the values for 3-alkyl and 4-*t*-butylcyclohexanone drop to ~0.4 and 0.18, respectively. These differences can be understood by examining the transition states for alkenal and ketene formation,[9]

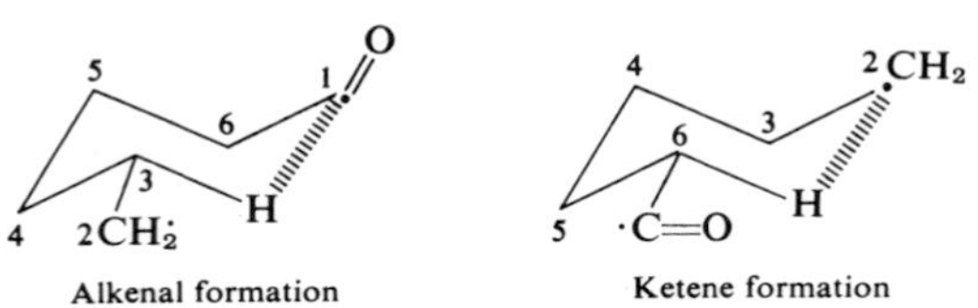

Alkenal formation Ketene formation

Introduction of an alkyl substituent into the α position will introduce unfavorable skew interactions in the transition state for ketene formation but should have little steric effect for alkenal formation. Hence the ratio of alkenal to ester should increase upon 2-substitution. With 3-substituents, the alkyl group must be axial in the transition state leading to alkenal while causing only further skew interactions in that leading to ketene. Assuming that an axial methyl group is more drastic than skew interactions, the alkenal:ketene ratio should drop relative to cyclohexane. A substituent in the 4 position should introduce skew interactions in both transition states, although in the alkenal transition state these are alkyl–alkyl interactions while in the ketene transition state they are alkyl–hydrogen. Hence again one would expect greater steric retardation for alkenal formation and the ratio should drop.

Dalton *et al.*[10] and Weiss, Turro, and Dalton[11] have attempted to assess the relative reactivity of the singlet and triplet states for cyclic alkanones toward α-cleavage. It has been found that certain bicyclic ketones and 2,2,5,5-tetramethylcyclopentanone undergo type I cleavage reactions which cannot be quenched by high concentrations of 1,3-pentadiene. This implies that in these molecules α-cleavage occurs (a) entirely by the singlet state, (b) by both singlet and fast (unquenchable) triplet states, or (c) by an extremely short-lived triplet. Assuming that (b) is correct, the triplet must have a lifetime less than 2×10^{-11} sec since it could not be quenched by 1,3-pentadiene in benzene ($k_q = 5 \times 10^9\ M^{-1}\ \text{sec}^{-1}$) and therefore k_r^t must have a lower limit of $5 \times 10^{10}\ \text{sec}^{-1}$. The singlet lifetimes of these derivatives, on the other hand, are in the range 4.2–8.7 × 10^{-9} sec; hence the upper limit for k_r^s is in the range 1.2–2.4 × $10^8\ \text{sec}^{-1}$. Thus it appears that the triplet states of these derivatives are at least two orders of magnitude more reactive than the singlet states toward type I cleavage. The reason for this great difference in reactivity is not clear at this time.

Barltrop and Coyle[12] have presented strong evidence for the participation of a discrete diradical intermediate in type I cleavage. In this study a mixture of the *cis* and *trans* isomers of 2,3-dimethylcyclohexanone was partially separated and the enriched isomers were photolyzed to yield the same mixture of *cis* and *trans* enals:

(4.9)

Further mechanistic information concerning type I cleavage reactions has been obtained by Yang in a study of alkyl-*t*-butyl ketones.[13] Irradiation of these ketones in hexane solution results in a mixture of products corresponding to both type I and type II cleavage, for example,

$$t\text{-}C_4H_9COR \xrightarrow{h\nu} i\text{-}C_4H_8 + i\text{-}C_4H_{10} + RCHO + t\text{-}C_4H_9COCH_3 + C_2H_5R$$
$$R = n\text{-}C_3H_7 \qquad (4.10)$$

In the presence of 1,3-pentadiene only part of the reaction was quenched. That is, a plot of Φ^0/Φ_q vs. [Q] is initially linear with a steep slope at low quencher concentrations. As [Q] is increased, a gradual decrease in the slope occurs until a final linear region is attained (see Figure 4.1). The initial slope at low quencher concentrations is attributed to normal triplet quenching. The final lower slope is attributed to singlet quenching at higher quencher concentrations. By extrapolating this slope back to zero quencher, one can obtain the limiting value of Φ from the singlet state. By difference Φ^t can be calculated. The data obtained in this way is seen in Table 4.3. These results show that type I cleavage occurs from both the singlet and triplet excited

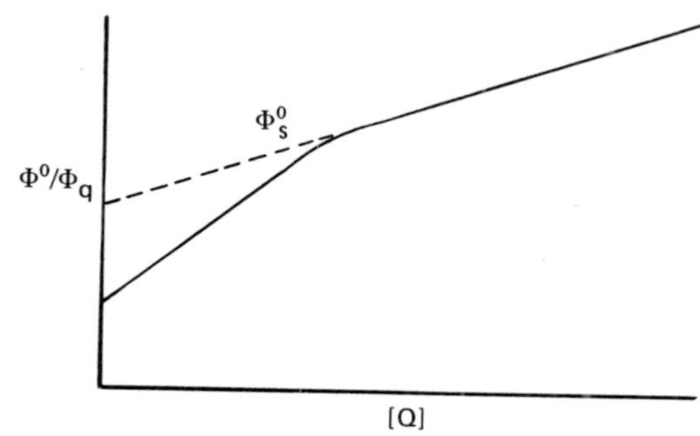

FIGURE 4.1. 1,3-Pentadiene quenching of alkyl-*t*-butyl ketones.

TABLE 4.3. *Quantum Yields of the Primary Processes in the Photolysis of Alkyl-t-butyl Ketones*[a]

Ketone	Φ_{dis}	Φ^s	Φ_I^s	Φ_{II}^s	Φ_{CB}^s	Φ^t	Φ_I^t	Φ_{II}^t	Φ_{CB}^t
Methyl-*t*-butyl	0.51	0.18	0.18	—	—	0.33	0.33	—	—
i-Propyl-*t*-butyl	0.59	0.24	0.22	0.02	—	0.35	0.35	—	—
n-Butyl-*t*-butyl	0.31	0.14	0.07	0.07	—	0.17	0.17	—	—
2-Hexanone	0.33	0.11	—	0.10	—	0.22	—	0.15	0.07

[a] Φ_{dis}, quantum yield for disappearance of ketone; Φ_I^s, Φ_I^t, quantum yields for type I cleavage; Φ_{II}^s, Φ_{II}^t, quantum yields for type II cleavage; Φ_{CB}^s, Φ_{CB}^t, quantum yields for cyclobutane formation.

states of these molecules, whereas type II cleavage occurs mainly from the singlet states. Furthermore, the predominant reaction leading to loss in the starting ketone for the *t*-butyl derivatives is the type I cleavage.

Di-*t*-butyl ketone reacts in pentane solution to yield the following products[14]:

$$(CH_3)_3CCOC(CH_3)_3 \xrightarrow{h\nu} (CH_3)_3CH + (CH_3)_2C{=}CH_2 + (CH_3)_2CC(CH_3)_3 + CO \qquad (4.11)$$

The quantum yield of ketone decomposition for this molecule was found to be $\Phi_{dis} = 0.71 \pm 0.04$. Quantum yields for carbon monoxide and 2,2,3,3-tetramethylbutane were found to be 0.62 ± 0.03 and 0.10 ± 0.01, respectively. As with the alkyl-*t*-butyl ketones, a nonlinear Stern–Volmer quenching plot was obtained with *cis*-1,3-pentadiene, indicating that both singlet and triplet states are involved in this reaction. From the quenching plot it was determined that $\Phi_s = 0.31$ for the disappearance of ketone ($\Phi_t = 0.40$). The singlet state lifetime of di-*t*-butyl ketone was obtained by a double quenching experiment using biacetyl as a singlet state quencher and *cis*-1,3-pentadiene as a triplet quencher (biacetyl will also quench triplets). From the slope of the linear Stern–Volmer plot obtained (Φ^0/Φ_q vs. [biacetyl]) and the assumed $k_q = 1 \times 10^{10}$ liters/mole-sec in hexane, a singlet lifetime of 4.45 nsec was obtained. By single-photon counting of fluorescence a value of 5.6 ± 0.5 nsec was obtained, in good agreement with the quenching study. The triplet lifetime of the ketone was calculated from the initial slope of the Stern–Volmer plot of piperylene quenching. The value obtained was 0.11 nsec. These results yield rate constants for reaction of $k_r^s = 6 \times 10^7$ and $k_r^t \approx 7\text{–}9 \times 10^9\ \text{sec}^{-1}$. Thus again we see that the triplet state is approximately two orders of magnitude more reactive than the singlet state toward type I cleavage.

A similar dissection of the excited states responsible for reaction by quenching has been carried out on *cis*- and *trans*-8-methyl-1-hydrindanone and *cis*- and *trans*-9-methyl-1-decalone with the purpose of determining the extent to which these biradical intermediates recombine in the solvent cage.[15] That solvent cage radical recombination may be an important process is suggested by the fact that in solution the sum of the quantum yields for all type I cleavage reactions falls short of unity whereas these processes occur with essentially a total unit efficiency in the gas phase.[13,14] With the *cis* and *trans* hydrindanones and decalones one should observe isomerization to the opposite isomer as well as product formation if cage recombination is important[16]:

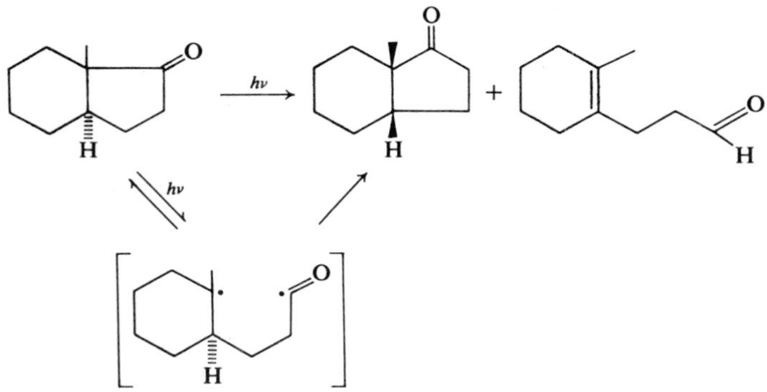

(4.12)

The data obtained in this study are presented in Table 4.4.

These results contain several interesting observations. First of all, it can be concluded that radical recombination is an important process, at least for these compounds. It can be seen that the *cis*-ketones undergo isomerization only from their first excited singlet states (linear zeroth-order Stern–Volmer plots were obtained with *cis*-1,3-pentadiene) while the *trans*-ketones undergo isomerization from both singlet and triplet states. Furthermore, the isomerization quantum yields are considerably higher for the *trans* compounds than for the *cis*. This latter result can be explained by examination of the stereochemistry of the *cis*- and *trans*-hydrindanones together with some known

TABLE 4.4. *Cleavage and Isomerization of 8-Methyl-1-hydrindanones and 9-Methyl-1-decalones*[15],[a]

Compound	Φ_{dis}	Φ_A	$\Phi_A{}^s$	$\Phi_A{}^t$	Φ_I	$\Phi_I{}^s$	$\Phi_I{}^t$	Φ_{iso}
cis-8-Methyl-1-hydrindanone	0.39	0.31	0.19	0.12	0.02	0.03	—	0.13
trans-8-Methyl-1-hydrindanone	0.83	0.27	0.22	0.05	0.46	0.37	0.09	0.13
cis-9-Methyl-1-decalone	0.40	0.25	0.07	0.18	0.05	0.05	—	0.75
trans-9-Methyl-1-decalone	0.56	0.25	0.15	0.10	0.16	0.10	0.06	0.39

[a] Φ_{dis}, quantum yield for disappearance of ketone; Φ_A, quantum yield for aldehyde formation; Φ_I, quantum yield for ketone isomerization; Φ_{iso}, intersystem crossing quantum yields determined by method of Lamola and Hammond.[17]

facts and some assumptions. The *cis*-hydrindanone can exist in the two conformations shown below, while the *trans* can exist only with a 1,2-diequatorial ring juncture (the diaxial juncture is too highly strained). It is known, furthermore, that cyclohexyl radicals prefer to react from the axial side rather than the equatorial;(18) thus axial ring closure of the biradical will be preferred. Assuming that the photochemical process is faster than ring flipping and that the stereochemistry at a reactive carbon is lost upon radical formation, one can see that the *trans*-ketone should isomerize more readily

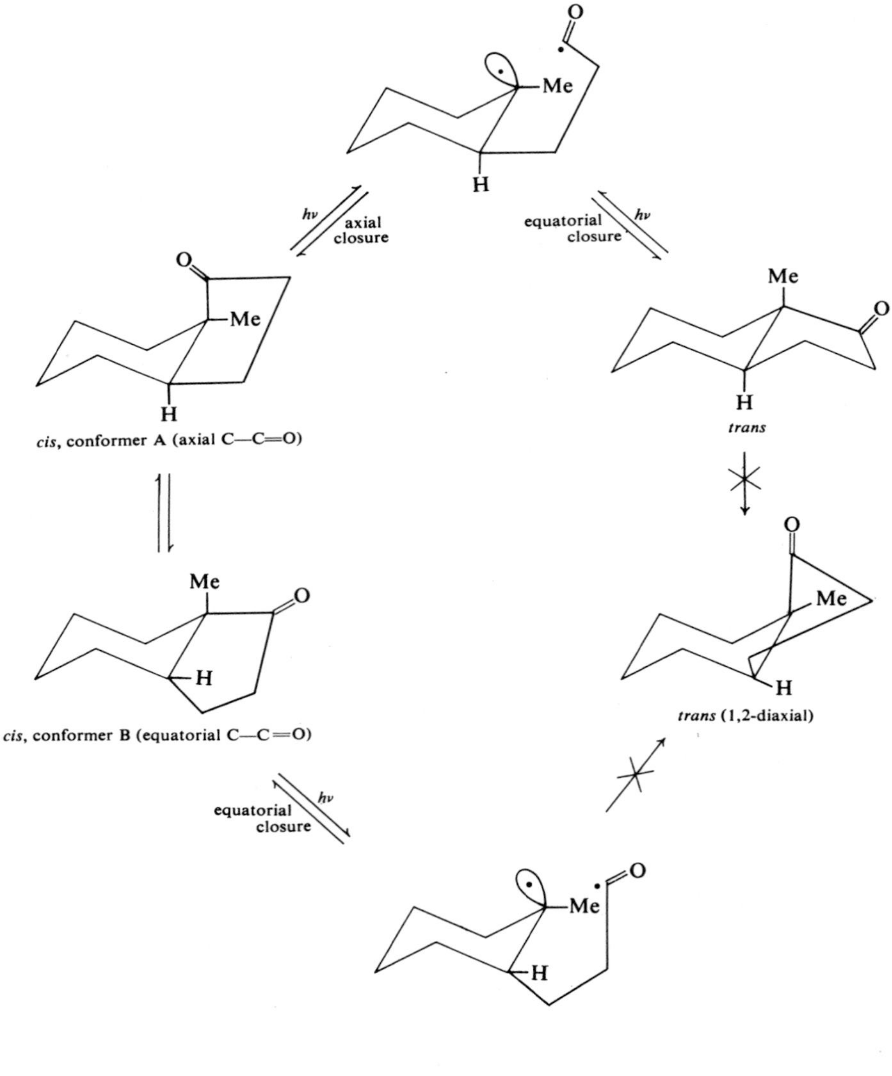

hν axial closure

equatorial closure hν

cis, conformer A (axial C—C=O)

trans

cis, conformer B (equatorial C—C=O)

trans (1,2-diaxial)

equatorial closure hν

than the *cis* since there are more pathways involving axial closure that lead to *cis* than lead to *trans*.

As can be seen in Table 4.4, the intersystem crossing quantum yields for these compounds increase as *cis*-hydrindanone $\approx$ *trans*-hydrindanone $<$ *trans*-decalone $<$ *cis*-decalone. This order also parallels the thermodynamic stability of these compounds[19] and is probably due to the varying amounts of internal strain. The data indicate that most of the photochemistry arises from the singlet excited states of these derivatives and is competitive with intersystem crossing even for the less favored cleavage of an equatorial C—C=O bond. In addition, the singlets of the hydrindanones are seen to be more reactive than those of the less strained decalones. Thus it appears that the amount of α-cleavage is proportional to the amount of internal strain in the excited singlet state.

Summary

The following is a short summary of the mechanistic aspects of type I cleavage discussed in this section.

(a) Ketones readily undergo type I cleavage from both the excited singlet and the excited triplet states.

(b) In some cases the triplet state may be as much as two orders of magnitude more reactive than the singlet.

(c) In cyclic ketones the efficiency of α-cleavage may reflect the amount of internal strain.

(d) Type I cleavage probably involves a discrete biradical intermediate, although concerted pathways may exist for some molecules.

(e) The low quantum yields for α-cleavage observed in many compounds in solution may result from cage recombination of the initially formed radicals. This process may occur from either the singlet or the triplet state.

(f) In general α-substituents increase the efficiency of α-cleavage while β- and γ-substituents decrease the efficiency.

4.1b. *Some Examples and Synthetic Applications of Type I Cleavage Reactions*

Reactions involving type I cleavage are especially prominent in gas-phase photolyses of carbonyl compounds. With molecules containing γ-hydrogens type II cleavage is also commonly observed[20]:

$$(CH_3)_2CH{-}CH_2{-}\overset{\overset{\displaystyle O}{\|}}{C}{-}H \xrightarrow[\text{gas phase}]{h\nu} \begin{cases} \xrightarrow{47\%} (CH_3)_3CH + CO \quad \text{(type I)} \\ \xrightarrow{52\%} CH_3{-}\overset{\overset{\displaystyle O}{\|}}{C}{-}H + CH_2{=}CHCH_3 \quad \text{(type II)} \end{cases} \tag{4.13}$$

In the liquid phase, α-cleavage is predominant among strained tricyclic and small ring ketones. The products of photolyses in the gas and liquid phases often differ markedly due to the possibility of reaction from higher vibrational levels of the excited state in the gas phase. In general the rapid cascade of vibrational excitation to the zero vibrational level of the excited state caused by solvent collisions prevents reaction from higher vibrational states in solution. In addition, the possibility of radical recombination in the solvent cage can drastically alter the amount and type of photochemistry observed in the liquid phase. For example, photolysis of cyclobutanone in the vapor phase at 100°C gives high yields of carbon monoxide, ethylenes, and presumably ketene,[21] while in ethanol solution only a low yield of the acetal is observed[22]:

vapor phase $h\nu$, 3130 Å; CO $\Phi = 0.40$; $\Phi = 0.14$; $\Phi = 0.014$; C_2H_4 + $H_2C{=}C{=}O$, $\Phi = 0.54$; $h\nu$ EtOH; OEt, H; 8%

(4.14)

In this section a brief review of the various reactions arising from type I cleavage and some synthetic applications of these reactions will be presented.

Type I Cleavage Reactions Proceeding through Carbene Intermediates

As stated previously, the photolysis of cyclobutanone in ethanol solution results in an 8% yield of an acetal, presumably formed from ethanol addition to a carbene intermediate. Alkylation of the α positions of cyclobutanone increases the yield of this rearrangement product,[22]

$h\nu$ / MeOH; O, OMe, H; 68%

(4.15)

On the other hand, tetramethylcyclobutadione proceeds by an entirely different pathway to yield tetramethylcyclopropanone,[23,24]

$h\nu$, $CHCl_3$ + CO + $>C{=}C{=}O$ (4.16)

Turro and McDaniel[(25)] have shown that this reaction probably occurs from the excited singlet state (reaction could not be quenched by 1,3-pentadiene) and is stereospecific,

Et, MeO, $h\nu$, $\Phi = 0.03$; $-CO$: Et, OMe, $\Phi = 0.01$; MeO, Et + $CH_2{=}C{=}O$, $\Phi = 0.11$; Et, OMe, MeO (4.17)

By studying the effect of various α-substituents, it has been shown that the bond to the most highly substituted α carbon is preferentially cleaved and that the more nucleophilic alkyl carbon migrates to the relatively electron-poor free radical to form the carbene intermediate,

ϕ, CF_3, CF_3 $\xrightarrow{h\nu}$ [CF_3, CF_3] $\xrightarrow{>95\%}$ $CH_2{=}C{=}O$ + $\phi CH{=}C(CF_3)_2$ (4.18)

ϕ, CH_3, CH_3 $\xrightarrow{h\nu}$ [CH_3, CH_3] $\xrightarrow{>85\%}$ [CH_3, CH_3, ϕ] $\xrightarrow{MeOH}$ ϕ, H_3C, H_3C, OMe (4.19)

Perhaps the first examples of this type of reaction resulting from type I cleavage were reported by Yates[(27)] in a study of the photochemistry of cyclocamphanone and nortricyclanone,

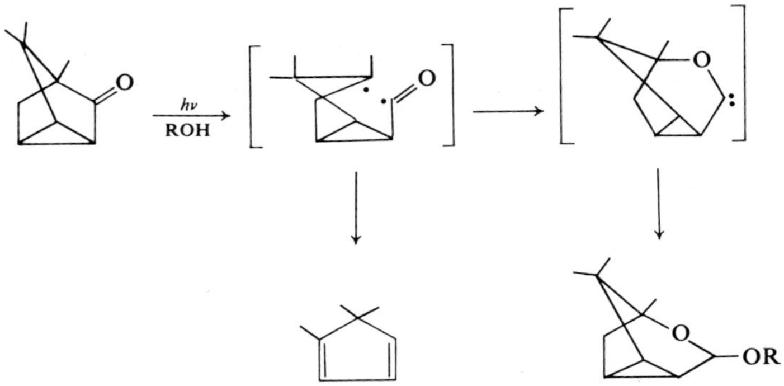

$h\nu$, ROH

(4.20)

That the carbene is actually an intermediate in the photolysis of cyclocamphanone was shown by trapping the carbene with cyclohexene,

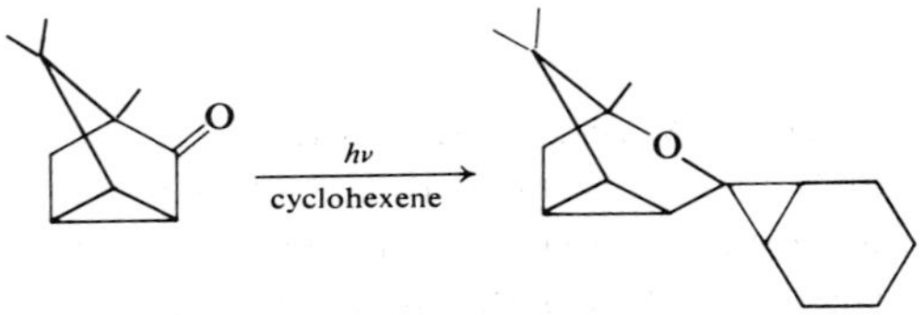

$h\nu$, cyclohexene

(4.21)

Irradiation of nortricyclanone in alcoholic solution has also been shown to yield cyclic acetals analogous to those observed with cyclocamphanone,

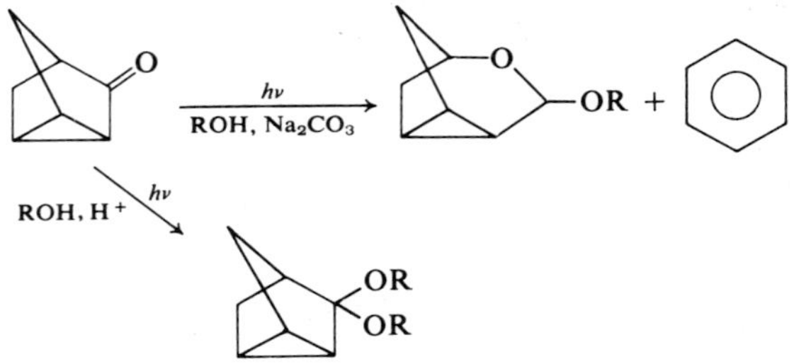

$h\nu$, ROH, Na_2CO_3 — OR +

$h\nu$, ROH, H^+ — OR, OR

(4.22)

In the presence of just a slight trace of acid, however, the ketal shown above is formed in preference to the acetal.

The formation of cyclic acetals has also been reported by Hostettler for a series of 3-substituted 2,2,4,4-tetramethylcyclobutanones,[28]

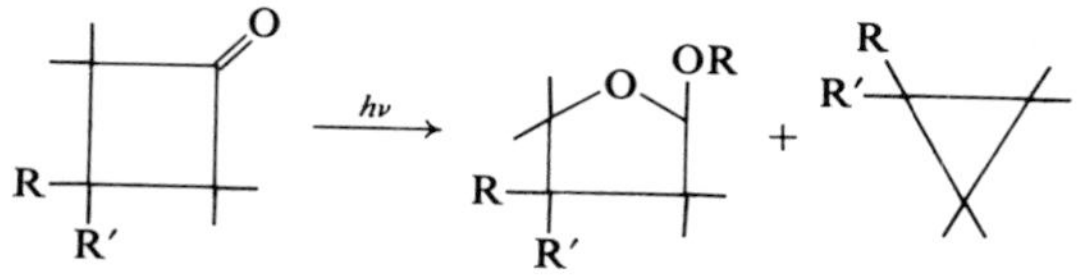

(4.23)

Interestingly, the compound shown below yields an acetal formed by bonding of oxygen at the cleavage site but a decarbonylation product corresponding to bonding at an allylic position relative to the site of original cleavage:

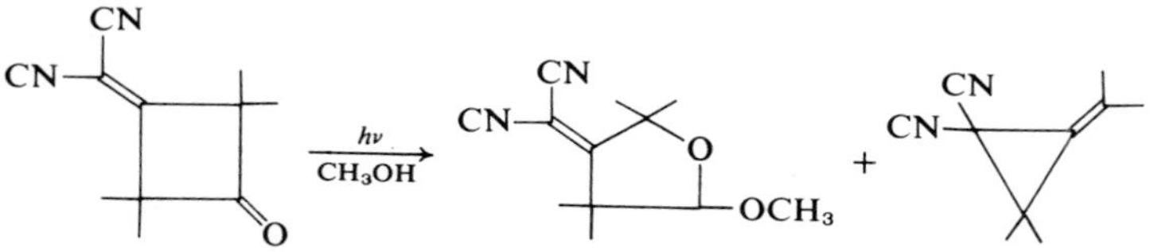

(4.24)

This may indicate that the reaction to form the carbene is concerted or at least indicates that if a diradical is involved, reaction must be too fast to permit rotation around the C-2–C-3 bond,

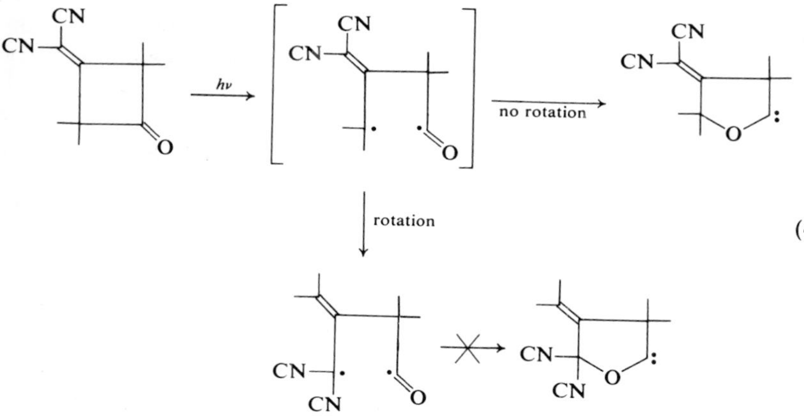

(4.25)

The photochemistry of some isomeric D-nor-16-keto steroids has been reported by Quinkert and co-workers,[29]

$$\xrightarrow[\text{EtOH}]{h\nu}$$ 54% + 12% + (1) + (2) (4.26)

$$\xrightarrow[\text{EtOH}]{h\nu}$$ 71% + (1) + (2) (4.27)

The *trans*-fused steroid produced 12% of a product corresponding to cleavage at the less substituted carbon atom, whereas the *cis*-fused isomer gave only products characteristic of cleavage at the more highly substituted carbon. The lack of discrimination shown by the *trans* isomer is thought to be due to the higher amount of strain in this isomer relative to the *cis*. The fact that the *cis*-fused isomer yields only *cis*-fused acetals and the *trans*-fused isomer yields only *trans*-fused acetals would tend to indicate again that carbene formation must be either concerted or sufficiently rapid to prevent loss of stereochemistry in the intermediate diradical.

In ethanol solution 1,2-benzocyclobutenedione undergoes reaction to produce a lactol ether, the analog of the products produced upon photolysis of the tricyclic ketones and cyclobutanones discussed above,[30]

$$\xrightarrow{h\nu}$$ (4.28)

However, in pentane–dichloromethane solution other products are produced,[30,31]

(4.29)

Migration to form the carbene intermediate occurs even when carbon is replaced by silicon, as shown by Brook and Pierce,[32]

(4.30)

It has been stated that bridged bicyclic systems fail to yield any cyclic acetal upon photolysis,[27,33]

(Ref. 33)

(4.31)

However, more recent work indicates that some bicyclic systems do indeed yield cyclic acetals from addition to carbene intermediates,[34,35]

70–80%

15%

(4.32)

In contrast to the report by Srinivasan[33] no ketone product was isolated.[34]

From the reactions presented in this section one can conclude that cyclic acetal formation via addition to a carbene intermediate is a general reaction for type I cleavage of cyclobutanones, tricyclic compounds, and certain bridged bicyclics as minor products. No acetal has been isolated from photolyses of cyclopentanones or cyclohexanones except for the special case of an α-sila ketone previously discussed.

Type I Cleavage Reactions Proceeding through Ketene Intermediates

The following reactions are believed to result from addition to ketene intermediates produced from type I cleavage reactions:

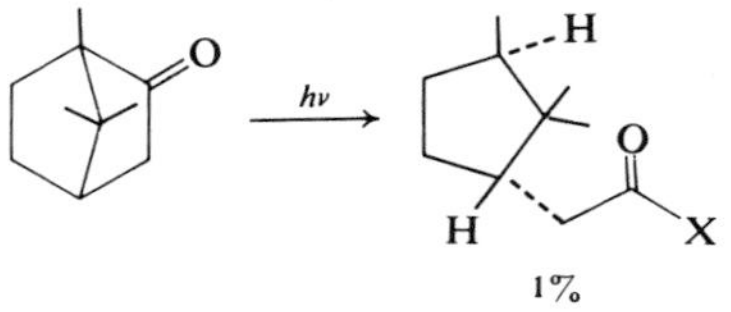

$h\nu$ 1% (4.33)

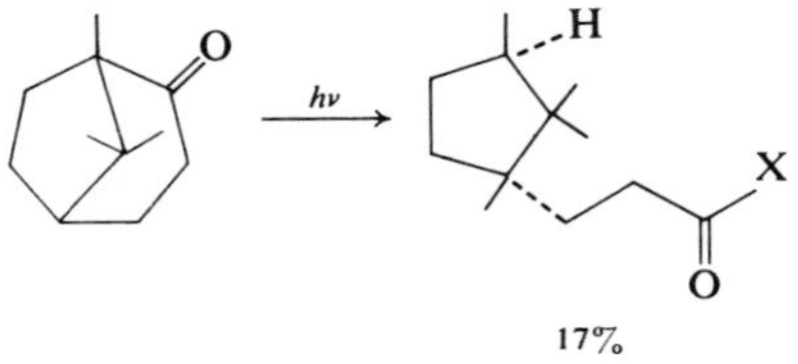

$h\nu$ 17% (4.34)

The importance of the six-membered transition state for hydrogen transfer to form the ketene is clearly demonstrated by the 17-fold increase in the ketene addition product (X = NHR) in proceeding from a five-membered transition state to a six-membered transition state above.[36,37]

A ketene intermediate is also proposed in the following photolysis of the oxosteroid[38]:

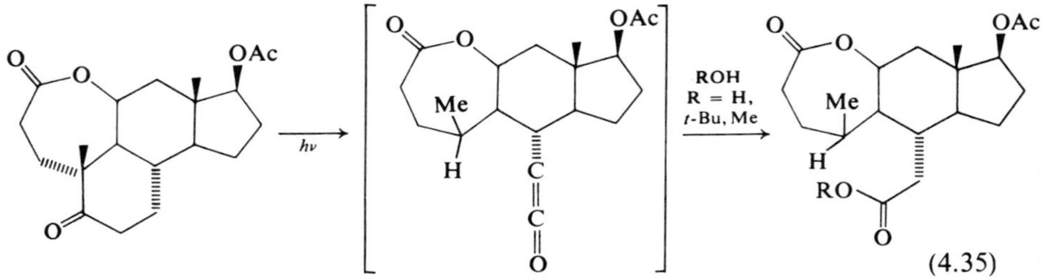

ROH R = H, *t*-Bu, Me (4.35)

Photolysis of the following 5-α-hydroxycholestan-6-one has been reported to produce the lactone stereospecifically. The 5-β isomer yields the lactone of opposite stereochemistry:

RO OH O → RO H O O (4.36)

R = H or OH

The stereospecificity of this reaction is thought to arise from retention of the stereochemistry about the alkyl radical center due to hydrogen bonding of the OH group with the ketone,

RO OH---O → RO H OH C O → RO H O O (4.37)

Although the following examples involve ketene intermediates, no hydrogen transfer is necessary[40]:

O ϕ ϕ $\xrightarrow[H_2O,\ CH_3CN]{h\nu}$ ϕ ϕ O → ϕ ϕ C=C=O $\xrightarrow{H_2O}$ ϕ ϕ COOH 78% (4.38)

O $\xrightarrow[MeOH]{h\nu}$ COOMe 62% (4.39)

Type I Cleavage Reactions Leading to Molecular Rearrangement

Cyclic β,γ-unsaturated ketones undergo a photochemical rearrangement thought to involve α-cleavage and radical rearrangement. The products of these cyclic ketones are β,γ-unsaturated ketones possessing exocyclic olefinic linkages,[41,42]

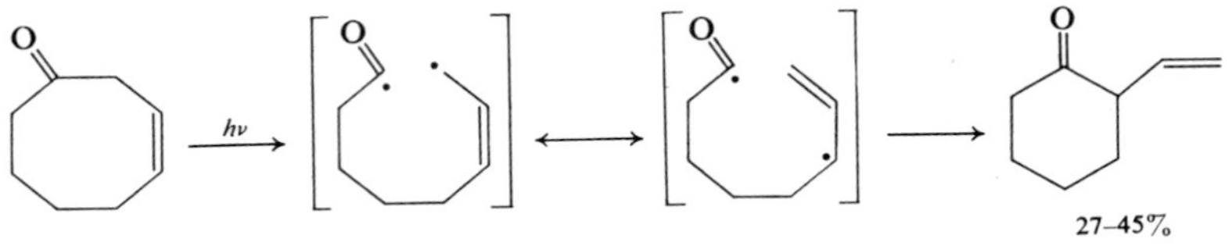

(4.40)

27–45%

The effect of substituents and ring size on this photochemical rearrangement was studied by Paquette and Meehan.[43] Irradiation of the following ketones resulted in steady state mixtures of starting ketone and photoketone whose ratios depended upon structure:

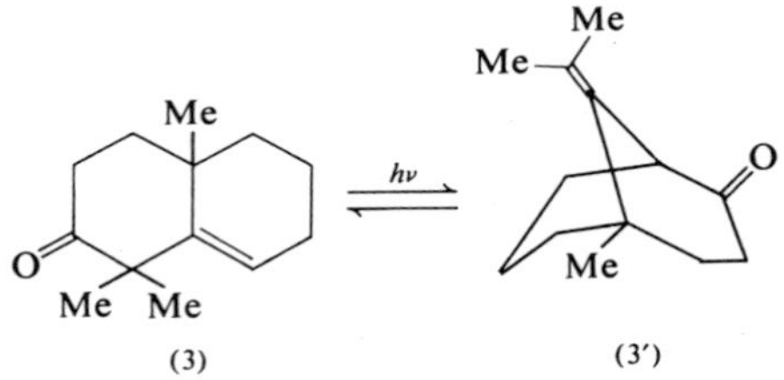

(3) ⇌ (3′) $h\nu$ (3)/(3′) = 10 (4.41)

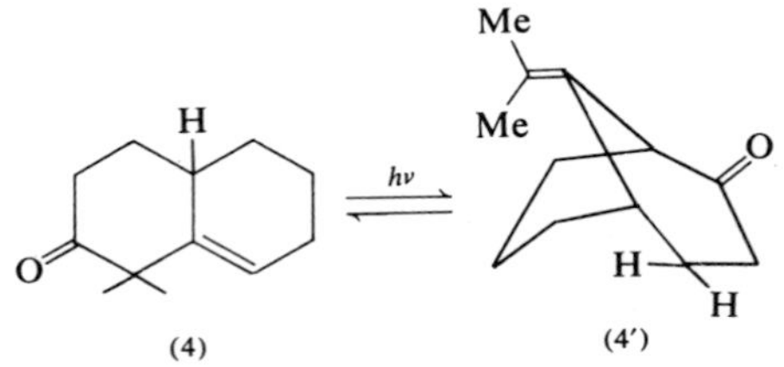

(4) ⇌ (4′) $h\nu$ (4)/(4′) = 4 (4.42)

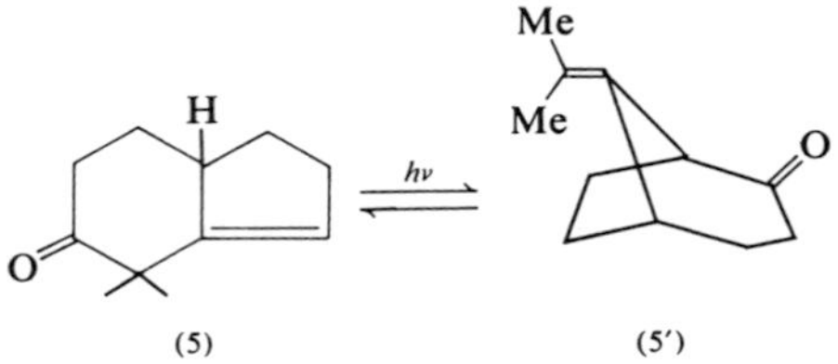

(5) ⇌ (5′) $h\nu$ (5)/(5′) = 7 (4.43)

However, examination of the spectral properties of the starting ketones and isolated photoproducts indicated that the product distribution was determined by the photochemical properties of the molecules rather than their relative thermodynamic stabilities. These workers proposed that these rearrangements may proceed concertedly by 1,3-suprafacial sigmatropic shifts rather than discrete biradical intermediates.

A similar rearrangement but where the allylic radical closes to yield a cyclopropyl radical can be seen in the following examples[44,45]:

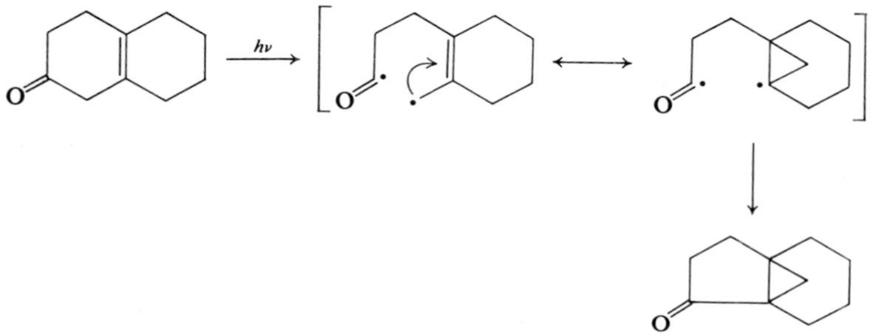

(4.44)

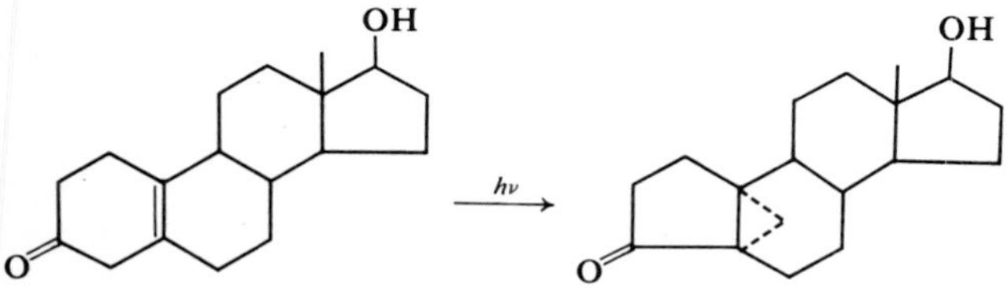

(4.45)

The fact that only one cyclopropanone is produced in the photolysis of the above enone may indicate that the rearrangement occurs by a concerted pathway.[45,46]

This rearrangement of β,γ-unsaturated ketones was utilized to achieve a photochemical synthesis of homocubanone,[47]

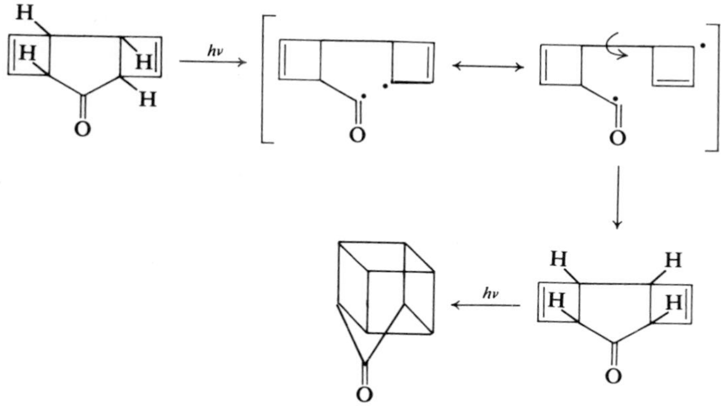

(4.46)

A further type of rearrangement occurs in 1,3-dicarbonyl compounds,[48]

(4.47)

Cleavage of Keto-Sulfides

γ-Keto sulfides undergo a cleavage reaction which can be viewed as a formal α-cleavage followed by elimination,[49]

(4.48)

27%

However, an alternative mechanism involving electron transfer can be proposed,

(4.49)

A further example of this type of reaction follows:

13% 1% 2%

(4.50)

15% 6%

Recently Padwa and Battisti reported the photolysis of a 9-heterobicyclo-[3.3.1]nonenone system.[50] A mechanism postulating α-cleavage was proposed,

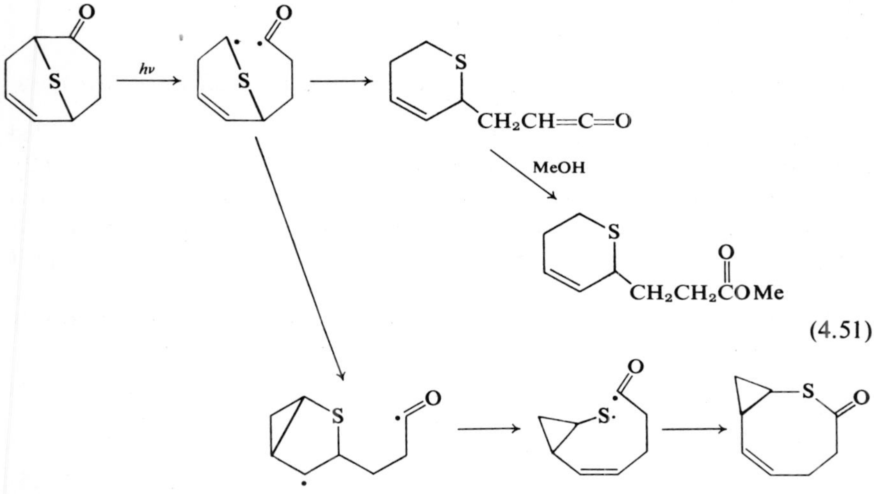

$CH_2CH{=}C{=}O$

MeOH

$CH_2CH_2\overset{O}{\overset{\|}{C}}OMe$

(4.51)

However, a pathway involving electron transfer could also be proposed to account for the products,

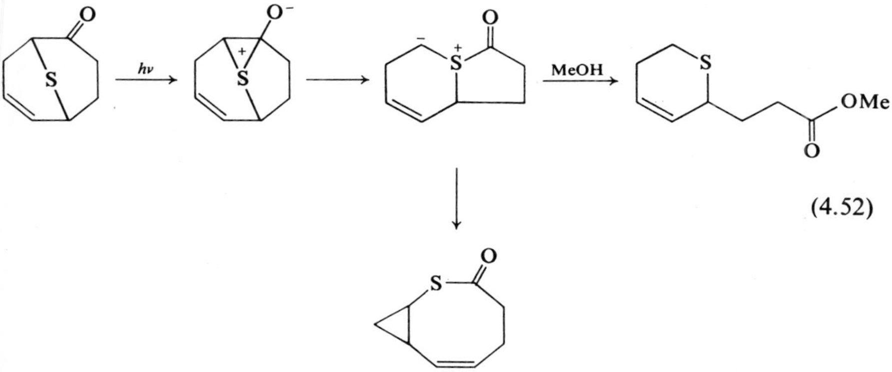

$h\nu$ MeOH OMe

(4.52)

To determine which of these possible pathways was actually operative, the photolysis was carried out in CH_3OD. Under these conditions the product ester contained one deuterium atom in the C-6 position:

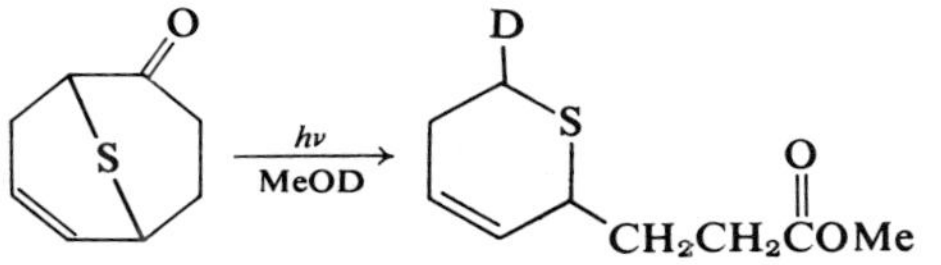

$\xrightarrow[\text{MeOD}]{h\nu}$ D $CH_2CH_2\overset{O}{\overset{\|}{C}}OMe$ (4.52′)

This result eliminates an α-cleavage pathway to form ketene since deuterium in this case should have appeared α to the carbonyl group of the ester. This result is, however, compatible with the charge-transfer mechanism.

Low-Temperature Observation of Intermediates in Type I Cleavage Reactions

By carrying out photolyses in liquid nitrogen- or liquid helium-cooled infrared cells using a special low-temperature apparatus (see Figure 4.2), one is often able to obtain direct spectroscopic evidence for intermediates of photochemical reactions. In this section we will briefly review how low-temperature techniques have been used to observe intermediates in type I cleavage reactions.

In 1959 Wheeler and Eastman reported that umbellulone quantitatively produced thymol upon photolysis at room temperature[51]:

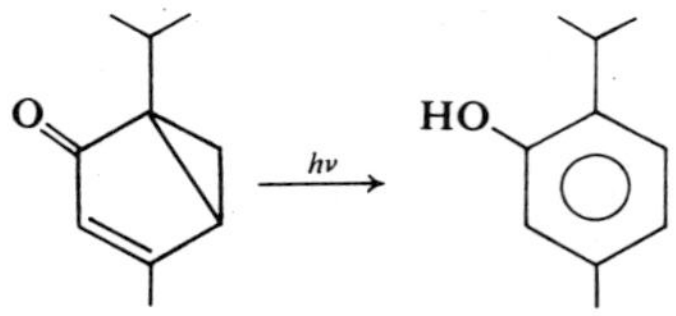

$$\xrightarrow{h\nu} \qquad (4.53)$$

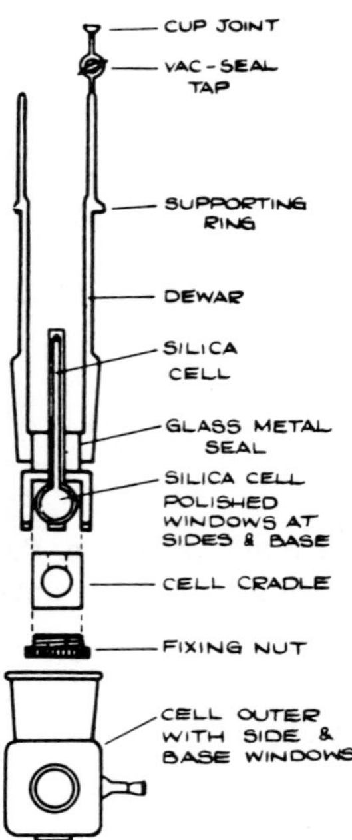

FIGURE 4.2. Spectrodewar for low-temperature spectroscopy (redrawn after a dewar available from Rho Scientific, New York).

This reaction has been reinvestigated by Barbers, Chapman, and Lasila using low-temperature techniques to observe intermediates.[52] These workers found that irradiation of umbellulone in an infrared cell at liquid nitrogen temperatures yielded two primary photoproducts, a ketene derivative ($\nu_{C=O} = 2113$ cm^{-1}) and another photoproduct with absorption at 1670 and 1630 cm^{-1}. Thymol was not observed to be a primary photoproduct. Upon warming to -90°C, the photoproduct with absorption bands at 1670 and 1630 cm^{-1} was rapidly converted to thymol. The ketene was also converted to thymol at temperatures above -70°C, but could be trapped by formation in a methanol ether glass and warming to produce the ester. The following mechanistic scheme was proposed:

Dihydrocoumarin produces the following product upon photolysis:

(4.54)

Two mechanisms have been proposed to account for this product, one via a ketene intermediate[53] and another via a spirodienone intermediate:

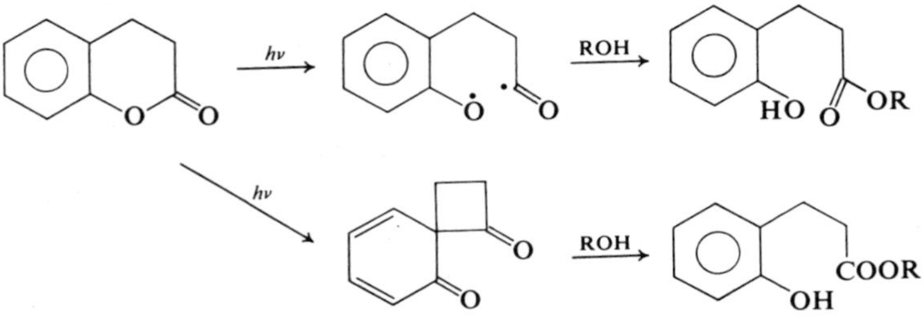

(4.55)

To determine which mechanism was correct, the photolysis was studied at low temperature.[55] An intermediate with a strong ketene band at 2115 cm^{-1} was observed. No bands which could be attributed to the spirodienone could be detected. Warming of the matrix to $-70°C$ caused the loss of the ketene absorption and the appearance of new bands characteristic of the ester (ROH = MeOH), indicating the ketene mechanism to be correct:

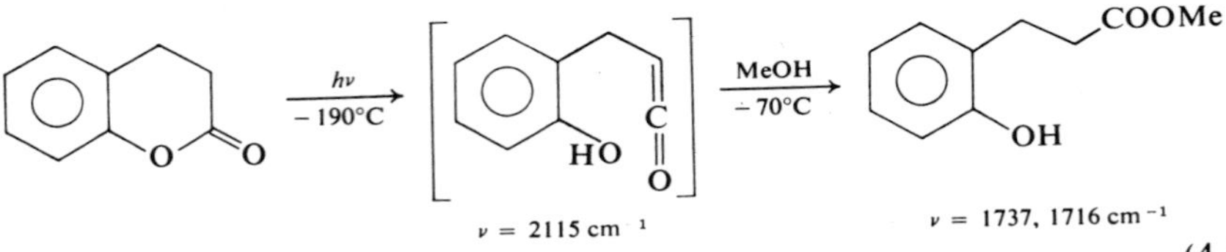

$\nu = 2115\ cm^{-1}$ $\nu = 1737,\ 1716\ cm^{-1}$

(4.56)

Irradiation of 2-phenyl-3,1-benzoxathian-4-one at room temperature has been reported to yield the following product[55]:

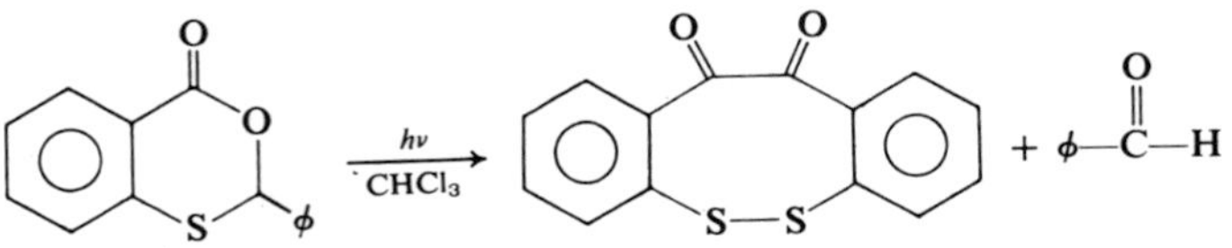

$$+\ \phi\text{—}\overset{\overset{\displaystyle O}{\|}}{C}\text{—}H \qquad (4.57)$$

Irradiation of this compound at 77°K produced bands characteristic of benzaldehyde ($\nu_{CO} = 1699\ cm^{-1}$) and a new species with $\nu_{CO} = 1803\ cm^{-1}$. Upon warming the irradiated sample to $-40°C$, new bands ($\nu = 900\ cm^{-1}$) appeared, characteristic of the following compound:

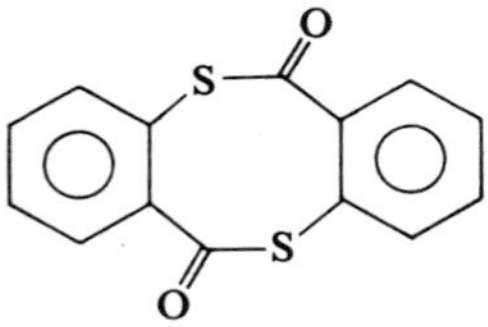

A mechanism involving 2-thiobenzopropiolactone as an intermediate was proposed[57]:

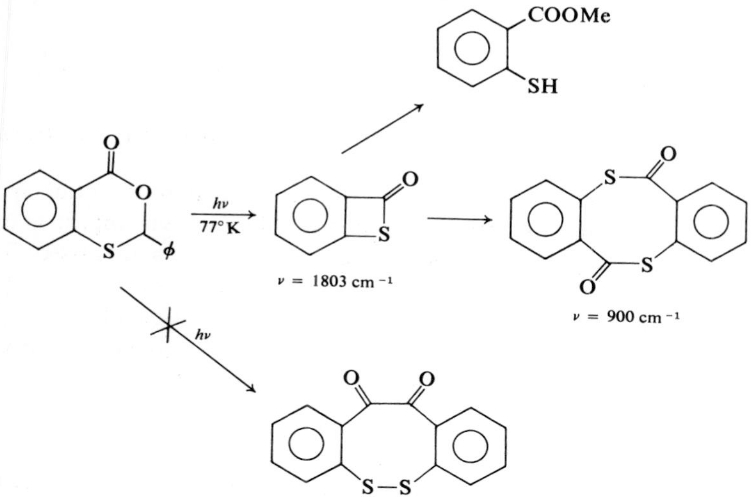

COOMe
SH
O
O
S
φ
$h\nu$
77°K
O
S
ν = 1803 cm^{-1}
S
O
O
S
ν = 900 cm^{-1}
$h\nu$
O O
S—S

(4.58)

Ketene products have also been observed in the following reactions at liquid nitrogen temperatures[52,56,57]:

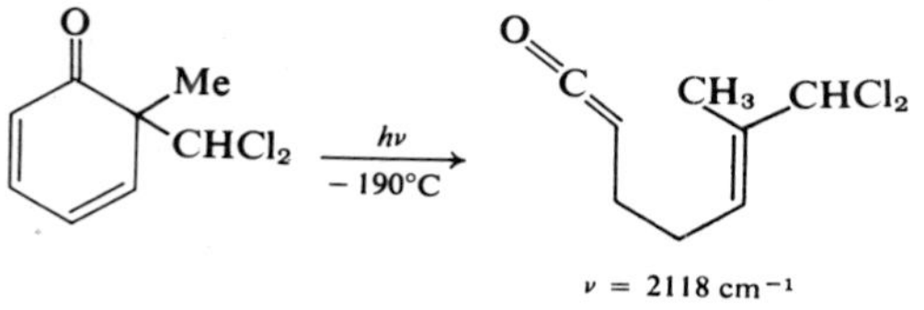

O
Me
$CHCl_2$
$h\nu$
− 190°C
O
C
CH_3 $CHCl_2$
ν = 2118 cm^{-1}

(4.59)

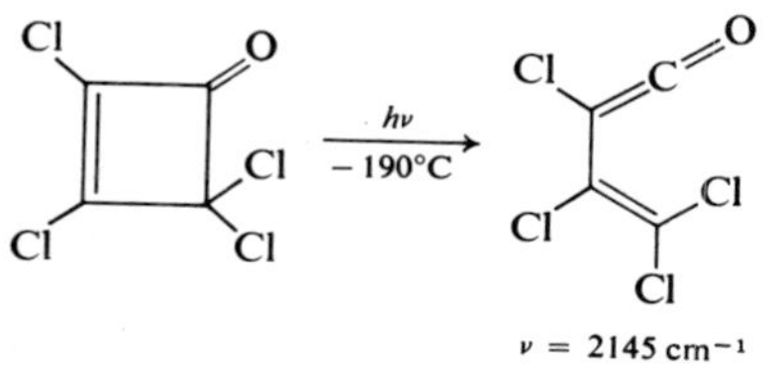

Cl
O
Cl
Cl
Cl
$h\nu$
− 190°C
Cl
C
O
Cl
Cl
Cl
ν = 2145 cm^{-1}

(4.60)

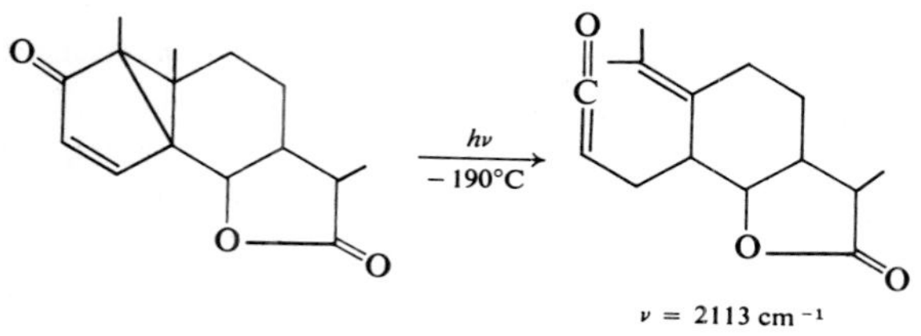

O
O
O
$h\nu$
− 190°C
O
C
O
O
ν = 2113 cm^{-1}

(4.61)

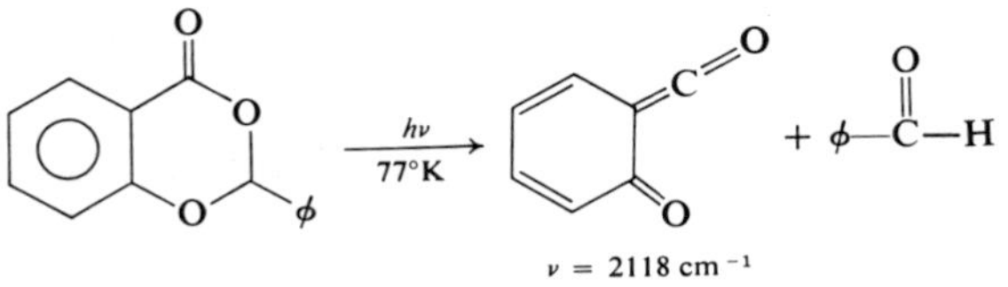

$$\xrightarrow[77°K]{h\nu} \quad + \ \phi\text{—}\overset{O}{\overset{\|}{C}}\text{—H} \qquad (4.62)$$

$\nu = 2118\ cm^{-1}$

4.1c. *Type I Cleavage Reactions Resulting in Loss of Carbon Monoxide*

Two of the three general types of secondary reactions resulting from photochemical α-cleavage of carbonyls, namely molecular rearrangement and hydrogen transfer to yield aldehydes or ketenes, have been discussed. The third type of reaction observed, decarbonylation, will be discussed in this section. The discussion will begin with the decarbonylation of small ring carbonyls. By way of example of this type of reaction, diphenylcyclopropenone decarbonylates upon photolysis to yield diphenylacetylene[57]:

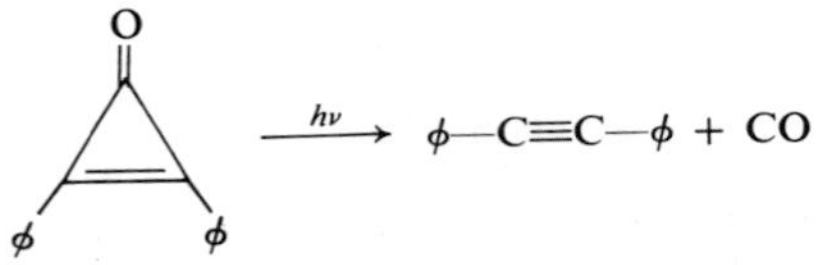

$$\xrightarrow{h\nu} \phi\text{—C}\equiv\text{C—}\phi + CO \qquad (4.63)$$

Tetramethylcyclobutadione yields several initial products, among them tetramethylcyclopropanone, corresponding to loss of carbon monoxide[23,24,58]:

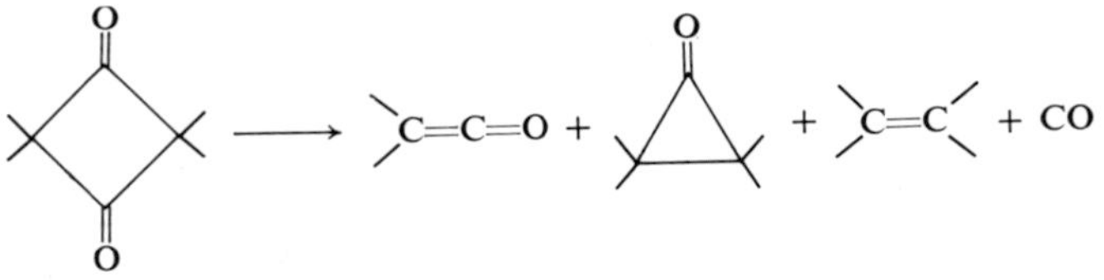

$$\longrightarrow \ \text{C=C=O} + \quad + \ \text{C=C} \ + CO \qquad (4.64)$$

The formation of a cyclopropanone derivative (originally determined by the isolation of degradation products from this unstable species) stimulated considerable interest in this reaction. Tetramethylcyclopropanone, however, cannot be isolated from the reaction mixture under normal photolysis conditions even with the use of an inert solvent. That it is indeed formed as an initial product of α-cleavage results from various trapping experiments in which chemical agents present in the reaction mixture were used to produce stable derivatives of the cyclopropanone [see equation (4.65)].

Using low-temperature techniques such as those described in the previous section, Haller and Srinivasan[24b] obtained direct spectroscopic evidence for the formation of the cyclopropanone. At 4°K infrared bands corresponding to carbon monoxide, dimethylketene, tetramethylethylene, and a compound

hν, + CO, MeOH, LiAlH$_4$, HO OMe, OH

(4.65)

with an intense absorption at 1840 cm^{-1}, which was assigned to tetramethylcyclopropanone, were observed.

The presence of oxygen in the reaction mixture drastically changes the course of the reaction.[23] Under these conditions acetone, carbon monoxide, carbon dioxide, and tetramethylethylene oxide are produced. Presumably tetramethylcyclopropanone is still produced as an initial product and the products observed result from oxygen addition to this species:

hν, O_2, O_2, + CO, + CO_2

(4.66)

Other tetrasubstituted diones react in a similar fashion.[23] For example,

hν, , O_2

(4.67)

Some five- and six-membered ring ketones undergo efficient decarbonylation upon photolysis:

$h\nu$, Ref. 59; + CO (4.68)

$h\nu$, Ref. 60; + + CO (4.69)

$h\nu$, Ref. 61; + CO (4.70)

100%

$h\nu$, Ref. 61; + + CO (4.71)

13% 28% 19–22%

$h\nu$, Ref. 61; + CO (4.72)

95%

$h\nu$, Ref. 61; + + CO (4.73)

55% 45%

H H; $h\nu$, Ref. 61; H H + H H + (4.74)

53% 18% 27%

In previous sections of this chapter we have seen many examples of type I cleavage reactions in which loss of carbon monoxide was not an important process. In the examples given above, however, decarbonylation is important, as evidenced by the high yields of decarbonylated products. Factors which facilitate decarbonylation include the presence of a suitably located cyclopropyl ring, β,γ unsaturation, and alkyl substitution at α positions. Presumably these conditions allow greater stabilization of intermediate biradicals formed upon loss of carbon monoxide.[61]

That resonance stabilization of intermediate biradicals is important in determining the efficiency of decarbonylation follows from the following examples yielding benzyl radicals upon loss of carbon monoxide[57]:

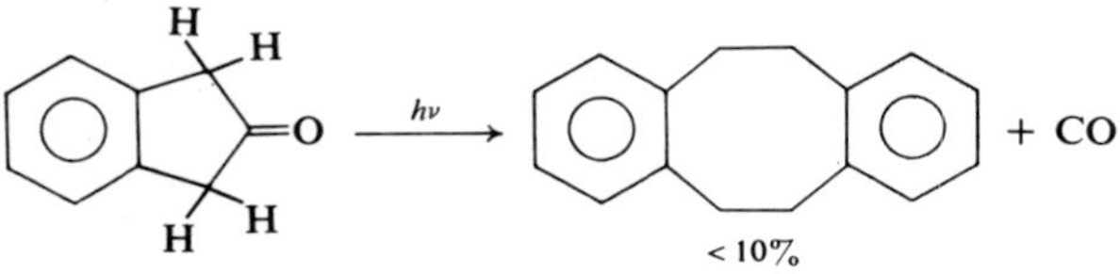

$\xrightarrow{h\nu}$ + CO (4.75)

< 10%

$\xrightarrow{h\nu}$ (4.76)

> 50%

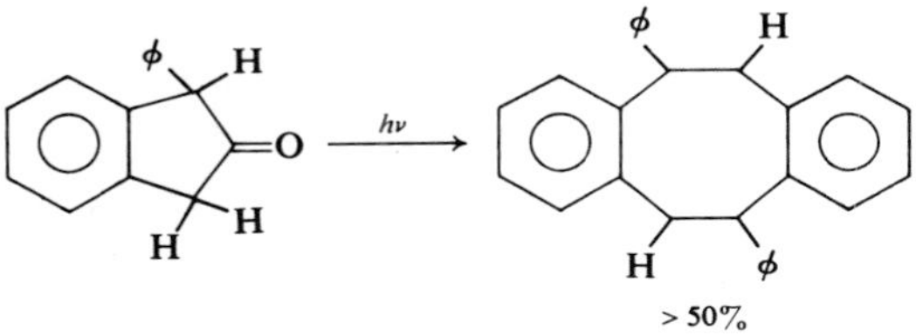

$\xrightarrow{h\nu}$ + (4.77)

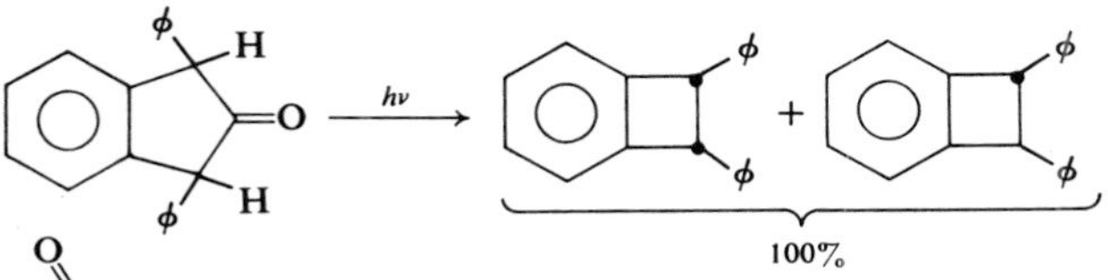

100%

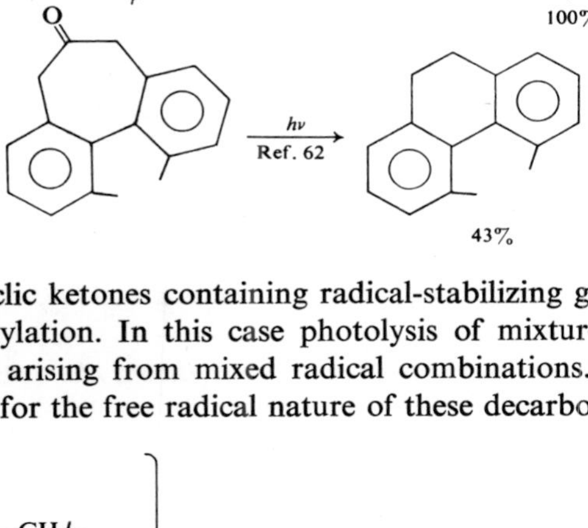

$\xrightarrow[\text{Ref. 62}]{h\nu}$ (4.78)

43%

Acyclic ketones containing radical-stabilizing groups similarly undergo decarbonylation. In this case photolysis of mixtures of ketones results in products arising from mixed radical combinations.[57] This result is direct evidence for the free radical nature of these decarbonylations:

$$\left.\begin{array}{c}\phi_2CH\text{—}\overset{\overset{\large O}{\|}}{C}\text{—}CH\phi_2 \\ + \\ \phi\text{—}CH_2\text{—}\underset{\underset{\large O}{\|}}{C}\text{—}CH_2\text{—}\phi\end{array}\right\} \xrightarrow{h\nu} \phi_2CHCH\phi_2 + \phi CH_2CH_2\phi + \phi_2CHCH_2\phi + CO \qquad (4.79)$$

The photodecarbonylation of a series of dibenzyl ketones was studied by Robbins and Eastman.[63] The results of this study are presented in Table 4.5. The data in Table 4.5 indicate that the presence of a *p*-methyl or a *p*-methoxy group has little effect on the quantum yield for this reaction. *p*-Cyano groups, on the other hand, essentially totally eliminated the decarbonylation. Since the reaction could also be quenched (inefficiently) by benzonitrile or biphenyl, it was concluded that the decarbonylation occurs from a short-lived triplet state. The effect of the *p*-cyano groups then could result from internal triplet quenching.

We have seen that the observation of "mixed" radical combination products from mixtures of $\phi_2CHCOCH\phi_2$ and $\phi CH_2COCH_2\phi$ upon photolysis indicates the decarbonylation to be stepwise rather than concerted, at least for these molecules. Further evidence in support of a stepwise process was reported by Robbins and Eastman in a study of *p*-methoxybenzyl ketone.[64] Three products are isolated from this reaction:

$$CH_3O\text{-}C_6H_4\text{-}CH_2\overset{O}{\overset{\|}{C}}CH_2\phi \xrightarrow[-CO]{h\nu} \phi CH_2CH_2\phi + CH_3O\text{-}C_6H_4\text{-}CH_2CH_2\phi + CH_3O\text{-}C_6H_4\text{-}CH_2CH_2\text{-}C_6H_4\text{-}OCH_3 \quad (4.80)$$

Quantum yields for the formation of symmetrical and unsymmetrical (mixed) products were determined as a function of solvent viscosity. If perchance expulsion of CO were concerted, yielding two benzyl radicals, formation of mixed combination products may reflect the ability of the radicals to escape from the solvent cage. If this were true, variation of the solvent viscosity should alter the rate of escape of these radicals and the ratio of symmetrical to unsymmetrical products should change. The results found in this study are presented in Table 4.6. The data in Table 4.6 indicate that the reaction is

TABLE 4.5. *The Photodecarbonylation of Some Dibenzylketones*

Compound	Φ(diphenylethane)[a]	Solvent
Dibenzyl ketone	0.70 ± 0.10	Benzene
p,p'-Dimethyldibenzyl ketone	0.71 ± 0.11	Benzene
p-Methoxydibenzyl ketone	0.66 ± 0.10[b]	Benzene
p,p'-Dicyanodibenzyl ketone	<0.02	Acetonitrile
Dibenzyl ketone	0.71 ± 0.11	Acetonitrile

[a] $\lambda = 313$ nm.
[b] Three products were obtained in a statistical ratio of 1:2:1.

TABLE 4.6. *The Effect of Solvent Viscosity on the Symmetrical and Unsymmetrical Product Quantum Yields for the Photolysis of p-Methoxydibenzyl Ketone*

Solvent	Viscosity, cP at 30°C	Φ_{sym}	Φ_{unsym}	Φ_{total} [a]
Benzene	0.56	0.12	0.24	0.48
Cyclohexane	0.80	0.12	0.25	0.49
Decane	0.80	0.12	0.27	0.51
Ethanol	1.00	0.08	0.16	0.32
2-Propanol	1.77	0.13	0.26	0.52
2-Methyl-2-propanol	3.32	0.11	0.22	0.44
Cyclohexanol	41.1	0.09	0.22	0.40

[a] $\Phi_{total} = 2\Phi_{sym} + \Phi_{unsym}$.

essentially unaffected by a change in solvent viscosity. This indicates that cleavage must be stepwise and that the resulting radicals must diffuse away,

$$\mathrm{R{-}\phi{-}CH_2\overset{O}{\overset{\|}{C}}{-}CH_2{-}\phi \xrightarrow{h\nu} [R{-}\phi{-}CH_2\overset{O}{\overset{\|}{C}}\cdot\ \cdot CH_2{-}\phi] \longrightarrow R{-}\phi{-}CH_2\overset{O}{\overset{\|}{C}}\cdot + \phi{-}CH_2\cdot;\quad R{-}\phi{-}CH_2\overset{O}{\overset{\|}{C}}\cdot \rightarrow R{-}\phi{-}CH_2\cdot + CO} \tag{4.81}$$

In addition, Robbins and Eastman were able to trap the intermediate phenylacetyl radicals by photolysis in the presence of the 2,2,6,6-tetramethylpiperidine-1-oxyl free radical,

$$\mathrm{R{-}\phi{-}CH_2\overset{O}{\overset{\|}{C}}{-}CH_2{-}\phi \xrightarrow[\text{(tetramethylpiperidinyl)N{-}O}\cdot]{h\nu} (\text{tetramethylpiperidinyl})N{-}O\overset{O}{\overset{\|}{C}}CH_2{-}\phi + (\text{tetramethylpiperidinyl})N{-}OCH_2{-}\phi} \tag{4.82}$$

The benzyl ether and the phenyl acetate were isolated in a 3:2 ratio.

The decarbonylation of dibenzyl ketone has been shown to result from the carbonyl triplet state by its ability to be quenched by 1,3-cyclohexadiene or 1,3-pentadiene.[65] Using 1,3-cyclohexadiene as quencher, photodimers of the cyclohexadiene were obtained. Since these are formed only by triplet sensitization,[66] the quenching of ketone triplet states, rather than their excited singlets, was assured. Further evidence for a triplet reaction follows from the fact that decarbonylation could be sensitized by acetone under conditions where the sensitizer absorbed 93% of the light.

Di-*t*-butyl ketone reacts to yield pivalaldehyde, isobutylene, and 2-methylpropane,

$$(CH_3)_3C{-}CO{-}C(CH_3)_3 \xrightarrow{h\nu} (CH_3)_3C\overset{O}{\overset{\|}{C}}H + CH_2{=}C(CH_3)_2 + (CH_3)_3CH \qquad (4.83)$$

These reactions can result from two sets of radical pairs:

$$[(CH_3)_3C\dot{C}{=}O \;\; \cdot C(CH_3)_3] \longrightarrow (CH_3)_3CCHO + CH_2{=}C(CH_3)_2 \qquad (4.84)$$

$$[(CH_3)_3C\cdot \;\; \cdot C(CH_3)_3] \longrightarrow (CH_3)_3CH + CH_2{=}C(CH_3)_2 \qquad (4.85)$$

Observation of spin-polarized products resulting from these radical pairs by the method of chemically induced dynamic nuclear polarization (CIDNP)[67] was accomplished by photolysis in the probe of an NMR spectrometer using perfluoromethylcyclohexane as solvent. The results obtained were consistent with nuclear spin polarization steps involving radical pairs formed from dissociated radicals and also directly from excited states, although the former could not be detected in carbon tetrachloride, probably due to radical scavenging by the solvent. It was not possible to determine the fraction of the reaction proceeding by singlet and triplet radical pairs.[68]

Decarbonylation is also observed upon photolysis of β,γ-unsaturated homoconjugated aldehydes and α-aryl aldehydes. For example, *R*-laurolenal produces two enantiomeric products in yields of $\sim 88\%$ and 12%, respectively:

$$\text{(R-laurolenal, CHO)} \xrightarrow[\Phi_{3130\text{Å}} = 0.61]{h\nu} \text{(product, H)} + \text{(enantiomer, H)} \qquad (4.86)$$

It was concluded in a study of this reaction that singlet state α-cleavage yields a solvent-caged radical-pair intermediate involving an allyl radical and a formyl radical. Carbon monoxide is released concurrent with a stereospecific transfer of the formyl hydrogen. The orientation of the partners in the radical pairs, determined by the ground state configuration, is thought to determine the position of the transferred hydrogen,[69]

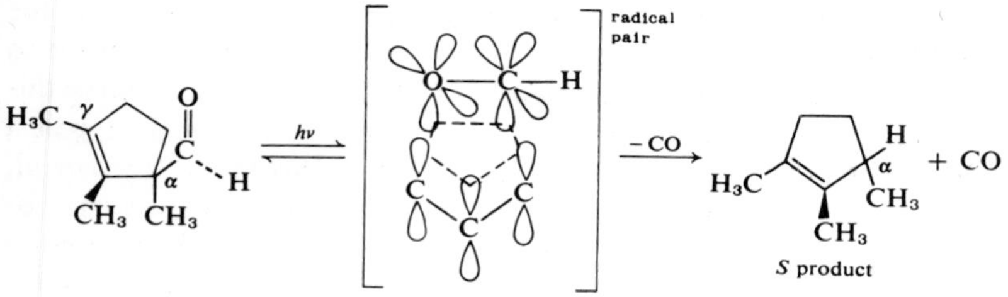

(4.87)

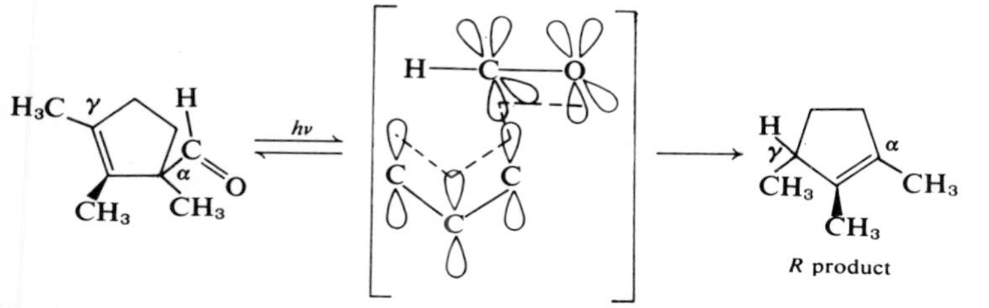

(4.88)

The photodecarbonylation of α-aryl aldehydes has been studied as a function of structure by Küntzel, Wolf, and Schaffner[70]:

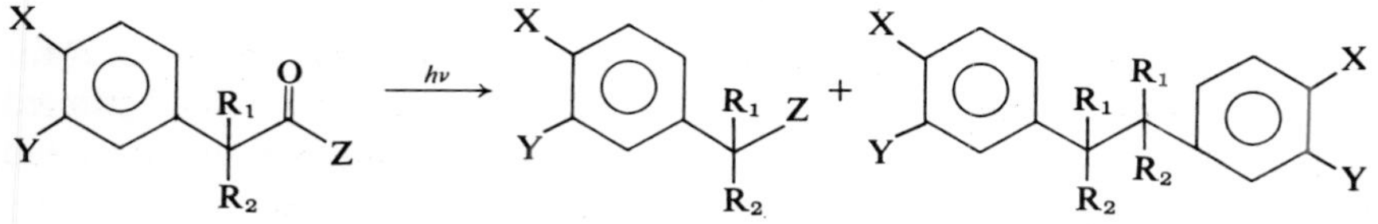

(4.89)

TABLE 4.7. *The Photolysis of α-Aryl Aldehydes*

X	Y	R_1	R_2	Z	[C]	λ, Å	A,[a] %	C,[b] %	*E*,[c] %
H	H	Me	Me	H	0.011	3130	87	11	2
H	H	Me	Me	H	0.040	3130	11	67	21
H	H	Me	Me	D	0.012	3130	16	81	3
H	H	Me	Me	D	0.042	3130	20	73	7
H	H	Me	Me	D	0.10	2537	47	50	3
Me	H	Me	Me	H	0.01	3130	63	27	0.2
Br	H	Me	Me	H	0.01	3130	46	38	1.2
H	Me	Me	Me	H	0.01	3130	20	65	7.2
H	OMe	Me	Me	H	0.01	3130	79	13	4
H	H	—CH_2—CH_2		H	0.01	3130	95.4	1.0	—
H	H	—$(CH_2$—$CH_2)_2$		H	0.01	3130	86	11	—
α-Naphthyl		Me	Me	H	0.1	2537	55	45	—
α-Naphthyl		Me	Me	H	0.1	3130	85	15	—

[a] Unreacted aldehyde.
[b] Product of hydrogen transfer.
[c] Substituted diphenylethane.

The results of this study are presented in Table 4.7. As can be seen from the data in Table 4.7, decarbonylation with hydrogen or deuterium transfer to the resulting radical is a relatively efficient process. The failure to observe this reaction using acetone or acetophenone as photosensitizer would suggest a singlet pathway for the direct photolysis of the aldehyde. In agreement, decarbonylation could not be quenched by naphthalene, piperylene, or 1,3-cyclohexadiene when the aldehyde was excited directly. The reaction could, however, be somewhat quenched by the addition of tri-*n*-butylstannane. The products in this case were

$$PhC(Me)_2C(=O)D\ (0.3\ M) \xrightarrow[R_3SnH\ (1.1\ M)]{h\nu} PhC(Me)_2R\ (R = H, 10\%;\ R = D, 24\%) + PhC(Me)_2CH(OH)D\ (66\%) \qquad (4.90)$$

corresponding to photoreduction of the aldehyde and hydrogen abstraction from the stannane by a dimethylbenzyl radical.

Some quantum yields for decarbonylation of these compounds are given in Table 4.8.

Rough Hammett correlations of Φ_{CO} with the resonance constants R for *para*-substituted isomers ($\rho_{para} = -0.25$) and with σ_m^+ for *meta*-substituted isomers ($\rho_{meta} \sim -0.53$) agreed with the α-cleavage mechanism proposed leading to an associated radical pair. If totally "free" free radicals were involved, a value of $\rho = -1.36$ might be expected for the hydrogen abstrac-

TABLE 4.8. *Some Photodecarbonylation Quantum Yields for α-Aryl Aldehydes*

$$X\text{-}C_6H_4C(Me)_2C(=O)H$$

[C], *M*	X	Φ_{CO}[a]
0.1	H	0.76
0.01	*p*-Me	0.80
0.01	*m*-Me	1.00
0.01	*p*-OMe	1.04
0.01	*m*-OMe	0.76
0.01	*p*-Br	1.25[b]
0.01	*p*-CF_3	0.71
0.01	*m*-CF_3	0.60

[a] Quantum yield at 3130 Å.
[b] Indicates a radical chain reaction.

tion.[71] The use of a Hammett plot for *excited state* reactions may be of questionable value, however.

4.1d. β-*Cleavage of Cyclopropyl Ketones*

Although not formally a type I cleavage process, β-cleavage of cyclopropyl ketones to form propenyl ketones and/or epimerized cyclopropyl groups bears such a close similarity to α-cleavage that it should be discussed in combination with the latter. The first report of a reaction thought to occur by β-cleavage was made by Pitts in 1954 for the formation of methyl propenyl ketone from methyl cyclopropyl ketone,[72]

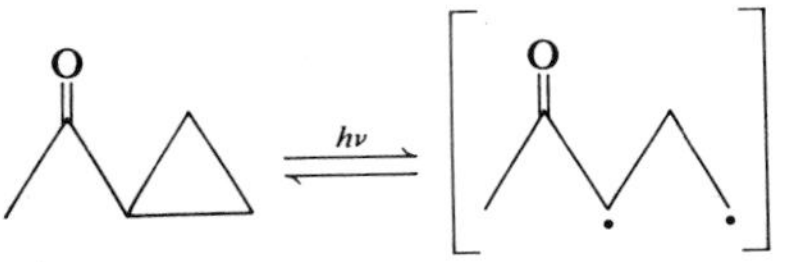

hν (4.91)

When ketones bearing substituted cyclopropyl groups are photolyzed, products having different stereochemistry result, indicating that the initial β-cleavage is reversible.

α-Cleavage and β-cleavage may often be competitive reactions. Which pathway predominates is dependent upon the structure of the ketone and the presence of substituents, as shown in the following examples[73]:

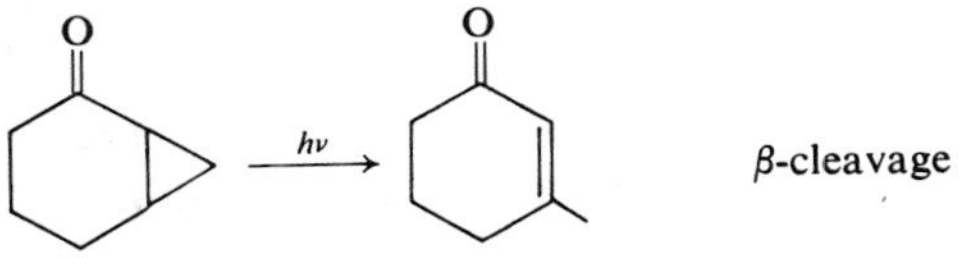

hν β-cleavage (4.92)

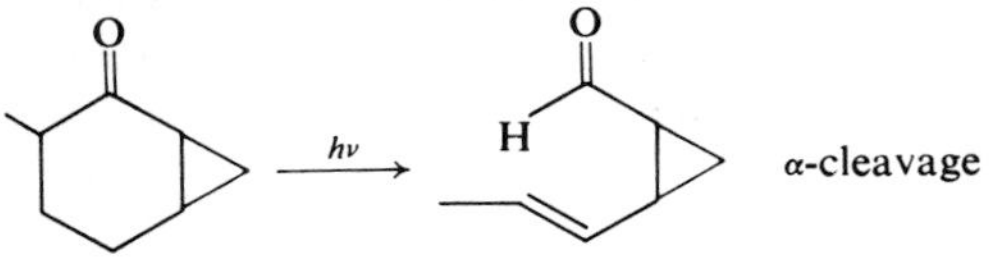

hν H α-cleavage (4.93)

The mode of bond cleavage is extremely sensitive to alkyl substitution,[73–75]

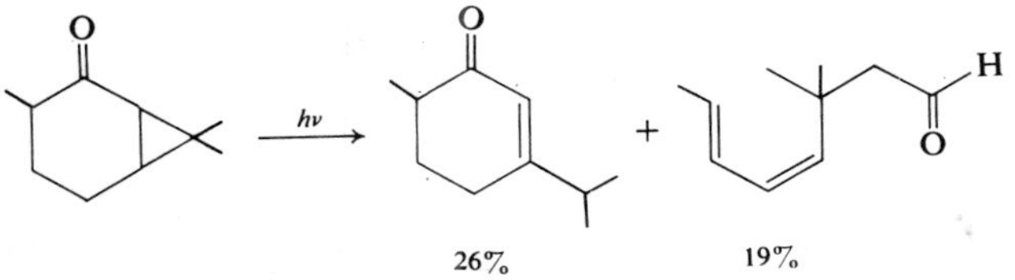

hν + H (4.94)

26% 19%

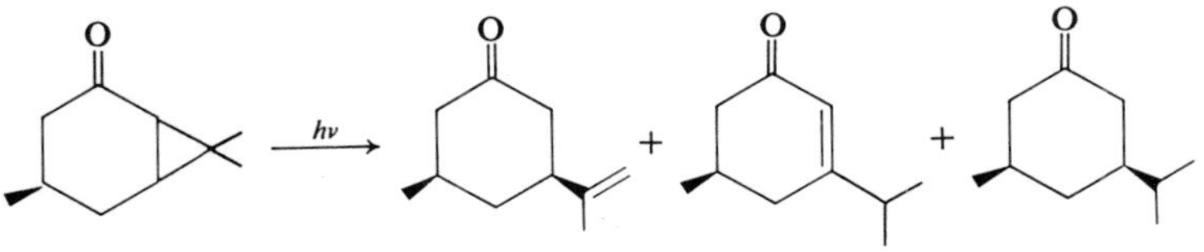

hν + + (4.95)

(4.96)

60–70% 10–20%

Irradiation of the following ketone in acetic acid-d_4 failed to produce product containing deuterium. Thus an enol intermediate cannot be involved in the ring opening[76]:

hν MeOH or AcOH

AcOH–d_4

(4.97)

The photochemical interconversion of *cis*- and *trans*-5,6-diphenylbicyclo[3.1.0]hexanones has been extensively studied by Zimmerman, Hancock, and Licke.[77] Using isotope dilution analysis, these workers obtained very accurate quantum yields for the various photochemical processes involved in this reaction:

hν, Φ = 0.94

hν, Φ = 0.093

trans *cis*

hν, Φ = 0.0101

hν, Φ = 0.0028

(4.98)

Cleavage at any of the three cyclopropyl bonds (*a*, *b*, or *c*) could potentially account for the isomerization.[78] However, cleavage at *a* or *b*, although both yielding *cis* product, have different stereochemical implications since the products obtained are enantiomers:

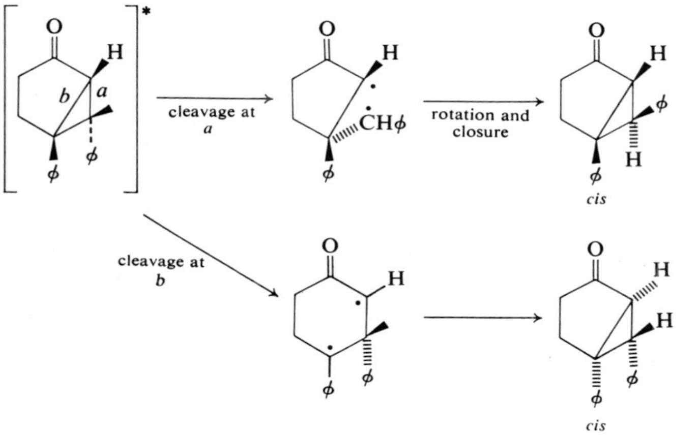

cleavage at *a* — rotation and closure — *cis*

cleavage at *b* — *cis*

(4.99)

By starting with a single enantiomer of the *trans* ketone, the stereochemical course of the reaction was determined:

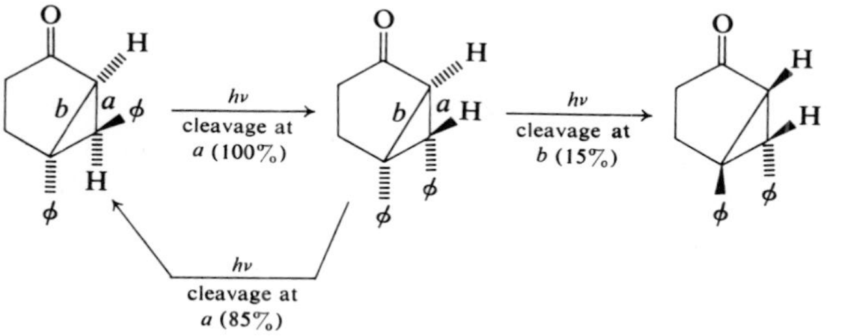

$h\nu$ cleavage at *a* (100%) — $h\nu$ cleavage at *b* (15%) — $h\nu$ cleavage at *a* (85%)

(4.100)

Thus isomerization of the *trans* to *cis* ketone is stereospecific while isomerization of *cis* to *trans* results in about 28% racemization. Postulating different pathways for *trans* to *cis* and *cis* to *trans* isomerizations would appear at first hand to violate the principle of microscopic reversibility. However, what we are dealing with here are the excited state intermediates as reactants and these are not necessarily in equilibrium for the two isomers. For example, taking the analogy

$$\mathrm{A} \xrightarrow{h\nu} \mathrm{B}$$

$$\mathrm{B} \overset{h\nu}{\not\longrightarrow} \mathrm{A}$$

if the pathway for A → B is

$$\mathrm{A} \xrightarrow{h\nu} \mathrm{A}^s \longrightarrow \mathrm{A}^t \longrightarrow \mathrm{B(ground\ state)} \qquad (4.101)$$

an analogous pathway for B is

$$\mathrm{B} \xrightarrow{h\nu} \mathrm{B}^s \longrightarrow \mathrm{B}^t \not\longrightarrow \mathrm{A} \qquad (4.102)$$

If B^t is unreactive or gives different products, no A results. Since different intermediates are involved, the principle of microscopic reversibility does not even enter the picture. If, however,

$$\mathrm{A}^t \longrightarrow \mathrm{B}^t \longrightarrow \mathrm{B(ground\ state)} \qquad (4.103)$$

a violation would result. The general feeling in reactions of this type is that once bond cleavage has occurred, electronic excitation is lost and the resulting biradical is a ground state species.

The predominance of cleavage at bond *a* over cleavage at bond *b* probably results from overlap control over the reaction pathway.[79] As shown below, there is greater overlap between the carbonyl orbitals and the cyclopropane *a* bond orbitals than between the carbonyl and *b* bond orbitals:

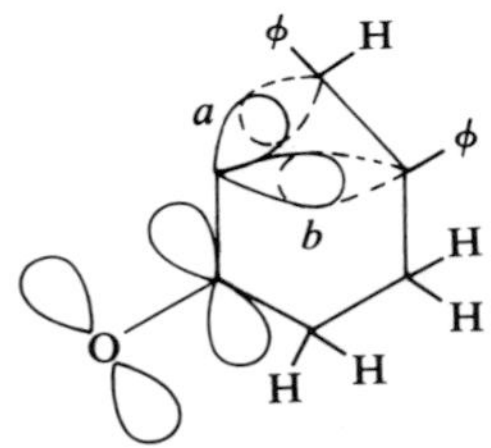

However, bond *b* must cleave for both the *cis* and *trans* isomers since both produce the ring-opened product 3,4-diphenylcyclohexenone although the quantum yields for this process are considerably lower than those for isomerization by cleavages at bond *a*. Finally, the possibility of cleavage at bond *c* cannot be ruled out by these results. We will return to this possibility later in this section.

Zimmerman and co-workers were also able to obtain some information regarding the multiplicities of the excited states responsible for the initial β-cleavage through quenching and sensitization studies. It was found that both *trans*-to-*cis* and *cis*-to-*trans* isomerizations could be sensitized by chlorobenzene under conditions where the latter absorbed over 95% of the light. The same product ratio was obtained under these conditions as in the direct irradiation of the ketones. With 1,3-cyclohexadiene or 2,5-dimethyl-2,4-hexadiene as quenchers nearly 90% of the reaction of the *trans* isomer could be quenched. Again the ratio of the quenched reaction products was the same as in the unquenched reaction. The reaction of the *cis* isomer, on the other hand, could not be quenched by 1,3-cyclohexadiene or 2,5-dimethyl-2,4-

hexadiene. These results indicate that both reactions result from excited triplet states. The *trans* isomer reacts from a quenchable triplet with a rate constant for isomerization k_r between 4.6×10^8 and 4.6×10^9 sec^{-1}. The fact that *cis* to *trans* isomerization could be sensitized by chlorobenzene indicates that it, too, results from a triplet state; however, since isomerization could not be quenched, the triplet must be extremely short lived. Alternatively, a singlet precursor could be involved in the direct photolysis.

Earlier in this section we represented β-cleavage as an opening of the cyclopropyl ring to yield a biradical. However, an alternative possibility exists.(58) For example, of the intermediates in the following,

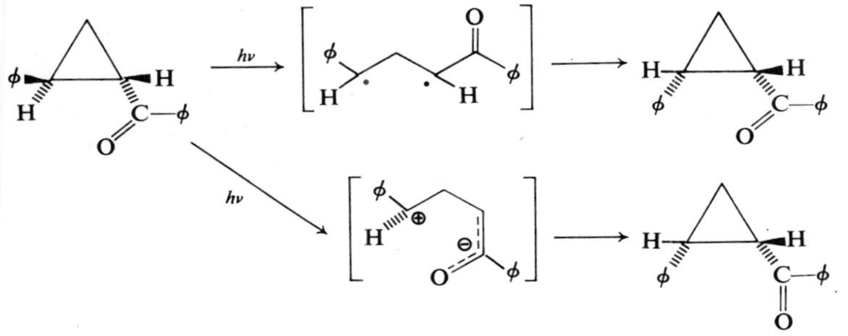

(4.104)

which of them is actually involved? In order to answer this question, Zimmerman and co-workers studied a *trans,trans*-2,3-diphenyl-1-benzoylcyclopropane system containing various *p*-substituents on the 3-phenyl group. The unsubstituted compound was found to yield the following isomerized product(80):

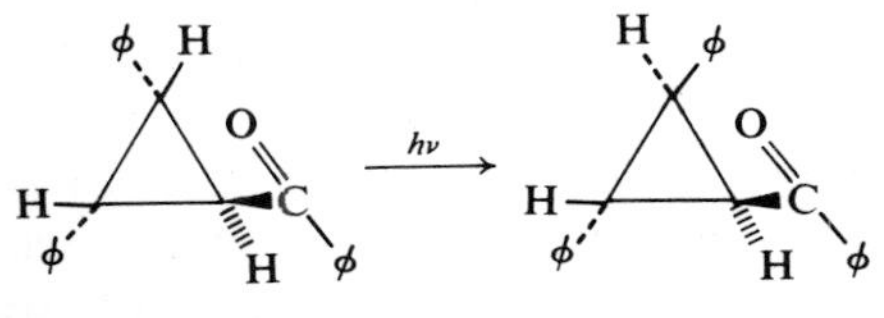

(4.105)

The quantum yield for isomerization in the direct photolysis was found to be $\Phi = 0.94$. The reaction could also be sensitized with acetophenone ($\Phi = 1.02$) and quenched with piperylene, indicating a reactive triplet species with a rate constant k_r of 3×10^{10} sec^{-1}. With a 3-(*p*-methoxyphenyl) derivative two products were obtained(81):

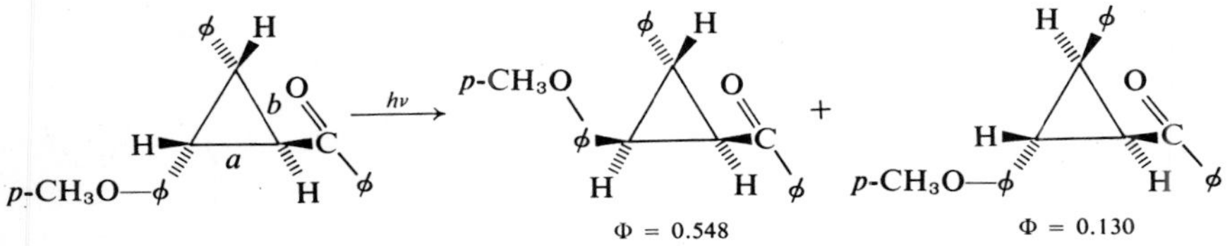

$\Phi = 0.548$ $\Phi = 0.130$

(4.106)

The product mixture contained 81% of the *cis*-3-*p*-methoxyphenyl-*trans*-2-phenyl derivative. Again a reactive triplet state was involved, as evidenced by sensitization and quenching studies ($k_r = 1.5 \times 10^{10}\ \text{sec}^{-1}$).

The fact that two products were produced indicates that cleavage occurs at both bond *a* and bond *b*, although cleavage at bond *a* predominates assuming that single rotations are favored, as shown below:

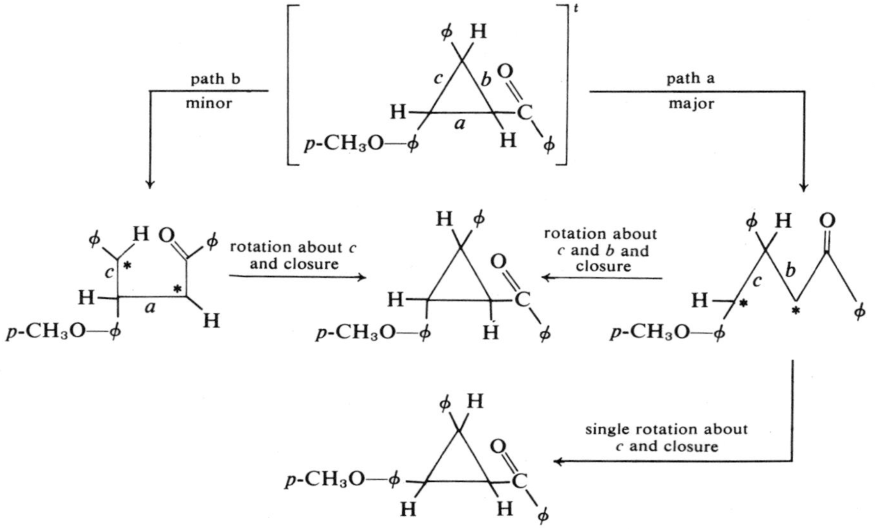

path b
minor
path a
major
rotation about *c* and closure
rotation about *c* and *b* and closure
single rotation about *c* and closure

A comparison of the rate constant for photoisomerization of the unsubstituted 3-phenyl derivative ($k_r = 3 \times 10^{10}\ \text{sec}^{-1}$) to that of the 3-(*p*-methoxy phenyl) derivative ($k_r = 1.5 \times 10^{10}\ \text{sec}^{-1}$) indicates that the presence of the *p*-methoxy groups imparts no special stability to the intermediate responsible for isomerization even though cleavage of a cyclopropane bond is predominant. Clearly these results are inconsistent with an intermediate possessing electron-poor or electron-rich species such as would be obtained from heterolytic cleavage of the cyclopropane. On the other hand, the results are consistent with a biradical species as intermediate. Further evidence consistent with this conclusion was obtained in a study of *trans*-3-*p*-cyanophenyl-*trans*-2-phenyl-1-benzoylcyclopropane,[82]

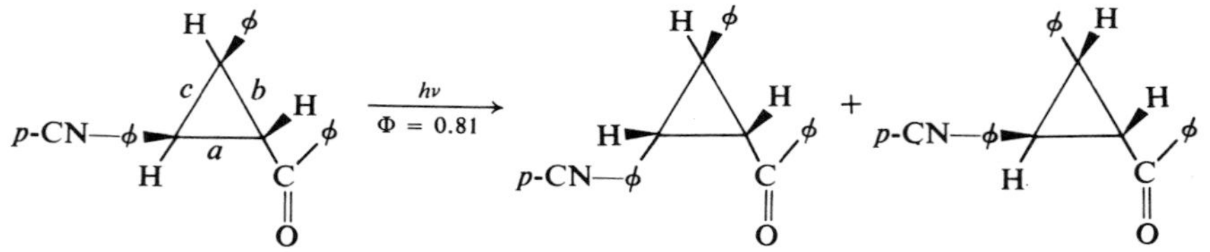

$h\nu$, $\Phi = 0.81$

(4.107)

Again two products were obtained from a reactive triplet state with a rate constant for photoisomerization of 2×10^{11}. This rate, however, is over six times greater than that obtained for the unsubstituted phenyl derivative and indicates that the *p*-cyano-phenyl group facilitates reaction by stabilization of the intermediate biradical.

Cleavage in the above reaction was found to occur primarily at the *a* bond (90% cleavage at this position). More detailed information concerning the various routes was obtained by studying an optically active derivative as shown in Figure 4.3. The results of this study are given in Table 4.9.

Returning now to the question of the relative importance of cleavage at bond *c* of the cyclopropyl group, if *c* cleavage in the above reaction were important, one would expect to find equal amounts of compounds CT (+) and TC (−). The data in Table 4.9 indicate, however, that a total of only 11% of CT (+) is formed potentially from two pathways, cleavage at bond *b* and at bond *a*. Therefore the maximum cleavage at bond *c* would correspond to 22%. Assuming therefore that cleavage at the *c* bond is relatively unimportant, the product ratio would indicate that single bond rotation is more likely than double rotation for cleavage at the *a* and *b* bonds. For cleavage at the *a* bond k_s^a/k_d^a (where k_s^c is the rate constant for single rotation and k_d^a is the rate constant for double rotation) is equal to 8. If *c* bond cleavage is again assumed to be unimportant, $k^a/k^b = 8$, confirming the results obtained earlier.

α,β-Epoxyketones undergo β-cleavage to yield 1,3-diketones and α-hydroxyketones through migration of methyl and hydrogen,[83]

$$\phi\text{—C(CH}_3\text{)—CH(—O—)—C(=O)—CH}_3 \xrightarrow{h\nu} \left[\phi\text{—C(CH}_3\text{)(O}\cdot\text{)—}\dot{\text{C}}\text{H—C(=O)—CH}_3\right] \longrightarrow \phi\text{—C(=O)—CH(CH}_3\text{)—C(=O)—CH}_3$$

$$\text{R—C(CH}_3\text{)—CH(—O—)—C(=O)}\phi \xrightarrow{h\nu} \left[\text{R—C(}\dot{\text{C}}\text{H}_2\text{)—CH(—O—)—}\dot{\text{C}}\text{(OH)}\phi\right] \longrightarrow \left[\text{R—C(=CH}_2\text{)—CH(O}\cdot\text{)—}\dot{\text{C}}\text{(OH)—}\phi\right] \longrightarrow \text{R—C(=CH}_2\text{)—CH(OH)—C(=O)—}\phi$$

(4.108)

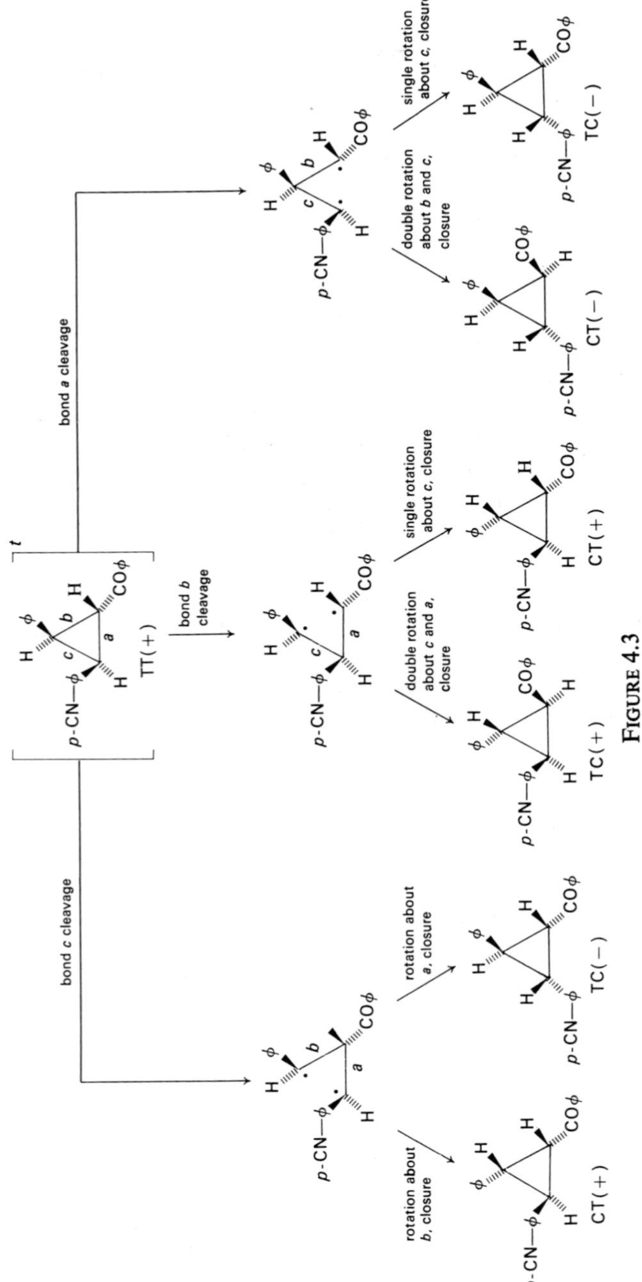
bond a cleavage
bond b cleavage
bond c cleavage
TT(+)
single rotation about c, closure
double rotation about b and c, closure
double rotation about c and a, closure
rotation about a, closure
rotation about b, closure
TC(−)
CT(−)
CT(+)
TC(+)
TC(−)
CT(+)

FIGURE 4.3

TABLE 4.9. *Photolysis of (+)-trans-3-p-Cyanophenyl-trans-2-phenyl-1-benzoylcyclopropane*

Compound	TC (+)	CT (+)	CT (−)	TC (−)
Relative yield	0	0.5	0.5	3.5

Cis–trans isomerizations also occur[84,85]:

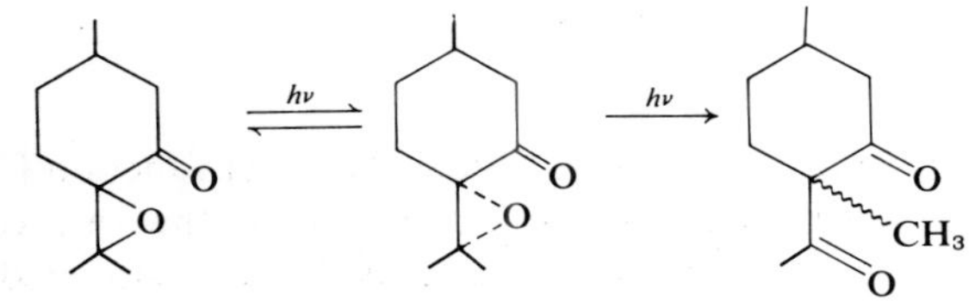

(4.109)

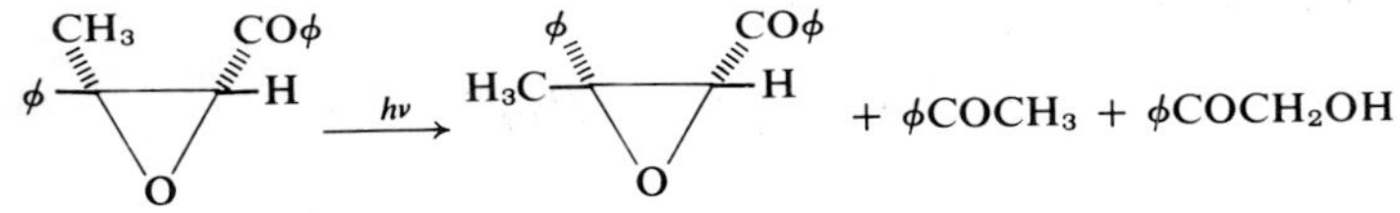

$\phi COCH_3$ + $\phi COCH_2OH$ (4.110)

4.1e. *The Nature of the Excited State: Part II*

As an aid in the qualitative understanding of why photocleavage occurs, it is interesting to consider the results of semiempirical extended Hückel theory (EHT)[86] for the following model compounds[87]:

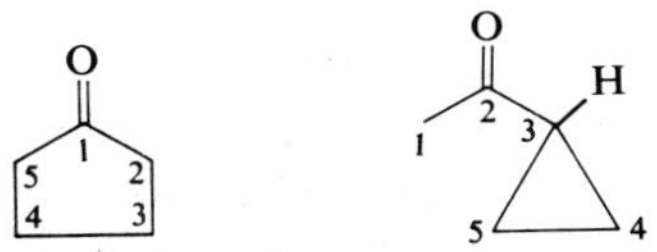

The calculations involve the usual secular determinant of the form

$$|H_{ij} - S_{ij}\varepsilon| = 0 \tag{4.111}$$

On expanding this determinant, we obtain a polynomial in ε whose order is the number of functions in the LCAO MO approximation:

$$\Psi_K = \sum_i c_{Ki}\chi_i \tag{4.112}$$

In (4.111), S_{ij} is the matrix of the overlap integrals

$$S_{ij} = \langle \chi_i | \chi_j \rangle \tag{4.113}$$

and H_{ij} is the Hamiltonian matrix

$$H_{ij} = \langle \chi_i | H | \chi_j \rangle \tag{4.114}$$

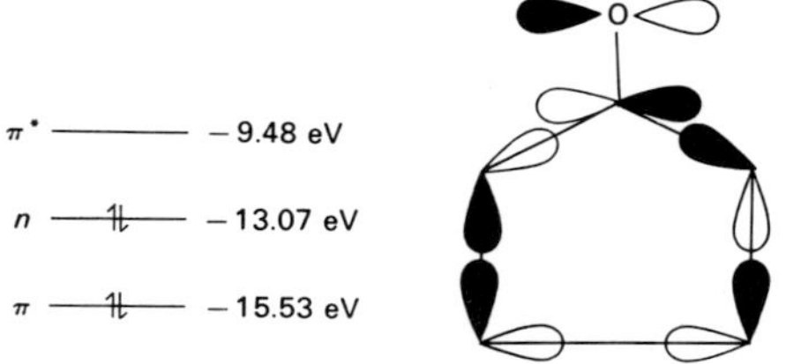

FIGURE 4.4

H is the one-electron operator and χ_i is the Slater basis set function ($2s$, $2p$). The diagonal elements of H_{ij} (H_{ii}) are approximated as the valence state ionization potentials and the off-diagonal elements H_{ij} are estimated using the Wolfsberg–Helmholtz approximation,

$$H_{ij} = 1.75S_{ij}(H_{ii} + H_{jj})/2 \tag{4.115}$$

The information obtainable upon solution of the eigenvalue problem includes the orbital energies ε_K and the corresponding wave function as a linear combination of the atomic basis set χ_i. The wave functions can then be subjected to a Mulliken population analysis[88] to provide the overlap populations P_{ij}:

$$P_{ij} = \sum 2c_{Ki}c_{Kj}S_{ij} \tag{4.116}$$

A decrease in the overlap population between the atoms upon excitation would suggest that the bond is weaker in the excited state than in the ground state. The EHT orbital energy levels and a sketch of the n orbital for cyclopentanone are shown in Figure 4.4. Thus we can see that excitation of a "lone-pair electron on oxygen" in fact alters the electron density between atoms 1 and 2 where α-cleavage occurs. The overlap population for the ground state and the $n \rightarrow \pi^*$ and $\pi \rightarrow \pi^*$ excited states are shown in Figure 4.5.

While $\pi \rightarrow \pi^*$ excitation has little effect on the C_1–C_2 overlap population (0.81 vs. 0.79), $n \rightarrow \pi^*$ excitation weakens the C_1–C_2 bond (0.81 vs. 0.70). Consequently we should expect photocleavage to occur only from the $n \rightarrow \pi^*$ excited state and only the α bond to be broken.

O 0.86 0.81 0.73 0.75 (a)

O 0.77 0.70 0.70 0.76 (b)

O 0.58 0.79 0.74 0.74 (c)

FIGURE 4.5. (a) Ground state; (b) $n \rightarrow \pi^*$ excited state; (c) $\pi \rightarrow \pi^*$ excited state.

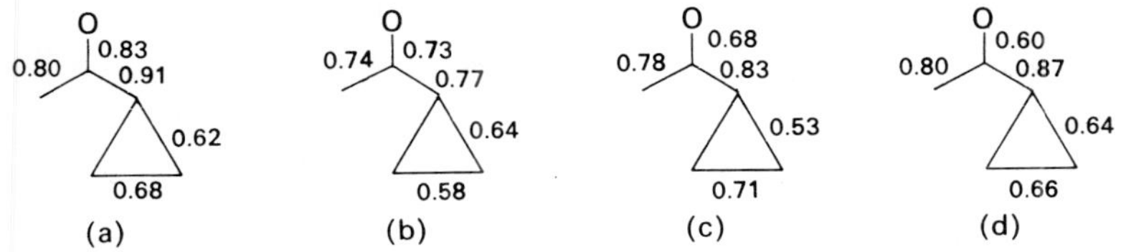

FIGURE 4.6. (a) Ground state; (b) $n \rightarrow \pi^*$ excited state: (c) $\Delta \rightarrow \pi^*$ excited state; (d) $\pi \rightarrow \pi^*$ excited state.

For methyl cyclopropyl ketone the question is more complex since in addition to the low-lying n and π levels, a cyclopropyl level is present also[89]:

π^* ——— -9.16

n ⇅ -12.84

Δ ⇅ -13.32

π ⇅ -15.16

The overlap populations for the ground and various excited states of methyl cyclopropyl ketone are shown in Figure 4.6. Only for the $\Delta \rightarrow \pi^*$ excited state would the EHT calculations suggest that cyclopropyl (β) cleavage (C_3–C_4) should be important. While this has not been experimentally confirmed, it should be subjected to test with the proper compound.

4.2. THE FORMATION OF OXETANES FROM CARBONYLS AND OLEFINS

We have seen that when carbonyl compounds are photolyzed in the presence of olefins, such as 1,3-pentadiene, which have lowest triplet energies below those of the carbonyls, the carbonyl triplet is quenched by energy transfer to the olefin. This behavior, as evidenced by the reduction or elimination of carbonyl photoproducts, is generally considered diagnostic of a triplet reaction. If, however, an olefin having a triplet energy higher than that of the carbonyl is used, energy transfer cannot occur and often addition of the carbonyl to the olefin, oxetane formation, results. The arrangement of triplet levels for this reaction to occur is diagrammed in Figure 4.7.

This reaction was first discovered by Paterno and Chieffi in 1909.[90] These workers isolated a compound from the photolysis of benzophenone in

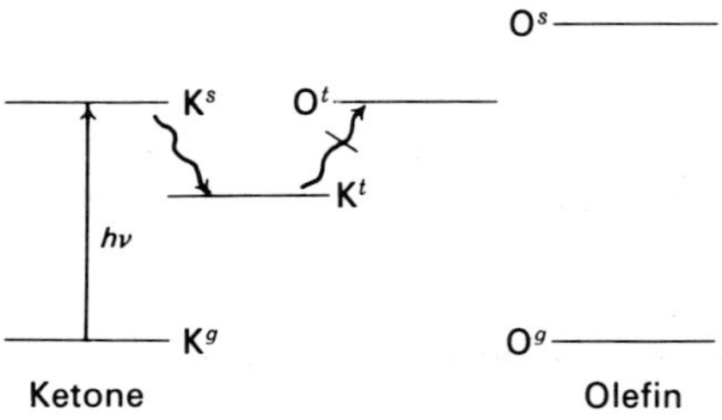

FIGURE 4.7

the presence of 2-methyl-2-butene. The oxetane structure shown below was proposed:

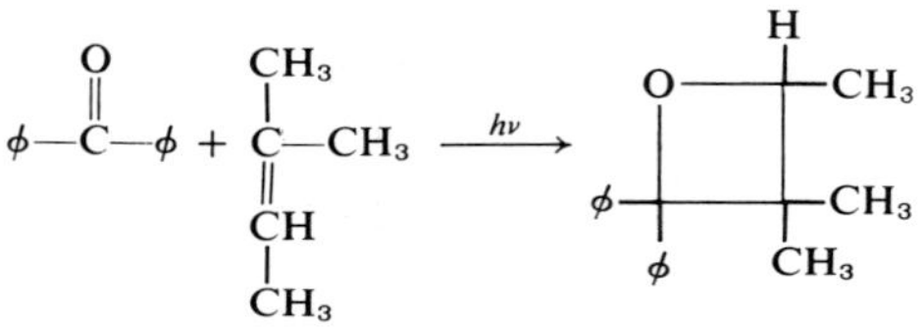

$$\phi\text{—}\overset{\overset{\text{O}}{\|}}{\text{C}}\text{—}\phi + (\text{CH}_3)_2\text{C}{=}\text{CH—CH}_3 \xrightarrow{h\nu} \text{oxetane: O—C(H)(CH}_3\text{)—C(CH}_3\text{)(CH}_3\text{)—C}(\phi)_2\text{—O} \tag{4.117}$$

In 1954 this reaction was more extensively investigated by Buchi, Inman, and Lipinsky,(91) who confirmed the oxetane structure of the photoproduct. To credit their initial work on this interesting reaction, it commonly is referred to as the Paterno–Buchi reaction.

4.2a. *Oxetane Formation from Olefins and Aryl Ketones and Aldehydes*

With aromatic carbonyls, oxetane formation appears to arise from the carbonyl triplet state, as evidenced by quenching studies. For example, benzaldehyde irradiated in the presence of cyclohexene yields products indicative of hydrogen abstraction reactions and an oxetane:

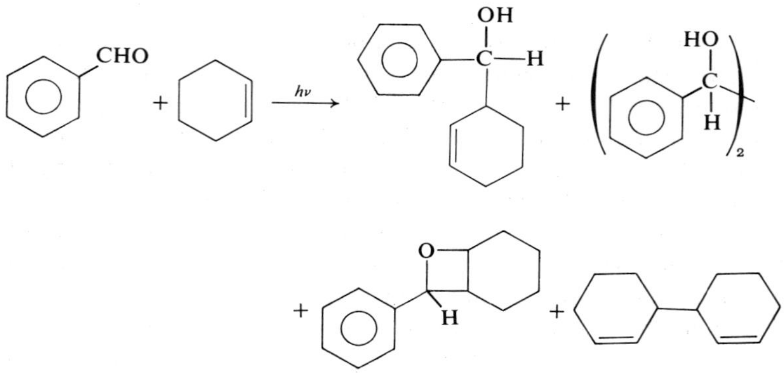

$$\text{C}_6\text{H}_5\text{CHO} + \text{cyclohexene} \xrightarrow{h\nu} \text{C}_6\text{H}_5\text{CH(OH)(cyclohexenyl)} + \left(\text{C}_6\text{H}_5\text{CH(OH)—}\right)_2 + \text{oxetane (phenyl, H)} + \text{bicyclohexenyl} \tag{4.118}$$

TABLE 4.10. *Reactivity of Various Aromatic Carbonyls toward Oxetane Formation and Photoreduction Compared to the Nature of the Lowest Triplet*[93,94–98]

Compound	2-Methyl-2-butene[a]	IPA[a]	Phosphorescent state
Benzaldehyde	Oxetane (0.45)	Pinacol (1.0)	(n, π^*)
Acetophenone	Oxetane (0.1)	Pinacol (0.5)	(n, π^*)
Benzophenone	Oxetane (0.5)	Pinacol (1.0)	(n, π^*)
1-Naphthaldehyde	Oxetane (0.05)	Pinacol (0.1)	(π, π^*)
2-Naphthaldehyde	Oxetane (0.05)	—	(π, π^*)
1-Acetonaphthone	No reaction	No reaction	(π, π^*)
2-Acetonaphthone	No reaction	No reaction	(π, π^*)
2-Benzoylnaphthalene	Oxetane (0.005)	Pinacol (0.1)	(π, π^*)

[a] Numbers in parentheses are quantum yields of reactions.

In the presence of naphthalene the product distribution remains the same although the product yield is reduced, indicating that all the products are derived from the same state, the carbonyl triplet.[92] A similar result was reported by Yang, Loeschen, and Mitchell[93] in a study of the photolysis of benzaldehyde in the presence of 2,3-dimethyl-2-butene and various concentrations of the triplet quencher piperylene. Linear Stern–Volmer plots were obtained for quenching of oxetane formation at two different olefin concentrations, indicating that only one reactive state is involved, this being the carbonyl triplet. The nature of the reactive carbonyl triplet would appear to be $n \rightarrow \pi^*$ since the reactivity of various carbonyls toward oxetane formation parallels that toward photoreduction, as can be seen in Table 4.10. Compounds that undergo facile photoreduction (i.e., those with $n \rightarrow \pi^*$ lowest triplets) also form oxetanes in the presence of olefins. Interestingly, some that do not photoreduce do form oxetanes, such as 2-naphthaldehyde. It should be recalled, however, that although inert in the presence of isopropyl alcohol, 2-naphthaldehyde can be efficiently photoreduced in the presence of a better hydrogen donor, such as tri-*n*-butyl stannane (Chapter 3).

In a recent study of intramolecular oxetane formation in naphthoylnorbornenes Sauers and Rousseau provided evidence that two distinct excited states are involved[99]:

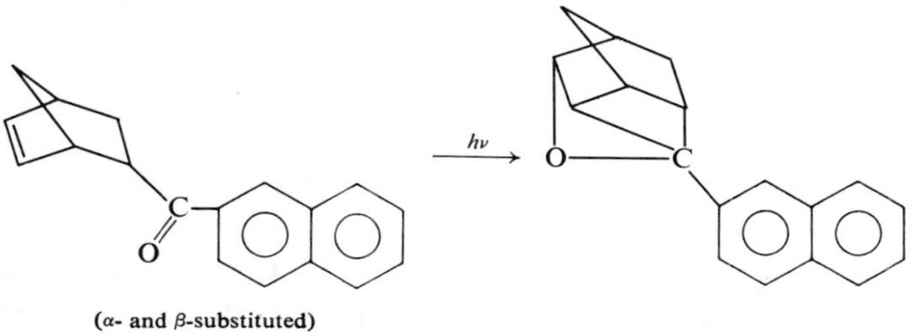

$$\xrightarrow{h\nu} \qquad (4.119)$$

(α- and β-substituted)

In the presence of 1,3-cyclohexadiene as quencher, linear Stern–Volmer plots were obtained between 2×10^{-5} and 10^{-4} *M* quencher. Above this latter concentration, however, the slope began to fall off and at about 10^{-3} *M* quencher became level. The quantum yield for this quenchable state was determined to be $\Phi = 0.65$ for both the α- and β-substituted naphthoylnorbornenes. Further quenching was achieved by addition of triethylamine to a solution containing sufficient diene to totally quench the reaction from the lowest triplet. From the quenching plots, lifetimes of 2.5×10^{-6} and 4.8×10^{-6} sec were obtained for the lowest triplets of the α- and β-naphthoylnorbornenes, respectively. From the slopes of triethylamine quenching a lifetime of 4×10^{-12} sec was obtained for the second state of both isomers. Although this does not preclude the existence of a higher reactive triplet state, the second state is most probably a singlet. From the intersystem crossing quantum yields and quantum yields of triplet state reaction one can determine that the efficiency for triplet reaction is 0.4. The efficiency of the singlet reaction is one-fifth of this value.

To explain the stereochemistry of the photoaddition, Buchi proposed that the reaction of electron-rich olefins and excited ketone involves an interaction of the electron-deficient carbonyl lone-pair orbital with the electron-rich π-olefin orbitals to form a diradical intermediate which could subsequently close to give the observed products. Indeed, reaction to yield the most stable diradical intermediate usually does nicely rationalize the observed product distribution. Examples of this are as follows[100]:

$h\nu$ or (4.120)

(most stable diradical)

(6) (7)

(6):(7) = 1:9
(yield 93%)

Reactions of aromatic ketones with electron-rich olefins are in general nonstereospecific, as evidenced by the same product distribution being obtained from reaction with both *cis*- and *trans*-2-butene[92]:

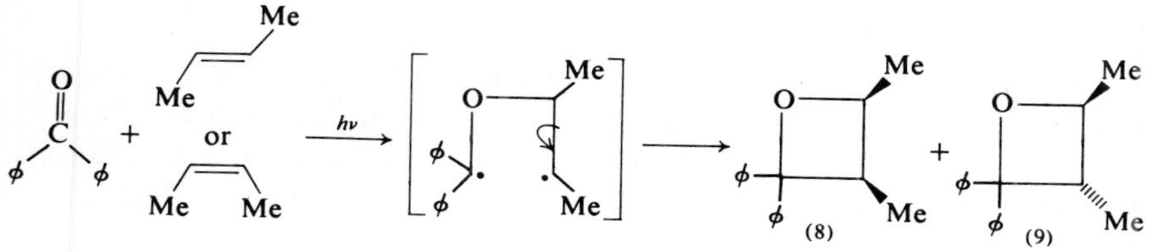

(8):(9) = 6:1
(yield 79%)

(4.121)

Norbornene reacts with benzophenone to yield predominantly the exo-oxetane, because approach from this side is less sterically hindered(100,101):

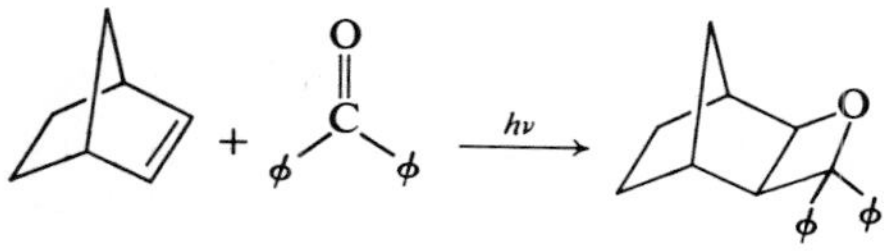

80% (*exo*)

(4.122)

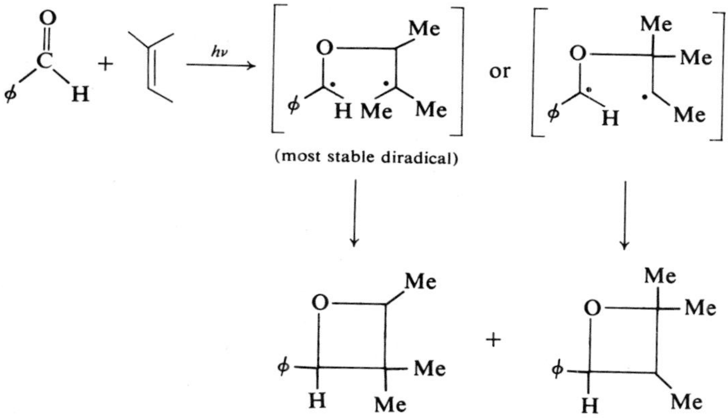

(most stable diradical)

(10) (11)

(10):(11) = 1.6:1
(yield 94%)

(4.123)

1,4-Benzoquinone, having a low triplet energy (E_t = 50 kcal/mole),(104) reacts with dienes to yield spiro-pyrans, presumably through the following allylic radical intermediate(105):

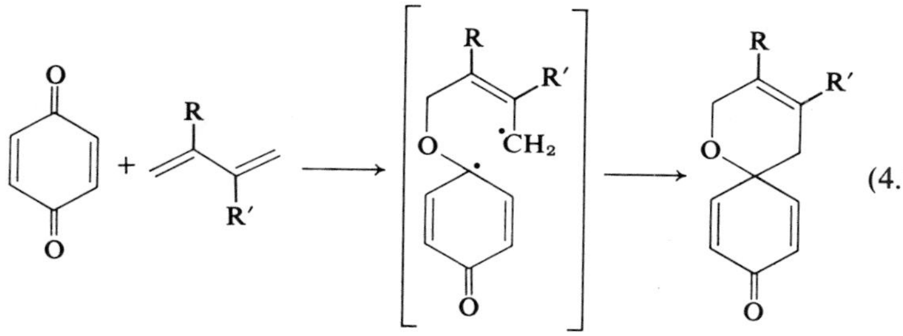

(4.124)

4.2b. *Synthetic Applications of Oxetane Formation*

The reaction of carbonyl compounds to olefins often yields products difficult to obtain synthetically by other routes. The excellent yields obtainable under proper conditions make this reaction of definite preparative interest. Examples of some synthetic applications of oxetane formation follow:

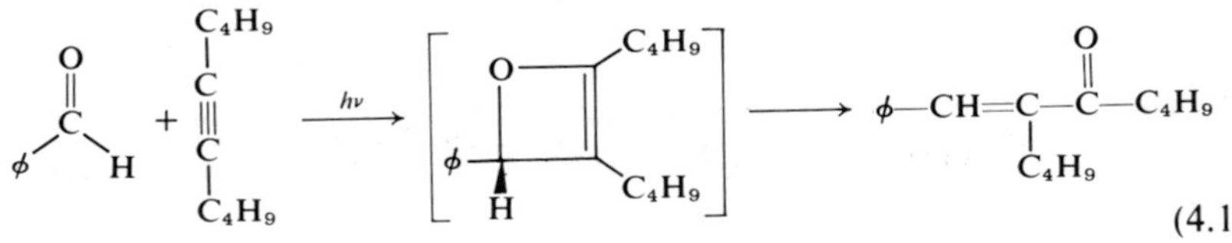

$$\phi CHO + C_4H_9C \equiv CC_4H_9 \xrightarrow{h\nu} [\text{oxetene}] \longrightarrow \phi{-}CH{=}C(C_4H_9){-}C(=O){-}C_4H_9 \quad (4.125)$$

Irradiation of benzaldehyde in the presence of 5-decyne results in the formation of the above α,β-unsaturated ketone, presumably from decomposition of an intermediate oxetene which may be formed in a vibrationally excited (hot) ground state.[106]

Acetylcyanide adds to olefins to form oxetanes having a useful functional group directly attached to the trimethylene oxide ring[107]:

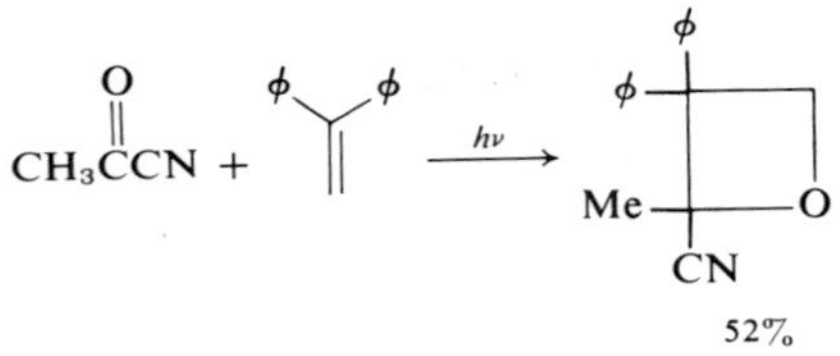

$$CH_3\overset{O}{\overset{\|}{C}}CN + \phi_2C{=}CH_2 \xrightarrow{h\nu} \text{oxetane } (\phi, \phi, Me, CN) \quad (4.126)$$

52%

Similarly to aldehydes, esters add to acetylenes to produce unstable oxetenes which decompose to β-alkoxy-α-benzoylstyrene derivatives when diphenylacetylene is used.[108] For example,

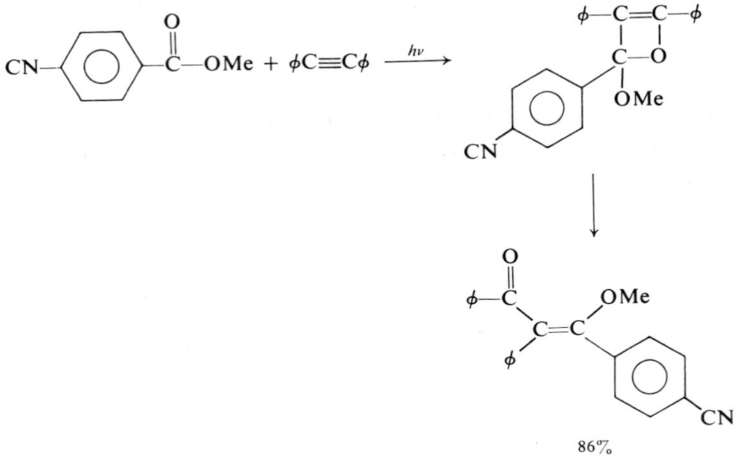

$$CN{-}C_6H_4{-}\overset{O}{\overset{\|}{C}}{-}OMe + \phi C \equiv C\phi \xrightarrow{h\nu} \text{oxetene} \longrightarrow \phi{-}CO{-}C(\phi){=}C(OMe){-}C_6H_4{-}CN \quad (4.127)$$

86%

Lange and Bosch[109] reported the interesting intramolecular oxetane formation shown below. Reduction with lithium aluminum hydride afforded *trans*-9-decalol (32%):

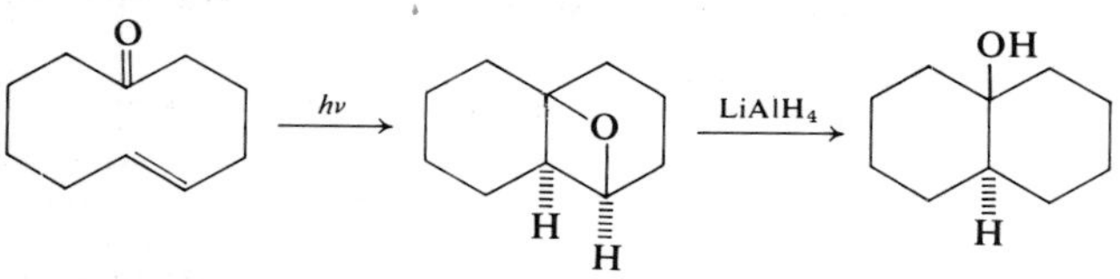

(4.128)

Aldehydes and ketones also add to allenes to form oxetanes.[110,111] Further reaction of the oxetanes produced with excited carbonyls results in dioxaspiro[3.3]heptane derivatives[111]:

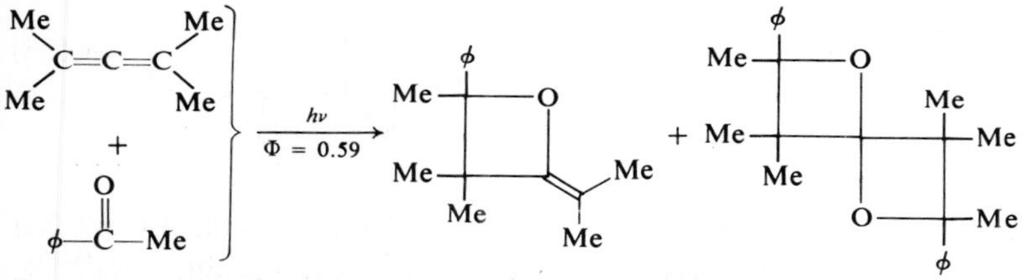

(4.129)

In some cases rearrangement to cyclobutanones occurs:

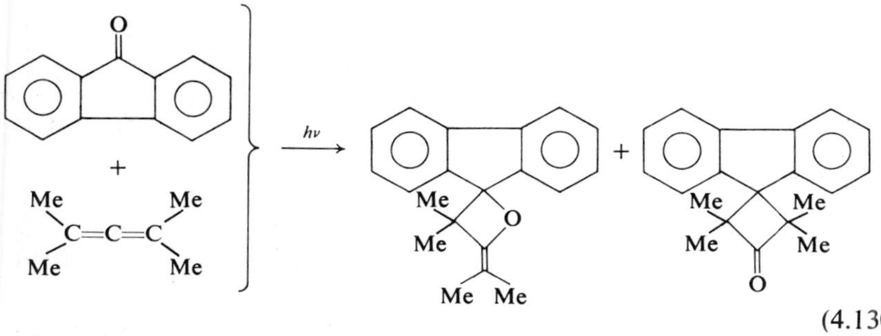

(4.130)

The addition of thioketones to olefins is very interesting indeed. Unsubstituted electron-rich olefins yield 1,4-dithianes as final products[112]:

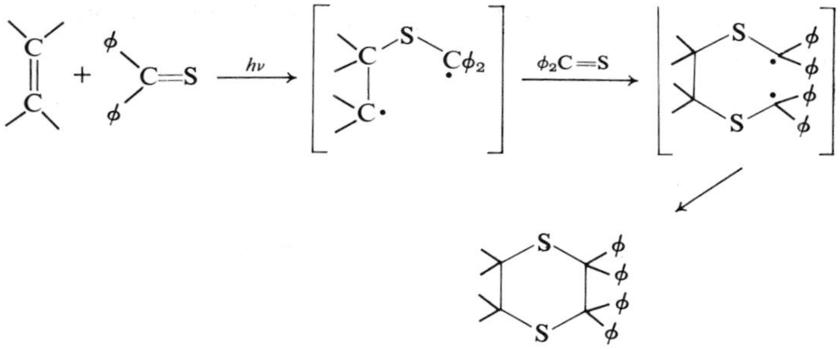

(4.131)

Olefins bearing electron-donating substituents yield thietane derivatives(113):

$$(\phi)_2C{=}S + H_3C{-}CH{=}CH\phi \xrightarrow[\lambda = 5890\ \text{Å}]{h\nu} \textbf{(12)} + \textbf{(13)}$$

cis and *trans*

(12) (13)

63% (12) (from *trans*)
79% (12) + 21% (13) (from *cis*)

(4.132)

Olefins bearing electron-withdrawing groups also yield thietanes(114):

$$(\phi)_2C{=}S + CH_2{=}CHCN \xrightarrow[\lambda = 3660\ \text{Å}]{h\nu} \text{thietane (CN, }\phi,\phi\text{)} \quad 93\%$$ (4.133)

When light from a mercury arc in the range 210–280 nm is used, other products presumably result from photodecomposition of the thietanes(115):

$$(\phi)_2C{=}S + CH_3CH{=}CHCH_3 \xrightarrow{210\text{–}280\ \text{nm}} [\text{thietane}] + [\text{thietane}]$$

$$\xrightarrow{h\nu} (\phi)_2C{=}CHCH_3 + CH_3CH{=}S \qquad \xrightarrow{h\nu} (\phi)_2C{=}CHCH_3 + CH_3CH{=}S$$

(4.134)

Benzophenone and other aryl ketones add readily to vinyl ethers to yield mono- and dialkoxyoxetanes(116,117):

$$(\phi)_2C{=}O + CH_2{=}CHOCH_2CH_3 \xrightarrow{h\nu} \text{oxetane (}OCH_2CH_3\text{)} + \text{oxetane (}OCH_2CH_3\text{)}$$

(99% yield)

(4.135)

$$\phi_2C{=}O + CH_2{=}C(OEt)_2 \xrightarrow{h\nu} \text{oxetane } (\phi, \phi, OEt, OEt) \qquad (4.136)$$

$$\phi_2C{=}O + \text{furan} \xrightarrow{h\nu} [\text{biradical}] \longrightarrow \text{oxetane} + \text{bis-adducts} \qquad (4.137)$$

Aryl ketones are often used to effect *cis* and *trans* isomerization of olefins.[118–120] Although this, in some cases, can be viewed as an energy transfer process where the ketone triplet transfers its energy to the olefin, which then isomerizes, the failure of noncarbonyl sensitizers of comparable triplet energy to isomerize the olefins suggests that a process other than energy transfer may be involved. Schenck and Steinmetz[118] suggested that isomerization results from decomposition of a biradical carbonyl-olefin adduct similar to that involved in oxetane formation:

$$\phi\overset{O}{\overset{\|}{C}}\phi + RHC{=}CHR' \xrightarrow{h\nu} [\text{biradical; rotation}] \longrightarrow \phi\overset{O}{\overset{\|}{C}}\phi + RHC{=}CHR' \qquad (4.138)$$

This concept will be explored more thoroughly in a later chapter on energy transfer.

4.2c. *Oxetane Formation from Olefins and Aliphatic Aldehydes and Ketones*

As is the case with aryl carbonyls previously studied, aliphatic carbonyls add to olefins to form oxetanes. The picture in this case is far more complicated, however, primarily due to the increased importance of singlet state carbonyl addition to the olefin. In Chapter 1 we saw that in an $n \rightarrow \pi^*$

excitation an electron is removed from the carbonyl oxygen lone pair and promoted to a vacant π^* orbital shared by carbon and oxygen:

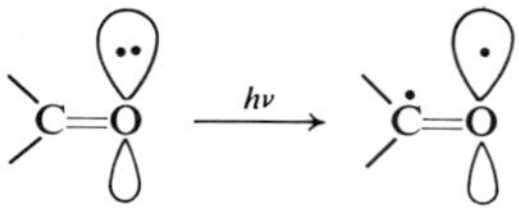

$$>C=\ddot{O} \xrightarrow{h\nu} >\dot{C}=\dot{O}$$

Thus we end up with an oxygen atom which is somewhat electron deficient and a carbon atom which is electron rich. The oxygen then would be expected to behave as an electrophilic reagent and the carbon (or rather the regions bounded by the π^* orbital) should behave as a nucleophilic reagent. The "amphoteric" nature of the carbonyl $n \rightarrow \pi^*$ singlet state is mirrored in its reactivity toward electron-rich and electron-deficient olefins.

Irradiation of mixtures of acetone and *cis*- or *trans*-1-methoxybutene yields the following oxetanes[121]:

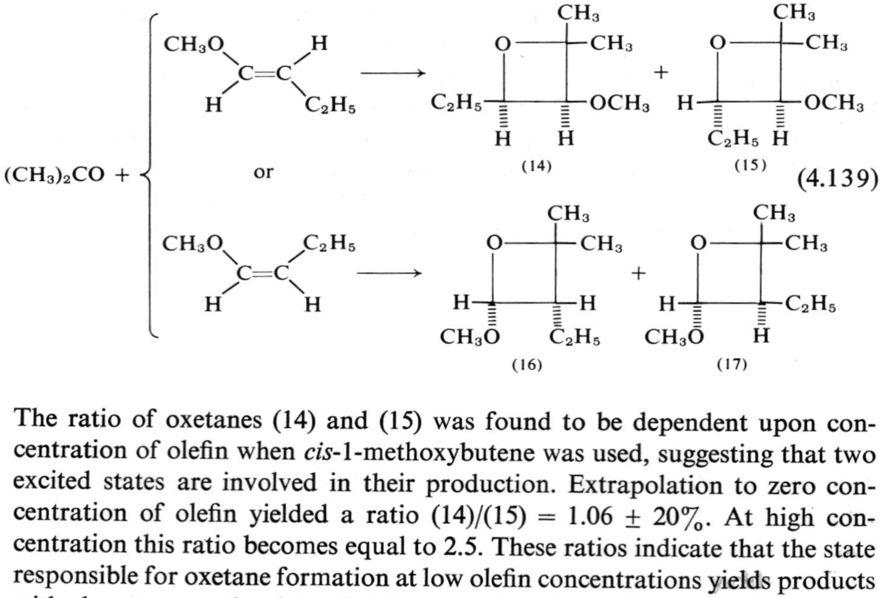

$$(CH_3)_2CO + \begin{cases} CH_3O(H)C{=}C(H)C_2H_5 \longrightarrow (14) + (15) \\ \text{or} \\ CH_3O(H)C{=}C(H)C_2H_5 \longrightarrow (16) + (17) \end{cases} \tag{4.139}$$

The ratio of oxetanes (14) and (15) was found to be dependent upon concentration of olefin when *cis*-1-methoxybutene was used, suggesting that two excited states are involved in their production. Extrapolation to zero concentration of olefin yielded a ratio (14)/(15) = 1.06 ± 20%. At high concentration this ratio becomes equal to 2.5. These ratios indicate that the state responsible for oxetane formation at low olefin concentrations yields products with almost a complete loss of olefin stereochemistry, whereas at higher olefin concentrations initial olefin stereochemistry is somewhat preserved. With the triplet quencher piperylene, the formation of both oxetanes (14) and (15) is quenched although nonlinear Stern–Volmer plots were obtained, again indicating the presence of two reactive states. At high piperylene concentrations quenching leveled off, providing evidence that only a singlet mechanism for oxetane formation is operative at high quencher concentrations. The limiting quantum yield for singlet reaction under these conditions was found to be 0.02 for *cis*-olefin and 0.005 for *trans*-olefin. It thus appears that the two states leading to oxetane formation are the acetone singlet and triplet $n \rightarrow \pi^*$ states. Furthermore, the degree of stereospecificity observed at 0.4 M piperylene (singlet reaction) is essentially that observed at high olefin concentrations. This indicates that the triplet reaction (predominant at low olefin concentrations) is essentially nonstereospecific, whereas a reasonable degree of stereospecificity is retained in the singlet reaction. These data are rationalized by the intermediacy of singlet and triplet biradical pairs produced by electrophilic attack of the excited carbonyl oxygen on the electron-rich olefin:

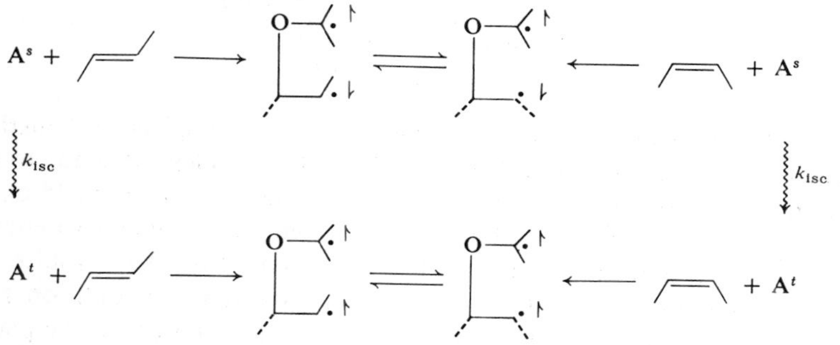

A^s + (olefin) ⟶ [biradical] ⇌ [biradical] ⟵ (olefin) + A^s

k_{isc} | | k_{isc}

A^t + (olefin) ⟶ [biradical] ⇌ [biradical] ⟵ (olefin) + A^t

The higher degree of stereospecificity in the singlet reaction relative to that of the triplet is not unexpected in light of the greater lifetime of the triplet biradical allowing bond rotation or bond breaking to yield acetone and an isomerized olefin.

Irradiation of mixtures of acetone and the electron-poor olefin *cis*- or *trans*-dicyanoethylene leads to the stereospecific formation of oxetane[122]:

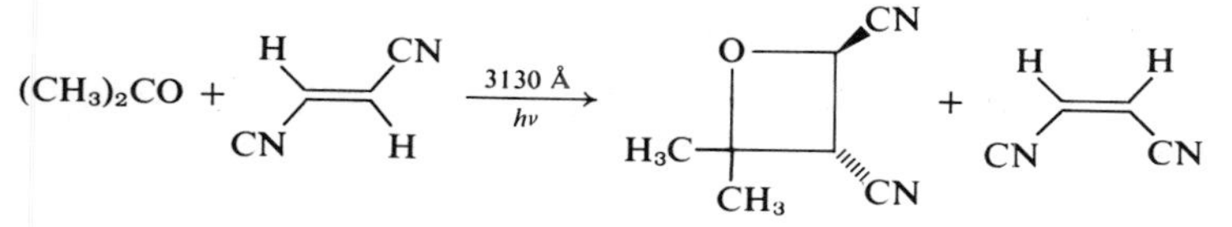

$$(CH_3)_2CO + \text{trans-NC–CH=CH–CN} \xrightarrow[h\nu]{3130\ \text{Å}} \text{oxetane (H}_3\text{C, CH}_3\text{; CN, CN trans)} + \text{cis-NC–CH=CH–CN} \quad (4.140)$$

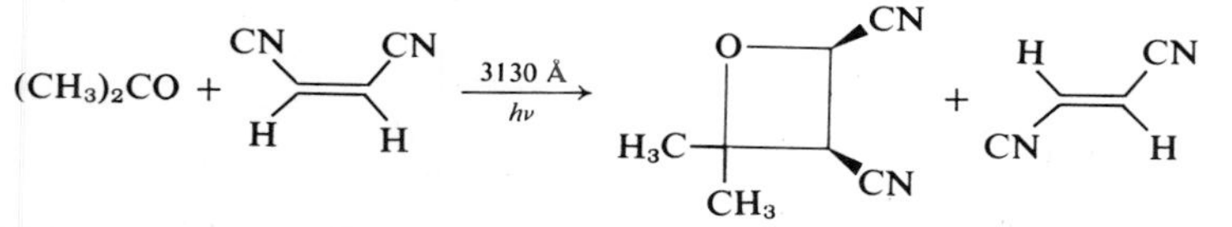

$$(CH_3)_2CO + \text{cis-NC–CH=CH–CN} \xrightarrow[h\nu]{3130\ \text{Å}} \text{oxetane (H}_3\text{C, CH}_3\text{; CN, CN cis)} + \text{trans-NC–CH=CH–CN} \quad (4.141)$$

That oxetane formation results from a singlet state reaction follows from the following evidence: (a) Acetone fluorescence is quenched by addition of the olefin, (b) oxetane formation is relatively insensitive to piperylene, and (c) *cis–trans* isomerization of the olefin is quenched at high olefin concentrations but oxetane formation is not affected. Since oxetane formation was found to be inefficient ($\Phi = 0.5$ at 0.5 M olefin) although quenching of acetone singlets was efficient, an intermediate singlet complex was proposed:

$$
\begin{aligned}
A &\longrightarrow A^s \\
A^s &\longrightarrow A + h\nu \\
A^s + t\text{-olefin} &\longrightarrow \text{complex} \\
\text{complex} &\longrightarrow \text{oxetane} \\
\text{complex} &\longrightarrow A + t\text{-olefin} \\
A^s &\longrightarrow A^t \\
A^t &\longrightarrow A \\
A^t + t\text{-olefin} &\longrightarrow A + [t\text{-olefin}]^t \\
[t\text{-olefin}]^t &\longrightarrow t\text{-olefin} \\
[t\text{-olefin}]^t &\longrightarrow c\text{-olefin}
\end{aligned}
$$

In view of the results obtained for attack of acetone singlets on 1-methoxybutene to yield singlet biradicals (partial loss of stereochemistry due to bond rotation in the biradical), acetone attack on dicyanoethylene to yield oxetane stereospecifically via a similar biradical intermediate is difficult to envision. Thus a new mechanism must be developed to account for these results.

As stated previously, the $n \rightarrow \pi^*$ singlet also has, in addition to the electrophilic oxygen, an electron-rich π system above and below the plane of the carbonyl group. This three-electron system should then undergo nucleophilic attack on electron-poor double bonds. Turro *et al.*[124] have proposed that addition of the singlet acetone occurs via a concerted or "quasi-concerted" attack of the nucleophilic π-system to yield an exciplex which can stereospecifically dissociate into ground state ketone and ground state olefin or collapse into oxetane:

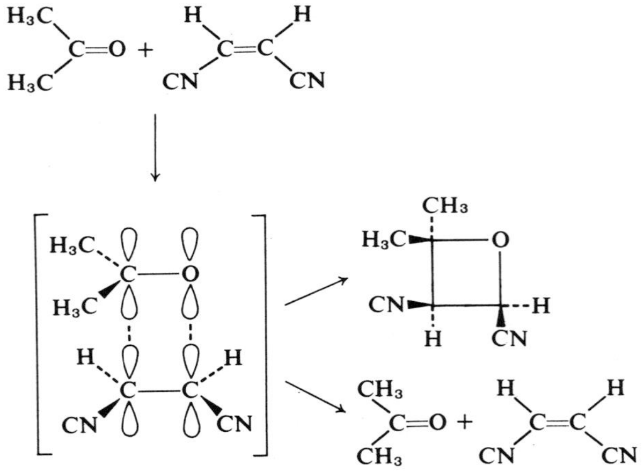

H3C, H3C C=O + H, CN C=C H, CN → [H3C, H3C C—O ; H, CN C—C H, CN] → oxetane (CH3, H3C, O, CN, H, H, CN) or CH3, CH3 C=O + H, CN C=C H, CN (4.142)

Similar conclusions were reached by Barltrop and Carless in a recent study of the photoaddition of aliphatic ketones to acrylonitrile and methacrylonitrile.[123]

Further examples of oxetane formation from aliphatic carbonyls are as follows[116b,124–129]:

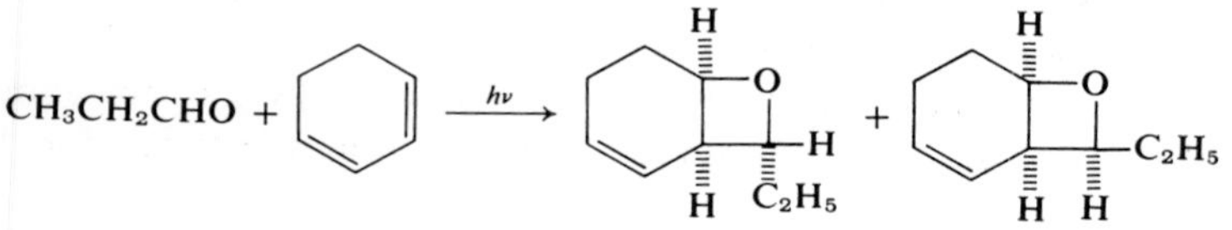

(4.143)

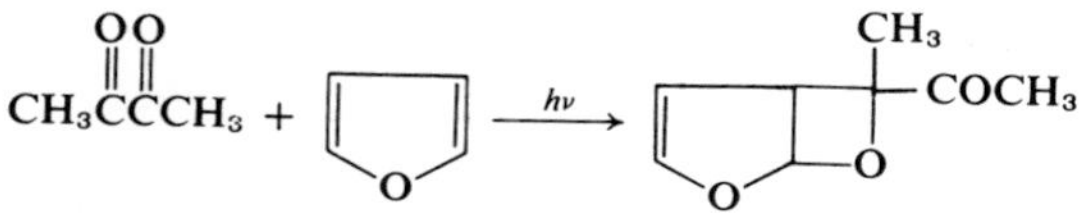

(4.144)

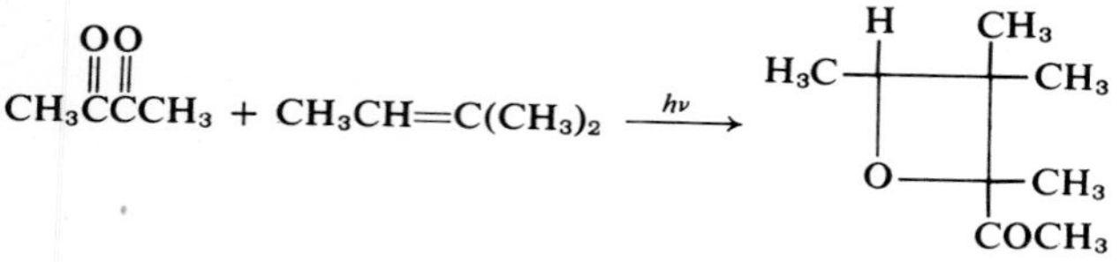

(4.145)

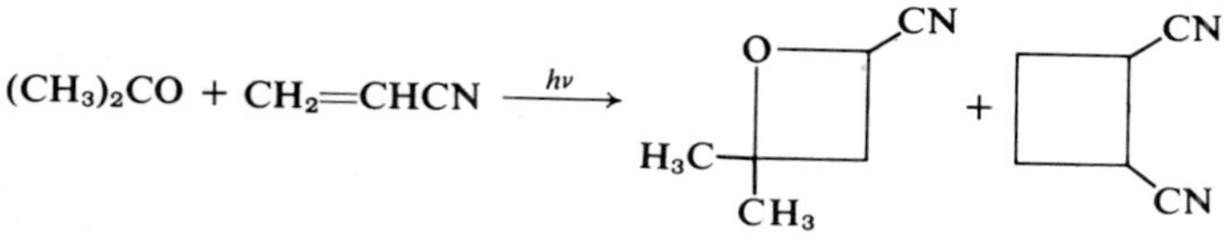

(4.146)

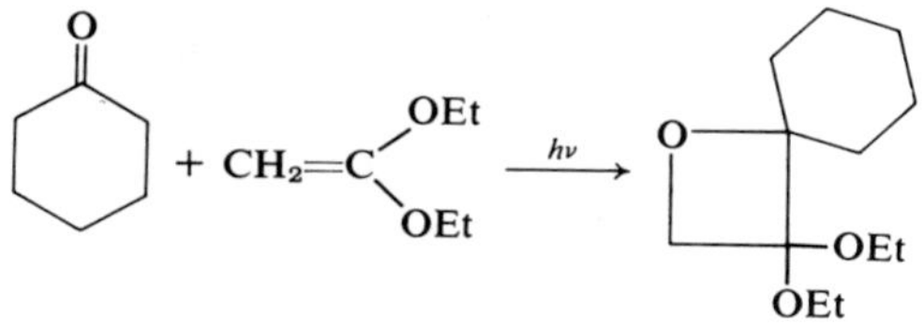

(4.147)

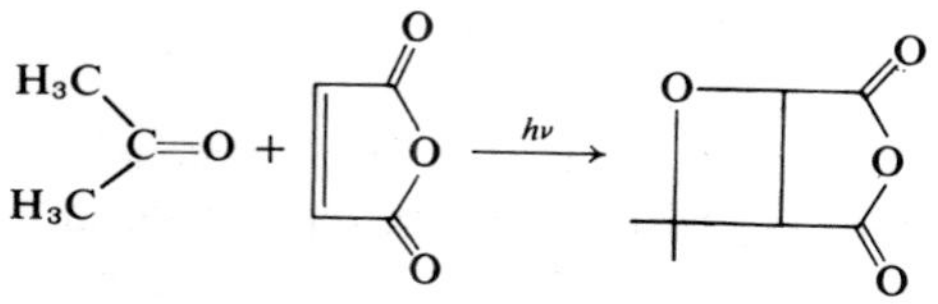

(4.148)

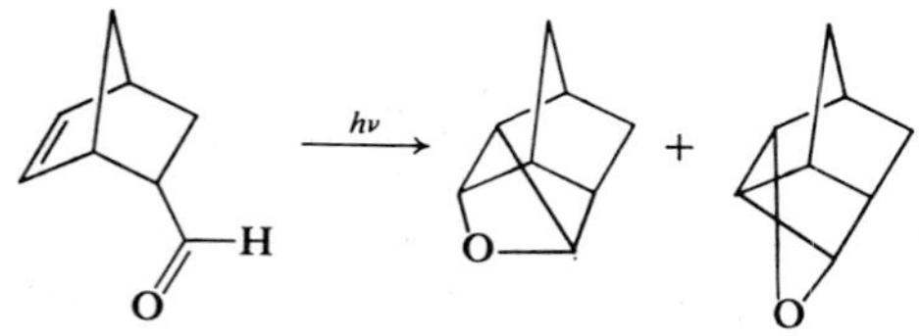

(4.149)

4.2d. *Perturbational Molecular Orbital Theory (PMO) Applied to Oxetane Formation*

Using PMO theory, it is possible to estimate the relative activation energy of reactions by assuming some reasonable structure for the transition state and then using perturbation theory to calculate the difference between the energy of the transition state and the energy of the reactants.[130]

For oxetane formation from formaldehyde and ethylene, we should consider the following four transition states and intermediates for the reaction[131]:

case 1 $r_1(n) + s_1(\pi)$ ⟶

case 2 $r_1(\pi) + s_1(\pi)$ ⟶

case 3 $r_2(\pi) + s_1(\pi)$ ⟶

case 4 $r_1r_2(\pi) + s_1s_2(\pi)$ ⟶

As a simple introduction to PMO theory suppose we consider the bond formation between the two-atom, two-orbital system shown in Figure 4.8. The energy gained on forming the bond A—B is given by the second-order perturbation equation

$$\Delta_2 = \frac{n(C_1C_2)^2}{E_2 - E_1}\gamma^2 \tag{4.150}$$

where n is the occupancy number (1 or 2 when the interaction involves one or two electrons), C_1 and C_2 are the orbital coefficients (HMO in the cases to

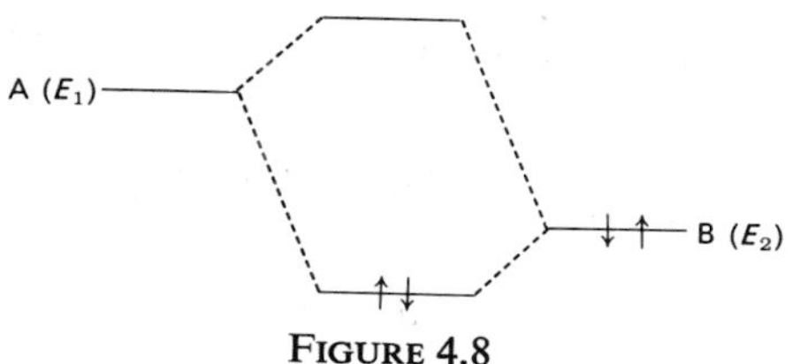

FIGURE 4.8

be considered), E_2 and E_1 are the energy levels (in units of β), and γ is the atomic orbital interaction exchange (resonance) integral. If the two orbitals have similar (ΔE is less than 0.6β) or identical energies, then the first-order perturbation term is dominant,

$$\Delta_1 = nC_1C_2\gamma \tag{4.151}$$

When more than two orbitals are involved, the energy change must take into account all important orbital interactions. This will be illustrated for the formaldehyde–ethylene case following the method of Herndon and Giles.(131) If we assume the bonds are half-formed in the transition state, then the exchange integral γ is just equal to $\frac{1}{2}\beta$. Since β has a value of about 40 kcal/mole, then

$$\gamma = 20 \text{ kcal/mole}, \qquad \gamma^2/\beta = 10 \text{ kcal/mole}$$

The Hückel energy levels and p basis orbital coefficients for formaldehyde and ethylene are shown in Figure 4.9. If we examine the orbital interactions for case 1, we can see that in the orientation such that the lone pair on oxygen is oriented in a way that a bond can begin to form between the oxygen lone pair and the carbon π orbital, there are two possible orbital interactions.

+0.707 −0.707
C——C $E(\pi^*) = \alpha - \beta$

$F(\pi^*) = \alpha - 0.62\beta$ +0.851 −0.526 C——O

$F(n) = \alpha$ 1.0 C——O

+0.707 +0.707
C——C $E(\pi) = \alpha + \beta$

$F(\pi) = \alpha + 1.62\beta$ +0.526 +0.851 C——O

FIGURE 4.9. Hückel orbitals for formaldehyde and ethylene.

$F(n)$–$E(\pi^*)$ and $E(\pi)$–$F(n)$. The change in energy will be the sum of these two interactions:

$$\Delta_{2,F(n)-E(\pi^*)} = \frac{(1)(1 \times 0.707)^2}{\beta}\gamma^2$$

$$= 0.50\gamma^2/\beta \tag{4.152}$$

For the $F(n)$–$E(\pi^*)$ interaction only one electron is involved. For the $E(\pi)$–$F(n)$ interaction there are three electrons involved: Two are lowered in energy and one is increased by the same amount. The others are lowered, so that there is a net one electron stabilized:

$$\Delta_{2,E(\pi)-F(n)} = \frac{(1)(1 \times 0.707)^2}{1\beta}\gamma^2$$

$$= 0.50\gamma^2/\beta \tag{4.153}$$

For case 2 we have four possible interactions to consider:

(a) $F(\pi^*)$–$E(\pi^*)$
(b) $F(\pi)$–$E(\pi^*)$
(c) $F(\pi)$–$E(\pi)$
(d) $E(\pi)$–$F(\pi^*)$

Interaction (c) is not important inasmuch as four electrons are involved, two stabilized and two destabilized. The energies for the other interactions can be easily calculated:

$$\Delta_{1,F(\pi^*)-E(\pi^*)} = (1)(-0.526)(-0.707)\gamma$$

$$= 0.372\gamma$$

$$\Delta_{2,F(\pi)-E(\pi^*)} = \frac{2(0.851 \times 0.707)^2}{2.62\beta}\gamma^2$$

$$= 0.276\gamma^2/\beta$$

$$\Delta_{2,E(\pi)-F(\pi^*)} = \frac{(1)(0.526 \times 0.707)^2}{1.62\beta}\gamma^2$$

$$= 0.0854\gamma^2/\beta$$

The results are tabulated in Table 4.11.

As can be seen from Table 4.11, the concerted $r_1r_2(\pi) + s_1s_2(\pi)$ reaction is favored. However, caution is required since this mode of reaction is possible only for the singlet excited formaldehyde. This also brings up the problem that the Hückel eigenvalues and eigenvectors represent some average between the singlet and triplet states. However, for the $n \rightarrow \pi^*$ excited state it is known that the S–T separation is small and the difference in charge distribution is

TABLE 4.11. *PMO Results for Formaldehyde–Ethylene Reactions*[a]

Reaction	PMO energy, kcal/mole	Interactions	$\Delta_1(\gamma)$	$\Delta_2(\gamma^2/\beta)$
$r_1(n) + s_1(\pi)$	10.00	$F(n)$–$E(\pi^*)$		0.5000
		$E(\pi)$–$F(n)$		0.5000
$r_1(\pi) + s_1(\pi)$	11.05	$F(\pi^*)$–$E(\pi^*)$	0.3718	
		$F(\pi)$–$E(\pi^*)$		0.2764
		$E(\pi)$–$F(\pi^*)$		0.0854
$r_2(\pi) + s_1(\pi)$	15.29	$F(\pi^*)$–$E(\pi^*)$	0.6015	
		$F(\pi)$–$E(\pi^*)$		0.1056
		$E(\pi)$–$F(\pi^*)$		0.2236
$r_1r_2(\pi) + s_1s_2(\pi)$	20.19	$F(\pi^*)$–$E(\pi^*)$	0.9233	
		$F(\pi)$–$E(\pi^*)$		0.0403
		$E(\pi)$–$F(\pi^*)$		0.0326

[a] From Ref. 131.

TABLE 4.12. *PMO Results for Benzaldehyde–Trimethylethylene Reaction*[a]

ϕ(H)C=O + (H)(CH_3)C=C(CH_3)(CH_3) $\xrightarrow{h\nu}$ oxetanes (O ← r_1, C ← r_2; s_1 → C, s_2 → C)

Reaction	PMO energy, kcal/mole
$r_1(n) + s_1(\pi)$	17.33
$r_1(n) + s_2(\pi)$	16.07
$r_1(\pi) + s_1(\pi)$	4.72
$r_1(\pi) + s_2(\pi)$	4.99
$r_2(\pi) + s_1(\pi)$	6.33
$r_2(\pi) + s_2(\pi)$	6.04
$r_1r_2(\pi) + s_1s_2(\pi)$	7.92
$r_1r_2(\pi) + s_2s_1(\pi)$	8.49

[a] From Ref. 131.

TABLE 4.13. *PMO Results for Acetone–Dicyanoethylene Reactions*[a]

$$\begin{array}{c} O \leftarrow r_1 \\ \| \\ C \leftarrow r_2 \end{array} + \begin{array}{c} s_1 \rightarrow C(H)(CN) \\ \| \\ s_2 \rightarrow C(H)(CN) \end{array} \xrightarrow{h\nu} \text{oxetanes}$$

Reaction	PMO energy, kcal/mole
$r_1(n) + s_1(\pi)$	14.89
$r_1(\pi) + s_1(\pi)$	8.66
$r_2(\pi) + s_1(\pi)$	13.18
$r_1r_2(\pi) + s_1s_2(\pi)$	19.09

[a] From Ref. 131.

negligible. (The charge distribution can be related to the squares of the AO coefficients.) Thus the approximate HMO wave functions and energies are probably equally good or equally bad for PMO calculations on both singlet and triplet excited states.

The PMO calculation for the benzaldehyde–trimethylene reaction is given in Table 4.12 and that for acetone and dicyanoethylene in Table 4.13. The predicted products based on the PMO calculations for these two reactions agree very well with the experimental results given in the previous sections.

It is easy to qualitatively understand the differences in the results for the three systems. If we use the formaldehyde–ethylene system as our reference, then electron-releasing substituents (alkyl groups) on formaldehyde and electronegative substituents (cyano groups) on ethylene will narrow the energy gap $F(\pi^*)$–$E(\pi^*)$ and thus lead to a concerted reaction. Electronegative substituents on formaldehyde and electron-releasing substituents on ethylene will decrease the $F(\pi^*)$–$E(\pi^*)$ and $E(n)$–$F(\pi^*)$ interaction but increase the $E(\pi)$–$F(n)$ interaction, with the result that the diradical intermediate reaction should predominate. Conjugation with the carbonyl portion of the system should decrease the $F(\pi^*)$–$F(\pi)$ interaction and increase the possibility of a diradical reaction.

PROBLEMS

1. Suggest products and/or mechanisms for the following reactions:

   (a) [J. R. Scheffer and R. A. Wostradowski, *Tetrahedron Lett.*, 677 (1972).]

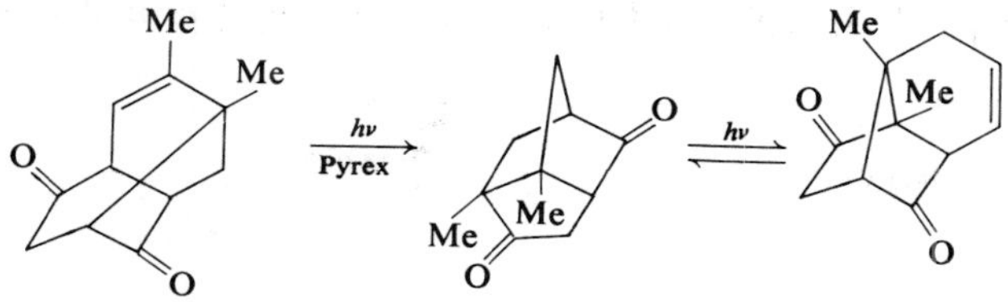

Me Me O O $\xrightarrow[\text{Pyrex}]{h\nu}$ O Me Me O $\underset{}{\overset{h\nu}{\rightleftharpoons}}$ Me O Me O

(b) [O. L. Chapman, P. W. Wojtkowski, W. Adam, O. Rodriguez, and R. Rucktäschel, *J. Amer. Chem. Soc.* **94**, 1367 (1972).]

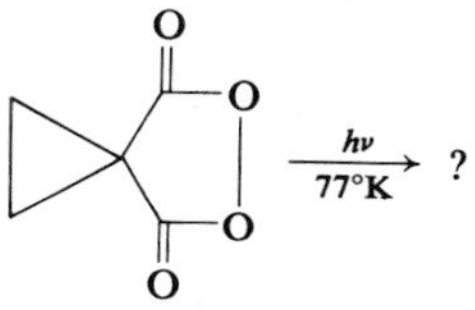

O O O O $\xrightarrow[77°K]{h\nu}$?

(c) [F. M. Beringer, R. E. K. Winter, and J. A. Castellano, *Tetrahedron Lett.*, 6183 (1968).]

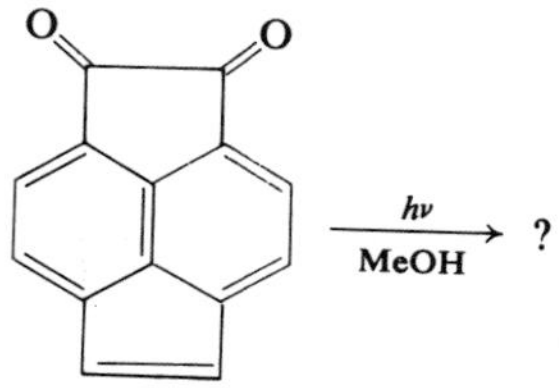

O O $\xrightarrow[\text{MeOH}]{h\nu}$?

(d) [G. W. Shaffer, A. B. Doer, and K. L. Purzycki, *J. Org. Chem.* **37**, 25 (1972).]

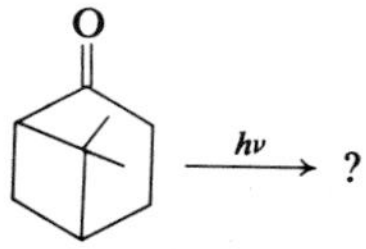

O $\xrightarrow{h\nu}$?

(e) [N. J. Turro and R. M. Southan, *Tetrahedron Lett.*, 545 (1967); P. Dowd and K. Sachder, *J. Amer. Chem. Soc.* **89**, 715 (1967).]

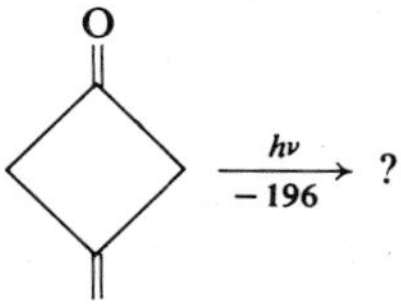

O $\xrightarrow[-196]{h\nu}$?

(f) [R. J. Spangler and J. C. Sutton, *J. Org. Chem.* **37**, 1462 (1972).]

CO_2Me H O $\xrightarrow[\text{MeOH}]{h\nu}$?

(g) [R. C. Cookson and N. R. Rogers, *Chem. Comm.*, 809 (1972).]

O $\xrightarrow{h\nu}$?

2. Compare the photochemical reactions of the following two compounds [A. A. Baum, *J. Amer. Chem. Soc.* **94**, 6866 (1972)]:

O —ϕ $\xrightarrow{h\nu}$?

O —ϕ $\xrightarrow{h\nu}$?

REFERENCES

1. G. Quinkert, *Angew. Chem. Int. Ed. Eng.* **4**, 211 (1965).
2. P. Yates, *Pure Appl. Chem.* **16**, 93 (1968).
3. R. L. Alumbaugh, G. O. Pritchard, and B. Rickborn, *J. Phys. Chem.* **69**, 3225 (1965).
4. P. J. Wagner and R. W. Spoerke, *J. Amer. Chem. Soc.* **91**, 4437 (1969).
5. P. Dunion and C. N. Trumbore, *J. Amer. Chem. Soc.* **87**, 4211 (1965).
6. R. Simonaitis, G. W. Cowell, and J. N. Pitts, Jr., *Tetrahedron Lett.* 3751 (1967).
7. P. J. Wagner and I. Kochevar, *J. Amer. Chem. Soc.* **90**, 2232 (1968); P. J. Wagner, *J. Amer. Chem. Soc.* **88**, 5672 (1966).
8. J. C. Dalton, K. Dawes, N. J. Turro, D. S. Weiss, J. A. Barltrop, and J. D. Coyle, *J. Amer. Chem. Soc.* **93**, 7213 (1971); J. D. Coyle, *J. Chem. Soc. B*, 1736 (1971).
9. J. C. Dalton and N. J. Turro, *Ann. Rev. Phys. Chem.* **21**, 499 (1970).
10. J. C. Dalton, D. M. Pond, D. S. Weiss, F. D. Lewis, and N. J. Turro, *J. Amer. Chem. Soc.* **92**, 2564 (1970).

11. D. S. Weiss, N. J. Turro, and J. C. Dalton, *Mol. Photochem.* **2**, 91 (1970).
12. J. A. Barltrop and J. D. Coyle, *Chem. Comm.*, 1081 (1969).
13. N. C. Yang and E. D. Feit, *J. Amer. Chem. Soc.* **90**, 504 (1968).
14. N. C. Yang, E. D. Feit, M. H. Hui, N. J. Turro, and J. C. Dalton, *J. Amer. Chem. Soc.* **92**, 6974 (1970).
15. N. C. Yang and R. H.-K. Chen, *J. Amer. Chem. Soc.* **93**, 530 (1971).
16. J. Franck and E. Rabinowitsch, *Trans. Faraday Soc.* **30**, 120 (1934).
17. A. A. Lamola and G. S. Hammond, *J. Chem. Phys.* **43**, 2129 (1965).
18. H. L. Goering, D. I. Delyea, and D. W. Larsen, *J. Amer. Chem. Soc.* **78**, 348 (1956); F. R. Jensen, L. H. Gale, and J. E. Rodgers, *J. Amer. Chem. Soc.* **90**, 5793 (1968); J. G. Traynham, A. G. Lane, and N. S. Bhacca, *J. Org. Chem.* **34**, 1302 (1969).
19. A. Ross, P. A. S. Smith, and A. S. Drieding, *J. Org. Chem.* **20**, 905 (1955).
20. C. H. Bamford and R. G. W. Norrish, *J. Chem. Soc.*, 1504 (1935); R. G. W. Norrish and R. P. Wayne, *Proc. Roy. Soc.* (*London*) *A* **284**, 1 (1965); J. N. Pitts, Jr., L. D. Hess, E. J. Baum, E. A. Schuck, and J. K. S. Wan, *Photochem. Photobiol.* **4**, 305 (1965).
21. F. E. Blacet and A. Miller, *J. Amer. Chem. Soc.* **79**, 4327 (1957); M. C. Flowers and H. M. Frey, *J. Chem. Soc.*, 2758, (1960); S. W. Benson and G. B. Kistiakowsky *J. Amer. Chem. Soc.* **64**, 80 (1942).
22. N. J. Turro and R. M. Southam, *Tetrahedron Lett.*, 545 (1967).
23. (a) N. J. Turro, P. A. Leermakers, H. R. Wilson, D. C. Neckers, G. W. Byers, and G. F. Vesley, *J. Amer. Chem. Soc.* **87**, 2613 (1965); (b) N. J. Turro, W. B. Hammond, and P. A. Leermakers, *J. Amer. Chem. Soc.* **87**, 2774 (1965).
24. (a) H. G. Richey, N. J. Richey, and D. C. Claggett, *J. Amer. Chem. Soc.* **87**, 1144 (1965); (b) I. Haller and R. Srinivasan, *Can. J. Chem.* **43**, 3165 (1965).
25. N. J. Turro and D. M. McDaniel, *J. Amer. Chem. Soc.* **92**, 5727 (1970).
26. D. R. Morton, Columbia University; results cited in N. J. Turro, J. C. Dalton, K. Dawes, G. Farrington, R. Hautala, D. Morton, M. Niemczyk, and N. Schore, *Acc. Chem. Res.* **5**, 92 (1972).
27. P. Yates, *Pure Appl. Chem.* **16**, 93 (1968); P. Yates and L. Kilmurry, *Tetrahedron Lett.*, 1739 (1964); *J. Amer. Chem. Soc.* **88**, 1563 (1966).
28. H. U. Hostettler, *Helv. Chim. Acta* **49**, 2417 (1966).
29. G. Quinkert, G. Cimbollek, and G. Buhr, *Tetrahedron Lett.*, 4573 (1966).
30. H. A. Staab and J. Ipaktschi, *Tetrahedron Lett.*, 583 (1966).
31. R. F. C. Brown and R. K. Jolly, *Tetrahedron Lett.*, 169 (1966).
32. A. G. Brook and J. B. Pierce, cited in: P. Yates, *Pure Appl. Chem.* **16**, 93 (1968).
33. R. Srinivasan, *J. Amer. Chem. Soc.* **81**, 2604 (1959).
34. W. C. Agosta and D. K. Herron, *J. Amer. Chem. Soc.* **90**, 7025 (1968).
35. H. Takeshita and Y. Fukazawa, *Tetrahedron Lett.*, 3395 (1968).
36. G. Quinkert, *Angew. Chem.* **77**, 229 (1965).
37. G. Quinkert, B. Wegemund, and E. Blanke, *Tetrahedron Lett.*, 221 (1962).
38. W. Koch, M. Carson, and R. W. Kierstead, *J. Org. Chem.* **33**, 1272 (1968).
39. R. C. Cookson, R. P. Ghandi, and R. M. Southam, *J. Chem. Soc. C*, 2494 (1968).
40. W. C. Agosta, A. B. Smith, A. S. Kende, R. G. Eilerman, and J. Renham, *Tetrahedron Lett.*, 4517 (1969).
41. L. A. Paquette and R. F. Eizember, *J. Amer. Chem. Soc.* **84**, 6205 (1967).
42. J. K. Crandell, J. P. Arrington, and J. Hen, *J. Amer. Chem. Soc.* **89**, 6208 (1967).
43. L. A. Paquette and G. V. Meehan, *J. Org. Chem.* **34**, 450 (1969).
44. J. R. Williams and H. Ziffer, *Chem. Comm.* **194**, 469 (1967).
45. J. R. Williams and H. Ziffer, *Tetrahedron* **24**, 6725 (1968).

46. C. P. Tenney, D. W. Boykin, and R. E. Lutz, *J. Amer. Chem. Soc.* **88**, 1835 (1966).
47. R. L. Cargill and T. Y. King, *Tetrahedron Lett.*, 409 (1970).
48. H. Nazaki, Z. Yamaguti, T. Okada, R. Noyori, and M. Kawanisi, *Tetrahedron* **23**, 3993 (1967).
49. P. Y. Johnson and G. A. Berchtold, *J. Org. Chem.* **35**, 584 (1970).
50. A. Padwa and A. Battisti, *J. Amer. Chem. Soc.* **94**, 521 (1972).
51. J. W. Wheeler and R. H. Eastman, *J. Amer. Chem. Soc.* **81**, 236 (1959).
52. L. Barber, O. L. Chapman, and J. D. Lassila, *J. Amer. Chem. Soc.* **90**, 5933 (1968).
53. D. A. Plank, Ph.D. Thesis, Purdue University (1966).
54. (a) C. D. Gutsche and B. A. M. Oude-Alink, *J. Amer. Chem. Soc.* **90**, 5855 (1968); (b) O. L. Chapman and C. L. McIntosh, *J. Amer. Chem. Soc.* **91**, 4309 (1969).
55. A. O. Pederson, S.-O. Lawesson, P. D. Klemmensen, and J. Kolc, *Tetrahedron* **26**, 1157 (1970).
56. O. L. Chapman and C. L. McIntosh, *J. Amer. Chem. Soc.* **92**, 7001 (1970).
57. G. Quinkert, K. Opitz, W. W. Wiersdorff, and J. Weinlich, *Tetrahedron Lett.*, 1863 (1963); G. Quinkert, *Pure Appl. Chem.* **9**, 607 (1964).
58. R. C. Cookson, M. J. Nye, and G. Subrahamanyam, *Proc. Chem. Soc.*, 144 (1964).
59. R. H. Eastman, J. E. Starr, R. St. Martin, and M. K. Sakata, *J. Org. Chem.*, **28**, 2162 (1963).
60. C. D. Gutsche and C. W. Armbruster, *Tetrahedron Lett.*, 1297 (1962); C. D. Gutsche and J. W. Baum, *Tetrahedron Lett.*, 2301 (1965).
61. J. E. Starr and R. H. Eastman, *J. Org. Chem.* **31**, 1393 (1966).
62. K. Mislow and A. J. Gordon, *J. Amer. Chem. Soc.* **85**, 3521 (1963).
63. W. K. Robbins and R. H. Eastman, *J. Amer. Chem. Soc.* **92**, 6076 (1970).
64. W. K. Robbins and R. H. Eastman, *J. Amer. Chem. Soc.* **92**, 6077 (1970).
65. P. S. Engel, *J. Amer. Chem. Soc.* **92**, 6074 (1970).
66. L. M. Stephenson and G. S. Hammond, *Pure Appl. Chem.* **16**, 125 (1968); D. I. Schuster and D. J. Patel, *J. Amer. Chem. Soc.* **90**, 5145 (1968).
67. R. Kaptein and L. J. Oosterhoff, *Chem. Phys. Lett.* **4**, 195, 214 (1969); F. L. Closs and A. D. Trifunac, *J. Amer. Chem. Soc.* **92**, 2183 (1970); R. E. Merrifield, *J. Chem. Phys.* **48**, 4318 (1968); R. C. Johnson and R. E. Merrifield, *Phys. Rev.* **31**, 896 (1970).
68. M. Tomkiewicz, A. Groen, and M. Cocivera, *Chem. Phys. Lett.* **10**, 39 (1971).
69. E. Baggiolini, H. P. Hamlow, and K. Schaffner, *J. Amer. Chem. Soc.* **92**, 4906 (1970); J. Hill, J. Iriarte, K. Schaffner, and O. Jeger, *Helv.* **49**, 292 (1966); K. Schaffner, *Chimia* **19**, 575 (1965).
70. H. Küntzel, H. Wolf, and K. Schaffner, *Helv.* **54**, 868 (1971).
71. R. S. Neale and E. Gross, *J. Amer. Chem. Soc.* **89**, 6579 (1967).
72. J. N. Pitts and I. Norman, *J. Amer. Chem. Soc.* **76**, 4815 (1954).
73. W. G. Dauben and G. W. Shaffer, *Tetrahedron Lett.*, 4415 (1967).
74. R. S. Carson, W. Cocker, S. M. Evans, and P. V. R. Shannon, *Chem. Comm.*, 726 (1969).
75. R. Beugelmans, *Bull. Soc. Chim. Fr.*, 244 (1967).
76. R. E. K. Winter and R. F. Landauer, *Tetrahedron Lett.*, 2345 (1967).
77. H. E. Zimmerman, K. G. Hancock, and G. C. Licke, *J. Amer. Chem. Soc.* **90**, 4892 (1968).
78. H. E. Zimmerman and J. W. Wilson, *J. Amer. Chem. Soc.* **86**, 4036 (1964).
79. T. Norin, *Acta Chim. Scand.* **19**, 1289 (1965); W. G. Dauben and E. J. Deviny, *J. Org. Chem.* **31**, 3794 (1966); H. E. Zimmerman, R. D. Rieke, and J. R. Scheffer, *J. Amer. Chem. Soc.* **89**, 2033 (1967); H. E. Zimmerman and R. L. Morse, *J. Amer. Chem. Soc.* **90**, 954 (1968).

80. H. E. Zimmerman and T. W. Flechtner, *J. Amer. Chem. Soc.* **92**, 6931 (1970).
81. H. E. Zimmerman and C. M. Moore, *J. Amer. Chem. Soc.* **92**, 2023 (1970).
82. H. E. Zimmerman, S. S. Hixson, and E. F. McBride, *J. Amer. Chem. Soc.* **92**, 2000 (1970).
83. H. E. Zimmerman, B. R. Cowley, C. Y. Tseng, and J. W. Wilson, *J. Amer. Chem. Soc.* **86**, 947 (1964).
84. H. Wehrli, C. Lehrmann, K. Schaffner, and O. Jeger, *Helv.* **47**, 1336 (1964); O. Jeger, K. Schaffner, and H. Wehrli, *Pure Appl. Chem.* **9**, 555 (1964); C. K. Johnson, B. Dominy, and W. Reusch, *J. Amer. Chem. Soc.* **85**, 3894 (1963).
85. H. E. Zimmerman, in *Abstract of Papers, 17th National Organic Chemistry Symposium, American Chemical Society, Bloomington, Indiana* (June 1961), p. 31.
86. R. Hoffmann and W. N. Lipscomb, *J. Chem. Phys.* **36**, 2179, 3487 (1962); R. Hoffmann, *J. Chem. Phys.* **39**, 1397 (1963); R. Hoffmann, *Tetrahedron Lett.*, 3819 (1965).
87. R. Gleiter and D. O. Cowan, unpublished calculations.
88. R. S. Mulliken, *J. Chem. Phys.* **23**, 1833, 1841, 2338, 2343 (1955).
89. A. Y. Meyer, B. Muel, and M. Kasha, *Chem. Comm.*, 401 (1972).
90. E. Paterno and C. Chieffi, *Gazz. Chim. Ital.* **39**, 341 (1909).
91. G. Buchi, C. G. Inman, and E. S. Lipinsky, *J. Amer. Chem. Soc.* **76**, 4327 (1954).
92. D. R. Arnold, R. L. Hinman, and A. H. Glick, work cited in D. R. Arnold, *Adv. Photochem.* **6**, 301 (1968).
93. N. C. Yang, R. Loeschen, and D. Mitchell, *J. Amer. Chem. Soc.* **89**, 5465 (1967).
94. N. C. Yang, N. Nussim, M. J. Jorgenson, and S. Murov, *Tetrahedron Lett.*, 3657 (1964); N. C. Yang, *Pure Appl. Chem.* **9**, 591 (1964).
95. G. S. Hammond and P. A. Leermakers, *J. Amer. Chem. Soc.* **84**, 207 (1962).
96. V. L. Ermolaev and N. A. Terenin, *J. Chim. Phys.* **55**, 698 (1958).
97. D. S. McClure, *J. Chem. Phys.* **17**, 905 (1949).
98. W. A. Bryce and C. H. J. Wells, *Can. J. Chem.* **41**, 2722 (1963).
99. R. R. Sauers and A. D. Rousseau, *J. Amer. Chem. Soc.* **94**, 1776 (1972).
100. D. R. Arnold, R. L. Hinman, and A. H. Glick, *Tetrahedron Lett.*, 1425 (1964).
101. D. Sharf and F. Korte, *Tetrahedron Lett.*, 821 (1963).
102. H. Kristinsson and G. W. Griffin, *J. Amer. Chem. Soc.* **88**, 1579 (1966).
103. I. P. Stepanov, O. A. Ikonopistseva, and T. I. Temnikova, *J. Org. Chem. USSR* **2**, 2216 (1966).
104. W. G. Herkstroeter and G. S. Hammond, *J. Amer. Chem. Soc.* **88**, 4769 (1966); A. Kuboyama, *Bull. Chem. Soc. Japan* **35**, 295 (1962).
105. J. A. Barltrop and B. Hesp, *J. Chem. Soc.*, 5182 (1965).
106. G. Buchi, J. T. Kofron, E. Koller, and D. Roesthal, *J. Amer. Chem. Soc.* **78**, 876 (1956).
107. Y. Shigemitsu, Y. Odaira, and S. Tsutsumi, *Tetrahedron Lett.*, 55 (1967).
108. T. Miyamoto, Y. Shigemitsu, and Y. Odaira, *Chem. Comm.*, 1410 (1969).
109. G. L. Lange and M. Bosch, *Tetrahedron Lett.*, 315 (1971).
110. D. A. Arnold and A. H. Glick, *Chem. Comm.*, 813 (1966).
111. H. Gotthardt, R. Steinmetz, and G. S. Hammond, *J. Org. Chem.* **33**, 2774 (1968); *Chem. Comm.*, 480 (1967).
112. G. Tsuchiahashi, M. Yamauchi, and M. Kukuyama, *Tetrahedron Lett.*, 1971 (1967); A. Ohno, Y. Ohnishi, M. Fukuyama, and G. Tsuchihashi, *J. Amer. Chem. Soc.* **90**, 7038 (1968).
113. A. Ohno, Y. Ohnishi, and G. Tsuchihashi, *Tetrahedron Lett.*, 283 (1969).
114. A. Ohno, Y. Ohnishi, and G. Tsuchihashi, *Tetrahedron Lett.*, 161 (1969).

115. E. T. Kaiser and T. F. Wulfers, *J. Amer. Chem. Soc.* **86**, 1897 (1964).
116. (a) S. H. Schroeter, *Chem. Comm.*, 12 (1969); (b) S. H. Schroeter and C. M. Orlando, *J. Org. Chem.* **34**, 1181 (1969).
117. G. R. Evanega and E. B. Whipple, *Tetrahedron Lett.*, 2163 (1967).
118. G. O. Schenck and R. Steinmetz, *Bull. Soc. Chim. Belges* **71**, 781 (1962).
119. N. C. Yang, *Photochem. Photobiol.* **7**, 767 (1968).
120. N. C. Yang, J. E. Cohen, and A. Shani, *J. Amer. Chem. Soc.* **90**, 3264 (1968); J. Saltiel, D. R. Neuberger, and M. Wrighton, *J. Amer. Chem. Soc.* **91**, 3658 (1969).
121. N. J. Turro and P. A. Wriede, *J. Amer. Chem. Soc.* **92**, 320 (1970).
122. J. C. Dalton, P. A. Wriede, and N. J. Turro, *J. Amer. Chem. Soc.* **92**, 1318 (1970); N. J. Turro, P. A. Wriede, and J. C. Dalton, *J. Amer. Chem. Soc.* **90**, 3274 (1968); N. J. Turro, P. A. Wriede, J. C. Dalton, D. Arnold, and A. Glick, *J. Amer. Chem. Soc.* **89**, 3950 (1967).
123. J. A. Barltrop and H. A. J. Carless, *J. Amer. Chem. Soc.* **94**, 1951 (1972).
124. N. J. Turro, J. C. Dalton, K. Dawes, G. Farrington, R. Hautala, D. Morton, M. Niemazyk, and W. Schore, *Acc. Chem. Res.* **5**, 92 (1972).
125. T. Kubota, K. Shima, S. Toki, and H. Sakurai, *Chem. Comm.*, 1462 (1969).
126. H.-S. Ryang, K. Shima, and H. Sakurai, *Tetrahedron Lett.*, 1091 (1970).
127. W. G. Bentrude and K. R. Darnall, *Chem. Comm.*, 862 (1969).
128. N. J. Turro and P. A. Wriede, *J. Org. Chem.* **34**, 3562 (1969).
129. R. R. Sauers and J. A. Whittle, *J. Org. Chem.* **34**, 3579 (1969); R. R. Sauers and K. W. Kelly, *J. Org. Chem.* **35**, 498 (1970).
130. M. J. S. Dewar, *The Molecular Orbital Theory of Organic Chemistry*, McGraw-Hill, New York (1969), pp. 191–247.
131. W. C. Herndon, *Tetrahedron Lett.*, 125 (1971); W. C. Herndon and W. B. Giles, *Mol. Photochem.* **2**, 277 (1970).

5

The Triplet State

5.1. INTRODUCTION

In our discussion in Chapter 1 of the various photophysical processes which can occur in an organic molecule upon absorption of light, we saw that two types of electronic excited states are possible. The first, initially reached upon excitation, is the singlet, in which the electronic spins remain paired. The second, obtained from the excited singlet through a process known as intersystem crossing, is the triplet state, in which the electron spins are antiparallel or unpaired. We have seen in subsequent chapters that a considerable amount of photochemistry is found to arise from the triplet state and in some cases it was suggested that the apparent greater reactivity of the excited triplet as compared to the excited singlet state arises from the greater lifetime of the former. However, we have not considered the questions: What exactly is a triplet state? Why should it be a longer lived state than the singlet? How are its properties different from the singlet? How are these properties determined? In this chapter we will attempt to answer these and other questions concerning the nature of the triplet state.

5.1a. *The Identity of the Phosphorescent State as a Triplet*

The afterglow or phosphorescence of many organic compounds when excited in viscous solutions or in the solid state has been recognized since the last two decades of the nineteenth century.[1,2] This afterglow, having a lifetime from 10^{-4} to 10^2 sec, is always observed at frequencies lower than a particular compound's fluorescence. However, it was not until the early

1930's that a scheme was developed to at least partially explain the origin of the afterglow. This scheme involved three distinct energy levels, the ground state, an excited state populated by light absorption, and a metastable state populated by degradation of the excited state. These states and the processes occurring between them are shown in Figure 5.1. In this figure solid lines are used to denote radiative transitions and broken lines to denote nonradiative transitions. Some energy loss was thought to occur by nonradiative decay of the excited and metastable states back to the ground state. Phosphorescence was postulated to arise by radiative decay of the metastable state, designated as a forbidden process to account for the relatively long lifetime of the process.[3,4]

In the early 1940's this radiative process from the metastable state to the ground state was suggested to be a transition from a triplet state to the ground state singlet, based on the efforts of Terenin,[5] Lewis, Lipkin, and Magel,[6] and Lewis and Kasha.[7] The experiment which demonstrated this hypothesis to be correct was carried out by Lewis and Calvin in 1945.[8] Since the proposed triplet state was known to be metastable, it should be possible to produce a relatively large population of this state by intense illumination. Since the triplet state has two unpaired electrons, it should be paramagnetic. Combining these two ideas, Lewis and Calvin were able to detect a light-induced paramagnetic susceptibility upon intense irradiation of fluorescein in a boric acid glass. It was concluded that about 85% of the fluorescein (cationic) molecules were in the metastable state and that this state possessed two unpaired electrons. It was later shown by Evans[9] that the rates of decay of fluorescein phosphorescence and of its magnetic susceptibility were identical and thus the states responsible for each behavior were identical (or at least intimately related). Failure to observe the expected EPR absorption which should be associated with a triplet state delayed general acceptance of this theory for some time, however. Finally, in 1964 the work of Lesclaux and Joussot-Dubien[10] removed all doubt of the validity of the triplet phosphorescent state. These workers obtained $\chi_m = (3280 \pm 300) \times 10^{-6}$ for fluorescein in boric acid at 25°C. The theoretical value for this triplet state is $\chi_m = 3340 \times 10^{-6}$ (cgs).

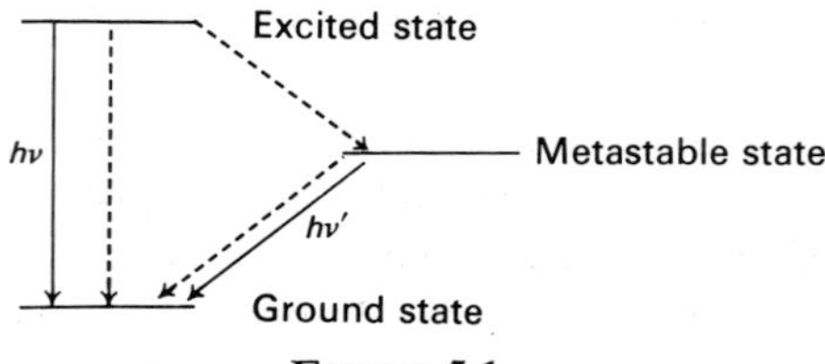

FIGURE 5.1

5.1b. *The Definition and Properties of a Triplet State*[11]

As seen in the previous section, one characteristic of the triplet state is its paramagnetism. This alone would of course not suffice as a definition of the triplet since there are many odd-electron species that also exhibit paramagnetism but do not exist as triplets. Thus we might state that a triplet is a paramagnetic even-electron species. This still does not constitute a limiting definition since compounds containing even numbers of electrons may exhibit two, three, or even five distinct electronic levels. For example, when in a biradical the radical centers are separated by several carbon atoms as below, no interaction between the electron spins occurs and the radicals appear as two doublet states:

$$H_2\dot{C}—(CH_2)_n—\dot{C}H_2$$

two doublets
(no spin interaction)

$$\underset{\cdot}{C}H_2 \xleftrightarrow[D]{} \underset{\cdot}{C}H_2 \quad \text{(bridged by } (CH_2)_n\text{)}$$

one triplet
(spin interaction)

However, if the molecular configuration is such as to bring the radical centers into close contact (say, to within some critical distance D), interaction will begin and the two doublet states will collapse since the spins of the two electrons are no longer independent of each other. Quantum mechanics tells us that under these conditions the total number of states is dependent upon the sum of the spin quantum numbers for the two electrons:

$$\text{number of states} = 2|S| + 1 \tag{5.1}$$

Since we have two unpaired electrons, $S = 1$, and the number of states is three. If we should have other even numbers of unpaired electrons, such as four or six, we could have three, five, or even as many as seven states. Thus we arrive at a definition for the triplet as a paramagnetic species possessing an even number of unpaired electrons and existing in a set of three energetically similar electronic levels which result from interaction of the electronic spin. Generally these three distinct electronic levels, between which transitions may be observed under certain conditions, are collectively referred to as the triplet state.

For a triplet to result, two electrons must have different orbitals and the same spin (the Pauli principle forbids having two unpaired electrons in the same orbital since all four quantum numbers would be the same). We saw in Chapter 1 that upon light absorption an electron is promoted from one orbital to a higher orbital. If we were to excite directly from the ground state singlet to the triplet state we would have to simultaneously change orbitals and electronic spins. Since this process is relatively improbable, direct absorption to the triplet state is seldom observed (later in this chapter, however, we will see that such a process can be observed under certain special conditions). Since direct absorption to the triplet is improbable, the reverse process, or return of the triplet to the singlet ground state, is also improbable. Since the triplet has few modes for deactivation (other than to react), it has a relatively long lifetime (on the order of seconds for some molecules). Decay to the ground state does occur, however, by two possible processes: (1) a non-radiative return similar to the intersystem crossing process that originally populated the triplet, and (2) the radiative process, phosphorescence.

As is the case with singlet states, absorption of light by the triplet can occur to produce higher triplet levels. We have previously seen how triplet–triplet absorption by flash photolysis is used as a tool in photochemistry.

The paramagnetism of the triplet state can be observed by electron spin resonance spectroscopy. This is perhaps the most reliable means of determining the existence of a triplet state since the ESR signals can be predicted using the following Hamiltonian operator:

$$H = g_0 H_0 S + D S_z^2 + E(S_x^2 - S_y^2) \tag{5.2}$$

where H_0 is the external magnetic field, S is the electronic spin, and the term $DS_z^2 + E(S_x^2 - S_y^2)$ describes spin–spin dipolar interactions along the x, y, and z axis of the molecule. The D and E are constants having the dimensions of energy and can frequently be easily determined from the simple triplet ESR spectrum. In qualitative terms, D is a measure of the interaction energy of the two unpaired spins. Thus D is a measure of the average separation of the two electrons in the molecule. Large values of D indicate extensive interaction and imply that the electrons are not greatly separated. The triplet excited state of naphthalene has $D = 0.1003\ \text{cm}^{-1}$, the triplet ground state of diphenylmethylene has $D = 0.405\ \text{cm}^{-1}$, and phenylmethylene has $D = 0.518\ \text{cm}^{-1}$. The constant E is a measure of the deviation of the molecule from threefold or higher axial symmetry.

Thus the properties of the triplet state are (a) paramagnetism, (b) emission of light (phosphorescence) to return to the ground state singlet, (c) absorption of light to produce higher triplet levels, and (d) absorption of electromagnetic (microwave) radiation to produce transitions among the triplet sublevels.

5.2. DETERMINATION OF TRIPLET ENERGY LEVELS

A number of experimental techniques are available for the determination of triplet energy levels. Those most commonly employed are phosphorescence spectroscopy, phosphorescence excitation spectroscopy, singlet to triplet absorption spectroscopy, flash photolysis, and electron excitation spectroscopy. Each of these techniques will be described in some detail in this section.

5.2a. *Phosphorescence Spectroscopy*

The simplest method for determination of the energy of the *lowest* triplet state for a particular compound is to determine the phosphorescence spectrum of the compound in a rigid glass at low temperature. The triplet energy is given directly as the energy of the $0 \rightarrow 0$ transition as shown in Figure 5.2. In principle this method can be applied to any compound provided its phosphorescence efficiency is sufficiently high to be detected using the available instrumentation. However, practical limitations arise since for many compounds the $0 \rightarrow 0$ band for phosphorescence emission lies in the red, a region of rather low sensitivity for most currently available photomultipliers. Other limitations include difficulty in determining the location of the $0 \rightarrow 0$ vibrational band for molecules whose $0 \rightarrow 0$ band is weak compared to the $0 \rightarrow 1$ band and the requirement of high purity, particularly for compounds that phosphoresce only weakly.

For compounds that are very weakly phosphorescent or that phosphoresce at wavelengths out of the normal range of sensitivity of the spectrometer this method of triplet energy determination cannot be applied. For these compounds triplet energies can sometimes be determined by measuring their E-type or P-type delayed fluorescence.

One of the first explanations for the afterglow of organic compounds in rigid solutions after exposure to UV light was offered by Perrin.[12] Perrin postulated that the excited molecules could undergo a transition to a metastable state of lower energy. Emission from this state was thought to be

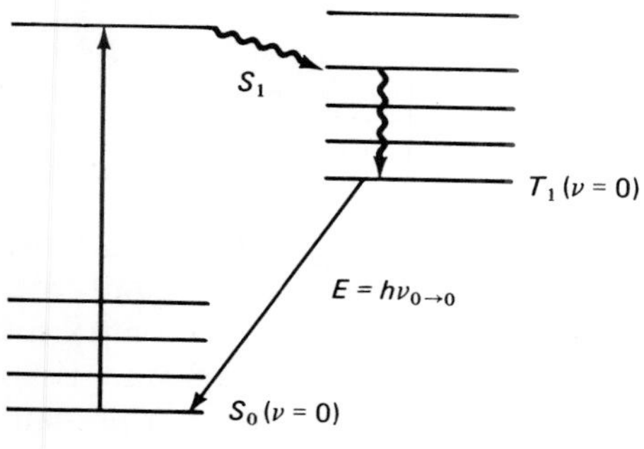

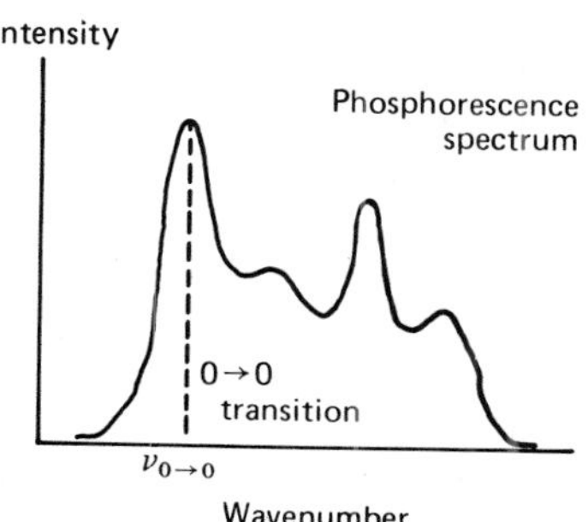

FIGURE 5.2

impossible but return to the fluorescent state could be achieved by thermal activation. The emission from the fluorescent state would thus be delayed by a time corresponding to the lifetime of the metastable state:

$$A \xrightarrow{h\nu} A' \rightsquigarrow M \overset{\Delta}{\rightsquigarrow} A' \longrightarrow A + h\nu' \tag{5.3}$$

Further extension of Perrin's idea by Jablonski led to the inclusion of the possibility of direct emission from the metastable state to the ground state as well as thermally activated return to the fluorescent state. Thus the observation of two bands of long-lived emission could readily be explained. One arises from direct emission from the metastable state to the ground state (observed at low temperature) and the other, observed at high temperatures and having a wavelength distribution identical to rapid fluorescence, arises from thermally activated return from the metastable state to the fluorescent state. Studies by Leoshin and Vinokurov[13] showed that the rates of decay of the delayed emission bands from fluorescein in boric acid glass at room temperature were identical, and thus a common state was indeed involved. Lewis, Lipkin, and Magel[6] termed these two emission bands α-phosphorescence (corresponding to delayed fluorescence) and β-phosphorescence. In 1961 Parker and Hatchard[14] reinvestigated the question of delayed luminescence, using a high-sensitivity spectrophosphorimeter. They found that the spectrum of eosin dissolved in deoxygenated glycerol and ethanol exhibited two long-lived emission bands at room temperature. One of these bands, having spectral characteristics identical to that of prompt fluorescence, was termed the eosin-type or E-type delayed fluorescence. The other band observed occurred in the extreme red and was assigned to a radiative $T_1 \rightarrow S_0$ transition (phosphorescence):

$$\begin{aligned}
&E \xrightarrow{h\nu} E^s \\
&E^s \longrightarrow E + h\nu' \text{ (prompt fluorescence)} \\
&E^s \rightsquigarrow E^t \\
&E^t \overset{\Delta}{\rightsquigarrow} E^s \\
&E^s \longrightarrow E + h\nu' \text{ (delayed fluorescence)} \\
&E^t \longrightarrow E + h\nu'' \text{ (phosphorescence)}
\end{aligned}$$

The first observations of P-type delayed fluorescence arose from the photoluminescence of organic vapors.[15] It was reported that phenanthrene, anthracene, perylene, and pyrene vapors all exhibited two-component emission spectra. One of these was found to have a short lifetime characteristic of prompt fluorescence while the other was much longer lived. For phenanthrene it was observed that the ratio of the intensity of the longer lived emission to that of the total emission increased with increasing phenanthrene vapor pressure. Excited dimers formed by reaction of singlet monomers with ground state molecules were proposed to explain the later emission:

$$P^s + P \rightarrow (P_2^s)^* \longrightarrow P^s + P \qquad (5.4)$$
$$\downarrow$$
$$P + h\nu$$

In a study of the fluorescence of concentrated pyrene solutions, Förster and Kasper[16] postulated that singlet dimers were responsible for a broad, structureless emission band occurring at longer wavelengths than the monomer emission. Stevens and Hutton[17] found that this emission was long lived and proposed the term excimer emission (from an excited dimer). However, Parker and Hatchard observed that the intensity of this delayed fluorescence was proportional to the *square* of the rate of light absorption and showed that it arose through collision of two triplet excited molecules, one being raised to the excited singlet level and emitting (fluorescence) at a rate equal to twice the rate of triplet decay. In this scheme singlet excited dimers may or may not be formed, although, with the exception of pyrene emission, the postulation of excimers was found not to be necessary to explain the long-lived emission. This emission has been termed the pyrene-type or P-type delayed fluorescence:

$$P^t + P^t \longrightarrow P^s + P \qquad (5.5)$$

$$P^s \longrightarrow P + h\nu \qquad (5.6)$$

Let us now return to the question of how E-type and P-type delayed fluorescence may be used to determine the triplet energy level. The efficiency of E-type delayed fluorescence is given by the following equation:

$$\Phi_e/\Phi_f = \tau_r \Phi_p A e^{-\Delta E/RT} \qquad (5.7)$$

where Φ_e is the quantum yield of delayed fluorescence, Φ_f is the quantum yield of prompt fluorescence, τ_r is the radiative lifetime of the triplet state, Φ_p is the quantum yield of phosphorescence, A is a frequency factor, and ΔE is the activation energy required to cause intersystem crossing from the triplet level back to the singlet. Since

$$\Phi_p = \Phi_{isc}\tau/\tau_r \qquad (5.8)$$

where τ is the actual triplet lifetime and Φ_{isc} is the quantum yield of triplet state population, this equation can be written as

$$\Phi_e/\tau\Phi_f = \Phi_{isc} A e^{-\Delta E/RT} \qquad (5.9)$$

Since Φ_{isc} should be independent of temperature in the temperature range used, a plot of $\ln(\Phi_e/\tau\Phi_f)$ vs. $1/T$ should be a straight line,

$$\text{slope} = -\Delta E/R \qquad (5.10)$$

Thus the triplet energy can be calculated from the singlet energy, determined from absorption and fluorescence measurements, and ΔE.

The P-type delayed fluorescence can also be used to determine triplet energy levels, although the results are considerably less accurate than those obtained using E-type delayed fluorescence. Generally the triplet energy of a particular compound is "bracketed" by using two compounds of known triplet energies—one to act as an energy donor to sensitize the P-type delayed fluorescence and the other to quench the delayed fluorescence. Thus the triplet energy of the compound of interest must lie between these two known triplet energies. Sensitization and quenching of the P-type delayed fluorescence has been used to determine the triplet energies of a number of aromatic hydrocarbons. Proflavine, acridine orange, and eosin were used as sensitizers.[19]

Instrumentation designed to measure fluorescence was discussed in Chapter 2. In order to distinguish between fluorescence and phosphorescence or delayed fluorescence, it is necessary to allow an interval between excitation and detection for prompt fluorescence to decay, leaving only the longer lived emission. This is accomplished by chopping the exciting and emission beams by a set of rotating disks with holes cut around the circumference or by using a rotating can with a slit cut in the side. A schematic diagram of a spectrophosphorimeter using chopper disks driven by synchronous motors is shown in Figure 5.3.[14,20] Here the two disks are rotated by separate synchronous

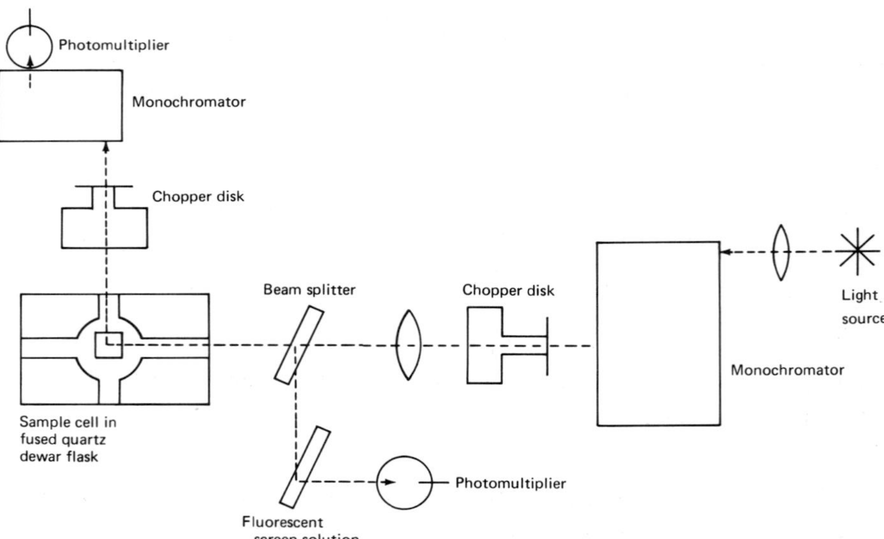

FIGURE 5.3. Diagram of a spectrophosphorimeter (after Parker and Hatchard[14]).

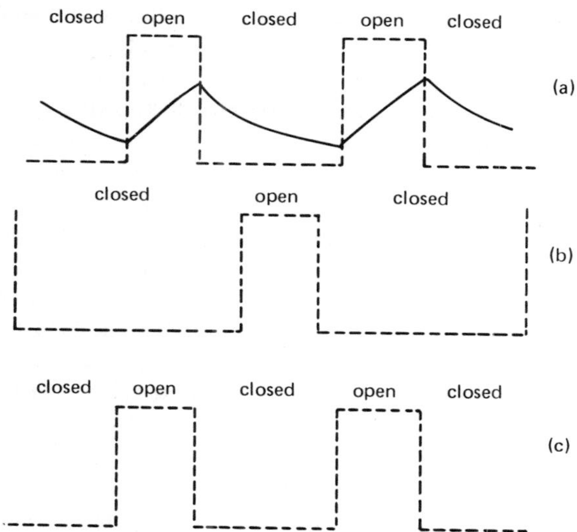

FIGURE 5.4. (a) First chopper, (b) second chopper (out of phase), (c) second chopper (in phase). (After Parker and Hatchard.[14])

motors, although both motors are run from the same power supply to maintain the phase relationship. The choppers can be run in or out of phase, allowing measurement of either total emission or of long-lived emission only, as shown in Figure 5.4.[20] In Figure 5.4(a) the solid line represents the rise and decay of the long-lived luminescence during light and dark periods, respectively. Rise and decay of prompt fluorescence would essentially correspond to the broken line in Figure 5.4(a). For applications in which the determination of the total phosphorescence intensity is important (to determine the relative intensities of fluorescence and phosphorescence, for example) it is necessary to know the phosphorimeter factor. This is given by the ratio of the observed intensity of phosphorescence with choppers out of phase to the total intensity of its emission (P_o/P). If the phosphorescence decay time is large compared to the light and dark periods of the two choppers, the intensity will be essentially constant during the entire cycle and the phosphorimeter factor is simply the ratio of "open" time for the second chopper (out of phase) to the time required for a complete cycle, $T_{2(o)}/T_c$. To obtain the total phosphorescence intensity, the observed intensity must be divided by the phosphorimeter factor. If, however, the luminescence lifetime is of the same order of magnitude as the light and dark periods, the emission will decay considerably during the dark cycle and the phosphorimeter factor will be less than $T_{2(o)}/T_c$. Parker has derived an equation (assuming the decay is exponential) and the interested reader is referred to his book.[20] Quantum efficiencies of phosphorescence or delayed fluorescence can be obtained in a manner identical to that discussed for prompt fluorescence in Chapter 2, by comparing the area under a corrected emission curve with that of a standard of known quantum yield (the phosphorimeter factor must of course be included). For the determination of absolute quantum yields of phosphorescence or for quantitative investigations of the efficiency of P-type delayed fluorescence (intensity dependent upon the square of the rate of light absorption) an accurate knowledge of the intensity of the exciting light is required. This is accomplished by focusing the light through an aperture onto the region of the sample cell from which the luminescence is observed. Ferrioxalate actinometry is used to measure the intensity of the exciting light, and from knowledge of the area of aperture and the optical density of the luminescent solution at the excitation wavelength the quantum yield of emission can be calculated.

Since triplet states are long lived and subject to quenching by impurities, sample purity standards for the determination of phosphorescence emission are very high. As we have seen, one of the most efficient triplet quenchers is oxygen. Therefore it is necessary to reduce the oxygen content of a sample to very low levels before attempting to measure its phosphorescence. This is achieved by several cycles of freezing, pumping under high vacuum, and thawing. In addition, special precautions must be taken to remove nonvolatile luminescent impurities or quenchers from all solvents to be used in a phosphorescence study. Methods for purification of some solvents commonly used in phosphorescence studies can be found in Ref. 20.

To minimize quenching of triplets due to bimolecular collisions, phosphorescence studies are generally carried out in rigid glasses at 77°K. Some materials which may be used to produce these glasses are listed in Table 5.1.[21,22]

5.2b. *Singlet → Triplet Absorption Spectra*

As stated in Chapter 1, transitions involving a change in multiplicity are spin forbidden. However, for reasons which we will consider later, such transitions do indeed occur although with very low transition probabilities in most cases. The intensity of an absorption corresponding to a transition from the ground state S_0 to the lowest triplet state T_1 is related to the triplet radiative lifetime $\tau_p{}^0$ by the following equation[23,24]:

$$\frac{1}{\tau_p{}^0} = \frac{g_l}{g_u} \frac{8\pi \times 2303 n_e{}^3}{c^2 N n_a} \frac{\int F(\nu_p)\, d\nu_p}{\int \nu_p^{-3} F(\nu_p)\, d\nu_p} \int \frac{\varepsilon_{st}\, d\nu_a}{\nu_a} \tag{5.11}$$

TABLE 5.1. *Solvents Yielding Clear, Rigid Glasses at 77°K*[a]

Solvent	Proportion, v/v	Viscosity, P
Pentane	—	—
2-Methylpentane	—	7×10^4 (96°K)
Ethyl ether	—	—
2-Bromobutane	—	—
Ethanol	—	1×10^{13} (78°K)
n-Propanol	—	1×10^{13} (100°K)
2-Methyltetrahydrofuran	—	4×10^7 (100°K)
1-Propanol/2-propanol	2:3	—
Ethanol/methanol[b]	4:1	—
Ethanol/methanol (+water)	1:1 (+4.5%)	—
Ethanol/methanol (+water)	1:1 (+9%)	—
Isooctane/isononane	1:1	—
Methylcyclohexane/decalin	1:1	1×10^{14} (122°K)
Methylcyclohexane/toluene	1:1	7×10^9 (114°K)
Methylcyclohexane/isohexanes	3:2	3×10^6 (102°K)
Methylcyclohexane/methylcyclopentane	1:1	2×10^5 (99°K)
Methylcyclohexane/isopentane[b]	4:1	—
Methylcyclohexane/isopentane	1:3	1×10^3 (90°K)
Ether/isopentane/ethanol[b]	5:5:2	9×10^3 (93°K)
Isopentane/methylcyclopentane/methylcyclohexane/ethyl bromide	7:7:4:1	—

[a] See Refs. 21, 22, and 159.
[b] These are the most frequently used mixtures.

where g_l and g_u are the multiplicities of the final and initial electronic states, respectively, c is the speed of light, N is Avogadro's number, n_a is the mean refractive index of the medium at the wavelength range of absorption, n_e is the mean refractive index of the medium at the wavelength range of emission, $\int F(\nu)\, d\nu$ is the integrated phosphorescence spectrum, ν_p is the frequency of the phosphorescence emission, ν_a is the frequency of the absorption, and ε_{st} is the molar extinction coefficient for the transition. Since the final state in this case is a triplet, $g_l/g_u = 1/3$. Abbreviating the third term in the equation by

$$\frac{\int F(\nu_p)\, d\nu_p}{\int \nu_p^{-3} F(\nu_p)\, d\nu_p} = \langle \nu_p^{-3} \rangle_{av}^{-1} \tag{5.12}$$

where $\langle \nu_p^{-3} \rangle_{av}^{-1}$ is the reciprocal of the mean value of ν_p^{-3} over the entire phosphorescence spectrum, the equation becomes

$$\frac{1}{\tau_p^0} = \frac{1}{3} \frac{8\pi \times 2303 n_e^3}{c^2 N n_a} \langle \nu_p^{-3} \rangle_{av}^{-1} \int \frac{\varepsilon_{st}\, d\nu_a}{\nu_a} \tag{5.13}$$

Since $k_{st} = 1/\tau_p{}^0$,

$$k_{st} = 9.6 \times 10^{-10} \frac{n_e{}^3}{n_a} \langle \nu_{\bar{p}}{}^{-3} \rangle_{av}^{-1} \int \frac{\varepsilon_{st}\, d\bar{\nu}_a}{\bar{\nu}_a} \tag{5.14}$$

where the equation has been changed from frequency units ν to wavenumbers $\bar{\nu}$. For a value of $k_{st} = 0.03\ \text{sec}^{-1}$, this equation yields $\varepsilon_{st} \sim 10^{-6}$. Thus it can be seen that attempts to measure this absorption without using fantastically long path lengths ($D = \varepsilon cd$) will be unsuccessful. However, various methods have been devised by which singlet–triplet absorption spectra can be determined. These methods will be the subject of this section.

One method for the observation of singlet–triplet absorption spectra, developed by Evans,[(25)] involves perturbation of the spin-forbidden transition by interaction in solution with certain paramagnetic molecules such as O_2 or NO under pressures of about 20 atm. Interactions between the solute and the paramagnetic molecules serve to decrease the degree of "forbiddenness" of spin-forbidden transitions, probably by a contact charge-transfer mechanism.[(26)] If an aromatic molecule (A) interacts with oxygen to yield a (contact) charge-transfer complex, several new energy levels are produced as shown in Figure 5.5.[(26a,27)] Enhanced singlet–triplet absorption of the aromatic molecule corresponds to a transition from the triplet ($^1A_0$, $^3O_2$) state to the triplet ($^3A_1$, $^3O_2$) state, a spin-allowed process since the states are of the same multiplicity. This transition gains intensity by interaction between the triplet ($^3A^+$, $^3O_2{}^-$), which in turn is interacting with the triplet ($^1A_1$, $^3O_2$) state. Thus what results is an indirect mixing of the triplet ($^1A_1$, $^3O_2$) and the ($^3A_1$, $^3O_2$) state through the intermediate ^{3}CT state.

Since the transition is thus mixed with the ^{3}CT state, the resulting $S_0 \rightarrow T_1$ absorption appears somewhat diffuse and is slightly red-shifted

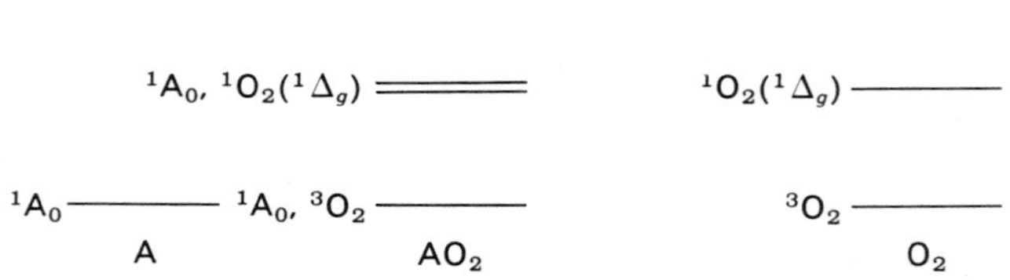

$^1A_1$ — $^1A_1$, $^3O_2$
(CT) $^2A^+$, $^2O_2{}^-$
$^3A_1$ — $^3A_1$, $^3O_2$
Σ
$^1A_0$, $^1O_2(^1\Sigma_g{}^+)$ — $^1O_2(^1\Sigma_g{}^+)$
$^1A_0$, $^1O_2(^1\Delta_g)$ — $^1O_2(^1\Delta_g)$
$^1A_0$ — $^1A_0$, $^3O_2$ — $^3O_2$
A — AO_2 — O_2

FIGURE 5.5. (Adapted from Kahn and Kearns.[(27b)])

relative to the unperturbed $S_0 \rightarrow T_1$ absorption. The magnitude of the shift depends upon the strength of the charge-transfer interaction and the degree of mixing of the CT energy levels with the other levels. For pyrazine a red shift of 40 cm^{-1} has been observed.[27a] Of course this shift should actually be taken into consideration when obtaining a triplet energy from the red edge (corresponding to the $0 \rightarrow 0$ transition) of an oxygen perturbed singlet–triplet absorption spectrum, but the difficulty involved in determining the position of the $0 \rightarrow 0$ transition from the diffuse spectrum probably involves more error than the slight red shift.

A second method of relaxing the spin restrictions on singlet–triplet transitions is by using heavy atoms (atoms of high atomic number). Substitution of a heavy atom directly onto an aromatic nucleus greatly enhances the intensity of the $S_0 \rightarrow T_1$ absorption by the *internal heavy-atom effect*. A less potent, but more desirable, method is to use a solvent containing heavy atoms (e.g., ethylene bromide, ethyl iodide). The enhancement of singlet–triplet absorption intensities using a heavy-atom solvent arises from the *external heavy-atom effect*. The mechanism of heavy-atom effects will be discussed in detail in a later section of this chapter.

The use of heavy atoms to obtain measurable singlet $\rightarrow$ triplet absorption spectra of aromatic molecules has been extensively studied by McGlynn and co-workers.[28] Absorption spectra of chloronaphthalene, anthracene, 9,10-dichloroanthracene, 9,10-dibromoanthracene, and tetracene were obtained using heavy-atom solvents. Some of the compounds studied contained heavy atoms substituted directly onto the aromatic nucleus. These compounds showed a significant singlet $\rightarrow$ triplet absorption even in light-atom solvents, such as hexane, due to the internal heavy-atom effect of the halogen substituent. A further enhancement of the absorption could be obtained, however, by using a solvent also containing heavy atoms such as ethyl iodide. The degree of enhancement of the singlet–triplet absorptions was found to depend upon the atomic number of the heavy atom present in the solvent. Those with high atomic numbers (e.g., iodine) were found to be more effective than chlorine, which has a lower atomic number. For example, the enhancements were found to be in the ratio of 97:9.7:1 for ethyl iodide, ethylene bromide, and carbon tetrachloride, respectively, relative to the magnitude of the singlet–triplet absorption in hexane. Similarly, the ratio of the singlet $\rightarrow$ triplet absorption intensity for anthracene determined in ethyl iodide and carbon disulfide was found to be 17:1. Similar results can also be obtained by using solid matrices of aromatic molecules in argon, krypton, or xenon at 4°K.[29]

In contrast to the technique of oxygen perturbation to obtain enhanced singlet–triplet absorption spectra, the use of heavy-atom perturbation results in no significant changes in the position or energy of the singlet $\rightarrow$ singlet absorption bands or in the position of the singlet → triplet absorption.[30,31] This is true regardless of how "heavy" the heavy atom may be, although the intensity of the transition is dependent upon the atomic number of the perturbing heavy atom. The failure of the absorption to shift in position as the degree of perturbation is increased (for example, in proceeding from carbon tetrachloride to ethyl iodide as solvent) would tend to indicate that charge-transfer interactions, similar to those proposed for oxygen perturbation, are not important in heavy-atom perturbation. However, as we shall see, other experimental results indicate that charge-transfer interactions may be important for heavy-atom effects as well.[32]

5.2c. *Phosphorescence Excitation Spectroscopy*

A third method suitable for determining the triplet energy of molecules is that using phosphorescence excitation spectroscopy.[33] In this technique the same apparatus is employed as is used for the determination of the phosphorescence spectrum (a phosphorimeter) except that the analytical monochromator (that through which the emission is passed) is set at a fixed wavelength (usually corresponding to the wavelength of maximum phosphorescence intensity) and the exciting wavelength is scanned in the region of singlet–triplet absorption via the excitation monochromator. Using an intense light source, the spectrum thus obtained corresponds to the direct singlet → triplet absorption. Since the instrumentation available for the detection of emission is much more sensitive than that for absorption, the phosphorescence excitation technique may be used for a wider range of compounds exhibiting only weak singlet → triplet absorption than is possible using singlet → triplet absorption. In addition, this technique is much less sensitive to impurities than the absorption technique.

For materials exhibiting extremely weak singlet → triplet absorption, external heavy-atom perturbation may be used to enhance the phosphorescence intensity (more light is absorbed, and thus a higher triplet population is obtained). Phosphorescence excitation may also be used for compounds with low phosphorescence quantum yields. Some compounds do not phosphoresce appreciably even at low temperatures although substantial triplet populations are known to be present upon excitation. By doping crystals with a phosphorescence material of lower triplet energy, such as naphthalene, emission is observed by transfer of triplet energy. Since, under the proper conditions, light is absorbed only by the compound of interest in the region of its singlet → triplet absorption the spectrum of the nonphosphorescent triplet energy donor can be obtained by monitoring the emission of the phosphorescent triplet energy acceptor as a function of the excitation wavelength:

$$\begin{array}{rcll} \mathrm{C} & \xrightarrow{h\nu} & \mathrm{C}^t & \text{excitation} \\ \mathrm{C}^t + \mathrm{N} & \longrightarrow & \mathrm{C} + \mathrm{N}^t & \text{energy transfer} \\ \mathrm{N}^t & \longrightarrow & \mathrm{N} + h\nu' & \text{emission} \end{array}$$

5.2d. *Flash Photolysis*

It should be noted that the preceding techniques for determining triplet energies can be applied only to the lowest triplet (T_1) or to any higher lying triplet states between T_1 and the first excited singlet state S_1, since (a) the higher triplet levels do not phosphoresce and (b) absorption measurements are complicated in the region of S_1 and above due to the much higher intensities of singlet–singlet absorptions. The energies of higher triplet states can be determined, however, by the technique of flash photolysis. This technique and some results which can be obtained by flash photolysis were discussed previously in Chapter 3.

As seen in Chapter 3, the flash photolysis technique involves the rapid discharge of a condenser bank through a discharge lamp to produce an intense flash of light of a few microseconds duration. The intense flash converts a large proportion of the ground state molecules into the first excited singlet level, from which these molecules can either return to the ground state by various processes (fluorescence or internal conversion) or cross over to the triplet manifold. The number of molecules undergoing this latter process depends upon the intersystem crossing efficiency of the particular molecule in question Φ_{isc}. A relatively high concentration of molecules frequently end up in the lowest triplet level. A second flash, called the spectroscopic flash, fired at a preset interval after the initial flash, serves to excite those molecules in the triplet level to higher triplet levels through spin-allowed transitions (this involves no change in spin multiplicity). The transient absorption spectra are recorded photographically with a spectrograph[(24)]:

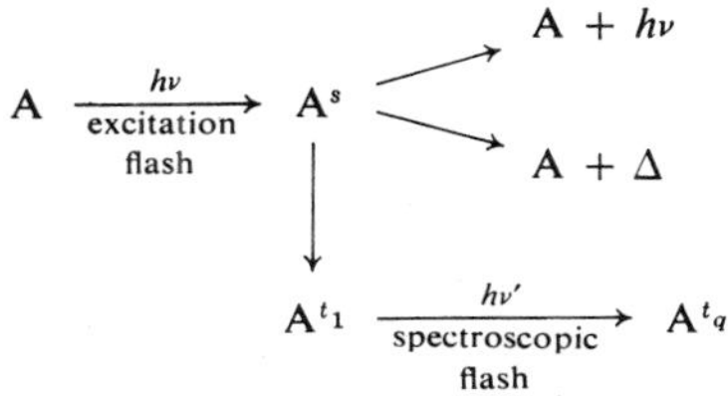

Flash photolysis has been used to study the triplet–triplet absorption spectra ($T^1 \rightarrow T^q$) of a number of aromatic molecules both in solution and in the gas phase.[(34)] A disadvantage associated with the flash technique for obtaining triplet–triplet absorption spectra is that the transient absorption can occur for only a short time after the initial flash (determined by the lifetime of triplet level) and the time available for recording of the spectra is limited to the length of the spectroscopic flash. Thus the spectra can be recorded photographically but not with a scanning photoelectric detector. A second method of observing triplet–triplet absorption spectra, developed by McClure,[35] involves continuous irradiation of the sample dissolved in a glass at 77°K to produce a photostationary triplet population. The absorption spectrum (triplet–triplet) is monitored by a light beam at right angles to the other source. Although much higher triplet populations can be obtained by flash photolysis, the photostationary technique offers the advantage that the strength of the triplet–triplet absorption remains constant with time and the spectrum can be recorded with a scanning photoelectric detector. Both techniques, however, until recently have been subject to instrumental limitations due to the relative insensitivity of photographic and photoelectric detection devices in the near infrared. The development of red-sensitive photomultipliers has extended the photoelectric detection limit down to $\bar{\nu} = 11{,}000\ \text{cm}^{-1}$, permitting the observation of triplet–triplet transitions between 11,000 and 16,000 cm^{-1} for a number of molecules.

Once the triplet–triplet absorption spectra have been recorded and the transition energies determined, the problem is to determine the transition to which the absorption actually corresponds ($T_1 \rightarrow T_2$ or T_3 or T_4, etc). Although triplet–triplet absorptions are not subject to multiplicity selection rules, they are subject to parity and sign selection rules. The energies of the π-electronic states of a number of polyacenes have been calculated by Pariser using the LCAO (linear combination of atomic orbitals) method including configuration interaction.[36] Since triplet–triplet absorptions are subject to some selection rules, all transitions do not occur with equal intensity. By comparing calculations of the transition energies and oscillator strengths with the experimentally determined data, it is often possible to assign the observed transitions.[34,37]

5.2e. *Electron Excitation*

A further technique exists for the determination of triplet energy levels. This technique, called electron impact spectroscopy, involves the use of inelastic scattering of low-energy electrons by collision with molecules. The inelastic collisions of the electrons with the molecules result in transfer of the electron energy to the molecule and the consequent excitation of the latter. Unlike electronic excitation by photons, excitation by electron impact is subject to no spin selection rule. Thus transitions that are spin and/or orbitally "forbidden" for photon excitation are totally allowed for electron impact excitation.

Inelastic scattering of low-energy electrons by molecules has been recognized as a tool for studying electronic transitions in atoms since 1914.[38] However, it was not until 1959 that the principle was applied to determine the electronic energy levels of molecules.[39] In this early work, however, the high incident electron energies used prohibited the observation of (photon) spin-forbidden transitions.[40] The first workers to develop the technique of low-energy electron impact spectroscopy specially for the observation of spin-forbidden transitions were Kuppermann and Raff, who reported obtaining an energy of 4.6 eV for the $S_0 \rightarrow T_1$ transition in ethylene.

An apparatus designed by Kuppermann, Rice, and Trajmar[42] for investigating optically allowed and forbidden transitions is diagrammed in Figure 5.6. The apparatus includes an electron gun, two hemispherical electrostatic energy selectors, a scattering chamber, and a detector and recorder system. The electron gun, consisting of a thermionic cathode and a two-stage lens system, produces a well-collimated beam of electrons of about 20 eV. This electron beam enters into the monochromator where the electrons are dispersed according to their kinetic energy. The electrons emerge from the monochromator with a narrow energy range and are accelerated to the proper impact energy and focused into the scattering chamber. The scattering chamber consists of two concentric, flexible bellows which permit a variation

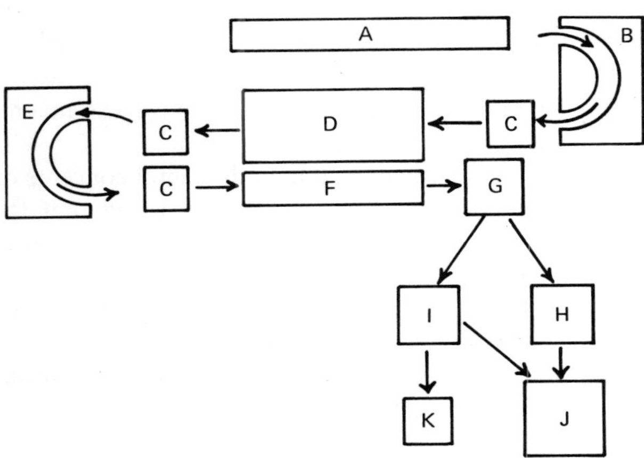

FIGURE 5.6. Diagram of a low-energy, high-angle electron-impact spectrometer. (A) Electron gun; (B) monochromator (180° spherical electrostatic energy selector); (C) electron optics; (D) scattering chamber; (E) analyzer (180° spherical electrostatic energy selector); (F) electron multiplier; (G) amplifier and pulse discriminator; (H) count-rate meter; (I) multichannel scaler; (J) *X–Y* recorder; (K) digital recorder. (After Kupperman *et al.*[42])

of the scattering angle from $-30°$ to $+90°$. The chamber can be temperature-controlled by passing a heating or cooling fluid into the space between the concentric bellows. The gas sample to be studied is introduced into the inner bellows at a pressure of about 10^{-3} Torr. The intensity of the incident electron beam is measured by a faraday cup. The entire scattering apparatus (including the scattering chamber) is housed in a surrounding vacuum chamber maintained at about 10^{-7} Torr. The scattered electrons enter into the analyzer monochromator, which can be tuned to transmit electrons of a particular kinetic energy and are detected by an electron multiplier. Signals from the electron multiplier are fed into a pulse amplifier and discriminator, which rejects low-intensity pulses (noise) and high-intensity pulses due to cosmic rays. The scattered signal intensity is then measured either with the count-rate meter or the multichannel scaler and recorded. The electron gun, monochromator, and scattering chamber optics are mounted on a gear wheel which can be rotated from outside the vacuum chamber to permit scanning of the scattering angle. The spectrum of a material can be determined by the energy lost from the electrons through inelastic collisions with the molecules. The lens potentials at the entrance and exit of the analyzer monochromator and the analyzer itself ride on top of a sweep voltage V applied between these elements and the scattering chamber. When this sweep voltage is zero, electrons having the same energy as the incident beam (from the monochromator) are transmitted to the detector. As the sweep voltage is increased (V_1 volts) electrons that have lost kinetic energy through inelastic collisions with the molecules ($\Delta E = eV_1$) are passed through the analyzer. Thus the signal output of the detector system as a function of the magnitude of the sweep voltage yields the energy loss spectrum.

Although a detailed treatment of the fundamental concepts of collisional processes is outside the scope of this text, it is hoped that the following discussion will suffice to give the reader an elementary understanding of what is occurring in electron impact spectroscopy.

When an electron beam of initial intensity I_0 passes through a gas, a loss of electrons due to elastic and inelastic collisions with the gas molecules occurs. For a concentration of gas particles c the remaining beam intensity I having traveled a distance l through the gas is

$$I = I_0 e^{-Qlc} \tag{5.15}$$

where Q is the total collision cross section (Q can be compared to ε, the absorption cross section in Beer's law). Collision of an electron with a molecule can cause the molecule to become excited or ionized. The total cross section is the sum of the cross sections for each of these processes. Electrons are scattered out of the beam anisotropically and the cross section for

TABLE 5.2. *Calculated[a] and Experimental Singlet–Singlet and Singlet–Triplet Transitions[b] for Benzene*

Transition	$S_0 \rightarrow S_x$			$S_0 \rightarrow T_x$	
x	ES[c]	Optical	Theory	ES[c]	Theory
1	5.0	4.74	4.76	3.9	3.92
2	6.2	6.02	6.05	4.7	4.67
3	6.9	6.85	6.79	5.6	5.12
4	—	—	—	—	6.72

[a] Unpublished calculations of D. O. Cowan using the Pariser–Parr–Pople (PPP) method. Parameters were those suggested by Zahradnik[(100)] and are not the same for $S \rightarrow S$ and $S \rightarrow T$ transitions.
[b] All values given in eV.
[c] ES, electron scattering.

excitation of the molecule to upper electronic states can be expressed as

$$Q_n = \int \sigma_n \sin \theta \, d\theta \, d\phi \qquad (5.16)$$

where σ_n is the differential cross section for excitation to the nth excited state and $\sin \theta \, d\theta \, d\phi$ is the solid angle into which the electrons are scattered. By comparison of expressions for the differential cross sections of a spin-allowed transition and spin-forbidden transition,[(43)] it can be shown that the scattered current at an energy loss corresponding to the latter transition falls off much less rapidly with angle than the former. In addition, the scattered current at energy loss corresponding to a spin-forbidden transition increases dramatically with a decrease in the incident electron energy. Thus at large scattering angles and low incident intensities singlet → triplet transitions are favored as compared to optically allowed transitions. By determining the intensity of the scattered electron signal as a function of the scattering angle, it is possible to determine the multiplicity of the transition.

The energies of the six lowest excited states of benzene have been determined by electron impact spectroscopy by Doering[(44)] using incident electron energies of 13.6 and 20.0 eV and scattering angles from 9° to 80°. The results of this study together with the theoretical energies of the triplet levels are presented in Table 5.2.

5.2f. *The Lowest Triplet Levels of Organic Molecules*

The lowest triplet energies of a number of organic molecules, determined by the various techniques described in this section, are presented in Tables 5.3–5.5.

TABLE 5.3. *Lowest Triplet Energies of Some Carbonyl Compounds*(30,45a,b,51)

Compound[a]	Triplet energy,[b] kcal/mole		
	Hydrocarbon[c]	Alcohol[d]	Other
Propiophenone	74.6	—	—
Cyclopropyl phenyl ketone	74.3	—	—
Xanthone	74.2	—	—
Acetophenone	73.6	74.1	74.2[e]
1,3,5-Triacetylbenzene	73.3	—	—
4-Methylphenyl cyclopropyl ketone	73.2	—	—
Isobutyrophenone	73.1	—	—
4-Methylacetophenone	73.0	72.8	72.8
1,3-Diphenyl-2-propanone	72.2	—	—
4-Chloroacetophenone	72.1	71.9	—
Benzaldehyde	71.9	—	—
3-Trifluoromethylacetophenone	71.6	—	—
4-Methoxyacetophenone	—	71.7	71.8[e]
4-Bromoacetophenone	71.2	71.4	—
4-Hydroxyacetophenone	—	71.3	—
3-Acetylpyridine	71.1	73.0	73.1[e]
4-Trifluoromethylacetophenone	70.9	71.4	—
Triphenylmethyl phenyl ketone	70.8	—	—
α,α,α-Trifluoroacetophenone	70.7	—	—
3-Cyanoacetophenone	70.4	73.3	73.8[e]
α-Chloroacetophenone	70.1	—	—
2-Acetylpyridine	70.0	70.9	—
4,4′-Dimethoxybenzophenone	69.4	69.7	69.8[e]
4-Acetylpyridine	69.5	70.3	—
4-Cyanoacetophenone	—	69.5	69.6[e]
2-Methylbenzophenone	69.2	—	—
4,4′-Dimethylbenzophenone	68.9	69.3	—
4,4′-Diphenylcyclohexadienone	—	—	68.8[e]
2-Benzoylbenzophenone	68.7	—	—
4-Methoxybenzophenone	68.6	69.0	—
4-Methylbenzophenone	68.6	69.1	—
Benzophenone	68.6	69.1	69.2[e]
3-Benzoylpyridine	68.4	69.0	—
3-Methoxybenzophenone	68.4	—	—
3-Chlorobenzophenone	68.4	68.9	—
3,3′-Bis(trifluoromethyl)-benzophenone	68.3	69.2	—
4-Chlorobenzophenone	68.3	68.8	—
4-Fluorobenzophenone	68.9	69.7	—
4,4′-Dichlorobenzophenone	68.0	—	—
4-Acetylacetophenone	67.7	—	—
4-Trifluoromethylbenzophenone	67.6	68.1	—
4,4′-Bis(trifluoromethyl)-benzophenone	67.2	67.9	—
4-Benzoylpyridine	67.1	67.2	—
9-Benzoylfluorene	66.8	—	—

TABLE 5.3.—*Continued*

Compound[a]	Triplet energy,[b] kcal/mole		
	Hydrocarbon[c]	Alcohol[d]	Other
2-Benzoylpyridine	66.7	67.4	—
4-Cyanobenzophenone	66.4	—	—
Phenylglyoxal	62.5	—	—
Benzoylformamide	62.1	63.1	—
Ethyl phenylglyoxalate	61.9	63.0[e]	—
Anthraquinone	62.4	63.3	—
α-Naphthoflavone	62.2	—	—
Flavone	62.0	—	—
Benzoylformic acid	61.3	60.8	—
Michler's ketone	61.0	62.0	—
β-Naphthylphenyl ketone	59.6	—	—
β-Naphthaldehyde	59.5	—	—
Benzoylformylchloride	59.4	—	—
β-Acetonaphthone	59.3	59.5	—
α-Naphthoic acid	58.2	—	—
α-Naphthyl phenyl ketone	57.5	57.7	—
α-Acetonaphthone	56.4	58.0	—
α-Naphthaldehyde	56.3	56.3	—
Biacetyl	54.9	—	57.2[e]
2,3-Pentanedione	54.7	—	—
Benzil	53.7	—	—
Fluorenone	53.3	—	—

[a] A more extensive table can be found in the review by Arnold.[(45b)]
[b] Defined as the maximum of the 0–0 band when possible.
[c] Methylcyclohexane/isopentane (4:1).
[d] Ethanol/methanol (4:1).
[e] Ether/pentane/alcohol (5:5:2).

5.3. *Determination of the Efficiency of Intersystem Crossing*

As can be seen in Figure 5.7, there are in principle two modes for the population of the triplet manifold. The first, direct singlet → triplet absorption, is a multiplicity forbidden process and hence is not very productive for most organic molecules. We saw in the previous section, however, that under certain special conditions, such as oxygen or heavy-atom perturbation, respectable triplet populations can result from this process. The second process for populating the triplet state is by intersystem crossing from the singlet manifold. For most molecules this process is thought to occur primarily from the lowest excited singlet state. Although intersystem crossing involves a change in multiplicity and is formally also a forbidden process, for reasons discussed in a later section of this chapter, the efficiency of this process

TABLE 5.4. *Lowest Triplet Energies of Various Hydrocarbons*

Compound	E_t, kcal/mole	Ref.
Benzene	83.9	30
Toluene	82.2	30
Biphenyl	~70, 65.6[a]	50, 45*a*
Fluorene	67.6[a]	45*a*
Triphenylene	66.6,[a] 67.2[b]	45*a*
Diphenylacetylene	62.5	47
Phenanthrene	62.2,[a] 61.8[b]	45*a*
Styrene	61.5[a]	45*a*, 47
Naphthalene	60.9,[a] 61[b]	45*a*
1-Methylnaphthalene	60.1	45*a*
Isoprene	60.1	46
trans-1,3-Pentadiene	58.8	46
Cyclopentadiene	58.3	46
cis-Stilbene	57	48
cis-1,3-Pentadiene	56.9	46
Chrysene	56.6[a]	45*a*
Coronene	54.5	30
1,3-Cyclohexadiene	53.5, 52.5	46
1,2,5,6-Dibenzanthracene	52.2[b]	45*a*
1,2,3,4-Dibenzanthracene	50.8[b]	45*a*
trans-Stilbene	49	48
Pyrene	48.7[a]	45*a*
trans,trans-1,3,5-Hexatriene	47.5	46
Anthracene	42.0,[a] 42.6[b]	45*a*, 46
	42.5	
1,3,5,7-Octatetraene (all *trans*)	39.0	46
Azulene	31–39	49
Naphthacene	29.3	49

[a] Determined in a hydrocarbon glass (methylcyclohexane–isopentane) at 77°K.
[b] Determined in EPA at 77°K.

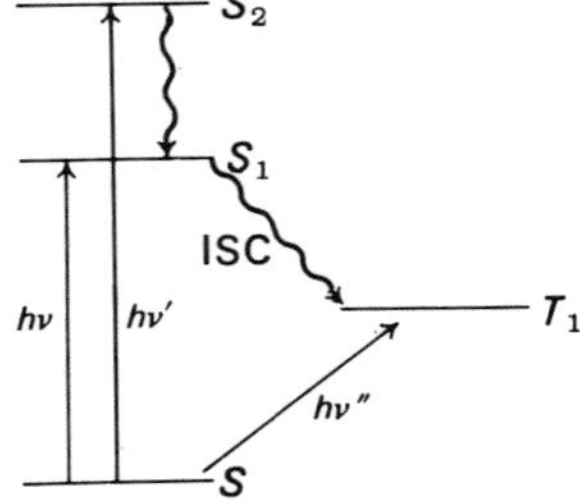

FIGURE 5.7. Modes for triplet state population.

TABLE 5.5. *Lowest Triplet Energies of Other Organic Molecules*

Compound	E_t, kcal/mole	Ref.
Phenol	81.6	30
Anisole	80.5	30
Aniline	76.5	30
Carbazole	70.1,[a] 70[b]	45
Diphenylene oxide	70.1[a]	45
Triphenylamine	70.1,[a] 70.1[b]	45
Dibenzothiophene	69.7,[a] 69.3[b]	45
Thioxanthone	65.5[a]	45
2-Triphenylene sulfonic acid sodium salt	65.0	45
Quinoline	62.8	51
Quinolinium ion	61.6	51
β-Naphthylammonium ion	60.8	51
N,*N*-Dimethyl-β-naphthyl-ammonium ion	60.8	51
β-Naphthol	60.2	51
β-Bromonaphthalene	60.2	7
2,6-Naphthalene disulfonic acid disodium salt	60.0[b]	45
α-Fluoronaphthalene	59.9	30
β-Chloronaphthalene	59.9	7
β-Idonaphthalene	59.9	7
β-Naphthylamine	59.6	51
2,4-Hexadien-1-ol	59.5	47
α-Naphthoate ion	59.3	51
α-Chloronaphthalene	59.1	7
α-Bromonaphthalene	59.1	7
Chloroprene	58.6	46
α-Iodonaphthalene	58.5	52
α-Naphthol	58.5	7
β-Naphtholate	58.2	51
N,*N*-Dimethyl-β-Naphthyl-amine	57.9	51
β-Naphthoic acid	57.6	51
β-Naphthoate ion	57.6	51
1-Chlorobutadiene	57.4	47
1-Methoxybutadiene	56.6	47
α-Nitronaphthalene	54.8	7
α-Naphthylamine	54.2	30
Acridine	45.3[a]	45

[a] Determined in a hydrocarbon glass (methylcyclohexane–isopentane) at 77°K.

[b] Determined in EPA at 77°K.

is considerably greater than direct singlet → triplet absorption. Hence for most molecules this is the primary mechanism for triplet population.

The efficiency or quantum yield of intersystem crossing is expressed by the following equation:

$$\Phi_{isc} = \frac{k_{isc}}{(k_{isc} + k_f + k_r)} \tag{5.17}$$

where k_{isc} is the rate constant for intersystem crossing, k_f is the fluorescence rate constant, and k_r is the rate constant for radiationless internal conversion from the lowest excited singlet state to the ground state. In this section we will discuss various methods for the experimental determination of Φ_{isc}.

5.3a. *Flash Photolysis*

Intersystem crossing quantum yields can be determined by the flash photolysis technique in several different ways. The accuracy of each method depends to a large extent upon how much is known about the particular molecule under study.

If a very dilute solution ($[A] \approx 10^{-7}\ M$) is exposed to a flash of high intensity (~1000 J) from a flash lamp, *all* of the ground state molecules are excited to their upper singlet levels. After a time τ corresponding to the lifetime of the lowest excited singlet state some of the molecules will have returned to the ground state via fluorescence and internal conversion processes. Others, however, will be in the lowest triplet state through intersystem crossing. The fraction of molecules in the triplet state can be determined by the loss in optical density of the solution as shown below. The optical density of the solution before the flash is given by

$$OD_i = \varepsilon_s[A]_i d \tag{5.18}$$

where ε_s is the molar absorptivity for singlet → singlet transitions ($S_0 \to S_x$) and d is the path length of the photolysis cell. At a time T (greater than the singlet lifetime τ) after the flash the OD of the solution is given by

$$OD_T = \varepsilon_s[A]_T d \tag{5.19}$$

Since the change in OD is due to some molecules being in the triplet state,

$$\Delta OD = \varepsilon_s \Delta[A]d = \varepsilon_s[A^t]_T d \tag{5.20}$$

Thus from the experimentally determined ΔOD and a knowledge of ε_s at the wavelength of excitation the concentration of triplet molecules $[A^t]_T$ can be determined. By measuring ΔOD as a function of time T after the flash and extrapolating the data to $T = 0$ one obtains $[A^t]$. Since all the ground state molecules were initially excited, the quantum yield of triplet formation is simply

$$\Phi_t = \Phi_{isc} = [A^t]/[A]_i \tag{5.21}$$

In practice, however, the situation is more complicated since some of the triplets may be excited to higher triplet levels ($T_1 \rightarrow T_x$) by the same wavelengths of light used to determine ΔOD. This results in a $[A^t]_T$ less than the actual concentration depending upon the magnitude of the triplet molar absorptivity ε_t. If ε_t is known, a correction for $T_1 \rightarrow T_x$ absorption can be made[53]:

$$\Delta OD = (\varepsilon_s - \varepsilon_t)[A^t]_T d \tag{5.22}$$

Alternatively, one can determine the strength of the $T_0 \rightarrow T_x$ absorption in a wavelength region free of $S_0 \rightarrow S_x$ absorption as a function of time after the flash and extrapolate to $T = 0$. This yields the quantity $\varepsilon_t[A^t]$. Therefore

$$\varepsilon_t \Phi_{isc} = \varepsilon_t [A^t]/[A]_i \tag{5.23}$$

and Φ_{isc} can be determined from a knowledge of ε_t.

If ε_t is not known, it is possible to obtain more accurate values of Φ_{isc} than above by using a lower flash intensity such that all the molecules are not excited during the flash ($I_0 \simeq$ 20–100 J). For this method the intensity of the light absorbed I_a must be accurately determined from the absorption spectrum and the incident light intensity I_0 determined by actinometry. The concentration of triplet molecules $[A^t]$ can be determined from $\Delta[A^s]$ as above. Since I_a and $[A^t]$ are smaller than in the previous case, errors due to the underlying $T_0 \rightarrow T_x$ absorption are reduced. The quantum yield of triplet formation is now

$$\Phi_t = \Phi_{isc} = [A^t]/I_a \tag{5.24}$$

In both cases discussed above determinations should be made at several different flash intensities and the results extrapolated to $I = 0$ to eliminate errors due to bimolecular triplet processes arising from the high triplet populations.

Midinger and Wilkinson[54] have used flash photolysis and fluorescence quenching by heavy atoms to determine the intersystem crossing efficiencies of anthracene and a number of its derivatives. As discussed in Section 5.2b, heavy atoms present as molecular substituents or in the solvent serve to promote multiplicity forbidden transitions. When anthracene is excited the following processes can occur:

	rate
$A \xrightarrow{h\nu} A^s$	I_a
$A^s \longrightarrow A + h\nu'$	$k_f[A^s]$
$A^s \longrightarrow A$	$k_{ic}[A^s]$
$A^s \longrightarrow A^t$	$k_{isc}[A^s]$

Thus we see that we have three processes which can compete for deactivation of the excited singlet: fluorescence, internal conversion, and intersystem crossing. If we increase the rate of the latter by adding a heavy atom, this should result in a decrease or quenching of the fluorescence intensity:

$$A^s + Q \longrightarrow A^t + Q \qquad k_q[A^s][Q] \tag{5.25}$$

$$A^s + Q \longrightarrow A + Q \qquad k_q'[A^s][Q] \tag{5.26}$$

The ratio of the fluorescence quantum efficiency in the absence of the heavy atom to that in the presence of the heavy atom is given by

$$\Phi_f{}^0/\Phi_f = 1 + (k_q + k_q')[Q]\tau_s \tag{5.27}$$

where τ_s is the singlet state lifetime $[=1/(k_f + k_{ic} + k_{isc})]$ in the absence of quencher. The ratio of the triplet population (just after a short flash of light) in the presence and absence of heavy atoms, however, is given by

$$\frac{[A^t]}{[A^t]^0} = \frac{1 + k_q/k_{isc}[Q]}{1 + (k_q + k_q')[Q]\tau_s} \tag{5.28}$$

Combining these two equations, one obtains

$$\frac{\Phi_f{}^0}{\Phi_f} = \left(\frac{[A^t]\Phi_f{}^0}{[A^t]^0\Phi_f} - 1\right)\left(1 + \frac{k_q'}{k_q}\right)\Phi_{isc}^0 + 1 \tag{5.29}$$

where $\Phi_{isc}^0 = k_{isc}\tau$ $[=k_{isc}/(k_{isc} + k_f + k_{ic})]$. This equation states that plots of $\Phi_f{}^0/\Phi_f$ vs. $([A^t]\Phi_f{}^0/[A^t]^0\Phi_f) - 1$ should be linear with a slope equal to $(1 + k_q'/k_q)\Phi_{isc}^0$. Assuming that k_q' is negligible in comparison to k_q, the slope is simply equal to Φ_{isc}^0. As we have seen, relative fluorescence quantum yields can be obtained fairly easily using a standard fluorimeter and relative triplet populations can be determined by flash photolysis by measuring $T_1 \rightarrow T_x$ absorption. By determining these values at several different concentrations of heavy atoms (bromobenzene was used as heavy-atom quencher), Medinger and Wilkinson were able to show that the quenching of fluorescence observed in the presence of heavy atoms was indeed due to an increased efficiency of intersystem crossing and were able to obtain the intersystem crossing efficiencies for anthracene and several of its derivatives listed in Table 5.7.

5.3b. *Triplet-Sensitized Isomerization*

A method for the determination of intersystem crossing quantum yields involving triplet-sensitized isomerization has been developed by Lamola and Hammond.[55] In this method the compound whose intersystem crossing quantum yield is desired is used as a sensitizer (energy supplier) for a photochemical reaction which is known to occur via the energy acceptor's triplet state. The process by which energy is transferred from one triplet state (donor) to another compound (acceptor) causing the latter to be raised to its lowest triplet level is called *energy transfer*. Mechanisms of energy transfer and factors which determine the efficiency of this process will be discussed in detail in Chapter 6. For our purposes in this section it will suffice to state that under certain conditions the efficiency of transfer from the donor to the acceptor molecule can be 100%. By measuring the quantum yield of the triplet reaction of the acceptor under conditions where energy transfer is 100% efficient (every donor triplet transfers its energy to an acceptor molecule), the intersystem quantum yield of the donor can be determined since

$$\Phi_a = \alpha\Phi_{isc} \tag{5.30}$$

where Φ_a is the measured quantum yield of the induced acceptor photochemistry, α is the efficiency of product formation from the acceptor triplet, and Φ_{isc} is the intersystem crossing efficiency of the donor molecule. The quantity α can be determined by using a donor molecule whose intersystem crossing efficiency is known to be unity. The processes involved in this method are shown schematically in Figure 5.8.

Although essentially any acceptor triplet reaction can potentially be used for determining Φ_{isc} of the donor if its quantum yield is accurately known, some precautions must be observed in order to assure success. First, the lowest excited singlet level of the acceptor should be above that of the donor to eliminate the possibility of singlet energy transfer. Second, the acceptor should be chosen such that the *donor* molecules absorb all of the incident light.

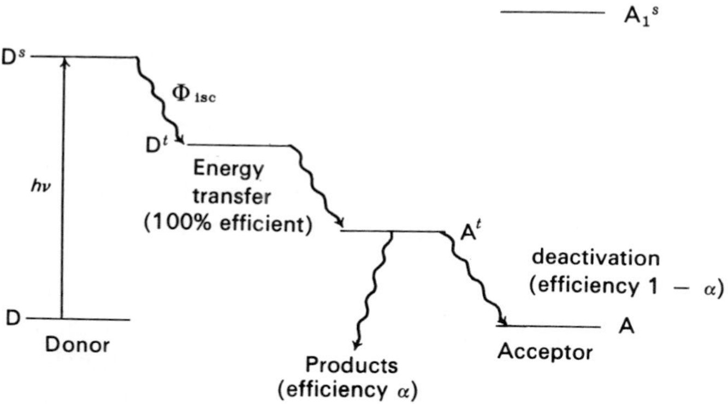

FIGURE 5.8. Processes involved in the determination of intersystem crossing efficiencies by energy transfer.

The acceptor originally chosen by Hammond and Lamola was *cis*-piperylene since the quantum yield of *cis–trans* isomerization from the triplet was well studied and the isomerization could be conveniently followed using vapor-phase chromatographic techniques. Subsequent work showed, however, that conjugated dienes such as piperylene can quench lower energy donor singlet states. Although no chemistry arises from this quenching process, the results obtained for the donor Φ_{isc} under conditions of singlet quenching are necessarily low. At concentrations of piperylene commonly used in these determinations (0.05 M) only very long-lived singlets are significantly quenched. Singlet quenching at higher concentrations can be corrected for by measuring the intersystem crossing efficiency of the donor at several different acceptor concentrations and extrapolating the data to zero piperylene concentration.[56]

A similar method for determining intersystem crossing efficiencies has been developed by Parker and Joyce[57] using acceptor delayed fluorescence (P-type, see Section 5.2a). The processes involved in this method are

$$
\begin{array}{rcll}
\mathrm{D} & \xrightarrow{h\nu} & \mathrm{D}^s & \text{efficiency} = I_a \\
\mathrm{D}^s & \longrightarrow & \mathrm{D}^t & \text{efficiency} = \Phi_{\text{isc}} \\
\mathrm{D}^t + \mathrm{A} & \longrightarrow & \mathrm{A}^t + \mathrm{D} & \text{efficiency} = \Phi_{\text{et}} \\
\mathrm{A}^t + \mathrm{A}^t & \longrightarrow & \mathrm{A}^s + \mathrm{A} & \text{efficiency} = \Phi_{\text{TT}} \\
\mathrm{A}^s & \longrightarrow & \mathrm{A} + h\nu' & \text{efficiency} = \Phi_f
\end{array}
$$

For two solutions containing the same acceptor but different donor molecules the ratio of the delayed fluorescence intensities is given by

$$\frac{(I_{\text{DF}})_1}{(I_{\text{DF}})_2} = \frac{(\Phi_{\text{et}}\Phi_{\text{isc}})_1}{(\Phi_{\text{et}}\Phi_{\text{isc}})_2} \tag{5.31}$$

If conditions are such that energy transfer from both donors occurs with 100% efficiency, the relative Φ_{isc} values can be obtained by comparing the intensities of the delayed fluorescence of the acceptor. If one Φ_{isc} is known, the other can be directly determined.

A second method involving luminescence measurements was developed by Sandros.[58] In this technique donor triplet states are used to populate biacetyl triplets, which then phosphoresce in an inert, degassed fluid environment. If the lowest singlet and triplet levels of the acceptor (biacetyl) are lower than the corresponding singlet and triplet levels of the donor, both singlet and triplet energy transfer will occur. The biacetyl fluorescence quantum yield, however, is very small and nearly 100% of the singlet excited molecules cross over to the triplet state. Thus energy received by a biacetyl singlet state will also appear as phosphorescence:

(a)	$D + h\nu \longrightarrow D^s$		I_a
(b)	$D^s \longrightarrow D + h\nu'$		k_f
(c)	$D^s \longrightarrow D$		k_{ic}
(d)	$D^s \longrightarrow D^t$		k_{isc}
(e)	$D^s + A \longrightarrow D + A^s$		k_s
(f)	$D^t + A \longrightarrow D + A^t$		k_t
(g)	$D^t \longrightarrow D$		k_d
(h)	$A^s \longrightarrow A^t$		k'_{isc}
(i)	$A^t \longrightarrow A + h\nu''$		k_p
(j)	$A^t \longrightarrow A$		k_d'

D = sensitizer, A = biacetyl

The biacetyl phosphorescence intensity I is given by the following equation:

$$I_p = \alpha k_p[A^t] \tag{5.32}$$

where α is an instrumental factor. The biacetyl triplets are derived from two processes: (e) followed by (h), and (d) followed by (f):

$$\text{(e), (h) mechanism:} \qquad [A^t] = I_a \Phi^s_{et} \Phi'_{isc} \tau_p \tag{5.33}$$

where Φ^s_{et} is the quantum yield for singlet energy transfer, τ_p is the lifetime of biacetyl triplet states, and Φ'_{isc} is the intersystem crossing quantum yield of biacetyl, $\Phi'_{isc} = 1$. We have

$$\text{(e), (h):} \qquad [A^t] = I_a \tau_p \frac{k_s[A]}{k_f + k_{ic} + k_{isc} + k_s[A]} \tag{5.34}$$

$$\text{(d), (f):} \qquad [A^t] = I_a \Phi_{isc} \Phi^t_{et} \tau_p \tag{5.35}$$

$$[A^t] = I_a \tau_p \left(\frac{k_{isc}}{k_f + k_{ic} + k_{isc} + k_s[A]}\right)\left(\frac{k_t[A]}{k_t[A] + k_d}\right) \tag{5.36}$$

The total biacetyl concentration $[A^t]$ is given by the sum of these two results:

$$[A^t] = \frac{I_a \tau_p [A]}{k_f + k_{ic} + k_{isc} + k_s[A]}\left(k_s + \frac{k_{isc} k_t}{k_t[A] + k_d}\right) \tag{5.37}$$

Substituting this expression for $[A^t]$ into Eq. (5.32) and recognizing that $\tau_p = 1/k_p$, we find

$$\frac{I_p}{\tau_p} = \frac{\alpha k_p I_a [A]}{k_f + k_{ic} + k_{isc} + k_s[A]}\left(k_s + \frac{k_{isc} k_t}{k_t[A] + k_d}\right) \tag{5.38}$$

If we choose conditions such that energy transfer from the singlet is not important, i.e., $(k_f + k_{ic} + k_{isc}) \gg k_s[A]$ and Φ_{isc} is sufficiently large so that

$k_{isc} \gg k_d$, then the above equation reduces to the following:

$$\frac{I_p}{\tau_p} = \alpha I_a k_p \frac{k_{isc} k_t[A]}{(k_f + k_{ic} + k_{isc})(k_d + k_t[A])} \tag{5.39}$$

Inversion of this equation yields

$$\frac{\tau_p}{I_p} = \frac{k_f + k_{ic} + k_{isc}}{\alpha I_a k_{isc} k_p}\left(1 + \frac{k_d}{k_t[A]}\right) \tag{5.40}$$

If the quantities that are constant are combined into one term denoted by K, equation (5.40) becomes

$$\frac{\tau_p}{I_p} = \frac{k_f + k_{ic} + k_{isc}}{K k_{isc}}\left(1 + \frac{k_d}{k_t[A]}\right) \tag{5.41}$$

Thus a plot of τ_p/I_p vs. $1/[A]$ should yield a straight line with intercept $(k_f + k_{ic} + k_{isc})/Kk_{isc}$ or $1/K\Phi_{isc}$. Thus Φ_{isc} can be determined if the constant K is known (e.g., by evaluation using a compound of known Φ_{isc} as donor). We see then that by determining the phosphorescence intensity and lifetime as a function of [A] (within the limits of our assumptions), the intersystem crossing efficiency of the donor can be obtained.

5.3c. *Photooxidation*

As discussed in Chapter 2, triplet excited anthracene transfers its energy to oxygen to produce singlet excited oxygen ($^1\Delta_g$). The singlet oxygen in turn attacks a ground state anthracene to form a 9,10-endoperoxide.

If the light is absorbed by a sensitizer and then after intersystem crossing is transferred to oxygen and the singlet oxygen forms a 9,10-endoperoxide with anthracene, the intersystem crossing quantum yield Φ_{isc} of the sensitizer can be determined. The following set of equations were used by Stevens and Algar[(59)] for the Φ_{isc} determinations:

$$h\nu + D \longrightarrow D^s \qquad I_a$$

$$D^s \longrightarrow D + h\nu' \qquad k_f$$

$$D^s \longrightarrow D^t \qquad k_{isc}$$

$$D^s \longrightarrow D \qquad k_{ic}$$

$$O_2 + D^s \longrightarrow D^t + O_2 \qquad k'_{isc}$$

$$D^t \longrightarrow D \qquad k_d'$$

$$D^t + O_2 \longrightarrow D + O_2^* \qquad k_{et}$$

$$A + O_2^* \longrightarrow AO_2 \qquad k_r$$

$$O_2^* \longrightarrow O_2 \qquad k_d$$

The quantum yield of endoperoxide is given by the expression

$$\Phi_{AO_2} = \Phi_A \Phi_{et} \Phi'_{isc} \tag{5.42}$$

where Φ_A describes the reactivity of the singlet oxygen once formed,

$$\Phi_A = k_r[O_2^*]/(k_r[O_2^*] + k_d) \tag{5.43}$$

For energy transfer from the triplet state

$$\Phi_{et} = k_{et}[O_2]/(k_{et}[O_2] + k_d')$$

Since the conditions can be easily adjusted such that $k_{et}[O_2] \gg k_d'$, then $\Phi_{et} = 1$. The intersystem crossing quantum yield in the presence of oxygen is

$$\Phi_{isc} = \frac{k_{isc} + k'_{isc}[O_2]}{k_{isc} + k_{ic} + k_f + k'_{isc}[O_2]}$$

Consequently the quantum yield for the production of AO_2 is

$$\Phi_{AO_2} = \Phi_A \frac{k_{isc} + k'_{isc}[O_2]}{k_{isc} + k_{ic} + k_f + k'_{isc}[O_2]} \tag{5.44}$$

This can be transformed into a more useful equation by noting the following relationship for the ratio of fluorescence intensity in the absence (F) and in the presence (F_O) of oxygen:

$$\frac{F_O}{F} = 1 + \frac{k'_{isc}[O_2]}{k_{isc} + k_{ic} + k_f} \tag{5.45}$$

Combining this equation with the previous one, we obtain the following result:

$$\frac{F_O}{F}\Phi_{AO_2} = \Phi_A \frac{k_{isc} + k'_{isc}[O_2]}{k_{isc} + k_{ic} + k_f + k'_{isc}[O_2]} \cdot \frac{k_{isc} + k_{ic} + k_f + k'_{isc}[O_2]}{k_{isc} + k_{ic} + k_f} \tag{5.46}$$

$$= \Phi_A \frac{k_{isc} + k'_{isc}[O_2]}{k_{isc} + k_{ic} + k_f} \tag{5.47}$$

or

$$\frac{F_O}{F}\Phi_{AO_2} = \Phi_A\left(\Phi_{isc} + \frac{F_O}{F} - 1\right) \tag{5.48}$$

Thus a plot of the quantity $F_O/F\Phi_{AO_2}$ vs. the quantity $(F_O/F) - 1$ should yield a straight line with a slope of Φ_A and an intercept of $\Phi_A\Phi_{isc}$. Dividing the intercept by the slope should yield Φ_{isc}:

$$\Phi_{isc} = \text{intercept/slope} \tag{5.49}$$

This method can be successfully applied to any aromatic hydrocarbon whose lowest excited singlet and triplet states are lower in energy than the substrate

(A), providing of course that the donor (singlet) has sufficient energy to excite oxygen to its $^1\Delta g$ level ($\sim$23 kcal/mole). A useful variant of this procedure is to eliminate the sensitizer for molecules that form endoperoxides themselves.

5.3d. *Delayed Fluorescence*

The subject of delayed fluorescence was discussed in Section 5.2a. It was seen that there are two common types of delayed fluorescence, that arising from thermally activated return from the triplet state to the lowest excited singlet (E-type delayed fluorescence) and that arising from collision of two excited triplet molecules resulting in a singlet excited molecule and a ground state molecule (P-type delayed fluorescence). The P-type delayed fluorescence can be used as a convenient tool for the determination of intersystem crossing efficiencies[57,63]:

$$\begin{aligned} A^t + A^t &\longrightarrow A^s + A \\ A^s &\longrightarrow A + h\nu \end{aligned} \qquad \text{P-type delayed fluorescence} \tag{5.50}$$

In the presence of a molecule D capable of transferring its triplet energy to A to produce acceptor triplets, the following processes can occur:

$$D \xrightarrow{h\nu} D^s \longrightarrow D^t \tag{5.51}$$

$$D^t + A \longrightarrow D + A^t \tag{5.52}$$

$$D^t + A \longrightarrow D + A \tag{5.53}$$

At low light intensities the probability of collision of two acceptor triplets is small; thus the main process for deactivation of the acceptor triplet is

$$A^t \longrightarrow A$$

In comparing two solutions containing the same acceptor but different donors, the ratio of the intensities of the delayed fluorescence is

$$\frac{(I_{\text{DF}})_1}{(I_{\text{DF}})_2} = \left[\frac{(I_a)_1(\Phi_{\text{et}}\Phi^{\text{D}}_{\text{isc}})_1\tau_1}{(I_a)_2(\Phi_{\text{et}}\Phi^{\text{D}}_{\text{isc}})_2\tau_2}\right]^2 \tag{5.54}$$

By measuring the intensities and lifetimes of delayed fluorescence of both solutions in the same apparatus $[(I_a)_1 = (I_a)_2]$ we obtain

$$\frac{(I_{\text{DF}})_1}{(I_{\text{DF}})_2} = \frac{(\Phi_{\text{et}}\Phi^{\text{D}}_{\text{isc}})_1{}^2\ \tau_1{}^2}{(\Phi_{\text{et}}\Phi^{\text{D}}_{\text{isc}})_2{}^2\ \tau_2{}^2} \tag{5.55}$$

Since Φ_{et} for all donors is essentially the same (generally assumed to be unity), the ratio of delayed fluorescence intensities is simply

$$\frac{(I_{\text{DF}})_1}{(I_{\text{DF}})_2} = \left[\frac{(\Phi^{\text{D}}_{\text{isc}})_1}{(\Phi^{\text{D}}_{\text{isc}})_2}\right]^2 \frac{\tau_1{}^2}{\tau_2{}^2} \tag{5.56}$$

If Φ_{isc} is known for one of the donors, the intersystem crossing efficiency of the other can be determined by measuring the ratio of the delayed fluorescence intensities and lifetimes of the two solutions. The experimental procedure involved in this method for the determination of Φ_{isc} is simple. Solutions containing equal optical densities of the two donors are prepared (OD ≈ 0.20) and equal amounts of acceptor are added to both (perylene is generally used at a concentration of about 4×10^{-5} M) and the solutions are degassed. Fluorescence intensities and lifetimes are measured separately using monochromatic light of the same wavelength used to determine the optical densities of the donors. This method may be applied to all donors having a triplet level higher than that of the acceptor.

5.3e. *Electron Spin Resonance Spectroscopy*

We saw in Section 5.1b that the triplet state is actually composed of three distinct states with quantum numbers $m = +1$, 0, and -1. If there were no interactions between the two electrons in the triplet state, these three states would be degenerate, that is, of equal energy. However, spin–spin dipolar interaction between the electrons does exist, removing the degeneracy of the three states. In the presence of an external magnetic field H_0, as in an ESR spectrometer, the magnitude of the splitting between the levels is increased (Figure 5.9) such that transitions between the levels occur in the microwave frequency range. Transitions occurring with $\Delta m = 1$ are allowed. Those with $\Delta m = 2$ are forbidden.

Thus under continuous electronic excitation in the cavity of an ESR spectrometer an ESR signal due to the triplet excited molecules can be obtained. The magnitude of this signal is directly proportional to the number of triplets produced by intersystem crossing (steady state triplet concentration). By comparing the magnitude of the integrated ESR signal of the steady state population of triplets produced upon continuous UV irradiation in a

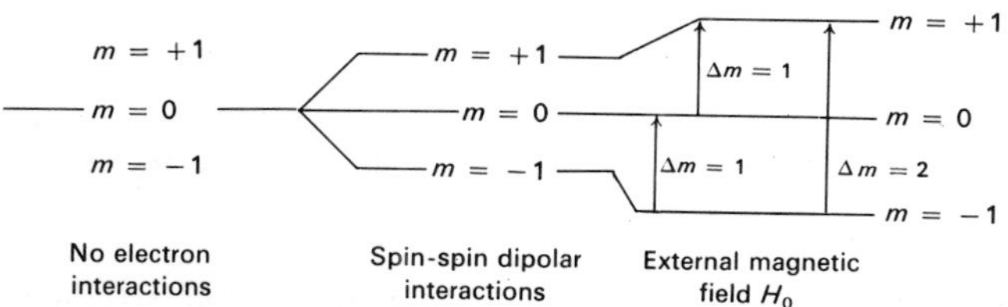

FIGURE 5.9. Effect of spin–spin dipolar interaction and an external magnetic field on triplet levels.

rigid glass at low temperatures ($\Delta m = 2$) with that obtained from a standard free radical of known concentration ($\Delta m = 1$) the steady state concentration of triplet molecules can be calculated[64]:

$$[A_1{}^t] = \frac{[R]15(h\omega)^2 I_t}{8(D^2 + 3E^2)I_R} \tag{5.57}$$

where [R] is the molar concentration of the standard free radical (a 10^{-2} *M* solution of di-*t*-butyl nitroxide in toluene is used), I_t and I_R are the integrated areas of the ESR absorption from the triplet and free radical standard solutions, respectively, $h\omega$ is the microwave energy, and D and E are the zero-field splitting parameters for the triplet molecules. The ratio of the integrated areas must be corrected for the difference in transition probability for the allowed ($\Delta m = 1$) free radical transition and the forbidden ($\Delta m = 2$) triplet transition.

If the rate of light absorption is known from the incident intensity I_0 and the optical density of the solution, Φ_{isc} can be determined:

$$\Phi_{\text{isc}} = [A^t]/I_a \tag{5.58}$$

5.3f. *Intersystem Crossing Quantum Yields of Organic Molecules*

Intersystem crossing quantum yields determined by the methods discussed in this section and other methods[24] are presented in Table 5.6.

5.4. DETERMINATION OF TRIPLET LIFETIMES

It will be recalled from previous discussions that the lifetime of the triplet state is given by

$$\tau_t = 1/(k_p + k_d + k_q[Q]) \tag{5.59}$$

where the rate constants refer to the processes

$$A^t \xrightarrow{k_p} A + h\nu \qquad \text{phosphorescence}$$
$$A^t \xrightarrow{k_d} A + \text{heat} \qquad \text{intersystem crossing}$$
$$A^t + Q \xrightarrow{k_q} A + Q \qquad \text{triplet quenching}$$

Since the first two processes are spin-forbidden, it can clearly be seen that in the absence of triplet quenchers (e.g., oxygen) the triplet will be long lived. Consequently the experimental determination of the lifetime of triplet states

TABLE 5.6. *Intersystem Crossing Quantum Yields for Various Organic Molecules*

Compound	Solvent	Φ_{isc}	Ref.
Naphthalene	Benzene	0.40, 0.82	55, 65
	Ethanol	0.80, 0.71	65, 20
Naphthalene-d_8	Benzene	0.38	55
	Cellulose acetate[a]	0.53	66
1-Methylnaphthalene	Benzene	0.48	55
2-Methylnaphthalene	Benzene	0.51	55
1-Fluoronaphthalene	Benzene	0.63	55
1-Methoxynaphthalene	Benzene	0.26	55
	Ethanol	0.50, 0.46	65, 20
Acenaphthene	Benzene	0.47	55
	Ethanol	0.45, 0.58	20, 65
Anthracene	Ethanol	0.72, 0.70	65, 20, 57
	Liquid paraffin	0.58, 0.75	53, 54, 68
Anthracene-d_{10}	EPA[a]	0.53	69
9-Methylanthracene	Liquid paraffin	0.48	68
	Ethanol	0.67	20
9,10-Dimethylanthracene	Ethanol	0.03	20
9,10-Diphenylanthracene	Liquid paraffin	0.12	54, 68
	Ethanol	0.02	20
9-Phenylanthracene	Liquid paraffin	0.37, 0.35	54, 68
	Ethanol	0.47, 0.51	20, 71
	Isopropanol	0.51	54
	Ethylene glycol	0.24	71
9-Ethylanthracene	Liquid paraffin	0.43	68
9,10-Dichloroanthracene	Ethanol	0.48	70
Phenanthrene	Benzene	0.76	55
	Ethanol	0.85, 0.80	65, 20
	3-Methylpentane	0.70	53
	Cellulose acetate[a]	0.88	66
Phenanthrene-d_{10}	Cellulose acetate[a]	0.90, 0.84	66
9,10-Dihydrophenanthrene	*n*-Hexane	0.13	76
Pyrene	Ethanol	0.28, 0.38, 0.27	57, 65, 20, 72
	n-Hexane	0.38	76
	Propylene glycol[b]	0.02	73
Tetracene	Benzene	0.63	59
1,2-Benzanthracene	Hexane	0.55, 0.77, 0.79	65, 74, 76
	Ethanol[a]	0.82	65
Chrysene	Benzene	0.67	55
	n-Hexane	0.81	76
	Ethanol	0.85, 0.82	65, 20
	EPA[a]	0.70	69
Chrysene-d_{12}	Cellulose acetate	0.60	66
Triphenylene	Benzene	0.95	55
	Ethanol	0.89	20
	Cellulose acetate[a]	0.85	66
Triphenylene-d_{12}	Cellulose acetate[a]	0.85, 0.85, 0.89	66

TABLE 5.6.—*Continued*

Compound	Solvent	Φ_{isc}	Ref.
Perylene	Ethanol	0.01	62
	n-Hexane	0.015	62
	Cyclohexane	0.014	62
	Benzene	0.06	60
1,2,5,6-Dibenzanthracene	EPA[a]	0.98	69
	3-Methylpentane	1.03	53
Picene	EPA[a]	0.36	69
Anthanthrene	Benzene	0.23	59
1,12-Benzoperylene	EPA[a]	0.59	69
Coronene	Ethanol	0.56	65
1,2-Benzocoronene	EPA[a]	0.64	75
Fluorene	Benzene	0.31	55
	n-Hexane	0.10	76
	Ethanol	0.32	65
p-Terphenyl	Cellulose acetate	0.07	66
	n-Hexane	0.11	76
p-Terphenyl-d_{14}	Cellulose acetate	0.09	66
Fluoranthene	Ethanol	0.79	20
Biphenyl	*n*-Hexane	0.51	76
Eosin (dianion)	Water	0.76	77
Fluorescein (dianion)	Water	0.032	77
	Ethanol	0.035	77
Acridine Orange	Ethanol	0.37	77
Acridine Orange (cation)	Ethanol	0.10	77
Proflavine (cation)	Ethanol	0.22	77
Pentacene	1-Chloronaphthalene	0.16	77

[a] Determined at 77°K.
[b] Determined at 180°K.

is considerably simpler than that of the lifetime of singlet states discussed in Chapter 2.

Two principal methods are employed for the determination of triplet lifetimes: flash photolysis and triplet luminescence decay. Since both of these techniques have been discussed previously in some detail, only a short treatment of the application of these techniques to the determination of triplet lifetimes will be given in this section.

5.4a. *Flash Photolysis*[(78)]

In the flash photolysis technique a large population of ground state molecules are raised to an excited singlet state by the initial photolysis flash. In a time τ (singlet lifetime) after the photolysis flash a certain proportion of

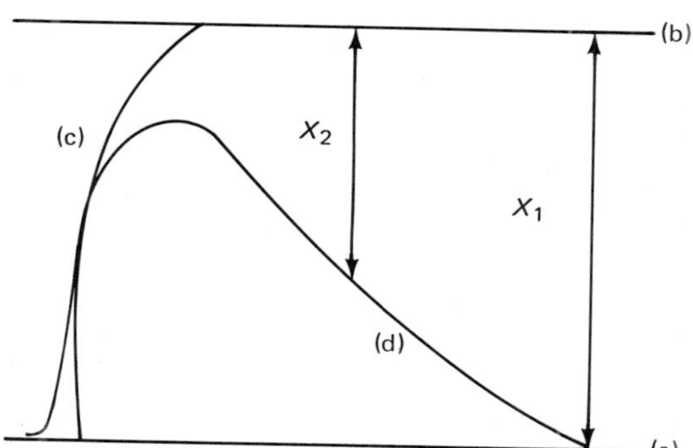

FIGURE 5.10. Oscilloscope traces obtained by the flash spectrophotometric technique. (After Porter.[160])

the singlet molecules have crossed over into the lowest triplet state (Φ_{isc}). A second, continuous light source is used to excite the triplet molecules into higher triplet levels. These triplet–triplet absorptions are detected by a photomultiplier and the decay of photomultiplier response is displayed on an oscilloscope or, if the lifetime is sufficiently long, on a fast-response recorder. A spectrum taken before the photolysis flash (no triplet species) shows only singlet–singlet absorption. By scanning the solution with the slit open and the continuous monitoring light on, trace (a) in Figure 5.10 is obtained. By scanning with the slit closed, trace (b) is obtained. By triggering the photolysis flash with the monitoring light off, a trace yielding the scattered light from the photolysis flash is obtained [trace (c)]. Finally the photolysis flash is triggered with the monitoring light on and the decay of the transient triplet absorption is measured [trace (d)]. The optical density of the transient species may be obtained from these oscilloscope traces as a function of time. If the oscilloscope response is proportional to the light intensity,

$$I_0 = k_E(\text{light on deflection} - \text{light off deflection}) = k_E X_1$$

$$I = k_E(\text{triplet} \rightarrow \text{triplet deflection} - \text{light off deflection}) = k_E X_2$$

where k_E is a constant depending upon the detector sensitivity. Since

$$\text{OD} = \log(I_0/I) = \varepsilon_\lambda c_\lambda d \tag{5.60}$$

$$= \log(X_1/X_2) \tag{5.61}$$

for unimolecular triplet decay we have

$$d[\text{OD}]/dT = -k_t[\text{OD}] \tag{5.62}$$

$$d[\varepsilon cd]/dT = -k_t[\varepsilon cd] \tag{5.63}$$

$$[\varepsilon cd] = [\varepsilon cd]_0 e^{-k_t T} \tag{5.64}$$

where $[\varepsilon cd]$ is the optical density at time T after the flash, $[\varepsilon cd]_0$ is the initial optical density, and k_t is the sum of all unimolecular rate constants leading to triplet deactivation. Thus a plot of $\log_e[\varepsilon cd]$ versus time T after the photolysis flash should be a straight line with a slope equal to k_t:

$$\log_e[\varepsilon cd] = \log_e[\varepsilon cd]_0 - k_t T \tag{5.65}$$

The triplet lifetime (in the absence of triplet quenchers) is given by

$$\tau_t = 1/k_t \tag{5.66}$$

A nonlinear plot of $\log_e[\varepsilon cd]$ vs. T indicates that bimolecular processes such as triplet–triplet annihilation or triplet quenching are contributing to triplet state deactivation.

5.4b. *Luminescence Decay*[20]

Phosphorescence and delayed fluorescence lifetimes can be determined with the apparatus shown in Figure 5.3 with the choppers out of phase by blocking the exciting light beam with a shutter and recording the decay of the luminescence intensity (photomultiplier output) with an oscilloscope or, if the lifetime is longer than about 5 sec, with a fast pen recorder. The results are plotted as $\log_e I_f$ against time T, since

$$dI/dT = -k_t I \tag{5.67}$$

$$I_T = I_0 e^{-k_t T} \tag{5.68}$$

$$\log_e I_T = \log_e I_0 - k_t T \tag{5.69}$$

A linear plot indicates that the luminescence decay is exponential. The slope of the line gives k_t, and τ_t can be calculated as above. The lifetime obtained by measuring the decay of P-type delayed fluorescence is equal to one-half the lifetime of the triplet state (see Section 5.2). Since in fluid solution at room temperature phosphorescence is generally much weaker than delayed fluorescence, the measurement of delayed fluorescence decay offers a convenient method for determining the lifetime of triplets at room temperature.

5.4c. *The Effect of Deuteration on Triplet Lifetime*[79]

In the absence of triplet quenchers two processes compete for triplet deactivation: radiative (phosphorescence) and nonradiative decay of the triplet to the ground state:

$$A_1{}^t \longrightarrow A_0{}^1 + h\nu \tag{5.70}$$

$$A_1{}^t \longrightarrow A_0{}^1 + \text{heat} \tag{5.71}$$

As discussed in Chapter 1, the probability of a nonradiative transition is proportional to the square of the vibrational overlap integral $\int \chi_1\chi_2 \, d\tau_v$:

$$\Pr \propto \left[\int \chi_1\chi_2 \, d\tau_v\right]^2 \tag{5.72}$$

For the nonradiative transition of interest, this can be written

$$\Pr \propto \left[\int \chi_{T_1}\chi_{S_0} \, d\tau_v\right]^2 \tag{5.73}$$

For the triplet state χ_{T_1} will generally correspond to the $j = 0$ vibrational level. In order for the triplet $\rightarrow$ singlet transition to occur isoenergetically, crossover must occur from the $j = 0$ vibrational level of the triplet to a higher j level of the ground state. The probability of the transition then depends upon the degree of spatial overlap of the vibrational wave functions of the $j = 0$ (triplet) and $j = x$ (singlet) vibrational levels. The vibrational wave functions for χ_{S_0} at high j's oscillate rapidly from positive to negative values in the region where χ_{T_1} is always positive. Hence the higher the j value in S_0 at the crossing point, the smaller will be the vibrational overlap integral and the smaller the probability of the transition. To illustrate the effect of deuteration on this nonradiative transition, let us consider the case of benzene and perdeuterobenzene. The triplet energy levels of benzene and perdeuterobenzene are both 85 kcal/mole. We see that energy and multiplicity effects on this transition will be essentially the same for both molecules. In order for intersystem crossing to occur from triplet benzene to ground state benzene, crossover must take place from the $j = 0$ level of the benzene triplet to a j level of the benzene ground state whose vibrational energy equals 85 kcal. This corresponds to $j = 10$ for C—H vibrations. However, C—D vibrations occur at a lower amplitude than C—H vibrations; consequently, a much larger j value will be necessary to equal 85 kcal in perdeuterobenzene. This means a smaller vibrational overlap integral and hence a lower probability for the transition.

Since we have inhibited one pathway leading to triplet depopulation by deuteration, it is clear that it will take longer for the triplet to decay by the radiative pathway and the lifetime of the triplet is increased. If phosphorescence were the sole pathway leading to triplet decay, the measured triplet lifetime would correspond to the radiative lifetime and would be equal to

$$\tau_r^0 = 1/k_p \tag{5.74}$$

where k_p is the rate constant for phosphorescence. Under conditions where both processes lead to triplet decay the measured phosphorescence lifetime is related to the natural or radiative lifetime by

$$\tau_r^0 = \tau_p(1 - \Phi_f)/\Phi_p \tag{5.75}$$

Thus the extent to which radiationless decay leads to triplet depopulation can be evaluated by comparison of the measured triplet lifetime τ_p with the calculated radiative lifetime if Φ_f and Φ_p are known. Alternatively, one can obtain a qualitative idea of the importance of triplet $\rightarrow$ singlet intersystem crossing by measuring the triplet lifetime of the deuterated analog since in these molecules the measured phosphorescence lifetime approaches the natural radiative lifetime. It is generally felt that deuteration has no effect on the triplet radiative lifetime τ_r^0, although this has been recently questioned for several aromatic polycyclic hydrocarbons.[80] The effect of deuteration on the triplet lifetime of several organic compounds can be seen in Table 5.7. We see that deuteration serves to lengthen the triplet state lifetime. Heavy-atom substitution or using a heavy-atom solvent, on the other hand, *decreases* the triplet lifetime, as we shall see in Section 5.6.

TABLE 5.7. *Effect of Deuteration on the Triplet Lifetime of Organic Molecules*[30,80–86]

Compound	τ_t, sec
Benzene	7.0, 6.3[a]
Benzene-d_6	26.0, 9.4[a]
Triphenylene	4.9, 14.0[a]
Triphenylene-d_{12}	23.0, 21.4[a]
Biphenyl	3.1, 4.6[a]
Biphenyl-d_{10}	11.3, 9.9[a]
Phenanthrene	3.3, 3.7,[a] 3.3[b]
Phenanthrene-d_{10}	16.4, 15.2[a]
Naphthalene	2.3,[c] 2.3,[a] 3.3[b]
Naphthalene-d_8	22.0, 18.4[a]
Pyrene	0.2,[a] 0.2, 0.7[b]
Pyrene-d_{10}	3.2[a]
Anthracene	0.06,[a] 0.1[c]
Anthracene-d_{10}	0.10[a]
Chrysene	2.7[a]
Chrysene-d_{10}	13.2[a]
Quinoxaline	0.27[a]
Quinoxaline-d_6	0.62[a]
p-Methoxybenzaldehyde	0.10[a]
p-Methoxybenzaldehyde-d_1	0.21[a]
α,α,α-h_3-*p*-Methoxyacetophenone	0.19[a]
α,α,α-d_3-*p*-Methoxyacetophenone	0.25[a]

[a] Data of Li and Lim,[80] determinations at 77°K.

[b] Determined by flash photolysis in isopentane or EPA at 77°K.

[c] Determined by phosphorescence decay in EPA at 77°K.

TABLE 5.8A. *Triplet Lifetimes of Aromatic Hydrocarbons and Derivatives*

Compound	τ_t,[a] sec	Ref.
Benzene	7.0	101
Toluene	8.8, 8[b]	30, 102
Durene	5.7, 6.4	103, 30
Hexamethylbenzene	5.7, 8.55	103, 30
Aniline	4.7	30
Phenol	2.9	30
Anisole	3.0	30
Chlorobenzene	0.004	30
Bromobenzene	0.0001	30
p-Dichlorobenzene	0.016	30
p-Dibromobenzene	0.0003	30
1,3,5-Trichlorobenzene	0.022	30
1,3,5-Tribromobenzene	0.00074	30
1,2,4,5-Tetrachlorobenzene	0.018	30
Benzoic acid	2.5	30
Naphthalene	2.6, 2.3	30, 104
1-Methylnaphthalene	2.5, 2.1	30, 104
1-Aminonaphthalene	1.5	30
1-Hydroxynaphthalene	1.9	30
2-Carboxynaphthalene	2.5	30
1-Nitronaphthalene	0.049	30
1-Fluoronaphthalene	1.5	30
1-Chloronaphthalene	0.30, 0.29	30, 104
1-Bromonaphthalene	0.018	30
1-Iodonaphthalene	0.0025, 0.002	30, 104
Biphenyl	3.6	30
Anthracene	0.04	105
2-Methylanthracene	0.021[c]	106

[a] Determined by phosphorescence decay at 77°K unless otherwise noted.
[b] Pure toluene at 4°K.
[c] In Lucite at room temperature.

5.4d. *Triplet Lifetimes of Various Organic Molecules*

The triplet lifetimes of various aromatic, carbonyl, and heterocyclic compounds are listed in Tables 5.8A–5.8C.

5.5. EXCITED STATE GEOMETRY[87]

Many techniques are available for the determination of the geometry of organic molecules in their ground states, such as microwave spectroscopy, infrared and Raman spectroscopy, and electron and x-ray diffraction.

TABLE 5.8B. *Triplet Lifetimes of Carbonyl Compounds*

Compound	τ_t,[a] sec	Ref.
Acetone	0.0004, 0.0006	107, 30
Methyl ethyl ketone	0.00085	30
Diethyl ketone	0.00126	30
Diisopropyl ketone	0.0037	30
Di-*t*-butyl ketone	0.0086	30
Benzophenone	0.0077, 0.006	101
1-Naphthaldehyde	0.08	108
2-Acetonaphthone	0.97, 0.95	108, 30
p,p'-Bis(dimethylamino)benzophenone	0.27	109
p-Aminobenzophenone	0.40	109
p-Dimethylaminobenzophenone	0.35	109
p,p'-Dimethoxybenzophenone	0.065	109
p-Methylbenzophenone	4.5	110
p-Bromobenzophenone	3.6	110
o-Chlorobenzophenone	2.8	110
o-Bromobenzophenone	2.0	110
p,p'-Dibromobenzophenone	3.8	110
Acetophenone	0.008, 0.0023, 0.004	30, 111, 112
p-Methylacetophenone	0.084	112
3,5-Dimethylacetophenone	0.11	112
p-Methoxyacetophenone	0.25	112
m-Methylacetophenone	0.074	112
m-Methoxyacetophenone	0.25	112
p-Bromoacetophenone	0.0066	110
Butyrophenone	0.002[b]	113
p-Chlorobutyrophenone	0.014[b]	113
p-Methylbutyrophenone	0.009[b]	113
p-Hydroxybutyrophenone	0.084[b]	113
p-Aminobutyrophenone	0.084[b]	113
p-Methoxybutyrophenone	0.051[b]	113
Benzaldehyde	0.0015, 0.0023	112, 114
p-Methylbenzaldehyde	0.043	114
p-Methoxybenzaldehyde	0.15	114
Ethylphenylketone	0.0037	111
Benzoin	2.4	111
Biacetyl	0.00225	30
Anthraquinone	4.0	115
2-Methylanthraquinone	4.4	115

[a] Determined by phosphorescence decay at 77°K unless otherwise noted.
[b] Determined at −190°C.

However, one of the few sources of information regarding the configurations of molecules in their electronically excited states is electronic spectra. Consequently the excited state geometries of relatively few molecules are known to date. The most accurate method for the determination of excited

TABLE 5.8C. *Triplet Lifetimes of Heterocyclic Compounds*

Compound	τ_t,[a] sec	Ref.
Pyrimidine	0.01	116
Pyrazine	0.02	30
sym-Triazine	0.4	117
2-Phenyl-*sym*-triazine	1.4	118
2,4-Diphenyl-*sym*-triazine	1.2	118
2-Phenyl-4,6-dimethyl-*sym*-triazine	1.9	118
2,4,6-trimethyl-*sym*-triazine	0.65	118
Quinoline	1.3, 13,[b] 0.9[c]	119, 120
Isoquinoline	0.9,[b] 0.7[c]	120
Quinoxaline	0.25	121
Phenanthridine	1.2,[b] 0.9[c]	120
1,10-Diazaphenanthrene	1.6	123
9,10-Diazanthracene	0.02	122
5,6-Benzoquinoline	2.9,[b] 2.6[c]	120
Dibenzofuran	5	124
Carbazole	10, 7.6	124, 125
Thianaphthene	0.5	124
Indole	5, 6.3	124, 125
9,10-Dihydroacridine	3.5	124
Adenine	2.2	126
Guanine	1.4	126
Purine	1.6, 1.6	127

[a] Determined by phosphorescence decay at 77°K.
[b] In EPA.
[c] In a hydrocarbon glass.

state geometries involves analysis of the rotational fine structure of the electronic bands. Since the spectra of polyatomic molecules are complex, most analyses of this sort have been carried out on the triatomic or tetraatomic molecules. For molecules whose spectra contain unresolved fine structure, estimates of the excited state geometry can be obtained by vibrational analyses.

The first polyatomic molecule whose excited state geometry was investigated was formaldehyde. Electronic excitation of formaldehyde to its lowest excited state involves the excitation of an electron from a $2p$ atomic orbital localized on the oxygen atom to a π^* orbital between the carbon and oxygen atoms ($n \rightarrow \pi^*$). This transition therefore should be accompanied by a lengthening of the C=O bond in the excited state. Rotational analyses of the singlet → singlet and singlet → triplet spectra of formaldehyde are in agreement with this picture and show that the C—O bond length in the lowest singlet excited state is 1.32 Å as compared to a bond length of 1.22 Å in the ground state.[88] The bond length in the lowest triplet state is 1.31 Å.[88,89] Vibrational analyses revealed that singlet excited formaldehyde exists in a

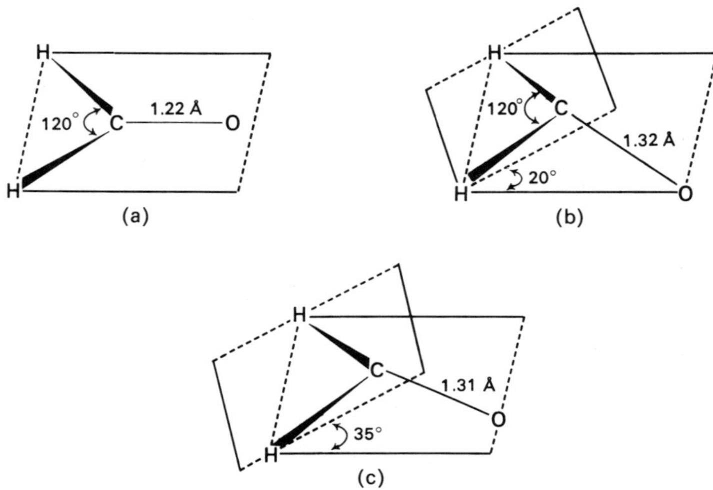

FIGURE 5.11. Geometry of formaldehyde in its (a) ground, (b) first excited singlet, and (c) first excited triplet electronic states.

nonplanar geometry with two equivalent configurations with a low-energy barrier with respect to inversion.[90–92] Robinson and DiGiorgio[88] determined that the angle between the CH_2 plane and the C—O axis in the singlet excited state is 20°. Similarly, vibrational and rotational analyses of the triplet formaldehyde spectra showed that it also has a nonplanar structure with an out-of-plane angle 35°.[88] These geometries are shown in Figure 5.11.

Upon light absorption by ethylene, an electron is promoted from the π orbital between the two carbons to a π^* orbital. This transition, therefore, should be accompanied by a large change in the C—C bond length. From the electronic spectrum of ethylene it has been deduced that the C—C bond length in the excited state is 1.69 Å as opposed to 1.34 Å in the ground state.[93] Theoretical calculations indicate, furthermore, that the two CH_2 groups in the excited state are twisted with their planes at right angles.[94] This is shown in Figure 5.12.

Using LCAO MO theory, one can calculate the state energies of ethylene as a function of the angle of twist between the two planes in the excited state and in the ground state.[94–96] From Figure 5.13 it can be seen that a molecule of ethylene excited to its $^1S_1$ state may twist around the double bond to achieve the minimum energy of the state ($\theta = 90°$). Similarly a triplet ethylene produced by intersystem crossing from the excited singlet or by direct singlet → triplet absorption will be at minimum energy when $\theta = 90°$. For the singlet molecule, return to the ground state will occur with equal

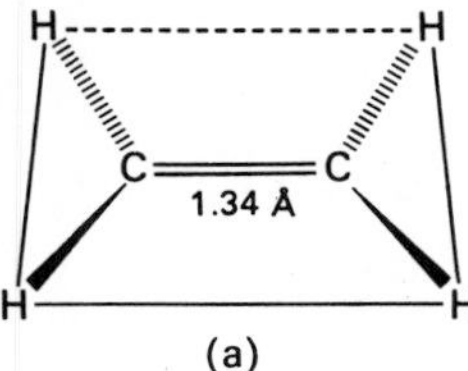

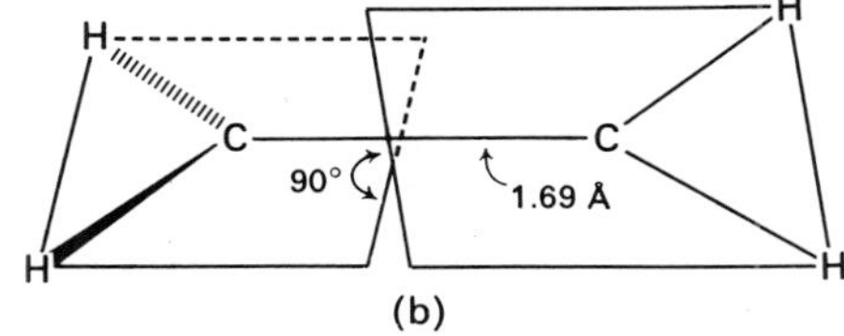

FIGURE 5.12. Geometry of ethylene in (a) the ground and (b) the excited state.

probability at $\theta = 0°$ or 180°. The latter possibility leads to isomerization of a *cis-* or *trans*-substituted ethylene to the opposite isomer. A triplet ethylene can pass through values of θ where the triplet and ground state energies overlap (60 or 120°). Intersystem crossing at these points and vibrational relaxation to the ground state result in the original molecule if crossing occurs at 60° and result in its *cis–trans* isomer if crossing occurs at 120°. Thus we see that *cis–trans* isomerization (to be discussed in detail in a later chapter) follows naturally from the excited state geometry of ethylenes.

The absorption spectrum of acetylene provided the first example of an electronic transition involving a change in molecular shape in the excited state. While the acetylene molecule is linear in the ground state it has a *trans*-bent configuration in the excited state.[97,98] Rotational analysis reveals that the bond length (C—C) increases from 1.207 Å in the ground state to 1.38 Å in the excited state. In addition, the C—H bond lengths increase from 1.06 Å in the ground state ($\angle$CCH = 180°) to 1.08 Å in the excited state ($\angle$CCH = 120 ± 2°). These configurations are shown in Figure 5.14.[97–99]

It is expected that as the ability of spectrometers to resolve closely spaced rotational and vibrational bands from molecular spectra improves, the number of molecules whose excited state geometries are known will increase. However, it must be realized that the task of determining excited

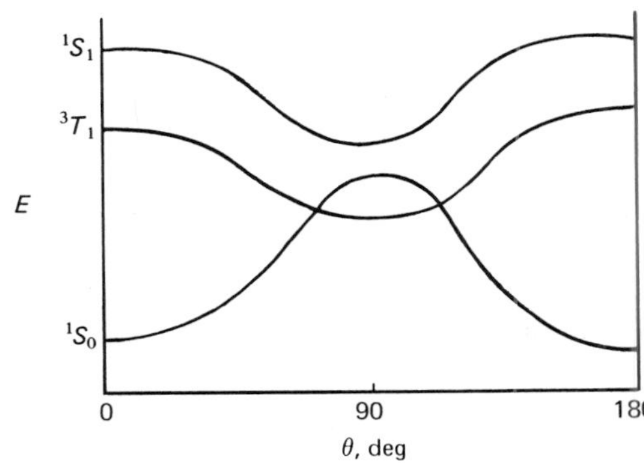

FIGURE 5.13. State energies of ethylene as a function of twist.

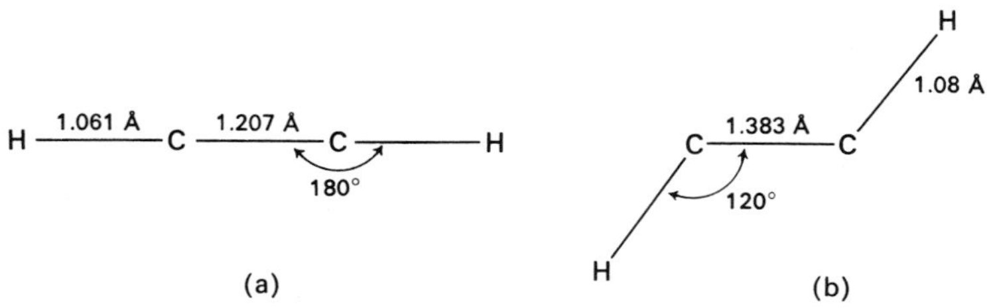

FIGURE 5.14. Geometry of (a) ground and (b) first excited singlet states of acetylene.

state geometries of the complex molecules of interest to most organic photochemists would nevertheless be immense.

5.6. SPIN–ORBIT COUPLING AND INTERSYSTEM CROSSING

In this section the effect of spin–orbit coupling on radiative and radiationless intercombinational transitions (transitions occurring between states of different multiplicity) will be investigated. We will be particularly concerned with the use of internal and external heavy atoms to induce spin–orbit coupling. The effect of heavy atoms on intercombinational processes occurring in aromatic hydrocarbons, carbonyl compounds, and heterocyclic compounds will be discussed.

5.6a. *The Nature of Spin–Orbit Coupling*[130,135]

It has been seen that the probability (Pr) of radiationless decay of a higher energy state to a lower energy state is given by

$$\mathrm{Pr} = \left| \int \psi_1 P \psi_2 \, d\tau \right|^2 \tag{5.76}$$

where the ψ's represent the total wave functions for the two states and P is a suitable perturbation operator. The total wave function for each state ψ contains an electronic portion ϕ and a vibrational eigenfunction χ. The probability thus becomes

$$\mathrm{Pr} = \left| \int \phi_1 P \phi_2 \, d\tau \int \chi_1 \chi_2 \, d\tau_v \right|^2 \tag{5.77}$$

In addition, one can factor out a spin component from the electronic wave function,

$$\mathrm{Pr} = \left| \int \phi_1 P \phi_2 \, d\tau_e \int \chi_1 \chi_2 \, d\tau_v \int S_1 S_2 \, d\tau \right|^2 \tag{5.78}$$

If the electron spins remain paired during the transition, $\int S_1 S_2 \, d\tau$ is unity. If the spins become unpaired, as they must in crossing from a singlet to a triplet state, this integral becomes zero.(79) Thus intercombinational transitions are formally forbidden due to spin restrictions. This would be strictly true if the spin and orbital motions of the electrons were independent. In actuality, however, this is not true since the orbital motion of the electron produces a magnetic field which interacts with its spin magnetic moment. This interaction causes a change in the direction of the spin-angular momentum of the electron. The probability of this change is dependent upon the charge of the nucleus.(131) This can perhaps be best understood by reference to Figure 5.15.(11) We tend to think that the electron orbits about the nucleus, but the electron sees itself as the center of the atom with the nucleus and core electrons orbiting about it. Since the nucleus and core electrons are charged, from the electron's point of view it is surrounded by an electric circuit which is carrying a current. This current causes a magnetic field which is perpendicular to the plane of the nuclear orbit. The electron itself, however, has a magnetic moment parallel to its axis of spin. The energy of the electron will vary depending upon the relative directions of this spin magnetic moment and the magnetic field from the orbital motion of the nucleus.(132)

Thus the change in the direction of the spin angular momentum of the electron effectively imparts some singlet character to a triplet state and, conversely, triplet character to a singlet state. This relaxes the spin selection rule since $\int S_s S_t \, d\tau$ is no longer strictly zero. The greater the nuclear charge,

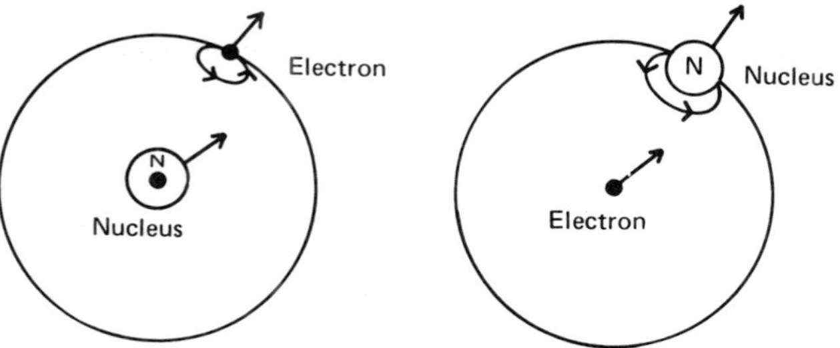

FIGURE 5.15. Diagrams representing a primitive model of spin–orbit coupling.

the greater the amount of mixing of the states and the transition becomes increasingly more allowed.

Robinson and Frosch[84,133] have developed a theory in which the molecular environment is considered to provide many energy levels which can be in near resonance with the excited molecules. The environment can also serve as a perturbation, coupling with the electronic system of the excited molecule and providing a means of energy dissipation. This perturbation can mix the excited states through spin–orbit interaction. Their expression for the intercombinational radiationless transition probability is

$$K_{s,t} = \frac{8\pi^2\tau}{h^2}\left|\sum_{ij} \langle \chi_{S_i}|\chi_{T_j}\rangle^2 \langle \phi_s|H'|\phi_t\rangle^2\right| \tag{5.79}$$

where χ_i and χ_j are vibrational wave functions belonging to the two electronic states ϕ_s and ϕ_t, and τ is the vibrational relaxation time. In this case (intersystem crossing) H' contains a contribution due to spin–orbit coupling. The vibrational overlap integral in the above equation is a product over all vibrational modes and is dependent upon differences in geometry and vibrational frequencies in the electronic states as well as the energy gap between the initial and final states. El-Sayed[131,134] has suggested that large differences in the internuclear distances between two states result if the states are of different character. Such changes in the internuclear distance contribute significantly to the vibrational overlap integral, so that $^1(n, \pi^*) \leftrightarrow {}^3(\pi, \pi^*)$ or $^1(\pi, \pi^*) \leftrightarrow {}^3(n, \pi^*)$ transitions can be about 10^2 times faster than similar transitions between states of like character.

As seen in the radiationless process, intercombinational radiative transitions can also be affected by spin–orbit interaction. As stated previously, spin–orbit coupling serves to "mix" singlet and triplet states. Although this mixing is of a highly complex nature, some insight can be gained by first-order perturbation theory. From first-order perturbation theory one can write a total wave function for the triplet state as

$$^3\Psi_1 = {}^3\Psi_1{}^0 + \sum_K \delta_{1K}\, {}^1\Psi_K{}^0 \tag{5.80}$$

where $^3\Psi_1{}^0$ is the wave function for the unperturbed triplet and the summation is carried over all singlet states. A similar wave function can be written for the singlet ground state:

$$^1\Psi_0 = {}^1\Psi_0{}^0 + \sum_j \delta_{0j}\, {}^3\Psi_j{}^0 \tag{5.81}$$

According to perturbation theory, δ_{1K} is given by

$$\delta_{1K} = \frac{\langle {}^3\Psi_1{}^0|H_{\mathrm{so}}|{}^1\Psi_K{}^0\rangle}{|{}^3E_1 - {}^1E_K|} \tag{5.82}$$

where H_{so} is the spin–orbit Hamiltonian and the denominator represents the difference in energy between the triplet state and the perturbing singlet state. The radiative transition moment is therefore

$$\mathbf{M}_{T_1,S_0} = \left< {}^3\Psi'_1 \middle| \sum_i e\mathbf{r}_i \middle| {}^1\Psi'_0 \right>$$

$$= \sum_K \delta_{1K}\mathbf{M}_{S_K,S_0} + \sum_j \delta_{0j}\mathbf{M}_{Tj,T_1} \qquad (5.83)$$

and the summation is carried over all electrons i ($\mathbf{M}_{T_1{}^0,S_0{}^0}$ and $\mathbf{M}_{T_j{}^0,S_i{}^0}$ are not included because of spin selection rules).[134] It should be noted that since $\mathbf{M}_{T_1,S_0}$ depends upon $\mathbf{M}_{S_K,S_0}$ and $\mathbf{M}_{T_j,T_1}$, the resulting emission (phosphorescence) must show the polarization characteristic of the latter transitions.

The spin–orbit Hamiltonian (H_{so}) requires some explanation. The energy of interaction between the magnetic moment $\mathbf{M}$ and the magnetic field caused by the orbital motion of an electron can be derived as[134]

$$H' = \frac{1}{2m^2c^2}\sum_i \frac{1}{r_i}\frac{\partial V(r_i)}{\partial r_i}\mathbf{L}_i\cdot\mathbf{S}_i = \sum_i \zeta_{r_i}\mathbf{L}_i\cdot\mathbf{S}_i \qquad (5.84)$$

where V is the potential energy of the electron, r_i is the distance of the electron from the nucleus, and ζ is the spin–orbit coupling constant. The summation is over all electrons i. Including terms for the spin–other orbit interaction (interaction between the spin of one electron and the orbital motion of another) and coupling of the spin and orbital motions of one electron with the repulsive field of another, H_{so} can be written

$$H_{so} = \alpha^2\left[\sum_\mu\sum_i \frac{Z_\mu}{r_{i\mu}^3}\mathbf{L}_i\cdot\mathbf{S}_i + 2\sum_{i,j}\frac{\mathbf{P}_j\times\mathbf{r}_{i,j}}{r_{i,j}^3}\cdot\mathbf{S}_i - \sum_{i,j}\frac{\mathbf{P}_i\times\mathbf{r}_{i,j}}{r_{i,j}^3}\cdot\mathbf{S}_i\right] \qquad (5.85)$$

where α is the fine structure constant, Z_μ is the effective nuclear charge of the μth nucleus, and $\mathbf{P}$ is the linear momentum operator. The first term in this equation represents the interaction between the spin and orbital momentum of an electron, the second the spin–other orbit interaction, and the third the coupling of the spin and orbital motions of one electron with the repulsive field of another. In the presence of atoms of high atomic number (i.e., heavy atoms) the first term contributes most strongly to H_{so}. It is clear, therefore, that substitution of an atom such as bromine or iodine onto an organic molecule should lead to enhancement of multiplicity forbidden intercombinational transitions due to the contribution of the large effective nuclear charge of these atoms Z_μ to the value of H_{so} (heavy-atom effect).

Alternatively, rather than to substitute a heavy atom onto a molecule directly, a solvent containing heavy atoms can be used to promote intercombinational transitions (external heavy-atom effect). Robinson and

Frosch[84,133] have explained the external heavy-atom effect in intersystem crossing by postulating that the singlet and triplet states of the solute, which cannot interact directly, couple with the solvent singlet and triplet states, which themselves are strongly coupled through spin–orbit interaction. Thus the transition integral becomes[134]

$$\langle T|H'|S\rangle = \frac{\langle T|H'|{}^3\Psi^b_{\text{solv}}\rangle\langle{}^3\Psi^b_{\text{solv}}|H_{\text{so}}|{}^1\Psi^a_{\text{solv}}\rangle\langle{}^1\Psi^a_{\text{solv}}|H''|S\rangle}{\Delta E_b\,\Delta E_a} \tag{5.86}$$

where ΔE_b and ΔE_a are the energy differences between the solute triplet state T and the perturbing solvent triplet state (${}^3\Psi^b{}_{\text{solv}}$) and between the solute singlet state S and the perturbing solvent singlet state (${}^1\Psi^a{}_{\text{solv}}$), respectively. The "allowedness" of the singlet–triplet transition in the heavy-atom solvent (due to the strong spin–orbit coupling) therefore serves to relax the "forbiddenness" of intersystem crossing in the solute.

This brings us to the question of which intercombinational process is more strongly affected by the presence of an internal or external heavy-atom-intersystem crossing from the lowest excited singlet to the triplet state or from the triplet level to the ground state. In other words, is the steady state triplet population greater or smaller in the presence of heavy-atom perturbation? Since the probability of a singlet–triplet transition resulting from spin–orbit coupling depends inversely upon the energy separation between the two states, one would expect a higher probability of crossing from a perturbed excited singlet to the triplet state than from the perturbed triplet to the ground state[84,133,135]:

$$A^s \xrightarrow{k_{\text{isc}}} A^t \tag{5.87}$$

$$A^t \xrightarrow{k_d} A \tag{5.88}$$

Hence the steady-state population of triplets should increase under heavy-atom perturbation. However, this conclusion is valid only if unimolecular decay is the main route leading to triplet state depopulation. If bimolecular triplet quenching as shown below is more important than unimolecular decay by several orders of magnitude, k_d could be increased as much or more than k_{isc} without decreasing the steady state triplet population[136]:

$$A^t + A \longrightarrow 2A \quad \text{(triplet quenching)} \tag{5.89}$$

The third process sensitive to heavy-atom perturbation is the radiative decay from the triplet to the ground state (phosphorescence). Since phosphorescence is commonly not observed in fluid solution at room temperature, the rate of phosphorescence in the presence of heavy-atom perturbation relative to the rate of intersystem crossing and nonradiative decay need not be considered. At low temperatures in a rigid glass, however, phosphorescence does occur. In general, as can be seen in Table 5.9, heavy-atom substitution serves to decrease the lifetime of phosphorescence, the magnitude of the reduction in lifetime depending upon the atomic number of the substituent. This potentially reflects an increase in the phosphorescence rate constant k_p due to spin–orbit coupling. The relative degrees of change of k_{isc}, k_d, and k_p with heavy-atom substitution, however, must be determined by an examination of Φ_f, Φ_p, and τ_p in the absence and presence of heavy-atom perturbation.

5.6b. *Effect of Heavy Atoms on Intercombinational Transitions in Aromatic Compounds*

The effect of the substitution of a heavy-atom directly onto the nucleus of aromatic compounds (internal heavy-atom effect) on intercombinational radiative and nonradiative processes can be seen by examination of experimental data obtained for naphthalene and its derivatives. The data obtained by Ermolaev and Svitashev[(104)] and analyzed by Birks[(24)] to obtain individual rate constants for the various processes are collected in Table 5.9.

The effect of the heavy-atom substituents is best illustrated by comparing the data obtained with the heavy-atom substituents with that obtained for the 1-substituted derivative containing only light atoms (1-methylnaphthalene). It is seen first of all that the fluorescence quantum yields of the naphthalene derivatives decrease as the atomic number of the halogen substituent increases. Assuming that the only process competing with fluorescence is intersystem crossing (i.e., radiationless internal conversion of the excited singlet to the ground state is unimportant), one can conclude that the observed quenching of fluorescence is due to an increased rate of intersystem crossing in the presence of the heavy atom. Indeed this is what we would expect from a previous discussion of spin–orbit coupling. A glance at the intersystem crossing rate constant k_{isc} indicates that k_{isc} increases by a factor of more

TABLE 5.9. *Internal Heavy-Atom Effect on Naphthalene*[(24,104),a]

Compound	Φ_f	Φ_p	τ_t, sec	k_p, sec^{-1}	k_d, sec^{-1}	k_{isc}, 10^6 sec^{-1}
Naphthalene	0.55	0.051	2.3	0.048	0.39	1.60
1-Methylnaphthalene	0.85	0.044	2.1	0.14	0.34	0.53
1-Fluoronaphthalene	0.84	0.056	1.5	0.23	0.44	0.57
1-Chloronaphthalene	0.058	0.30	0.29	1.10	2.35	49
1-Bromonaphthalene	0.0016	0.27	0.02	14	36	1850
1-Iodonaphthalene	<0.0005	0.38	0.002	190	310	>6000

[a] Determined in solid ethanol/ether solutions at 77°K; k_p, k_d, and k_{isc} determined assuming $\Phi_f + \Phi_p = 1$, $k_f = 3 \times 10^6$ sec^{-1} for 1-methylnaphthalene.

TABLE 5.10. *External Heavy-Atom Effect on Naphthalene*[24,32b],[a]

Solvent	Φ_f	Φ_p	τ_t, sec^{-1}	k_p, sec^{-1}	k_d, sec^{-1}	k_{isc}, 10^6 sec^{-1}
Ethanol/methanol	0.55	0.055	2.5	0.048	0.35	1.60
Propyl chloride	0.44	0.080	2.27	0.061	0.38	2.54
Propyl bromide	0.13	0.24	1.73	0.162	0.42	13.4
Propyl iodide	0.026	0.35	1.33	0.271	0.48	75

[a] Solid solutions at 77°K; k_p, k_d, and k_{isc} calculated assuming $\Phi_f + \Phi_p = 1$ and $k_f = 2 \times 10^6$ sec^{-1}.

than 10^4 in proceeding from 1-methylnaphthalene to 1-iodonaphthalene. Second, the lifetime of the triplet decreases with increasing atomic number of the halogen substituent. This indicates that the presence of the heavy atom also promotes processes leading to decay of the naphthalene triplet state. These processes are of course phosphorescence and radiationless decay to the singlet ground state. The data show that the rates of both of these processes (k_d and k_p) are increased, and by approximately the same factor ($\sim 10^3$), in proceeding from 1-methylnaphthalene to 1-iodonaphthalene. Thus we conclude that in aromatic hydrocarbons the presence of internal heavy atoms results in increases in the rates of all intercombinational transitions, although the effect on singlet → triplet intersystem crossing is greater than that on phosphorescence and radiationless triplet decay.

The external heavy-atom effect in naphthalene has been studied by McGlynn and co-workers.[32b]* A similar analysis of these data by Birks[24] is presented in Table 5.10. These data show that qualitatively the effect of an external heavy atom is the same as that of an internal heavy atom: Fluorescence is quenched by the presence of the heavy atom in a degree depending upon the atomic number of the perturber, the phosphorescence quantum yield is increased by the heavy-atom perturber, and the rates of all intercombinational processes are increased to a degree dependent upon the atomic number of the heavy atom. We note that in this case, however, k_{isc} is increased by a factor of only about 50 in proceeding from the light-atom solvent to the solvent containing propyl iodide (in contrast to a factor of 10^4 increase using direct substitution of iodine). Thus we see that the magnitude of the heavy-atom effect is greater for internal heavy atoms than for external heavy atoms. This results from the greater proximity of the heavy atom when substituted directly onto the aromatic nucleus. We note in addition that k_d is only slightly affected by the external heavy atom whereas k_p is increased by a factor of six

* For a recent study on the sensitivity of $A^s \rightarrow A^t$ and $A^t \rightarrow A$ to external heavy-atom perturbation in fluid solution, see R. P. Detoma and D. O. Cowan, *J. Amer. Chem. Soc.* **97**, 3283, 3291 (1975), where A is anthracene and 9,10-dibromoanthracene.

in proceeding from the light-atom solvent to that containing propyl iodide. Thus we conclude that in the presence of external heavy atoms k_p in aromatic hydrocarbons is increased more than k_d although k_{isc} is the most sensitive to the heavy-atom perturbation. Similar conclusions were drawn by Giachino and Kearns[137] in a study on the external heavy-atom effect on radiative and nonradiative singlet → triplet transitions in naphthalene, phenanthrene, and triphenylene.

Cowan and Drisko have studied the photodimerization of acenaphthylene[130,136] in detail and have concluded that the *cis* dimer is derived primarily from a singlet excimer state of acenaphthylene while the *trans* dimer is derived entirely from the acenaphthylene triplet state:

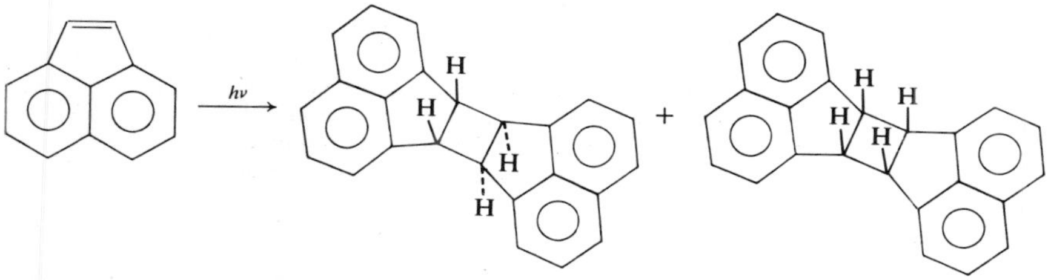

$h\nu$ → H H H H + H H H H

The amount of *trans* dimer formed upon the photolysis of acenaphthylene in heavy-atom solvents could be correlated with the square of the spin–orbit coupling constant for the heaviest atom in the solvent (see Table 5.11):

$$\frac{p_1}{p_2} = \frac{\zeta_1^2}{\zeta_2^2} = \frac{trans(1)}{trans(2)} \tag{5.90}$$

Thus it can be seen that the heavy-atom effect, so useful in spectroscopy, can also be an important tool in photochemistry, not only to facilitate the study of a reaction mechanism, but also to control the major reaction product. This general system and other related $2\pi + 2\pi$ photoaddition reactions will be considered in more detail in Chapter 10.

5.6c. *Effect of Heavy Atoms on Intercombinational Transitions in Carbonyl and Heterocyclic Compounds*

In contrast to aromatic hydrocarbons, heavy-atom substitution onto carbonyl and heterocyclic molecules appears to have little effect on radiative and nonradiative intercombinational transitions. Wagner[138] has shown that as determined by the type II photoelimination, aliphatic ketones ($n \rightarrow \pi^*$ excited states) are not sensitive to external heavy-atom perturbation. As seen previously in our discussion of type II photoelimination, aliphatic ketones undergo this cleavage from both the excited singlet and triplet states (in

TABLE 5.11. *Comparison of the Trans Dimer Ratio with the Ratio of the Squares of the Spin–Orbit Coupling Parameters*[a]

Solvent	ζ_1^2/ζ_2^2	$\frac{trans(1)}{trans(2)}$
Ethyl iodide[b]	1	1
n-Propyl bromide[b]	0.24	0.24
Neopentyl bromide[b]	0.24	0.29
Neopentyl bromide[c]	0.24	0.20
Bromobenzene[b]	0.24	0.17
n-Butyl chloride[b]	0.01	0.01

[a] It is assumed that the presence of 5–10 mole % of the heavy-atom solvent exerts no effect other than that through spin–orbit coupling. Therefore the *trans* dimer yield in these solvents has been corrected for the amount of *trans* formed in pure cyclohexane. Here ζ_1 is the spin-orbit coupling constant for I, Br, or Cl, and ζ_2 is the spin-orbit coupling constant for I.

[b] 10 mole % heavy-atom solvent in cyclohexane.

[c] 5 mole % heavy-atom solvent in cyclohexane.

2-hexanone, 42% of the reaction occurs from the singlet state and 58% from the triplet[(139)]). The photoelimination of three ketones (2-hexanone, 2-heptanone, and 2-octanone) in both hexane and *n*-propyl bromide, was investigated under conditions where the triplet reaction was eliminated by quenching with 2,5-dimethyl-2,4-hexadiene. Thus a comparison of the singlet reaction in propyl bromide as solvent with that in hexane as solvent should demonstrate the effect of the heavy atom on the rate of intersystem crossing of the carbonyl. After correction for the slight absorption of the propyl bromide (16% at 313 nm) the ratios of $\Phi_{\text{hexane}}/\Phi_{\text{propyl bromide}}$ were unity within experimental error. Similar conclusions concerning the internal heavy-atom effect on carbonyl possessing lowest $n \rightarrow \pi^*$ triplet states were made by Borkman and Kearns[(140)] through a spectroscopic study of various halogenated carbonyls.

Through a theoretical study of spin–orbit and vibronic coupling in the formaldehyde molecule, El-Sayed[(141)] concluded that spin–orbit perturbation due to the carbonyl moiety itself is comparable to or larger than intramolecular heavy-atom perturbation due to halogen substitution. This conclusion has, however, been questioned by Carroll *et al.*,[(144)] who, including terms neglected by El-Sayed, calculated triplet lifetimes varying as $H_2CO \sim 10^2$ sec, $F_2CO \sim 10^{-2}$ sec, $Cl_2CO \sim 10^{-2}$ sec, $Br_2CO \sim 10^{-4}$ sec, and $I_2CO \sim 10^{-5}$ sec. Heavy-atom effects on $S_1 \rightarrow T_1$ intersystem crossing, however, are probably unimportant since this process is already efficient in most carbonyls.

Since external heavy-atom perturbation is expected to be considerably less important than internal heavy-atom perturbation, one can conclude that use of a heavy-atom solvent should have a relatively small effect on intercombinational processes involving $n \rightarrow \pi^*$ transitions.

Thus we see that in molecules possessing $n \rightarrow \pi^*$ excited states intercombinational transitions (intersystem crossing, phosphorescence, and nonradiative triplet decay) should be efficient compared to the same processes in aromatic hydrocarbons. This conclusion is consistent with the high phosphorescence efficiencies and low fluorescence efficiencies exhibited by most carbonyl and heterocyclic compounds.

In carbonyls and heterocycles possessing both $n \rightarrow \pi^*$ and $\pi \rightarrow \pi^*$ excited states one would expect $S \rightarrow T_{n \rightarrow \pi^*}$ transitions to be 10^2–10^3 times more efficient than $S \rightarrow T_{\pi \rightarrow \pi^*}$ transitions.[142] We saw in Section 5.6a that the contribution of the overlap integral $\left(\int \chi_1 \chi_2 \, d\tau_0\right)$ in the transition probability is greater for transitions occurring between states of different type ($n, \pi^* \rightsquigarrow \pi, \pi^*$) than for transitions between states of the same type ($\pi, \pi^* \rightsquigarrow \pi, \pi^*$). Similarly, spin–orbit coupling between singlet states of different configurations is expected to be about 10^2 times greater than that between singlet and triplet states of the same configuration.[134,143] Thus for substituted carbonyls and heterocycles having lowest singlet and triplet states of $n \rightarrow \pi^*$ character (e.g., benzophenone) the pathway of intersystem crossing should be

$$S_{n,\pi^*} \rightsquigarrow T_{\pi,\pi^*} \rightsquigarrow T_{n,\pi^*} \longrightarrow h\nu \qquad (5.91)$$

while for compounds possessing lowest $n \rightarrow \pi^*$ singlets but lowest $\pi \rightarrow \pi^*$ triplets (e.g., substituted acetophenones):

$$S_{n,\pi^*} \rightsquigarrow T_{\pi,\pi^*} \longrightarrow h\nu \qquad (5.92)$$

Thus we would expect the phosphorescence efficiency to be greater for the first case than the second. In agreement with this conclusion, Φ_p for p-methylbenzophenone has been determined[110] to be 1.04 while Φ_p for p-methylacetophenone (lowest π, π^* triplet) is 0.61.[112] Similar effects are observed for heterocycles; for example, the phosphorescence quantum yield for pyrazine (lowest n, π^* triplet) is 0.30[119] while that for quinoline in a hydroxylic solvent (lowest π, π^* triplet) is 0.19.[30]

5.6d. *External Heavy-Atom Effects and Charge Transfer*

Although some workers suggest that exchange interactions are dominant in determining the effects of external heavy atoms,[145] experimental evidence at this point tends to favor charge transfer as the mechanism for external heavy atoms. This evidence[135] is summarized as follows: (a) The magnitude of the enhancement of intercombinational processes in aromatic compounds increases as the electron donor ability of the aromatic increases and as the electron acceptor ability of the perturber increases[32a]; (b) stoichiometries, as determined by changes in optical density of solutions of aromatic compounds and heavy-atom perturbers as compared to the optical densities of the individual components, indicate 1:1 complexes in general[146]; (c) Benesi–Hildebrand[147] plots for various aromatic–heavy-atom systems indicate 1:1 complexes with small but finite stability; (d) the optical density of equimolar solutions of 1-chloronaphthalene and ethyl iodide is temperature dependent; (e) the infrared stretching band (C—H stretch) for 1-chloronaphthalene in various heavy-atom solvents is sensitive to the perturbing medium.[146] The effects observed appear to suggest C—H compression due to the heavy-atom solvent. This would eliminate hydrogen bonding effects. The sum of this evidence would also appear to eliminate collisional complexes from being of dominant importance.

The best evidence for a charge-transfer exciplex (hetero excimer) has been provided by Thomaz and Stevens.[148,149] They note a reduction in fluorescence yield of pyrene with increasing heavy-atom concentration and proposed the following set of reactions to explain their results:

$$
\begin{array}{ccccccccc}
 & & A^t & & A + Q + h\nu_f & & & & \\
 & & \uparrow k_{isc} & & \uparrow k_f' & & & \nearrow & A + Q + h\nu_p \\
h\nu + A & \underset{k_f}{\rightleftharpoons} & A^s + Q & \underset{k_{-e}}{\overset{k_e}{\rightleftharpoons}} & (AQ)^s & \xrightarrow{k_{isc}'} & (AQ)^t & & \\
 & & \downarrow k_{ic} & & \downarrow k_{ic}' & & & \searrow\!\!\!\nwarrow & A^t + Q \\
 & & A & & A + Q & & & &
\end{array}
$$

where A is pyrene, Q is a heavy-atom compound, and $(AQ)^s$ is a singlet exciplex.

A similar kinetic scheme with $Q \equiv A$ can be used to account for concentration quenching, where $(AQ)^s = (AA)^s$, an excimer. Leonhardt and Weller[150] and Matoga, Okada, and Ezumi[151] first proposed such a scheme to account for the quenching of fluorescence from aromatic hydrocarbons by *N*-alkyl anilines. In addition, they observed a long-wavelength structureless emission from the excited hydrocarbon–aniline complex. The key step in the above scheme is the unimolecular reaction $(AQ)^s \xrightarrow{k_{-e}} A^s + Q$. Inasmuch as the activation energy required for this process is about 8 kcal/mole, a negative temperature coefficient for the rate constant k_q for the dimethylaniline quenching of perylene fluorescence is observed. Table 5.12 lists various

TABLE 5.12. *Properties of Exciplexes and Excimers in Hexane Solution*[a]

A	D	k, M^{-1}	$-\Delta H_a$, kcal/mole	$-\Delta S_a$, cal/mole-°C	E_R, kcal/mole	μ, D	τ_f^c, nsec	τ_f, nsec	Ref.
Anthracene*	$C_6H_5N(C_2H_5)_2$	>1400	10.2	18.7	5.5	10	105	4.7	152, 153, 154
Pyrene*	$C_6H_5N(C_2H_5)_2$	>130	8.4	18.3	5.0	11	130	30	152, 153, 155
Biphenyl	$C_6H_5N(C_2H_5)_2$*	—	7.3	17.3	8.0	13.5	—	—	152, 153
Pyrene*	Pyrene	—	9.4	17.6	7.7	0.0	67	300	153, 156, 157

[a] A and D are the exciplex or excimer components, * denotes the primarily excited species, k is the limiting photoassociation equilibrium constant, ΔH_a, ΔS_a, and E_R are the thermodynamic parameters for the exciplex–excimer, and μ is the excited state dipole moment of the complex. Note that the large dipole moment for the exciplex indicates almost complete charge transfer in the excited state, $(D^+, A^-)^*$. τ_f^c and τ_f are the fluorescence lifetimes for the complex and the * component.

measured quantities for the best studied excimers and exciplexes. Figure 2.27 helps explain the nature of exciplex formation (Chapter 2).

Some evidence for the process $(AQ)^s \rightarrow (AQ)^t$ when $Q \equiv A$, or for photoassociation of the triplet $A^t + A \rightarrow (AA)^t$, has been provided by Hoytink and co-workers,[158] who reported excimer phosphorescence from cooled ethanolic solutions of phenanthrene and naphthalene.

Based on analogies we have cited, the kinetic scheme proposed for heavy-atom fluorescence quenching is reasonable and would predict the following relationship for fluorescence quenching:

$$\begin{aligned}\frac{\Phi_f^0}{\Phi_f} &= 1 + k_e[Q]\tau_f^0 \frac{k_f' + k_{ic}' + k_{isc}'}{k_{-e} + k_f' + k_{ic}' + k_{isc}'} \\ &= 1 + k_q\tau_f^0[Q] \qquad (5.93)\end{aligned}$$

where

$$\begin{aligned}k_q &= k_e(k_f' + k_{ic}' + k_{isc}')/(k_{-e} + k_{ic}' + k_f' + k_{isc}') \\ \tau_f^0 &= 1/(k_f + k_{isc} + k_{ic})\end{aligned} \qquad (5.94)$$

Thus it is possible to obtain k_q from a graph of Φ_f^0/Φ_f vs. [Q] or from a fluorescence lifetime study,

$$1/\tau_f = (1/\tau_f^0) + k_q[Q] \qquad (5.95)$$

From measurements of this type Thomaz and Stevens found a linear relationship for a graph of $\log(k_q/n^2\zeta^2)$ vs. $E_{1/2}$, where n is the number of halogen atoms in the molecule, ζ is the spin–orbit coupling constant, and $E_{1/2}$ is the polarographic half-wave reduction potential of the heavy-atom quencher (Figure 5.16). This correlation suggests that an exciplex is formed by partial

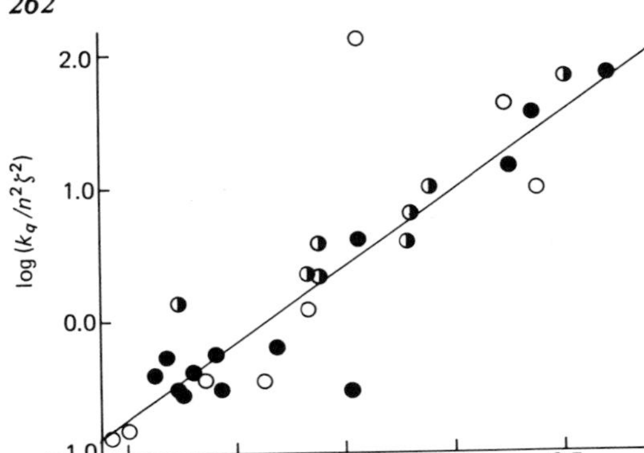

FIGURE 5.16. Plot of data for the external heavy-atom quenching of pyrene fluorescence in benzene at 20°C. Polarographic half-wave reduction potentials $E_{1/2}$ are used as a measure of the electron affinity of the quencher containing chlorine (○), bromine (●), or iodine (◑). From Thomaz and Stevens[148] with permission of W. A. Benjamin, New York.

electron transfer from pyrene singlet state to the heavy-atom quencher. The higher is the electron affinity of the quencher, the smaller will be k_{-e} and the larger will be the value of k_q; [see Eq. (5.94).]

REFERENCES

1. J. Dewar, *Proc. Roy. Soc. (London)* **36**, 164 (1880); *Proc. Roy. Inst. G. B.* **12**, 557 (1888); *Proc. Roy. Soc. (London)* **43**, 1078 (1888).
2. E. Wiedemann, *Ann. Phys.* **34**, 446 (1888).
3. H. Kautsky and A. Hirsch, *Ber.* **64**, 2677 (1931); **65**, 401 (1932); H. Kautsky, A. Hirsch, and F. Davidshofer, *Ber.* **65**, 1762 (1932); H. Kautsky, H. deBruijn, R. Neuwirth, and W. Baumeister, *Ber.* **66**, 1588 (1933).
4. A. Jablonski, (a) *Nature* **131**, 839 (1933); (b) *Z. Physik.* **94**, 38 (1935).
5. A. Terenin, *Acta Physicochim. URSS* **18**, 210 (1943); *Zh. Fiz. Khim.* **18**, 1 (1944).
6. G. N. Lewis, D. Lipkin, and T. T. Magel, *J. Amer. Chem. Soc.* **63**, 3005 (1941).
7. G. N. Lewis and M. Kasha, *J. Amer. Chem. Soc.* **66**, 2100 (1944); **67**, 994 (1945).
8. G. N. Lewis and M. Calvin, *J. Amer. Chem. Soc.* **67**, 1232 (1945).
9. D. F. Evans, *Nature* **176**, 777 (1955).
10. R. Lesclaux and J. Joussot-Dubien, *J. Chim. Phys.* **61**, 1147 (1964).
11. N. J. Turro, *J. Chem. Ed.* **46**, 2 (1969).
12. F. Perrin, *Ann. Phys. (Paris)* **12**, 169 (1929).
13. W. L. Leoshin and L. A. Vinokurov, *Physik. Z. Sowjet U.* **10**, 10 (1936).
14. C. A. Parker and C. G. Hatchard, *Trans. Faraday Soc.* **57**, 1894 (1961).
15. P. P. Dikun, *Zh. Eksperim. i Teor. Fiz.* **20**, 193 (1950); R. Williams, *J. Chem. Phys* **28**, 577 (1958).
16. Th. Förster and K. Kasper, *Z. Elektrochem.* **59**, 977 (1955).
17. B. Stevens and E. Hutton, *Nature* **186**, 1045 (1960); *Spectrochim. Acta* **18**, 425 (1962).
18. C. A. Parker and C. G. Hatchard, *Trans. Faraday Soc.* **59**, 284 (1963); *Proc. Chem. Soc.*, 147 (1962); *Proc. Roy. Soc. (London) A* **269**, 574 (1962).
19. C. A. Parker, C. G. Hatchard, and T. A. Joyce, *Nature* **205**, 1282 (1965).
20. C. A. Parker, *Photoluminescence of Solutions*, Elsevier Publishing Company, Amsterdam (1968).

21. J. D. Winefordner and P. A. St. John, *Anal. Chem.* **35**, 2211 (1963).
22. H. Greenspan and E. Fischer, *J. Phys. Chem.* **69**, 2466 (1965).
23. J. B. Birks and D. J. Dyson, *Proc. Roy. Soc. A* **275**, 135 (1963).
24. J. B. Birks, *Photophysics of Aromatic Molecules*, Wiley–Interscience, London (1970).
25. D. F. Evans, *J. Chem. Soc.*, 1351 (1957).
26. (a) H. Tsubomura and R. S. Mulliken, *J. Amer. Chem. Soc.* **82**, 5966 (1960); (b) J. N. Murrell, *Molec. Phys.* **3**, 319 (1960).
27. (a) K. Kawaoka, A. U. Khan, and D. R. Kearns, *J. Chem. Phys.* **46**, 1842 (1967); (b) A. U. Khan and D. R. Kearns, *J. Chem. Phys.* **48**, 3272 (1968).
28. S. P. McGlynn, T. Azumi, and M. Kasha, *J. Chem. Phys.* **40**, 507 (1964).
29. G. W. Robinson, *J. Mol. Spectrosc.* **6**, 58 (1961).
30. D. S. McClure, *J. Chem. Phys.* **17**, 905 (1949).
31. M. Kasha, *J. Chem. Phys.* **20**, 71 (1952).
32. (a) S. P. McGlynn, R. Sunseri, and N. Christodouleas, *J. Chem. Phys.* **37**, 1818 (1962); (b) S. P. McGlynn, M. J. Reynolds, G. W. Daigre, and N. D. Christodouleas, *J. Phys. Chem.* **66**, 2499 (1962); (c) N. D. Christodouleas and S. P. McGlynn, *J. Chem. Phys.* **40**, 166 (1964).
33. A. P. Marchetti and D. R. Kearns, *J. Amer. Chem. Soc.* **89**, 768 (1967).
34. G. Porter and M. W. Windsor, *Proc. Roy. Soc. A* **245**, 238 (1958).
35. D. S. McClure, *J. Chem. Phys.* **19**, 670 (1951).
36. R. Pariser, *J. Chem. Phys.* **24**, 250 (1956).
37. S. D. Colson and E. R. Bernstein, *J. Chem. Phys.* **43**, 2661 (1965).
38. J. Franck and G. Hertz, *Verhandl. Deut. Physik. Ges.* **16**, 457 (1914); *Physik. Z.* **17**, 409 (1916).
39. E. N. Lassetre, *Radiation Res., Suppl.* **1**, 530 (1959).
40. A. Kuppermann and L. M. Raff, *J. Chem. Phys.* **39**, 1607 (1963); E. N. Lassitre, *J. Chem. Phys.* **42**, 2971 (1966); J. R. Oppenheimer, *Phys. Rev.* **32**, 361 (1928).
41. A. Kuppermann and L. M. Raff, *Disc. Faraday Soc.* **35**, 30 (1963); *J. Chem. Phys.* **37**, 2497 (1962).
42. A. Kuppermann, J. K. Rice, and S. Trajmar, *J. Phys. Chem.* **72**, 3894 (1968).
43. M. A. Dillon, *Creation and Detection of the Excited State*, Vol. 1, A. A. Lamola, ed., Marcel Dekker, New York (1971), Chapter 8.
44. J. P. Doering, *J. Chem. Phys.* **51**, 2866 (1969).
45. (a) W. G. Herkstroeter, A. A. Lamola, and G. S. Hammond, *J. Amer. Chem. Soc.* **86**, 4537 (1964); (b) D. R. Arnold, R. L. Hinman, and A. H. Glick, reported in *Advances in Photochemistry*, Vol. 6 (1968), p. 328.
46. D. F. Evans, *J. Chem. Soc.*, 1351 (1957); 3885 (1957); 2753 (1959); 1735 (1960); 1987 (1961).
47. R. E. Kellogg and W. T. Simpson, *J. Amer. Chem. Soc.* **87**, 4230 (1965).
48. W. G. Herkstroeter and G. S. Hammond, *J. Amer. Chem. Soc.* **88**, 4769 (1966).
49. A. A. Lamola, W. G. Herstroeter, J. C. Dalton, and G. S. Hammond, *J. Chem. Phys.* **42**, 1715 (1965).
50. P. Wagner, *J. Amer. Chem. Soc.* **89**, 2980 (1967).
51. G. Jackson and G. Porter, *Proc. Roy. Soc. (London) A* **260**, 13 (1961).
52. A. Terenin and V. L. Ermolaev, *J. Chim. Phys.* **55**, 698 (1958).
53. P. G. Bowers and G. Porter, *Proc. Roy. Soc. (London) A* **296**, 435 (1967).
54. T. Medinger and F. Wilkinson, *Trans. Faraday Soc.* **61**, 620 (1965).
55. A. A. Lamola and G. S. Hammond, *J. Chem. Phys.* **43**, 2129 (1965).
56. L. M. Stephenson, Ph.D. Thesis, California Institute of Technology, Pasadena, California (1967).

57. C. A. Parker and T. A. Joyce, *Chem. Comm.*, 234 (1966).
58. K. Sandros, *Acta Chim. Scand.* **23**, 2815 (1969).
59. B. Stevens and B. E. Algar, *Chem. Phys. Lett.* **1**, 58 (1967).
60. B. Stevens and B. E. Algar, *Chem. Phys. Lett.* **1**, 219 (1967).
61. C. A. Parker and T. A. Joyce, *Photochem. Photobiol.* **6**, 395 (1967).
62. C. A. Parker and T. A. Joyce, *Chem. Comm.*, 108 (1966).
63. C. A. Parker and T. A. Joyce, *Trans. Faraday Soc.* **62**, 2785 (1966).
64. I. V. Aleksandrov and K. K. Pukhov, *Opt. Spectrosc.* **17**, 513 (1964); J. S. Brinen, *J. Chem. Phys.* **49**, 586 (1968).
65. A. R. Horrocks and F. Wilkinson, *Proc. Roy. Soc. A* **306**, 257 (1968).
66. R. E. Kellogg and R. G. Bennett, *J. Chem. Phys.* **41**, 3042 (1964).
67. I. B. Berlman, *Handbook of Fluorescence Spectra of Aromatic Molecules*, Academic Press, New York (1965).
68. A. R. Horrocks, T. Medinger, and F. Wilkinson, in *International Symposium on Luminescence. The Physics and Chemistry of Scintillators*, N. Ruhl and H. Kallmann, eds., Verlag Karl Thiemig, Munich (1966).
69. M. W. Windsor and W. R. Dawson, *Mol. Cryst.* **4**, 253 (1968).
70. R. G. Bennett and P. J. McCartin, *J. Chem. Phys.* **44**, 1969 (1966).
71. A. R. Horrocks, T. Medinger, and F. Wilkinson, *Trans. Faraday Soc.* **62**, 1785 (1966).
72. T. Medinger and F. Wilkinson, *Trans. Faraday Soc.* **62**, 1785 (1966).
73. J. B. Birks, B. N. Srinivasan, and S. P. McGlynn, *J. Mol. Spectrosc.* **27**, 266 (1968).
74. H. Labhart, *Helv. Chim. Acta* **47**, 2279 (1964).
75. W. R. Dawson, *J. Opt. Soc. Am.* **58**, 222 (1968).
76. W. Heinzelmann and H. Labhart, *Chem. Phys. Lett.* **4**, 20 (1969).
77. B. Soep, A. Kellmann, M. Martin, and L. Lindquist, *Chem. Phys. Lett.* **13**, 241 (1972).
78. G. Porter, *Tech. Org. Chem.* **8**, 1081 (1961).
79. N. J. Turro, *Molecular Photochemistry*, Benjamin, New York (1965).
80. R. Li and E. C. Lim, *J. Chem. Phys.* **57**, 605 (1972).
81. M. R. Wright, R. P. Frosch, and G. W. Robinson, *J. Chem. Phys.* **33**, 934 (1960).
82. M. R. Wright, R. P. Frosch, and G. W. Robinson, *J. Chem. Phys.* **38**, 1187 (1963).
83. E. C. Lim, *J. Chem. Phys.* **36**, 3497 (1962).
84. G. W. Robinson and R. P. Frosch, *J. Chem. Phys.* **37**, 1962 (1962).
85. R. E. Kellogg and R. P. Schwenker, *J. Chem. Phys.* **41**, 2860 (1964).
86. D. P. Craig and I. G. Ross, *J. Chem. Soc.*, 1589 (1954).
87. D. Ramsey, in *Determination of Organic Structure by Physical Methods*, F. C. Nachod and W. D. Phillips, eds., Academic Press, New York (1962), Vol. 2.
88. G. W. Robinson and V. E. DiGiorgio, *Can. J. Chem.* **36**, 31 (1968); *J. Chem. Phys.* **31**, 1678 (1959).
89. S. E. Hodges, J. R. Henderson, and J. B. Coon, *J. Mol. Spectrosc.* **2**, 99 (1958).
90. A. D. Walsh, *J. Chem. Soc.*, 2260 (1953).
91. J. C. D. Brand, *J. Chem. Soc.*, 858 (1956).
92. G. W. Robinson, *Can. J. Phys.* **34**, 699 (1956).
93. P. G. Wilkinson and R. S. Mulliken, *J. Chem. Phys.* **23**, 1895 (1955).
94. R. S. Mulliken and C. C. J. Roothaan, *Chem. Revs.* **41**, 219 (1947).
95. H. C. Longuet-Higgins, *J. Chem. Phys.* **18**, 265 (1950).
96. S. P. McGlynn, T. Azumi, and M. Kinoshita, *Molecular Spectroscopy of the Triplet State*, Prentice-Hall, Englewood Cliffs, New Jersey (1969).
97. C. K. Ingold and G. W. King, *J. Chem. Soc.*, 2702 (1953).
98. K. K. Innes, *J. Chem. Phys.* **22**, 863 (1954).

99. J. H. Callomon and B. P. Stoicheff, *Can. J. Phys.* **35**, 373 (1957).
100. J. Pancir and R. Zahradnik, *Theoret. Chim. Acta* (*Berl.*) **14**, 426 (1969).
101. E. Gilmore, G. Gibson, and D. McClure, *J. Chem. Phys.* **20**, 829 (1952); **23**, 399 (1955).
102. Y. Konda and H. Sponer, *J. Chem. Phys.* **28**, 798 (1958).
103. D. Olness and H. Sponer, *J. Chem. Phys.* **38**, 1799 (1963).
104. V. Ermolaev and K. Svitashev, *Opt. Spectrosc.* **7**, 399 (1959).
105. R. Kellogg and N. Wyeth, *J. Chem. Phys.* **45**, 3156 (1966).
106. R. Bennett and J. McCartin, *J. Chem. Phys.* **44**, 1969 (1966).
107. R. Borkman and D. Kearns, *J. Chem. Phys.* **44**, 945 (1966).
108. V. Ermolaev and A. Terenin, *Soviet Phys.—Uspekhi* **3**, 423 (1960).
109. J. Pitts, H. Johnson, and T. Kuwana, *J. Phys. Chem.* **66**, 245 (1962).
110. R. Borkman and D. Kearns, *J. Chem. Phys.* **46**, 2333 (1966).
111. A. Terenin and V. Ermolaev, *Trans. Faraday Soc.* **57**, 1042 (1956).
112. N. C. Yang, P. McClure, S. Murov, J. Hauser, and R. Dusenbery, *J. Amer. Chem. Soc.* **89**, 5466 (1967).
113. E. Baum, J. Wan, and J. Pitts, *J. Amer. Chem. Soc.* **88**, 2652 (1966).
114. D. Murov, Ph.D. Dissertation, University of Chicago (1967).
115. W. Neeley and H. Dearman, *J. Chem. Phys.* **44**, 1302 (1966).
116. B. Cohen and L. Goodman, *J. Chem. Phys.* **46**, 548 (1962).
117. J. Paris, R. Hirt, and R. Schmitt, *J. Chem. Phys.* **34**, 1851 (1961).
118. J. Brinen, J. Koren, and W. Hodgson, *J. Chem. Phys.* **44**, 3095 (1966).
119. V. Ermolaev and I. Kotlyar, *Opt. Spectrosc.* **9**, 183 (1960).
120. E. Lim and J. Yu, *J. Chem. Phys.* **45**, 4742 (1966); **49**, 3878 (1968).
121. J. Vincent and A. Maki, *J. Chem. Phys.* **39**, 3088 (1963).
122. R. Harrell, Ph.D. Dissertation, Florida State University (1959).
123. J. Brinen, D. Rosebrook, and R. Hirt, *J. Phys. Chem.* **67**, 2651 (1963).
124. R. Heckman, *J. Mol. Spec.* **2**, 27 (1968).
125. V. Ermolaev, *Opt. Spectrosc.* **11**, 266 (1961).
126. J. Longworth, R. Rahn, and R. Shulman, *J. Chem. Phys.* **45**, 2930 (1966).
127. J. Drobnik and L. Augenstein, *Photochem. Photobiol.* **5**, 13 (1966).
128. C. Helene, R. Santus, and P. Douzou, *Photochem. Photobiol.* **5**, 127 (1966).
129. R. S. Becker, *Theory and Interpretation of Fluorescence and Phosphorescence*, Wiley–Interscience, New York (1969).
130. R. L. E. Drisko, Ph.D. Dissertation, The Johns Hopkins University (1968).
131. M. A. El-Sayed, *Acc. Chem. Res.* **1**, 8 (1968).
132. W. Kauzmann, *Quantum Chemistry*, Academic Press, New York (1957).
133. G. W. Robinson and R. Frosch, *J. Chem. Phys.* **38**, 1187 (1963).
134. S. K. Lower and M. A. El-Sayed, *Chem. Rev.* **66**, 199 (1966).
135. S. P. McGlynn, T. Azumi, and M. Kinoshita, *The Triplet State*, Prentice-Hall, Englewood Cliffs, New Jersey (1969).
136. D. O. Cowan and R. L. E. Drisko, *J. Amer. Chem. Soc.* **92**, 6281 (1970).
137. G. G. Giachino and D. R. Kearns, *J. Chem. Phys.* **52**, 2964 (1970).
138. P. J. Wagner, *J. Chem. Phys.* **45**, 2335 (1966).
139. P. J. Wagner and G. S. Hammond, *J. Amer. Chem. Soc.* **87**, 4009 (1965); **88**, 1245 (1966).
140. R. F. Borkman and D. R. Kearns, *J. Chem. Phys.* **46**, 2333 (1967).
141. M. A. El-Sayed, *J. Chem. Phys.* **41**, 2462 (1964).
142. D. Kearns and W. Case, *J. Amer. Chem. Soc.* **88**, 5087 (1966).
143. M. A. El-Sayed, *J. Chem. Phys.* **38**, 2834 (1963).

144. D. G. Carroll, L. Vanquickenborne, and S. P. McGlynn, *J. Chem. Phys.* **45**, 2777 (1966).
145. C. Dijkgraaf and G. J. Hoijtink, *Tetrahedron Suppl.* **2**, 179 (1963).
146. V. Ramakrishnan, R. Sunseri, and S. P. McGlynn, *J. Chem. Phys.* **45**, 1365 (1966).
147. H. A. Benesi and J. H. Hildebrand, *J. Amer. Chem. Soc.* **71**, 2703 (1949).
148. M. F. Thomaz and B. Stevens, in *Molecular Luminescence*, E. C. Lim, ed., Benjamin, New York (1969).
149. B. Stevens, in *Advances in Photochemistry*, Vol. 8, Wiley Interscience, New York (1971), p. 161.
150. H. Leonhardt and A. Weller, *Ber. Bunsenges. Phys. Chem.* **67**, 791 (1963).
151. N. Matago, T. Okada, and K. Ezumi, *Mol. Phys.* **10**, 203 (1966).
152. A. Weller, in *Fast Reactions and Primary Processes in Chemical Kinetics*, S. Claesson, ed., Wiley–Interscience, New York (1967), p. 413.
153. H. Beens, H. Knibbe, and A. Weller, *J. Chem. Phys.* **47**, 1183 (1967).
154. H. Knibbe, K. Röllig, F. P. Schafer, and A. Weller, *J. Chem. Phys.* **47**, 1184 (1967).
155. N. Matago, T. Okada, and N. Yamamoto, *Chem. Phys. Lett.* **1**, 119 (1967).
156. J. B. Birks, M. P. Lamb, and I. H. Munro, *Proc. Roy. Soc.* (*London*) *A* **280**, 289 (1964).
157. E. Döller and Th. Förster, *Z. Phys. Chem. N.F.* **34**, 132 (1962).
158. J. Langelaar, R. P. H. Rettschnick, A. M. F. Lamboy, and G. J. Hoytink, *Chem. Phys. Lett.* **1**, 609 (1967).
159. D. R. Scott and J. B. Allison, *J. Phys. Chem.* **66**, 561 (1962).
160. G. Porter, *Tech. Org. Chem.* **8**, 1055 (1961).

6

Electronic Energy Transfer

Intramolecular and intermolecular electronic energy transfers from singlet and triplet states are of great interest both theoretically and experimentally because of their importance in radiation chemistry, molecular spectroscopy, photochemistry, and photobiology. A number of reviews concerning the experimental and theoretical aspects of energy transfer have appeared recently.[1–6] In this chapter we will discuss the theories for electronic energy transfer and the role that energy transfer plays in organic photochemistry.

6.1. EXCITATION TRANSFER WITHIN A CHROMOPHORE SYSTEM

6.1a. *Internal Conversion and Intersystem Crossing Theory*

In Chapters 4 and 5 we made use of the theory of radiationless transitions developed by Robinson and Frosch.[7] In this theory the transition is considered to be due to a time-dependent intramolecular perturbation on nonstationary Born–Oppenheimer states. Henry and Kasha[8] and Jortner and co-workers[9–12] have pointed out that the Born–Oppenheimer (BO) approximation is only valid if the energy difference between the BO states is large relative to the vibronic matrix element connecting these states. When there are near-degenerate or degenerate zeroth-order vibronic states belonging to different configurations the BO approximation fails.

In the more recent theory[8–13] the BO zeroth-order states are not considered to be pure, because of configuration interaction between the nearly degenerate zeroth-order states. That is, the system under investigation is in a compound state. The zeroth-order states of the system have no real physical significance; such properties of the total system as absorption coefficients, relaxation times, etc., must be derived from the compound states. While the theory is beyond the level of this text, the particular advantages of this approach as described in Ref. 12 are as follows.

(a) "A theory based on the use of compound states automatically focuses attention on the exact molecular eigenstates of the system. It is these states which are needed in the description of absorption line shapes and other spectroscopic properties of the system."[(12)]

(b) "An important implication of the break down of the BO approximation in the excited electronic states of medium and large molecules is the prediction of inhomogeneous line broadening."[(12)] This theory generates a scheme for extracting information about electronic relaxation processes from spectroscopic line shape.

(c) "If the widths of two or more inhomogeneously broadened molecular eigenstates exceed their separation, interference effects may be observed."[(12)]

(d) "In the general analysis based on the use of compound molecular states the irreversibility of the electronic relaxation process enters, without *ad hoc* assumptions, in a natural fashion."[(12)]

(e) "The present theory accounts properly for the properties of coupled radiative and radiationless decay processes and, for the first time, elucidates the nature of the phenomenological rate constants used to describe the gross kinetics of the electronic relaxation in large molecules."[(12)]

The empirical relations

$$1/\tau = 1/\tau_r + 1/\tau_n$$

$$\Phi_f = \tau/\tau_r = \tau_n/(\tau_r + \tau_n)$$

where r denotes radiative and n denotes nonradiative, have been justified.

(f) "The general theoretical approach used leads to a unified formal description of photochemical dissociation reactions and of simple photorearrangement processes."[(12)]

However, it has been pointed out[(13–16)] for large organic molecules ("statistical limit" case) that the decay times and quantum yields can legitimately be handled by the Fermi golden rule:

$$k_{sl} = \frac{2\pi}{\hbar} \sum_{i,j} \frac{\exp(-\beta E_{si})}{\sum_i \exp(-\beta E_{si})} \langle \Phi_{si} | v | \Phi_{lj} \rangle^2 \, \delta(E_{si} - E_{lj})$$

where Φ_{si} and Φ_{lj} are the Born–Oppenheimer wave functions of the initial and final vibronic states with energies E_{si} and E_{lj}. The indices i and j denote the vibronic states Φ_{s1}, Φ_{s2} and Φ_{l1}, Φ_{l2}. This equation of course can be factored into the Robinson–Frosch equation. Consequently, in the statistical limit (molecules the size of benzene and larger) both theories give similar expressions for the rate of radiationless transitions.

6.1b. *Radiationless Transitions: Phosphorescence Microwave Double Resonance*[17]

With the possible exception of nanosecond and picosecond flash photolysis, phosphorescence microwave double resonance is probably the most exciting new technique available to help characterize excited state processes. Most of the phosphorescence studies that have been made since Lewis and Kasha[18] in 1944 determined that the long-lived emission originated from the triplet state have used mixtures of organic solvents that form glasses at 77°K. We noted in Chapter 5 that even in the absence of a magnetic field, the triplet state of an organic molecule is in fact composed of three closely spaced "micro states." This splitting of states in the absence of a magnetic field results from the anisotropy of the spin–spin interaction in the molecule. At 77°K the thermal energy available to establish a Boltzmann distribution is about 50 cm^{-1} (kT). This is over two orders of magnitude larger than the zero-field splitting (zfs) between the three tiplet-state magnetic levels of T_1 for most organic molecules. Further, at 77°K the spin–lattice relaxation (slr) between these zf levels is much faster than phosphorescence from the levels. Consequently, the phosphorescence decay, polarization, and spectral distribution observed are a population-weighted average of the properties of individual zf states. Below 4.2°K the slr processes are slow and thermal equilibrium between the three zf levels is not established. This means that at temperatures below 4.2°K the triplet state retains memory of the path by which it was formed inasmuch as the probability of intersystem crossing to each zf level can be quite different. This is a result of the fact that "when a molecule changes its spin state from the lowest singlet (S_1) to one of the triplet states, T_i, it does so by exchanging spin angular momentum with orbital angular momentum. This leads not only to a change in the total spin angular momentum from $S = 0$ (singlet state) to $S = 1$ (triplet state), *but also to a preferential spin direction in the molecular framework*, depending on the symmetry of the spatial distribution of the excited electrons in S_1 and T_i."[17a] The molecule pyrazine is a good example.[19] Intersystem crossing from the lowest singlet state of pyrazine, $^1B_{3u}(n, \pi^*)$, to only one spin state (τ_y) of the $^3B_{1u}(\pi, \pi^*)$ is allowed by spin–orbit coupling

$$\Pr \propto \langle {}^1\Psi_{n,\pi^*} | H_{so} | {}^3\Psi_x \rangle^2$$

TABLE 6.1. *The Spin–Orbit Functions of Different Triplet States That Are Below the Lowest Singlet State* $[^1B_{3u}(n, \pi^*)]$ *of Pyrazine*[19]

Triplet state	Spin function	Spin–orbit function[a]
$^3B_{1u}(\pi, \pi^*)$	τ_x	B_{2u}
	τ_y	$\boxed{B_{3u}}$
	τ_z	A_u
$^3B_{3u}(n, \pi^*)$	τ_x	A_u
	τ_y	B_{1u}
	τ_z	B_{2u}

[a] Spin–orbit function for the lowest singlet is $B_{3u} \times a_g = B_{3u}$.

where H_{so} should be considered totally symmetric. This conclusion is based on the symmetries shown in Table 6.1,

$$\langle B_{3u}|\text{symmetric}|B_{3u}\rangle^2 \neq 0$$

In addition, it can be shown that second-order vibronic perturbation will make possible some intersystem crossing to the $^3B_{3u}(n, \pi^*)$ state. However, this second-order perturbation should be much less important than the first-order spin–orbit perturbation.[19] This will produce the unequal population of the spin states shown in Figure 6.1. In the absence of slr the ratios of population densities n are given by the following equations:

$$\begin{aligned} n_z/n_y &= K_z k_y/K_y k_z \\ n_z/n_x &= K_z k_x/K_x k_z \\ n_x/n_y &= K_x k_y/K_y k_x \end{aligned} \tag{6.1}$$

where K_z and k_z are the rate constants for intersystem crossing ($S_1 \rightsquigarrow T_1[\tau_z]$) and for the $T_1(\tau_z) \rightsquigarrow S_0$ radiative and nonradiative processes, respectively.

Possibly the most easily observable process is the phosphorescence decay.[17] While at 77°K the decay is exponential, at 1.6°K the observed decay for compounds like pyrazine is nonexponential and is composed of three first-order decays. Thus one can determine the values of k_x, k_y, and k_z from this decay. Typical values are shown in Table 6.2.

TABLE 6.2. *Comparison of Phosphorescence Lifetimes* (sec^{-1}) *of the Individual Zero-Field Levels*[20]

zf level	Quinoxaline-d_6	2,3-Dichloroquinoxaline
k_z	6.7 ± 0.7	3.0 ± 0.3
k_y	≤ 0.28	7.1 ± 0.5
k_x	≤ 0.24	≤ 0.4

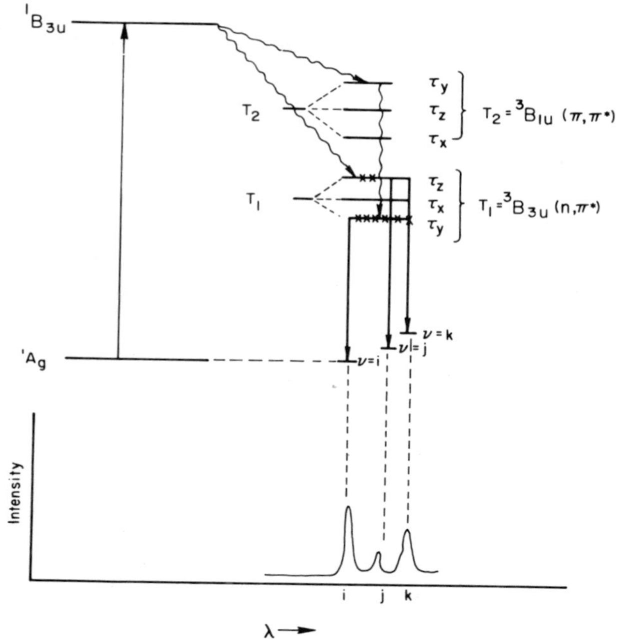

FIGURE 6.1. Jablonski-type diagram for pyrazine. The zero-field splittings (between τ_x, τ_y, τ_z) are not drawn to scale. Spin polarization (× × ×) resulting from the most probable intersystem crossing routes and part of the emission spectrum where different vibronic bands ($\nu = i, j, k$) have different zf origins are schematically indicated. (After El-Sayed.[17])

For pyrazine three exponential decays of lifetime 6, 130, and 400 msec can be extracted. From these three values the 77°K rate constant for phosphorescence decay can be calculated:

$$k^{77} = \tfrac{1}{3}k_z + \tfrac{1}{3}k_y + \tfrac{1}{3}k_x$$

$$= \tfrac{1}{3}\cdot\tfrac{1}{6} + \tfrac{1}{3}\cdot\tfrac{1}{130} + \tfrac{1}{3}\cdot\tfrac{1}{400} \approx \tfrac{1}{3}\cdot\tfrac{1}{6} = \tfrac{1}{18}$$

The corresponding lifetime of 18 msec is in excellent agreement with the observed triplet lifetime of pyrazine at 77°K.[19]

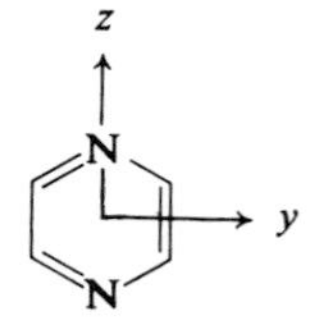

pyrazine

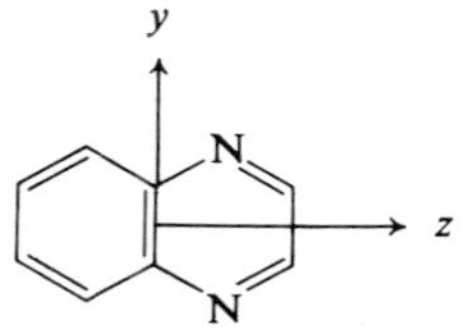

quinoxaline

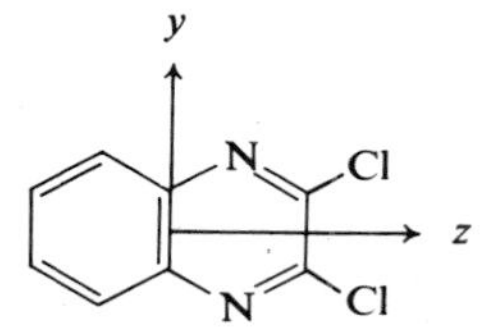

2,3-dichloroquinoxaline

Tinti and El-Sayed[20] have pointed out that even though the effect of the heavy atoms on the overall lifetime of 2,3-dichloroquinoxaline is not very large (see Table 6.2), its effect on the decay of the τ_y zf level is quite large, <0.28 to 7 sec^{-1}. For the τ_z levels there is a small "negative" heavy-atom effect. The large change in k_y is due to an increase in the radiative rate of decay from that spin state. In addition, a number of other conclusions regarding this system have been extracted from the lifetimes by El-Sayed. While it is possible to obtain the three rate constants (k_x, k_y, k_z) from the nonexponential decay, it is more accurate to observe the exponential decay of individual vibronic bands. As indicated in Figure 6.1, the different vibronic bands can arise from different zf states. If we know which bands are associated with which zf states, it is possible to monitor the decay of one vibronic band and more directly determine k_x, k_y, or k_z.

It is relatively easy to decide which vibronic bands have a common origin. This is accomplished by observing the phosphorescence intensity change of each band upon microwave saturation at a frequency that corresponds to transitions between τ_z and τ_x. This is known as phosphorescence–microwave double resonance (PMDR) spectroscopy. These frequencies for 2,3-dichloroquinoxaline are given in Table 6.3.

From Figure 6.1 it is easy to see that microwave saturation of the τ_y–τ_z transition of pyrazine should decrease the $\nu = i$ transition intensity and increase the $\nu = j + k$ transition intensity. Microwave saturation of the τ_x–τ_y transition should have no effect on $\nu = i$, but decrease the intensity of $\nu = j + k$.

To determine not only which bands have a common zf origin but also to assign the origin (τ_x, τ_y, τ_z) requires more information. If the microwave power is modulated[22] at a frequency not very different from $1/\tau$, where τ is the emission lifetime, to which the detection system is locked, then all bands with a common origin will have the same sign (see Figure 6.2). Figure 6.2(I) gives the phosphorescence spectrum of 2,3-dichloroquinoxaline oriented in a durene crystal at 1.6°K. Figure 6.2(II) is the amplitude-modulated phosphorescence–microwave double resonance spectrum (am-PMDR). Bands that originate from different zf levels appear with opposite signs in

TABLE 6.3. *Zero-Field Energy Levels*[20,21] *for 2,3-Dichloroquinoxaline*

Transition	Frequency, GHz
$\tau_z \rightarrow \tau_y$	1.055
$\tau_y \rightarrow \tau_x$	2.457
$\tau_z \rightarrow \tau_x$	3.512

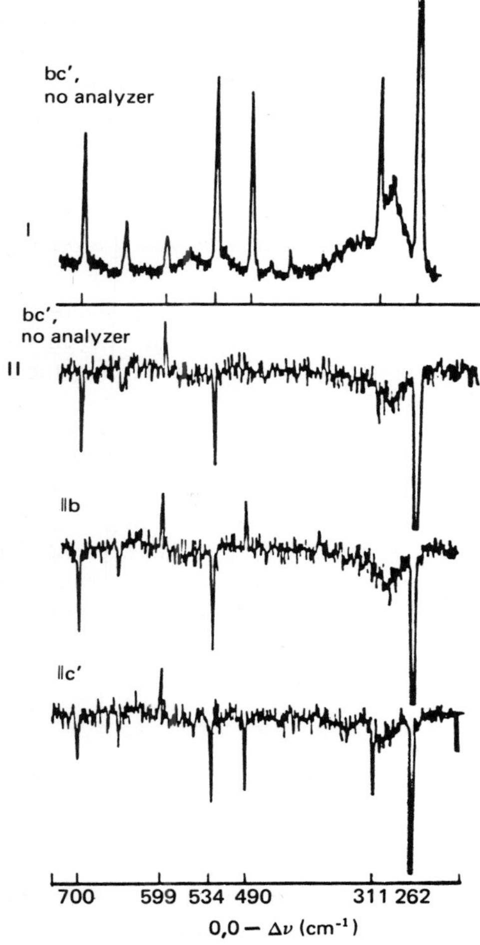

FIGURE 6.2. (I) Conventional phosphorescence spectrum of 2,3-dichloroquinoxaline in durene at 1.6°K. (II) am-PMDR spectrum, obtained by amplitude modulation of microwave radiation that pumps the τ_y–τ_z (1.055 GHz) zf transition with the detection at the modulation frequency. Only bands whose intensities change upon microwave radiation (1.055 GHz) and thus originate from τ_y or τ_z appear in the am-PMDR spectrum. Transitions from τ_x and τ_y appear with opposite sign (phase-shifted by 180°). (IIb, IIc′) Polarization of the am-PMDR spectral transitions, relative to the crystal axes. The band at 0,0-490 cm^{-1} originates from both the τ_y and τ_z spin states; its intensity does not change upon the 1.055-GHz saturation (no band in II); however, its polarization does change (bands in IIb and IIc′). (Reproduced with permission from M. A. El-Sayed.[17b])

the am-PMDR spectrum. The effect of inserting a polarizer parallel to the two crystal axes is shown in Figure 6.2(IIb, IIc). From the polarization spectra obtained when the different zf transitions are saturated and a knowledge of how the guest molecules are oriented in the host crystal (often durene) a complete assignment of the zf origins of the different vibronic bands in the phosphorescence spectrum can be made.

6.1c. *Zero-Field Optically Detected Magnetic Resonance* (*ODMR*)

It is possible to determine the energy differences between τ_x, τ_y, τ_z (for example, those given in Table 6.3) by monitoring the change in phosphorescence intensity of the sample (1.6°K) while varying the frequency of the microwave field. The phosphorescence intensity from a particular zf level will change when the microwave frequency corresponds to the zf transition involving the emitting level. This was first observed by Schmidt and van der Waals and others.[23] If the effect of magnetic field orientation on the ODMR spectra of guest molecules oriented in a host single crystal is studied, it is possible to assign the zf transitions, that is, $\tau_x \leftrightarrow \tau_y$. There are various modifications, such as optically detected electron–electron double resonance at zero field and optically detected electron nuclear double resonance at zero field, that we will not consider here.

Intersystem Crossing[24]

The intensity of phosphorescence from the *i* level is given by

$$I_i = \gamma_1 k_i^r n_i \tag{6.2}$$

where k_i^r is the radiative rate constant of the *i*th zf level; n_i is the population density; and γ_1 is the fraction of the total intensity emitted from the *i* level that appears in the vibronic band monitored. If we compare the intensity emitted from two states τ_1 and τ_2 using two vibronic phosphorescence bands, one band from τ_1 and one from τ_2, we obtain the following equation:

$$I_1/I_2 = \gamma_1 k_1^r n_1/\gamma_2 k_2^r n_2 \tag{6.3a}$$

If we saturate the $\tau_1 \leftrightarrow \tau_2$ transition, then $n_1 = n_2$, and we have

$$I_1^v/I_2^v = \gamma_1 k_1^r/\gamma_2 k_2^r \tag{6.3b}$$

A ratio of Eqs. (6.3a) and (6.3b) gives

$$\frac{I_1^v/I_2^v}{I_1/I_2} = \frac{n_2}{n_1} \tag{6.4}$$

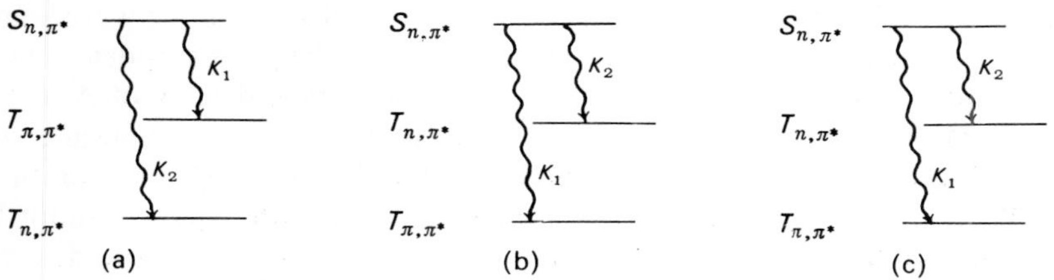

FIGURE 6.3. Experimentally determined ratio for the intersystem crossing rate constants. (a) Pyrazine,[19] $K_1/K_2 > 20$. (b) Quinoxaline,[25] $K_1/K_2 > 10$. (c) 2,3-Dichloroquinoxaline,[25] $K_1/K_2 > 15$.

However, from Eq. (6.1)

$$\frac{n_2}{n_1} = \frac{K_2 k_1}{K_1 k_2} = \frac{I_1{}^{\nu}/I_2{}^{\nu}}{I_1/I_2} \tag{6.5}$$

Since $(I_1{}^{\nu}/I_2{}^{\nu})/(I_1/I_2)$ can be measured and k_1 and k_2 are just the observed reciprocal decay constants of levels τ_1 and τ_2, it is possible to determine the ratio of the intersystem crossing rate constants K_2/K_1. This information is difficult if not impossible to obtain in other ways. The results for several molecules are shown in Figure 6.3.

This work verifies El-Sayed's earlier suggestion[26] that $S_{\pi,\pi^*} \rightsquigarrow T_{n,\pi^*}$ and $S_{n,\pi} \rightsquigarrow T_{\pi,\pi^*}$ are at least one order of magnitude more probable than $S_{\pi,\pi^*} \rightsquigarrow T_{\pi,\pi^*}$ or $S_{n,\pi^*} \rightsquigarrow T_{n,\pi^*}$.

Work on intermolecular energy transfer at low temperature from a donor (D) to an acceptor (A) has indicated that if a donor is prepared such that its triplet state has a unique spin direction, the acceptor will also have a unique and predictable direction when both donor and acceptor are oriented in a single crystal.[27]

Phosphorescence microwave multiple resonance studies are likely to grow in importance in the next ten years as more systems are studied.

6.2. THEORY OF EXCITATION TRANSFER BETWEEN TWO CHROMOPHORES

6.2a. *Radiative Transfer (Trivial Mechanism)*

The radiative mechanism for energy transfer can be represented by

$$\mathrm{D} + h\nu' \longrightarrow \mathrm{D}^* \tag{6.6}$$

$$\mathrm{D}^* \longrightarrow \mathrm{D} + h\nu'' \tag{6.7}$$

$$\mathrm{A} + h\nu'' \longrightarrow \mathrm{A}^* \tag{6.8}$$

This can be an important process if the acceptor A absorbs in the wavelength region in which the donor D emits. The efficiency of the process is determined by the quantum yield of D emission and by the optical density of A at the donor emission wavelength. The probability that an acceptor molecule will reabsorb the light varies as R^{-2}, where R is the donor–acceptor separation.

When D and A are similar molecules emission–reabsorption cannot be very important due to the usually small overlap of the emission and absorption spectra. Also, this mechanism should not be important for triplet–triplet energy transfer because of (a) low phosphorescence quantum yields in fluid solutions and (b) the low oscillator strengths for singlet–triplet absorption.

Energy transfer by the trivial mechanism is characterized by (a) change in the donor emission spectrum (inner filter effect), (b) invariance of the donor emission lifetime, and (c) lack of dependence upon viscosity of the medium.

6.2b. *Resonance Transfer (Long-Range Transfer)*[1,4]

Resonance transfer can occur over distances as great as 50–100 Å. While the trivial mechanism for energy transfer involves no donor–acceptor interaction and no change in donor lifetime, the resonance transfer mechanism is dependent upon a long-range donor–acceptor interaction (often dipole–dipole) and the donor lifetime is decreased. In a qualitative manner the resonance energy transfer is similar to the behavior of two coupled oscillators. That is, the energy can be readily transferred from one oscillator to the other. Just as the two oscillators should have the same frequency for optimum coupling, the transition in the donor ($D^* \rightarrow D$) and the transition in the acceptor ($A \rightarrow A^*$) should involve nearly the same energy. If there is a small difference in the electronic energy gaps for the two transitions, it can be adjusted by including vibrational levels of the donor and acceptor. This type of coupled transition is shown in Figure 6.4. This coupling interaction between the initial (Ψ_i) and final (Ψ_f) states can be expressed as follows:

$$u = \langle \Psi_i | H | \Psi_f \rangle \tag{6.9}$$

If only two electrons are considered, the antisymmetrized initial and final state wave functions are given by

$$\Psi_i = (1/\sqrt{2})[\Phi_{D^*}(1)\Phi_A(2) - \Phi_{D^*}(2)\Phi_A(1)] \tag{6.10}$$

$$\Psi_f = (1/\sqrt{2})[\Phi_D(1)\Phi_{A^*}(2) - \Phi_D(2)\Phi_{A^*}(1)] \tag{6.11}$$

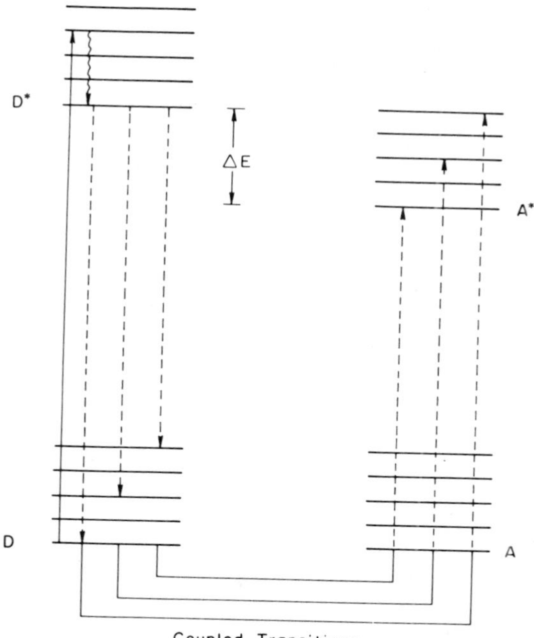

FIGURE 6.4. Energy level diagram for a donor **D** and an acceptor **A** pair with an electronic energy difference ΔE. Possible coupled transitions are indicated.

From Eq. (6.9) we obtain four terms:

$$+\tfrac{1}{2}\langle \Phi_{D^*}(1)\Phi_A(2)|H|\Phi_D(1)\Phi_{A^*}(2)\rangle \tag{6.12}$$

$$-\tfrac{1}{2}\langle \Phi_{D^*}(2)\Phi_A(1)|H|\Phi_D(1)\Phi_{A^*}(2)\rangle \tag{6.13}$$

$$-\tfrac{1}{2}\langle \Phi_{D^*}(1)\Phi_A(2)|H|\Phi_D(2)\Phi_{A^*}(1)\rangle \tag{6.14}$$

$$+\tfrac{1}{2}\langle \Phi_{D^*}(2)\Phi_A(1)|H|\Phi_D(2)\Phi_{A^*}(1)\rangle \tag{6.15}$$

In (6.12) and (6.15) there has been no exchange of electrons, so these terms can be added; similarly, (6.13) and (6.14) can be combined:

$$u = \langle \Phi_{D^*}(1)\Phi_A(2)|H|\Phi_D(1)\Phi_{A^*}(2)\rangle - \langle \Phi_{D^*}(1)\Phi_A(2)|H|\Phi_D(2)\Phi_{A^*}(1)\rangle \tag{6.16}$$

The first term is a Coulomb term and the second is an exchange term. The exchange term, as we will see in the following section on exchange transfer, is a short-range interaction.

If the interaction Hamiltonian in the Coulomb term is expanded in a series about the separation vector, the first term of the expansion is a dipole–dipole interaction, the second a dipole–quadrupole interaction, etc.[(4)] Again reverting to a classical analog (dipole oscillators), the energy of interaction between the two dipoles is inversely proportional to the third power of the distance separating the dipoles (R^{-3}) and is directly proportional to the oscillator strengths, to the extent of spectral overlap, and a dipole–dipole orientation factor. The probability that energy will be transferred from one chromophore to another is proportional to the square of the interaction energy, and thus is inversely proportional to the sixth power of the distance between the two oscillators:

$$k_{D^* \to A} = (2\pi/\hbar)\, u^2 \rho_E \tag{6.17}$$

where ρ_E is the density of states.

Inasmuch as the interaction energy can be related to the transition moments, Förster has been able to develop a quantitative expression for the rate of energy transfer due to dipole–dipole interactions in terms of experimental parameters[4,28–30]:

$$k_{D^* \to A} = \frac{9000\,(\ln 10) K^2 \Phi_D}{128 \pi^5 n^4 N \tau_D R^6} \int_0^\infty F_D(\nu) \varepsilon_A(\nu) \frac{d\nu}{\nu^4} \tag{6.18}$$

where ν is the wavenumber, $\varepsilon_A(\nu)$ is the molar absorptivity of the acceptor, $F_D(\nu)$ is the spectral distribution of the fluorescence of the donor, measured in quanta and normalized to unity on a wavenumber scale, N is Avogadro's number, R is the distance between the donor and acceptor, Φ_D is the quantum yield of fluorescence of the donor, n is the refractive index of the solvent, τ_D is the mean lifetime of the excited state, and K is an orientation factor, and is equal to $\cos \Theta_{DA} - 3 \cos \Theta_D \cos \Theta_A$, where Θ_{DA} is the angle between the dipole vectors of D and A and Θ_D and Θ_A are the angles between these vectors and the direction $D \to A$. For a random distribution $K^2 = \frac{2}{3}$.

It is important to note that the emission yield of the donor is normalized to unity. Consequently, resonance energy transfer can occur not only when neither donor nor acceptor changes its multiplicity, but also when the donor changes multiplicity (weak or forbidden transition) but the acceptor does not. This suggests that singlet–singlet and triplet–singlet energy transfer may occur by the resonance mechanism:

$$D^s + A \longrightarrow D + A^s$$
$$D^t + A \longrightarrow D + A^s$$

However, the very important triplet–triplet energy transfer should not occur by this mechanism:

$$D^t + A \not\longrightarrow A^t + D$$

The efficiency of resonance transfer is often given in terms of a critical radius R_0. If R_0 is the distance such that the rate of energy transfer is equal to the sum of all other donor deactivation rates

$$k_{D^* \to A} = 1/\tau_D$$

then

$$R_0^6 = \frac{9000\,(\ln 10) K^2 \Phi_D}{128\pi^5 n^4 N} \int_0^\infty F_D(\nu)\varepsilon_A(\nu)\,\frac{d\nu}{\nu^4} \tag{6.19}$$

Thus the "theoretical" critical distance can be calculated entirely on the basis of spectroscopic properties of the donor and acceptor.

This value can be compared with the value calculated from quenching or lifetime studies using the following equation:

$$R_0 = \left(\frac{3000}{4\pi N[A]_{1/2}}\right)^{1/3} = \frac{7.35}{\sqrt[3]{[A]_{1/2}}} \tag{6.20}$$

where $[A]_{1/2}$ is the critical concentration such that transfer is 50% efficient, that is, donor fluorescence is half-quenched. It is also possible to extract energy transfer rate constants from the quenching or lifetime studies and to compare these values with the "theoretical" values obtained from Eq. (6.18). A frequently used expression to relate experimental resonance transfer efficiency Φ_{et} with the critical distance R_0 is

$$\Phi_{et} = \frac{(R_0/R)^6}{1 + (R_0/R)^6} \tag{6.21}$$

6.2c. *Energy Transfer via Exchange Interaction*

The electron exchange interaction [see Eq. (6.16)] requires overlap of the electronic wave functions of the donor and acceptor. Dexter[31] evaluated the exchange expectation value equation and obtained the following result:

$$k_{D^* \to A} = \frac{2\pi K^2}{\hbar}\, e^{-2R/L} \int_0^\infty F_D(\nu)\varepsilon_A(\nu)\, d\nu \tag{6.22}$$

Inasmuch as K and L are constants not readily available from experimental data, only the form of Eq. (6.22) is of interest. For example, the rate constant should decrease exponentially with increasing separation R between the donor and acceptor. Also, because donor and acceptor multiplicities can change during the transfer, the overlap integral is calculated with both the donor emission and acceptor absorption normalized to unity.

When energy transfer takes place via an exchange mechanism, the system obeys Wigner's spin rule.[32] If χ_1 and χ_2 are the initial spin quantum numbers of the species, the resulting spin quantum number of the two systems taken together must have *one* of the values

$$\chi_1 + \chi_2, \quad \chi_1 + \chi_2 - 1, \quad \chi_1 + \chi_2 - 2, \ldots, \chi_1 - \chi_2 \tag{6.23}$$

The triplet–triplet energy transfer, which is doubly forbidden by a resonance mechanism, is allowed by an exchange mechanism:

$$\mathrm{D}^t + \mathrm{A} \longrightarrow \mathrm{D} + \mathrm{A}^t$$

Energy transfer by the following route is forbidden both by exchange (spin) and resonance transfer:

$$\mathrm{D}^s + \mathrm{A} \longrightarrow \mathrm{D} + \mathrm{A}^t$$

Several of the spin-allowed processes are listed below:

$$\mathrm{D}^s + \mathrm{A} \longrightarrow \mathrm{D} + \mathrm{A}^s$$

$$\mathrm{D}^t + \mathrm{A} \longrightarrow \mathrm{D} + \mathrm{A}^s$$

$$\mathrm{D}^s + \mathrm{A}^t \longrightarrow \mathrm{D} + \mathrm{A}^t_x$$

$$\mathrm{D}^t + \mathrm{A}^t \longrightarrow \mathrm{D} + \mathrm{A}^{s,t,q} \qquad \text{triplet–triplet annihilation}$$

6.2d. *Exciton Transfer* (*Strong Coupling*)[4]

For resonance transfer the interaction integral u [Eq. (6.9)] is much smaller than the energy separating the states ΔE (see Figure 6.4), which is also much larger than α,

$$\Delta E \gg \alpha \gg u \qquad \text{very weak coupling resonance transfer}$$

where α is a measure of the efficiency of vibrational relaxation. Thus for resonance transfer the rate of electronic energy transfer is slow with respect to vibrational relaxation after the energy has been transferred. When the interaction energy u is large, quite dramatic changes can be found in the emission and absorption spectra:

$$u \gg \Delta E \gg \alpha \qquad \text{strong coupling exciton transfer}$$

For dilute solutions of essentially independent donor and acceptor molecules the Förster or resonance interaction is quite important; in molecular aggregrates and in molecular crystals exciton interactions are likely to be important. When the interaction is strong the excitation is *not localized* on the donor or acceptor but is spread over both. If larger aggregrates are involved, the excitation can be spread over many molecules.[33–38] This can easily be seen for the case of a "dimer" where the donor and acceptor are

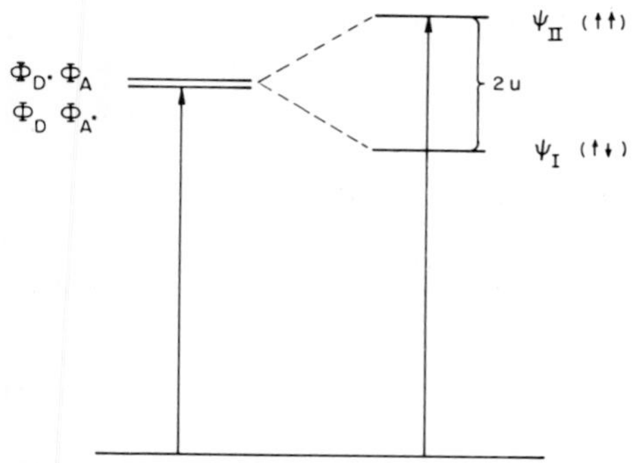

FIGURE 6.5. Exciton model for the coupling of dimers. When the transition dipoles are aligned (↑↑, ↓↓) in a card pack fashion only the transition to Ψ_{II} is allowed; thus a blue shift in the spectrum is expected.

identical. The excited states of this system are described by Ψ_I and Ψ_{II} with an exciton splitting of $2u_{el}$ (Figure 6.5):

$$\begin{aligned} \Psi_I &= (1/\sqrt{2})(\Phi_{D^\bullet}\Phi_A + \Phi_D\Phi_{A^\bullet}) \\ \Psi_{II} &= (1/\sqrt{2})(\Phi_{D^\bullet}\Phi_A - \Phi_D\Phi_{A^\bullet}) \end{aligned} \tag{6.24}$$

The arrangement of the individual transition dipoles in an aggregrate or polymer determines the selection rule for exciton transitions. Several possible dipole arrangements are shown in Figure 6.6

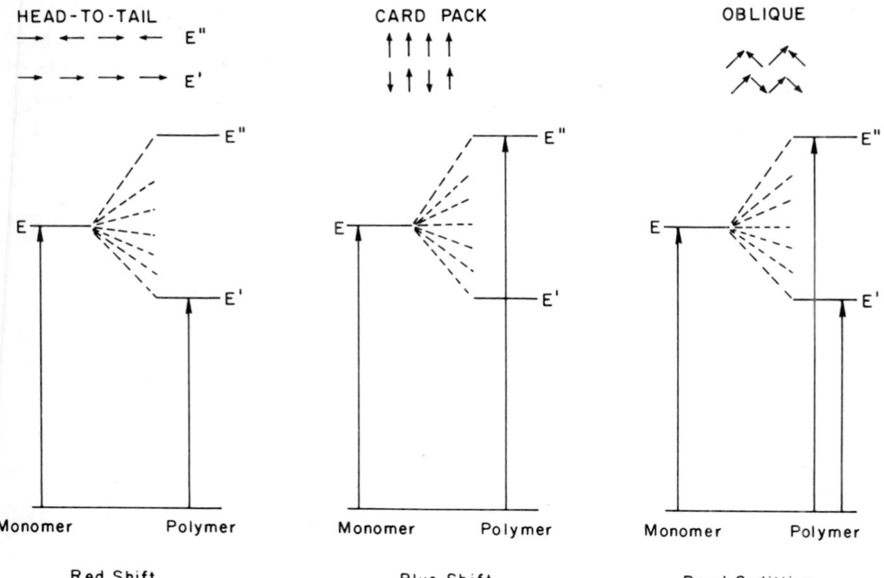

FIGURE 6.6. Energy level diagram for exciton interaction with different geometric arrangements of molecules in a linear polymer.

6.3. EXCITATION TRANSFER BETWEEN TWO CHROMOPHORES

6.3a. *Singlet–Singlet Energy Transfer (Förster Type)*

The most important mechanism for singlet energy transfer is the Förster resonance transfer:

$$D^s + A \longrightarrow D + A^s \tag{6.25}$$

It will not be possible in the limited space available to survey all the experimental literature for either singlet–singlet or triplet–triplet energy transfer. Instead, we will try to give several representative examples.

Intermolecular Transfer

Ware[39,40] investigated singlet–singlet energy transfer by measuring the shortening of the donor fluorescence lifetime. His data, along with the value of R_0 (theoretical) and $k_{D^* \to A}$ (theoretical) from Förster equations, are given in Table 6.4. The observed R_0 and $k_{D^* \to A}$ are in reasonable agreement with the calculated values. Forster's equations are valid only for rigid systems. Consequently, we would expect diffusion in mobile systems to introduce some error in the comparison of calculated and observed values of R_0 and $k_{D^* \to A}$. Bennett[41] has studied intermolecular singlet–singlet energy transfer in rigid plastic matrices where the donor–acceptor distance is fixed.

TABLE 6.4. *Typical Values for the Critical Distance R_0 Associated with Long-Range Energy Transfer*[a]

Pair[b]	ϕ_F	$k_{D^* \to A}$, mole^{-1} sec^{-1}		R_0, Å	
		Theor.	Expt.	Theor.	Expt.
Anthracene–perylene	0.19[c]	2.3×10^{10}	1.2×10^{11}	31	54
Perylene–rubrene	0.89	2.8×10^{10}	1.3×10^{11}	38	65
9,10-Dichloro-anthracene–perylene	0.65	1.7×10^{10}	8.0×10^{10}	40	67
Anthracene–rubrene	0.265	7.7×10^{9}	3.7×10^{10}	23	39
9,10-Dichloro-anthracene–rubrene	0.65	8.5×10^{9}	3.1×10^{10}	32	49

[a] Taken from Ware.[53]
[b] First member of pair initially excited.
[c] Reduced from 0.265 by self-quenching.

In these systems the theoretical and experimental R_0 values agree very well [R_0(theor) = 26, R_0(expt) = 27 ± 1; R_0(theor) = 39, R_0(expt) = 42 ± 2].

Intramolecular Transfer

Schnepp and Levy[42] have studied the fluorescence quantum yield of the following system as a function of the wavelength of the exciting light:

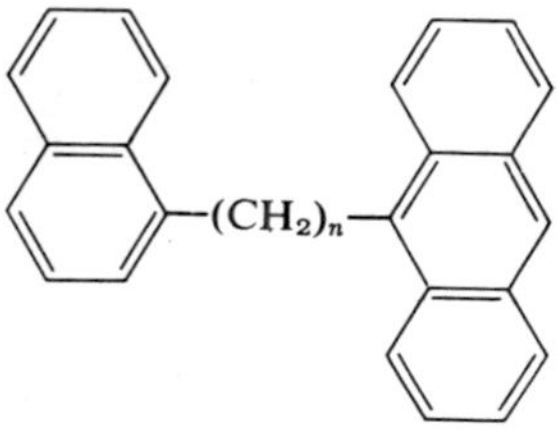

$(CH_2)_n$ $n = 1–3$

They observed a constant quantum yield of fluorescence ($\Phi_f = 0.3$) for all members of the series independent of whether the anthracene moiety absorbed and emitted the energy or the naphthalene moiety absorbed the energy and transferred it to the anthracene moiety. Thus at these short distances singlet energy transfer is 100% efficient.

Latt, Cheung, and Blout[43,44] have made use of the following bisteroid as a rigid framework on which to attach two donor–acceptor pairs:

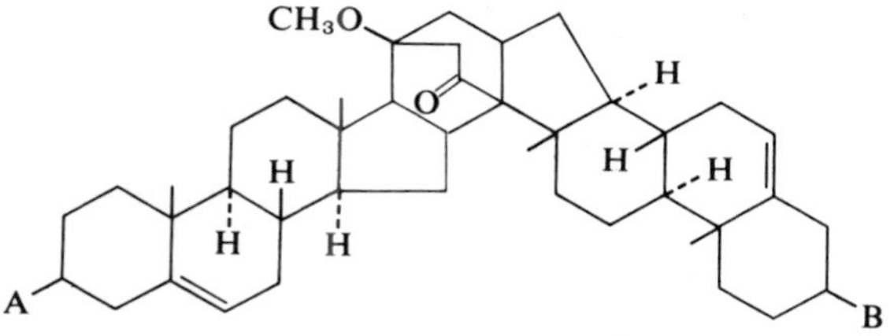

CH_3O O H H H H H H A B

where

(1) A = 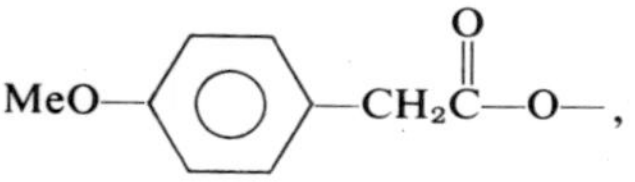 MeO—C_6H_4—$CH_2C(=O)$—O—, B = $C_{10}H_7$—C(=O)—O—

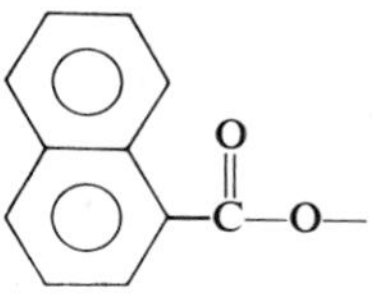

and

(2) A = $C_{10}H_7$—C(=O)—O—, B = $C_{14}H_9$—C(=O)—O—

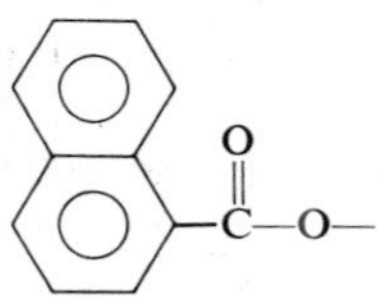

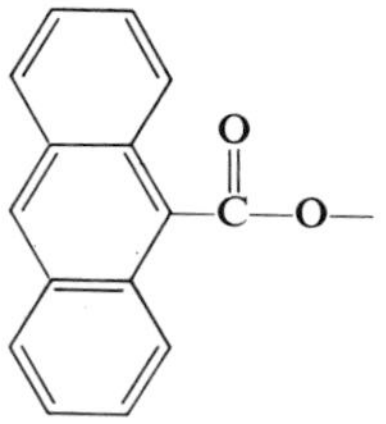

Energy transfer measurements were used, together with fluorescence and absorption spectral data of the donor and acceptor moieties, to calculate the donor–acceptor separation via the Förster equation. The average values of R obtained assuming random donor–acceptor orientations were 21.3 ± 1.6 for (1) and 16.7 ± 1.4 for (2). The average separation obtained from molecular models is 21.8 ± 2.0 for (1) and 21.5 ± 2.0 for (2). The somewhat low calculated separation between the groups of (2) may be due to nonrandom donor–acceptor orientations.

A very nice example of Förster-type energy transfer has been provided by Stryer and Haugland,[45] who studied a series of long-chain compounds (dansyl-(L-prolyl)$_n$-α-naphthyl):

(3)
$n = 1$–12

The transfer efficiency decreases from a value close to 100% for short oligomers to 16% for the $n = 12$ oligomer. The efficiency of energy transfer as a function of distance is shown in Figure 6.7.

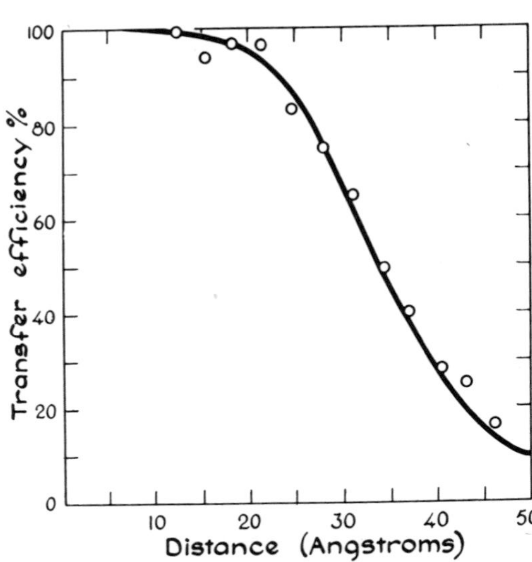

FIGURE 6.7. Efficiency of energy transfer as a function of distance for compounds (3), $n = 1$–12. From Stryer and Haughland.[45] Reprinted by permission of *Proc. Nat. Acad. Sci. U.S.*

The efficiency of energy transfer as a function of distance is given by [see also Eq. (6.21)]

$$\Phi_{\text{et}} = \frac{(R_0/R)^j}{1 + (R_0/R)^j} \tag{6.26}$$

R_0 and j can be obtained from the experimental data by plotting $\log(\Phi_{\text{et}}^{-1} - 1)$ vs. $\log R$. The slope of the line is j, while R_0 is given by the value of R at $\Phi_{\text{et}} = 0.5$. Their data, shown in Figure 6.8, give a straight line $j = 5.9 \pm 0.3$. Thus this result is in excellent agreement with Förster's R^{-6} dependence.

These and other workers[46] have proposed that the energy transfer process can serve as a spectroscopic ruler (10–60 Å) in biological systems.

6.3b. *Singlet–Singlet Energy Transfer via Collisions*

All of the examples of singlet energy transfer we have considered take place via the long-range resonance mechanism. When the oscillator strength of the acceptor is very small (for example, $n \rightarrow \pi^*$ transitions) so that the Förster critical distance R_0 approaches or is less than the collision diameter of the donor–acceptor pair, then all evidence indicates that the transfer takes place at a diffusion-controlled rate. Consequently, the transfer mechanism should involve exchange as well as Coulomb interaction. Good examples of this type of transfer have been provided by Dubois and co-workers.[47–49]

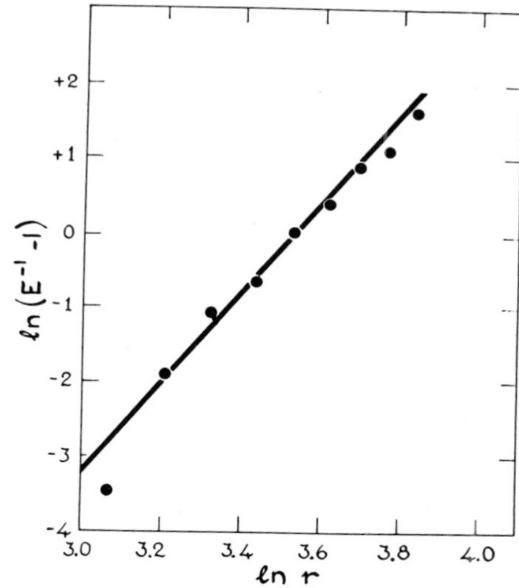

FIGURE 6.8. Dependence of the energy transfer efficiency (E $= \Phi_{\text{et}}$ in the figure) on distance. The slope of 5.9 is in excellent agreement with the r^{-6} dependence for Förster-type transfer. From Stryer and Haugland.[45] Reprinted by permission of *Proc. Nat. Acad. Sci. U.S.*

They have shown that singlet energy transfer from many donors to biacetyl takes place by a diffusion-controlled process.

6.3c. *Intermolecular Triplet–Triplet Energy Transfer*

Sensitized phosphorescence due to triplet–triplet energy transfer in low-temperature glasses was first reported by Terenin and Ermolaev. The unambiguous demonstration of this type of energy transfer was obtained by showing the phosphorescence of naphthalene in a rigid glass at 77°K could be excited by 3660-Å light, which napthalene does not absorb, if the glass also contains benzophenone. The energy level diagram for this type of system is shown in Figure 6.9. Aromatic ketones are ideal donors inasmuch as they have high intersystem crossing efficiencies and small singlet–triplet energy separation. Terenin and Ermolaev[50,51] have been able to show that the quenching process can be described by the equation

$$\frac{I_0}{I_A} = e^{\alpha C_A} \tag{6.27}$$

where I_0 is the phosphorescence intensity in the absence of an acceptor, I_A is the phosphorescence intensity in the presence of a quencher at concentration C_A (moles/liter), and α is a constant. This equation has also been derived based on a spheres of quenching action model by Perrin.[52] The critical radius R_c based on the spheres of quenching action model is given by

$$R_c = \left(\frac{3000 \ln(I_0/I_A)}{4\pi N C_A}\right)^{1/3} \tag{6.28}$$

where N is Avogadro's number. In this model the donor is quenched if the acceptor is within the sphere of radius R_c. The value of R_c for a number of donor–acceptor pairs is in the range of 11–15 Å (see Ref. 1, p. 45).

Bäckstrom and Sandros[54,55] found that the phosphorescence of biacetyl in benzene solution at room temperature was quenched at a diffusion-controlled rate by aromatic hydrocarbons when the triplet energy of the hydrocarbon was sufficiently below that of biacetyl.

Porter and Wilkinson[56] measured the rates of quenching for a variety of triplet donors with triplet acceptors at room temperature in fluid solution by flash photolysis. The appearance of the triplet–triplet absorption spectrum of the acceptor and the simultaneous disappearance of the donor triplet–triplet absorption spectrum provided unequivocal evidence for the triplet–triplet energy transfer process. Table 6.5 provides some of the quenching rate constants reported in this classic paper.

When the triplet energy of the donor is 3 kcal/mole or more higher than the acceptor triplet energy, the energy transfer rate is about the diffusion-

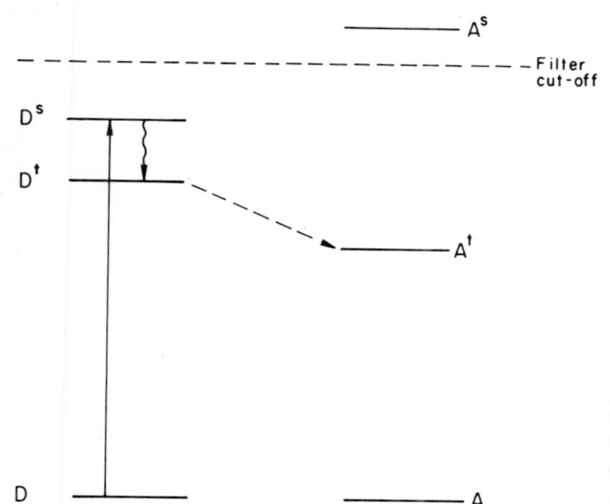

FIGURE 6.9. Ideal donor–acceptor energy level arrangement for triplet–triplet energy transfer.

controlled rate. When the triplet energy level of the donor is not 3 kcal/mole above the acceptor level, energy transfer becomes rather inefficient.

This dependence on energy has been nicely demonstrated by Herkstroeter and Hammond,[57,58] who studied the energy transfer for a series of 18 triplet donors to *cis*- and *trans*-stilbene and *cis*- and *trans*-1,2-diphenylpropene by flash photolysis. The results for *cis*- and *trans*-stilbene are summarized graphically in Figure 6.10.

It has been proposed that when a triplet sensitizer has insufficient excitation energy to promote an acceptor to its triplet state, this deficiency can be supplied as an activation energy:

$$\frac{\Delta(\log k)}{\Delta E_t} = -\frac{1}{2.303RT} \tag{6.29}$$

TABLE 6.5. *Rate Constants for Triplet–Triplet Energy Transfer*[56]

Donor	Acceptor	Solvent	ΔE_t, kcal/mole	k_{et}, M^{-1} sec^{-1}
Triphenylene	Naphthalene	Hexane	6.30	$1.3 \pm 0.8 \times 10^9$
Phenanthrene	1-Iodonaphthalene	Hexane	3.15	$7 \pm 2 \times 10^9$
Phenanthrene	1-Iodonaphthalene	Ethylene glycol	3.15	$2.1 \pm 0.2 \times 10^8$
Phenanthrene	1-Bromonaphthalene	Hexane	2.57	$1.5 \pm 0.8 \times 10^8$
Naphthalene	1-Iodonaphthalene	Ethylene glycol	2.29	$2.8 \pm 0.3 \times 10^8$
Phenanthrene	Naphthalene	Hexane	0.86	$2.9 \pm 0.7 \times 10^6$
Naphthalene	Phenanthrene	Hexane	−0.86	$\leq 2 \times 10^4$
Naphthalene	Benzophenone	Benzene	−8.90	$\leq 1 \times 10^4$

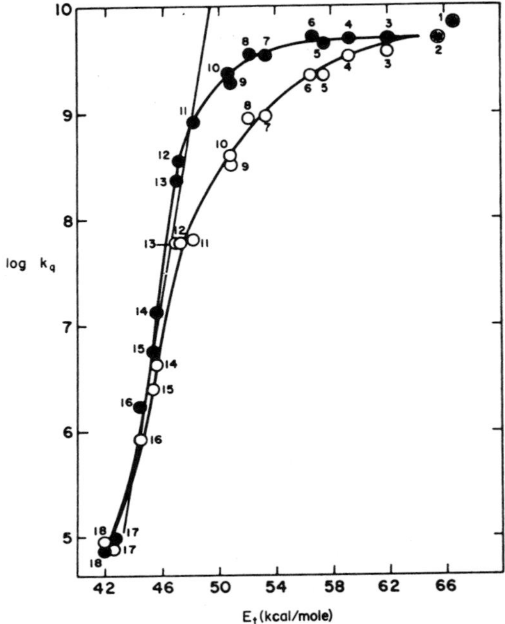

FIGURE 6.10. Rate constants for quenching of sensitizers by *cis*- and *trans*-stilbenes (open and filled circles, respectively). Sensitizers are as follows: (1) triphenylene, (2) thioxanthone, (3) phenanthrene, (4) 2-acetonaphthone, (5) 1-naphthyl phenyl ketone, (6) crysene, (7) fluorenone, (8) 1,2,5,6-dibenzanthracene, (9) benzil, (10) 1,2,3,4-dibenzanthracene, (11) pyrene, (12) 1,2-benzanthracene, (13) benzanthrone, (14) 3-acetyl pyrene, (15) acridine, (16) 9,10-dimethyl-1,2-benzanthracene, (17) anthracene, (18) 3,4-benzpyrene.[57] Reprinted by permission of the American Chemical Society.

The line drawn through the data for *trans*-stilbene has the slope predicted by Eq. (6.29) of -0.73 kcal^{-1} (for room-temperature studies).

6.3d. *Application of Triplet–Triplet Energy Transfer*

When Hammond and co-workers[59] found that the intersystem crossing quantum yield for aromatic ketones was unity (see Chapter 3) it was a short but very important step to realize that these compounds should be ideal triplet sensitizers. Thus one can excite the triplet state of molecules that otherwise would be formed inefficiently, if at all, by intersystem crossing. This idea resulted in a number of papers in the early 1960's from the Hammond group on this topic. It is not possible in this short section to survey this area, but a few of the early studies are indicated by the following reactions:

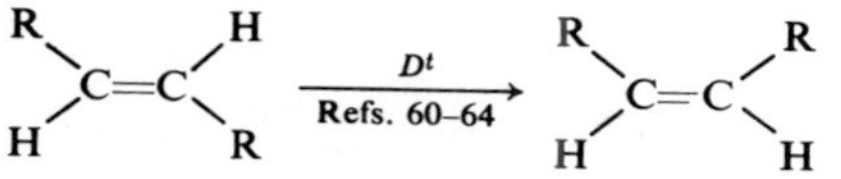

R, H, H, R $\xrightarrow[\text{Refs. 60–64}]{D^t}$ R, R, H, H (6.30)

$\xrightarrow[\text{Refs. 65–68}]{D^t}$ H, H + H, H + (6.31)

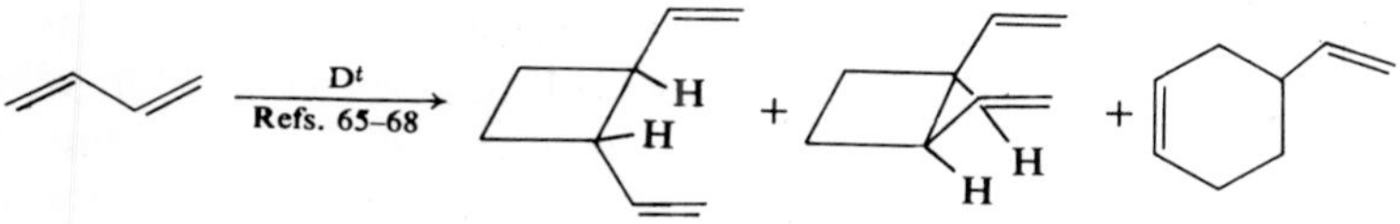

$$CH_2N_2 \xrightarrow[\text{Refs. 69, 70}]{D^t} \underset{\text{triplet}}{CH_2:\uparrow\uparrow} + N_2 \tag{6.32}$$

$$CH_3\overset{O}{\overset{\|}{C}}-\overset{O}{\overset{\|}{C}}-OC_2H_5 \xrightarrow[\text{Ref. 71}]{D^t} 2CH_3CHO + CO \tag{6.33}$$

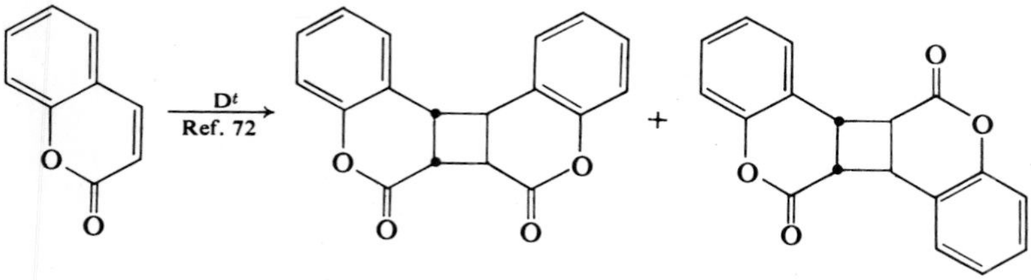

$\xrightarrow[\text{Ref. 72}]{D^t}$ + (6.34)

trace

$\xrightarrow[\text{Ref. 73}]{D^t}$ (6.35)

$\xrightarrow[\text{Ref. 74}]{D^t}$ (6.36)

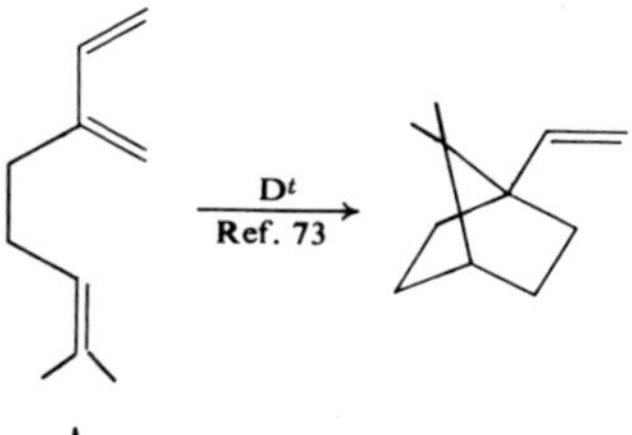

$\xrightarrow[\text{Ref. 75}]{D^t}$ (6.37)

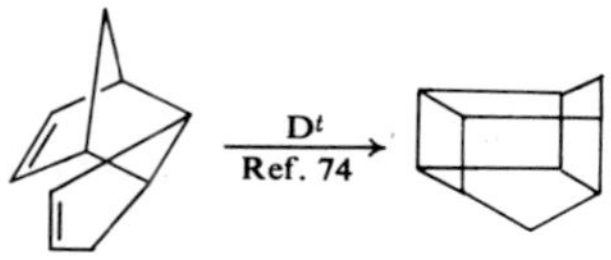

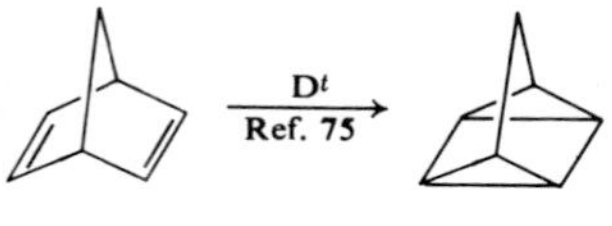

$\xrightarrow[\text{Ref. 76}]{D^t}$ (6.38)

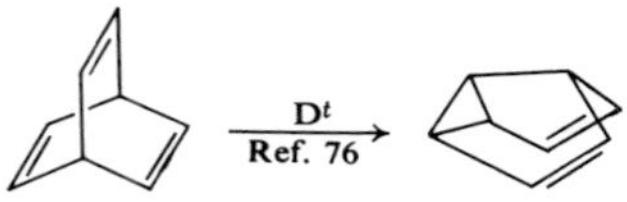

6.3e. *Schenck Mechanism*

Not all sensitized photochemical reactions occur by electronic energy transfer. Schenck[77,78] has proposed that many sensitized photoreactions involve a sensitizer–substrate complex. The nature of this interaction could vary from case to case. At one extreme this interaction could involve σ-bond formation and at the other extreme involve loose charge transfer or exciton interaction (exciplex formation). The σ-Schenck mechanism for a photosensitized reaction is illustrated by the following hypothetical reaction:

$$D + h\nu \longrightarrow D^s \rightsquigarrow (\cdot D \cdot)^t$$

$$(\cdot D \cdot)^t + \text{RHC=CHR} \longrightarrow \cdot D\text{–C(R)(H)–}\dot{\text{C}}\text{(R)(H)} \xrightarrow{-D} \text{RHC=CHR}$$

6.3f. *Intramolecular Triplet Energy Transfer*

Leermakers, Byers, Lamola, and Hammond[79,80] have studied both singlet and triplet energy transfer in the following three compounds:

$$C_6H_5\text{–CO–}C_6H_4\text{–}(CH_2)_n\text{–}C_{10}H_7$$

(4)

$n = 1\text{–}3$

Figure 6.11 is the energy level scheme for this system. Excitation at 3130 Å excites the naphthalene moiety; singlet energy transfer to the benzophenone moiety takes place with high efficiency ($n = 1$, 98%; $n = 2$, 80%; $n = 3$, 94%); intersystem crossing then takes place with unit efficiency to form the benzophenone triplet; triplet energy transfer then occurs ($n = 1, 2, 3$) with 100% efficiency to form the naphthalene triplet, which can phosphoresce (if in a low-temperature glass). Excitation at 3660 Å starts the above scheme at the benzophenone singlet state.

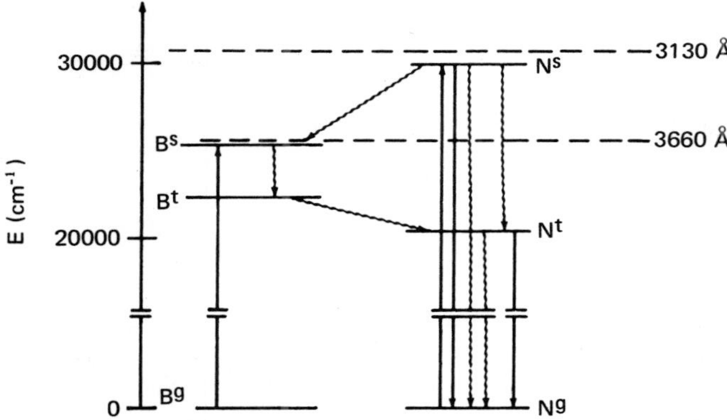

FIGURE 6.11. Electronic transitions to the first excited singlet (*s*) and lowest triplet (*t*) states from the ground states (*g*) of benzophenone (B) and naphthalene (N) moieties in compounds (4), $n = 1$–3. Possible radiative transitions are represented by straight arrows, radiationless transitions by wavy arrows.[80] Reprinted by permission of the American Chemical Society.

Keller and co-workers have studied the following three related systems:

Again 100% efficient singlet–singlet and triplet–triplet energy transfer was observed. However, Keller and Dolby[83] found triplet–triplet energy transfer to be 12% efficient in compound (5) and 20% efficient in compound (6), shown on page 292. These authors suggested a mechanism based on a singlet–triplet mixed wave function and a dipole–dipole coupling interaction.

Filipescu and co-workers[84–86] have also examined intramolecular singlet and triplet energy transfer using fluorescence and phosphorescence as a gauge for energy transfer. Lamola[87] has repeated one of the studies

O—C(=O)—R

(5) R = $-C_6H_4-C(=O)-C_6H_5$ $k_{et} = 25\ sec^{-1}$

(6) R = $-CH_2-N$ (carbazolyl) $k_{et} = 0.040\ sec^{-1}$

that reportedly verified the orientation requirement for electric dipole energy transfer and found an error which negates the conclusion.

The intramolecular triplet energy transfer studies described so far generally have used phosphorescence emission measurements to follow the

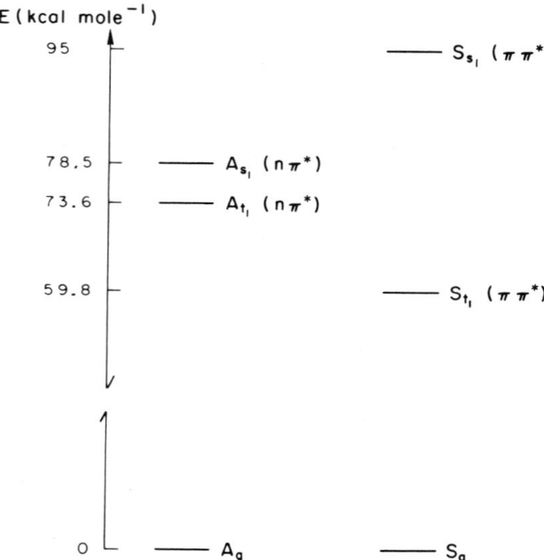

FIGURE 6.12. Energy level diagram for acetophenone and *trans*-β-methylstyrene. From Cowan and Baum.[89] Reprinted by permission of the American Chemical Society.

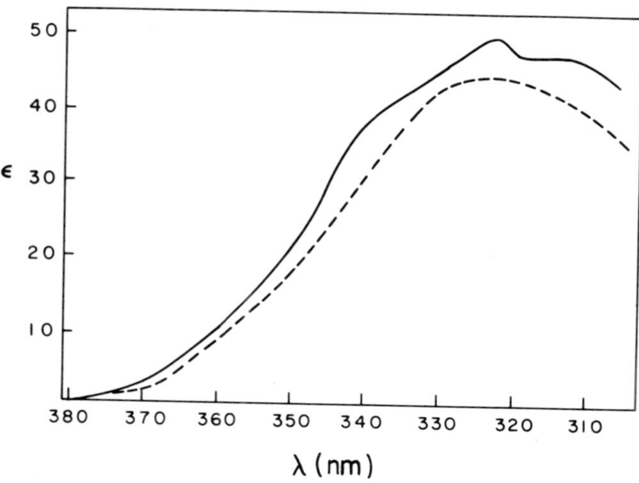

FIGURE 6.13. The n–π^* region of the ultraviolet absorption spectrum of (a) compound (7), $n = 4$ (solid line) and (b) equal molar mixture of acetophenone and *trans*-β-methylstyrene (dotted line).[89] Reprinted by permission of the American Chemical Society.

energy transfer. The following series of compounds was prepared and studied by Cowan and Baum,[88,89] who used *cis–trans* isomerization to follow intramolecular triplet energy transfer:

C_6H_5–C(=O)–$(CH_2)_n$–CH=CH–C_6H_5 $\quad n = 1–4$

A (acetophenone moiety), T (styrene moiety)

(7)

Because of our obvious prejudices we will describe this study in some detail. The energy levels of the acetophenone and β-methylstyrene moieties are ideally situated for intramolecular triplet energy transfer (see Figure 6.12). If these two chromophores are not interacting, then the absorption spectra for these compounds should be the composite of the acetophenone–β-methylstyrene spectra. Figure 6.13 indicates that this is true for $n = 4$ (and also 2 and 3); however, it is not correct for $n = 1$ (Figure 6.14). In the case of $n = 1$ the increased intensity of the vibronic structure of the $n \rightarrow \pi^*$

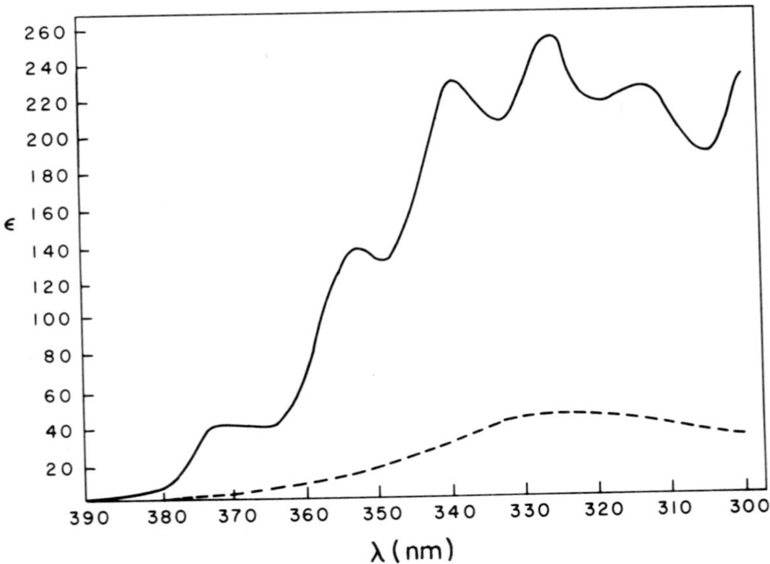

FIGURE 6.14. The n–π^* region of the ultraviolet absorption spectrum of (a) compound (7), $n = 1$ (solid line) and (b) equal molar mixture of acetophenone and *trans*-β-methylstyrene.[89] Reprinted by permission of the American Chemical Society.

absorption strongly indicates some type of interaction between the two chromophores.

Selective excitation of the acetophenone portion (A) of these compounds with 3660-Å light results in efficient isomerization of the *trans*-β-methylstyryl group (T) as shown in Table 6.6.

For each of compounds (7), n = 1–4, the photostationary isomer ratio ([A–T]/[A–C]) is about 0.91. Radical equilibration of n = 1–4 and equilibration with low-energy sensitizers which are thought to operate via Schenck

TABLE 6.6. *Isomerization Quantum Yield Data for Compounds* (7), n = *1–4*

n	$\Phi_{T \to C}$	% *Cis* at photo-equilibrium	$\Phi_{iso}\Phi_{et}$
1	0.53 ± 0.02	53	1.01
2	0.52 ± 0.02	52	0.99
3	0.52 ± 0.02	59[a]	0.99
4	0.51 ± 0.01	58[a]	0.98

[a] These are only approximate values inasmuch as type II reactions become important with long irradiation times.

intermediates produce mixtures where the *trans* isomer strongly predominates. This indicates that isomerization is the result of electronic energy transfer.

Irradiation of solutions containing equimolar amounts of $n = 1$–4 and *trans*-β-methylstyrene as a quencher at concentrations comparable to those of the experiments reported in Table 6.6 indicated that intermolecular energy transfer cannot compete with intramolecular energy transfer under these conditions.

A possible kinetic scheme for this system is shown below, where A–T and A–C are the *trans* and *cis* enones, superscripts s and t refer to singlet and triplet states, and A–P* corresponds to the common triplet intermediate (twisted triplet) produced after triplet energy transfer:

$$\text{A–T} \xrightarrow{I_a} \text{A}^s\text{–T} \qquad \text{A–C} \xrightarrow{I_a} \text{A}^s\text{–C}$$

$$\text{A}^s\text{–T} \xrightarrow{\Phi_{\text{isc}}} \text{A}^t\text{–T} \qquad \text{A}^s\text{–C} \xrightarrow{\Phi_{\text{isc}}} \text{A}^t\text{–C}$$

$$\text{A}^t\text{–T} \xrightarrow{k_d} \text{A–T} \qquad \text{A}^t\text{–C} \xrightarrow{k_d} \text{A–C}$$

$$\text{A}^t\text{–T} \xrightarrow{k_{\text{et}}} \text{A–P*} \qquad \text{A}^t\text{–C} \xrightarrow{k_{\text{et}}} \text{A–P*}$$

$$\text{A–P*} \xrightarrow{k_t} \text{A–T}$$

$$\text{A–P*} \xrightarrow{k_c} \text{A–C}$$

$$\text{A}^t\text{–T} + \text{Q} \xrightarrow{k_q} \text{A–T} + \text{Q}^t$$

Selective excitation of the acetophenone portion of these compounds results in formation of the $n \rightarrow \pi^*$ singlet state of this chromophore (A^s). Singlet energy transfer from A^s to the olefin portion should be endothermic and very slow. However, intersystem crossing should be fast and efficient. Triplet energy transfer can then occur at an appreciable rate since the process is exothermic. The rates of processes in competition with energy transfer are given by the rate constant k_d. Excitation of the styryl group due to energy transfer allows for *cis–trans* isomerization of this portion of the molecule. This kinetic scheme leads to the following equations:

$$\Phi_{\text{T}\rightarrow\text{C}} = \Phi_{\text{isc}}\Phi_{\text{et}}\frac{k_c}{k_t + k_c} \tag{6.39}$$

$$\Phi_{\text{C}\rightarrow\text{T}} = \Phi_{\text{isc}}\Phi_{\text{et}}\frac{k_t}{k_t + k_c} \tag{6.40}$$

Adding (6.39) and (6.40) yields

$$\Phi_{\text{T}\rightarrow\text{C}} + \Phi_{\text{C}\rightarrow\text{T}} = \Phi_{\text{isc}}\Phi_{\text{et}} \tag{6.41}$$

Thus if one measures $\Phi_{\text{T}\rightarrow\text{C}}$ and $\Phi_{\text{C}\rightarrow\text{T}}$, a value for $\Phi_{\text{isc}}\Phi_{\text{et}}$ can be calculated. In this study only $\Phi_{\text{T}\rightarrow\text{C}}$ was measured, along with β:

$$\beta = ([\text{A–T}]/[\text{A–C}])_{\text{PSS}} = \Phi_{\text{C}\rightarrow\text{T}}/\Phi_{\text{T}\rightarrow\text{C}} = k_t/k_c \tag{6.42}$$

This allows one to calculate $\Phi_{C\to T}$ $(=\Phi_{T\to C}\beta)$ and thus evaluate $\Phi_{isc}\Phi_{et}$. For compound (7), $n = 2$, we have $\Phi_{T\to C} = 0.52$ and $\beta = 0.91$. Thus $\Phi_{C\to T} = 0.52 \times 0.91 = 0.47$, and we obtain

$$\Phi_{isc}\Phi_{et} = \Phi_{T\to C} + \Phi_{C\to T} = 0.52 + 0.47 = 0.99$$

As we can see from Table 6.6, both energy transfer and intersystem crossing take place with unit efficiency in all four cases.

The results of external quenching studies using *trans*-β-methylstyrene as a quencher are shown in Figures 6.15 and 6.16.

The above mechanism leads to the following Stern–Volmer equation:

$$\frac{1}{\Phi_{T\to C}} = \frac{1}{F\Phi_{isc}}\left(\frac{1}{\Phi_{et}} + \frac{k_q}{k_{et}}[Q]\right) \tag{6.43}$$

where $F = k_c/(k_c + k_t)$.

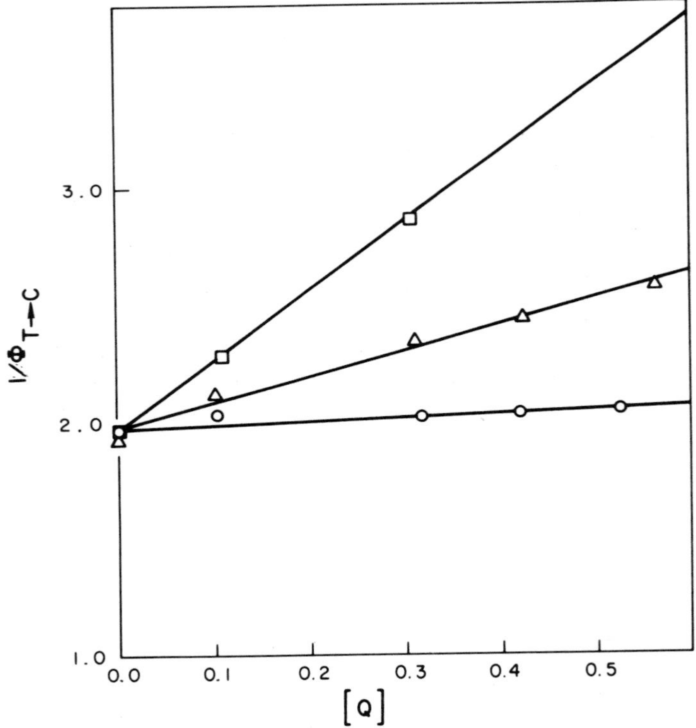

FIGURE 6.15. Quenching of *trans* → *cis* isomerization of compounds (7), $n = 2$ (○), $n = 3$ (△), and $n = 4$ (□) by *trans*-β-methylstyrene (benzene).[89] Reprinted by permission of the American Chemical Society.

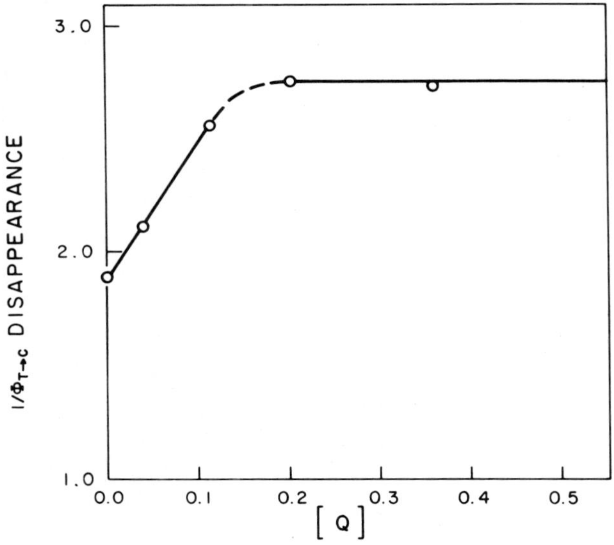

FIGURE 6.16. Quenching of *trans* → *cis* isomerization of compound (7), $n = 1$, by *trans*-β-methylstyrene (benzene).[89] Reprinted by permission of the American Chemical Society.

Recognition of the fact that $\Phi_{et}\Phi_{isc} = 1$ makes the interpretation of the quenching data straightforward:

$$\frac{1}{\Phi_{T\to C}} = \frac{k_c + k_t}{k_c}\left(1 + \frac{k_q}{k_{et}}[Q]\right) \tag{6.44}$$

The slope divided by the intercept gives the rate ratio k_q/k_{et}. Table 6.7 gives the results for compounds (7), $n = 2$–4 in benzene solution. The detailed mechanism of energy transfer must be able to account for the 25-fold decrease in k_{et} in going from $n = 2$ to $n = 4$. While inspection of molecular models shows that the average distance between chromophores

TABLE 6.7. *Rates and Rate Ratios for Intramolecular Triplet Energy Transfer in Compounds (7), n = 2,3,4*

n	k_q/k_{et}, M^{-1}	k_{et}, sec^{-1}
2	0.069	7.2×10^{10}
3	0.55	1.0×10^{10}
4	1.5	3.3×10^{9}

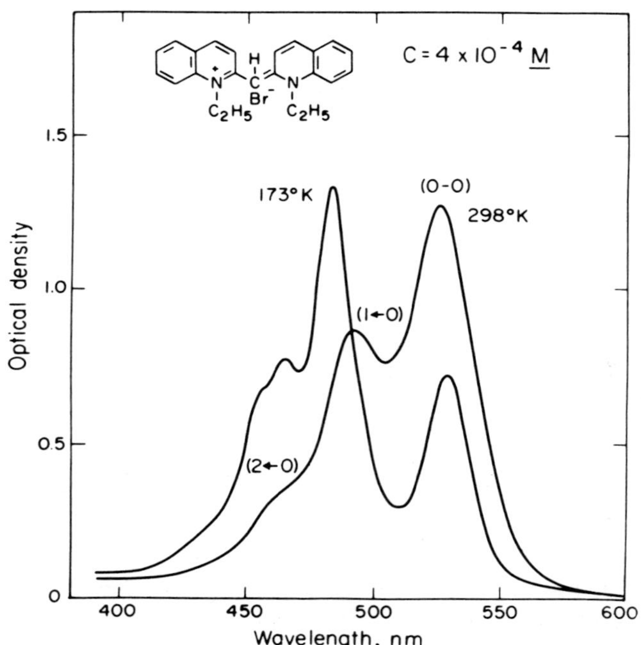

FIGURE 6.17. Absorption spectra of 1,1′-diethyl-2,2′-cyanine bromide in ethylene glycol:water (1:1) at room temperature and at 173°K, showing the *H*-aggregate states.[94] Reprinted by permission of *Chemical Physics Letters*.

remains nearly constant in the compounds, the number of conformations in which the two ends of the molecule are nearly in contact decreases greatly with increasing number of methylene groups. In addition, the rate ratio k_q/k_{et} for compound (7), $n = 3$, in *t*-butyl alcohol is 0.75 M^{-1}. Since the solvent properties of *t*-butyl alcohol are quite different from those of benzene (viscosities are a factor of ten different) this result suggests that k_{et} depends on the same properties that influence k_q. The most reasonable conclusion to be drawn from both of these observations is that intramolecular triplet energy transfer in these compounds takes place by the same mechanism as bimolecular quenching, that is, the exchange mechanism that requires a collision between the donor and acceptor.

The quenching results for compound (7), $n = 1$, imply that energy transfer occurs from two different excited states.

A number of intramolecular singlet and triplet energy transfer studies have been reported by Morrison and co-workers[90–92] and an extensive review of this general area has been published recently.[93]

6.3g. *Exciton Interaction*

While we will not give an extensive survey of this area, it is interesting to examine one recent study of 1,1′-diethyl-2,2′-cyanine bromide.[94] At a dye concentration of 4×10^{-4} *M* (in 1:1 mixture of ethylene glycol and water) at room temperature the absorption spectrum (see Figure 6.17) is essentially that of the unperturbed molecule. At 173°K the intensity of the strongest monomer band (525 nm) is diminished and a high-intensity band appears at 483 nm (*H* band). This blue-shifted absorption was assigned by the author as the allowed exciton dimer transition (see Figure 6.17). Using the method of McCrae and Kasha[33] and the card-pack exciton model, it is possible to predict the exciton shifts of the trimer, tetramer, pentamer, and infinite polymer. There is excellent agreement between theory and experiment (see Figure 6.17 and Table 6.8). From Figure 6.17 the trimer absorbs at 465 and the tetramer at 458 nm. Spectra at a higher concentration and lower temperature are shown in Figure 6.18.

If the transition dipoles are aligned in a head-to-tail formation, then a red shift is expected. This is the reported explanation for the sharp bands at 573 and 578 (*J* bands). The narrow half-bandwidths of the split *J* aggregate absorption suggest that the exciton states are not strongly coupled with external perturbations. The two distinct electronic transitions were proposed to arise from two structural modifications of the aggregates.

Energy Transfer[95–98]

Strong exciton splitting, e.g., 1000 cm^{-1}, indicates extremely high rates of energy migration in a lattice. Even exciton splittings of 10 cm^{-1}

TABLE 6.8. *Spectral Shifts of Aggregated States of 1,1′-Diethyl-2,2′-cyanine Bromide in Ethylene Glycol–Water Matrix at 173°K*[a]

N[b]	λ_{max}, nm	ν_{max}, cm^{-1} Expt.	ν_{max}, cm^{-1} Calcd.
1	530	19,048	—
2	483	20,704	20,713
3	465	21,504	21,256
4	458	21,834	21,532
5[c]	455	21,978	21,696
∞	447(calcd.)	—	22,360

[a] This table is from Cooper.[94] $C = 4 \times 10^{-4}$ *M*.
[b] *N* is the number of molecules in the aggregated state.
[c] Data taken from absorption spectrum of dye at $C = 4 \times 10^{-3}$ *M*.

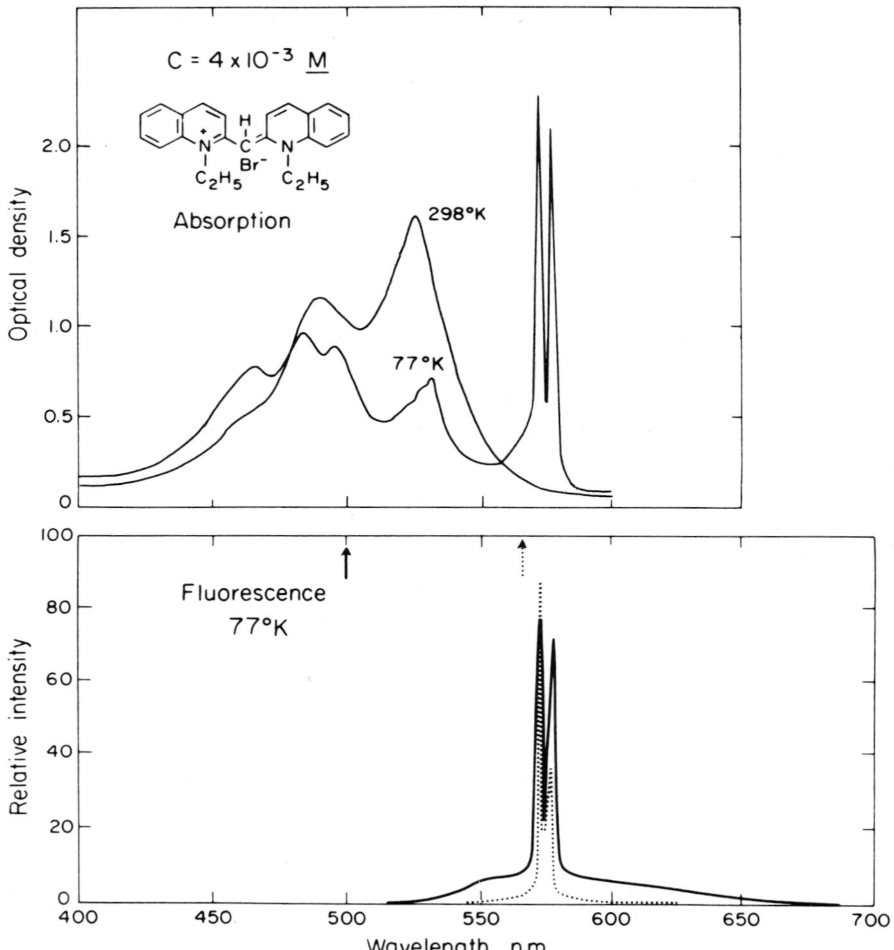

FIGURE 6.18. Absorption and fluorescence spectra of 1,1′-diethyl-2,2′-cyanine bromide at 298 and 77°K. Fluorescence excitation at 498 nm and 565 nm, as indicated by arrows.[94] Reprinted by permission of *Chemical Physics Letters*.

indicate high transfer rate constants calculated as the Heisenberg resonance lifetime of the localized excited state.

It has been suggested that trapping of this delocalized excitation by low-energy "impurity" molecules could play an important role in the photophysics of chloroplasts. After trapping the excitation energy, it could be transferred to a reactant or an electron transfer reaction could occur.[35]

Birks[99] gives an excellent description of the two modes for exciton migration, the exciton band model [where exciton–phonon (lattice vibration) scattering is small and the mean free path of an exciton is large compared with the lattice spacing] and the exciton hopping model (where exciton–phonon scattering is large).

6.4. EXCIPLEX QUENCHING[100,105]

The fact that quadricyclene and dienes quench the fluorescence of aromatic hydrocarbons despite the fact that the energetics for classical energy transfer are very unfavorable has been rationalized by the formation of an exciplex. A general mechanism is as follows:

$$\begin{aligned} D &\longrightarrow D^s \\ D^s &\longrightarrow D \\ D^s &\longrightarrow D^t \\ D^s + A &\rightleftharpoons (DA)^s \text{ exciplex} \\ (DA)^s &\longrightarrow D + A \\ (DA)^s &\longrightarrow D + \text{products} \\ (DA)^s &\longrightarrow \text{photoadduct} \end{aligned}$$

Hammond and co-workers have suggested that the exciplex should be described in terms of the four contributing structures

$$D^*A \longleftrightarrow DA^* \longleftrightarrow D^+A^- \longleftrightarrow D^-A^+$$

Quenching rate constants for dienes and quadricyclenes have similar sensitivities to the electronic and structural features of the excited aromatic hydrocarbon. However, during this process quadricyclene isomerizes to norbornadiene with a quantum yield of 0.52, whereas dienes usually remain unchanged.[103] Hammond has suggested that vibrational energy which is partitioned to the acceptor upon internal conversion of the exciplex can lead to isomerization[102,103]:

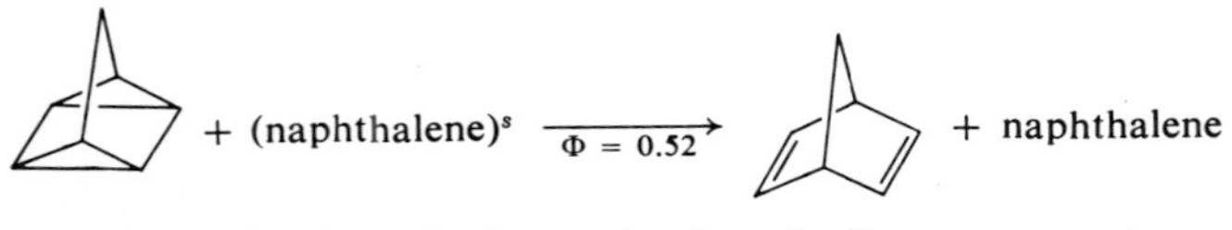

$$\text{[quadricyclene]} + (\text{naphthalene})^s \xrightarrow[\Phi\,=\,0.52]{} \text{[norbornadiene]} + \text{naphthalene}$$

The fact that quadricyclene is isomerized and dienes are not could be a result of (a) factors that govern how the vibrational energy is partitioned and (b) the large difference in activation energies for isomerization.

Another point of view is that exciplex formation can in some cases involve bonding changes in the acceptor. Solomon, Steel, and Weller[(104)] proposed that the quenching by quadricyclene leads to an exciplex where the quadricyclene is distorted.

A recent study of Murov and co-workers[(105)] indicates that the activation energies for isomerization are not the controlling factors. Thus the fluorescence of naphthalene is quenched ($5 \times 10^8\ M^{-1}\ \text{sec}^{-1}$) by *cis-trans*-1,3-cyclooctadiene with isomerization to form *cis-cis*-1,3-cyclooctadiene. However, the compound bicyclo[4,2.0]oct-7-ene is not formed despite the low activation energy for this process:

$E^* = 27.8$ kcal/mole

1,5-H shift

$E^* = 33$ kcal/mole

cis–trans isomerization

$E^* \approx 45$ kcal/mole

Similarly, 2,3-dimethyl-1,3-butadiene quenches naphthalene fluorescence ($1.7 \times 10^7\ M^{-1}\ \text{sec}^{-1}$) but does not rearrange:

$E^* = 36.0$ kcal/mole

Although these reactions do not involve the usual concept of energy transfer, it is clear that the physics and chemistry characterizing this class of reactions are of growing importance.

REFERENCES

1. A. A. Lamola, Electronic Energy Transfer in Solution: Theory and Applications, in *Techniques of Organic Chemistry*, XIV, P. A. Leermakers and A. Weissberger, eds., Interscience, New York (1969), pp. 17–132.
2. F. Wilkinson, in *Advances in Photochemistry*, Vol. 3, Wiley Interscience, New York (1964), p. 241.

3. R. G. Bennett and R. Kellogg, Mechanisms and Rates of Radiationless Energy Transfer, in *Progress in Reaction Kinetics*, Vol. 4, G. Porter, ed., Pergamon Press, London (1966).
4. Th. Förster, Delocalized Excitation and Excitation Transfer, in *Modern Quantum Chemistry*, Vol. 3, O. Sinnanoğlu, ed., Academic Press (1965).
5. R. S. Knox, *Theory of Excitons*, Academic Press, New York (1959).
6. F. Wilkinson, in *Luminescence in Chemistry*, E. J. Bowen, ed., D. Van Nostrand, London (1968), pp. 155–182.
7. G. W. Robinson and R. P. Frosch, *J. Chem. Phys.* **37**, 1962 (1962); **38**, 1187 (1963).
8. B. R. Henry and M. Kasha, *Ann. Rev. Phys. Chem.* **19**, 161 (1968).
9. M. Bixon and J. Jortner, *J. Chem. Phys.* **48**, 715 (1968).
10. J. Jortner and R. S. Berry, *J. Chem. Phys.* **48**, 2757 (1968).
11. D. P. Chock, J. Jortner, and S. A. Rice, *J. Chem. Phys.* **49**, 610 (1968).
12. J. Jortner, S. A. Rice, and R. M. Hochstrasser, in *Advances in Photochemistry*, Vol. 7, Wiley Interscience, New York (1969), p. 149.
13. K. F. Freed and J. Jortner, *J. Chem. Phys.* **52**, 6272 (1970).
14. S. H. Lin, *J. Chem. Phys.* **53**, 3766 (1970).
15. A. Nitzan and J. Jortner, *J. Chem. Phys.* **55**, 1355 (1971).
16. W. Siebrand, *J. Chem. Phys.* **55**, 5843 (1971).
17. (a) M. A. El-Sayed, *Acc. Chem. Res.* **4**, 23 (1971); (b) M. A. El-Sayed, *Pure Appl. Chem.* **24**, 475 (1970).
18. G. N. Lewis and M. Kasha, *J. Amer. Chem. Soc.* **66**, 2100 (1944).
19. M. A. El-Sayed, in *Proceedings, International Conference on Molecular Luminescence, Loyola University*, E. Lim, ed., Benjamin, New York (1969).
20. D. S. Tinti and M. A. El-Sayed, *J. Chem. Phys.* **54**, 2529 (1971).
21. M. A. El-Sayed, *J. Chem. Phys.* **54**, 680 (1971).
22. M. A. El-Sayed, D. S. Owens, and D. S. Tinti, *Chem. Phys. Lett.* **6**, 395 (1970).
23. M. Sharnoff, *J. Chem. Phys.* **46**, 3263 (1967); A. Kwiram, *Chem. Phys. Lett.* **1**, 272 (1967); J. Schmidt, I. A. M. Hesselmann, M. S. de Groot, and J. H. van der Waals, *Chem. Phys. Lett.* **1**, 434 (1967); J. Schmidt and J. H. van der Waals, *Chem. Phys. Lett.* **2**, 640 (1968).
24. M. A. El-Sayed, *J. Chem. Phys.* **52**, 6438 (1970).
25. D. S. Tinti, M. A. El-Sayed, A. H. Maki, and C. B. Harris, *Chem. Phys. Lett.* **3**, 343 (1969); M. A. El-Sayed, D. S. Tinti, and D. Owens, *Chem. Phys. Lett.* **3**, 339 (1969).
26. M. A. El-Sayed, *Acc. Chem. Res.* **1**, 8 (1968).
27. M. A. El-Sayed, D. S. Tinti, and E. M. Yee, *J. Chem. Phys.* **51**, 5721 (1969).
28. Th. Förster, *Ann. Physik.* **2**, 55 (1948).
29. Th. Förster, *Disc. Faraday Soc.* **27**, 7 (1959).
30. Th. Förster, in *Comparative Effects of Radiation*, M. Burton, J. S. Kirby-Smith, and J. L. Magee, eds., Wiley, New York (1960).
31. D. L. Dexter, *J. Chem. Phys.* **21**, 836 (1953).
32. E. Wigner, *Nachr. Ges. Wiss. Gottingen Math. Physik. Kl.*, 375 (1927).
33. E. G. McRae and M. Kasha, in *Physical Processes in Radiation Biology*, L. Augenstein, R. Mason, and B. Rosenberg, eds., Academic Press (1964), pp. 23–42.
34. M. Kasha, *Rev. Mod. Phys.* **31**, 162 (1959).
35. R. M. Hochstrasser and M. Kasha, *Photochem. Photobiol.* **3**, 317 (1964).
36. P. Avakian and R. E. Merrifield, *Mol. Cryst.* **5**, 37 (1968).
37. W. T. Simpson and D. L. Peterson, *J. Chem. Phys.* **26**, 588 (1957).
38. A. S. Davydov, *Theory of Molecular Excitons*, McGraw-Hill, New York (1962).
39. W. Ware, *J. Phys. Chem.* **66**, 455 (1962).

40. W. Ware, *J. Amer. Chem. Soc.* **83**, 4374 (1961).
41. R. G. Bennett, *J. Chem. Phys.* **41**, 3037 (1964).
42. O. Schnepp and M. Levy, *J. Amer. Chem. Soc.* **84**, 172 (1962).
43. S. A. Latt, H. T. Cheung, and E. R. Blout, *J. Amer. Chem. Soc.* **87**, 995 (1965).
44. D. Rauh, T. R. Evans, and P. A. Leermakers, *J. Amer. Chem. Soc.* **90**, 6897 (1968).
45. L. Stryer and R. P. Haugland, *Proc. Nat. Acad. Sci. U.S.* **58**, 719 (1967).
46. R. H. Conrad and L. Brand, *Biochemistry* **7**, 777 (1968).
47. J. T. Dubois and M. Cox, *J. Chem. Phys.* **38**, 2536 (1963).
48. J. T. Dubois and R. L. Van Hemert, *J. Chem. Phys.* **40**, 923 (1964).
49. F. Wilkinson and J. T. Dubois, *J. Chem. Phys.* **39**, 377 (1963).
50. A. N. Terenin and V. Ermolaev, *Dokl. Akad. Nauk SSSR* **85**, 547 (1952).
51. A. N. Terenin and V. Ermolaev, *Trans. Faraday Soc.* **52**, 1042 (1956).
52. F. Perrin, *Compt. Rend.* **178**, 1978 (1924).
53. W. R. Ware, *Survey Prog. Chem.* **4**, 205 (1968).
54. H. L. J. Bäckstrom and K. Sandros, *Acta Chem. Scand.* **12**, 823 (1958).
55. H. L. J. Bäckstrom and K. Sandros, *Acta Chem. Scand.* **18**, 48 (1960).
56. G. Porter and F. Wilkinson, *Proc. Roy. Soc. A* **264**, 1 (1961).
57. W. G. Herkstroeter and G. S. Hammond, *J. Amer. Chem. Soc.* **88**, 4769 (1966).
58. W. G. Herkstroeter, L. B. Jones, and G. S. Hammond, *J. Amer. Chem. Soc.* **88**, 4777 (1966).
59. W. M. Moore, G. S. Hammond, and R. P. Foss, *J. Amer. Chem. Soc.* **83**, 2789 (1961).
60. G. S. Hammond, J. Saltiel, A. A. Lamola, N. J. Turro, J. S. Bradshaw, D. O. Cowan, R. C. Counsell, V. Vogt, and C. Dalton, *J. Amer. Chem. Soc.* **86**, 3197 (1964).
61. J. Saltiel and G. S. Hammond, *J. Amer. Chem. Soc.* **85**, 2515 (1963).
62. G. S. Hammond and J. Saltiel, *J. Amer. Chem. Soc.* **85**, 2516 (1963).
63. G. S. Hammond and J. Saltiel, *J. Amer. Chem. Soc.* **84**, 4983 (1962).
64. G. S. Hammond, N. J. Turro, and P. A. Leermakers, *J. Phys. Chem.* **66**, 1144 (1962).
65. R. S. H. Liu, N. J. Turro, and G. S. Hammond, *J. Amer. Chem. Soc.* **87**, 3406 (1965).
66. G. S. Hammond, N. J. Turro, and A. Fischer, *J. Amer. Chem. Soc.* **83**, 4674 (1961).
67. N. J. Turro and G. S. Hammond, *J. Amer. Chem. Soc.* **84**, 2841 (1962).
68. G. S. Hammond, N. J. Turro, and R. S. H. Lui, *J. Org. Chem.* **28**, 3297 (1963).
69. K. R. Kopecky, G. S. Hammond, and P. A. Leermakers, *J. Amer. Chem. Soc.* **84**, 1015 (1962).
70. D. O. Cowan, M. M. Couch, K. R. Kopecky, and G. S. Hammond, *J. Org. Chem.* **29**, 1922 (1964).
71. G. S. Hammond, P. A. Leermakers, and N. J. Turro, *J. Amer. Chem. Soc.* **83**, 2395 (1961).
72. G. S. Hammond, C. A. Stout, and A. A. Lamola, *J. Amer. Chem. Soc.* **86**, 3103 (1964).
73. R. S. H. Liu, *J. Amer. Chem. Soc.* **86**, 1892 (1964).
74. G. O. Schenck and R. Steinmetz, *Chem. Ber.* **96**, 520 (1963).
75. G. S. Hammond, N. J. Turro, and A. Fischer, *J. Amer. Chem. Soc.* **83**, 4674 (1961).
76. H. E. Zimmerman and G. L. Grunewald, *J. Amer. Chem. Soc.* **88**, 183 (1966).
77. K. Gollnick and G. O. Schenck, *Pure Appl. Chem.* **9**, 507 (1964).
78. G. O. Schenck and R. Steinmetz, *Tetrahedron Lett.*, *1* (1960).
79. P. A. Leermakers, G. W. Byers, A. A. Lamola, and G. S. Hammond, *J. Amer. Chem. Soc.* **85**, 2670 (1963).
80. A. A. Lamola, P. A. Leermakers, G. W. Byers, and G. S. Hammond, *J. Amer. Chem. Soc.* **87**, 2322 (1965).
81. D. E. Breen and R. A. Keller, *J. Amer. Chem. Soc.* **90**, 1935 (1968).
82. R. A. Keller, *J. Amer. Chem. Soc.* **90**, 1940 (1968).

83. R. A. Keller and L. J. Dolby, *J. Amer. Chem. Soc.* **91**, 1293 (1969).
84. N. Filipescu, J. DeMember, and F. L. Minn, *J. Amer. Chem. Soc.* **91**, 4169 (1969).
85. N. Filipescu and J. R. Bunting, *J. Chem. Soc. B*, 1498 (1970).
86. J. R. DeMember and N. Filipescu, *J. Amer. Chem. Soc.* **90**, 6425 (1968).
87. A. A. Lamola, *J. Amer. Chem. Soc.* **91**, 4786 (1969).
88. D. O. Cowan and A. A. Baum, *J. Amer. Chem. Soc.* **92**, 2153 (1970).
89. D. O. Cowan and A. A. Baum, *J. Amer. Chem. Soc.* **93**, 1153 (1971).
90. H. Morrison and R. Peiffer, *J. Amer. Chem. Soc.* **90**, 3428 (1968).
91. H. Morrison, *J. Amer. Chem. Soc.* **87**, 932 (1965).
92. S. R. Kurowsky and H. Morrison, *J. Amer. Chem. Soc.* **94**, 507 (1972), and reference cited therein.
93. F. C. DeSchryver and J. Put, *Ind. Chim. Belg.* **37**, 1107 (1972).
94. W. Cooper, *Chem. Phys. Lett.* **7**, 73 (1970).
95. M. Kasha, *Rev. Mod. Phys.* **31**, 162 (1959).
96. M. Trlifaj, *Czech. J. Phys.* **6**, 533 (1956); **8**, 510 (1958).
97. S. A. Rice and J. Jortner, in *Physics and Chemistry of the Organic Solid State*, Vol. 3, D. Fox, M. M. Labes, and A. Weissberger, eds., Interscience, New York (1967).
98. G. C. Nieman and G. W. Robinson, *J. Chem. Phys.* **39**, 1298 (1963).
99. J. B. Birks, *Photophysics of Aromatic Molecules*, Interscience, New York (1970).
100. L. M. Stephenson, D. G. Whitten, G. F. Vesley, and G. S. Hammond, *J. Amer. Chem. Soc.* **88**, 3665 (1966).
101. L. M. Stephenson and G. S. Hammond, *Pure Appl. Chem.* **16**, 125 (1968).
102. G. N. Taylor and G. S. Hammond, *J. Amer. Chem. Soc.* **94**, 3684, 3687 (1972).
103. (a) S. L. Murov, R. S. Cole, and G. S. Hammond, *J. Amer. Chem. Soc.* **90**, 2957 (1968); (b) S. L. Murov and G. S. Hammond, *J. Phys. Chem.* **72**, 3797 (1968).
104. B. S. Solomon, C. Steel, and A. Weller, *Chem. Comm.*, 927 (1969).
105. S. L. Murov, L. Yu, and L. P. Giering, *J. Amer. Chem. Soc.* **95**, 4329 (1973).

7

Dienone and Enone Photochemistry[1-3]

7.1. DIENONE PHOTOREACTIONS

The photorearrangement of a dienone was noted[4] as early as 1830 in a study of the sesquiterpene α-santonin (1). However, the structure and stereochemistry of the various photoproducts were not conclusively established until 1965.[5] Upon irradiation in neutral media, α-santonin (1) undergoes rapid rearrangement to the cyclopropyl ketone, lumisantonin (2). However, if the irradiation is not terminated after a short period of time the lumisantonin itself rearranges into a linearly conjugated dienone (3). The dienone (3) can be isolated from the photolysis of either (1) or (2) in benzene or ether. In nucleophilic solvents (alcohol or water) the dienone (3) is also photochemically active and is further converted into an ester or an acid (photosantonic acid) (4).

In acidic solution the rearrangement of α-santonin to lumisantonin and then ultimately to photosantonic acid is not as efficient as rearrangement to the hydroxy ketone, isophotosantonic lactone (5), shown on page 308. Fisch and Richards[6] found that the photorearrangements of α-santonin could be sensitized with benzophenone or Michler's ketone. Moreover, if the irradiation of α-santonin (3660 Å) is carried out in piperylene as solvent, the photoreaction is completely quenched. This suggests that the rearrangements proceed via triplet states.

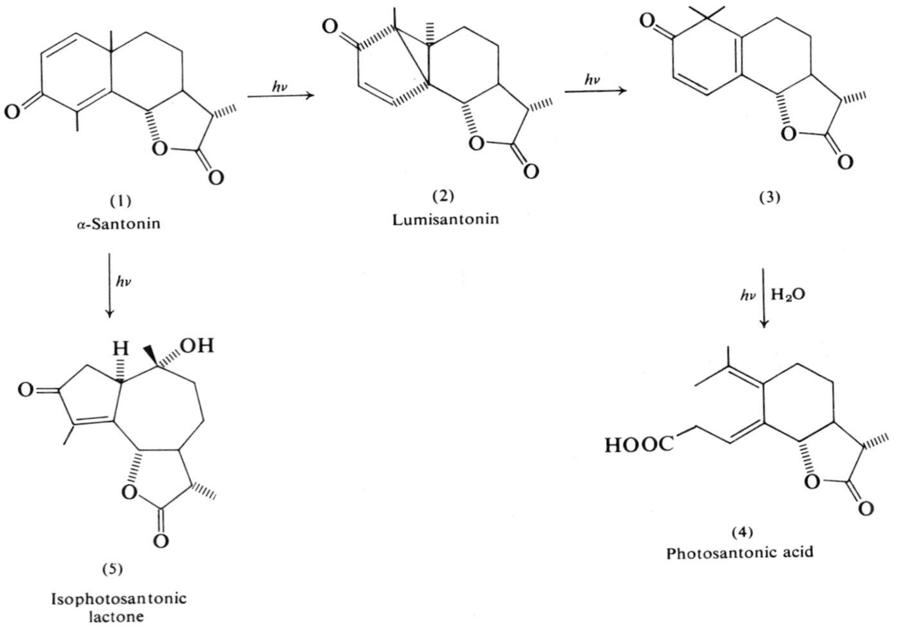

hν
hν
hν
hν
H₂O
(1)
α-Santonin
(2)
Lumisantonin
(3)
(4)
Photosantonic acid
(5)
Isophotosantonic lactone
HOOC
OH
H

7.2. DIENONE TO CYCLOPROPYL KETONE FORMATION

In order to defer stereochemical questions we will consider first the photoreaction of 6,6-diphenyl-cyclohexa-2,5-dienone (6). Either direct or sensitized photolysis[3,7,8] of compound (6) in aqueous dioxane follows the following course:

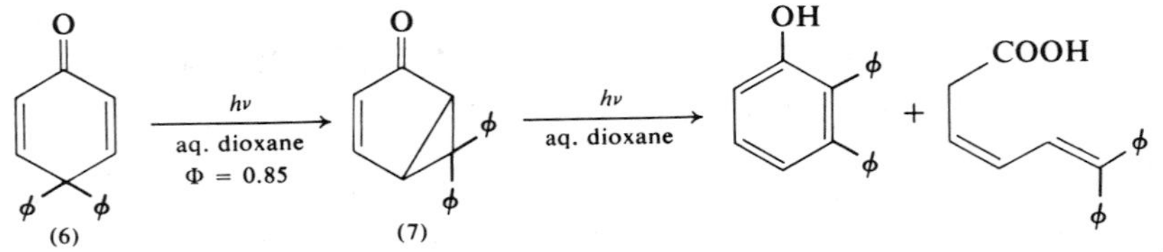

hν
aq. dioxane
Φ = 0.85
hν
aq. dioxane
COOH
OH
(6)
(7)

(7.1)

Since the phosphorescence emission from (6) (68.8 kcal/mole) is very similar in energy and vibrational structure to benzophenone, and has a short lifetime (0.5 msec), it was proposed that the photorearrangement takes place via the $n \rightarrow \pi^*$ triplet state. A Zimmerman-like mechanism is as follows for the formation of the cyclopropyl ketone (7) from dienone (6):

(6) $\xrightarrow[n-\pi^*]{h\nu}$ (8)

(8) $\xrightarrow{\text{excited state bond formation}}$ (9)

(9) $\xrightarrow{\text{intersystem crossing}}$ (10)

(10) $\xrightarrow{\text{ground state rearrangement}}$ (7) (7.2)

Several questions should be asked about this mechanism: (a) Is excited state bond formation [(8) → (9)] reasonable? (b) Could the rearrangement and intersystem crossing take place simultaneously [(9) → (7) or even (8) → (7)]? (c) If the ground state species (10) is formed, will it rearrange to (7)?

7.2a. *3–5 Bond Orders*

The bond orders for both $n \rightarrow \pi^*$ and $\pi \rightarrow \pi^*$ singlet and triplet states (PPP CI calculation) indicate (Table 7.1) that the one excited state which is not β,β bonding is the $\pi \rightarrow \pi^*$ triplet state. Such bonding is predicted to be energetically favorable for the $\pi \rightarrow \pi^*$ singlet, the $n \rightarrow \pi^*$ singlet, and the $n \rightarrow \pi^*$ triplet. Consequently, it was concluded that the excited bond formation step [(8) → (9)] is reasonable.[9] However, the authors did not report other bond orders that would have been equally interesting. For example, if the process is a concerted [(8) → (7)] transformation, we would also expect to observe a decrease in the 4–5 bond order (P_{45}) and an increase in the 2–4 bond order (P_{24}) in the $n \rightarrow \pi^*$ triplet state:

8–7 type transformation

TABLE 7.1. *The 3–5 Bond Orders*

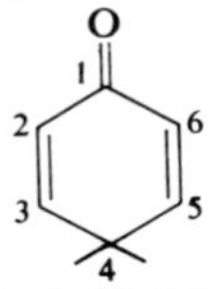

State	Bond order, P_{35}
Ground	−0.0678
$n \rightarrow \pi^*$ Singlet	+0.1107
$n \rightarrow \pi^*$ Triplet	+0.1005
$\pi \rightarrow \pi^*$ Singlet	+0.3338
$\pi \rightarrow \pi^*$ Triplet	−0.0567

7.2b. *Zwitterionic vs. Diradical Intermediates*

Swenton *et al.*[10] have shown that in the gas phase the cyclohexadienone (11) gives the bicyclic ketone (12) as the initial product ($\Phi = 0.40$, 3660 Å) and that this ketone is photoconverted to the dienone (13):

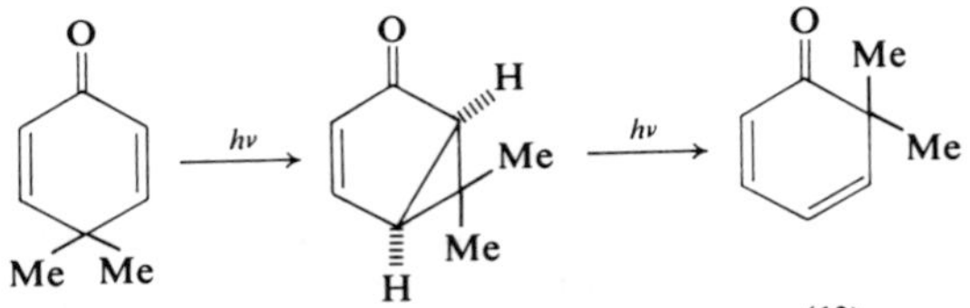

(11) (12) (13) (7.3)

In cyclohexane the same two ketones (12) and (13) are obtained from the photolysis of (11) but in aqueous dioxane two phenols are isolated as well as the bicyclic ketone (12). Swenton[10] suggested that the gas-phase reaction involves diradical species, whereas in polar solvents zwitterionic intermediates are favored:

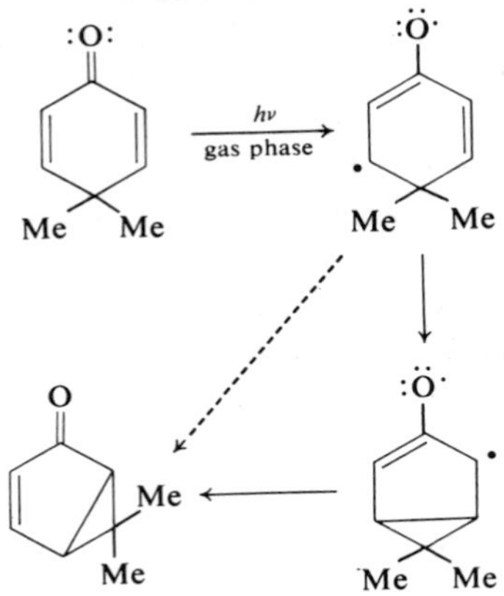

(7.4)

:Ö: Me CCl_3 (14) — $h\nu$ → :Ö· Me CCl_3 — $(CH_3)_2C{=}CH(CH_3)$ → O Me CCl_3 (15)

$CH_3CH_2OCH_2CH_3$

OH Me CCl_3 (16a) — $-\cdot CCl_3$, $\Phi = 0.72$ → OH Me (16b)

:Ö· Me CCl_3 — benzene, $\Phi = 0.75$ → O H Me CCl_3 H (17)

polar solvents

:Ö:⊖ ⊕ Me CCl_3 — MeOH → O OMe Me CCl_3 (18)

O ⊕ Cl Cl Me — MeOH (R = OMe) / *t*BuOH/LiCl (R = Cl) → O R Cl Cl H Me (19)

Schuster and co-workers[11–16] have provided one of the best examples where *both* diradicals and zwitterionic species are probably involved. The dienone (14), when photolyzed in 1,1,2-trimethylethylene, forms only the oxetane (15); all other photoreactions are eliminated. This finding confirms that $n \rightarrow \pi^*$ and not $\pi \rightarrow \pi^*$ states are involved in the dienone photorearrangements. Irradiation of (14) in a good hydrogen (H·) donating solvent like ether results in hydrogen abstraction by the carbonyl group and expulsion of the C-4 (CCl_3) substituent. This type of expulsion is frequently observed when heterolytic cleavage would lead to a very stable free radical.[17,18] In this case *p*-cresol is formed with a quantum yield of 0.72 along with chloroform and hexachloroethane. In benzene the photolysis of (14) gives the cyclopropylketone (17) with a quantum efficiency of 0.75. However, if polar solvents (MeOH) are used in place of benzene, in addition to the usual dienone rearrangement product (17), several new products (18) and (19) are formed by trapping of the ionic intermediates. (See structure on previous page).

Another interesting and relevant series of dienones are the spirodienones. Schuster and co-workers[19–21] found that photolysis of the parent dienone (20) gives exclusively reduction products (21)–(24):

O

$\xrightarrow[\text{ether}]{h\nu}$

O·

$CH_2CH_2\cdot$

(20)

OH

CH_2CH_3

(21)

+

OH

CH_3

CH_2CH_2—$CHOCH_2CH_3$

(22)

(7.5)

+

OH

$CH_3CHCHOCH_2CH_3$

CH_3

(23)

+

CH_3

$OCHOCH_2CH_3$

CH_2CH_3

(24)

On the basis of the observed products, these workers favor the diradical spirodienone excited states.

The photochemistry of the dienones (25) and (26) is of special interest inasmuch as these compounds not only undergo stereospecific photo-rearrangements, but the type of reaction observed is dependent upon the wavelength of the absorbed radiation. Photolysis of (25) or (26) at 350 nm in cyclohexane brings about *cis–trans* isomerization via opening of the cyclopropane ring to produce a 4:1 (*trans*:*cis*) photostationary mixture. Thermal equilibration produced a 1.8:1 equilibrium mixture with the *trans* isomer predominating. Photolysis at 254 nm of (25) or (26) leads not only to *cis–trans* isomerization but also produces the quinone methides (27) and (28) via methyl or hydrogen migration. The relative values for Φ_1 to Φ_6 are 25, 19.6, 9.0, 3.73, 0.90, 0.0. Thus *cis*-dienone (25) affords essentially only methyl migration, whereas *trans*-dienone (26) affords only hydrogen migration.

$\Phi_1 = 25$ $h\nu$ $\Phi_2 = 19.6$

(25) (26)

H H Me Me H Me Me H

$\Phi_3 = 9.0$ $\Phi_6 = 0.0$ $\Phi_5 = 0.90$ $\Phi_4 = 3.73$

H Me H Me Me H H Me

(27) (28)

From the photoreactions presented in this section it appears that both diradical intermediates and ionic intermediates can be involved in the photochemistry of dienones.

7.2c. *Pivot vs. Slither Mechanism*

Zimmerman[24,25] has provided strong circumstantial evidence that zwitterionic intermediates can be involved in formation of cyclopropyl ketones from dienones. His approach was to generate the dipolar species via ground state chemistry and show that these rearrange to the photoproduct:

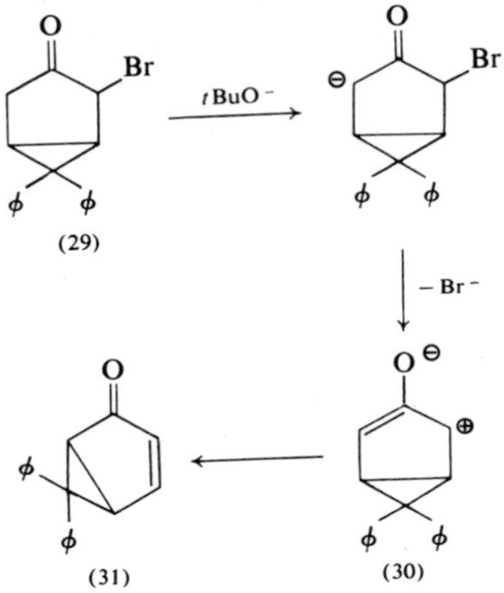

(29) $\xrightarrow{t\text{BuO}^-}$ $\xrightarrow{-\text{Br}^-}$ (30) $\longrightarrow$ (31) (7.6)

The rearrangement from (30) to (31) can occur via a "slither" or "pivot" mechanism:

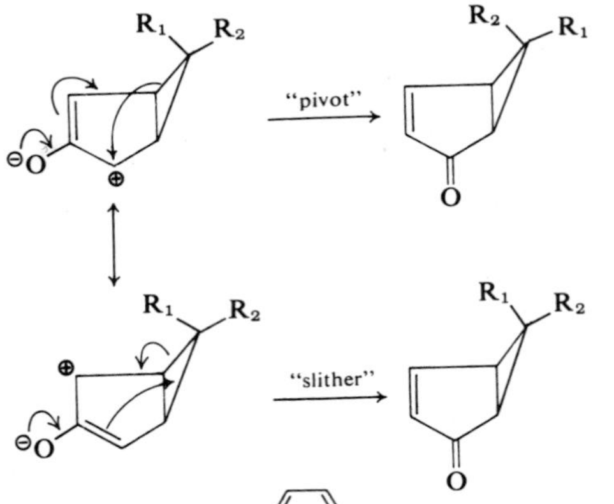

"pivot" (7.7)

"slither" (7.8)

$R_1 = \phi$; $R_2 = $ —C₆H₄—Br

$R_1 = $ —C₆H₄—Br; $R_2 = \phi$

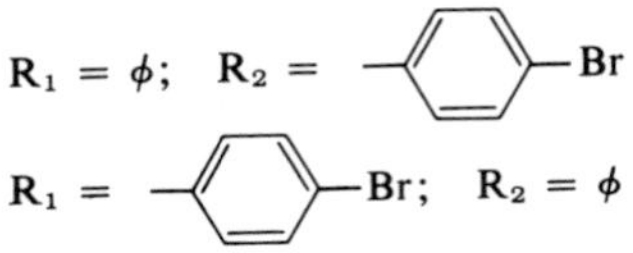

Since the rearrangement afforded *endo* product from *endo* zwitterion and *exo* product from *exo* zwitterion, the rearrangement follows the "slither" mechanism.

In light of these results, the study of Rodgers and Hart[27] is particularly interesting. They found that as one increases substitution at C-3 and C-5, there is a greater stereospecificity in cyclopropyl ketone formation:

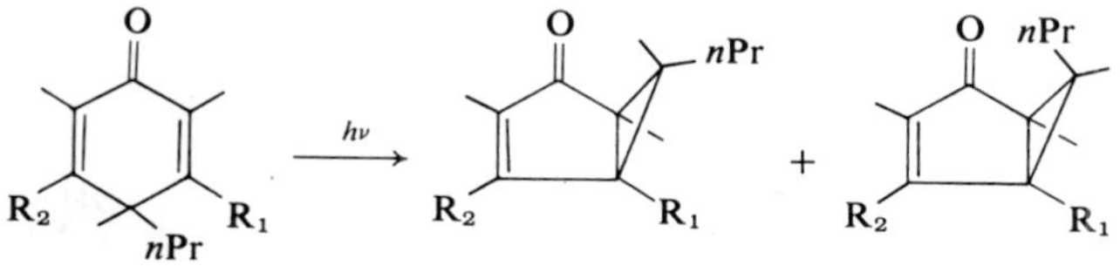

(32a,b,c) $\xrightarrow{h\nu}$ (33) + (34)

	(33)	(34)
(a) $R_1 = R_2 = H$	56%	44%
(b) $R_1 = H$; $R_2 = Me$	32%	68%
(c) $R_1 = R_2 = Me$	9%	91%

The preferred ground state conformation of (32c) is

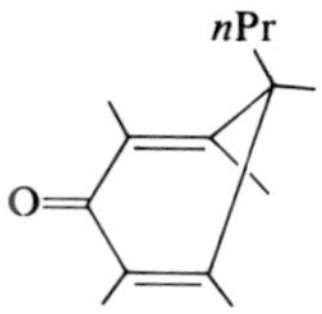

This preferred ground state conformation, coupled with the "slither" mechanism, would account for the observed stereospecificity [formation of (34c)]. The spirodienone[28,29] (35) undergoes photorearrangement to form

TABLE 7.2. *Solvent Effect on the Photoproducts of Compound (35)*

Solvent	Ratio (37):(36)	Ratio [(37) + (36)]:[(38) + (39)]
MeOH	2.5	0.14
t-BuOH	2.0	0.34
Dioxane	1.2	1.28
Et_2O	1.1	1.24
Hexane	0.91	0.35
Cyclohexane	0.80	0.52
Benzene	1.0	—

the isomeric cyclopropyl ketones [(36) and (37), $\Phi = 0.59$] and the phenols [(38) and (39)]:

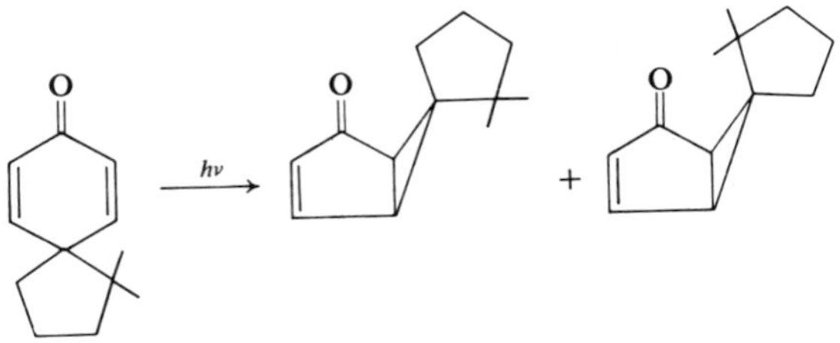

(35) (36) (37)

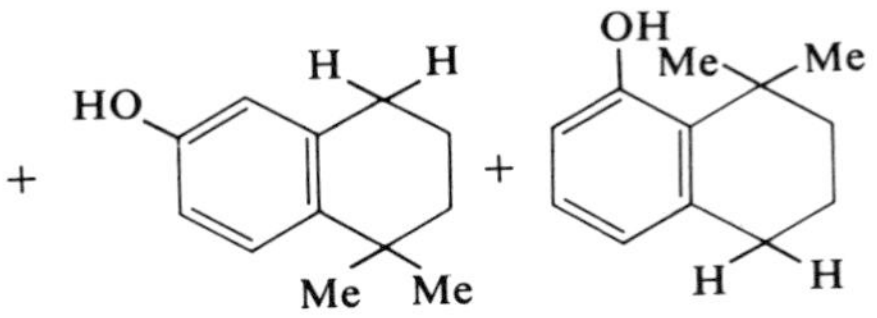

(38) (39)

(7.9)

In polar solvents the reaction is more stereoselective (Table 7.2). This is explained by preferential solvation of the diastereoisomeric zwitterionic intermediates which lead to the cyclopropyl ketones (36) and (37).

Ogura and Matsuura[30] argued that which mechanism (the slither or pivot) takes place depends on whether the rearrangement takes place from the ground state zwitterionic species or from the excited state diradical species. They claim that the following transformation occurs via the "diradical pivot" mechanism:

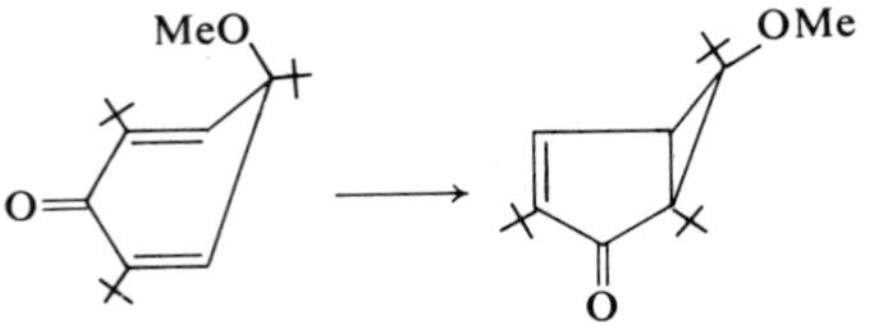

(7.10)

However, this explanation is rather speculative. Steric effects seem to be important but not entirely understood at the present time.[31]

7.3. DIENONE TO HYDROXY KETONE

In aqueous acid cross-conjugated cyclohexadienones are principally photoconverted to one or more hydroxy ketones. In the case of α-santonin (1), isophotosantonic lactone (5) is formed in about 50% yield. A series of papers by Kropp and co-workers has aided in understanding this reaction.[32,39–41] They have shown that the presence of a 4-methyl group (steroid numbering) results in the preferential formation of the 5–7 fused ring system (isophotosantonic lactone). An analogous cross-conjugated cyclohexadienone with 2-methyl substituent yields only the spiro compound under the same conditions:

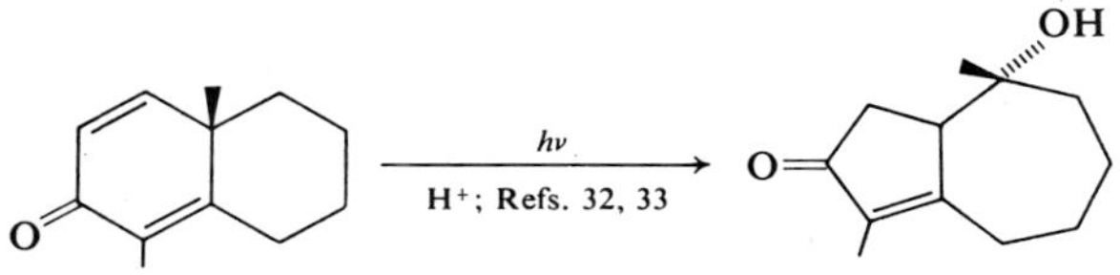

$h\nu$ / H^+; Refs. 32, 33 (7.11)

$h\nu$ / H^+; Ref. 34 (7.12)

$h\nu$ / H^+; Refs. 35–38 (7.13)

$h\nu$ / H^+; Ref. 39 (7.14)

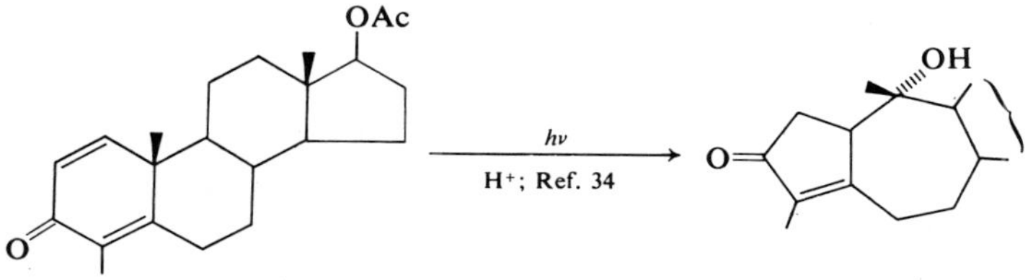

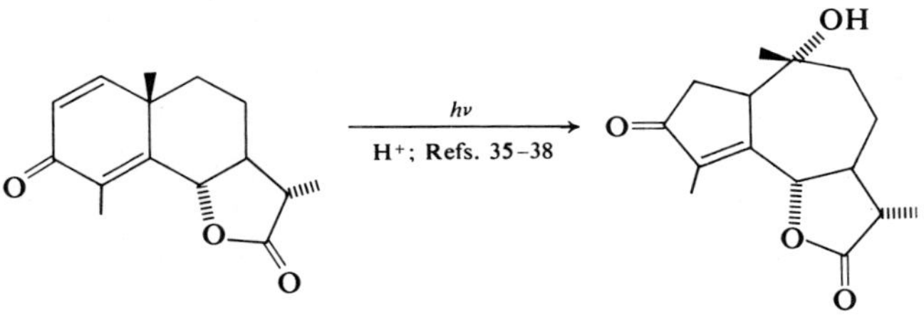

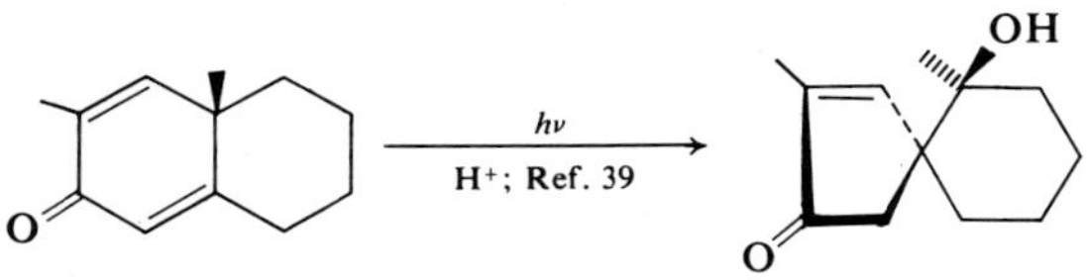

Kropp has explained these results via the following mechanism:

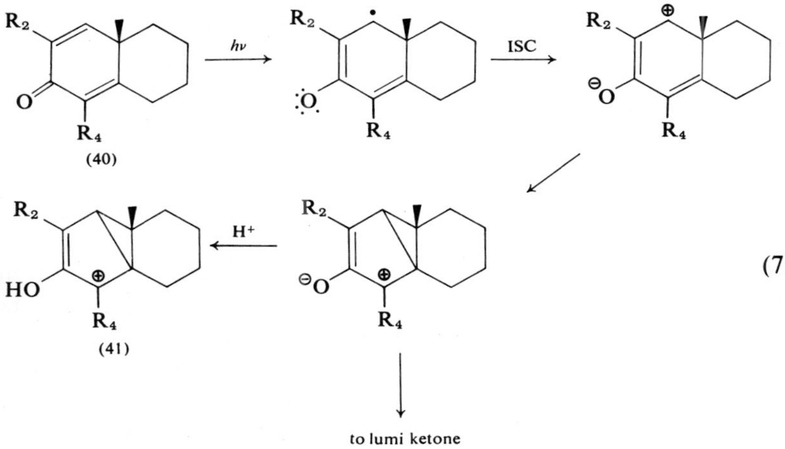

$h\nu$ ISC H^+ (40) (41) to lumi ketone (7.15)

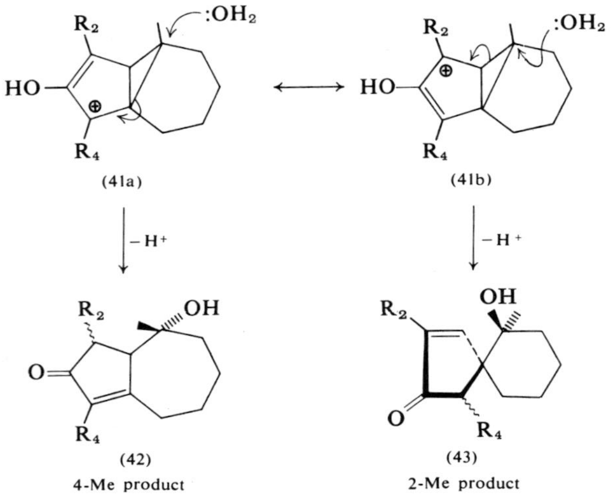

(41a) (41b) (7.16)

$-H^+$ $-H^+$

(42) 4-Me product (43) 2-Me product

The 4 or 2 methyl group tends to cause localization of the positive charge at the more highly substituted position via an inductive effect.

When the dienone does not have either a 2 or 4 substituent a mixture of the two types of products is obtained:

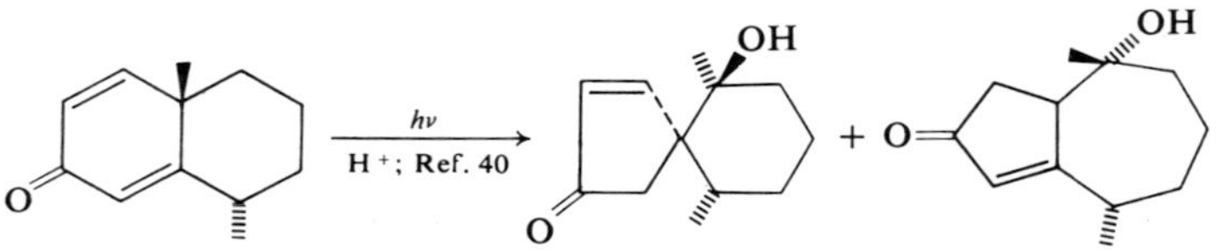

$h\nu$ H^+; Ref. 40 (7.17)

$h\nu$ H^+; Ref. 41 (7.18)

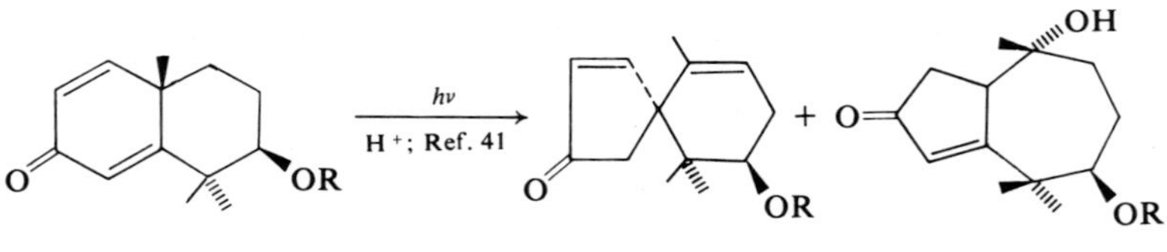

7.4. CYCLOPROPYL KETONES

The initially formed cyclopropyl ketones from the photolysis of cross-conjugated cyclohexadienones are also photolabile. Usually the photochemical rearrangements of cyclopropyl ketones involve cleavage of the cyclopropyl bond which forms part of the cyclopentenone ring followed by either (a) a substituent migration or (b) rearrangement through a spiro intermediate. While the literature in this area is too voluminous to review in detail, we will give a number of examples. Lumisantonin (2) photorearranges via a 1,2 shift of the angular methyl substituent[37]:

(7.19)

Products from the photolysis of the cyclopropyl ketone (44) are dependent on the *p*H of the solvent.[42] In aqueous dioxane only the 2,3-diphenylphenol (45) is formed along with the photoacid (46). Bond cleavage at *c* takes place via both the singlet and triplet states, whereas bond cleavage at *a* with phenol formation takes place in the triplet state:

(44)
$\Phi = 0.160$

(45)
30% yield; $\Phi = 0.066$

(46)
60% yield; $\Phi = 0.094$

TABLE 7.3

Compound	Solvent	% (56)	% (57)
(52)	MeOH	50	Trace
(53)	MeOH	62	Trace
(54)	MeOH	62	Trace
(55)	MeOH	57	Trace
(53)	45% HOAc	33	28
(55)	45% HOAc	47	33

While only the 2,3-diphenylphenol is formed upon photolysis of (44) in aqueous dioxane, in 50% acetic acid the 2,3- and 3,4-diphenylphenols are formed in a 1:1 ratio. This has been explained on the basis of the intermediates (47)–(50):

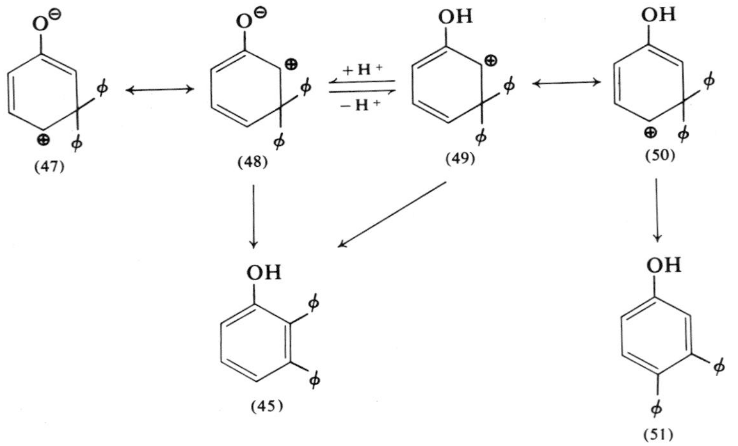

(47) (48) (49) (50)

(45) (51)

Zimmerman *et al.*[42] pointed out that the bridged phenonium ion from (48) is more stable (lower energy route) than is the phenonium ion from (47), while the phenonium ions from (49) and (50) should be of about equal stability. An alternate explanation has been proposed by Kropp,[1] who suggests that preferred migration to the position adjacent to the electron-rich oxygen in aprotic solvents arises from a preferred minimization of charge separation in in the dipolar species.

The photolysis of dienone (52) to the linearly conjugated dienone (56) involves a four-photon sequence and nicely illustrates the involvement of spiro intermediates[39,43] (Table 7.3).

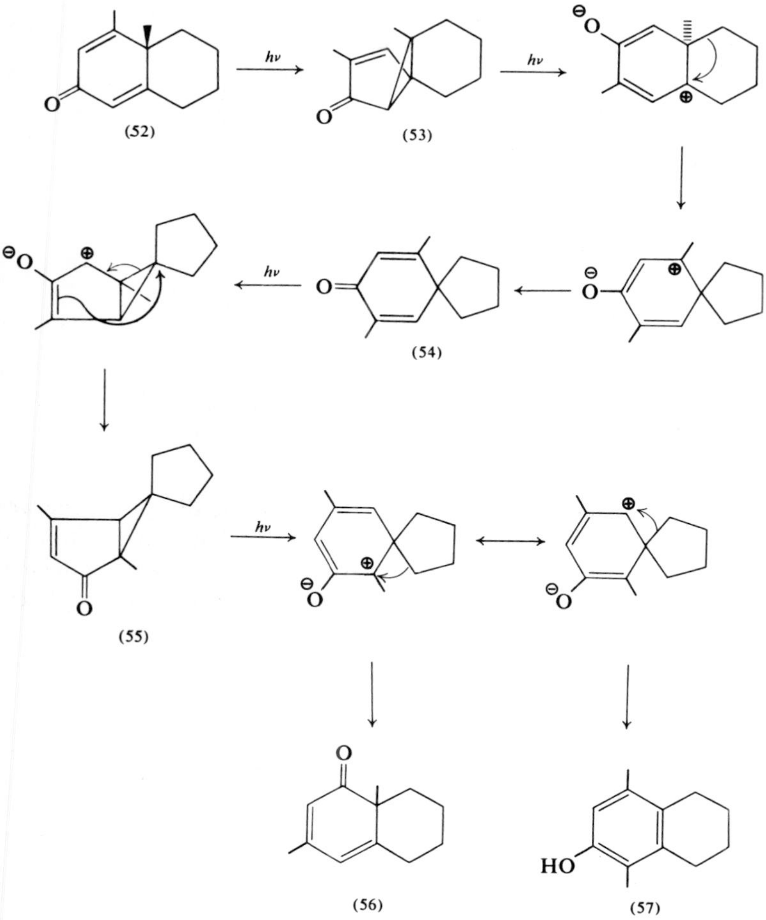

(52) (53)

(54)

(7.20)

(55)

(56) (57)

7.5. 2,4-CYCLOHEXADIENONES

Linearly conjugated cyclohexadienones usually photorearrange with ring fusion to a *cis*-diene-ketene. The reaction is reversible, so that in the absence of a nucleophile little change is observed. A good example of this type of transformation is the formation of photosantonic acid:

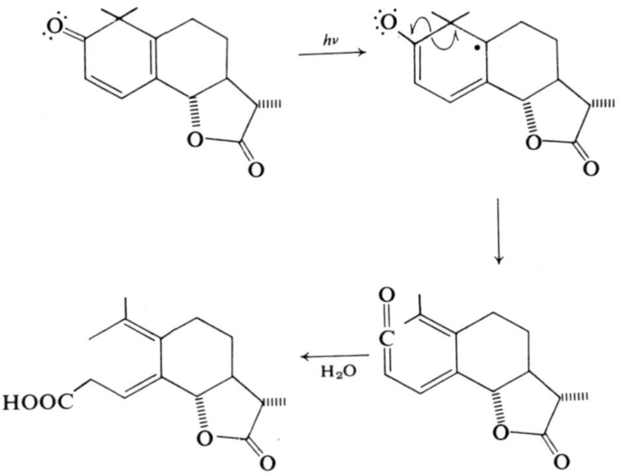

(7.21)

(7.22)

(4)

In the presence of cyclohexyl amine the photolysis of the dienone (58) gives the cyclohexyl amide expected from dienone ring fission and reaction of cyclohexyl amine with the ketene. However, in the presence of weaker nucleophiles the ring opened product is *not* trapped and instead there is a relatively slow formation of phenols[36,44]:

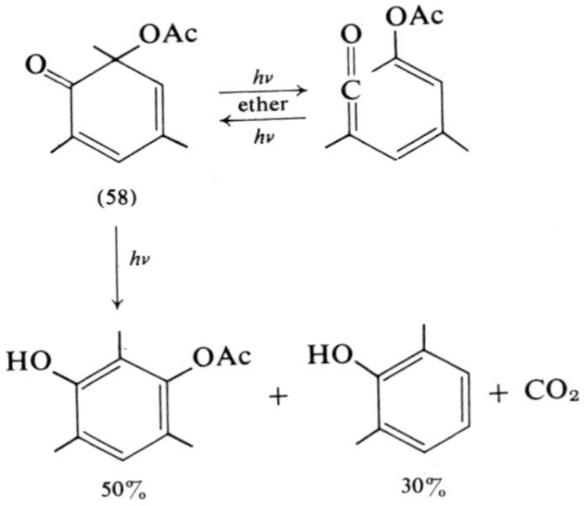

(58)

(7.23)

50% 30%

The acetoxyl migration is thought to go through intermediates like (59) and (60):

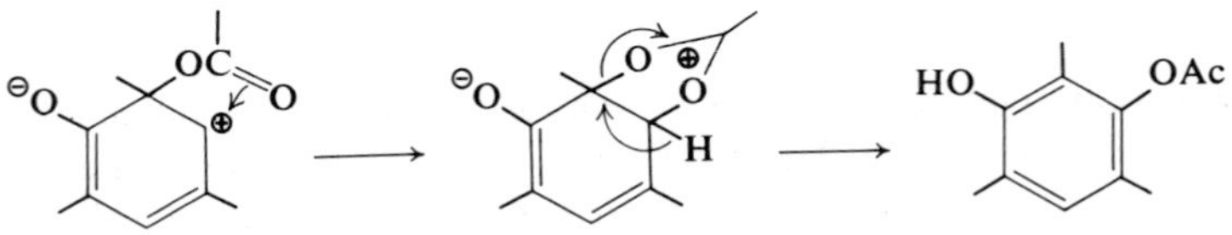

(7.24)

(59) (60)

Heavily substituted *o*-dienones, such as 2,3,4,5,6,6-hexamethyl and 2,3,4,6,6-pentamethyl cyclohexadienone, upon irradiation give not the ketene but a substituted bicyclic compound of the following type (62)[45–54]:

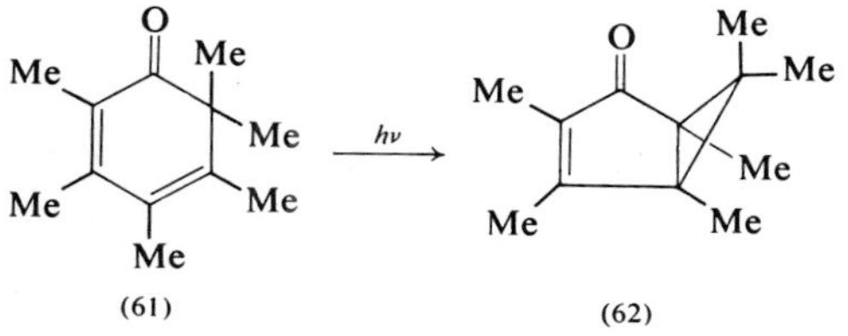

(61) $\xrightarrow{h\nu}$ (62) (7.25)

The use of trifluoroethanol as solvent or absorption of the dienone on silica gel promotes the photoconversion of dienones into bicyclic ketenes.[47] For the photolysis[48–50] of (63) it has been shown by low-temperature infrared and ultraviolet spectroscopy that the initial photolysis gives a ketene which can be efficiently trapped by cyclohexylamine or, in the absence of a good nucleophile, thermally rearranges by a $({}_{\pi}4_a + {}_{\pi}2_a)$ allowed process to a bicyclic ketone (64):

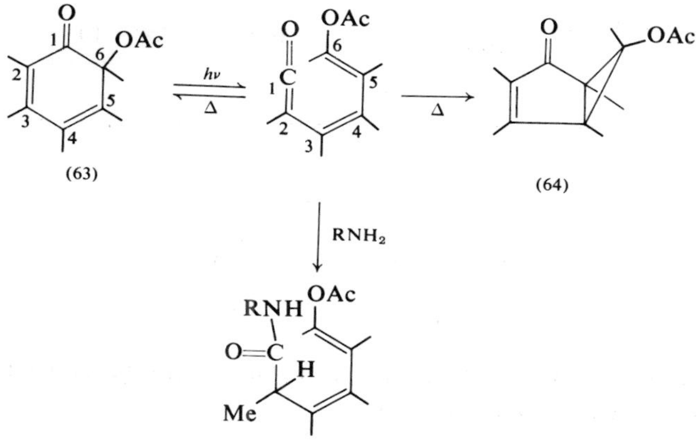

(63) $\underset{\Delta}{\overset{h\nu}{\rightleftharpoons}}$ ketene $\xrightarrow{\Delta}$ (64); ketene $\xrightarrow{RNH_2}$ amide (7.26)

Morris and Waring[49] suggested that to effect the necessary overlap to form (64), the π-orbital lobes at C-4, C-6 and C-1 and C-5 must be twisted from planarity. This could explain the finding that only heavily substituted dienones lead to the bicyclo[3.1.0]hex-3-en-2-one photolysis.

7.6. CYCLOHEXENONE PHOTOREARRANGEMENTS

The photorearrangement of cyclohexenones to bicyclo[3.1.0]hexanones (type A process) has been extensively studied[55–65]:

TABLE 7.4

Compound	Rearrangement process	Φ	k_r, sec^{-1}	k_d, sec^{-1}
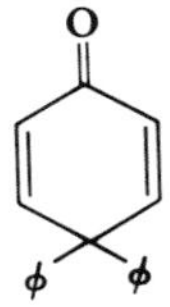 O; ϕ ϕ	Type A	0.85	$>2.0 \times 10^{10}$	$>3.5 \times 10^{9}$
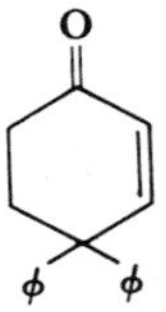 O; ϕ ϕ	ϕ migration	0.043	2.4×10^{7}	8.4×10^{8}
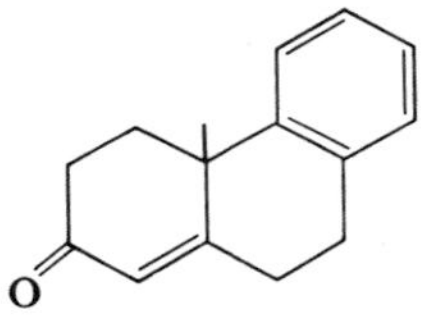 O	Type A	0.0084	2.9×10^{5}	3.9×10^{7}

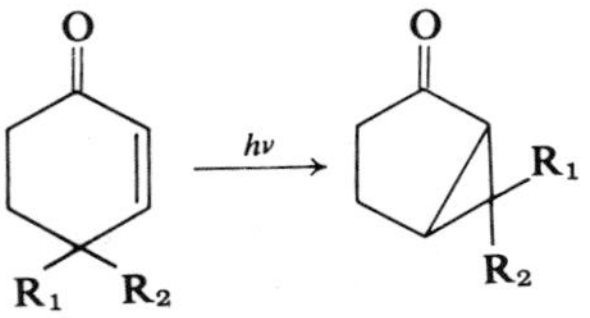

O; R_1 R_2 $\xrightarrow{h\nu}$ O; R_1; R_2 (7.27)

There are also numerous examples of a second rearrangement which occurs with 4-aryl or diarylcyclohexenones[66–75]:

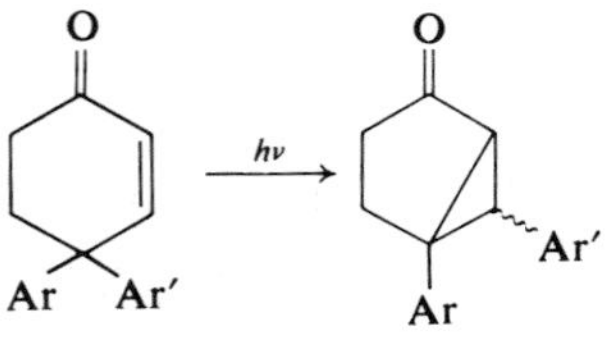

O; Ar Ar′ $\xrightarrow{h\nu}$ O; Ar′; Ar (7.28)

7.6a. *Aryl-Substituted Cyclohexenones*

Zimmerman and co-workers[66,72,73,75] have shown that photolysis of 4,4-diphenylcyclohex-2-en-1-one gives the following three products:

$h\nu$, 366 nm, benzene

(65) $\Phi = 0.043$ → (66) $\Phi = 0.043$ + (67) $\Phi = 0.0003$ + (68) $\Phi = 0.0002$ (7.29)

In addition, it was observed that the sensitized photolysis produced the same distribution of products with the same efficiency (fingerprint characteristic of the triplet state). From quenching studies the specific rate constant for the rearrangement could be obtained. Phenyl migration rearrangement is of intermediate efficiency, interposed between the more efficient and less efficient type A processes (Table 7.4). The type of mechanism proposed for this transformation is as follows:

$h\nu$

concerted or stepwise

(trace) (7.30)

While this reaction is insensitive to wavelength (254, 313, 366 nm), a 50° temperature increase led to a 16-fold rate enhancement for triplet rearrangement k_r and a twofold increase in the rate of triplet decay k_d. Zimmerman[75] concluded that an activation energy of about 10 kcal/mole is required for the rearrangement and suggested that the barrier is intuitively reasonable since the rearrangement does involve disruption of the electronics of the phenyl ring. This means that despite the fact that an electronically excited state is involved, there is appreciable loss in energy incurred in phenyl bridging. If the Eyring equation can be applied to excited state reactions, the results suggest that the stereoselectivity arises only partially from a concerted pathway being

TABLE 7.5. *Activation Energies, Enthalpies, and Entropies*[a]

Compound	ΔE_a	$\Delta H^{\ddagger}$	$\Delta S^{\ddagger}$
trans-(66)	10.53	9.86	6.95
cis-(67)	11.27	10.7	−0.65

[a] Results from two temperatures only. From Ref. 75.

energetically favored and mainly from an entropy effect favoring the concerted process (Table 7.5).

Photorearrangement of 4-*p*-cyanophenyl-4-phenylcyclohexene (69) took place mainly by *p*-cyanophenyl migration.[67,70,74] The conclusion could then be drawn that the rearranging excited state is not electron deficient at the β-carbon atom, since one would not expect a cyanophenyl group to migrate to a positive carbon. The β-carbon was proposed to have odd electron character:

(69) $\xrightarrow{h\nu}$ [bicyclic ketone, PhCN, H, ϕ] $\Phi = 0.177$ + [bicyclic ketone, ϕ, H, PhCN] $\Phi = 0.160$ (7.31)

$$k_r(\text{PhCN, migration}) = 3.37 \times 10^8\ \text{sec}^{-1}$$
$$k_r(\text{Ph, migration}) = 3.10 \times 10^7\ \text{sec}^{-1}$$
$$k_d = 15.2 \times 10^8\ \text{sec}^{-1}$$

This also indicates that loss of electronic excitation cannot precede the rate-limiting stage of the reaction; otherwise, an electron-rich oxygen and electron-deficient β-carbon (70) would lead to reversal of the selectivity:

(70) [$O^{\ominus}$, $\oplus$, ϕ, PhCN] ⟶ phenyl migration

The triplet decay rates k_d for (65) and (69) are so similar that it is not reasonable to suggest that decay is dominated by "unsuccessful" migration, but must be due to a radiationless transition.

It is interesting that the 4,5-diphenylcyclohexenone (71) photorearrangement takes place 98.6% via a type A route and only 1.4% via a phenyl type migration.[68] How this was shown by ^{14}C labeling is described by the following equations:

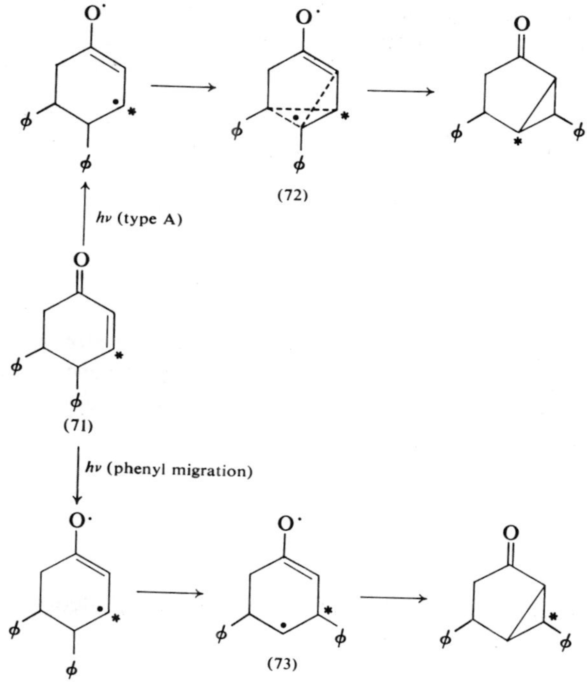

(72)

$h\nu$ (type A)

(71)

$h\nu$ (phenyl migration)

(73)

(7.32)

$* \equiv C^{14}$ label

The phenyl migration route leads, after migration, to (73), in which the odd electron is localized on C-4. In the 4,4-diphenylcyclohexenone system a phenyl group remains behind to delocalize the odd electron. Consequently the 4,4,5-triphenyl-cyclohexenone rearranges via phenyl migration.[71] Other factors that favor the migration of phenyl in the 4,4-diphenyl enones are (a) more relief of strain by migration of one of the phenyl groups, and (b) at least one phenyl group is in the proper conformation to migrate (axial).

The photorearrangement of 4-methyl-4-phenyl-2-cyclohexenone was shown by Dauben and co-workers to be very solvent dependent.[76] In aprotic, nonpolar solvents only two products were reported:

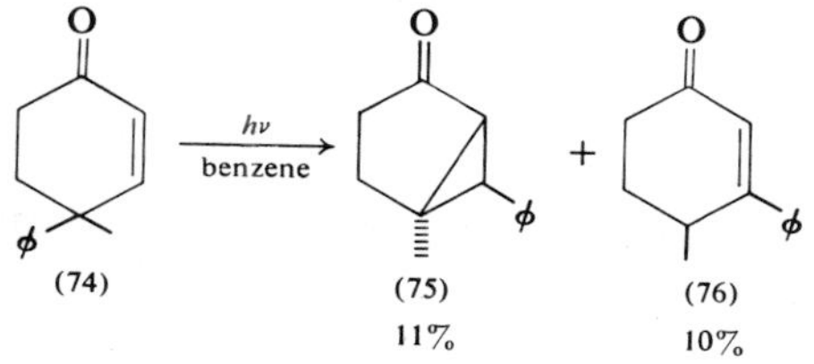

$h\nu$ benzene

(74) (75) 11% (76) 10%

(7.33)

However, when 78% of the starting ketone had reacted, only 22% of low molecular weight products was isolated. In 90% methanol–water solution under the same conditions five products were formed:

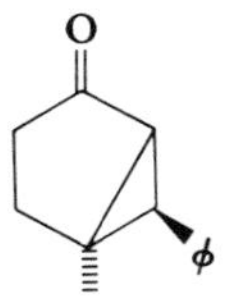

(75) 6% (76) 4% (77) 10% (78) 2.5% (79) 20%

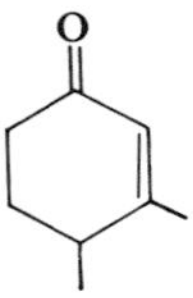

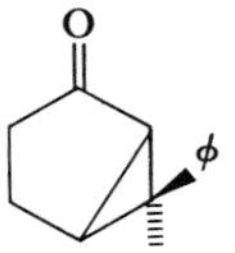

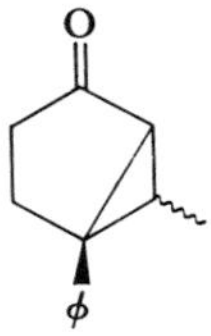

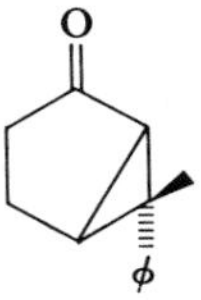

This result and the fact that a Stern–Volmer quenching plot for (75) and (77) had somewhat different slopes (no statistical analysis given) led the author to propose that the (75) and (76) were $n \rightarrow \pi^*$ triplet products and (77), (78), and (79) were $\pi \rightarrow \pi^*$ triplet products.

7.6b. *Alkyl-Substituted Cyclohexenones*

Photolysis of 4,4-dimethylcyclohexenone in *t*-butyl alcohol gives ketones from the type A rearrangement[56]:

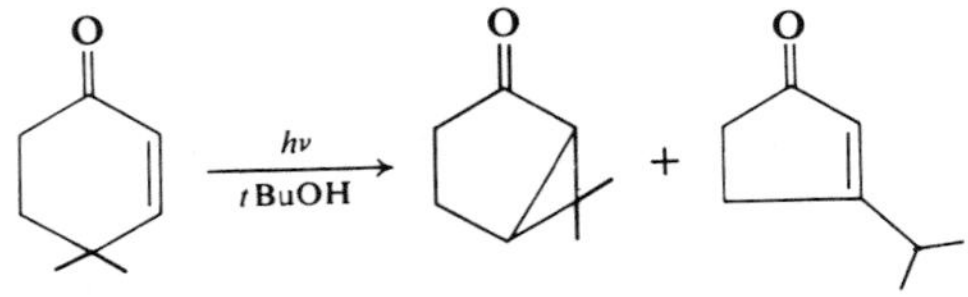

$h\nu$, *t*BuOH

(80) → (81) 60% + (82) 5%

$h\nu$, CH_3COOH

(81) (5%) + (82)(5–10%) + (83) (30–40%) + (84) (20–25%)

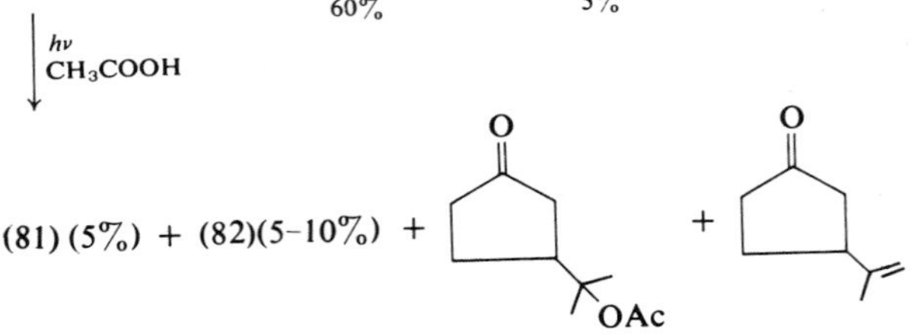

(7.34)

In acetic acid these products are accompanied by products (83) and (84).

Dauben *et al.* have investigated the scope of the photochemical type A rearrangement.[60] They conclude that the rearrangement occurs only if the fourth carbon atom of the 2-cyclohexenone ring is fully alkyl-substituted. If this requirement is not met, photodimers are the major products. This substituent requirement is necessary but not sufficient to ensure rearrangement since the presence of other groups can inhibit the reaction.

Compounds that do not give type A rearrangement are as follows:

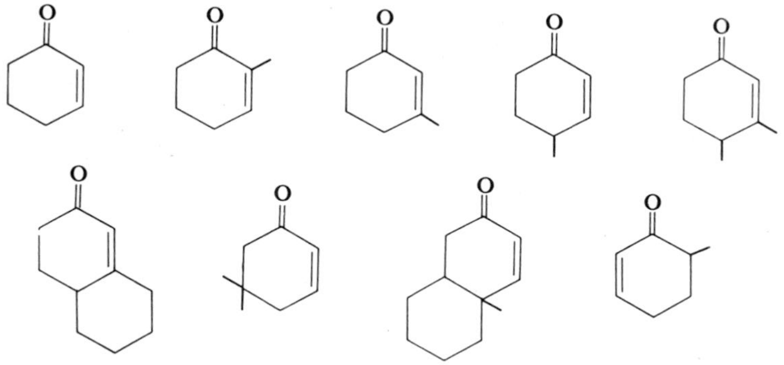

Compounds that do give type A rearrangement are as follows:

C_8H_{17}

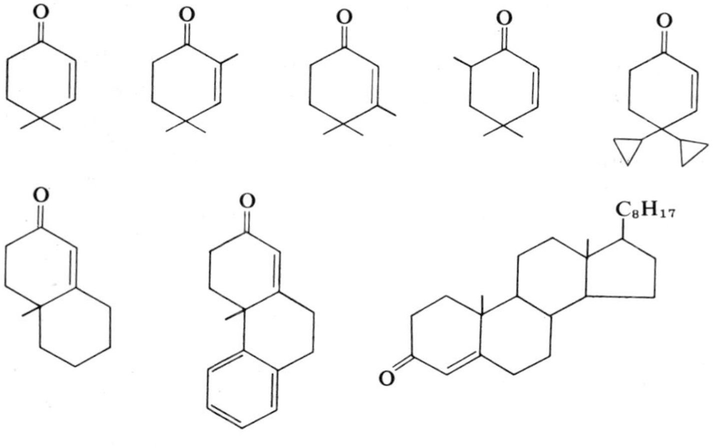

The type A rearrangement could take place via either a concerted or a stepwise route:

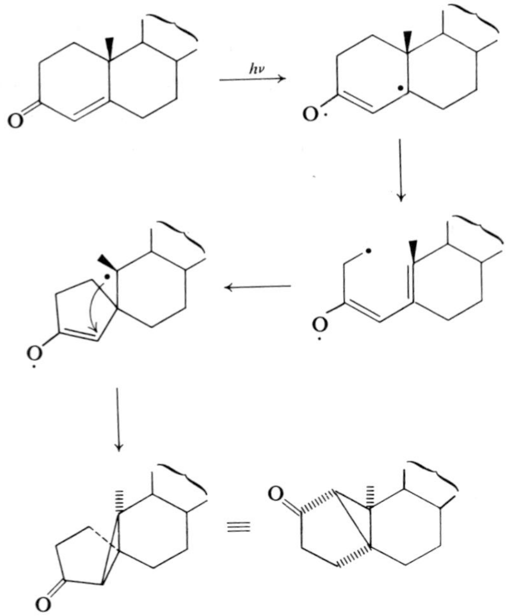

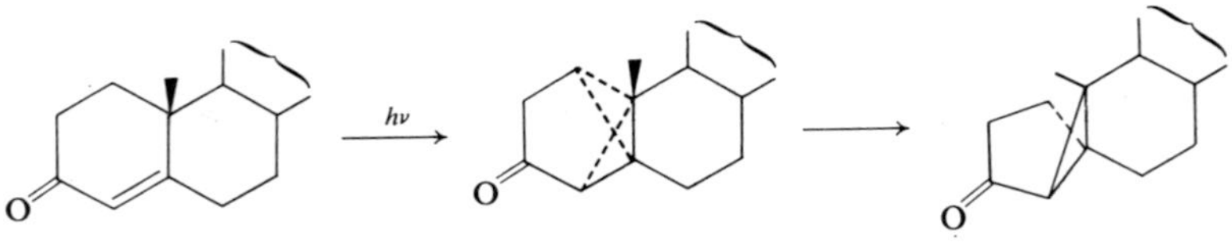

hν

hν

≡

The fact that the conversion of (85) to (86) proceeds with greater than 90% retention of optical activity would seem to favor the concerted mechanism.[59] However, Schuster and Brizzolara[61] have pointed out that the examination of molecular models clearly indicates that the angular methyl (or other group) would strongly inhibit the rotation that is necessary for formation of the new 1,5 bond in the lumiketone from other than the bottom side of the molecule:

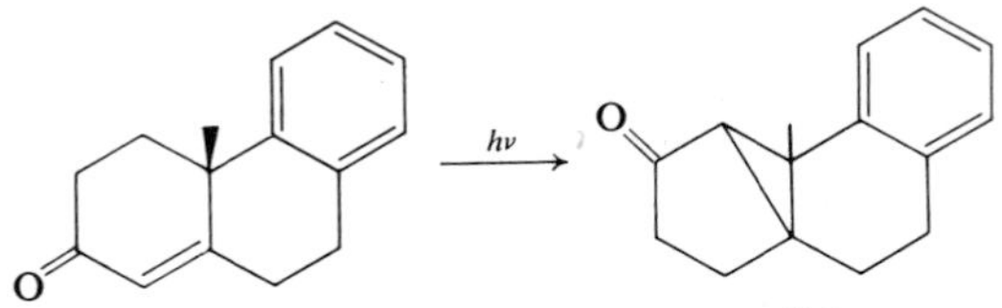

hν (7.35)

(85) −322° +322°

(86) −40.5° +41.0°

This reaction is sensitized with acetophenone and quenched by naphthalene, di-*t*-butylnitroxide, or piperylene.

TABLE 7.6. *Photolysis Products of (87)*

Solvent	Time, hr	(87)	(88)	(89)	(90)	(91)	(92)	Total
Chloroform	4	26	4	2	20	6	5	63
Toluene	10	17	15	8	0	0	3	43
Cumene	10	25	9	6	0	0	4	44
Benzene	10	23	19	11	0	0	3	56
t-Butyl alcohol	600	25	45	0	0	0	0	70

Schuster and Brizzolara[61] have provided a very nice study of the photochemistry of 10-hydroxymethyl-$\Delta^{1,9}$-2-octalone (87). Schuster and Patel[13] previously used radical fragmentation reactions as probes for the nature of the intermediates in the cyclohexadienone rearrangement. This compound (87) was designed so that it could undergo a radical fragmentation reaction in competition with the usual type A rearrangement if the intermediate involved has radical character ($n \rightarrow \pi^*$ triplet). Photolysis produced (88)–(92):

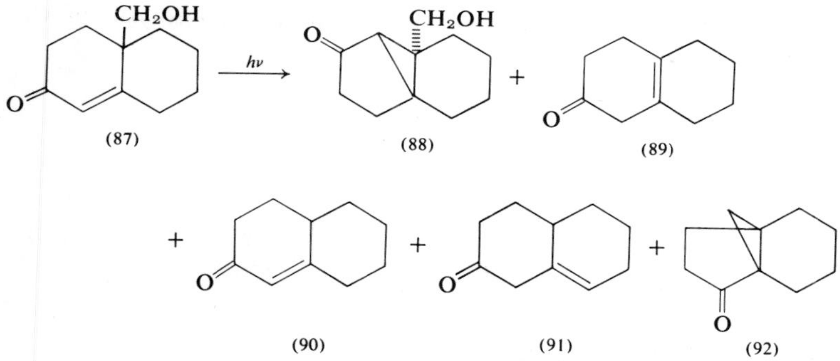

CH_2OH (87) $\xrightarrow{h\nu}$ CH_2OH (88) + (89) + (90) + (91) + (92) (7.36)

Table 7.6 summarizes the solvent effect on this photoreaction. The fact that fragmentation does not occur in *t*-butyl alcohol suggests that hydrogen abstraction precedes fragmentation, although the reverse was a possibility. Stern–Volmer quenching plots for the formation of the lumiketone (88) and the octalone (89) in benzene with piperylene as the quencher indicate that the lifetimes of the triplet state leading to each product are the same. They concluded that all the reactions in this system proceed from one excited triplet state or from two nearly isoenergetic states which are in thermal equilibrium.

Similar systems have been studied by Bellus, Kearns, and Schaffner,[62] who proposed two reactive triplets, one responsible for the formation of (95) and the other responsible for the remaining products [(94), (96), other]:

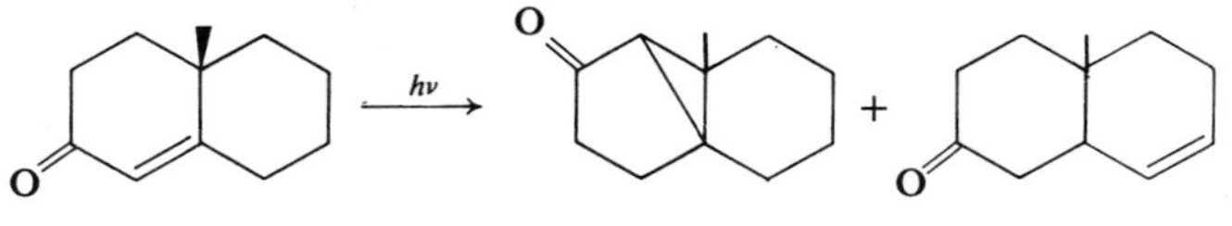

(93) $\xrightarrow{h\nu}$ (94) + (95)

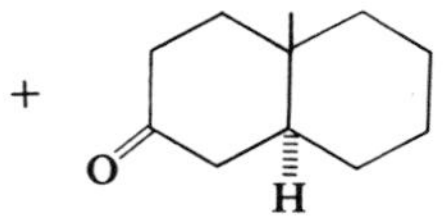

+ (96) + other (7.37)

PROBLEMS

1. Show that the stereochemistry of lumisantonin and isophotosantonic lactone formed from α-santonin are consistent with the mechanisms presented in this chapter.
2. Give the product (with stereochemistry) from the photolysis of the following compound[77]:

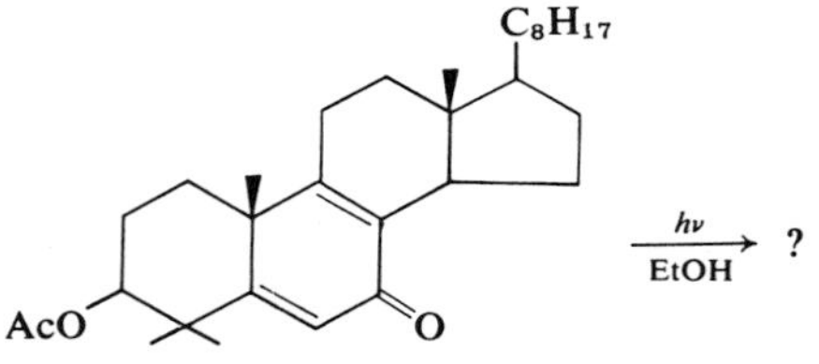

C_8H_{17}

AcO, O $\xrightarrow[EtOH]{h\nu}$?

3. Give a mechanism for the following reaction[46]:

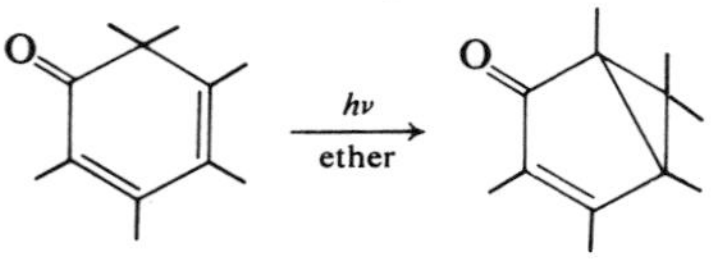

$\xrightarrow[ether]{h\nu}$

80%

4. Predict the product from the photolysis of the following compound[17]:

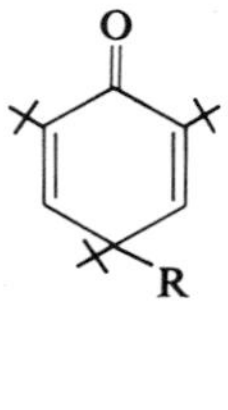

$R = Br, NO_2$

5. Provide a mechanism for the following transformations:

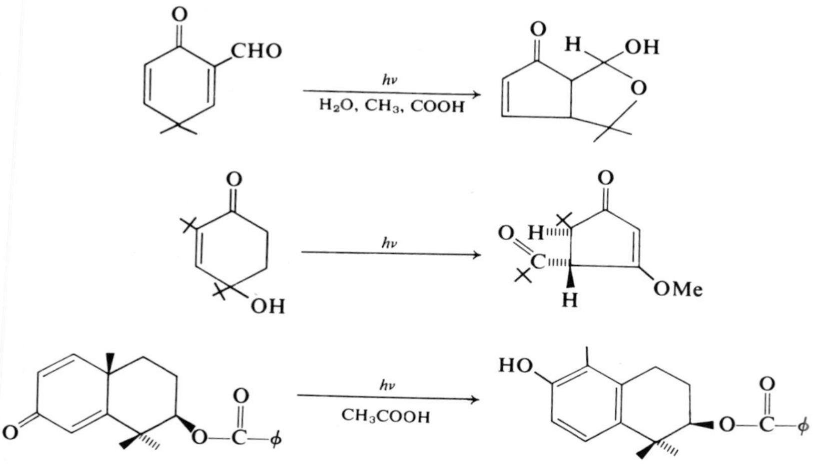

$h\nu$, H_2O, CH_3, COOH

$h\nu$

$h\nu$, CH_3COOH

REFERENCES

1. P. J. Kropp, in *Organic Photochemistry*, Vol. I, O. L. Chapman, ed., Marcel Dekker, New York (1967).
2. K. Schaffner, in *Advances in Photochemistry*, Vol. 4, W. A. Noyes, Jr., G. S. Hammond, and J. N. Pitts, Jr., eds., Interscience, New York (1966).
3. H. E. Zimmerman, in *Advances in Photochemistry*, Vol. 1, W. A. Noyes, Jr., G. S. Hammond, and J. N. Pitts, Jr., eds., Interscience, New York (1963).
4. Kahler, *Arch. Pharm.* **34**, 318 (1830).
5. J. D. M. Asher and G. A. Sim, *J. Chem. Soc.*, 1584 (1965).
6. M. H. Fisch and J. H. Richards, *J. Amer. Chem. Soc.* **85**, 3029 (1963).
7. H. E. Zimmerman and D. I. Schuster, *J. Amer. Chem. Soc.* **84**, 4527 (1962).
8. H. E. Zimmerman and J. S. Swenton, *J. Amer. Chem. Soc.* **89**, 906 (1967).
9. H. E. Zimmerman, R. W. Binkley, J. J. McCullough, and G. A. Zimmerman, *J. Amer. Chem. Soc.* **89**, 6589 (1967).
10. J. S. Swenton, E. Saurborn, R. Srinivasan, and F. I. Sontag, *J. Amer. Chem. Soc.* **90**, 2990 (1968).
11. D. I. Schuster and D. J. Patel, *J. Amer. Chem. Soc.* **87**, 2515 (1965); **88**, 1825 (1966).
12. D. J. Patel and D. I. Schuster, *J. Amer. Chem. Soc.* **89**, 184 (1967).
13. D. I. Schuster and D. J. Patel, *J. Amer. Chem. Soc.* **90**, 5137, 5145 (1968).
14. D. I. Schuster and V. Y. Abraitys, *Chem. Comm.*, 419 (1969).
15. D. I. Schuster, K. V. Prabhu, S. Adcock, J. van der Veen, and H. Fujiwara, *J. Amer. Chem. Soc.* **93**, 1557 (1971).
16. D. I. Schuster and K. Liu, *J. Amer. Chem. Soc.* **93**, 6711 (1971).
17. K. Ogura and T. Matsuura, *Bull. Chem. Soc. Japan* **43**, 3205 (1970).
18. D. G. Hewitt and R. F. Taylor, *Chem. Comm.*, 493 (1972).

19. D. I. Schuster and C. J. Polowczyk, *J. Amer. Chem. Soc.* **86**, 4502 (1964).
20. D. I. Schuster and C. J. Polowczyk, *J. Amer. Chem. Soc.* **88**, 1722 (1966).
21. D. I. Schuster and I. S. Krull, *J. Amer. Chem. Soc.* **88**, 3456 (1966).
22. W. H. Pirkle, S. G. Smith, and G. F. Koser, *J. Amer. Chem. Soc.* **91**, 1580 (1969).
23. W. H. Pirkle and G. F. Koser, *Tetrahedron Lett.*, 129 (1968).
24. H. E. Zimmerman and D. S. Crumrine, *J. Amer. Chem. Soc.* **90**, 5612 (1968).
25. H. E. Zimmerman, D. S. Crumrine, D. Dopp, and P. S. Huyffer, *J. Amer. Chem. Soc.* **91**, 434 (1969).
26. H. E. Zimmerman, D. Dopp, and P. S. Huyffer, *J. Amer. Chem. Soc.* **88**, 5352 (1966).
27. T. R. Rodgers and H. Hart, *Tetrahedron Lett.*, 4845 (1969).
28. W. V. Curran and D. I. Schuster, *Chem. Comm.*, 699 (1968).
29. D. I. Schuster and W. V. Curran, *J. Org. Chem.* **35**, 4192 (1970).
30. K. Ogura and T. Matsuura, *Bull. Chem. Soc. Japan* **43**, 2891 (1970).
31. D. I. Schuster and W. C. Barringer, *J. Amer. Chem. Soc.* **93**, 731 (1971).
32. P. J. Kropp, *J. Org. Chem.* **29**, 3110 (1964).
33. D. Caine and J. B. Dawson, *J. Org. Chem.* **29**, 3108 (1964).
34. K. Weinberg, E. C. Utzinger, D. Arigoni, and O. Jeger, *Helv. Chim. Acta* **43**, 236 (1960).
35. D. H. R. Barton, *Proc. Chem. Soc.*, *61* (1958).
36. D. H. R. Barton, *Helv. Chem. Acta* **42**, 2604 (1959).
37. D. H. R. Barton, J. E. D. Levisalles, and J. T. Pinhey, *J. Chem. Soc.*, 3472 (1962).
38. D. H. R. Barton, T. Miki, J. T. Pinhey, and R. J. Wells, *Proc. Chem. Soc.*, 112 (1962).
39. P. J. Kropp, *J. Amer. Chem. Soc.* **86**, 4053 (1964).
40. P. J. Kropp and W. F. Erman, *J. Amer. Chem. Soc.* **85**, 2456 (1963).
41. P. J. Kropp, *J. Amer. Chem. Soc.* **85**, 3779 (1963).
42. H. E. Zimmerman, R. Keese, J. Nasielski, and J. S. Swenton, *J. Amer. Chem. Soc.* **88**, 4895 (1966).
43. P. J. Kropp, *Tetrahedron* **21**, 2183 (1965).
44. D. H. R. Barton and G. Quinkert, *J. Chem. Soc.*, 1 (1960).
45. H. Hart and A. J. Waring, *Tetrahedron Lett.*, 325 (1965).
46. H. Hart, P. M. Collins, and A. J. Waring, *J. Amer. Chem. Soc.* **88**, 1005 (1966).
47. J. Griffiths and H. Hart, *J. Amer. Chem. Soc.* **90**, 3297, 5296 (1968).
48. M. R. Morris and A. J. Waring, *Chem. Comm.*, 526 (1969).
49. M. R. Morris and A. J. Waring, *J. Chem. Soc. C*, 3266, 3269 (1971).
50. A. J. Waring, M. R. Morris, and M. M. Islam, *J. Chem. Soc. C*, 3274 (1971).
51. R. J. Bastiani, D. Hart, and H. Hart, *Tetrahedron Lett.*, 4841 (1969).
52. H. Hart and R. K. Murray, *J. Org. Chem.* **35**, 1535 (1970).
53. H. Perst, *Tetrahedron Lett.*, 3601 (1970).
54. H. Perst and I. Wesemeier, *Tetrahedron Lett.*, 4189 (1970).
55. W. W. Kwie, B. A. Shoulders, and P. D. Gardner, *J. Amer. Chem. Soc.* **84**, 2268 (1962).
56. O. L. Chapman, T. A. Rettig, A. A. Griswold, A. I. Dutton, and P. Fitton, *Tetrahedron Lett.*, 2049 (1963).
57. B. Nann, D. Gravel, R. Schorta, H. Wehrli, K. Schaffner, and O. Jeger, *Helv. Chim. Acta* **46**, 2473 (1963).
58. H. E. Zimmerman, R. G. Lewis, J. J. McCullough, A. Padwa, S. W. Staley, and M. Semmelhack, *J. Amer. Chem. Soc.* **88**, 1965 (1966).
59. O. L. Chapman, J. B. Sieja, and W. J. Welstead, Jr., *J. Amer. Chem. Soc.* **88**, 161 (1966).

60. W. G. Dauben, G. W. Shaffer, and N. D. Vietmeyer, *J. Org. Chem.* **33**, 4060 (1968).
61. D. I. Schuster and D. F. Brizzolara, *J. Amer. Chem. Soc.* **92**, 4357 (1970).
62. D. Bellus, D. R. Kearns, and K. Schaffner, *Helv. Chim. Acta* **52**, 971 (1969).
63. O. Jeger and K. Schaffner, *Pure Appl. Chem.* **21**, 247 (1970).
64. R. C. Hahn and G. W. Jones, *J. Amer. Chem. Soc.* **93**, 4232 (1971).
65. J. Gloor, K. Schaffner, and O. Jeger, *Helv. Chim. Acta* **54**, 1864 (1971).
66. H. E. Zimmerman and J. W. Wilson, *J. Amer. Chem. Soc.* **86**, 4036 (1964).
67. H. E. Zimmerman, R. C. Hahn, H. Morrison, and M. C. Wani, *J. Amer. Chem. Soc.* **87**, 1138 (1965).
68. H. E. Zimmerman and D. J. Sam, *J. Amer. Chem. Soc.* **88**, 4905 (1966).
69. T. Matsurra and K. Ogura, *J. Amer. Chem. Soc.* **88**, 2602 (1966).
70. H. E. Zimmerman, R. D. Rieke, and J. R. Scheffer, *J. Amer. Chem. Soc.* **89**, 2033 (1967).
71. H. E. Zimmerman and R. L. Morse, *J. Amer. Chem. Soc.* **90**, 954 (1968).
72. H. E. Zimmerman and K. G. Hancock, *J. Amer. Chem. Soc.* **90**, 3749 (1968).
73. H. E. Zimmerman, K. G. Hancock, and G. C. Licke, *J. Amer. Chem. Soc.* **90**, 4892 (1968).
74. H. E. Zimmerman and N. Lewin, *J. Amer. Chem. Soc.* **91**, 879 (1969).
75. H. E. Zimmerman and W. R. Elser, *J. Amer. Chem. Soc.* **91**, 887 (1969).
76. W. G. Dauben, W. A. Spitzer, and M. S. Kellogg, *J. Amer. Chem. Soc.* **93**, 3674 (1971).
77. D. H. R. Barton, J. F. McGhie, and M. Rosenberger, *J. Chem. Soc.*, 1215, 1961.

8

The Di-π-Methane Photorearrangement

Early examples of the class of reactions now known as di-π-methane photorearrangements were reported by Griffin *et al.*,[1,2] Zimmerman and coworkers,[3,4] Meinwald and Smith,[5] and Srinivasan and Carlough.[6] This rearrangement occurs upon photolysis of molecules having two unsaturated moieties bonded to a single saturated carbon atom. A typical example is as follows[6]:

$$\text{di-}\pi\text{-methane} \xrightarrow[\text{Hg, gas phase}]{h\nu} \text{vinylcyclopropane} \tag{8.1}$$

This type of reaction has been observed for a large number of acyclic and cyclic molecules.

8.1. ACYCLIC DI-π-METHANE PHOTOREARRANGEMENT

The photochemistry of 1,1,5,5-tetraphenyl-3,3-dimethyl-1,4 pentadiene (1) has been studied in detail by Zimmerman and Mariano.[7] Photolysis of compound (1) gave 1,1-diphenyl-2,2-dimethyl-3-(2,2-diphenylvinyl) cyclopropane:

$$(1) \xrightarrow[t\text{-BuOH}]{\lambda = 250 \text{ nm}} (2) \qquad \Phi = 0.080 \tag{8.2}$$

TABLE 8.1. *Quantum Yield for Sensitized Reactions of Compound (1)*

Sensitizer	λ, nm	Percent light absorbed	$\Phi_{(2)}$
Benzophenone	313	99%	0.0047
Propiophenone	313	99%	0.0103
Chlorobenzene	262	93%	0.0096

Attempts to sensitize the rearrangement with benzophenone, propiophenone, and chlorobenzene failed, as indicated in Table 8.1. Although the reaction could not be sensitized, triplet energy transfer was taking place inasmuch as compound (1) quenched the photoreduction of benzophenone without the formation of any new products (Table 8.2).

Zimmerman has formally depicted the rearrangement in the following manner:

$$\xrightarrow{h\nu} \longrightarrow \longrightarrow \qquad (8.3)$$

While this scheme is useful in helping to predict products from di-π-methane rearrangements, all evidence indicates that the structures drawn are not intermediates in the reaction. That is, they do not represent energy minima on the potential energy surface leading from the excited state of the reactant to the ground state of the product.

The reaction is presumably concerted and can be indicated by the following simplified scheme:

Me Me 2 3 4 1 5 ϕ ϕϕ ϕ ⟶ [Me Me ϕ ϕϕ ϕ] ⟶ Me Me ϕ ϕϕ ϕ (8.4)

TABLE 8.2. *Test of Energy Transfer from Benzophenone to Compound (1)*[a]

Wt. ϕ—C(=O)—ϕ	Wt. ϕ—CH(OH)—ϕ	Wt. diene (1)	Wt. benzpinacol
6.0 g	2.0 g	0	5.81 g
6.0 g	2.0 g	0.5 g	0.89 g

[a] Solvent, *t*-butyl alcohol, Pyrex filter.

TABLE 8.3. *Electrocyclic Reaction Rules*[a]

Transition state	4n + 2		4n	
	Photo	Thermal	Photo	Thermal
Hückel (no sign change, or even number of sign changes)	Forbidden	Allowed	Allowed	Forbidden
Möbius (odd number of sign changes)	Allowed	Forbidden	Forbidden	Allowed

[a] Forbidden ≡ antiaromatic; allowed ≡ aromatic.

This pathway would involve simultaneous cleavage of bond 2–3 and formation of bonds 2–4 and 3–5. This picture is misleading in that, unlike the more refined pictures, it does not indicate inversion at C-3.

Two possible molecular orbital schemes, (3) and (4), are shown in the accompanying diagram. Orbital scheme (3) involves a cyclic array of six

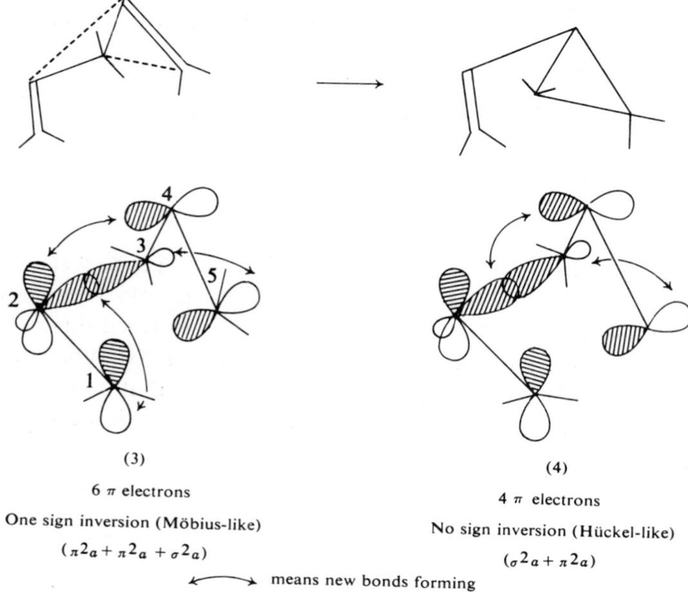

(3)
6 π electrons
One sign inversion (Möbius-like)
$(_{\pi}2_a + _{\pi}2_a + _{\sigma}2_a)$

(4)
4 π electrons
No sign inversion (Hückel-like)
$(_{\sigma}2_a + _{\pi}2_a)$

⟷ means new bonds forming

π electrons with one sign inversion (Möbius-like). A Möbius array in the ground state is antiaromatic with six electrons but is aromatic and "stable" in the excited state.[8] Thus the di-π-methane rearrangement via transition state (3) is photoallowed. The orbital scheme (4) involves a cyclic array of four π electrons with no sign inversions (Hückel-like). Again the ground state is antiaromatic but it is aromatic in the excited state and the reaction is photochemically allowed.

The aromatic–antiaromatic transition state rules are another formulation of the Woodward–Hoffmann type rules (Table 8.3).

The process described by transition state (4) is just the *trans* addition of a single bond 2–3 to π bond 4–5. Both transition states suggest inversion at C-3 since the 5–3 bond is formed from the back as the 2–3 bond is broken.

Zimmerman prefers transition state (3) inasmuch as transition state (4) suggests that the second double bond does not play an important role. Photorearrangements of this type without the second double bond are very inefficient[9]:

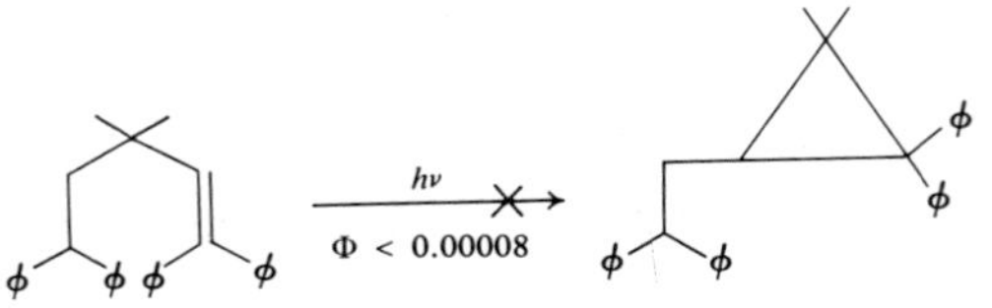

$h\nu$, $\Phi < 0.00008$ (8.5)

8.1a. *Regiospecificity and Stereochemistry*

Compound (5) undergoes a singlet state photorearrangement to (6) but not (7)[10,11]:

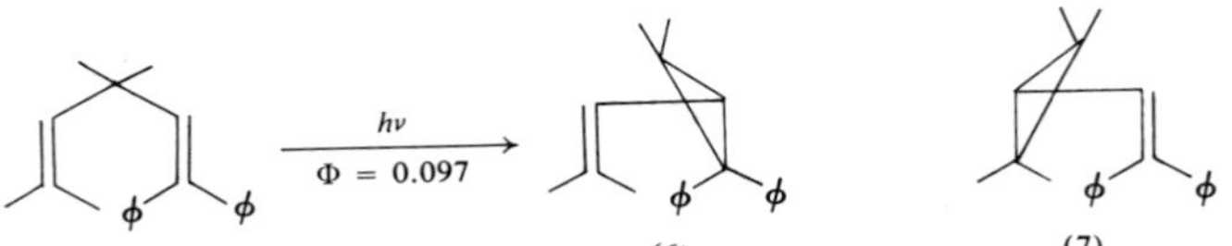

$h\nu$, $\Phi = 0.097$ (8.6)

(5) (6) (7)

In this and all other cases, we observe that the major product results from a pathway involving migration of the less conjugated π-system to the more conjugated one. Zimmerman and co-workers have described this as being due to the demand of the excited state for retention of maximum odd election delocalization during rearrangement or alternately that bonding between orbitals on C-5 and C-3 is minimal early along the reaction coordinate [see schemes (3) and (4)]:

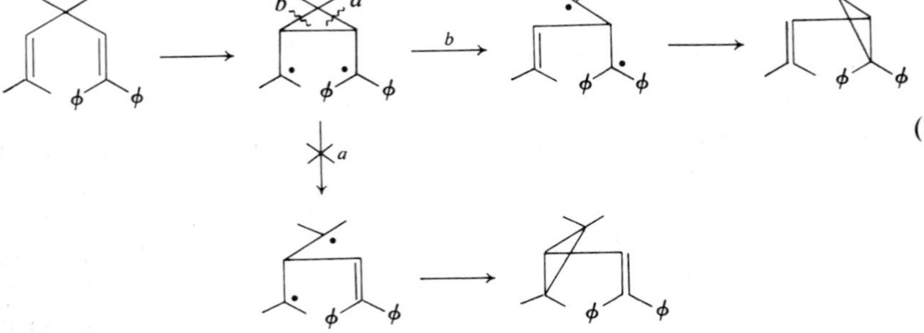

(8.7)

Zimmerman and Baum[12] observed only the formation of (9) from (8):

$$(8) \xrightarrow[\Phi\, =\, 0.008]{h\nu} (9) \qquad (8.8)$$

This regioselective example of a di-π-methane photorearrangement is of particular importance because in this case both π moieties have very similar singlet energy levels. This suggests that excitation localization does not control the reaction.

The di-π-methane reaction is not only regioselective but is also stereoselective, as indicated by the following examples[13,14]:

$$(10) \xrightarrow{h\nu} (12) \qquad (8.9)$$

$$(11) \xrightarrow{h\nu} (13) \qquad (8.10)$$

Upon low conversion direct photolysis the *cis* isomer (10) gave only the *cis* isomer (12) and the *trans* isomer (11) gave only the *trans* isomer (13). The triplet sensitized reaction of (10) and (11) gave rise only to *cis–trans* isomerization. Thus the di-π-methane photorearrangement from the triplet state cannot compete with triplet state deactivation via *cis–trans* isomerization (Zimmerman has termed this the free rotor effect). Several other examples of regiospecificity and stereospecificity in di-π-methane photoreactions are as follows[15,23]:

$$\xrightarrow[\Phi\, =\, 0.050;\ \text{Ref. 15}]{h\nu} \quad + \text{ other} \qquad (8.11)$$

$$\xrightarrow[\Phi\, =\, 0.050;\ \text{Ref. 15}]{h\nu} \qquad (8.12)$$

hν
Ref. 16

hν
sensitized

(8.13)

hν
$PhCOCH_3$;
Ref. 17

(8.14)

hν
Ref. 18

(8.15)

Ref. 18

(8.16)

254 nm hν;
Ref. 19

Minor Major

(8.17)

254 nm hν;
Ref. 19

Minor Major

(8.18)

Me Me

$h\nu$

i-PrOH; Ref. 20

No di-π-methane

(8.19)

$h\nu$

Refs. 21, 22

(8.20)

$h\nu$

Ref. 23

(8.21)

8.1b. *Substitution at the Central sp^3 Carbon Atom and Di-π-Methane Reactivity*

In most of the acyclic examples of the di-π-methane rearrangement studied, there has been methyl substitution on the central (C-3) atom. We should expect that electron-withdrawing substituents (relative to $—CH_3$) on C-3 would lower the energy of the basis orbitals on C-3 and slow the transformation shown in the diagrams for schemes (3) and (4).

The pentadiene (14) gave almost equal amounts of (15) and (16) upon preparative photolysis[24]:

$h\nu$

t-butylalcohol

(8.22)

(14) (15) (16)

$\Phi = 0.0024$ $\Phi = 0.0020$

However, Zimmerman has shown that (15) is not formed via a di-π-methane rearrangement by the photolysis of the dideuterio compound (17):

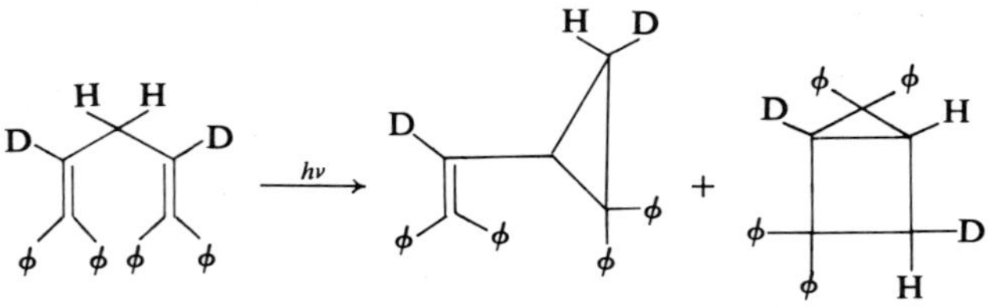

(17) (18) (19) (8.23)

The vinylcyclopropane formed (18) is not the isomer that one would predict for the di-π-methane reaction (20):

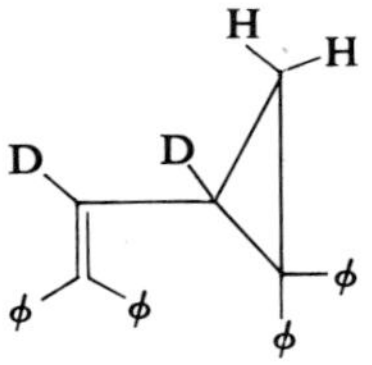

(20)

The unobserved di-π-methane rearrangement and the observed $\sigma + \pi$ hydrogen migration route are as follows:

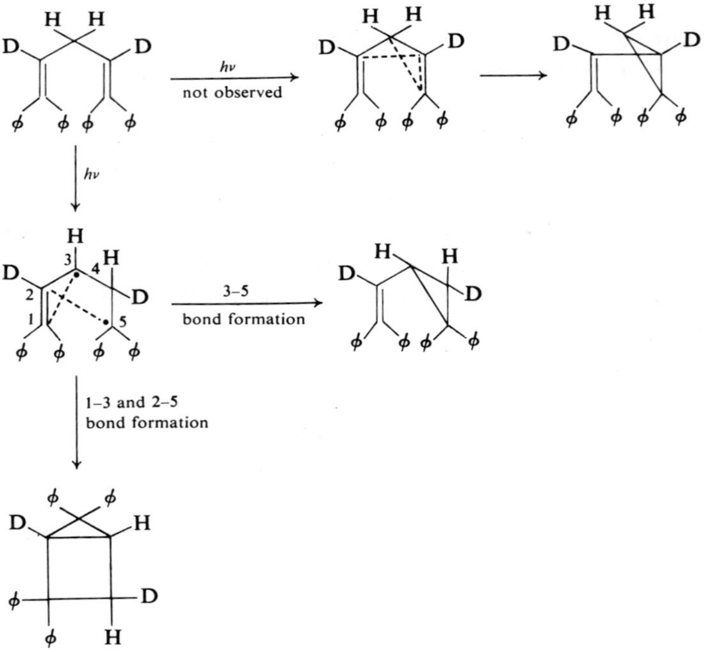

(8.24)

This example illustrates the effect of rather small electronic changes on the di-π-methane rearrangement.

When the central substituents are phenyl the di-π-methane rearrangement does take place,[25]

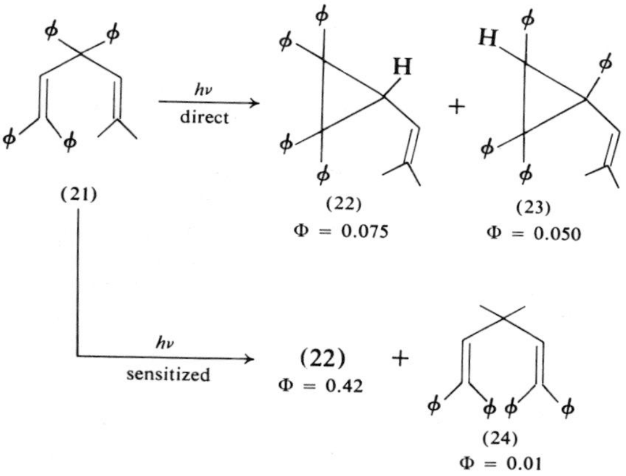

(21) —hν, direct→ (22) Φ = 0.075 + (23) Φ = 0.050

(21) —hν, sensitized→ (22) Φ = 0.42 + (24) Φ = 0.01 (8.25)

Again the di-π-methane reaction is regiospecific but this time it is competing with phenyl migration,

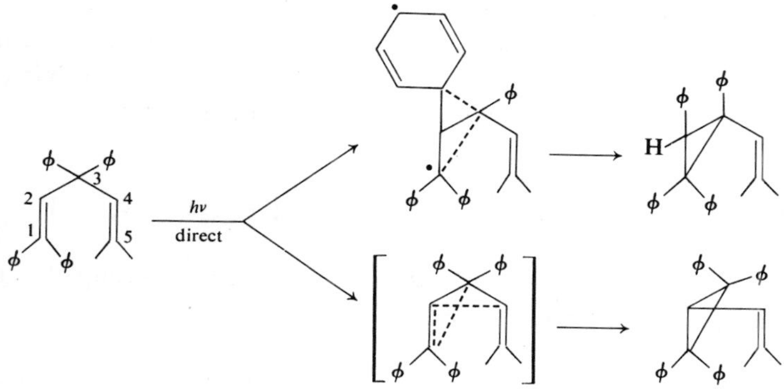

hν, direct (8.26)

The very efficient di-π-methane rearrangement from the triplet state in this case is probably due to two factors: (a) the energetics is favorable for cleavage of bond 3–4 and formation of bond 2–4 [see Eq. (8.26)] and (b) rotation about bond 1–2, "*cis–trans* isomerization," should be restricted because of steric interactions.

TABLE 8.4. *Reaction Rate Constants for Singlet Di-π-Methane Reactions*

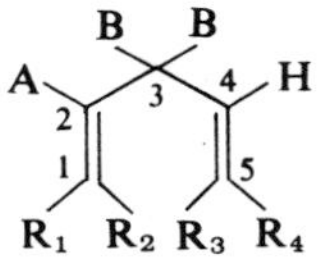

No.	R_1	R_2	R_3	R_4	A	B	Φ_r	τ_s,[a] sec	k_r, sec^{-1}
25	H	H	Ph	Ph	Ph	Me	0.008	1.4×10^{-11}	5.8×10^{8}
26	Me	Me	Ph	Ph	H	Me	0.097	1.4×10^{-11}	6.9×10^{9}
27	Me	Me	Ph	Ph	H	Ph	0.076	8.85×10^{-12}	8.5×10^{9}
28	Ph	Ph	Ph	Ph	H	Me	0.080	5.6×10^{-13}	1.4×10^{11}
29	Ph	Ph	Ph	Ph	H	H	0.0024[b]	9.1×10^{-13}	2.6×10^{9}

[a] Lifetimes were measured at 77°K and adjusted to room temperature using Φ_F^{298}/Φ_F^{77}.
[b] Not a di-π-methane reaction.

8.1c. *Reaction Rate Constants*

The rate constant for the singlet di-π-methane photorearrangement can be determined from the quantum yield for the reaction and the experimentally measured singlet lifetime:

$$k_r = \Phi_r/\tau_s \tag{8.27}$$

It is interesting to note that the reactivity of the excited states of (25), (26), (27), and (28) in Table 8.4 increases in this order as stabilizing terminal substitution is increased. Zimmerman suggests that vinyl–vinyl bridging (the start of bond formation between 2 and 4) controls the reaction rate.

8.2. ARYL DI-π-METHANE PHOTOREARRANGEMENT

A phenyl group may be substituted for one of the vinyl groups 1,2 in the di-π-methane rearrangement as follows:

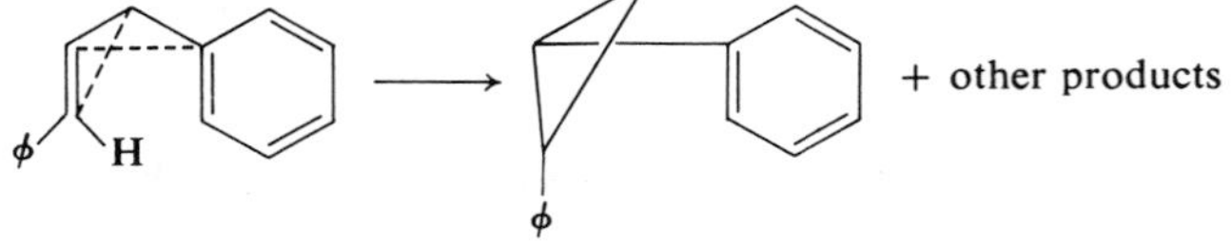

+ other products (8.28)

(30)

Hixson[27] prepared the deuterio compound (31) and found that photolysis produced only compound (32) resulting from a di-π-methane rearrangement and not from a hydrogen migration:

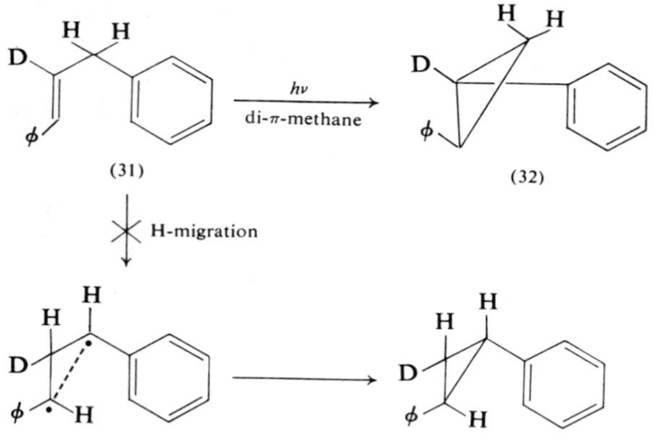
D
H H
ϕ
hν
di-π-methane
H-migration

(31) (32)

(8.29)

(33) Not formed

The generality of this reaction has been explored by Griffin[1,2,38] and Hammond[28] and their co-workers:

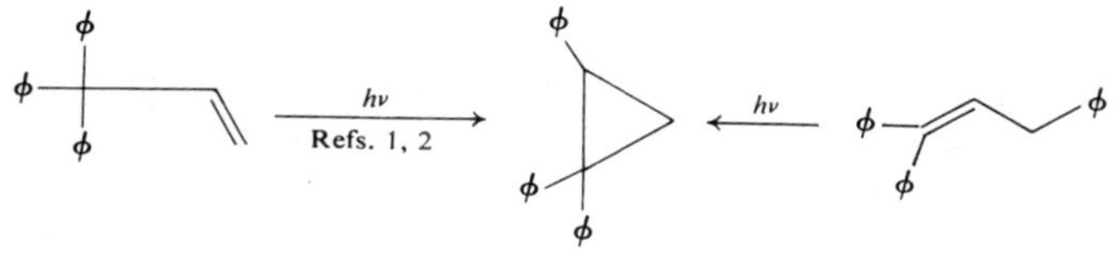
hν
Refs. 1, 2
hν

(8.30)

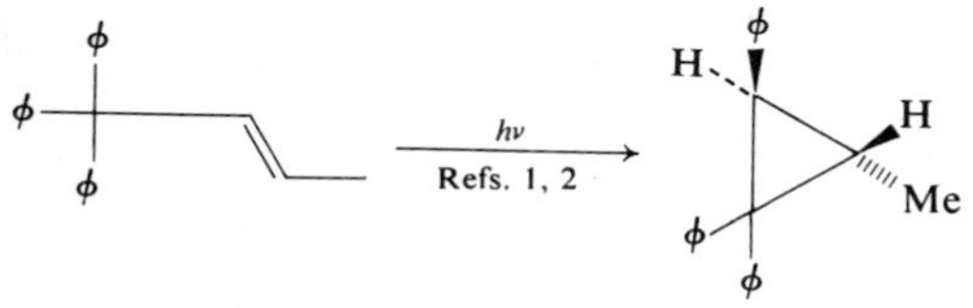
hν
Refs. 1, 2
H
Me

(8.31)

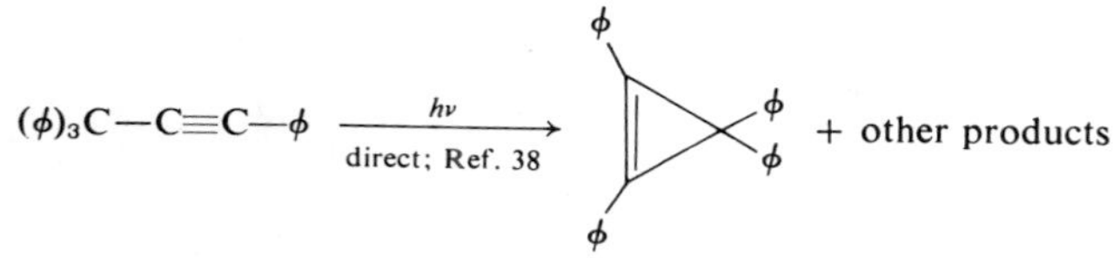
(ϕ)₃C—C≡C—ϕ
hν
direct; Ref. 38
+ other products

(8.32)

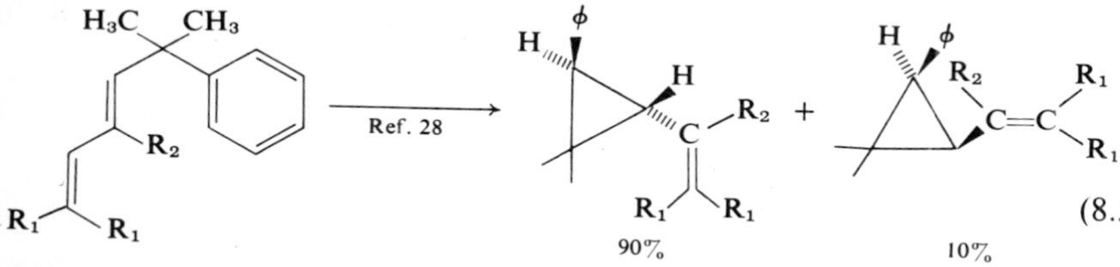
H₃C CH₃
R₂
R₁
Ref. 28
90%
10%

(8.33)

(a) $R_1 = R_2 = H$
(b) $R_1 = H$; $R_2 = CH_3$
(c) $R_1 = CH_3$; $R_2 = H$

Other examples where one of the two π-moieties is in a six-member ring are as follows:

$h\nu$ direct $\Phi = 0.112$ Ref. 29

(34) $\Phi = 0.00088$ (35) $\Phi = 0.112$ (8.34)

$h\nu$ direct $\Phi = 0.12$ Ref. 30 (8.35)

$h\nu$ direct Ref. 31 (8.36)

$h\nu$ direct $\Phi = 0.0030$ Ref. 32 (8.37)

$h\nu$ sensitized Ref. 33 (8.38)

$h\nu$ sensitized Refs. 34, 35 (8.39)

direct $h\nu$

TABLE 8.5. *Relative Quantum Yields and Rate Constants for Compounds (36)*

Compound				
No.	X	Y	Φ_r(rel)	k_r(rel)
(36a)	CN	H	3.6	41
(36b)	H	OCH_3	1.0	13
(36c)	H	CN	0.26	3.8
(36d)	H	H	1.0	1
(36e)	OCH_3	H	<0.05	<0.04

The absence of exocyclic π bonds in the reactants [(8.38) and (8.39)] eliminates the possibility of efficient *cis–trans* isomerization about that bond and allows the triplet di-π-methane rearrangement to become important.

Hixson[36] (also see Ref. 37) has determined the relative rate constants for the di-π-methane rearrangement of a series of *p*-substituted 1,3-diphenylpropenes (Table 8.5):

X—C₆H₄—CH=CH—CH₂—C₆H₄—Y

(36a–e)

$h\nu$

X—C₆H₄—(cyclopropane)—C₆H₄—Y

Decreasing the electron density at the migrating terminus [compare (36a,d,e); see Table 8.5] greatly enhances the rate of migration. Both the methoxy and the *p*-cyano group attached to the migrating ring increase the rate of rearrangement relative to the unsubstituted compound (36d). Hixson suggests that the results are in accord with an odd-electron mechanism.

An interesting gas-phase study of the photolysis of *trans*-1-phenyl-2-butene has been published by Comtet.[39,40] He has found that (a) it was not possible to quench the formation of the cyclopropane product under conditions that reduced the fluorescence quantum yield, (b) sensitization by acetophenone only gave *cis–trans* isomerization, and (c) the quantum yield of cyclopropane formation in the direct photolysis decreases as *n*-butane is added to the reaction mixture. Comtet suggests that the data are consistent with a reaction from the *second* triplet state.

8.3. BICYCLIC DI-π-METHANE PHOTOREARRANGEMENT

8.3a. *Barrelene*

A classic example of the di-π-methane photorearrangement is afforded by the reaction of barrelene (37)[41]:

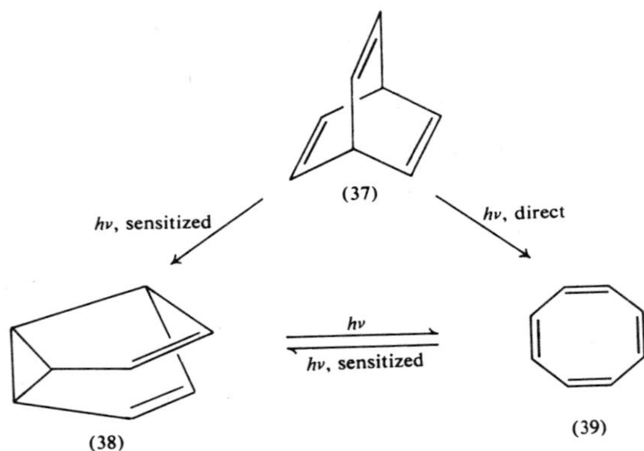

Acetone sensitization provided semibullvalene (38) while direct photolysis gave cyclooctatetraene (39). Several structural representations of semibullvalene are shown below:

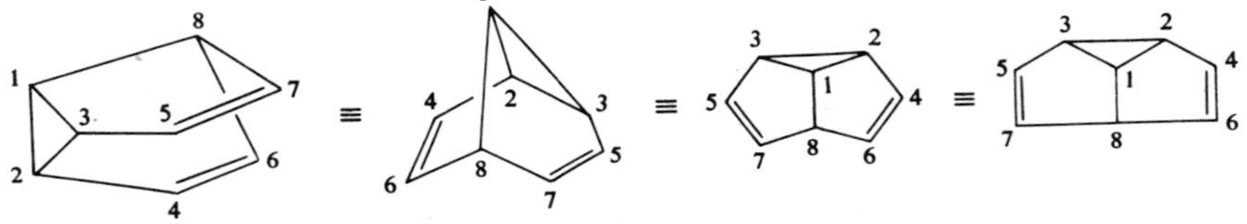

Based on the structure drawn for semibullvalene (38), it appears that there should be five different types of protons. However, because of the following Cope rearrangement only three types are observed (−110 to +117°) by NMR:

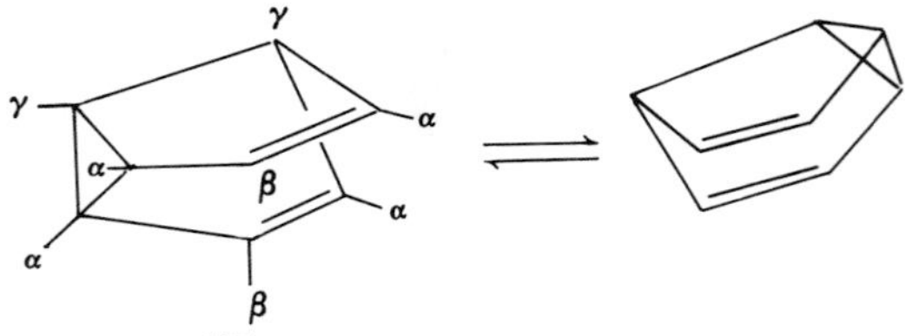

(40) (8.40)

The direct photolysis of barrelene to cyclooctatetraene can be pictured as follows:

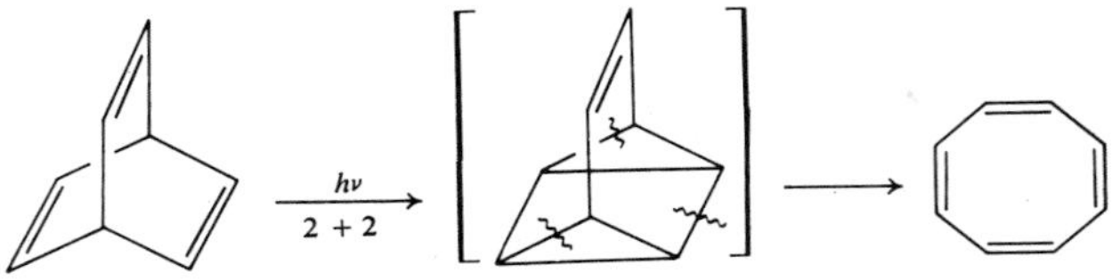

(8.41)

A precedent[42] for this type of reaction can be found in the formation of benz[*f*]oxepin (42) from the epoxide (41):

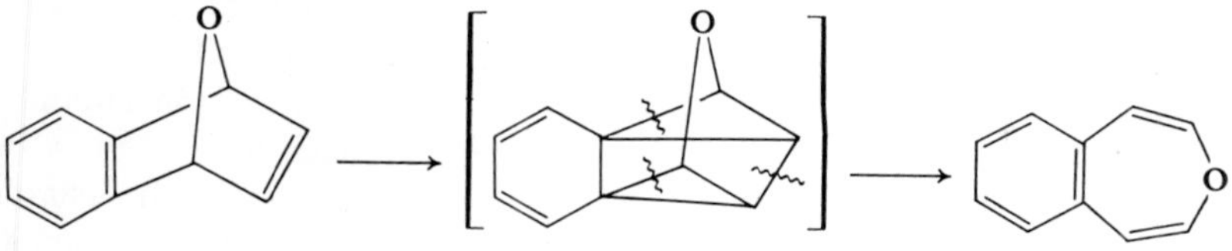

(41) (42) (8.42)

Two possible reaction schemes for the photosensitized formation of semibullvalene are as follows, where the large dot indicates a proton label (all other protons were exchanged for deuterium):

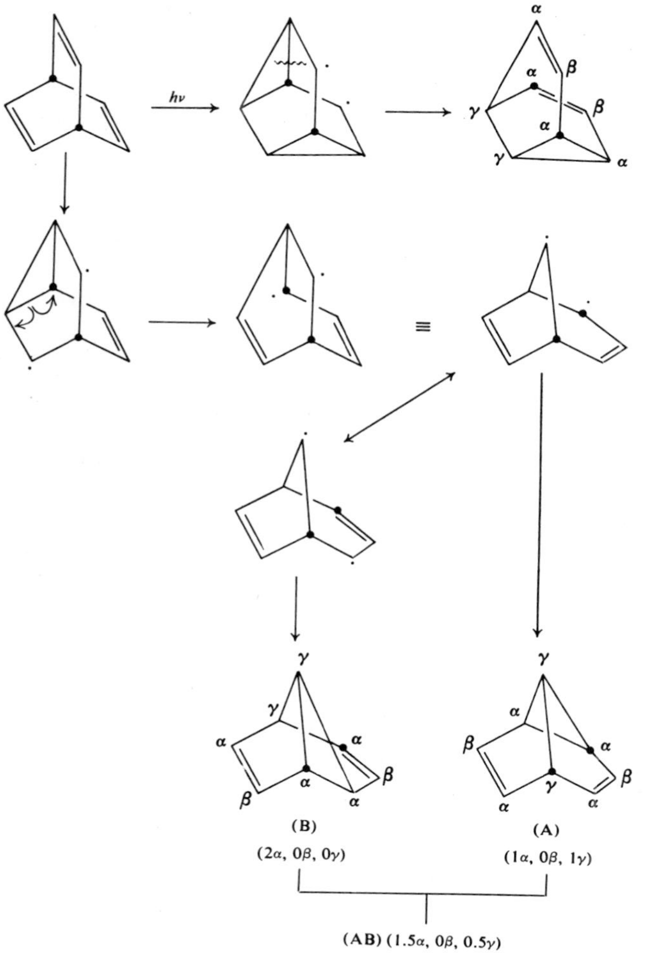

(8.43)

(8.44)

(B)
(2α, 0β, 0γ)

(A)
(1α, 0β, 1γ)

(AB) (1.5α, 0β, 0.5γ)

In mechanism (8.43) the bridgehead hydrogens of barrelene should be found at the α positions of semibullvalene ($2\alpha, 0\beta, 0\gamma$). Mechanism (8.44) can give three different hydrogen-label distributions. If the final bond formation is concerted with bond fission, and bond fission and formation take place at the same carbon atom [mechanism (8.44A)], the label distribution should be ($1\alpha, 0\beta, 1\gamma$). If bond formation is concerted with bond fission but with a preference for bond formation at the carbon allylic to bond fission [mechanism (8.44B)], the label distribution should be ($2\alpha, 0\beta, 0\gamma$). If there is a symmetric allylic biradical which has a finite existence [mechanism (8.44AB)], then the hydrogen-label distribution should be ($1.5\alpha, 0\beta, 0.5\gamma$).

The observed hydrogen distribution is consistent with mechanism (8.44AB).

Liu has found that 2,3-*bis*trifluoromethylbarrelene undergoes sensitized photorearrangement to three semibullvalenes[43]:

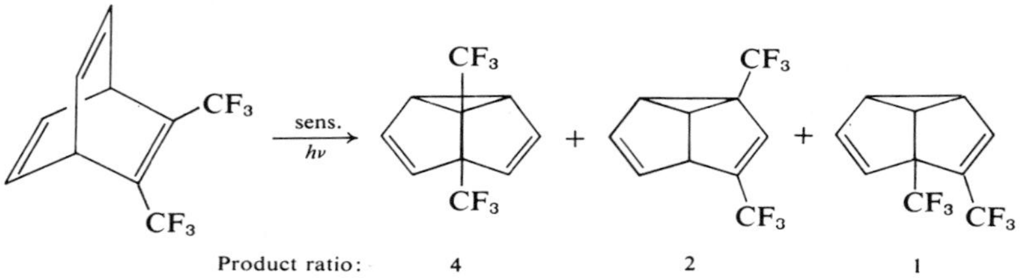

CF_3 CF_3 sens. $h\nu$ CF_3 CF_3 + CF_3 CF_3 + CF_3 CF_3 (8.45)

Product ratio: 4 2 1

8.3b. *Benzobarrelene*

Benzobarrelene is similar to barrelene in that rearrangement to benzocyclooctatetraene takes place from the singlet state while benzosemibullvalene is formed from the triplet state[44]:

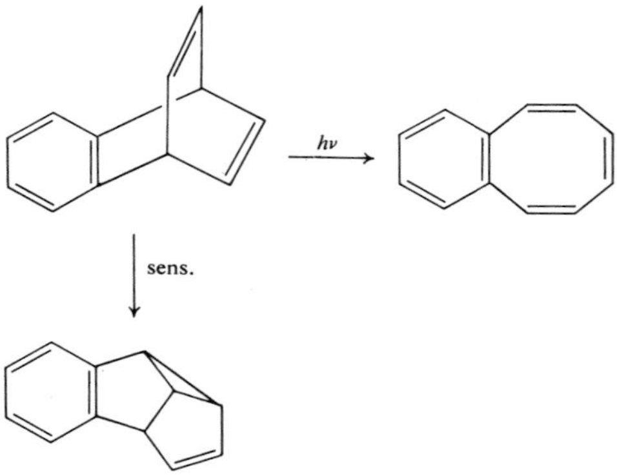

$h\nu$ sens. (8.46)

Brewer and Heaney[45] observed the same type of rearrangement with tetrafluorobenzobarrelene, while Friedman found that dibenzobarrelene formed dibenzosemibullvalene upon sensitized photolysis[46,47]:

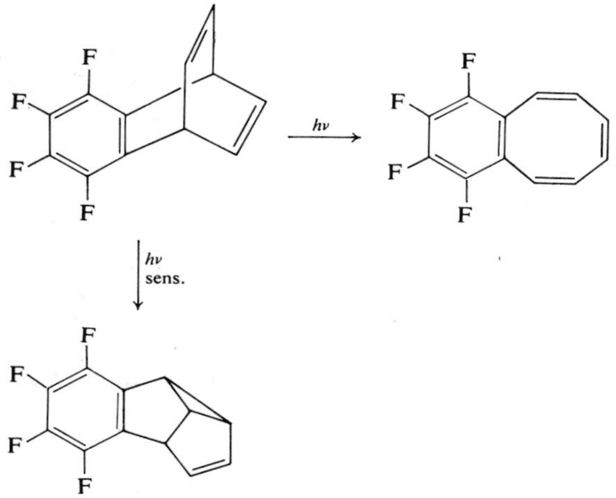

(8.47)

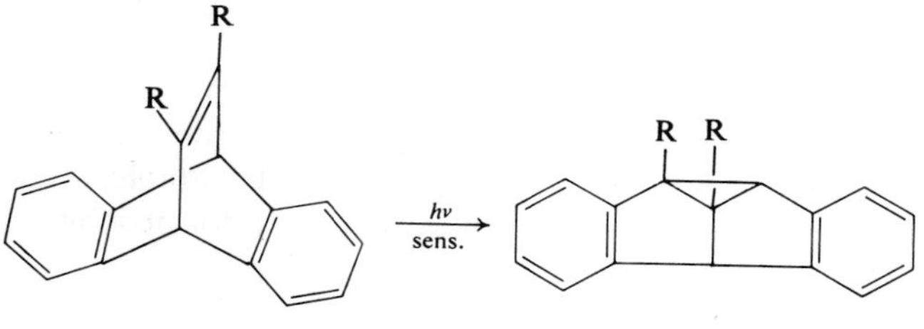

(8.48)

The hydrogen-labeled benzobarrelene indicates that the formation of benzocyclooctatetraene from the direct photolysis takes place predominantly via benzovinyl bridging[44]:

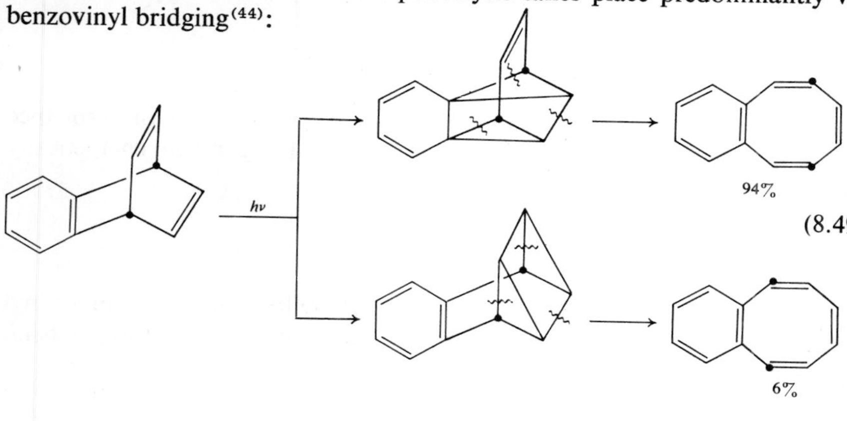

(8.49)

where the large dot represents hydrogen; deuterium elsewhere. The triplet reaction takes place via vinyl–vinyl bridging:

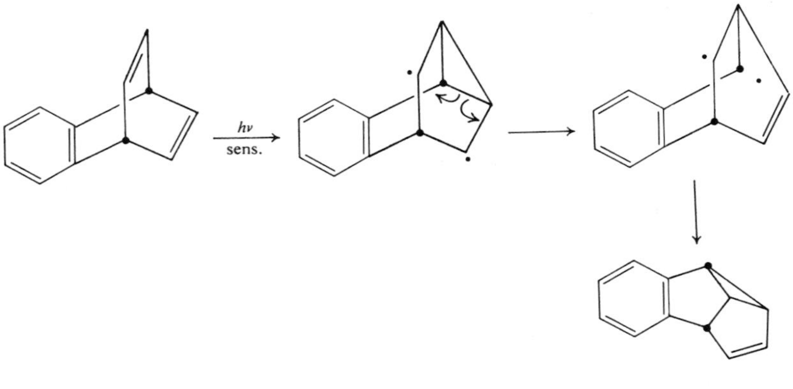

$h\nu$ sens. (8.50)

None of the products that would correspond to benzo–vinyl bridging [(43), (44)] were observed.

(43) (44)

Zimmerman suggests that initial interaction between the vinyl moieties leads to a "butadiene type" array of p orbitals ($E_t \approx$ 58 kcal/mole, butadiene) and that this array is of lower energy than the array formed by benzo–vinyl interaction ($E_t \approx$ 64 kcal/mole, styrene):

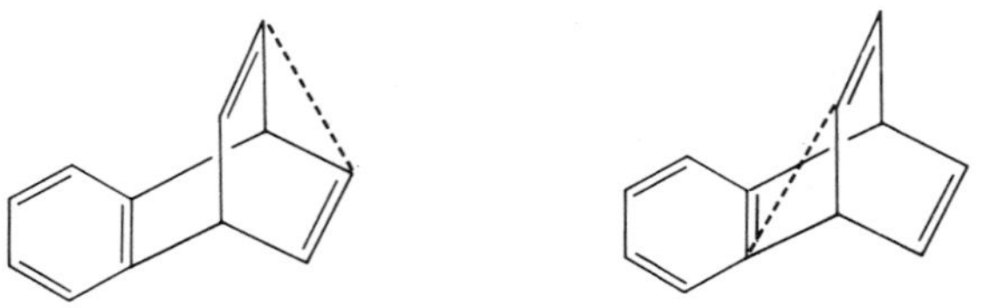

vinyl–vinyl benzo–vinyl

Another, possibly more attractive rationale is that the loss of resonance energy on benzo–vinyl bridging makes this bonding mode energetically unattractive to the triplet state.

8.3c. *Naphthobarrelenes*

2,3-Naphthobarrelene gives naphthosemibullvalene via vinyl–vinyl bridging upon either direct or sensitized photolysis[48] (Φ_{dir} = 0.45, Φ_{sen} = 0.47):

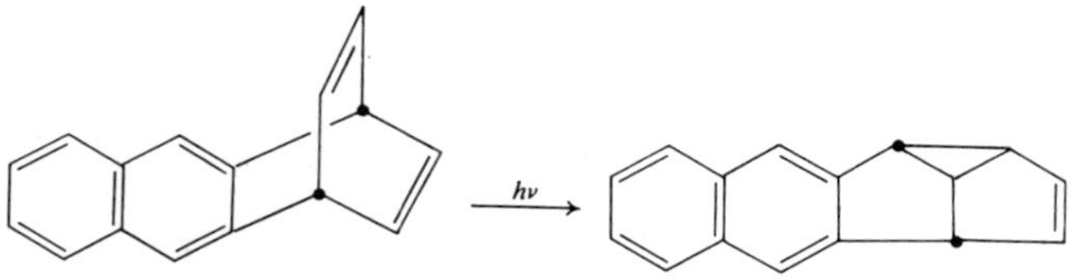

(8.51)

where the dot represents hydrogen; deuterium elsewhere. However, the sensitized reaction of 1,2-naphthobarrelene gives the *syn*- and *anti*-naphthosemibullvalenes by α-naphtho–vinyl bridging:

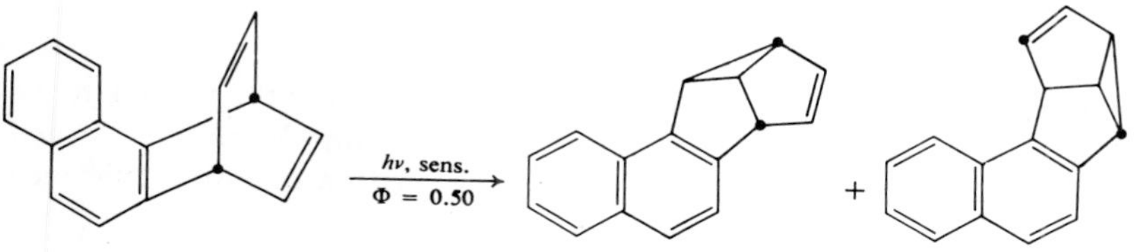

(8.52)

Zimmerman suggests that this bonding preference is consistent with approximate triplet energies of his isoconjugate models for bridge formation:

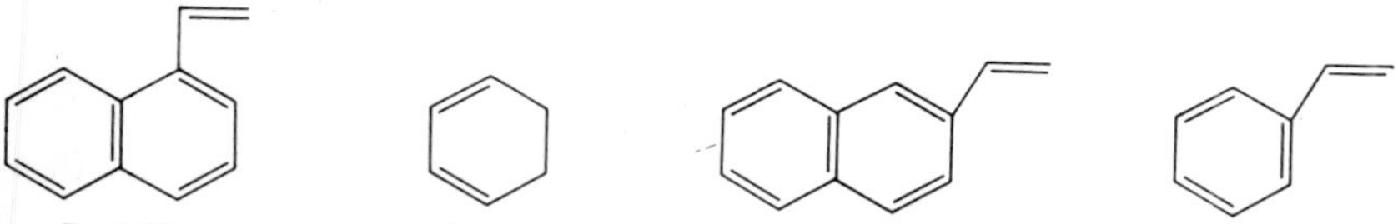

The pattern of bridging reactivity is as follows: α-naphtho–vinyl > vinyl–vinyl > β-naphtho–vinyl > benzo–vinyl.

When benzo-2,3-naphthobarrelene is photolyzed, formation of the semibullvalene takes place by (50%) β-naphtho–vinyl and (50%) benzo–vinyl bridging.[49] However, the quantum efficiency for this reaction (Φ = 0.077) is much lower than for the photolysis of 2,3-naphthobarrelene (Φ = 0.45), where vinyl–vinyl bridging takes place:

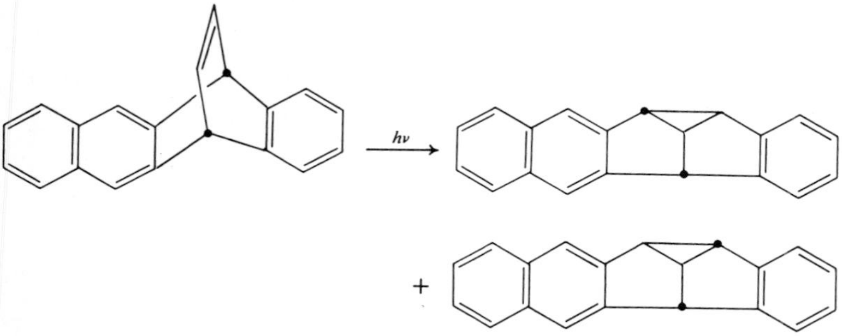

(8.53)

8.3d. *Anthrabarrelene*

Anthrabarrelene was prepared by generating anthracyne, from 2-amino-3-anthracenecarboxylic acid, in the presence of benzene[50]:

(45)

(8.54)

In sensitization studies, the T_1 state of anthrabarrelene (45) proved to be totally unreactive. However, direct photolysis of 2,3-anthrabarrelene afforded 2,3-anthrasemibullvalene via vinyl–vinyl bridging with a quantum efficiency of 0.25:

$h\nu$

$\Phi = 0.25$

(8.55)

With the exception of the α-naphtho position, bridging to aromatic sites is very unfavorable in the barrelene–semibullvalene rearrangements. If we exclude anthraceno–vinyl bridging, there is insufficient energy available (43 kcal/mole) for vinyl–vinyl bridging (58 kcal/mole) in the lowest anthrabarrelene triplet state. Consequently, it is not surprising that the T_1 state is unreactive. In contrast, S_1 (76 kcal/mole) and T_2 (74 kcal/mole) are not subject to this limitation. Evidence in favor of one or the other of these two possible electronic states is not available.

8.3e. *Other Selected Examples*

$h\nu$

sens.; Ref. 51

(8.56)

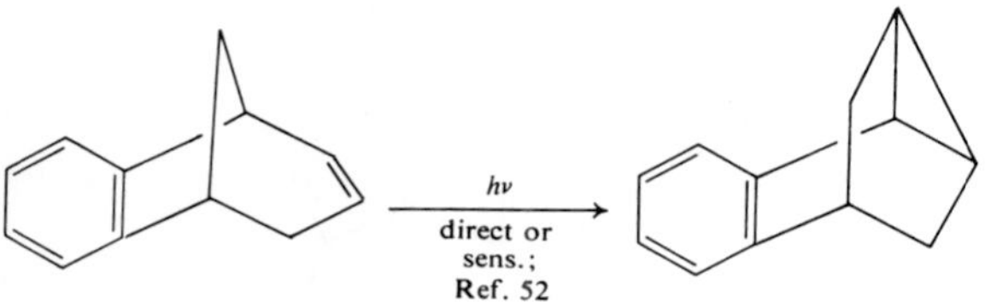
hν
direct or
sens.;
Ref. 52

(8.57)

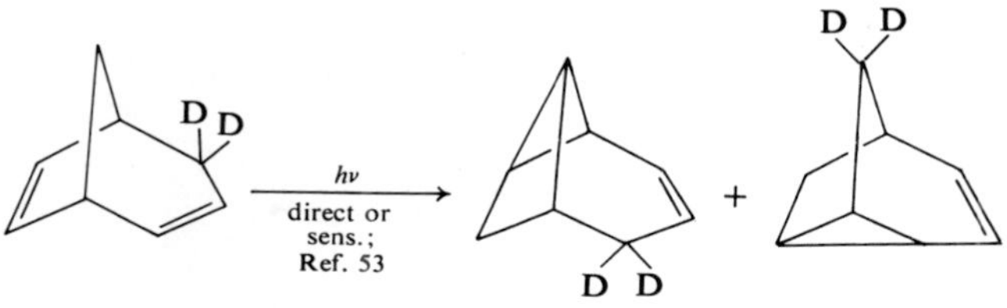
D D
hν
direct or
sens.;
Ref. 53
D D
+
D D

(8.58)

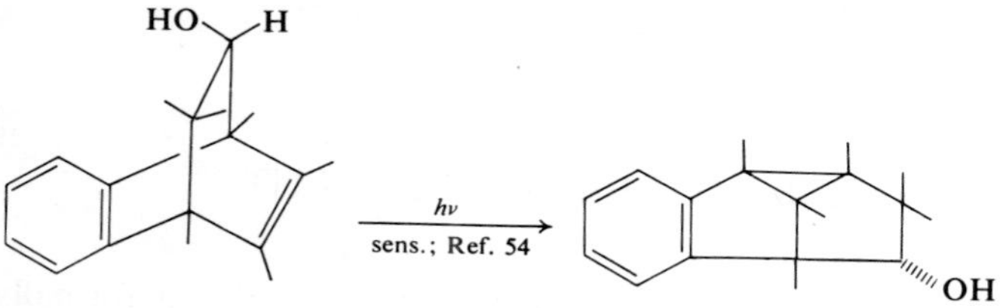
HO H
hν
sens.; Ref. 54
OH

(8.59)

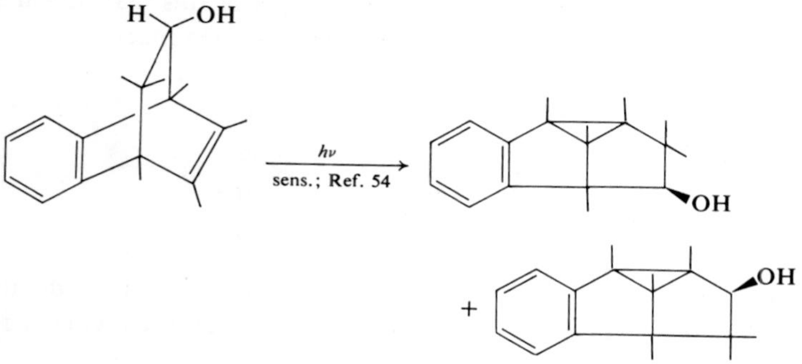
H OH
hν
sens.; Ref. 54
OH
+
OH

(8.60)

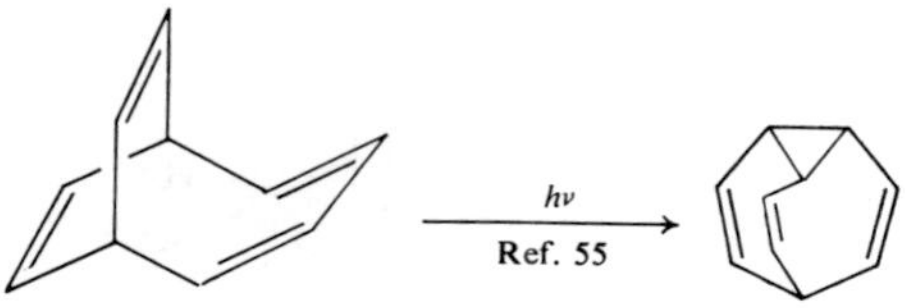
hν
Ref. 55

(8.61)

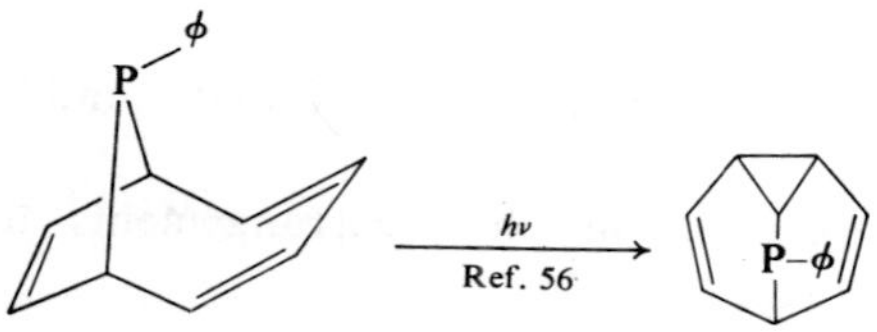
P
ϕ
hν
Ref. 56
P–ϕ

(8.62)

$h\nu$; Ref. 57 — $\Phi = 0.36$ direct and sens. (8.63)

8.4. OXA-DI-π-METHANE REARRANGEMENT

β,γ-Unsaturated ketones undergo a rearrangement that is formally like the di-π-methane photorearrangement. An example of this rearrangement is provided by the photolysis of 1,2,4,4-tetraphenyl-3-butenone:

$h\nu$; ϕ–H (8.64)

Tenney, Boykin, and Lutz[58] labeled carbon 1 with either ^{14}C or deuterium and showed that a hydrogen or phenyl migration was not involved in the rearrangement:

(8.65)

They also labeled one of the C-4 phenyl groups with ^{14}C and found that there was no migration of this group.

Another example of acyclic enone photorearrangements has been provided by Dauben and co-workers[59]:

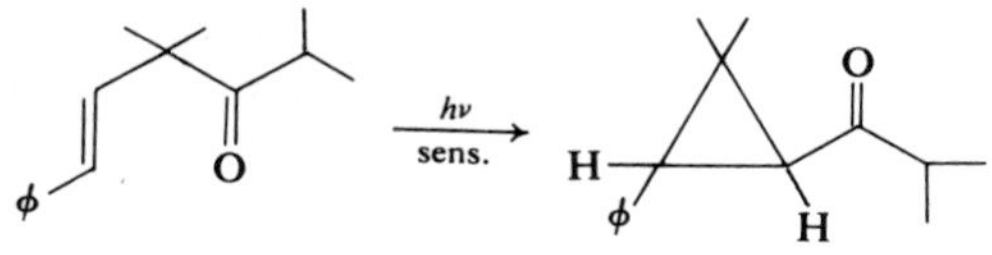

(8.66)

They argue that this and related enone photorearrangements do not take place via an α-cleavage (8.67), inasmuch as bonding to the central carbon atom of the allylic radical (47) or the rearrangement of an allyl radical to a cyclopropyl radical is the least favorable path available to the radical species:

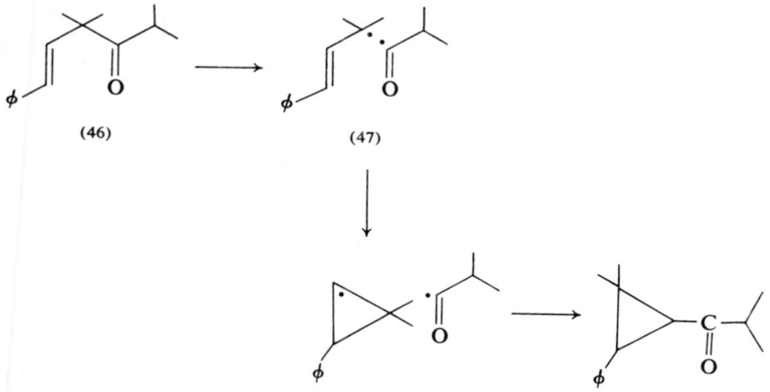

(46) (47)

(8.67)

Closure of the radical species (47) should form the starting materials or compound (48):

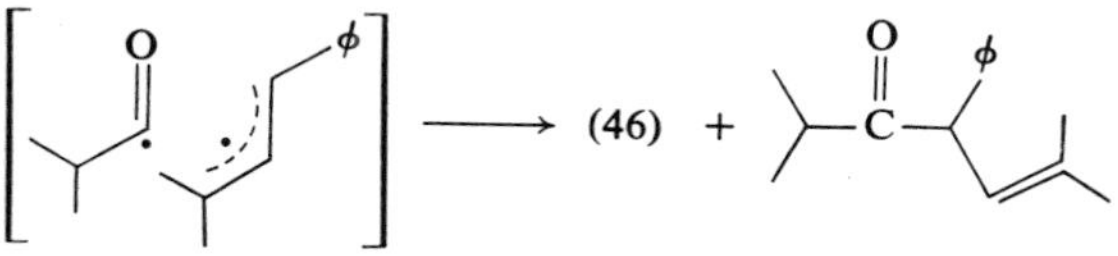

(8.68)

(47) (48)

While direct photolysis of (46) gave (48), the sensitized photolysis of (46) with acetophenone gave only the cyclopropane product (>93%). These authors favor a stepwise oxa-di-π-methane reaction mechanism (8.69) inasmuch as a concerted reaction should produce the product without a change in multiplicity. That is, the product would have to be formed in the triplet state and not enough energy is available for the product to be formed in an excited state:

(8.69)

A number of examples of this type of photorearrangement have been observed with cyclic and bicyclic molecules. Some photoreactions involving cyclohexenones are as follows:

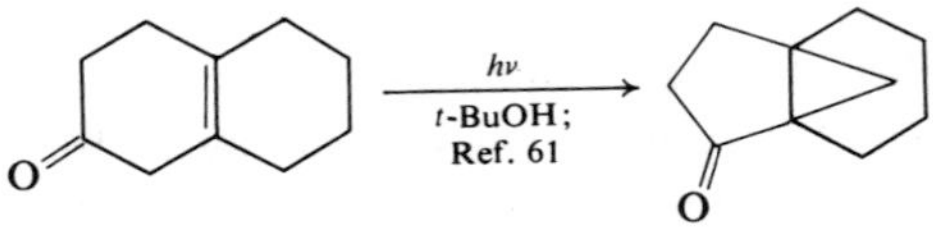
hν
t-BuOH;
Ref. 61

50%

(8.70)

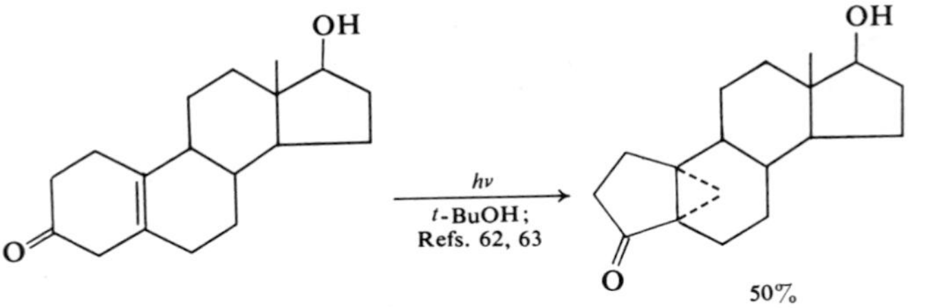
OH
hν
t-BuOH;
Refs. 62, 63
OH

50%

(8.71)

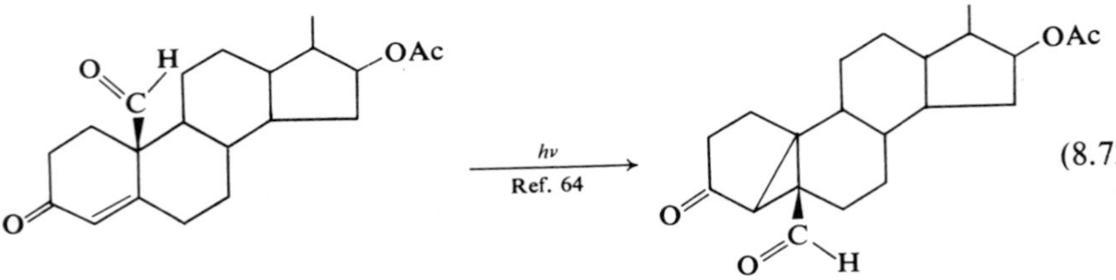
OAc
hν
Ref. 64
OAc

(8.72)

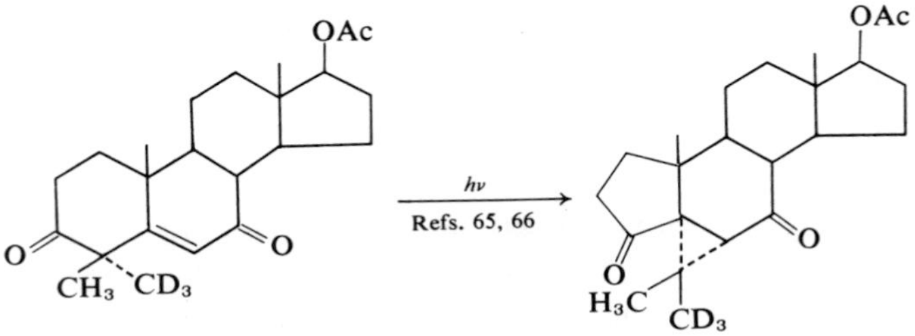
OAc
hν
Refs. 65, 66
OAc

(8.73)

(49) (50)

Major product;
—CD_3 and —CH_3 groups scrambled

Schaffner and co-workers found that steroid (49) upon photolysis gave products (50) in which the —CD_3 and —CH_3 groups were scrambled. This would be consistent with a stepwise oxa-di-π-methane type mechanism.

The photolysis of bicyclic enones is frequently a high-yield route to new tricyclic compounds:

$h\nu$ acetone, sens. — 75% (8.74)

$h\nu$ direct; Refs. 67, 68 (8.75)

$h\nu$ acetone, sens. Refs. 67, 68 — 60% (8.76)

$h\nu$ acetone, sens. Refs. 69, 70 — 70% (8.77)

$h\nu$ Ph–H; Refs. 71, 72 — 40% (8.78)

$h\nu$ sens.; Refs. 73–75 — (27%) + (8.79)

$h\nu$ sens.; Refs. 73–75 — (70%) (8.80)

hν sens.; Refs. 76, 77 → (26%) + other products (8.81)

All of the oxa-di-π-methane reactions studied to date occur from the triplet state. This is in contrast to the di-π-methane photorearrangement, which occurs from both singlet and triplet states. The oxa-di-π-methane rearrangement must compete with other triplet enone reactions: photoreduction (Chapter 3), type II cleavage (Chapter 3), type I or α-cleavage (Chapter 4), photocycloaddition reactions (Chapter 10), and *cis–trans* isomerization (Chapters 6 and 9). While more work needs to be done, it appears that the configuration of the reactive triplet state in the oxa-di-π-methane rearrangements is the $\pi \rightarrow \pi^*$ state.[78–80] This is based on the lifetime of the triplet states at 77°K in a rigid glass and upon the assumption that the emitting triplet is also the reacting one. All of the triplet enones found to undergo oxa-di-π-methane rearrangement had relatively long triplet lifetimes (30–200 msec) characteristic of the $\pi \rightarrow \pi^*$ excited state.*

PROBLEMS

1. Show that the product formed in the reaction described by Eq. (8.21) can be predicted by the di-π-methane mechanism.
2. Give a mechanism for the formation of compound (24), Eq. (8.25).
3. When τ_s is shorter than a vibration, what does this suggest? (See Table 8.4.)
4. Propose an explanation of why the photorearrangement of compound (14) takes place via a hydrogen migration while the rearrangement of compound (30) takes place via a di-π-methane reaction.
5. Show that the trans isomer (35) is the expected photoproduct from compound (33). How can you explain the formation of (34) at all?
6. Can a fortuitous combination of mechanisms (8.43) and (8.44A) account for the observed hydrogen-label distribution in the formation of semibullvalene from barrelene? If so, how can you eliminate this possibility?

* For a recent review of di-π-methane and oxa-di-π-methane rearrangements see Hixson *et al.*[81]

7. Give the products for the following reaction[51]

$\xrightarrow[\text{sens.}]{h\nu}$?

8. (a) Give a synthesis for the following compound:

CF$_3$

CF$_3$

(b) What are the expected photoproducts upon sensitized photolysis of this compound?

REFERENCES

1. G. W. Griffin, J. Covell, R. C. Petterson, R. M. Dodson, and G. Klose, *J. Amer. Chem. Soc.* **87**, 1410 (1965).
2. G. W. Griffin, A. F. Marcantonio, H. Kristinsson, R. C. Petterson, and C. S. Irving, *Tetrahedron Lett.*, 2951 (1965).
3. H. E. Zimmerman and G. L. Grunewald, *J. Amer. Chem. Soc.* **88**, 183 (1966).
4. H. E. Zimmerman, B. W. Binkley, R. S. Givens, and M. A. Sherwin, *J. Amer. Chem. Soc.* **89**, 3932 (1967).
5. J. Meinwald and G. W. Smith, *J. Amer. Chem. Soc.* **89**, 4923 (1967).
6. R. Srinivasan and K. H. Carlough, *J. Amer. Chem. Soc.* **89**, 4932 (1967).
7. H. E. Zimmerman and P. S. Mariano, *J. Amer. Chem. Soc.* **91**, 1718 (1969).
8. H. E. Zimmerman, *Acc. Chem. Res.* **4**, 272 (1971).
9. H. E. Zimmerman and R. D. Little, *J. Amer. Chem. Soc.* **94**, 8256 (1972).
10. H. E. Zimmerman and A. C. Pratt, *J. Amer. Chem. Soc.* **92**, 1407 (1970).
11. H. E. Zimmerman and A. C. Pratt, *J. Amer. Chem. Soc.* **92**, 6259 (1970).
12. H. E. Zimmerman and A. A. Baum, *J. Amer. Chem. Soc.* **93**, 3646 (1971).
13. H. E. Zimmerman and A. C. Pratt, *J. Amer. Chem. Soc.* **92**, 1410 (1970).
14. H. E. Zimmerman and A. C. Pratt, *J. Amer. Chem. Soc.* **92**, 6267 (1970).
15. H. E. Zimmerman, P. Baeckstrom, T. Johnson, and D. W. Kurtz, *J. Amer. Chem. Soc.* **94**, 5504 (1972).
16. P. S. Mariano and J. Ko, *J. Amer. Chem. Soc.* **94**, 1766 (1972).
17. M. J. Bullivant and G. Pattenden, *Chem. Comm.*, 864 (1972).
18. J. Perreten, D. M. Chihal, G. W. Griffin, and N. S. Bhacca, *J. Amer. Chem. Soc.* **95**, 3427 (1973).
19. D. C. Lankin, D. M. Chihal, G. W. Griffin, and N. S. Bhacca, *Tetrahedron Lett.*, 4009 (1973).

20. H. E. Zimmerman and J. A. Pincock, *J. Amer. Chem. Soc.* **95**, 3246 (1973).
21. W. R. Roth and B. Peltzer, *Ann. Chem.* **685**, 56 (1965).
22. T. Sasaki, S. Eguchi, M. Ohno, and T. Umemura, *Tetrahedron Lett.*, 3895 (1970).
23. P. S. Mariano and R. B. Steitle, *J. Amer. Chem. Soc.* **95**, 6115 (1973).
24. H. E. Zimmerman and J. A. Pincock, *J. Amer. Chem. Soc.* **95**, 2957 (1973).
25. H. E. Zimmerman, R. J. Boettcher, and W. Braig, *J. Amer. Chem. Soc.* **95**, 2155 (1973).
26. H. E. Zimmerman, D. P. Werthemann, and K. S. Kamm, *J. Amer. Chem. Soc.* **95**, 5094 (1973).
27. S. Hixson, *Tetrahedron Lett.*, 1155 (1972).
28. H. Kristinsson and G. S. Hammond, *J. Amer. Chem. Soc.* **89**, 5968 (1967).
29. H. E. Zimmerman and G. E. Samuelson, *J. Amer. Chem. Soc.* **91**, 5307 (1969); **89**, 2971 (1967).
30. H. E. Zimmerman, P. Hackett, D. F. Juers, and B. Schroder, *J. Amer. Chem. Soc.* **93**, 3653 (1971); **89**, 5973 (1967).
31. H. Hark, J. D. DeVrieze, R. M. Lange, and A. Sheller, *Chem. Comm.*, 1650 (1968).
32. H. E. Zimmerman, D. F. Juers, J. M. McCall, and B. Schroder, *J. Amer. Chem. Soc.* **93**, 3662 (1971).
33. W. G. Dauben and W. A. Spitzer, *J. Amer. Chem. Soc.* **92**, 5817 (1970).
34. J. S. Swenton, A. R. Crumrine, and T. J. Walker, *J. Amer. Chem. Soc.* **92**, 1406 (1970).
35. H. E. Zimmerman and G. A. Epling, *J. Amer. Chem. Soc.* **92**, 1411 (1970).
36. S. Hixson, *J. Amer. Chem. Soc.* **94**, 2507 (1972).
37. D. Kumari and S. K. Mukeyie, *Tetrahedron Lett.*, 4169 (1967).
38. B. Halton, M. Kulig, M. A. Battiste, J. Perreten, D. M. Gibson, and G. W. Griffin, *J. Amer. Chem. Soc.* **93**, 2327 (1971).
39. M. Comtet, *J. Amer. Chem. Soc.* **91**, 7761 (1969).
40. M. Comtet, *J. Amer. Chem. Soc.* **92**, 5308 (1970).
41. H. E. Zimmerman, R. N. Binkley, R. S. Givens, G. L. Grunewald, and M. A. Sherwin, *J. Amer. Chem. Soc.* **91**, 3316 (1969).
42. G. R. Ziegler and G. S. Hammond, *J. Amer. Chem. Soc.* **90**, 513 (1968).
43. R. Liu, *J. Amer. Chem. Soc.* **90**, 215 (1968).
44. H. E. Zimmerman, R. S. Givens, and R. M. Pagni, *J. Amer. Chem. Soc.* **90**, 6096 (1968).
45. J. Brewer and H. Heaney, *Chem. Comm.*, 811 (1967).
46. E. Ciganek, *J. Amer. Chem. Soc.* **88**, 2882 (1966).
47. P. W. Rabideau, J. B. Hamilton, and L. Friedman, *J. Amer. Chem. Soc.* **90**, 4465 (1968).
48. H. E. Zimmerman and C. O. Bender, *J. Amer. Chem. Soc.* **92**, 4366 (1970).
49. H. E. Zimmerman and M. Viriot-Villaume, *J. Amer. Chem. Soc.* **95**, 1274 (1973).
50. H. E. Zimmerman and D. R. Amick, *J. Amer. Chem. Soc.* **95**, 3977 (1973).
51. J. R. Edman, *J. Amer. Chem. Soc.* **91**, 7103 (1969).
52. R. C. Hahn and L. J. Rothman, *J. Amer. Chem. Soc.* **91**, 2409 (1969).
53. R. R. Sauers and A. Shurpik, *J. Org. Chem.* **33**, 799 (1968).
54. H. Hart and R. K. Murray, *J. Amer. Chem. Soc.* **91**, 2183 (1969).
55. M. Jones, S. D. Reich, and L. I. Scott, *J. Amer. Chem. Soc.* **92**, 3118 (1970).
56. T. J. Katz, J. C. Carnahan, G. M. Clarke, and N. Acton, *J. Amer. Chem. Soc.* **92**, 734 (1970).
57. H. E. Zimmerman, D. R. Amick, and H. Hemetsberger, *J. Amer. Chem. Soc.* **95**, 4606 (1973).

58. L. P. Tenney, D. W. Boykin, Jr., and R. E. Lutz, *J. Amer. Chem. Soc.* **88**, 1835 (1966).
59. W. G. Dauben, M. S. Kellog, J. I. Seeman, and W. A. Spitzer, *J. Amer. Chem. Soc.*, **92**, 1786 (1970).
60. W. G. Dauben and W. A. Spitzer, *J. Amer. Chem. Soc.* **90**, 802 (1968).
61. J. R. Williams and H. Ziffer, *Chem. Comm.*, 194 (1967).
62. J. R. Williams and H. Ziffer, *Chem. Comm.*, 469 (1967).
63. J. R. Williams and H. Ziffer, *Tetrahedron* **24**, 6725 (1968).
64. E. Pfenninger, D. E. Poel, C. Berse, H. Wehrli, K. Schaffner, and O. Jeger, *Helv. Chim. Acta* **51**, 772 (1968).
65. S. Bornb, G. Bozzato, J. A. Saboz, and K. Schaffner, *Helv. Chim. Acta* **52**, 2436 (1968).
66. S. Domb and K. Schaffner, *Helv. Chim. Acta* **53**, 677 (1970).
67. J. Ipaktschi, *Tetrahedron Lett.*, 2153 (1969).
68. J. Ipaktschi, *Chem. Ber.* **105**, 1840 (1972).
69. J. Ipaktschi, *Tetrahedron Lett.*, 3179 (1970).
70. D. I. Schuster and D. H. Sussman, *Tetrahedron Lett.*, 1661 (1970).
71. P. A. Knott and J. M. Mellor, *Tetrahedron Lett.*, 1829 (1970).
72. P. A. Knott and J. M. Mellor, *J. Chem. Soc.* **1**, 1030 (1972).
73. J. Ipaktschi, *Tetrahedron Lett.*, 215 (1969).
74. R. S. Givens, W. F. Oettle, R. L. Coffin, and R. G. Carlson, *J. Amer. Chem. Soc.* **93**, 3957 (1971).
75. R. S. Givens and W. F. Oettle, *J. Amer. Chem. Soc.* **93**, 3963 (1971).
76. R. K. Murray and H. Hart, *Tetrahedron Lett.*, 4995 (1968).
77. H. Hart and R. K. Murray, *Tetrahedron Lett.*, 379 (1969).
78. K. G. Hancock and R. O. Grider, *Chem. Comm.*, 580 (1972).
79. G. Marsh, D. R. Kearns, and K. Schaffner, *J. Amer. Chem. Soc.* **93**, 3129 (1971).
80. K. N. Houk and D. J. Northington, *J. Amer. Chem. Soc.* **94**, 1387 (1972).
81. S. S. Hixson, P. S. Mariano, H. E. Zimmerman, *Chem. Rev.* **73**, 531 (1973).

9

Photochemical Cis-Trans and Valence Isomerization of Olefins

9.1. INTRODUCTION: CIS–TRANS ISOMERIZATION OF STILBENE

The direct photolysis of *cis*- or *trans*-stilbene in solution gives rise to photoisomerization. In addition, dihydrophenanthrene is formed from the *cis* isomer:

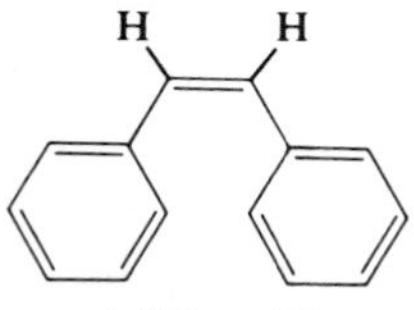

H H

cis-Stilbene (C)

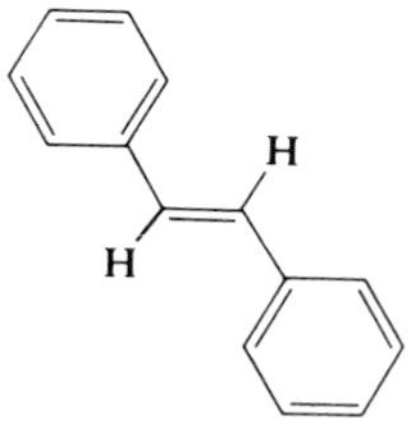

H

H

trans-Stilbene (T)

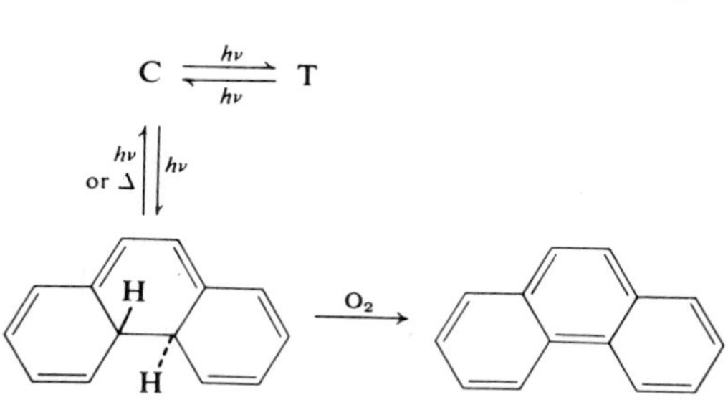

$$C \underset{h\nu}{\overset{h\nu}{\rightleftharpoons}} T$$

$h\nu$ or Δ ; $h\nu$

H

H

$\xrightarrow{O_2}$

The equilibrium concentration of dihydrophenanthrene under oxygen-free conditions is very small. However, this reaction, in the presence of O_2 and I_2, has been developed to the point where it is a good method for ring formation.

The mechanism for the direct photochemical *cis–trans* isomerization of stilbene has been a highly controversial subject. However, a recent review by Saltiel and co-workers greatly helps to clarify this area of research by painting a detailed and beautifully consistent picture. We will make extensive reference to this review.[(1)]

9.2. POTENTIAL ENERGY DIAGRAMS

Molecular orbital calculations on ethylene indicate that the lowest energy excited singlet and triplet states have a twisted geometry.[(2)] This geometry helps minimize electron–electron repulsion. Figure 9.1 gives the calculated

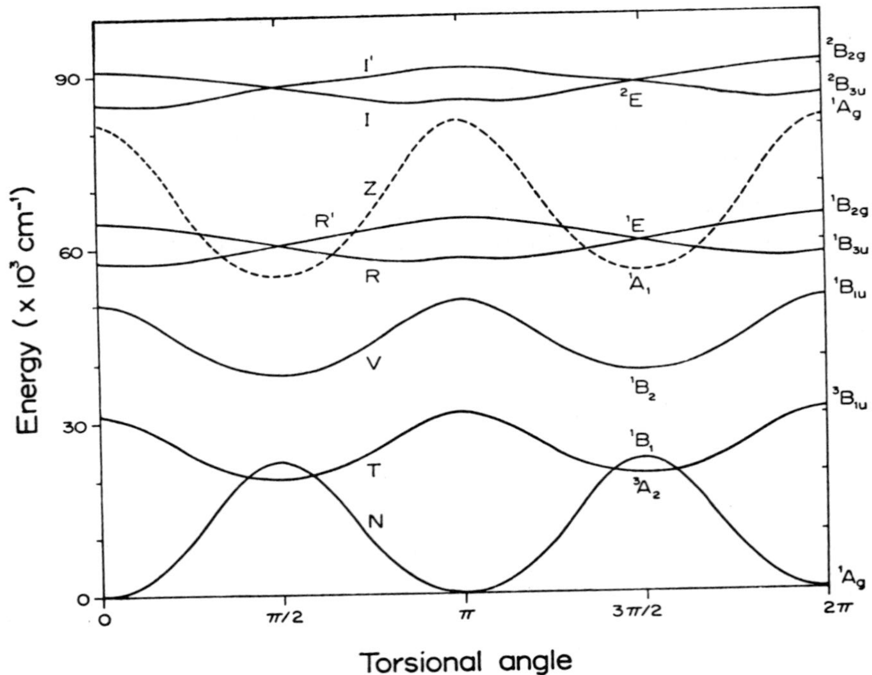

FIGURE 9.1. Potential energy diagram for the electronic states of ethylene: *N*, ground state $(\pi)^2$; *T* $(^3B_{1u})$ first excited triplet state $(\pi\pi^*)$; *V*, $(^1B_{1u})$ first excited singlet state $(\pi\pi^*)$; *Z*, two-electron excitation $(\pi^*)^2$. For the ion $C_2H_4^+$: R and *R′*, Rydberg states, *I*, *I′* ground and excited states. [From Ref. 2(c).]

(1)

potential energy curve for ethylene. Examination of this figure suggests that internal conversion from the twisted singlet or intersystem crossing from the twisted triplet could give either the *cis* or *trans* product. That is, isomerization could take place from either or both of the lowest excited states:

$$cis \xrightarrow{h\nu} {}^1(cis)$$

$${}^1(cis) \longrightarrow {}^1[\text{phantom (P) or twist form}]$$

$${}^1(\text{P}) \longrightarrow cis + trans$$

Note that the excited singlet state that should be produced upon optical excitation is a highly energetic vibronic form of the common twist singlet. This picture is probably valid for ethylene and simple alkyl-substituted olefins. The potential energy curves for stilbene have been calculated[3] and are reproduced in Figure 9.2. Again the indication is that isomerization should occur in either or both of the two lowest excited states. The experimental evidence concerning this prediction will be discussed in the next sections.

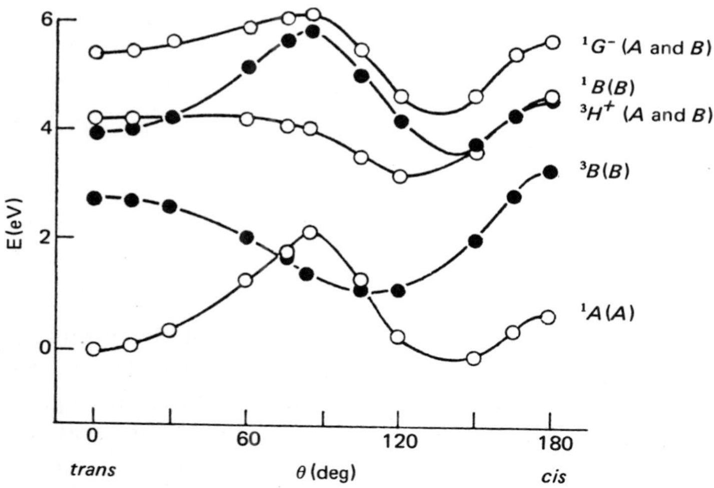

FIGURE 9.2. The energies of the ground ¹A state and electronically excited states shown as a function of θ, the angle of twist. (○) Singlet states: (●) triplet states. The G and H curves represent two states that nearly coincide. (From Ref. 3.)

9.3. PHOTOSENSITIZED STILBENE ISOMERIZATION

The *cis–trans* isomerization of stilbene is observed upon irradiation of solutions containing stilbene and a variety of triplet sensitizers, under conditions where only the sensitizer absorbs energy.[4–12] The photoequilibrium concentration ratio of *cis*/*trans* stilbene (pss, photostationary state; $([C]/[T])_s$) as a function of sensitizer triplet energy is given in Figure 9.3. We can start to understand this very important figure if we know the triplet energies of *cis*- and *trans*-stilbene. The *vertical* excitation energies, determined from singlet–triplet absorption spectra, are 57 kcal/mole for *cis*-stilbene and 49 kcal/mole for *trans*-stilbene. From this information Hammond and co-workers suggested that (a) triplet donors with energies $E_t > 62$ kcal/mole (3–5 kcal/mole above the triplet energy of *cis*-stilbene) transfer triplet excitation to both *cis*- and *trans*-stilbene with equal efficiencies at a diffusion-controlled rate, (b) triplet donors with energies above 48 kcal/mole but below 62 kcal/mole transfer triplet energy to the *trans* isomer at a diffusion-controlled rate but at a much lower rate to the *cis* isomer, and (c) triplet donors with energies below 48 kcal/mole produce a photostationary state rich in *trans*-stilbene. To account for this observation, energy transfer to the *cis* isomer to produce the twisted triplet was proposed and this process termed a nonvertical excitation [Eq. (9.5)]:

$$D + h\nu \longrightarrow D^s \qquad (9.1)$$

$$D^s \rightsquigarrow D^t \qquad (9.2)$$

$$D^t + C \longrightarrow D + C^t \quad (k_3) \qquad (9.3)$$

$$D^t + T \longrightarrow D + T^t \quad (k_4) \qquad (9.4)$$

$$D^t + C \longrightarrow D + P^t \quad (k_5) \qquad (9.5)$$

$$D^t \rightsquigarrow D \quad (k_6) \qquad (9.6)$$

where D = sensitizer or donor.

Herkstroeter and Hammond found support for this postulate from a flash photolysis study. They were able to measure directly the rate of sensitizer quenching (energy transfer) by *cis*- and *trans*-stilbene. When a sensitizer triplet had insufficient excitation energy to promote *trans*-stilbene to its triplet state, the energy deficiency could be supplied as an activation energy. The decrease in transfer rate as a function of excitation energy of the sensitizer is given by

$$\Delta(\log k)/\Delta E_t = -1/2.303RT \qquad (9.7)$$

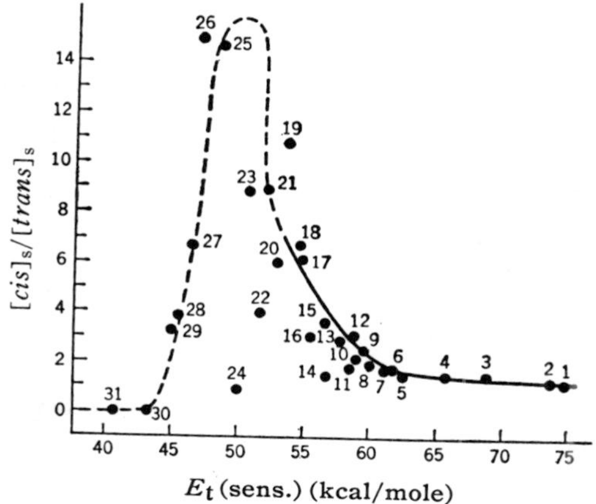

FIGURE 9.3. Variation of the stationary states in photosensitized isomerization of stilbenes with various sensitizers. [From G. S. Hammond, Kayaku to Kôgyô (Tokyo) **18**, 1464 (1965).]

However, *cis*-stilbene does not behave as a classical triplet energy acceptor inasmuch as the rate of energy transfer is much larger than would be predicted from Eq. (9.7) for donors with triplet energies less than 57 kcal/mole. This is then additional information consistent with Eq. (9.5).

The initial quantum yields for *cis*- to *trans*-stilbene isomerization ($\Phi_{C\to T}$) and for *trans* to *cis* isomerization ($\Phi_{T\to C}$) are consistent with Hammond's postulate that isomerization takes place from a common state, most likely the twisted or phantom triplet state:

$$\mathrm{C}^t \longrightarrow \mathrm{P}^t \qquad (k_8) \tag{9.8}$$

$$\mathrm{T}^t \longrightarrow \mathrm{P}^t \qquad (k_9) \tag{9.9}$$

$$\mathrm{P}^t \longrightarrow \alpha T + (1 - \alpha)\mathrm{C} \qquad (k_{10}) \tag{9.10}$$

The following generalizations were made for donors that transfer their energy at a diffusion-controlled rate (high-energy sensitizers):

$$\Phi_{C\to T} + \Phi_{T\to C} = 1 \qquad (\text{when } \Phi_{et}\Phi_{isc} \equiv 1) \tag{9.11}$$

$$\Phi_{T\to C}/\Phi_{C\to T} = [\mathrm{C}]_s/[\mathrm{T}]_s \tag{9.12}$$

A short summary of quantum yield data is given in Table 9.1. Equations (9.11) and (9.12) can be easily derived from the suggested mechanism:

$$\Phi_{C\to T} = \Phi_{isc}\Phi_{et}\alpha \tag{9.13}$$

where Φ_{isc} is the efficiency of intersystem crossing of the donor, Φ_{et} is the efficiency of energy transfer from the donor to *cis*-stilbene, and α is the fraction of phantom triplets that decay to *trans*-stilbene. We also have

$$\Phi_{T\to C} = \Phi_{isc}\Phi'_{et}(1 - \alpha) \tag{9.14}$$

TABLE 9.1. *Quantum Yields for Sensitized Stilbene Isomerization*

Sensitizer	$\Phi_{C \to T}$	$\Phi_{T \to C}$	$\Phi_{C \to T} + \Phi_{T \to C}$	$\Phi_{T \to C}/\Phi_{C \to T}$	$[C]_s/[T]_s$
Acetophenone[10]	0.37	0.52	0.89	1.40	1.27
Benzophenone[10]	0.41	0.55	0.96	1.34	1.4
Benzophenone[11]	0.42	0.55	0.97	1.31	1.4
(2,4,6-Triethyl-benzoyl)naphthalene[11]	0.41	0.57	0.98	1.39	—
9-Fluorenone[a]	0.19	0.47	0.66	2.47	6.2

[a] $E_t \approx 55$ kcal/mole, energy transfer is not diffusion controlled.

where Φ'_{et} is the efficiency of energy transfer from donor to *trans*-stilbene. If $\Phi_{et} = \Phi'_{et}$, then

$$\Phi_{C \to T} + \Phi_{T \to C} = \Phi_{isc}\Phi_{et} \tag{9.15}$$

and

$$\Phi_{T \to C}/\Phi_{C \to T} = (1 - \alpha)/\alpha \tag{9.16}$$

where $(1 - \alpha)/\alpha$ indicates how the phantom triplet decays and is equal to $[C]_s/[T]_s$ (see Problem 1).

If a common intermediate were not involved in the triplet state isomerization of stilbene, the sum of $\Phi_{C \to T}$ and $\Phi_{T \to C}$ could have any value from zero to two.

Before we cite another experiment that supports the idea that isomerization occurs from the phantom triplet state, it will be helpful to complete our set of stilbene reactions and construct an "experimental" potential energy diagram.

The observations that the pss depends upon the sensitizer concentration (for low-energy sensitizers $E_t < 53$ kcal/mole) and that added azulene ($E_t = 30$ kcal/mole) alters the $[C]_s/[T]_s$ ratio so that the pss contains more *trans*-stilbene requires the addition of the following reactions:

$$T^t + D \longrightarrow D^t + T \tag{9.17}$$

$$P^t \longrightarrow T^t \tag{9.18}$$

$$T^t + A \longrightarrow T + A^t \tag{9.19}$$

These reactions, while not very important for high-energy donors in the absence of azulene, do help define what the potential energy diagrams should look like, that is, the *trans*-stilbene triplet is a discrete intermediate while the *cis*-stilbene triplet is a high vibrational form of the phantom triplet. Herkstroeter and McClure[13] have been able to observe the *trans*-stilbene triplet via flash photolysis using low-temperature glasses; however, no similar triplet was observed for *cis*-stilbene. Only when they studied *cis*-stilbene-like

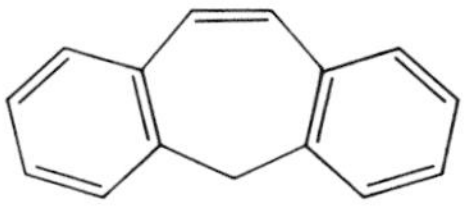

(2)

molecules (2) incapable of isomerization could they observe a *cis*-stilbene-like triplet. Additional information available to help construct the potential energy diagram is as follows:

(a) The enthalpy difference between *cis*- and *trans*-stilbene is 2.3 kcal/mole. This value was obtained from thermal equilibration studies in toluene.[14]
(b) The transition state (or activation energy) for thermal *cis–trans* isomerization is about 46 kcal/mole above the potential energy of *cis*-stilbene.[15] This means that the twist ground state has an energy of about 48 kcal/mole.
(c) The 0–0 bands from the absorption spectra of *trans*- and *cis*-stilbene give the singlet energies[16] 88 and 90 kcal/mole, respectively.
(d) The *vertical* excitation energies determined from singlet–triplet absorption spectra[16–19] in the presence of ethyl iodide[16,19] or high pressures of oxygen[17,18] are 57 and 49 kcal/mole, respectively, for *cis*- and *trans*-stilbene.

The potential energy diagram shown in Figure 9.4 is an attempt to incorporate what we know about the stilbene system. The twisted triplet is shown at lower energy than the *trans* triplet (49 kcal/mole). This would allow either crossing or near crossing with the ground state potential energy surface and account for the short lifetime of stilbene triplets[7,20–23] ($\tau \approx 8 \times 10^{-8}$ sec). This lifetime was estimated from the effect of azulene on the sensitized photostationary states of stilbene (see Problem 3).

Saltiel[1,21,22] has shown that the photostationary state is identical for perhydro and perdeuteriostilbene (with benzophenone and 0–2 $\times$ 10^{-3} mole/liter azulene). This is a very important result that is again consistent with isomerization only from the twist triplet state. If the isomerization resulted from two states (*trans* triplet and *cis* triplet or *trans* triplet and phantom triplet) the triplet decay ratio $(1 - \alpha)/\alpha$ should be affected by deuteration. The theory of radiationless decay[24–26] predicts that the effect of deuteration will diminish as the energy separation between electronic states decreases. Thus the effect on the phantom triplet should be relatively small (Figure 9.4, energy of the phantom triplet about equal to the energy of the ground electronic state at $\pi/2$), while the effect on either the *trans* or *cis* triplet should be quite large. Certainly the triplet decay ratio should be altered by deuteration if isomerization resulted from two states.

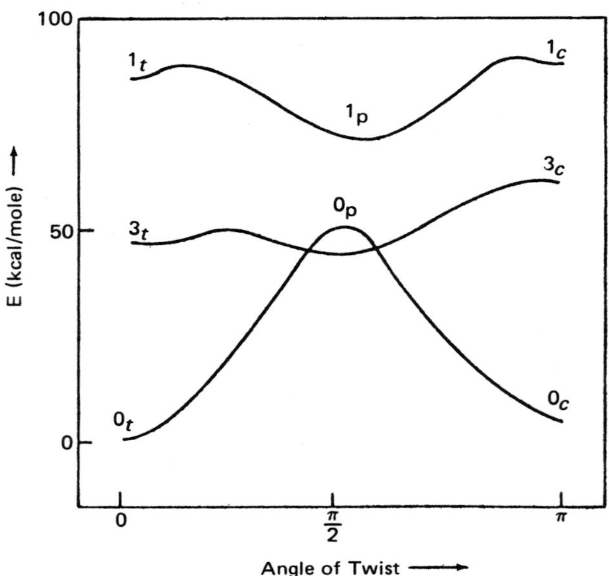

FIGURE 9.4. Potential energy diagram for stilbene as a function of angle of twist. (From Ref. 22.)

Saltiel[1,21,22,27] has also calculated, from the azulene studies, that perdeuteration increases the lifetime of the stilbene triplets by 30%. However, it is known from flash photolysis studies in rigid media at 77°K that perdeuteration results in a 500% increase in the *trans* triplet lifetime. Perdeuteration also increases the lifetime of pyrene triplets (S_0–T_1 = 49 kcal/mole) at 25°C by 600%. Thus the reasonable conclusion is that isomerization occurs from a state with a very small S_0–T_1 energy separation, the phantom triplet state. It is interesting to note that by a comparison[27,28] of vinyl-substituted stilbene-d_2, aromatic-substituted stilbene-d_{10}, and perdeuterated stilbene-d_{12} it can be concluded that deuteration of the vinyl positions increases the lifetime of stilbene triplets by 30% while ring deuteration has no effect. The lifetime increase is attributed to a change in the radiationless decay rate of twisted stilbene triplets.

9.4. NONVERTICAL ENERGY TRANSFER

Hammond and Saltiel[1,5,6,12] proposed the concept of nonvertical energy transfer to account for the nonclassical behavior of *cis*-stilbene and *cis*- and *trans*-α-methylstilbene.

They visualized the removal or relaxation of the Franck–Condon restrictions (vertical excitation) via a long-lived donor–acceptor complex where the donor–acceptor interaction alters the singlet and or triplet potential energy surface of the acceptor. They state that,[6] "Even though the donor–acceptor complex does not become equilibrated with its surroundings, the overall change may be effected by passage through many 'internal equilibrium' states which effectively mix vibrational and electronic states of the donor–acceptor system. In other words, the electronic systems of the two components must interact and some vibrational motions of the two molecules must become strongly coupled." (See Figure 9.5 mode b.) Two other explanations have been proposed for the nonvertical energy transfer process.[1,12,29,30] The "hot band" explanation involves vertical excitations from higher vibrational levels (Figure 9.5, mode c) of the ground state to *lower* levels of the triplet state. The efficiency of energy transfer from the higher vibrational levels would be underestimated if calculated from ΔE where the triplet energy is assumed to be that of the spectroscopic triplet (59 kcal/mole).

The third explanation is based on vertical excitations from the vibration functions outside the potential energy curves of the *cis*-stilbene (Figure 9.5, mode d).

At the present time it is not possible to decide whether one or more of these mechanisms are operative.

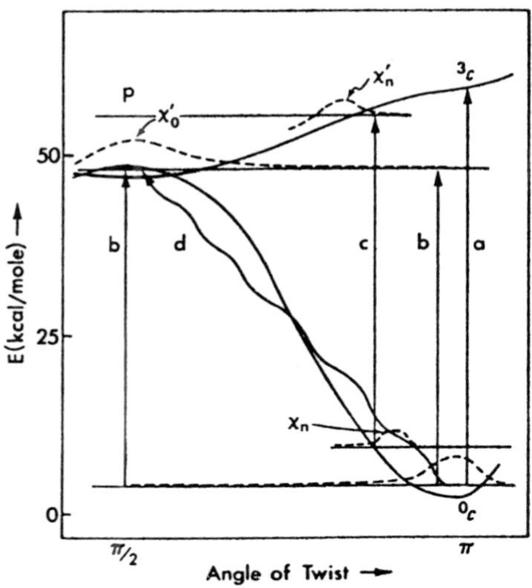

FIGURE 9.5. Nonvertical excitation of *cis*-stilbene. (From Ref. 1.)

9.5. STILBENE ISOMERIZATION VIA DIRECT PHOTOLYSIS

The previous sections have shown that *cis–trans* isomerization of stilbene can take place via the lowest triplet state of stilbene. The question to be considered now is whether the isomerization upon direct photolysis takes place via the singlet state, the triplet state, or a vibrationally excited ground state.[1,7,31–50]

9.5a. *Vibrationally Excited Ground State*

Any mechanism which involves isoenergetic, radiationless internal conversion from C^s, P^s, or T^s to a *high* vibrational level of the ground state would be expected to show a large deuterium isotope effect on the rate of internal conversion. In the direct photolysis of perdeuterio and perhydrostilbene, Saltiel[22] found no isotope effect on the photostationary state or upon the quantum yields of *cis*-to-*trans* and *trans*-to-*cis* conversion.

In addition, this mechanism was criticized on the basis of the Kassel–Rice theory of vibrationally excited ground states some time earlier by Zimmerman *et al.*[34]

9.5b. *Triplet State Mechanism*

A recently popular mechanism involves the intersystem crossing of the *cis*- or *trans*-stilbene singlet state, produced upon direct photolysis, to its corresponding triplet states, which would then undergo the type of reactions given in Eqs. (9.8)–(9.10) and (9.17)–(9.19):

$$T \xrightarrow{h\nu} T^s \tag{9.20}$$

$$C \xrightarrow{h\nu} C^s \tag{9.21}$$

$$T^s \longrightarrow T^t \tag{9.22}$$

$$C^s \longrightarrow C^t \tag{9.23}$$

$$C^t \longrightarrow P^t \qquad [(9.8)]$$

$$T^t \longrightarrow P^t \qquad [(9.9)]$$

$$P^t \longrightarrow T^t \qquad [(9.18)]$$

$$P^t \longrightarrow \alpha T + (1 - \alpha)C \qquad [(9.10)]$$

$$T^t + A \longrightarrow T + A^t \qquad [(9.19)]$$

If we know the extinction coefficients for the two isomers at the wavelength of photolysis (3130), it should be possible to calculate the pss for the direct photolysis from

$$\frac{[C]_s}{[T]_s} = \frac{1 - \alpha}{\alpha} \frac{\varepsilon_T}{\varepsilon_C} \tag{9.24}$$

The value of the pss calculated from Eq. (9.24) is 2.74 and the measured value[7] is 2.64. While this agreement could be taken as evidence in favor of a triplet mechanism, it could also mean that the twist singlet and twist triplet have similar geometries (see Figure 9.2), and therefore similar decay ratios. In this mechanism we have neglected fluorescence from *trans*-stilbene and dihydrophenanthrene formation from *cis*-stilbene. (See Problem 4.)

Evidence that eliminates the triplet mechanism as the mode for the *cis–trans* isomerization of stilbene upon direct photolysis has been provided by azulene quenching studies.[48] Using the experimentally determined decay ratio $\alpha/(1 - \alpha)$ and the triplet mechanism, it is possible to calculate what the effect of azulene is upon the pss. The predicted and observed azulene effects on the direct photoisomerization are shown in Figure 9.6. The failure of the triplet mechanism in predicting the very small changes observed in the pss provides a crucial test that is the basis for rejecting the triplet mechanism.

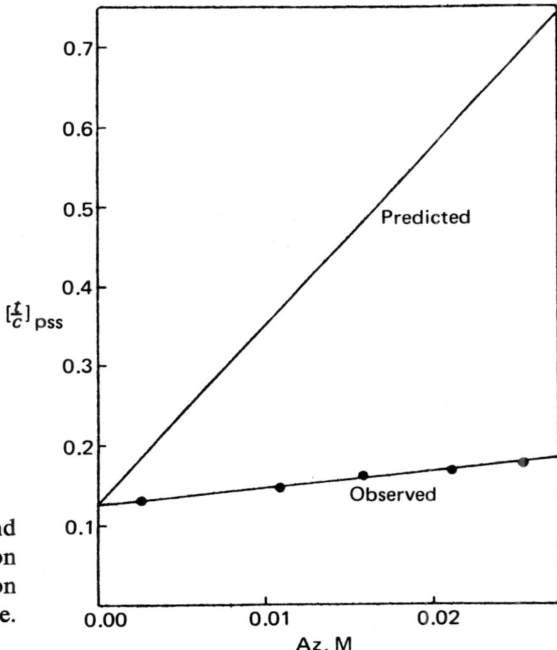

FIGURE 9.6. Predicted and observed azulene effects on the direct photoisomerization of the stilbenes in *n*-pentane. (From Ref. 1.)

The small effect that has been observed can be quantitatively accounted for via Förster-type long-range singlet energy transfer,[1,46,48]

$$T^s + A \longrightarrow T + A^s \tag{9.25}$$

In addition,[1,46,48] it was noted that whereas the azulene effect on the sensitized reaction is sensitive to changes in solvent viscosity, the azulene effect on the direct photoreaction was independent of solvent viscosity, as would be predicted for Förster-type energy transfer. The inescapable conclusion is that *cis–trans* isomerization upon direct irradiation of stilbene takes place in the singlet manifold.

9.5c. *Singlet State Mechanism*[1]

This mechanism involves rotation about the central bond from the *trans* singlet and the *cis* singlet to the twisted singlet state P^s (see Figures 9.2 and 9.4):

$$T \xrightarrow{h\nu} T^s \qquad [(9.20)]$$

$$C \xrightarrow{h\nu} C^s \qquad [(9.21)]$$

$$T^s \longrightarrow P^s \tag{9.26}$$

$$C^s \longrightarrow P^s \tag{9.27}$$

$$P^s \longrightarrow \beta T + (1 - \beta)C \tag{9.28}$$

$$T^s \longrightarrow T + h\nu \tag{9.29}$$

This mechanism is consistent with all of the experimental and theoretical information presented. The fact that no *trans*-stilbene fluorescence is observed when *cis*-stilbene is irradiated would indicate that the following reaction is not important, that is, the twist singlet is much lower in energy than is the *trans* singlet state:

$$P^s \longrightarrow T^s \tag{9.30}$$

Saltiel[1] has pointed out that the experimental study of Lewis, Dalton, and Turro[51] can be interpreted in terms of "nonvertical" singlet energy transfer to produce the twisted or phantom singlet state:

$$T + D^s \longrightarrow T^s \tag{9.31}$$

$$T + D^s \longrightarrow P^s \tag{9.32}$$

$$C + D^s \longrightarrow C^s \tag{9.33}$$

$$C + D^s \longrightarrow P^s \tag{9.34}$$

It is possible to estimate the rate of vertical singlet energy transfer (9.31) and (9.33) (when Förster-type energy transfer is negligible, spectral overlap integral is very small) from the relation

$$k_{et} = k_{diffn}\, e^{-\Delta E/RT} \tag{9.35}$$

where ΔE is the energy deficiency for endothermic energy transfer. When singlet energy transfer is exothermic under these conditions $k_{et} = k_{diffn}$. The calculated rate of singlet energy transfer[1] from chrysene to *cis*- or *trans*-stilbene using Eq. (9.35) with $\Delta E \approx 8.5$ kcal/mole and $k_{diffn} \approx 3.5 \times 10^{10}$ M^{-1} sec^{-1} is 2.5×10^4 M^{-1} sec^{-1}; the observed rates of energy transfer calculated from crysene fluorescence quenching by *cis*- or *trans*-stilbene are 2.5×10^7 and 8.6×10^8 M^{-1} sec^{-1}. Consequently, the tentative conclusion was drawn that *cis*- and *trans*-stilbene are "nonvertical" singlet excitation acceptors.[1,51]

If fluorescence and *cis–trans* isomerization (9.26)–(9.29) are the main competing reactions upon direct excitation, then inhibition of rotation about the central bond should produce an increase in the fluorescence quantum yield. The rigid systems (3) and (4) both have fluorescence quantum yields of 1.0 at room temperature.[44,52] While the fluorescence of *trans*-stilbene is a

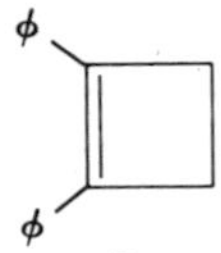

(3)

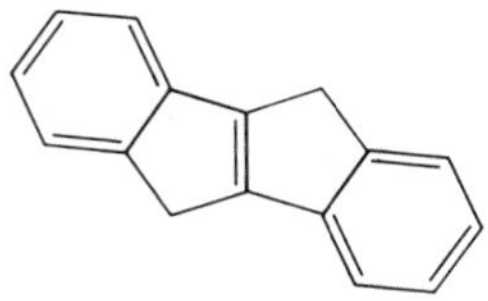

(4)

sensitive function of temperature (viscosity), the fluorescence of (4) in a (6:1) mixture of methylcyclohexane-3-methylpentane is temperature independent in the range 301–77°K.

A more direct test is to examine the fluorescence and *trans–cis* isomerization efficiency of *trans*-stilbene as a function of temperature. Several papers which claimed that uncoupling of fluorescence and *trans–cis* isomerization occurred at high viscosities[41,43] have been shown to be in error[44,47] and the early work of Malkin and Fischer[38] appears to be essentially correct. That is, Φ_f increases continuously as the temperature is decreased (from 0.05 to about 1) while the $\Phi_{T\to C}$ decreases (from about 0.50 to almost zero) (see Figure 9.7). If this had not been the case, it would have been good evidence against the singlet mechanism. As the fluorescence studies stand, they are consistent with the singlet mechanism but they are also consistent with a triplet mechanism where intersystem crossing takes place only from the phantom singlet to the phantom triplet. However, this last mechanism is not viable because of the azulene quenching studies.

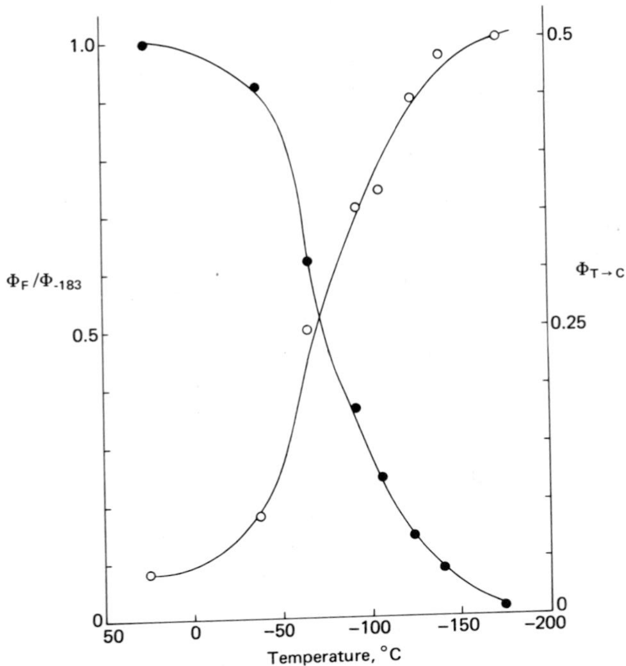

FIGURE 9.7. Temperature dependence of $\Phi_F/\Phi_{-183°C}$ and $\Phi_{T\to C}$ of *trans*-stilbene in methylcyclohexane and isohexane. (Data from Ref. 38; see also Ref. 1.)

9.6. SUBSTITUTED STILBENES

The photochemistry of α-methylstilbene (5) resembles stilbene photochemistry in many ways. However, as pointed out earlier, both the *cis* and *trans* isomers are nonclassical acceptors of triplet excitation. This suggests that both the *cis* and *trans* triplet states correspond to high-energy vibrational levels of the twisted or phantom triplet. Azulene does not alter the photo-

ϕ ϕ
H CH$_3$
(5)

Br
H H
(6)

stationary state ratio for the sensitized isomerization of the α-methylstilbenes. This result is consistent with the phantom triplet picture but it means that the most useful method for the comparison of the sensitized and direct photoreactions is not applicable to this system.

While a complete study of the photoreactions of 4-bromostilbene (6) has not been published yet, preliminary azulene quenching results[1] suggest that upon direct photoreaction about 35% of the *trans* singlets intersystem cross to the triplet state:

$$T^s \longrightarrow T^t \tag{9.36}$$

In addition, 20–30% of the phantom singlets intersystem cross to the phantom triplet state:

$$P^s \longrightarrow P^t \quad . \tag{9.37}$$

Thus, upon direct excitation of 4-bromostilbene, *cis–trans* isomerization takes place in both the singlet and triplet manifolds.[1,16,38,53]

Other related systems of some interest are the *cis*- and *trans*-β-styrylnaphthalenes (7),[54] the stilbazoles, and 1,2-bispyridylethylenes (8 and 9).[53,56]

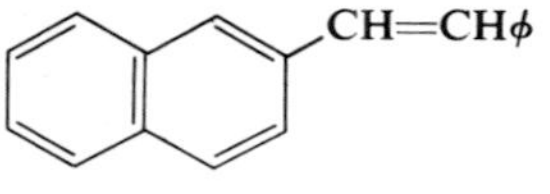

CH=CHϕ

(7)

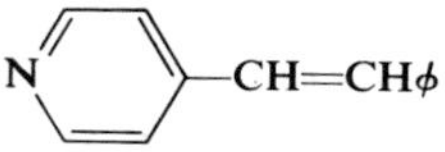

N CH=CHϕ

(8)

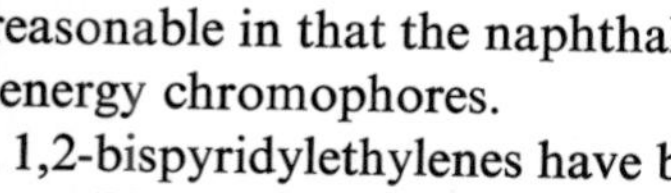

N CH=CH N

(9)

In the case of β-styrylnaphthalene, direct or sensitized excitation apparently does not involve exclusive formation of a single excited state from the two isomers.[54] This seems reasonable in that the naphthalene and styryl moieties can act as coupled low-energy chromophores.

The stilbazoles and 1,2-bispyridylethylenes have been extensively studied by Whitten and co-workers,[55] who find a reasonably efficient ($\Phi_{T\to C} = 0.44$, $\Phi_{C\to T} = 0.34$; 2-stilbazole) benzophenone-sensitized photoisomerization but a very inefficient direct photoisomerization ($\Phi_{T\to C} = 0.25$; 2-stilbazole; $\Phi_{T\to C} = 0.003$, 4,4′-bispyridylethylene). It was shown that internal conversion from the singlet to ground state not involving geometric change is important for the stilbazoles and bispyridylethylenes. In this respect the azastilbenes resemble pyridine and other *N*-heteroaromatics in which internal conversion is an important process. Azulene quenching experiments of direct and sensitized photoisomerization of two azastilbenes again indicate that direct *cis–trans* photoisomerization does not take place via the triplet state.[35]

9.7. PIPERYLENE PHOTOCHEMISTRY[7,57,58]

The direct excitation of *cis-* or *trans*-piperylene (1,3-pentadienes) at 254 nm inefficiently results in *cis–trans* isomerization,[59,60] $\Phi_{C\rightarrow T} = 0.09$ and $\Phi_{T\rightarrow C} = 0.010$; and very small amounts of 3-methylcyclobutene and 1,3-dimethyl-cyclopropene[60] [(DMCP); $\Phi_{T\rightarrow DMCP} = 0.002$; $\Phi_{C\rightarrow DMCP} = 0.009$] are formed.

These results were interpreted in terms of a cyclopropylmethylene mechanism[61]:

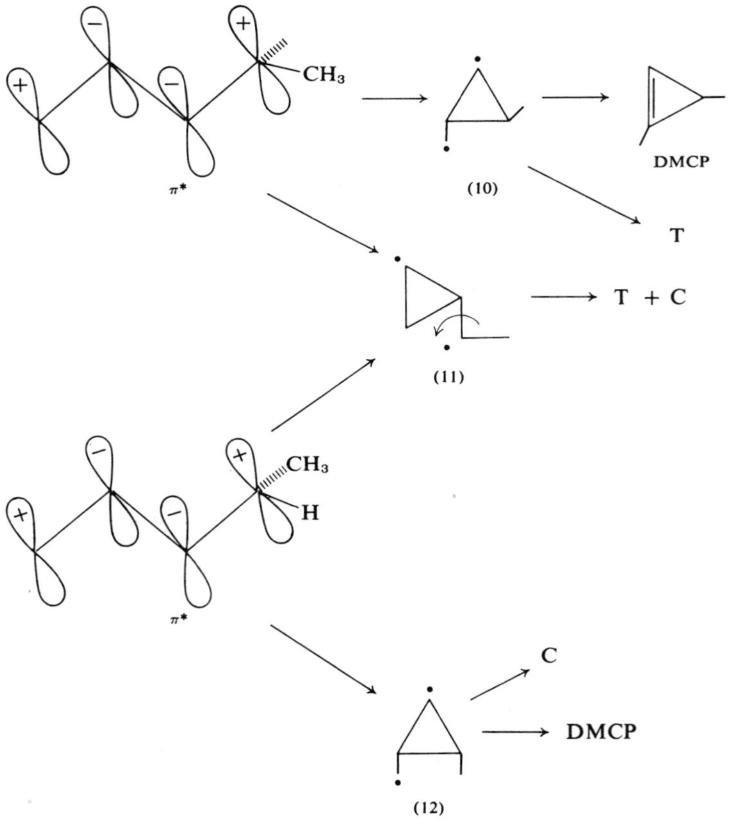

If the formation of (10) and (12) is more efficient than the formation of (11), then this mechanism accounts for the low quantum yield of stereo-isomerization.

While the direct *cis–trans* isomerization is inefficient, the sensitized reactions are very efficient[17]; $\Phi_{C\rightarrow T} = 0.55$ and $\Phi_{T\rightarrow C} = 0.44$, sensitizer is benzophenone. As for stilbene, the sum of the quantum yields for *cis*-to-*trans* isomerization and *trans*-to-*cis* isomerization is equal to unity:

$$\Phi_{C \to T} + \Phi_{T \to C} = 0.99 \pm 0.02 \tag{9.38}$$

Also, the photostationary state can be predicted from the ratio of the quantum yields:

$$\Phi_{C \to T}/\Phi_{T \to C} = 1.25 \tag{9.39}$$

$$[T]_s/[C]_s = 1.22 \pm 0.05 \tag{9.40}$$

To simplify the discussion, the following notation will be adopted:

t–T ≡ (13)

c–T ≡ (14)

t–C ≡ (15)

c–C ≡ (16)

The first letter represents the conformation about the carbon–carbon single bond and the second letter represents the configuration about the carbon–carbon double bond.

For sensitizers with triplet energies greater than 60 kcal/mole the following mechanism is consistent with the experimental observations:

$$D + h\nu \longrightarrow D^s \tag{9.41}$$

$$D^s \longrightarrow D^t \tag{9.42}$$

$$D^t + \text{t–T} \longrightarrow D + X \tag{9.43}$$

$$D^t + \text{t–C} \longrightarrow D + X \tag{9.44}$$

$$X \longrightarrow \text{t–T} \tag{9.45}$$

$$X \longrightarrow \text{t–C} \tag{9.46}$$

Since energy transfer from high triplet energy donors to both s–*trans* and s–*cis* conformations should be diffusion controlled, and the piperylene is more than 95% in the s–*trans* conformation at room temperature, the role of the c–T and c–C isomers can be neglected to a first approximation.

An important feature of the mechanism suggested by Hammond and co-workers was the increased bond order of the carbon–carbon single bond in the excited state. This is shown in the accompanying scheme.

c–C ⇌ t–C; c–T ⇌ t–T

triplet energy transfer

$\begin{Bmatrix} \text{X from Eqs.} \\ (9.43), (9.44) \end{Bmatrix}$

c–T + c–C

t–T + t–C

This system has been used to determine intersystem crossing quantum yields for high triplet energy sensitizers (E_t > 60 kcal/mole).[62] With sensitizers having lower triplet energies (50–61 kcal/mole) there is a change in the photo-equilibrium (see Figure 9.8). Two factors are involved: (a) preferential excitation of the *cis* isomer (t–C), which has a low triplet energy (~57 vs. 59), and (b) preferential excitation of the s–*cis* isomers (c–T and c–C). Even though the s–*cis* isomers are a minor component of the system at room temperature, their importance is obvious from Hammond's studies of butadiene dimerization discussed in Chapter 10.

The 2,4-hexadienes (17)–(19) are of interest because, if a common triplet intermediate (like X, a 1,4 diradical) is involved, the sum of all six quantum yields should be equal to 2, and the individual quantum yields should be predictable from the composition at the photostationary state[63,64]:

T–T (17) T–C (18) C–C (19)

$$\Phi_{TT\to CT} = \Phi_{CC\to CT} = [\%C\text{–}T]_s \times 10^{-2} \quad (9.47)$$

$$\Phi_{CT\to TT} = \Phi_{CC\to TT} = [\%T\text{–}T]_s \times 10^{-2} \quad (9.48)$$

$$\Phi_{CT\to CC} = \Phi_{TT\to CC} = [\%C\text{–}C]_s \times 10^{-2} \quad (9.49)$$

The photostationary state composition for the benzophenone-sensitized isomerization of 2,4-hexadienes is given in Table 9.2. Table 9.3 gives the measured quantum yields for benzophenone-sensitized isomerization of 2,4-hexadienes along with the calculated quantum yields based on Eqs. (9.47)–(9.49) and the pss values given in Table 9.2.

While the 1,4 biradical (20) mechanism adequately accounts for the observed quantum yields, it is not possible to eliminate a mechanism involving

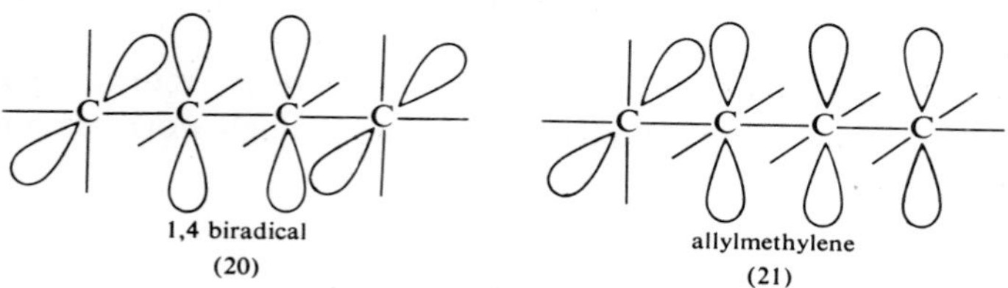

a pair of rapidly interconverting allylmethylene (21) triplets. In fact, molecular orbital theory at several levels of approximation suggests that (21) should be a lower energy state than (20).[66–68]

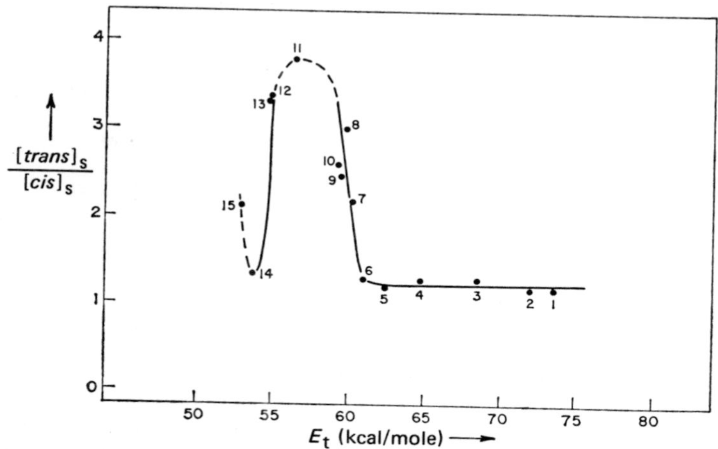

FIGURE 9.8. Piperylene photostationary states. (From Ref. 7.)

TABLE 9.2. *The pss for Benzophenone-Sensitized Isomerization of 2,4-Hexadienes*[64]

Diene (*M*)	% T–T	% C–T	% C–C
T–T (0.09)	31.3 ± 0.3	50.2 ± 0.1	18.5 ± 0.2
C–C (0.09)	31.3 ± 0.2	50.2 ± 0.4	18.7 ± 0.2

TABLE 9.3. *Quantum Yields of Sensitized Isomerization of 2,4-Hexadienes*[64]

	$\Phi_{TT\to CT}$	$\Phi_{CC\to CT}$	$\Phi_{CT\to TT}$	$\Phi_{CC\to TT}$	$\Phi_{CT\to CC}$	$\Phi_{TT\to CC}$	$\sum \Phi$
Observed	0.48	0.50	0.33	0.29	0.16	0.18	1.85
	±0.02	±0.01	±0.03	±0.01	±0.01	±0.02	±0.10
Calculated	0.50	0.50	0.31	0.31	0.19	0.19	2.00

Preliminary observations indicate that stilbenes can act as quenchers for piperylene triplets.[69] This, along with the effect of low-energy sensitizers upon piperylene or 2,4-hexadienes, is another area that needs more experimental work before a complete picture can be obtained.

Several recent studies on trienes,[70,71] vitamin A, retinal, and higher carotenoids[72–75] have been published; however, much work remains to be done in this area.

9.8. ALKENE PHOTOISOMERIZATION

The direct irradiation of *cis*- and *trans*-2-butenes in solution separately and in admixture gives stereospecific dimerization along with the more efficient *cis–trans* isomerization[76]:

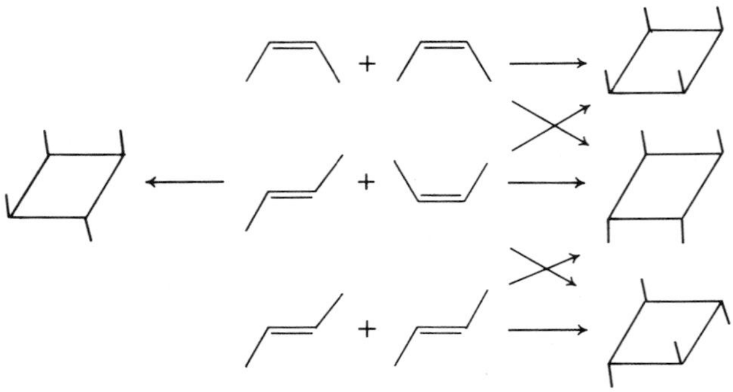

(9.50)

This is likely a $[_{\pi}2_s + _{\pi}2_s]$ cycloaddition from the first excited singlet state, and suggests that this state either does not have the twisted geometry noted for the singlet state of stilbene or that bimolecular reaction is faster than decay to a common twisted singlet state. Since the dimerization is most efficient in neat *cis*- or *trans*-2-butene a third possibility which we favor is that the dimerization takes place via a singlet excimer while isomerization is from the "free" singlet state.

It has been assumed[1,77] that dimerization and isomerization take place via the "free" singlet state, using the following scheme:

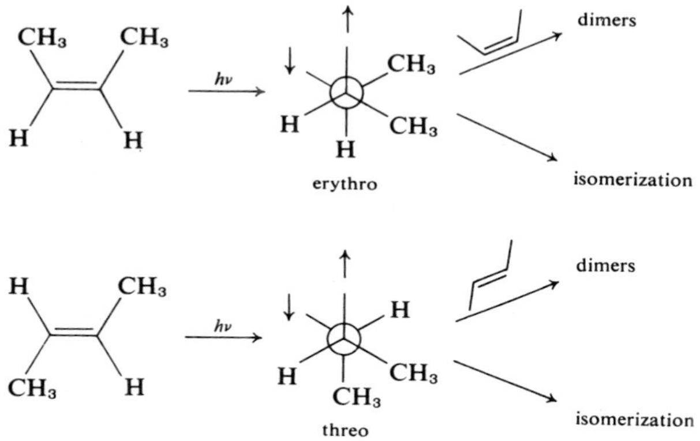

The sensitized photoisomerization of alkenes is complex, with a variety of sensitizer–olefin interactions operative.[78–83]

For simple alkenes, Saltiel has proposed that the deviation of the decay ratio from unity is a measure of the involvement of Schenck type intermediates (22) in the *cis–trans* photosensitized isomerization. Table 9.4 gives pss concentration ratios for the 2-pentenes at 30°C:

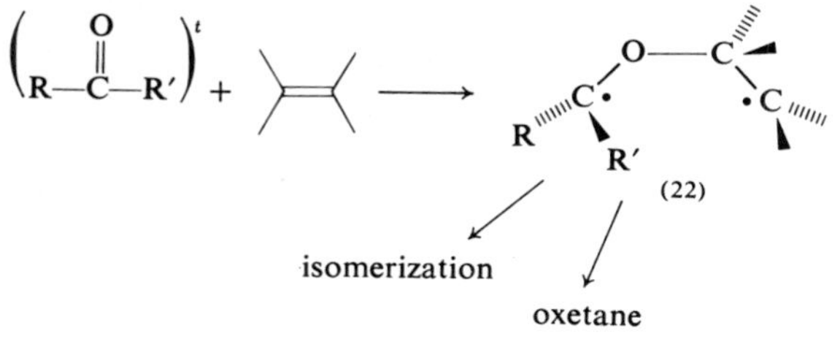

Thus Saltiel has concluded that the small increase in $[T]_s/[C]_s$ in going from benzene to acetone indicates that a mixed mechanism is operative for acetone-sensitized isomerization, that is, both triplet energy transfer and, to a minor extent, Schenck intermediates are involved. When acetophenone or benzophenone is used as a sensitizer the pss is close to the thermodynamic

TABLE 9.4. *The pss for 2-Pentenes*

Sensitizer	Triplet energy, kcal/mole	$[T]_s/[C]_s$
Benzene	84	1.0
Acetone	80	1.52
Acetophenone	72	5.4
Benzophenone	68	7.2

equilibrium distribution. Consequently, for these cases a Schenck intermediate is invoked.

Yang has observed that *cis–trans* isomerization of 3-methyl-2-pentenes is accompanied by oxetane formation and concluded that intermediates such as (22) are common to both isomerization and oxetane formation.[82] Deuterium isotope effects are also consistent with the involvement of this type of intermediate.[83,84]*

Another related mechanistic proposal is that oxetane cleavage can produce the excited olefin. This is based on the exchange results shown in the following scheme[88]:

9.9. INTRAMOLECULAR CYCLOADDITION REACTIONS

9.9a. *Theory*

The theory of electrocyclic reactions has had a dramatic impact on synthetic and physical organic chemistry. It will not be possible to develop the theory in detail here, and the interested reader is referred to

* For an elegant study of the secondary deuterium isotope effects on a non-Schenck sensitized *cis-trans* isomerization, see Refs. 85–87.

several good reviews.[(89–94)] We will examine the following simple ${}_{\pi}2_s + {}_{\pi}2_s$ cycloaddition:

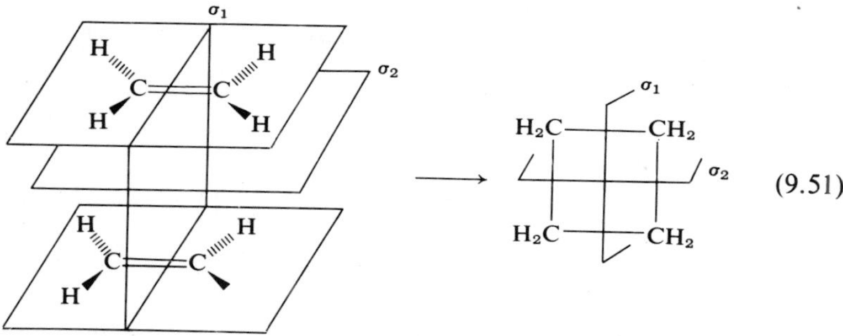

(9.51)

We can use the planes of symmetry σ_1 and σ_2 to characterize the π, π^*, σ, and σ^* orbitals taking part in the cycloaddition. The orbitals shown in Figure 9.9 are projections upon a plane passing through the four carbon atoms. Since we have four atomic orbitals making up the two π bonds, we can form four new molecular orbitals shown in Figure 9.9, where the symmetric–symmetric (SS) combination will be slightly lower in energy than the SA combination and the AS antibonding combination is lower than the AA combination. The relative energies are shown on the left-hand side of the correlation energy diagram in Figure 9.11. For the orbitals making up the σ bonds in cyclobutane shown in Figure 9.10 the SS combination is lower in energy than the AS and the SA is lower than the AA. The relative energies are shown on the right-hand side of the correlation diagram (Figure 9.11). The orbitals that undergo direct changes are the π_1, π_2, π_1^*, π_2^*, σ_1, σ_2, σ_1^*, and σ_2^*. The two planes of symmetry σ_1 and σ_2 are maintained throughout the course

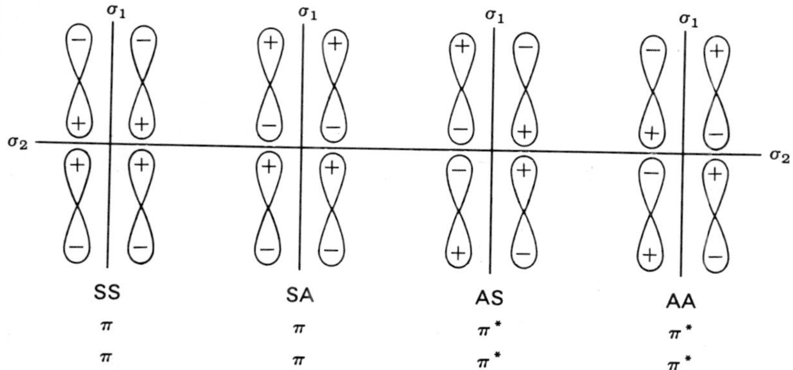

FIGURE 9.9. The π bases orbitals for a ${}_{\pi}2_s + {}_{\pi}2_s$ cycloaddition; S, A indicate symmetric or antisymmetric reflection through the symmetry plane σ_1 or σ_2.

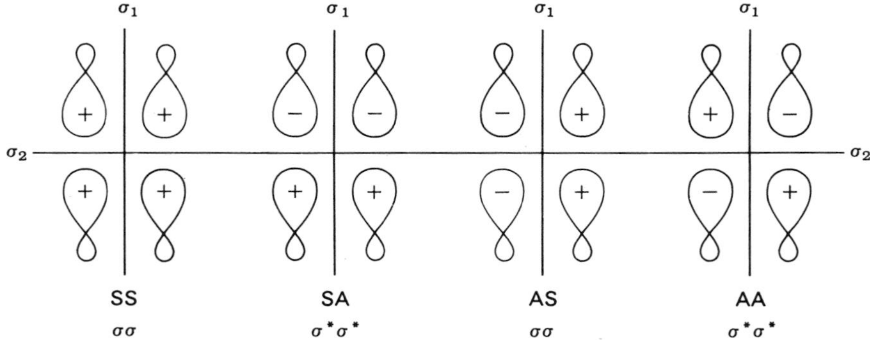

FIGURE 9.10. The σ bases orbitals for a ${}_{\pi}2_s + {}_{\pi}2_s$ cycloaddition.

of the reaction, so that orbital symmetry is conserved. Consequently, an SS transforms to an SS, an SA to an SA. If, for example, there had been four SS levels in the diagram, the correlation lines would be drawn so that there would be no crossing of lines with the same symmetry labels. Note that the ethylene ground state $\pi_1{}^2\pi_2{}^2$ correlates with a doubly excited state of cyclobutane $(\sigma_1)^2(\sigma_1{}^*)^2$. This is clearly an unfavorable or forbidden process. The first excited state $(\pi_1)^2(\pi_2)^1(\pi_1{}^*)^1$ correlates with the first excited state of the product $(\sigma_1)^2(\sigma_2)^1(\sigma_1{}^*)^1$ and to a first approximation there is no net gain or loss in energy. The photochemical ${}_{\pi}2_s + {}_{\pi}2_s$ cycloaddition is an allowed transformation.

It will be helpful to construct a state diagram from the orbital diagram.

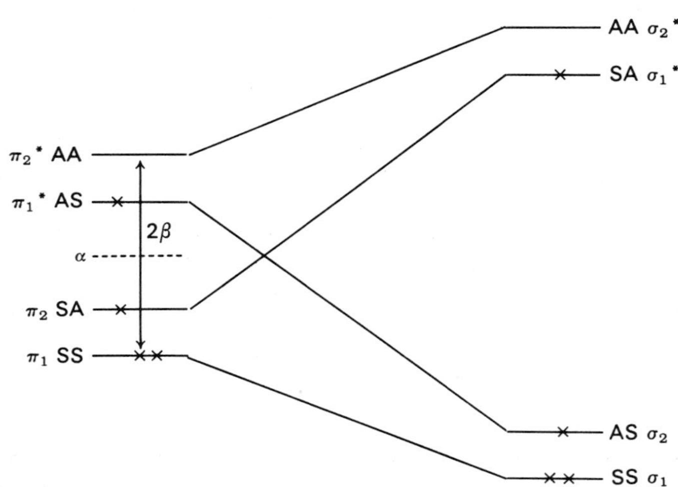

FIGURE 9.11. Correlation diagram for the photoallowed ${}_{\pi}2_s + {}_{\pi}2_s$ cycloaddition.

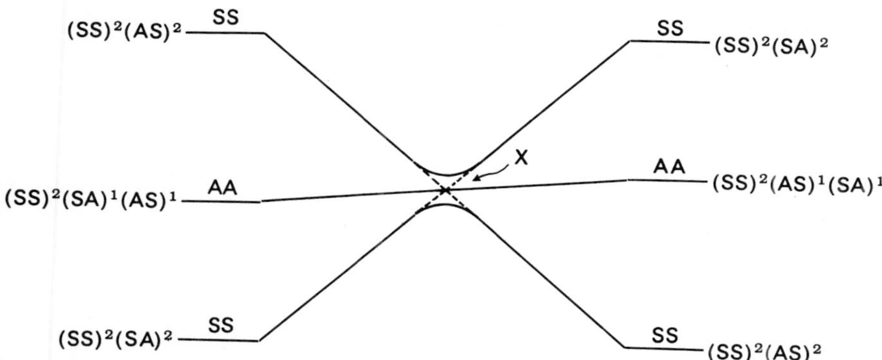

FIGURE 9.12. Electronic state diagram for the formation of cyclobutane from ethylene.

The state symmetries are derived from the product of the orbital symmetries using the following rules:

$$\mathrm{S \times S = S = A \times A, \qquad S \times A = A = A \times S}$$

Thus the ground state for the two ethylenes is

$$\mathrm{(SS)^2(SA)^2 \equiv SS}$$

The other states are shown in Figure 9.12. From Figure 9.12 it is seen that the ground state electron configuration of two ethylene molecules correlates with a doubly excited state of cyclobutane and the ground state of cyclobutane correlates with a doubly excited state of the two ethylenes. Since two states with the same symmetry cannot cross (noncrossing rule), electron interaction causes a correlation of one ground state with the other. However, there is a large activation energy for this process (that is, it is forbidden). The first excited state of the two ethylenes correlates with the first excited state of cyclobutane. This diagram, then, in many ways resembles the diagram for stilbene *cis–trans* isomerization. Thus it is not surprising that the photodimerization of tetramethylethylene does not produce the excited state of cyclobutane but gives directly the ground state. On proceeding from state $\mathrm{(SS)^2(SA)^1(AS)^1}$ to $\mathrm{(SS)^2(AS)^1(SA)^1}$ there is a high probability of internal conversion at point X (Figure 9.12) to produce a ground state species.

We have been discussing a cycloaddition where bonds are made or broken on the same face (suprafacial process). The alternative process is one where the bonds are made or broken on opposite faces of the reacting system (antarafacial):

suprafacial　　　antarafacial

Without going through the relevant correlation diagrams, two 2 + 2 and three 4 + 2 additions are shown below:

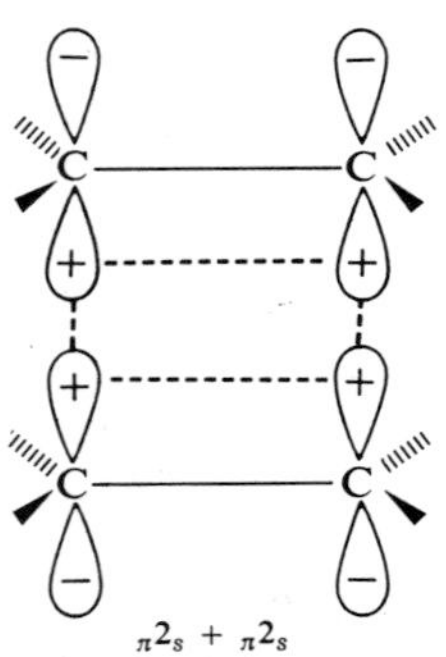

$_{\pi}2_s + _{\pi}2_s$
0 nodes, 4 electrons
$h\nu$ allowed, Δ forbidden

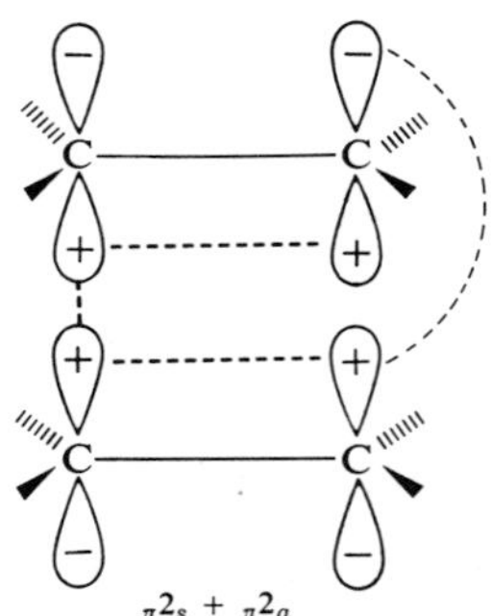

$_{\pi}2_s + _{\pi}2_a$
1 node, 4 electrons
Δ allowed, $h\nu$ forbidden

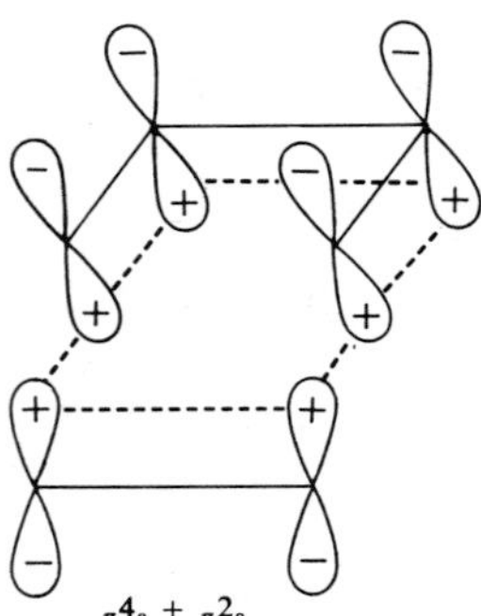

$_{\pi}4_s + _{\pi}2_s$
0 nodes, 6 electrons
Δ allowed, $h\nu$ forbidden

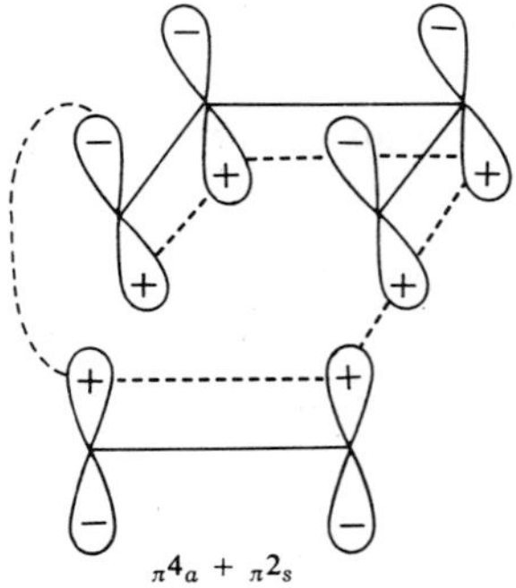

$_{\pi}4_a + _{\pi}2_s$
1 node, 6 electrons
$h\nu$ allowed, Δ forbidden

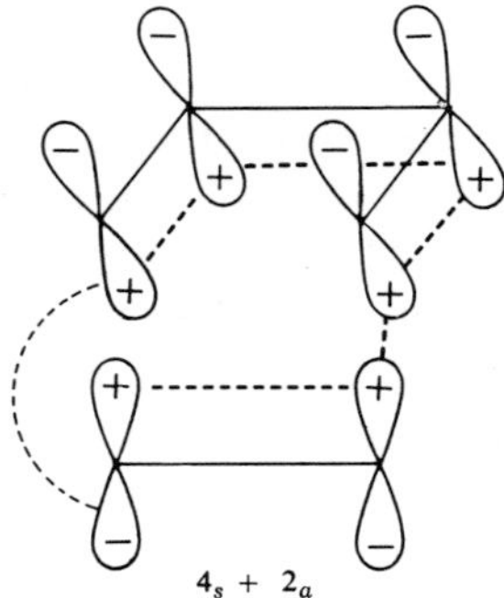

$4_s + 2_a$
1 node, 6 electrons
$h\nu$ allowed, Δ forbidden

TABLE 9.5. *Selection Rules for Cycloaddition Reactions*[a]

$m + n$	Thermal allowed, photo forbidden	Photo allowed, thermal forbidden
$4q$	$s + a$	$s + s$
	$a + s$	$a + a$
$4q + 2$	$s + s$	$s + a$
	$a + a$	$a + s$

[a] s = supra, a = antara, for addition of an m to an n electron system, where q is one of the integers 1, 2, 3.

Using the nomenclature of Dewar and Zimmerman, the transition state for the ${}_{\pi}2_s + {}_{\pi}2_s$ cycloaddition is a $4n$ Hückel system (zero nodes) and is antiaromatic in the ground state and aromatic in the excited state. The transition state for the ${}_{\pi}2_s + {}_{\pi}2_a$ cycloaddition is a $4n$ Möbius system (one node) and is aromatic in the ground state and antiaromatic in the excited state (see Chapter 8). The general cycloaddition rules are given in Table 9.5.

For an example of an orbital correlation diagram for a ${}_{\pi}4_s + {}_{\pi}4_s$ cycloaddition see Chapter 2 (Figure 2.17).

It is interesting to note that the most reasonable geometry for the ${}_{\pi}2_s + {}_{\pi}2_a$ cycloaddition is one where the two double bonds are orthogonal:

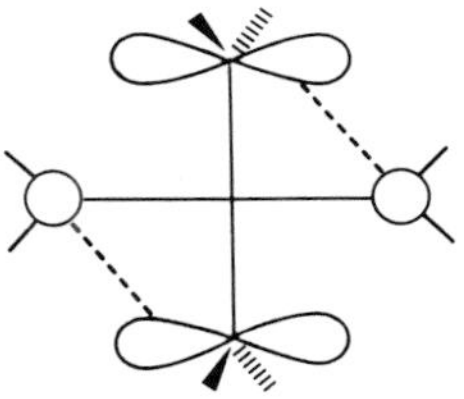

9.9b. *Intramolecular (2 + 2) Cycloadditions and Cycloreversion Reactions*

We have already discussed in this chapter [Eq. (9.50)] the 2 + 2 cycloaddition of *cis*- and *trans*-2-butenes. A number of intramolecular examples are shown below, in some of which cases the reactions may not be concerted:

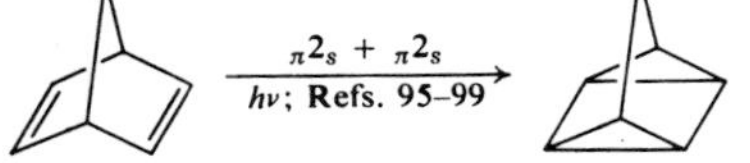

${}_{\pi}2_s + {}_{\pi}2_s$; $h\nu$; Refs. 95–99 ; 67% (9.52)

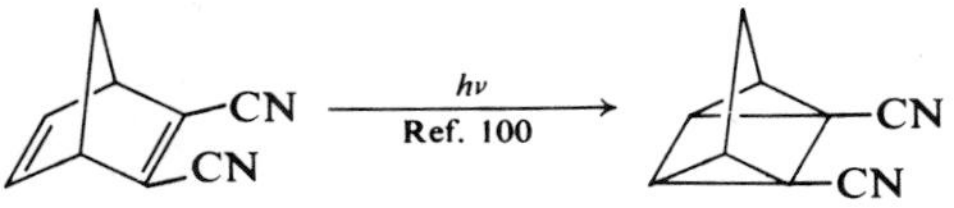

CN, CN ; $h\nu$; Ref. 100 ; CN, CN (9.53)

O COOMe COOMe $\xrightarrow[\text{Ref. 101}]{h\nu}$ O COOMe COOMe (9.54)

$\xrightarrow[\text{Ref. 102}]{h\nu}$ 100% (9.55)

H H $\xrightarrow[\text{acetone; 18 hr; Refs. 103, 104}]{h\nu}$ (9.56)

O O $\xrightarrow[\text{6 hr; Refs. 105, 106; 90\%}]{h\nu\text{, EtOAc}}$ O O (9.57)

R_2 R_1 COOMe COOMe $\underset{\Delta\text{; Ref. 98}}{\overset{h\nu}{\rightleftharpoons}}$ R_2 R_1 COOMe COOMe (9.58)

$R_1 = R_2 = \phi$
$R_1 = R_2 = \text{Me}$

H H O $\underset{\Delta\text{; Ref. 107}}{\overset{h\nu}{\rightleftharpoons}}$ O (9.59)

$h\nu$ Refs. 108, 109 31% (9.60)

HO H

H OH

$h\nu$ 29%

HO H

$h\nu$ Ref. 110 (9.61)

COOMe COOMe $h\nu$ Ref. 111 COOMe COOMe (9.62)

COOMe COOMe $h\nu$ Ref. 112 COOMe COOMe (9.63)

CF_3 CF_3 $h\nu$ Ref. 113 CF_3 CF_3 (9.64)

(9.65)

Kaupp and Prinzbach have shown that the sum of the quantum yields for the forward and reverse transformations of a number of norbornadienes and quadricyclanes is about unity(115–117):

(9.66)

Because of this they suggest that a common intermediate is involved in the reaction (23):

(23)

A concerted mechanism could give $\Phi_{\text{forward}} + \Phi_{\text{reverse}}$ values as large as 2.

A number of $2\pi + 2\sigma$ reactions have been recently reported. Some of these follow:

(9.67)

(9.68)

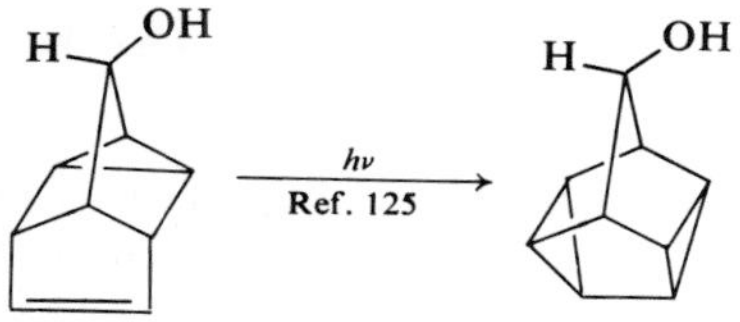
H
OH
hν
Ref. 125
H
OH
(9.69)

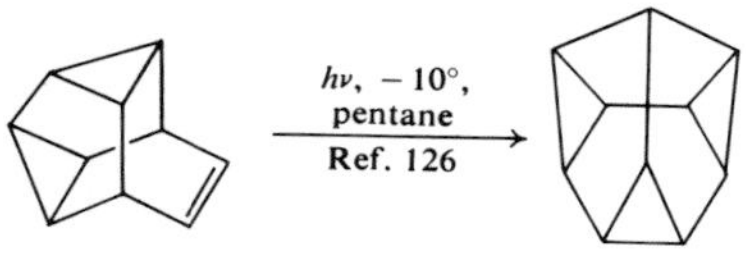
hν, − 10°,
pentane
Ref. 126
(9.70)

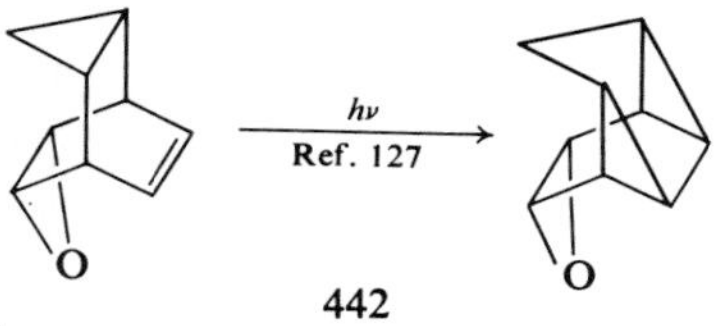
hν
Ref. 127
O
O
(9.71)

442

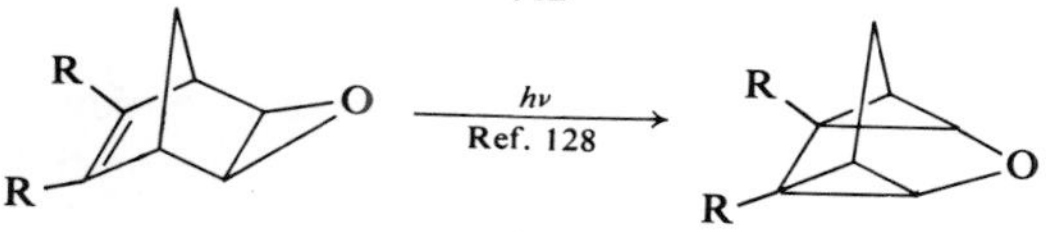
R
R
O
hν
Ref. 128
R
R
O
(9.72)

It is interesting to note that the following phototransformation does not take place:

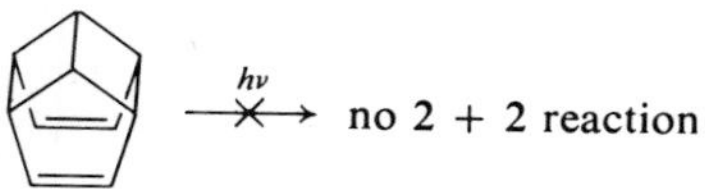
hν
no 2 + 2 reaction
(9.73)

Schmidt and Wilkins proposed that the reaction is rendered symmetry forbidden by through-bond coupling of the carbon–carbon π bonds with high-lying σ levels.[(129)]

Saltiel and Nghim have provided a nice example of the photochemical ${}_{\sigma}2_s + {}_{\sigma}2_s$ cycloreversion[(130)]:

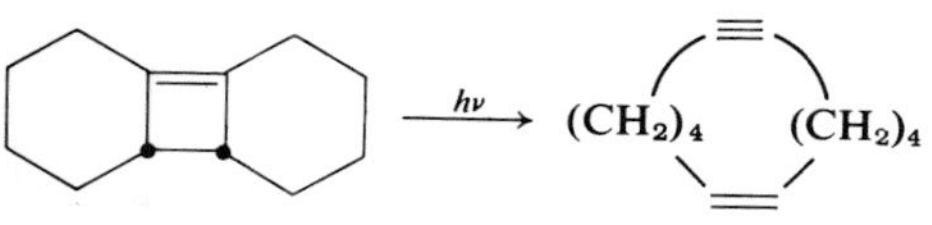
hν
(CH2)4
(CH2)4
(9.74)

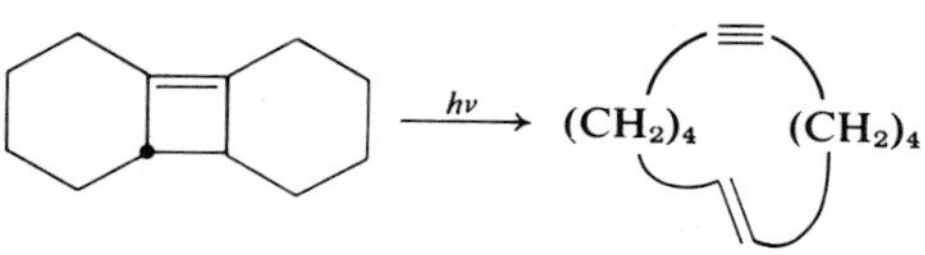
hν
(CH2)4
(CH2)4
(9.75)

9.9c. *Intramolecular (4 + 2) Photocycloaddition Reactions*

A number of photochemically induced ${}_{\pi}4 + {}_{\pi}2$ cycloadditions have been observed.[131–150] This reaction can be either a concerted $({}_{\pi}4_s + {}_{\pi}2_a)$ or a $({}_{\pi}4_a + {}_{\pi}2_s)$ addition. The stereochemical consequences of the allowed reactions are as follows:

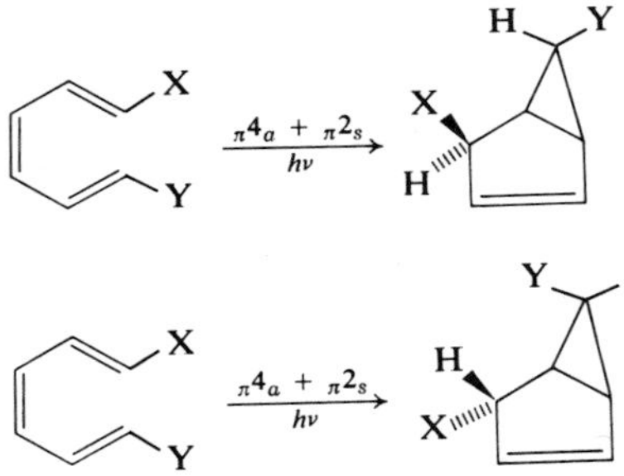

$\xrightarrow[h\nu]{{}_{\pi}4_a + {}_{\pi}2_s}$ (9.76)

One of the first examples of the formation of the bicyclo[3.1.0]hex-2-ene structure from an acyclic conjugated triene was the formation of suprasterol I (24) and suprasterol II (25) from vitamin D_2[131]:

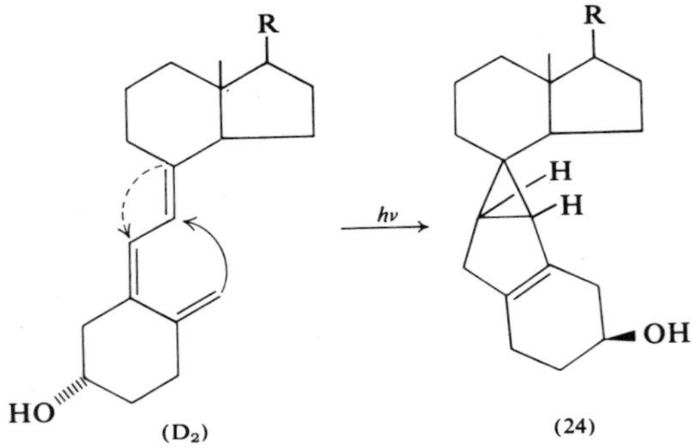

(D₂) $\xrightarrow{h\nu}$ (24) (9.77)

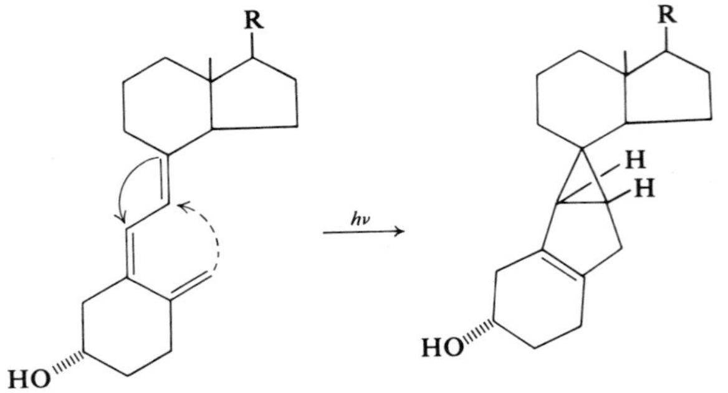

(D₂) $\xrightarrow{h\nu}$ (25) (9.78)

It is known from x-ray studies that the most stable conformation of vitamin D_2 is that shown in Eq. (9.77), s-*trans*, s-*cis* and not the s-*cis*, s-*cis* structure (26).

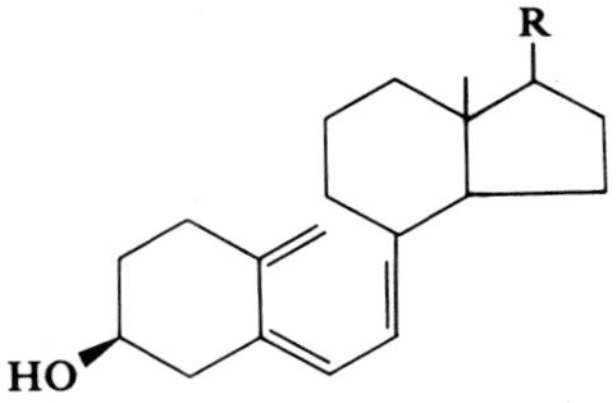

(26)

Dauben *et al.* have shown from the study of simple model trienes that this conformation favors the formation of the ($_{\pi}4$ + $_{\pi}2$) product[(135)] (see Table 9.6):

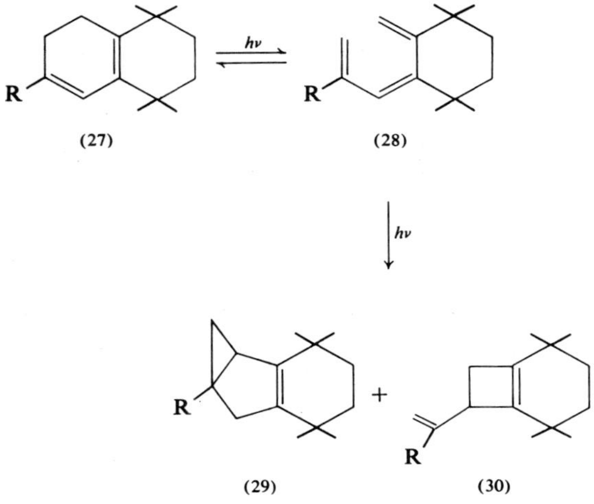

(27) (28)

(29) (30)

(9.79)

TABLE 9.6. *Effect of R on the Photoreaction of (28)*[(135)]

R	$t_{1/2}$, hr	(29)/(30)
H	0.4	2.3
Me	2.1	1.0
i-Pr	8.5	0.3
t-Bu	43.5	0.2

Several other examples are as follows[139,141,142,150]:

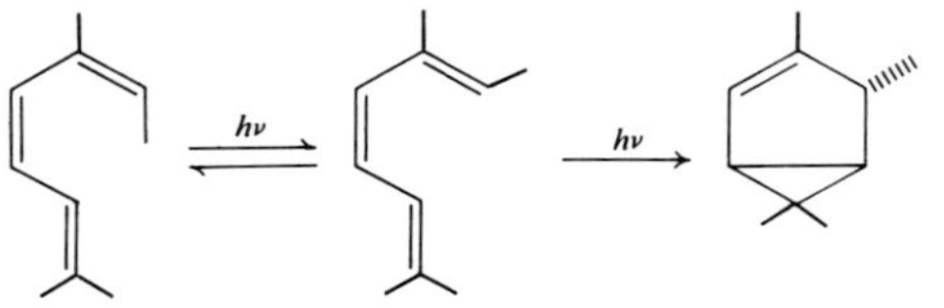

(9.80)

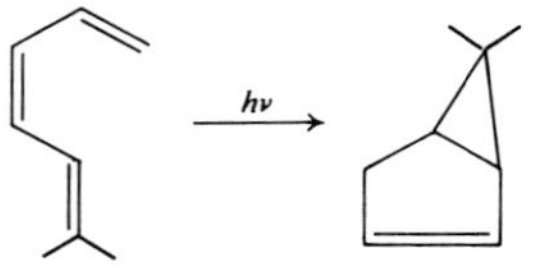

(9.81)

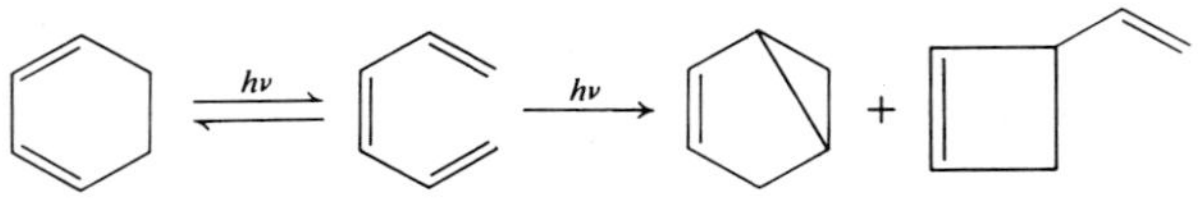

(9.82)

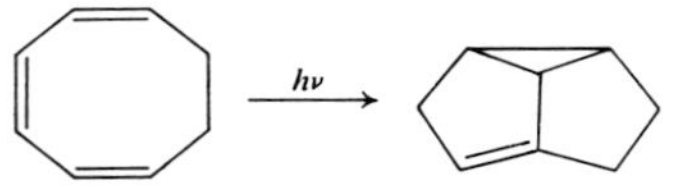

(9.83)

Padwa and Clough[145] examined the photochemistry of (31) and (32) and found that they did not give the expected products from either a simple $_{\pi}4_s + _{\pi}2_a$ or a $_{\pi}4_a + _{\pi}2_s$ cycloaddition:

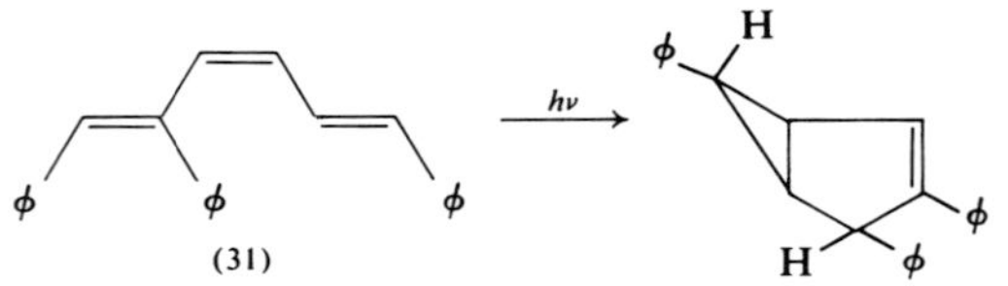

(31) (33) (9.84a)

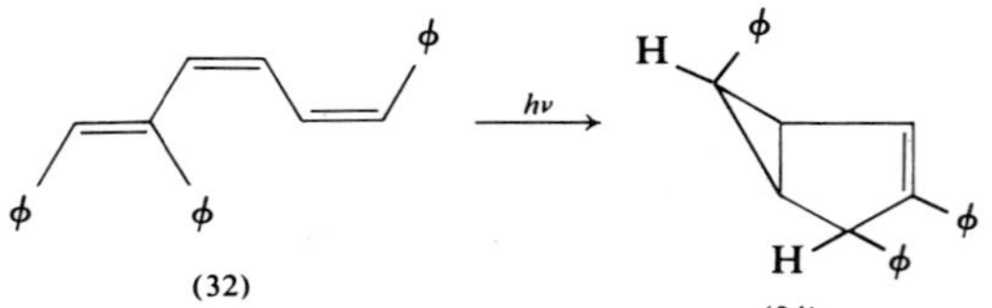

(32) (34) (9.84b)

Compound (31) forms the thermodynamically most favored isomer (33). It was for this reason that (32) was studied. Since (32) only formed (34), product stability does not govern the photoreaction.

The authors propose that the above observations can be rationalized if *cis–trans* isomerization precedes ${}_{\pi}4_s + {}_{\pi}2_a$ photoaddition.[145] It also seems reasonable to propose that a twisted or phantom singlet could be involved in this photoaddition reaction. In this case two photons would not be required. The nodal structure of the lowest π^* level is shown in structure (35)[145]:

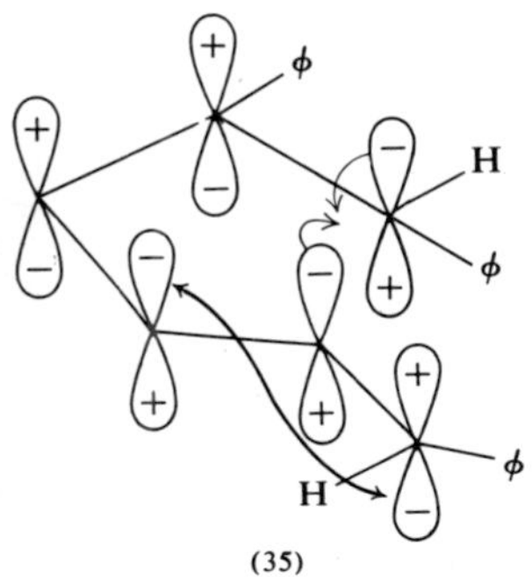

(35)

Dauben has provided a recent example where the ${}_{\pi}4_s + {}_{\pi}2_a$ or the ${}_{\pi}4_a + {}_{\pi}2_s$ mechanism fails to predict the observed products. Photolysis of (36) produces the following photoequilibrium with the triene:

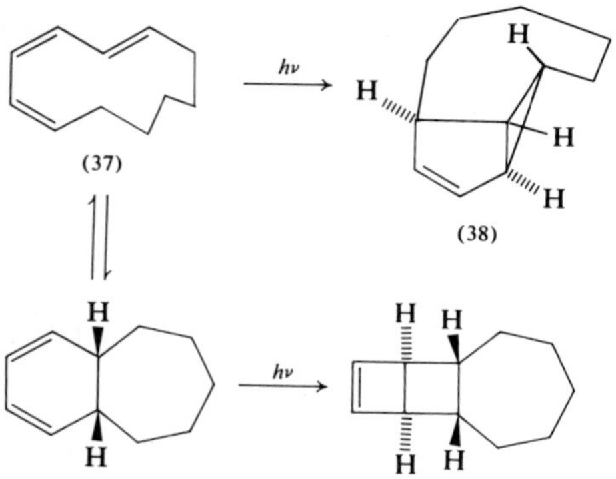

hν

(37)

(38)

(9.85)

hν

(36)

The expected product upon the photolysis of (37) is (39) and not the product observed (38) (see structure at top of page 402).

If prior *cis–trans* isomerization resulted in the formation of (40), compound (38) could be formed from (40) by a ${}_{\pi}4_s + {}_{\pi}2_a$ process. However, Dauben and co-workers found no prior buildup of this triene and since the ${}_{\pi}4_a + {}_{\pi}2_a$ is improbable on steric grounds, concluded that the reaction is not concerted. Another possible solution could be that the addition takes place via the twisted singlet of (37) to produce (38) directly in a ${}_{\pi}4_s + {}_{\pi}2_a$ process.

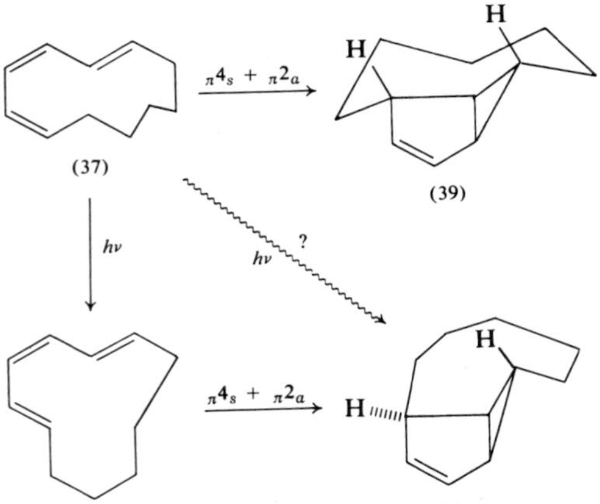

(37) $\xrightarrow{\pi 4_s + \pi 2_a}$ (39)

(37) $\xrightarrow{h\nu}$ (40); (37) $\xrightarrow{h\nu\ ?}$ (38)

(40) $\xrightarrow{\pi 4_s + \pi 2_a}$ (38)

9.10. PHOTOELECTROCYCLIC REACTIONS

9.10a. *Theory*(89–92)

Woodward and Hoffmann define as electrocyclic reactions the formation of a single bond between the termini of a system containing k π electrons, and the reverse process(89):

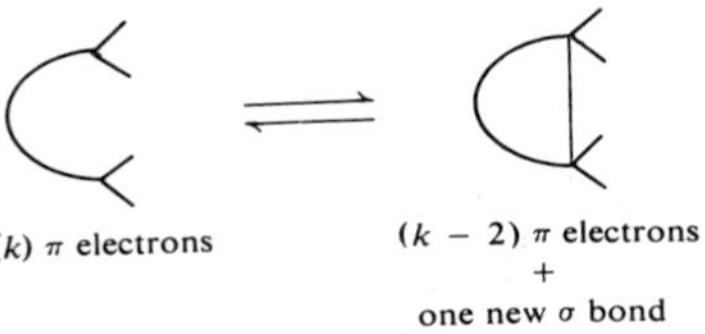

$$(k)\ \pi \text{ electrons} \rightleftharpoons (k-2)\ \pi \text{ electrons} + \text{one new } \sigma \text{ bond} \tag{9.86}$$

Examples of this type of process are the conrotatory and disrotatory reactions of butadiene:

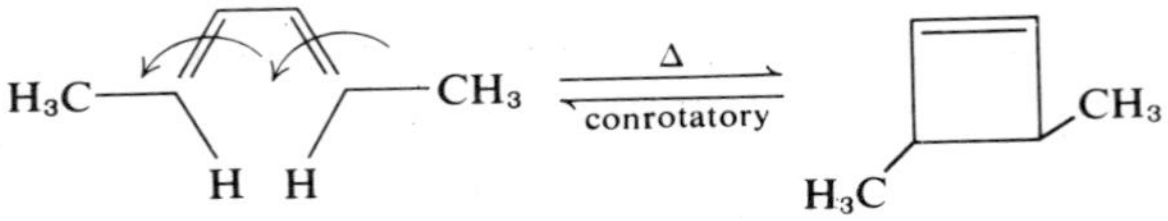

$$\text{H}_3\text{C–CH=CH–CH=CH–CH}_3 \underset{\text{conrotatory}}{\overset{\Delta}{\rightleftharpoons}} \text{3,4-dimethylcyclobutene } (trans) \tag{9.87}$$

$$\text{H}_3\text{C–CH=CH–CH=CH–CH}_3 \underset{\text{disrotatory}}{\overset{h\nu}{\rightleftharpoons}} \text{3,4-dimethylcyclobutene } (cis) \tag{9.88}$$

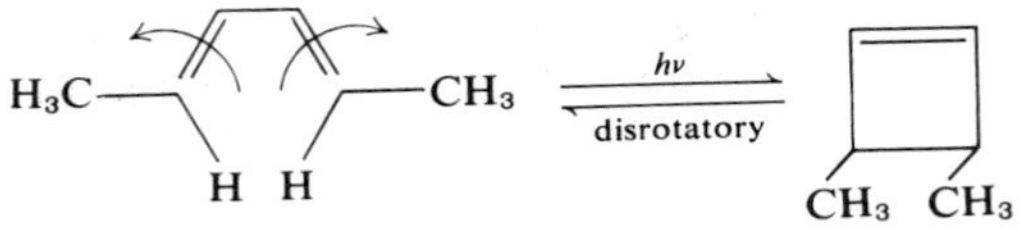

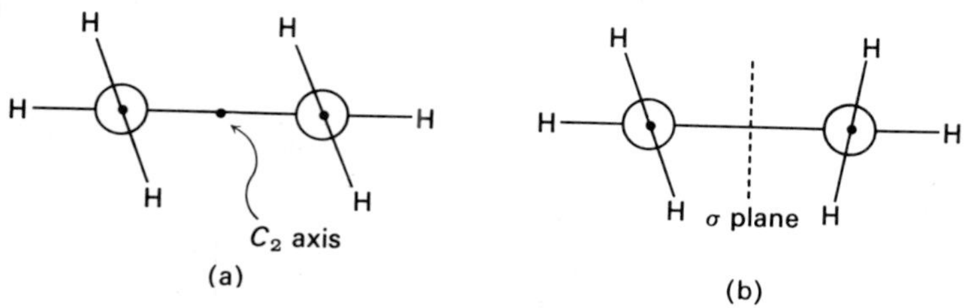

FIGURE 9.13. Transition states for (a) conrotatory and (b) disrotatory closure.

In the thermal conrotatory process the molecule maintains a C_2 axis of symmetry throughout the entire reaction, while the photochemical disrotatory process maintains a plane of symmetry as shown in Figure 9.13 for butadiene.

In a concerted reaction, orbital and state symmetry is conserved throughout the course of the reaction. Thus a symmetric orbital in butadiene must transform into a symmetric orbital in cyclobutene and an antisymmetric orbital must transform into an antisymmetric orbital. In drawing the correlation diagram, molecular orbitals of one symmetry on one side of the diagram are connected to orbitals of the same symmetry on the other side, while observing the noncrossing rule.

For the conrotatory process (Figure 9.14) the butadiene ground state ${\pi_1}^2{\pi_2}^2$ correlates with the ground state of the cyclobutene $\sigma^2\pi^2$. This is an allowed process. The first excited state of butadiene $(\pi_1)^2(\pi_2)^1({\pi_1}^*)^1$ for the conrotatory process correlates with a *very high*-energy excited state of cyclobutene $(\sigma)^2(\pi)^1(\sigma^*)^1$, (not the first excited state) and thus is *not* allowed. The disrotatory process is shown in Figure 9.15. Here the ground state of butadiene ${\pi_1}^2{\pi_2}^2$ correlates with a doubly excited state of cyclobutene $(\sigma)^2(\pi^*)^2$ and is forbidden, while the first excited state $(\pi_1)^2(\pi_2)^1({\pi_1}^*)^1$ correlates with the first excited state of cyclobutene $(\sigma)^2(\pi)^1(\pi^*)^1$, and is therefore an allowed process.

A *state* correlation diagram[151] is shown for the disrotatory process in Figure 9.16. Only those states derived from the transformations shown in the disrotatory orbital correlation diagram are included (Figure 9.15). From the state diagram for the disrotatory closure of butadiene to cyclobutene or the opening of cyclobutene to butadiene one can see that the photo disrotatory process is allowed while the thermal disrotatory process has a large activation energy. Again this diagram points out that the excited state of the product need not be formed directly but that the product can result from internal conversion at some point along the potential energy surface. If the conjugated system has more than two double bonds, the same basic treatment used for butadiene will give similar results that depend only upon the number of π electrons. The results are generalized in Table 9.7.

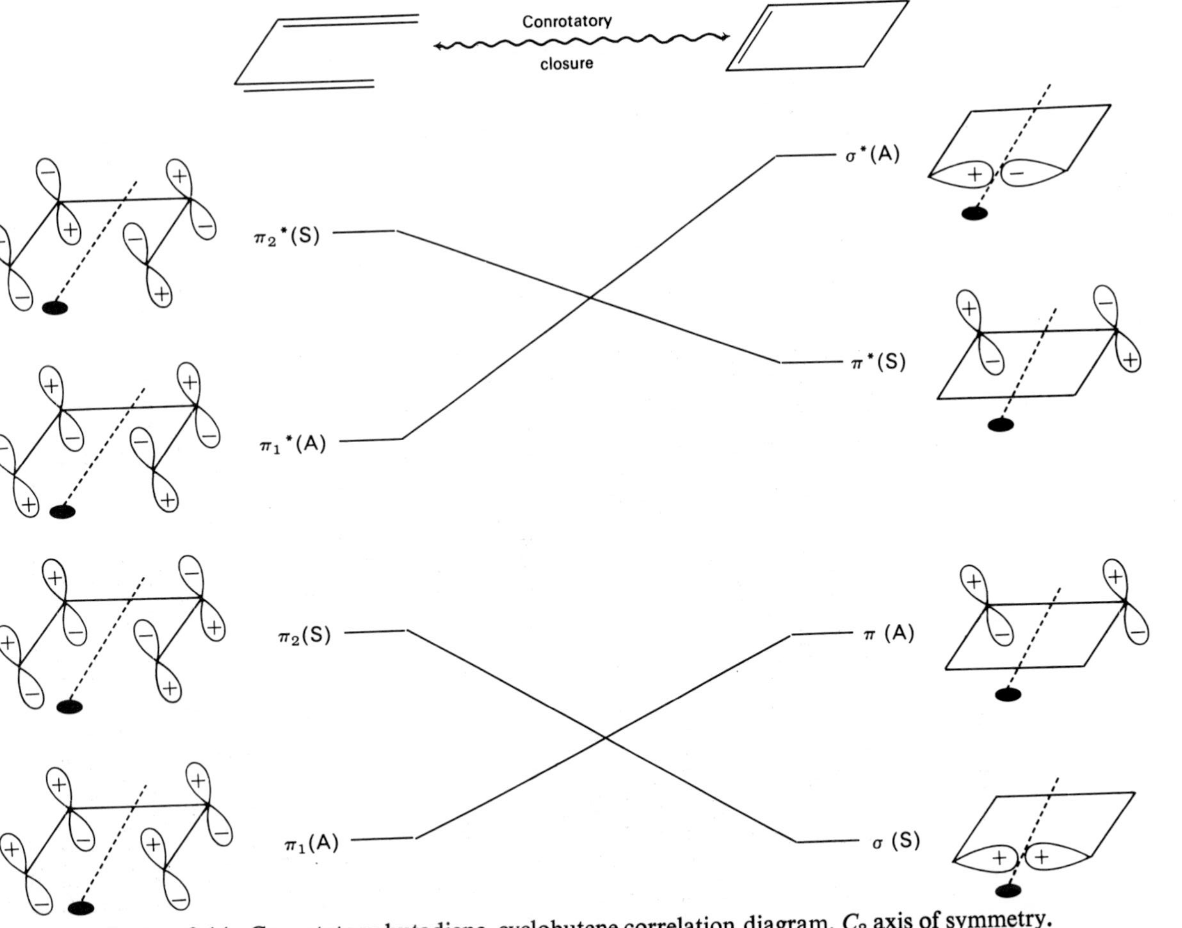

FIGURE 9.14. Conrotatory butadiene–cyclobutene correlation diagram. C_2 axis of symmetry.

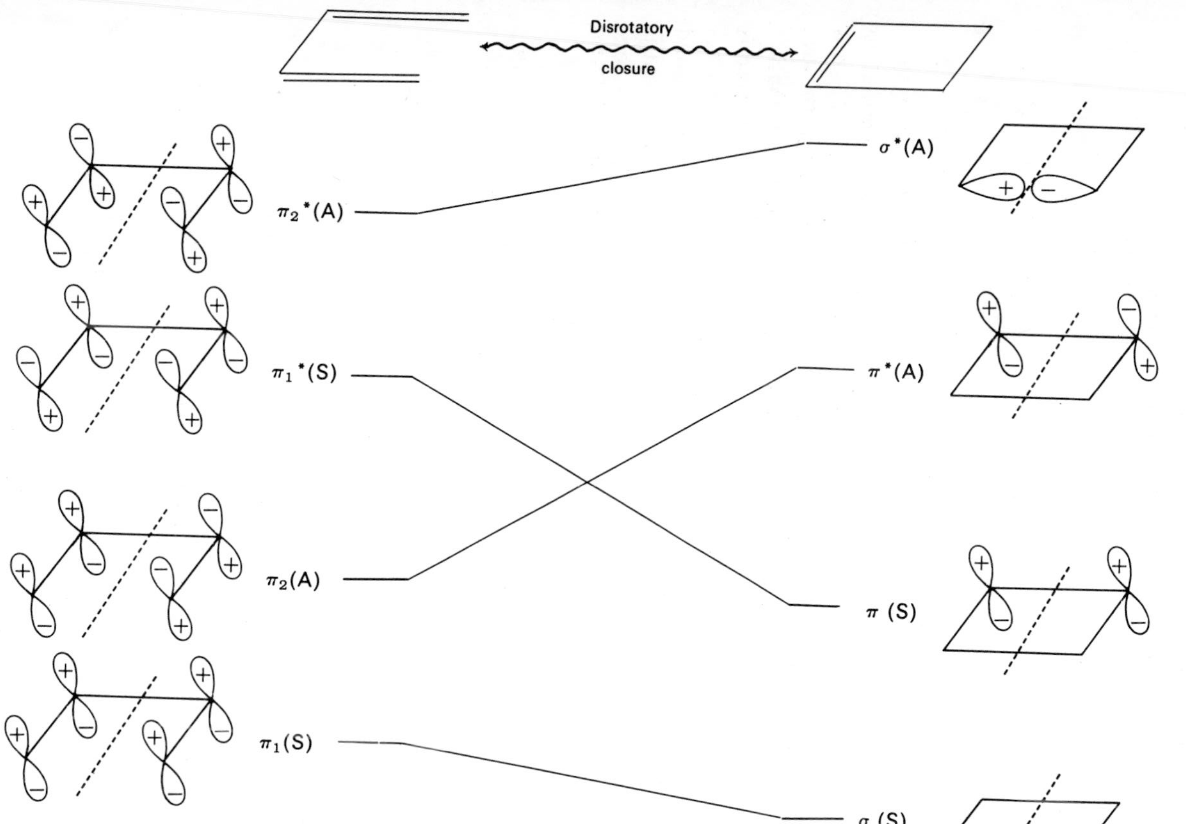

FIGURE 9.15. Disrotatory butadiene–cyclobutene correlation diagram. σ plane of symmetry.

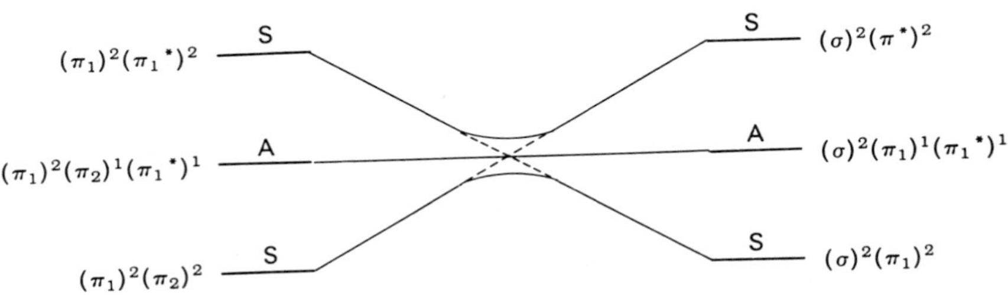

FIGURE 9.16. Disrotatory state diagram for butadiene–cyclobutene.

The alternate approach of Dewar and Zimmerman can be illustrated by an examination of the 1,3,5-hexatriene system.[91,92] The disrotatory closure has no sign discontinuity (Hückel system) and has $4n + 2$ (where $n = 1$) π electrons, so that the transition state for the thermal reaction is "aromatic" and the reaction is thermally allowed. For the conrotatory closure there is one sign discontinuity (Möbius system) and there are $4n + 2$ ($n = 1$) π electrons, so that the transition state for the thermal reaction is antiaromatic and forbidden but the transition state for the photochemical reaction is aromatic or allowed (see Chapter 8 and Table 9.8). If we reexamine the butadiene

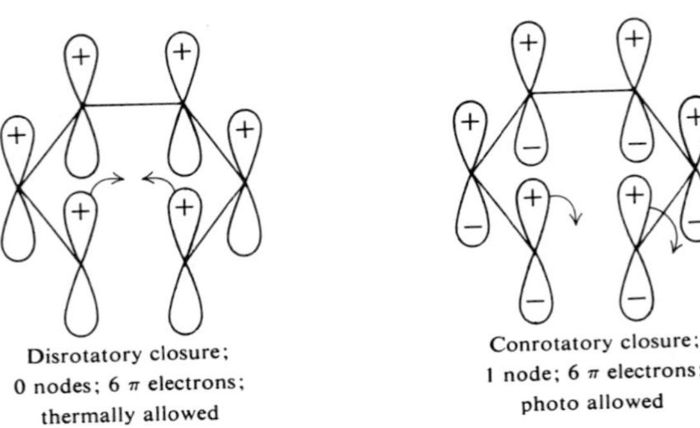

TABLE 9.7. *Selection Rules for Electrocyclic Reactions*

Number of π electrons	Reaction	Allowed mode
$4n$	Thermal	Conrotatory
$4n$	Photochemical	Disrotatory
$4n + 2$	Thermal	Disrotatory
$4n + 2$	Photochemical	Conrotatory

TABLE 9.8. *Characterization of the Transition State of Electrocyclic Reactions*[a]

System	Reaction	4*n*	4*n* + 2
Hückel	Thermal	−	+
	Photochemical	+	−
Möbius	Thermal	+	−
	Photochemical	−	+

[a] +, Aromatic transition state = allowed reaction; −, antiaromatic transition state = forbidden reaction.

system, it will be possible to evaluate the energy levels at the transition state. The basis set of orbitals can be drawn as they would interact in either a conrotatory or disrotatory manner (Figure 9.17). The disrotatory closure has no nodes and is therefore a Hückel system. The energy levels of the Hückel system can be derived by a simple device given by Frost and Musulin.[(152)] For this device the appropriate polygon (in this case a square) is inscribed with one vertex down in a circle of radius 2β with a center at the energy level of an isolated p orbital (α). Each vertex of the polygon then gives the energy of one molecular orbital from its vertical displacement. In the conrotatory process, there is one sign inversion, a Möbius system. The energy levels for the Möbius system are determined in the same manner except that the polygon is inscribed with *one side* down.[(153)]

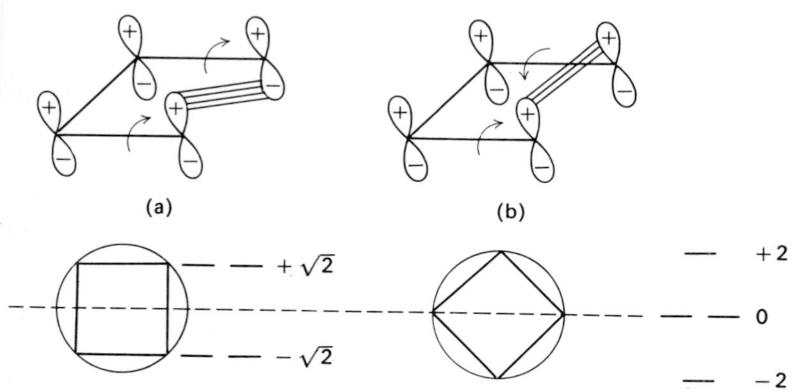

FIGURE 9.17. (a) Conrotatory, Möbius system, one sign inversion. (b) Disrotatory, Hückel system, no sign inversions.

9.10b. *Examples of Electrocyclic Reactions*

Again we must point out that many of the examples we will cite have not been shown to be concerted but are consistent with the electrocyclic rules. It is possible that some of these involve intermediates; however, in some cases the intermediates will be subject to the electrocyclic rules.

4n (n = 1) Examples

O, O —— $h\nu$ / Ref. 154 ——> O, O (9.89)

H_3C, H_3C —— $h\nu$, pentane, 10° / $\Phi = 0.12$ / Refs. 155–158 ——> H_3C, H_3C (9.90)

70% yield

—— $h\nu$, pentane / $\Phi = 0.025$; Ref. 157 ——> (9.91)

80% yield

—— $h\nu$ / $\Phi = 0.11$ / Refs. 157, 158 ——> (9.92)

90%

—— $h\nu$ / Refs. 159, 160 ——> (9.93)

—— $h\nu$ / Refs. 161, 162 ——> (9.94)

O, O, O —— $h\nu$ / Ref. 163 ——> O, O, O (9.95)

Me Me N—H Me O $\xrightarrow[\text{Refs. 164, 165}]{h\nu}$ Me Me O N Me (9.96)

H H [4*n*-electron system] $h\nu$ 300 nm; Ref. 166 (9.97)

[(4*n* + 2)-electron system] $h\nu$; 254 nm

H H $\xrightarrow[4n + 2]{\Delta}$ H H

O O O $\xrightarrow{h\nu}$ O O O $\xrightarrow{Pb(OAc)_4}$ (9.98)

t-Bu *t*-Bu *t*-Bu $\xrightarrow[\text{Ref. 167}]{h\nu}$ *t*-Bu *t*-Bu *t*-Bu (9.99)

70% of normal aromaticity

$h\nu$, EtOH 0°; Ref. 168 (9.100)

15% yield

TABLE 9.9. *Photolysis of cis,cis-1,3-Cyclooctadiene in Hexane at 248 nm*

Time, hr	Bicyclo[4.2.0]-oct-7-ene, mole %	*cis,trans*-1,3-Cyclooctadiene, mole %	*cis,cis*-1,4-Cyclooctadiene, mole %
0.50	0.9	12.1	—
1.0	2.0	20.4	—
3.0	7.4	31.5	2.4
4.0	12.4	31.1	3.2
9	26.3	24.6	5.6

In many cases the transformations may be more complex than indicated by Eqs. (9.89)–(9.100). An example of this is the photochemistry of *cis,cis*-1,3-cyclooctadiene [Eq. (9.94)].(169) A close examination of this reaction indicates that bicyclo[4.2.0]oct-7-ene is formed but in low relative yields during the initial reaction (see Table 9.9). In addition, the *cis,trans*-1,3-cyclooctadiene is formed and then consumed as the reaction proceeds. Fonken showed that the bicyclooctene initially formed, however, was *not* from thermal isomerization of the *cis,trans*-diene. Still a third reaction was the 1,3 sigmatropic hydrogen shift to form the *cis,cis*-1,4-cyclooctadiene:

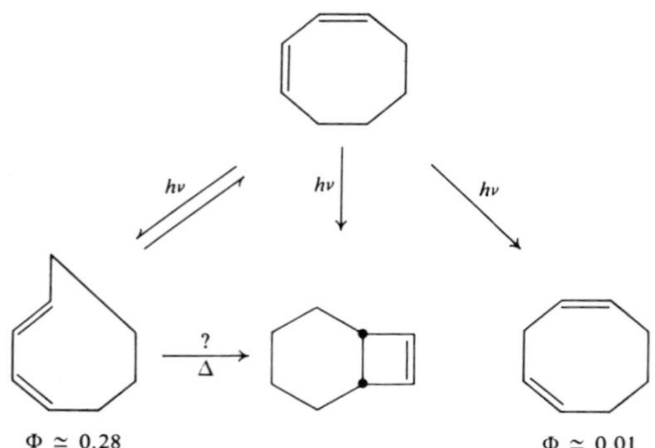

4n + 2 Examples

Vitamin D chemistry provided some of the first examples of both thermal and photo electrocyclic reactions(170–173):

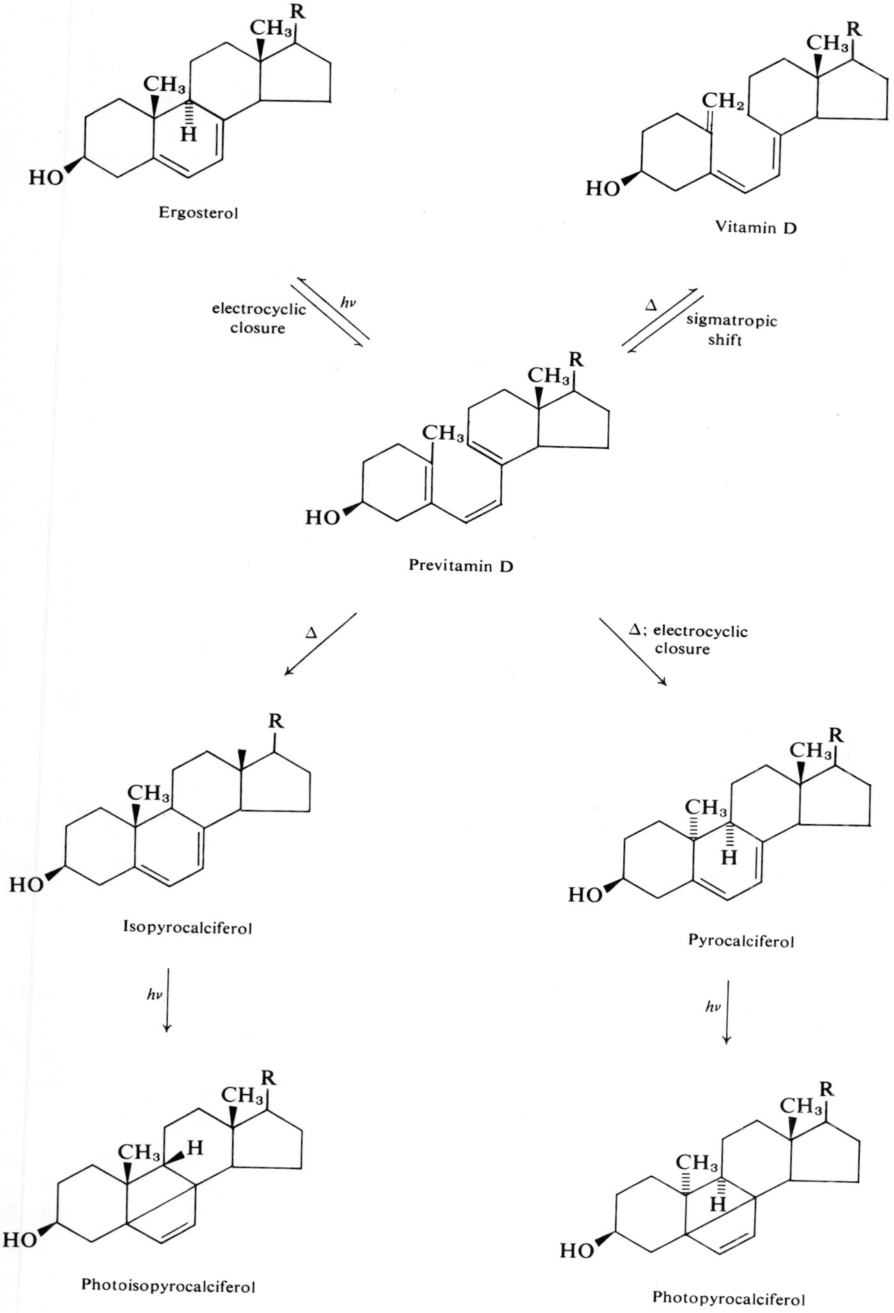

Ergosterol

Vitamin D

electrocyclic closure

$h\nu$

Δ

sigmatropic shift

Previtamin D

Δ

Δ; electrocyclic closure

Isopyrocalciferol

Pyrocalciferol

$h\nu$

$h\nu$

Photoisopyrocalciferol

Photopyrocalciferol

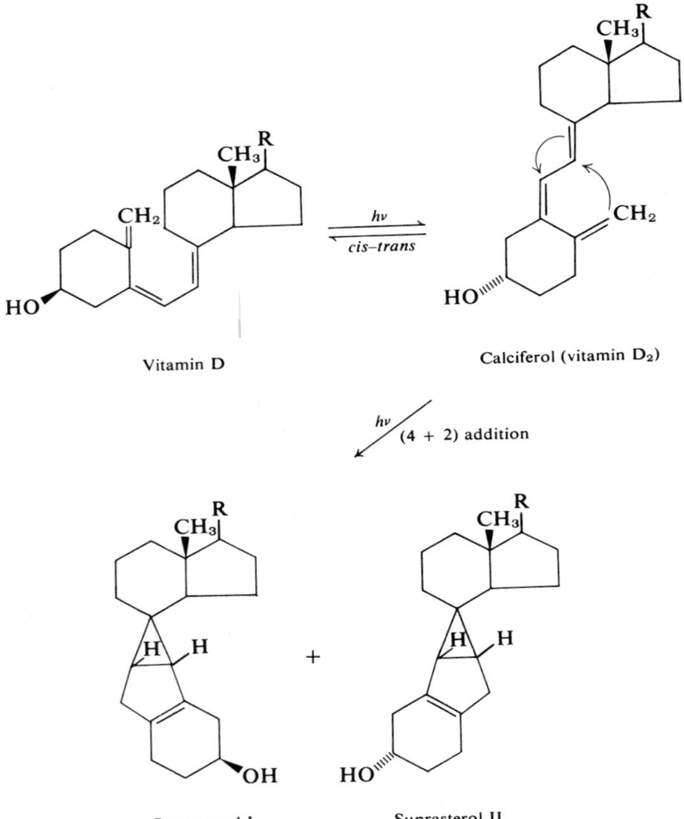
Vitamin D
hν
cis–trans
Calciferol (vitamin D₂)
hν
(4 + 2) addition
Suprasterol I
Suprasterol II

Other examples follow:

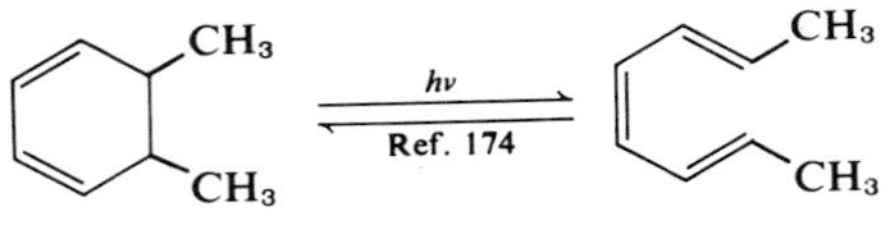
hν
Ref. 174

(9.101)

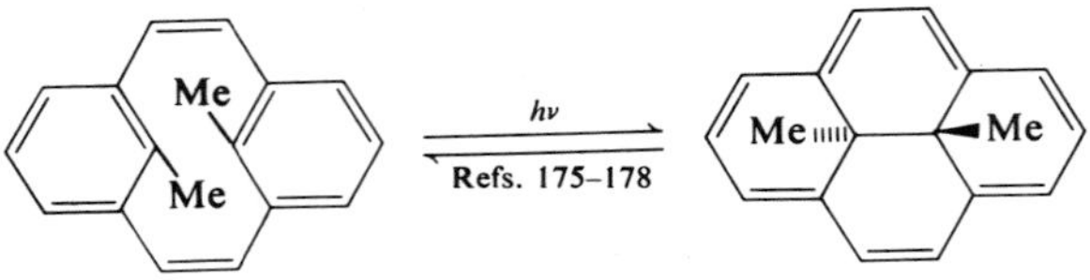
hν
Refs. 175–178

(9.102)

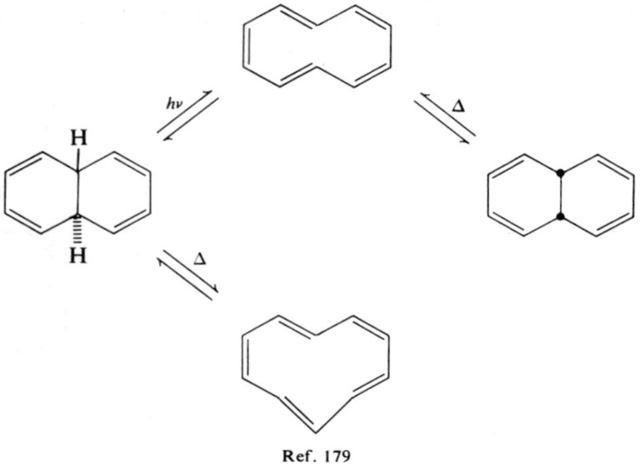

$h\nu$ Δ Δ

Ref. 179

(9.103)

$h\nu$ O_2, I_2

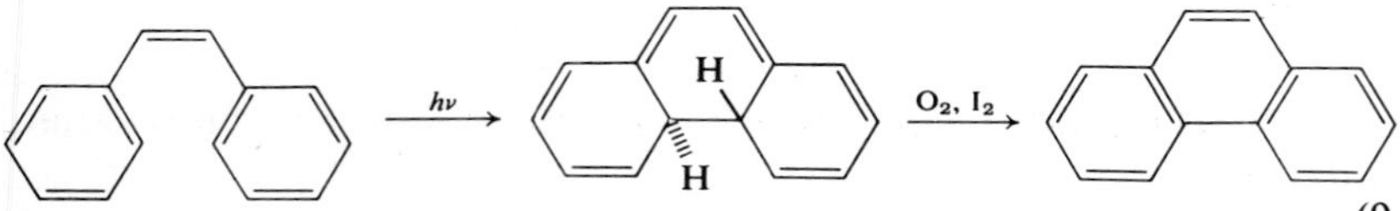

(9.104)

It is interesting to note that while the electrocyclic reaction shown in Eq. (9.104) has been developed into a very useful synthetic reaction, not all stilbene-type systems cyclize. For the reaction to occur, the sum of the free valence indices ($\sum F^*$) for the first excited state at atoms between which the new bond is formed must be greater than unity[183,184]:

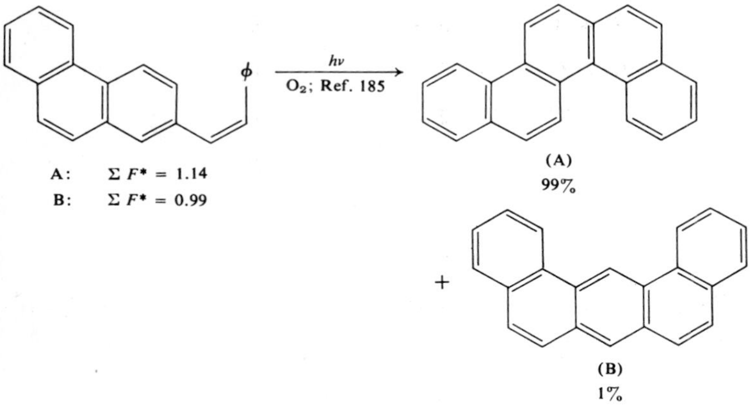

ϕ

$h\nu$

O_2; Ref. 185

A: $\Sigma F^* = 1.14$

B: $\Sigma F^* = 0.99$

(A)
99%

(9.105)

+

(B)
1%

An example of the synthetic utility of this reaction is the synthesis of tridecahelicene[186]:

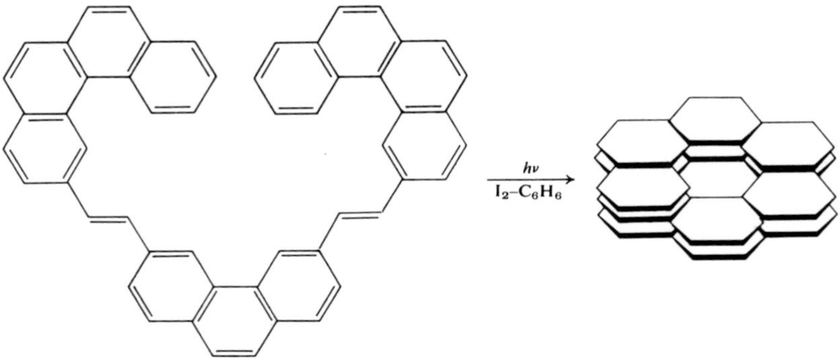

$$\xrightarrow[I_2\text{–}C_6H_6]{h\nu}$$

PROBLEMS

1. Using the suggested mechanism for sensitized stilbene isomerization, show that

$$[\mathrm{C}]_s/[\mathrm{T}]_s = (1 - \alpha)/\alpha$$

for high-energy sensitizers. At photoequilibrium $d\mathrm{T}/dt = d\mathrm{C}/dt$.

2. Calculate $[\mathrm{C}]_s/[\mathrm{T}]_s$, $\Phi_{\mathrm{T}\to\mathrm{C}}$, $\Phi_{\mathrm{C}\to\mathrm{T}}$, and the ratio $\Phi_{\mathrm{T}\to\mathrm{C}}/\Phi_{\mathrm{C}\to\mathrm{T}}$ for the following nonphantom mechanism for *cis–trans* isomerization. Assume $\Phi_{\mathrm{isc}}\Phi_{\mathrm{et}} = 1$ and $k_1 = k_2 = k_{\mathrm{diffn}}$.

$$\mathrm{S} + h\nu \longrightarrow \mathrm{S}^s \longrightarrow \mathrm{S}^t$$

$$\mathrm{S}^t + \mathrm{T} \longrightarrow \mathrm{S} + \mathrm{T}^t \qquad k_1$$

$$\mathrm{S}^t + \mathrm{C} \longrightarrow \mathrm{S} + \mathrm{C}^t \qquad k_2$$

$$\mathrm{T}^t \longrightarrow \mathrm{T} \qquad k_3$$

$$\mathrm{T}^t \longrightarrow \mathrm{C} \qquad k_4$$

$$\mathrm{C}^t \longrightarrow \mathrm{T} \qquad k_5$$

$$\mathrm{C}^t \longrightarrow \mathrm{C} \qquad k_6$$

3. Using equations (9.1)–(9.6), (9.8)–(9.10), and (9.17)–(9.19), derive the expression for the pss:

$$[\mathrm{C}]_s/[\mathrm{T}]_s = ?$$

4. Calculate the pss for the direct photolysis of stilbene [Eqs. (9.18)–(9.23), (9.8)–(9.10)] and include the following two reactions (DHP = dihydrophenanthrene):

$$T^s \longrightarrow T + h\nu \qquad k_f$$

$$C^s \longrightarrow \text{DHP} \qquad k_{\text{DHP}}$$

with $\Phi_f \approx 0.05$ and $\Phi_{\text{DHP}} \approx 0.1$.

5. Show that Eqs. (9.47)–(9.49) are consistent with the mechanism proposed.

6. What are the stereochemical consequences of ${}_{\pi}4_s + {}_{\pi}2_a$ and ${}_{\pi}4_a + {}_{\pi}4_s$ photoaddition reactions?

7. Construct orbital and state diagrams for the following processes:

   (a) ${}_{\pi}2_s + {}_{\pi}2_a$.
   (b) ${}_{\pi}4_s + {}_{\pi}2_s$.
   (c) ${}_{\pi}4_a + {}_{\pi}2_s$.

   See Ref. 153 if you have problems.

8. Does the Curtin–Hammett principle apply to photoelectrocyclic reactions when more than one conformation of reactant is present in solution? Why?

9. Construct a state correlation diagram for the conrotatory butadiene–cyclobutene system.

10. Explain the following transformation:

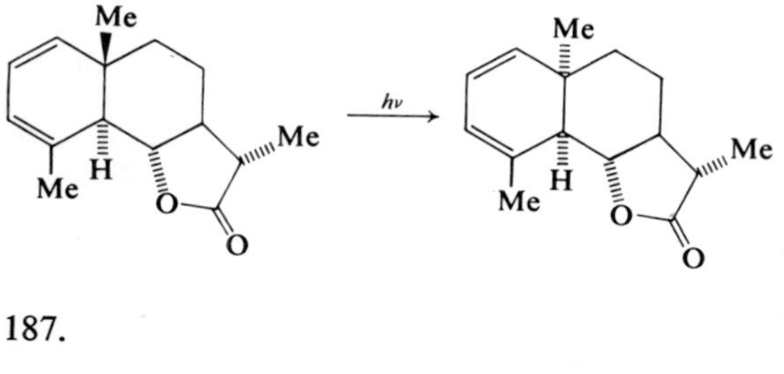

    See Ref. 187.

11. Construct a diagram like Figure 9.18 for (a) the Hückel closure of butadiene to cyclobutene and (b) the Hückel and Möbius cycloadditions of butadiene and ethylene.

12. Is the following reaction allowed? Explain.

$\xrightarrow{h\nu}$

See Refs. 188 and 189.

13. The following transformation can be carried out photochemically or thermally:

CN CN → CN CN ϕ ϕ

(A) (B)

The photoprocess could be an example of a ${}_{\sigma}2_s + {}_{\pi}2_s$ reaction. Explain the fact that (a) optically pure dinitrile (A) upon photolysis gives (B) with 83% retention and 17% inversion of configuration; (b) optically pure (A) upon thermal rearrangement gives (B) with 95.5% retention of configuration. See Refs. 190 and 191.

REFERENCES

1. J. Saltiel, J. D'Agostino, E. D. Megarity, L. Metts, K. R. Neuberger, M. Wrighton, and O. C. Zafiriou, *Org. Photochem.* **3**, 1 (1973).
2. (a) R. G. Parr and B. L. Crawford, Jr., *J. Chem. Phys.* **16**, 526 (1948); (b) R. S. Mulliken and C. C. J. Roothaan, *Chem. Rev.* **41**, 219 (1947); (c) A. J. Merer and R. S. Mulliken, *Chem. Rev.* **69**, 639 (1969).
3. P. Borrell and H. H. Greenwood, *Proc. Roy. Soc.* (*London*) *A* **298**, 453 (1967).
4. G. S. Hammond and J. Saltiel, *J. Amer. Chem. Soc.* **84**, 4983 (1962).
5. J. Saltiel and G. S. Hammond, *J. Amer. Chem. Soc.* **85**, 2515 (1963).
6. G. S. Hammond and J. Saltiel, *J. Amer. Chem. Soc.* **85**, 2516 (1963).
7. G. S. Hammond, J. Saltiel, A. A. Lamola, N. J. Turro, J. S. Bradshaw, D. O. Cowan, R. C. Counsell, V. Vogt, and C. Dalton, *J. Amer. Chem. Soc.* **86**, 3197 (1964).
8. W. G. Herkstroeter and G. S. Hammond, *J. Amer. Chem. Soc.* **88**, 4769 (1966).
9. W. G. Herkstroeter, L. B. Jones, and G. S. Hammond, *J. Amer. Chem. Soc.* **88**, 4777 (1966).
10. D. Valentine, Jr. and G. S. Hammond, *J. Amer. Chem. Soc.* **94**, 3449 (1972).
11. H. A. Hammond, D. E. DeMeyer, and J. L. R. Williams, *J. Amer. Chem. Soc.* **91**, 5180 (1969).
12. A. A. Lamola, in *Techniques of Organic Chemistry*, Vol. XIV, Wiley, New York (1969), p. 17.
13. W. G. Herkstroeter and D. S. McClure, *J. Amer. Chem. Soc.* **90**, 4522 (1968).
14. G. Fischer, K. A. Muszkat, and E. Fischer, *J. Chem. Soc. B*, 156 (1968).
15. W. W. Schmiegel, F. A. Litt, and D. O. Cowan, *J. Org. Chem.* **33**, 3334 (1968).

16. R. H. Dyck and D. S. McClure, *J. Chem. Phys.* **36**, 2326 (1962).
17. D. F. Evans, *J. Chem. Soc.*, 1351 (1957).
18. A. Bylina and Z. R. Grabowski, *Trans. Faraday Soc.* **65**, 458 (1969).
19. H. Stegemeyer, *Z. Physik. Chem.* **51**, 95 (1966).
20. W. J. Potts, Jr., *J. Chem. Phys.* **23**, 65 (1955).
21. J. Saltiel, *J. Amer. Chem. Soc.* **89**, 1036 (1967).
22. J. Saltiel, *J. Amer. Chem. Soc.* **90**, 6394 (1968).
23. H. Blume and D. Schulte-Frohlinde, *Tetrahedron Lett.*, 4693 (1967).
24. S. K. Lower and M. A. El-Sayed, *Chem. Rev.* **66**, 199 (1966).
25. G. W. Robinson and R. Frosch, *J. Chem. Phys.* **37**, 1962 (1962).
26. W. M. Gelbart, K. F. Freed, and S. A. Rice, *J. Chem. Phys.* **52**, 2460 (1970).
27. J. Saltiel, J. T. D'Agostino, W. G. Herkstroeter, G. Saint-Ruf, and N. P. Buu Hoi, *J. Amer. Chem. Soc.* **95**, 2543 (1973).
28. G. Heinrich, G. Holzer, H. Blume, and D. Schulte-Frohlinde, *Z. Naturforsch. B* **25**, 496 (1970).
29. A. Bylina, *Chem. Phys. Lett.* **1**, 509 (1968).
30. S. Yamauchi and T. Azumi, *J. Amer. Chem. Soc.* **95**, 2710 (1973).
31. A. R. Olson and F. L. Hudson, *J. Amer. Chem. Soc.* **56**, 1320 (1934).
32. G. N. Lewis, T. T. Magel, and D. Lipkin, *J. Amer. Chem. Soc.* **62**, 2973 (1940).
33. Th. Förster, *Z. Elektrochem.* **56**, 716 (1952).
34. G. Zimmerman, L. Chow, and V. Paik, *J. Amer. Chem. Soc.* **80**, 3528 (1958).
35. H. Stegemeyer, *J. Phys. Chem.* **66**, 2555 (1962).
36. S. Malkin and E. Fischer, *J. Phys. Chem.* **66**, 2482 (1962).
37. D. Schulte-Frohlinde, H. Blume, and H. Güsten, *J. Phys. Chem.* **66**, 2486 (1962).
38. S. Malkin and E. Fischer, *J. Phys. Chem.* **68**, 1153 (1964).
39. J. Saltiel, E. D. Megarity, and K. G. Kneipp, *J. Amer. Chem. Soc.* **88**, 2336 (1966).
40. K. A. Muszkat and E. Fischer, *J. Chem. Soc. B*, 662 (1967).
41. K. A. Muszkat, D. Gegiou, and E. Fischer, *J. Amer. Chem. Soc.* **89**, 4814 (1967).
42. D. Gegiou, K. A. Muszkat, and E. Fischer, *J. Amer. Chem. Soc.* **90**, 12 (1968).
43. D. Gegiou, K. A. Muszkat, and E. Fischer, *J. Amer. Chem. Soc.* **90**, 3907 (1968).
44. J. Saltiel, O. Zafiriou, E. D. Megarity, and A. A. Lamola, *J. Amer. Chem. Soc.* **90**, 4759 (1968).
45. L. D. Weis, T. R. Evans, and P. A. Leermakers, *J. Amer. Chem. Soc.* **90**, 6109 (1968).
46. J. Saltiel and E. D. Megarity, *J. Amer. Chem. Soc.* **91**, 1265 (1969).
47. S. Sharafy and K. A. Muszkat, *J. Amer. Chem. Soc.* **93**, 4119 (1971).
48. J. Saltiel and E. D. Megarity, *J. Amer. Chem. Soc.* **94**, 2742 (1972).
49. E. Fischer, *Mol. Photochem.* **5**, 227 (1973).
50. J. Saltiel, *Mol. Photochem.* **5**, 231 (1973).
51. F. D. Lewis, J. C. Dalton, and N. J. Turro, *Mol. Photochem.* **2**, 67 (1970).
52. C. D. DeBoer and R. H. Schlessinger, *J. Amer. Chem.* **90**, 803 (1968).
53. E. Lippert, *Z. Physik. Chem.* **42**, 125 (1964); *Acc. Chem. Res.* **3**, 74 (1970).
54. G. S. Hammond, S. C. Shim, and S. P. Van, *Mol. Photochem.* **1**, 89 (1969).
55. D. G. Whitten and M. T. McCall, *J. Amer. Chem. Soc.* **91**, 5097 (1969); Y. J. Lee, D. G. Whitten, and L. Pedersen, *J. Amer. Chem. Soc.* **93**, 6330 (1971).
56. P. Bortulus, G. Cauzzo, U. Mazzucato, and G. Caliazzo, *Z. Phys. Chem.* **51**, 264 (1966).
57. G. S. Hammond, N. J. Turro, and P. A. Leermakers, *J. Amer. Chem. Soc.* **83**, 2396 (1961).
58. G. S. Hammond, N. J. Turro, and P. A. Leermakers, *J. Phys. Chem.* **66**, 1144 (1962).

59. J. Saltiel, L. Metts, and M. Wrighton, *J. Amer. Chem. Soc.* **92**, 3227 (1970).
60. S. Boue' and R. Srinivasan, *J. Amer. Chem. Soc.* **92**, 3226 (1970).
61. R. Srinivasan and S. Boue', *Tetrahedron Lett.*, 203 (1970).
62. A. A. Lamola and G. S. Hammond, *J. Chem. Phys.* **43**, 2129 (1965).
63. H. L. Hyndman, B. M. Monroe, and G. S. Hammond, *J. Amer. Chem. Soc.* **91**, 2852 (1969).
64. J. Saltiel, L. Metts, and M. Wrighton, *J. Amer. Chem. Soc.* **91**, 5684 (1969).
65. J. Saltiel, L. Metts, A. Sykes, and M. Wrighton, *J. Amer. Chem. Soc.* **93**, 5302 (1971)
66. R. Hoffman, *Tetrahedron* **22**, 521 (1966).
67. E. M. Evleth, *Chem. Phys. Lett.* **3**, 122 (1969).
68. N. C. Baird and R. M. West, *J. Amer. Chem. Soc.* **93**, 4427 (1971).
69. R. A. Caldwell, *J. Amer. Chem. Soc.* **92**, 3229 (1970).
70. R. S. H. Liu, Y. Butt, and W. G. Herkstroeter, *Chem. Comm.*, 799 (1973).
71. R. S. H. Liu and Y. Butt, *J. Amer. Chem. Soc.* **93**, 1532 (1971).
72. A. Sykes and T. G. Truscott, *Chem. Comm.*, 274 (1969).
73. A. Sykes and T. G. Truscott, *Trans. Faraday Soc.* **67**, 679 (1971).
74. R. S. Becker, K. Inuzuka, and D. E. Balke, *J. Amer. Chem. Soc.* **93**, 38, 43 (1971).
75. W. H. Waddell, A. M. Schaffer, and R. S. Becker, *J. Amer. Chem. Soc.* **95**, 8223 (1973).
76. H. Yamazaki and R. J. Cvetanovic', *J. Amer. Chem. Soc.* **91**, 520 (1969).
77. D. R. Arnold and V. Y. Abraitys, *Mol. Photochem.* **2**, 27 (1970).
78. M. A. Golub and C. L. Stephens, *J. Phys. Chem.* **70**, 3576 (1966).
79. M. A. Golub, C. L. Stephens, and J. L. Brash, *J. Chem. Phys.* **45**, 1503 (1966).
80. J. Saltiel, K. R. Neuberger, and M. Wrighton, *J. Amer. Chem. Soc.* **91**, 3658 (1969).
81. R. F. Borkman and D. R. Kearns, *J. Chem. Phys.* **44**, 945 (1966); *J. Amer. Chem. Soc.* **91**, 3658 (1969).
82. N. C. Yang, J. I. Cohen, and A. Shani, *J. Amer. Chem. Soc.* **90**, 3264 (1968).
83. R. A. Caldwell and S. P. Jones, *J. Amer. Chem. Soc.* **91**, 5184 (1969).
84. R. A. Caldwell, *J. Amer. Chem. Soc.* **92**, 1439 (1970).
85. R. A. Caldwell and G. W. Sovocool, *J. Amer. Chem. Soc.* **90**, 7138 (1968).
86. R. A. Caldwell, G. W. Sovocool, and R. J. Perlsie, *J. Amer. Chem. Soc.* **93**, 779 (1971).
87. R. A. Caldwell, G. W. Sovocool, and R. J. Perlsie, *J. Amer. Chem. Soc.* **95**, 1496 (1973).
88. S. M. Japar, M. Pomerantz, and E. W. Abrahamson, *Chem. Phys. Lett.* **2**, 137 (1968).
89. R. B. Woodward and R. Hoffmann, *The Conservation of Orbital Symmetry*, Academic Press, New York (1970).
90. R. B. Woodward, *Aromaticity*, Special Publication No. 21, The Chemical Society, London (1967), p. 217.
91. (a) R. B. Woodward and R. Hoffmann, *Acc. Chem. Res.* **1**, 17 (1968); (b) H. E. Zimmerman, *Angew. Chem. Int. Ed.* **8**, 1 (1969).
92. M. J. S. Dewar, *Aromaticity*, Special Publication No. 21, The Chemical Society, London (1967), p. 177.
93. S. I. Miller, in *Advances in Physical Organic Chemistry*, Vol. 6, V. Gold, ed., Academic Press, New York (1970).
94. R. E. Lehr and A. P. Marchand, *Orbital Symmetry*, Academic Press, New York (1972).
95. W. G. Dauben and R. L. Cargill, *Tetrahedron* **15**, 197 (1961).
96. G. S. Hammond, N. J. Turro, and A. Fischer, *J. Amer. Chem. Soc.* **83**, 4674 (1961).

97. G. S. Hammond, P. Wyatt, C. D. DeBoer, and N. J. Turro, *J. Amer. Chem. Soc.* **86**, 2532 (1964).
98. H. Prinzbach, *Pure Appl. Chem.* **16**, 17 (1968).
99. W. Dilling, *Chem. Rev.* **66**, 373 (1966).
100. J. R. Edman, *J. Org. Chem.* **32**, 2920 (1967).
101. H. Prinzbach and J. Rivier, *Tetrahedron Lett.*, 3713 (1967).
102. H. D. Scharf, *Tetrahedron* **23**, 3057 (1967).
103. G. O. Schenck and R. Steinmetz, *Bull. Soc. Chim. Belges* **71**, 781 (1962).
104. G. O. Schenck and R. Steinmetz, *Chem. Ber.* **96**, 520 (1963).
105. R. C. Cookson, E. Crundwell, R. R. Hill, and J. Hudec, *J. Chem. Soc.*, 3062 (1964).
106. R. C. Cookson, R. R. Hill, and J. Hudec, *J. Chem. Soc.*, 3043 (1964).
107. G. W. Griffin and A. K. Price, *J. Org. Chem.* **29**, 3192 (1964).
108. W. L. Dilling and C. E. Reineke, *Tetrahedron Lett.*, 2547 (1969).
109. W. L. Dilling, C. E. Reineke, and A. Plepys, *J. Org. Chem.* **34**, 2605 (1969).
110. D. M. Lemal and J. P. Lokensgard, *J. Amer. Chem. Soc.* **88**, 5934 (1966).
111. W. G. Dauben and D. L. Whalen, *Tetrahedron Lett.*, 3743 (1966).
112. H. Prinzbach, W. Eberbach, and G. Philippossian, *Angew. Chem. Int. Ed.* **7**, 887 (1968).
113. R. S. Liu, *Tetrahedron Lett.*, 1409 (1969).
114. L. A. Paquette, *J. Amer. Chem. Soc.* **92**, 5765 (1970).
115. G. Kaupp and H. Prinzbach, *Helv. Chim. Acta* **52**, 956 (1969).
116. G. Kaupp and H. Prinzbach, *Chem. Ber.* **104**, 182 (1971).
117. G. Kaupp, *Angew. Chem. Int. Ed.* **10**, 340 (1971).
118. H. Prinzbach, W. Eberbach, and G. Veh, *Angew. Chem. Int. Ed.* **4**, 436 (1965).
119. P. K. Freeman, P. G. Kuper, and V. N. M. Rao, *Tetrahedron Lett.*, 3301 (1965).
120. H. Prinzbach and H. D. Martin, *Helv. Chim. Acta* **51**, 438 (1968).
121. C. F. Huebner, *Chem. Comm.* **13**, 419 (1966).
122. H. Prinzbach and D. Hunkler, *Angew. Chem. Int. Ed.* **6**, 247 (1967).
123. P. K. Freeman and D. M. Balls, *J. Org. Chem.* **32**, 2354 (1967).
124. R. M. Coates and J. L. Kirkpatrick, *J. Amer. Chem. Soc.* **90**, 4162 (1968).
125. R. M. Coates and J. L. Kirkpatrick, *J. Amer. Chem. Soc.* **92**, 4883 (1970).
126. A. deMeijere, D. Kaufmann, and O. Schallner, *Angew. Chem. Int. Ed.* **10**, 417 (1971).
127. H. Prinzbach, M. Klaus, and W. Mayer, *Angew. Chem. Int. Ed.* **8**, 883 (1969).
128. H. Prinzbach and M. Klaus, *Angew. Chem. Int. Ed.* **8**, 276 (1969).
129. W. Schmidt and B. T. Wilkins, *Tetrahedron* **28**, 5649 (1972).
130. J. Saltiel and L. S. Nghim, *J. Amer. Chem. Soc.* **91**, 5404 (1969).
131. W. G. Dauben, I. Bell, T. W. Hutton, G. F. Laws, A. Rheiner, and H. Urscheler, *J. Amer. Chem. Soc.* **80**, 4116 (1958).
132. W. G. Dauben and P. Baumann, *Tetrahedron Lett.*, 565 (1961).
133. W. G. Dauben and J. A. Smith, *J. Org. Chem.* **32**, 3244 (1967).
134. W. G. Dauben, J. Rabinowitz, N. D. Vietmeyer and P. H. Wendschuh, *J. Amer. Chem. Soc.* **94**, 4285 (1972).
135. W. G. Dauben, M. S. Kellogg, J. I. Seeman, N. D. Vietmeyer, and P. Wendschuh, in *IV International Symposium on Photochemistry*, Butterworths, London (1973).
136. G. R. Evanega, W. Bergmann, and J. English, *J. Org. Chem.* **27**, 13 (1962).
137. H. Prinzbach and E. Druckrey, *Tetrahedron Lett.*, 2959 (1965).
138. K. J. Crowley, *Tetrahedron Lett.*, 2863 (1965).
139. K. J. Crowley, *Photochem. Photobiol.* **7**, 775 (1968); *J. Org. Chem.* **33**, 3679 (1968).
140. R. C. Cookson and D. W. Jones, *J. Chem. Soc.*, 1881 (1965).
141. S. Meinwald, A. Eckell, and K. L. Erickson, *J. Amer. Chem. Soc.* **87**, 3532 (1965).

142. J. Meinwald and P. H. Mazzocchi, *J. Amer. Chem. Soc.* **88**, 2850 (1966); **89**, 1755 (1967).
143. R. C. Cookson, S. M. Be B. Costa, and J. Hudec, *Chem. Comm.*, 1272 (1969).
144. A. Padwa and S. Clough, *Chem. Comm.*, 417 (1971).
145. A. Padwa and S. Clough, *J. Amer. Chem. Soc.* **92**, 5803 (1970); A. Padwa, L. Brodsky, and S. Clough, *J. Amer. Chem. Soc.* **94**, 6767 (1972).
146. H. Heimgartner, L. Ulrich, H. J. Hansen, and H. Schmid, *Helv. Chim. Acta* **54**, 2313 (1971).
147. D. A. Seeley, *J. Amer. Chem. Soc.* **94**, 4378 (1972).
148. M. Pomerantz, *J. Amer. Chem. Soc.* **89**, 694 (1967).
149. K. R. Huffman, M. Burger, W. A. Henderson, M. Loy, and E. F. Ullman, *J. Org. Chem.* **34**, 2407 (1969), and references cited therein.
150. O. L. Chapman, G. W. Borden, R. W. King, and B. Winkler, *J. Amer. Chem. Soc.* **86**, 2660 (1964).
151. H. C. Longuet-Higgins and E. W. Abrahamson, *J. Amer. Chem. Soc.* **87**, 2045 (1965).
152. A. Frost and B. Musulin, *J. Chem. Phys.* **21**, 572 (1953).
153. H. E. Zimmerman, *J. Amer. Chem. Soc.* **88**, 1563 (1966).
154. E. J. Corey and J. Streith, *J. Amer. Chem. Soc.* **86**, 950 (1964).
155. R. Srinivasan, *J. Amer. Chem. Soc.* **84**, 4141 (1962).
156. K. J. Crowley, *Tetrahedron* **21**, 1001 (1965).
157. D. H. Aue and R. N. Reynolds, *J. Amer. Chem. Soc.* **95**, 2027 (1973).
158. J. M. Garrett and G. J. Fonken, *Tetrahedron Lett.*, 191 (1969).
159. W. G. Dauben and R. L. Cargill, *Tetrahedron* **12**, 186 (1961).
160. R. Srinivasan, *J. Amer. Chem. Soc.* **84**, 3432 (1962).
161. W. G. Dauben and R. C. Cargill, *J. Org. Chem.* **27**, 1910 (1962).
162. R. Srinivasan, *J. Amer. Chem. Soc.* **84**, 4143 (1962).
163. G. J. Fonken, *Chem. Ind.* (*London*), 1575 (1961).
164. O. L. Chapman and E. D. Hoganson, *J. Amer. Chem. Soc.* **86**, 498 (1964).
165. L. Paquette, *J. Amer. Chem. Soc.* **86**, 500 (1964).
166. W. G. Dauben and M. S. Kellogg, *J. Amer. Chem. Soc.* **93**, 3805 (1971).
167. E. E. van Tamelen, S. P. Pappas, and R. C. Kirk, *J. Amer. Chem. Soc.* **93**, 6092 (1971).
168. E. E. van Tamelen, J. I. Brauman, and L. E. Ellis, *J. Amer. Chem. Soc.* **93**, 6145 (1971).
169. W. J. Nebe and G. J. Fonken, *J. Amer. Chem. Soc.* **91**, 1249 (1969); R. S. H. Liu, *J. Amer. Chem. Soc.* **89**, 112 (1967).
170. W. G. Dauben and G. J. Fonken, *J. Amer. Chem. Soc.* **81**, 4060 (1959).
171. E. Havinga, R. J. de Kock and M. P. Rappold, *Tetrahedron* **11**, 278 (1960).
172. E. Havinga and J. L. Schlattmann, *Tetrahedron* **16**, 146 (1961).
173. G. M. Sanders, J. Pot, and E. Havinga, *Fortschr. Chem. Org. Naturstoffen* **29**, 131 (1969).
174. G. J. Fonken, *Tetrahedron Lett.*, 549 (1962).
175. V. Boekelheide and J. B. Phillips, *J. Amer. Chem. Soc.* **89**, 1695 (1967); **89**, 1704, 1709 (1967).
176. H. Blaschke and V. Boekelheide, *J. Amer. Chem. Soc.* **89**, 2747 (1969).
177. H. Blattmann and W. Schmidt, *Tetrahedron* **26**, 5885 (1970).
178. W. Schmidt, *Helv. Chim. Acta* **54**, 862 (1971).
179. E. E. van Tamelen, T. L. Burkoth, and R. A. Greeley, *J. Amer. Chem. Soc.* **93**, 6120 (1971), and references cited therein.

180. F. B. Mallory, C. S. Wood, J. T. Gordon, L. C. Lindquist, and M. Savitz, *J. Amer. Chem. Soc.* **84**, 4361 (1962).
181. F. B. Mallory, J. T. Gordon, and C. S. Wood, *J. Amer. Chem. Soc.* **85**, 823 (1963).
182. F. B. Mallory, C. S. Wood, and J. T. Gordon, *J. Amer. Chem. Soc.* **86**, 3094 (1964).
183. E. V. Blackburn and C. J. Timmons, *J. Chem. Soc. C*, 172 (1970).
184. W. H. Laarhoven, T. J. H. M. Cuppen, and R. J. F. Nivard, *Tetrahedron* **26**, 1069 (1970).
185. D. D. Morgan, S. W. Horgan, and M. Orchin, *Tetrahedron Lett.*, 4347 (1970); 1789 (1972).
186. R. A. Martin, G. Morren, and J. J. Schurter, *Tetrahedron Lett.*, 3683 (1969).
187. E. J. Corey and A. G. Hartmann, *J. Amer. Chem. Soc.* **87**, 5736 (1965).
188. W. von E. Doering and J. W. Rosenthal, *Tetrahedron Lett.*, 349 (1967).
189. S. Masamune, R. T. Seidner, H. Zenda, M. Wiesel, N. Nakatsuka, and G. Bigam, *J. Amer. Chem. Soc.* **90**, 5286 (1968).
190. R. C. Cookson, J. Hudec, and M. Sharma, *Chem. Comm.*, 107, 108 (1971).
191. R. C. Cookson and J. E. Kemp, *Chem. Comm.*, 385 (1971).

10

Photodimerization and Photocycloaddition Reactions Yielding Cyclobutanes

The formation of dimers and cycloaddition products upon irradiation of compounds containing olefinic bonds constitutes one of the first types of photochemical reactions observed. Early workers were dependent mainly upon irradiation in sunlight, often for periods of several months (varying as to the season) to obtain small yields of products. Laborious separation and purification often led to the elucidation of the gross structure of these products, but since the excellent physical tools commonly available to the present-day photochemist were lacking, the stereochemistry of the products often remained completely obscure. A review of the reactions discovered and investigated prior to the rapid acceleration in photochemical progress of the past twenty years is given by Mustafa.[(1)]

This chapter contains a review of some photodimerizations and photocondensations leading to the formation of substituted cyclobutanes. Since the literature in this field is indeed vast, it would be impossible to present here a comprehensive review of all the reactions reported to lead to such products. Instead, it is our goal to present a general survey of the various types of compounds known to produce cyclobutane compounds upon irradiation and to discuss mechanisms which have been proposed to account for some of these products.

It could be recognized at this point that dimerization of an unsymmetrically substituted olefinic bond can potentially lead to a total of 12 different isomeric dimers, depending upon whether the dimerization proceeds in a head-to-head or a head-to-tail stepwise fashion, or indeed, by both. Fortunately in very few photochemical reactions does the number of dimeric products formed in reasonable yield exceed more than four or five of the 12 possible isomers. In fact, as we shall see, many such photoreactions display an amazing degree of selectivity.

10.1. PHOTODIMERIZATION AND PHOTOCYCLOADDITION REACTIONS OF OLEFINS AND POLYENES

The subject of the first section of this chapter is the photodimerization and photocycloaddition reactions of olefins. In this category we include only those compounds in which the photoreactive olefin or polyene is not part of an aromatic system although it may bear an aromatic substituent, such as in styrene.

10.1a. *Photodimerization of Olefins and Polyenes*

The photodimerization of simple isolated olefinic bonds is rarely observed because of the absorption of these compounds in the high-energy or vacuum-ultraviolet region. One case reported is that of the photo-dimerization of 2-butene.[2] Irradiation of liquid *cis*-2-butene with light from a cadmium ($\lambda = 229, 227, 214$ nm) or zinc ($\lambda = 214$ nm) lamp was reported to lead to dimers (1) and (2):

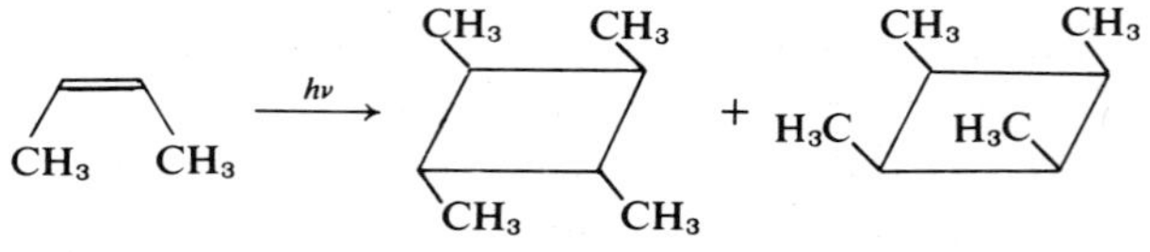

CH₃ CH₃ —hν→ (1) + (2) (10.1)

Similarly, irradiation of liquid *trans*-2-butene yielded dimers (1) and (3):

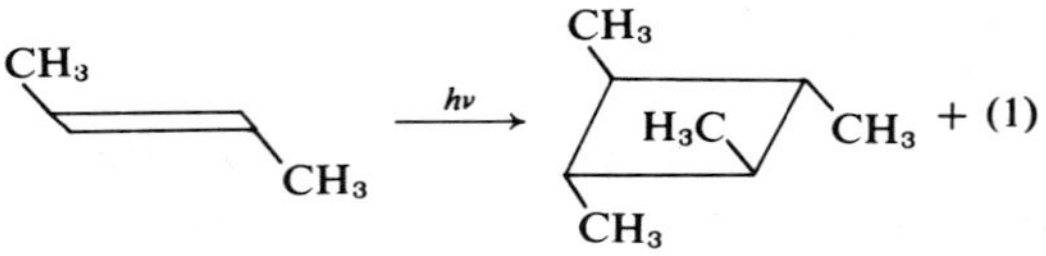

CH₃ CH₃ —hν→ (3) + (1) (10.2)

It can be seen that these products result from *stereospecific* dimerization of the corresponding 2-butenes. Irradiation of mixtures of *cis*- and *trans*-2-butene resulted in a fourth isomer (4) in addition to the above products:

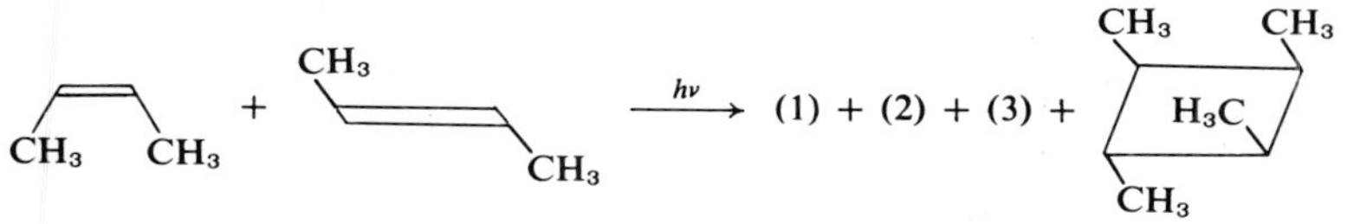

$$\text{cis-}CH_3CH{=}CHCH_3 + \text{trans-}CH_3CH{=}CHCH_3 \xrightarrow{h\nu} (1) + (2) + (3) + (4) \tag{10.3}$$

(4)

The stereospecificity of these reactions is surprising in light of the large energies absorbed by these molecules. Indeed, the major photochemical product of these photolyses was the alternate olefin isomer (1-butene was also observed). These results indicate that free rotation about the photo-excited double bond does not occur in those molecules that dimerize. This suggests the participation of ground state complexes or excimers in the photodimerization. This view is supported by the observations that dilution of *cis*-2-butene with neopentane (1:1) decreased the yield of dimers and a 1:4 dilution almost completely suppressed dimerization.

A photosensitized dimerization of an isolated olefin, norbornene, has been reported by Scharf and Korte.[3] Irradiation in acetone or in the presence of acetophenone (E_t = 74 kcal/mole) produced dimers (5) and (6) as major products. However, benzophenone (E_t = 69 kcal/mole) failed to sensitize the reaction to (5) and (6), but in ether solution led to the quantitative formation of benzpinacol and in benzene to the oxetane (7) in 80% yield. Sensitizers of intermediate energy, such as xanthone (E_t = 72 kcal/mole), demonstrated a competition between energy transfer to form triplet norbornene and cycloaddition to form the oxetane:

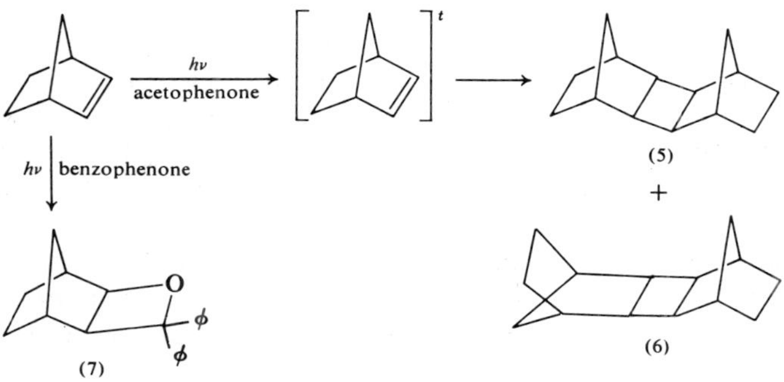

hν, acetophenone → [norbornene]^t → (5) + (6) (10.4)

hν, benzophenone → (7)

(5)

(6)

(7)

Irradiation of butadiene in isooctene solution is reported to yield cyclobutane dimers (8) and (9) in low yield[4]:

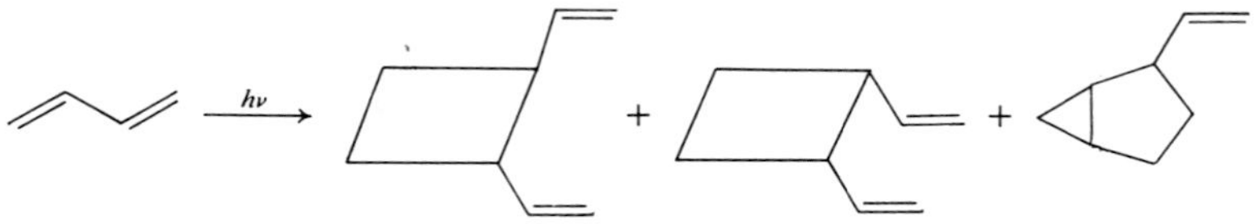

$h\nu$ (10.5)

(8) (9) (10)

In addition to (8) and (9), several other noncyclobutane dimers, such as (10) (50% yield), were isolated. Other reports[5–7] have identified dimers (11) and (12) as products of the direct photolysis of butadiene. It appears that the

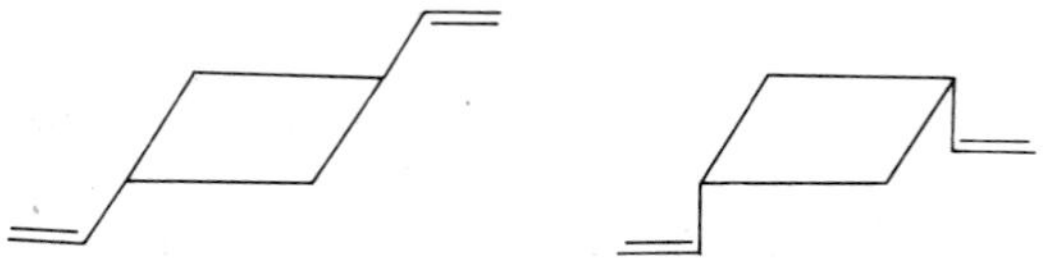

(11) (12)

products obtained from this photolysis are highly sensitive to reaction conditions.

The sensitized photolysis of butadiene has been extensively investigated by Hammond and co-workers.[8–12] Irradiation of butadiene in the presence of a sensitizer that absorbs all of the light yields dimers (8) and (9) and the cyclohexene dimer (13):

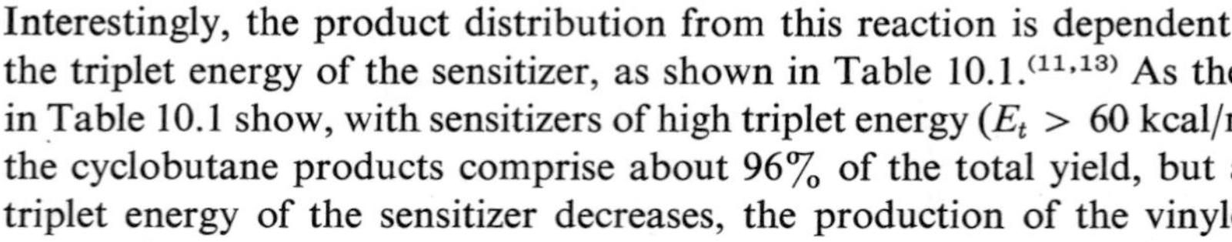

$h\nu$ sensitizer (10.6)

(8) (9) (13)

Interestingly, the product distribution from this reaction is dependent upon the triplet energy of the sensitizer, as shown in Table 10.1.[11,13] As the data in Table 10.1 show, with sensitizers of high triplet energy (E_t > 60 kcal/mole), the cyclobutane products comprise about 96% of the total yield, but as the triplet energy of the sensitizer decreases, the production of the vinylcyclohexene (13) becomes increasingly important until at triplet energies of ≈50 kcal/mole, this product comprises greater than 40% of the reaction products. For sensitizers of triplet energy < 50 kcal/mole the quantum yield for the reaction drastically decreases and the cyclobutane products again predominate.

TABLE 10.1. *Product Distributions for the Photosensitized Dimerization of Butadiene*(13)

Sensitizer	E_t, kcal/mole	% (8)	% (9)	% (13)
Xanthone	74.2	78	19	3
Acetophenone	73.6	78	19	3
Benzaldehyde	71.9	80	16	4
o-Dibenzoylbenzene	68.7	76	16	7
Benzophenone	68.5	80	18	2
2-Acetylfluorenone	62.5	78	18	4
Anthraquinone	62.4	77	19	4
Flavone	62.0	75	18	7
Michler's ketone	61.0	80	17	3
4-Acetylbiphenyl	60.6	77	17	6
β-Naphthylphenyl ketone	59.6	71	17	12
β-Naphthaldehyde	59.5	71	17	12
β-Acetonaphthone	59.3	76	16	8
α-Acetonaphthone	56.4	63	17	20
α-Naphthaldehyde	56.3	62	15	23
Biacetyl	54.9	52	13	35
Benzil	53.7	44	10	45
Fluorenone	53.3	44	13	43
Duroquinone	51.0	72	16	12
β-Naphthil	51.0	57	15	28
Benzoquinone	50	~51	~6	~43
Camphorquinone	50	30	7	63
Pyrene	48.7	~30	~10	~60
Benzanthrone	47	55	10	35
3-Acetylpyrene	45	43	12	45
Eosin	43.0	60	17	23
Anthracene	42.5	75	10	15
9,10-Dibromoanthracene	40.2	78	19	3

The dependence of the cyclobutane/cyclohexene ratio on sensitizer triplet energy is explained by assuming that energy transfer from the sensitizer triplet occurs to both the s-*cis* and s-*trans* forms of the butadiene. Since the s-*trans* form of the butadiene is strongly predominant at room temperature,(11,15) the product distribution with high sensitizer triplet energies (>60 kcal/mole), in which energy transfer to diene would be expected to be diffusion controlled, is thought to represent the ground state population of s-*trans* (E_t = 59.8 kcal/mole) and s-*cis* ($E_t \simeq$ 53 kcal/mole) dienes, with the *trans* form yielding the cyclobutanes. The respective butadiene triplets are not expected to be readily interconvertible since the electron being excited is promoted from a molecular orbital that is antibonding in the ground state (between carbon atoms 2 and 3) to one that is bonding in the excited state.(16)

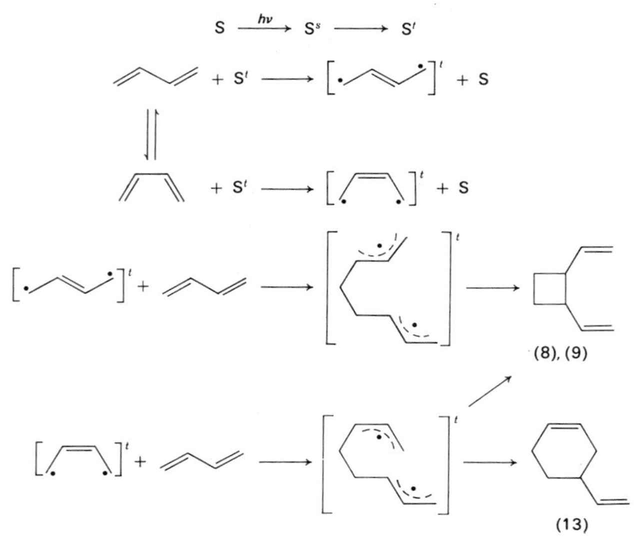

$$S \xrightarrow{h\nu} S^s \longrightarrow S^t$$

(8), (9)

(13)

FIGURE 10.1. Reaction scheme for the formation of cyclobutanes (8) and (9) and vinylcyclohexene (13) from butadiene.

As the triplet energy of the sensitizer becomes less than that necessary to excite the *trans*-diene triplet, energy transfer to the *cis* triplet becomes increasingly important and the product derived from this state (vinylcyclohexene) increases proportionally. This explanation is summarized in Figure 10.1.[11,14]

As the triplet energy of the sensitizer becomes less than that necessary to excite either form of the butadiene (E_t < 50 kcal/mole), it is proposed that energy is transferred via nonvertical excitation to lower energy twist forms of the diene triplets.[11] The product distribution again reflects the ground state population of s-*cis* and *s-trans* forms:

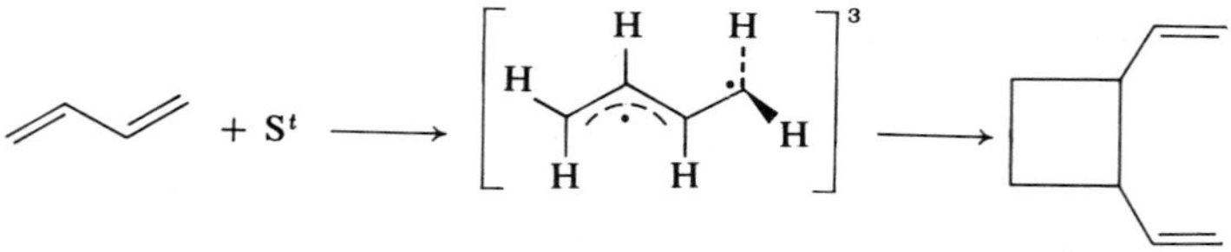

$$+ \; S^t \longrightarrow [\;]^3 \longrightarrow \qquad (10.7)$$

Another explanation has been offered to explain the large proportion of cyclobutane derivatives produced by low-energy sensitizers, especially for the anthracene derivatives.[17] This is that energy transfer to diene occurs from the second excited triplet state of the sensitizer rather than the first. Experiments using a large number of anthracene derivatives as sensitizers

TABLE 10.2. *Photosensitization of Butadiene Dimerization Using Anthracene Derivatives as Sensitizers*[17]

Sensitizer	E_t,[a] kcal/mole	E_t,[b] kcal/mole	% (8 + 9)	% (13)
9,9′-Bianthryl	—	—	96	4
Anthracene	74.4	42.5	95	5
2-Methylanthracene	73.5	40.6	95	5
9-Methylanthracene	—	40.6	95	5
9,10-Dichloroanthracene	—	40.2	95	5
9-Methyl-10-chloromethyl-anthracene	—	—	95	5
1,5-Dichloroanthracene	72.6	40.7	94	6
9,10-Dibromoanthracene	—	40.2	94	6
9-Methyl-10-chloroanthracene	—	—	92	8

[a] Energy for second excited triplet.
[b] Energy for lowest excited triplet.

have yielded product distributions characteristic of the high-energy sensitizers as shown in Table 10.2.

The fact that dimers (10)–(12), produced in the direct photolysis of butadiene, were not observed in the photosensitized experiments may indicate that these are derived from the excited singlet state of butadiene.[6]

The photosensitized dimerization of isoprene is considerably more complex than that of butadiene, yielding cyclobutanes (14)–(16) as well as four dimers of noncyclobutane types[9–11,18]:

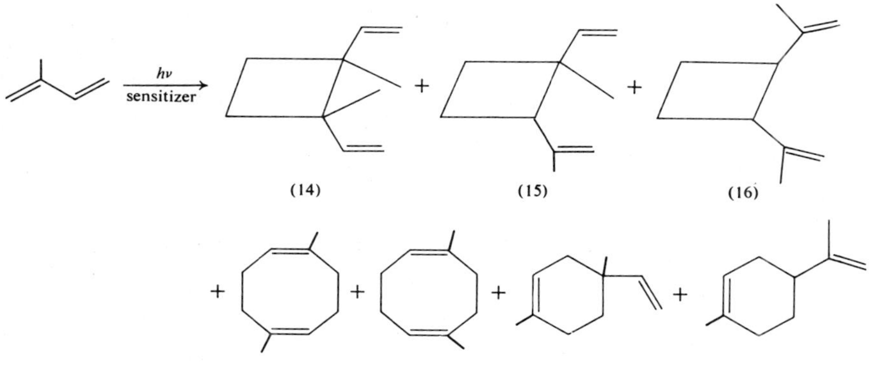

hv, sensitizer

(14) (15) (16)

(17) (18)

(10.8)

Dienes (17) and (18) may not actually be photoproducts but may arise from thermal rearrangement of *cis*-1,2-dialkenylcyclobutanes.[6,9] As in the photodimerization of butadiene, the product distributions of sensitized photolyses of isoprene were found to be dependent on the triplet energy of the photosensitizer and an explanation analogous to that proposed for butadiene photodimerization has been developed.[11,18] The quantum yield for the formation of all dimers in neat (10 *M*) isoprene was 0.40 with benzophenone (E_t = 69 kcal/mole), 0.25 with β-acetonaphthone (E_t = 59.3 kcal/mole), and 0.29 with fluorenone (E_t = 53.3 kcal/mole) as photosensitizer.[11,19] Stern–Volmer plots for these three sensitizers were linear with approximately the same slopes (indicating that k_d/k_r for the reaction of the diene triplets was nearly the same for all three sensitizers) although the yield of cyclobutanes fell from 68.1% (benzophenone) to 39.0% (fluorenone). This indicates that k_d/k_r for *cis* and *trans* isoprene triplets must be quite similar.[6]

Photodimerization of isoprene has also been observed using ferrocene as sensitizer[20] (E_t = 40.5 kcal/mole[21]). The product distribution in this case was similar to that observed using high-energy photosensitizers. To account for this behavior, the following mechanism was proposed,[20] where I^{t_n} indicates a triplet level of isoprene higher than the lowest triplet (I^{t_1}):

$$\begin{aligned} \text{ferrocene} + \text{I} &\rightleftharpoons \text{complex} \\ \text{complex} &\xrightarrow{h\nu} \text{complex}^{s_n} \\ \text{complex}^{s_n} &\longrightarrow \text{I}^{t_n} + \text{ferrocene}^{t_n} \\ \text{I}^{t_n} &\longrightarrow \text{I}^{t_1} \\ \text{I}^{t_1} + \text{I} &\longrightarrow \text{dimers} \end{aligned} \tag{10.9}$$

Although some evidence of the isoprene–ferrocene complex was obtained, a mechanism of this type may not be necessary to explain the product distribution since other sensitizers with triplet energies in this range also give distributions representative of high-energy sensitizers through nonvertical transfer or energy transfer from higher triplet levels.

The photosensitized dimerization of *cis*- and *trans*-piperylene has been found to lead to at least 15 products, six of which are cyclobutanes.[9]

Three products have been obtained from the photosensitized dimerization of cyclopentadiene,[9,22] in which the diene is constrained in the *cis* form:

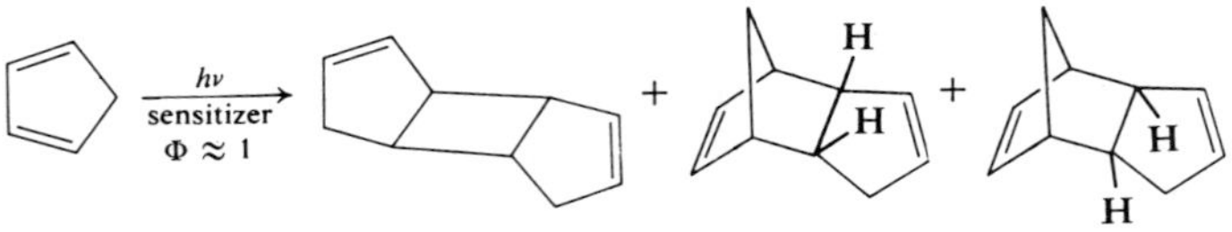

hν / sensitizer / Φ ≈ 1 (10.10)

H H H H

Significantly, there is no variation in the relative product distribution in this case as a function of the sensitizer triplet energy. 9-Anthraldehyde (E_t = 42 kcal/mole) was ineffective as sensitizer for this reaction, presumably because its lowest triplet is lower in energy than that of cyclopentadiene.[23]

Irradiation of 1,3-cyclohexadiene in the presence of a sensitizer affords dimers (19) and (20) as well as two other dimers[24,26]:

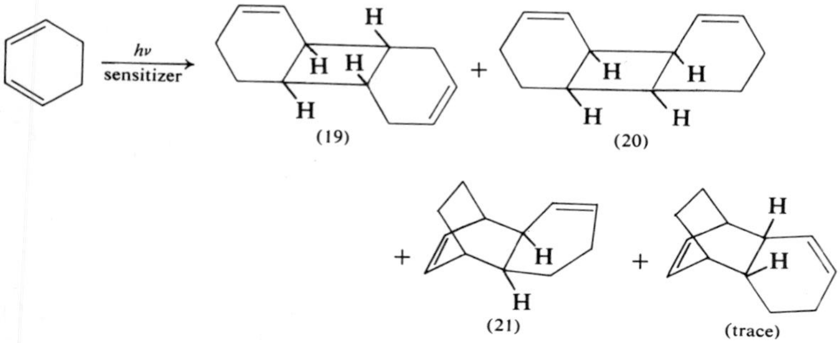

hν, sensitizer (19) (20) (21) (trace) (10.11)

As with cyclopentadiene, the relative yields of products (19)–(21) were relatively insensitive to the sensitizer triplet energy[27] although an effect of temperature on dimer distribution has been noted.[7,28] The direct photolysis of cyclohexadiene with wavelengths greater than 330 nm yielded products (19)–(21), although the product distribution in this case was more nearly statistical [(19):(20):(21) = 44%:24%:33%][28].

Photodimerization reactions of some other simple alkenes and dienes follow.[29,30,36,122] Although not a dimerization reaction, photochemical ring closures to yield cyclobutane derivatives are analogous and are included in this section[31–35]:

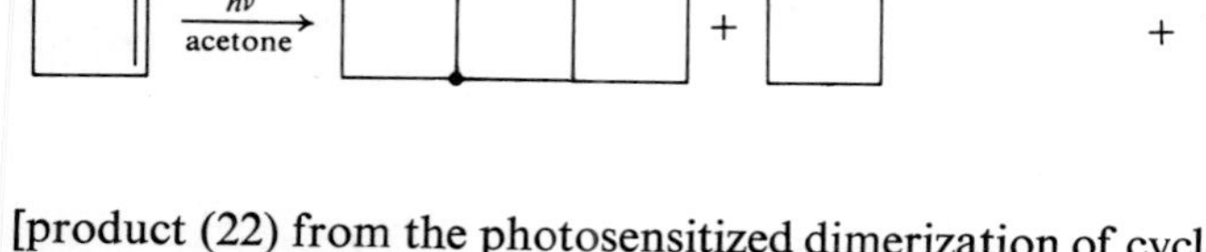

hν, acetone 80% 20% (10.12)

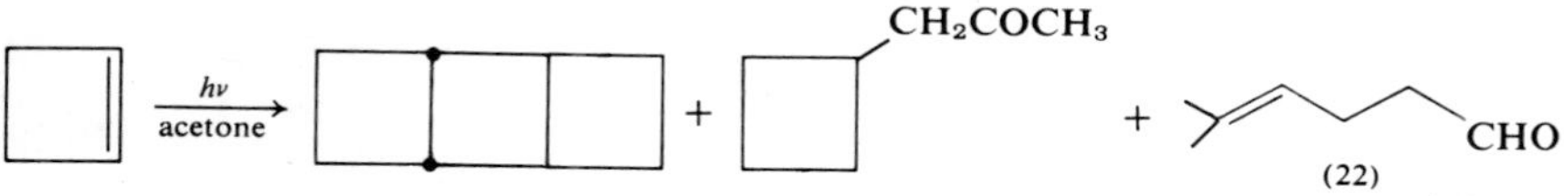

hν, acetone CH_2COCH_3 CHO (22) (10.13)

[product (22) from the photosensitized dimerization of cyclobutene is thought to be derived from ring opening of an intermediate oxetane. This indicates a competition between energy transfer from the acetone triplet to cyclobutene to lead to cyclobutene dimer and addition of the ketone triplet to ground state cyclobutene];

hν (10.14)

OH hν acetone OH (10.15)

hν acetone COOMe COOMe COOMe COOMe (10.16)

hν acetone 62% (10.17)

hν sensitizer (10.18)

H hν naphthalene H H H + H H + H H H H (10.19)

O O O hν sensitizer O O O O O O + O O O O O O (10.20)

The dimerization of *N*-vinylcarbazole and its derivatives proceeds with quantum yields well in excess of unity, indicating a chain mechanism such as shown below[123]:

N + sens.* ⟶ [R]$^{\dot{+}}$ + [sens.]$^{\dot{-}}$ $\xrightarrow{O_2}$ [R]$^{\dot{+}}$ $O_2^{\dot{-}}$ + sens.

(A) (B)

R + sens.

(10.21a)

(A) or (B) + R ⟶ R, +, H, H, •, R ⇌ [H, R, H, R]$^{\dot{+}}$

(C)

(10.21b)

(C) + R ⟶ H, R, H, R + (A) or (B) (10.21c)

(C) + $O_2^{\dot{-}}$ ⟶ H, R, H, R + O_2 (10.21d)

Me An-*p* $\xrightarrow{h\nu}$ Me, Me, An-*p*, An-*p* + Me An-*p*

(23)

$h\nu$ benzophenone

p-An Me ϕ O ϕ

(24)

(10.22)

Photodimerizations involving olefinic bonds conjugated to aromatic rings are common. Irradiation of *trans*-anethole produces the *cis–anti–cis* head-to-head dimer (23) as the sole dimeric product and *cis*-anethole[(37)], see page 433. In solutions containing benzophenone as sensitizer, in which the latter absorbs essentially all of the incident light, no dimeric product is formed but a high yield of the oxetane (24), which spontaneously decomposes in part to *p,p'*-dimethoxystilbene and acetaldehyde, is produced. Since the benzophenone triplet should have sufficient energy to excite the anethole to its triplet state, it appears that dimer (23) cannot result from the triplet level. Accordingly, a singlet excimer (I–I)* derived from the first excited singlet and a ground state molecule has been proposed to account for the remarkable stereospecificity of this dimerization.

Indene undergoes photosensitized dimerization to produce (25) in high yield[(38)]:

$$\text{indene} \xrightarrow[\substack{\text{benzophenone} \\ \Phi = 0.9}]{h\nu} \text{(25)} \qquad (10.23)$$

The following mechanism has been proposed to account for this photodimerization[(39)]:

$$\text{sens.} \xrightarrow{h\nu} \text{sens.}^{s} \longrightarrow \text{sens.}^{t} \qquad (10.24a)$$

$$\text{sens.}^{t} \longrightarrow \text{sens.} \qquad (10.24b)$$

$$\text{sens.}^{t} + \text{indene} \longrightarrow \text{sens.} + \text{indene}^{t} \qquad (10.24c)$$

$$\text{indene}^{t} \longrightarrow \text{indene} \qquad (10.24d)$$

$$\text{indene}^{t} + \text{indene} \longrightarrow 2\ \text{indene} \qquad (10.24e)$$

$$\text{indene}^{t} + \text{indene} \longrightarrow \text{dimer (25)} \qquad (10.24f)$$

Evidence for a step similar to (10.24e) was obtained by measuring the effect of indene concentration on the quantum yield of the photosensitized dimerization. That this energy-wasting step involved ground state indene was shown by irradiating an equimolar mixture of indene and *trans*-stilbene with benzophenone or Michler's ketone as sensitizer. In this case the quantum yield of stilbene isomerization was unchanged by the presence of indene although the dimerization of the latter was completely quenched. This effectively eliminates energy loss by a sensitizer–indene biradical as in the Schenck mechanism for photosensitized dimerization, shown below for indene[(39)]:

$$\text{sens.} \xrightarrow{h\nu} \text{sens.}^t$$

$$\text{sens.}^t + \text{indene} \longrightarrow \cdot\text{sens.–indene}\cdot$$

$$\cdot\text{sens.–indene}\cdot \longrightarrow \text{sens.} + \text{indene}$$

$$\cdot\text{sens.–indene}\cdot + \text{indene} \longrightarrow \text{sens.} + \text{dimer}$$

The photodimerizations of other indene derivatives are shown below[124]:

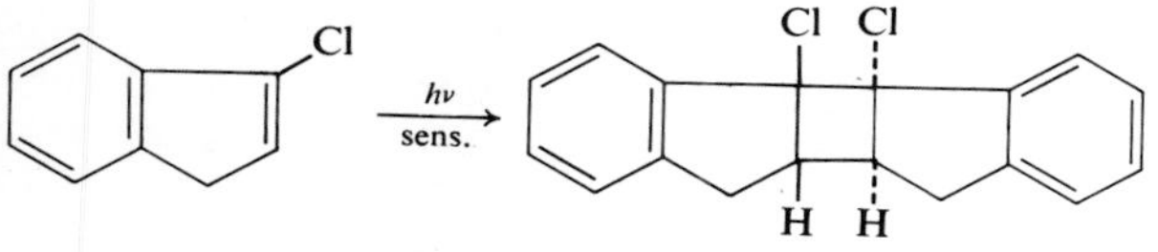

(10.25a)

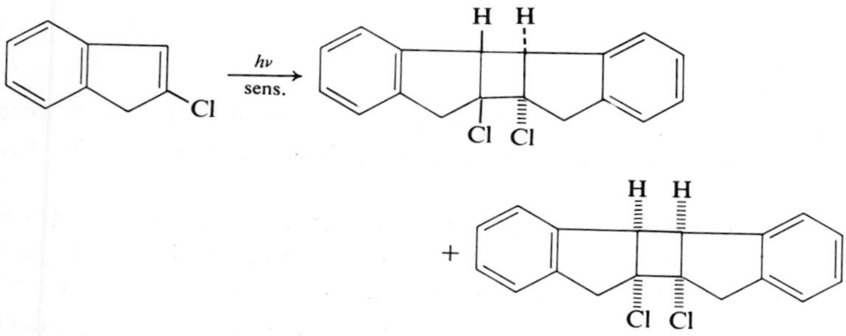

(10.25b)

The photodimerization of acenaphthylene has been extensively investigated[40–46]:

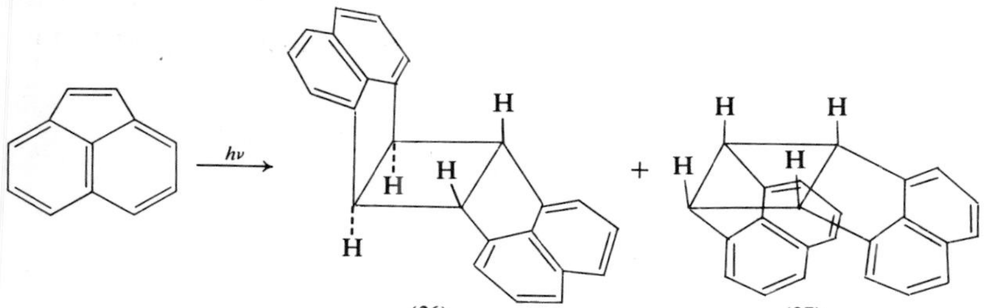

(26) (27) (10.26)

The ratio of *cis* (27) to *trans* (26) dimers produced in this reaction has been found to be solvent and concentration dependent, as can be seen in Table 10.3. The data in Table 10.3 show that the *cis* dimer predominates in all three solvents. It is also seen that as the acenaphthylene concentration is increased, proportionally more *cis* dimer is formed. This effect is slightly more pronounced in cyclohexane than in benzene (see Figure 10.2) Increases in the yield of dimer with increasing concentration would of course be expected

TABLE 10.3. *Solvent and Concentration Effects on the Photodimerization of Acenaphthylene*[41]

Solvent	Acenaphthylene[a]	*Cis/trans* dimers
Cyclohexane	5.0	3.69
	10.0	4.24
	15.2	4.97
	30.0	6.60
Benzene	5.0	1.74
	10.0	2.14
	15.2	2.41
	30.0	2.81
Methanol	15.2	5.74

[a] Grams of acenaphthylene per 150 ml of solvent.

since the dimerization is a bimolecular process. However, if both dimers were formed in a similar fashion, the concentration dependence should be the same for each isomer.

In order to determine the multiplicity of the reactive species, the photodimerization was carried out in the presence of the triplet quenchers oxygen and ferrocene. The results of these experiments are shown in Table 10.4.[41] It is obvious that the presence of oxygen exerts a large quenching effect on the production of the *trans* dimer and a smaller but significant effect on the formation of the *cis* dimer (the formation of *trans* dimer is decreased by oxygen by a factor of 25, while the *cis* dimer is decreased by a factor of 1.2). As with oxygen, the production of the *trans* dimer was quenched in the

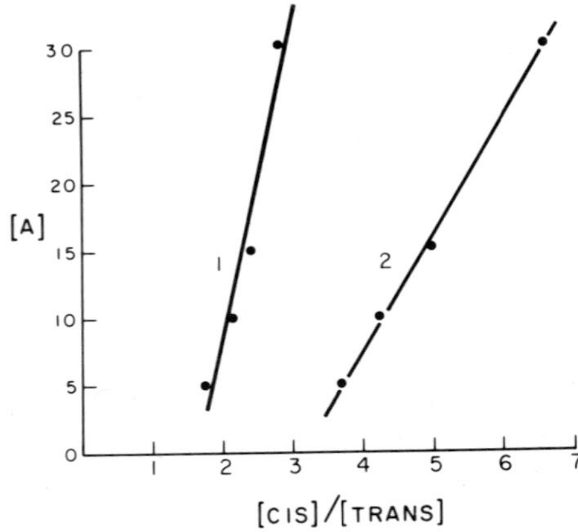

FIGURE 10.2. Ratio of *cis*-to-*trans* acenaphthylene dimers found in (1) benzene and (2) cyclohexane as a function of acenaphthylene concentration.

TABLE 10.4. *Quenching of Acenaphthylene Photodimerization*[a]

Quencher	[Quencher]	*Cis*(9)	*Trans*(9)	*Cis*/*trans*
—	—	5.42	1.09	4.97
O_2[b]	(sat.)	5.84	0.11	53.1
Ferrocene	5.67×10^{-5}	4.55	0.54	8.43
	8.03×10^{-5}	4.68	0.41	11.41
	1.06×10^{-4}	4.90	0.32	15.31
	2.10×10^{-4}	4.55	0.32	14.22
	3.49×10^{-4}	4.50	0.28	16.07
	5.31×10^{-4}	4.59	0.28	16.40

[a] All data for 15.2 g of acenaphthylene in 150 ml of cyclohexane, unless noted otherwise.
[b] Data for 15.2 g of acenaphthalene in 150 ml of benzene.

presence of ferrocene. A plot of Φ_0/Φ_q vs. the molar concentration of ferrocene (Stern–Volmer plot) is presented in Figure 10.3 for both *cis* and *trans* dimers. It is seen that for *trans*, quenching is linear for concentrations up to about 1×10^{-4} *M* ferrocene, whereupon a marked change in slope occurs and the quenching continues slowly but still in a linear manner. The *cis* dimer, however, is quenched only slightly and quenching is not increased as the concentration of ferrocene is increased. Further mechanistic information concerning this photodimerization was obtained by investigating the sensitized photolysis (Table 10.5).[(41)] Table 10.5 shows that the dimerization of acenaphthylene can be sensitized by molecules of triplet energy ≥44.6 kcal/mole. It appears therefore that the lowest triplet state of acenaphthylene must lie between 39 and 44.6 kcal/mole. It is noted in addition that ferrocene (E_t < 43 kcal/mole) served to quench the formation of the *trans* dimer. Thus the triplet energy of acenaphthylene must be equal to or higher than that of ferrocene.

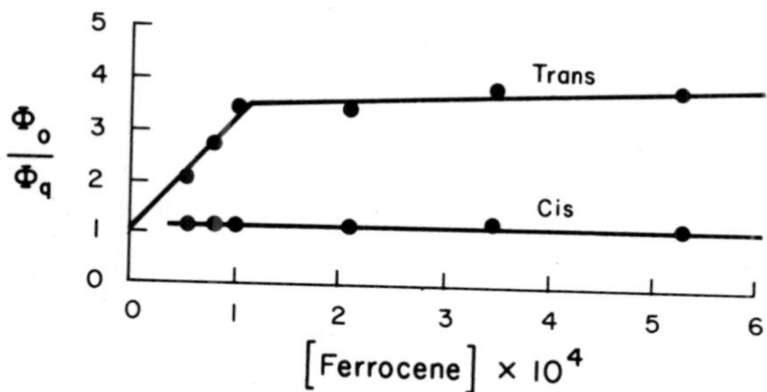

FIGURE 10.3. Stern–Volmer plot of the quenching of acenaphthylene dimer formation by ferrocene. (Reproduced with permission from Ref. 41.)

TABLE 10.5. *The Photosensitized Dimerization of Acenaphthylene*[a]

Sensitizer	E_t, kcal/mole	*Cis/trans*
Eosin-*y*	46.8	0.44
Rose Bengal	44.6	0.58
Crystal Violet	39	([b])
Hematoporphyrin	37.2	([b])
Methylene Blue	33.5	([b])

[a] All data for 11.5 g of acenaphthylene in 115 ml of methanol.
[b] No dimer was observed.

The quenching of the *trans* dimer with oxygen and ferrocene indicates that this product is formed almost entirely from the triplet state. It is possible to calculate the amount of triplet-derived product in benzene by subtracting the amount of product obtained in the presence of oxygen from the amount of product obtained in the absence of oxygen. Such a calculation indicates that acenaphthylene triplets in benzene give both *trans* and *cis* dimers in the ratio of 74:26. The triplet state accounts for almost all of the *trans* product and about 10% of the *cis* product. The break in the slope of the Stern–Volmer plot for the *trans* dimer (Figure 10.3) may be attributed to the presence of two excited species which are quenched at different rates. These two species could be (a) two different monomeric acenaphthylene triplet states T_1 and T_2 or (b) a monomeric acenaphthylene triplet state T_1 and a triplet excimer. This second triplet species is of relatively minor importance in the overall reaction since less than 5% of the total product in an unquenched reaction is due to this species.

Comparison of the dimer distribution from the direct photolysis of acenaphthylene in methanol to that for the same concentration (0.67 *M*) in the presence of a sensitizer indicates that the predominance of *cis* observed in the direct photolysis is reversed in the photosensitized process. Assuming that the presence of the sensitizer exerts no effect other than energy transfer, one would expect the dimer ratios to be identical in both cases if all products were triplet-derived. It appears that one of the products, the *cis* dimer, must be formed from two states of different multiplicity. Since the *cis* dimer is only slightly quenched by oxygen or ferrocene, a singlet reaction is indicated for this species. Since *trans* is almost totally eliminated by quenching, it appears that the singlet species exclusively forms the *cis* dimer; it is reasonable to propose that this singlet species is either formed from a ground state π complex (although no evidence of ground state complexes in acenaphthylene solutions has been obtained) or is a singlet excimer. Accordingly, the following mechanism was proposed[(41)]:

$$A + h\nu \longrightarrow A^s \tag{10.27a}$$

$$A^s + A \longrightarrow (A\cdots A)^s \tag{10.27b}$$

$$(A\cdots A)^s \longrightarrow 2A \tag{10.27c}$$

$$(A\cdots A)^s \longrightarrow A_2\ cis \tag{10.27d}$$

$$A^s \longrightarrow A \tag{10.27e}$$

$$A^s \longrightarrow A^t \tag{10.27f}$$

$$A^t + A \longrightarrow A_2\ trans\ \text{(major)} \tag{10.27g}$$

$$A^t + A \longrightarrow A_2\ cis\ \text{(minor)} \tag{10.27h}$$

$$A^t \longrightarrow A \tag{10.27i}$$

$$A^t + Q \longrightarrow A + Q \tag{10.27j}$$

$$A^t + A \longrightarrow 2A \tag{10.27k}$$

The symbols A, $(A\cdots A)^s$, and A_2 refer to acenaphthylene monomer, its singlet excimer, and its stable dimer(s), respectively. Experimental support for the singlet excimer mechanism [(10.27b)–(10.27d)] has been provided by Chu and Kearns,[47] who have shown that the *cis* (but not the *trans*) dimer in EPA (300–195°K) gives a broad fluorescence emission due to a singlet excimer. This excited state was also demonstrated to lead to dimer dissociation.

As mentioned briefly in Chapter 5, the photodimerization of acenaphthylene is subject to a very interesting heavy-atom solvent effect. The results of the photolysis of acenaphthylene in some heavy-atom solvents are given in Table 10.6.[42] The data in Table 10.6 show that the heavy-atom solvents *n*-propyl bromide and ethyl iodide yield product ratios similar to that obtained in the sensitized photolysis, indicating a greater role of the triplet state in

TABLE 10.6. *The Photodimerization of Acenaphthylene in the Presence of Some Heavy Atoms*

Solvent	*Cis*, g	*Trans*, g	*Cis/trans*
n-Butyl chloride	5.02	2.12	2.37
n-Propyl bromide	4.06	10.01	0.41
Ethyl iodide[a] (10 mole %)	2.78	10.96	0.25
Cyclohexane	5.42	1.09	4.97
Benzene	6.80	2.82	2.41

[a] In cyclohexane.

the presence of a heavy atom than in its absence. Processes that would be affected by spin–orbit coupling in acenaphthylene are as follows:

$$A^s \longrightarrow A^t \qquad [(10.27f)]$$

$$A^t + A \longrightarrow [\uparrow A\text{–}A \uparrow] \longrightarrow A_2\ (\textit{trans} \text{ or } \textit{cis}) \qquad [(10.27g, h)]$$

$$[\uparrow A\text{–}A \uparrow] \longrightarrow 2A \qquad [(10.27k)]$$

$$A^t \longrightarrow A \qquad [(10.27i)]$$

$$A + h\nu \longrightarrow A^t \qquad (10.27l)$$

$$A^t \longrightarrow A + h\nu \qquad (10.27m)$$

The importance of light absorption to directly populate the triplet state in the presence of heavy atoms [(10.27l)] and the effect of heavy atoms on the phosphorescence of acenaphthylene [(10.27m)] are easily ruled out. No new absorption bands or increased band intensities are noted in the presence of heavy atoms. Since no phosphorescence is observed, even at low temperature, there cannot be an important effect on this process. A heavy-atom effect on the partitioning of reactions (10.27g,h) and (10.27k) is eliminated by the data of Hartmann, Hartmann, and Schenck in their study of solvent effects on the photosensitized photolysis of acenaphthylene.[46] The influence of a number of solvents was described with the use of Kirkwood–Onsager solvent parameters (an empirical method for the correlation of reaction rate with the ability of the solvent to stabilize the change in dipole moment in proceeding to the transition state).[48] A linear plot of log(*trans*/*cis*) vs. the expression $[(D - 1)/(2D + 1)]\rho/M$, where D is the dielectric constant, ρ is the density, and M is the molecular weight of the solvent, was obtained for the product ratios of the Rose Bengal-photosensitized reaction. Since no heavy-atom effect on the sensitized reaction was observed (sensitized dimerizations in heavy-atom solvents followed the Kirkwood–Onsager relationship), the heavy-atom effect observed on the direct photolysis cannot be interpreted as an increased efficiency of triplet dimerization [(10.27g)]. This leaves two processes occurring from the excited acenaphthylene molecule which may be affected by heavy-atom perturbation [reactions (10.27f) and (10.27i)]. Although the observation of a greater proportion of triplet-derived product (*trans* dimer) in heavy-atom solvents would suggest that singlet-to-triplet intersystem crossing is more sensitive to heavy-atom perturbation than triplet decay, one cannot definitely decide which process is more sensitive since if bimolecular quenching [reaction (10.27h)] is more effective by several orders of magnitude than reaction (10.27i) in depopulating the triplet manifold, an increase in triplet-derived products would result regardless of a relative effect of the heavy-atom on intersystem crossing and unimolecular decay.[42]

Recent evidence of Koziar and Cowan indicates that the heavy-atom assisted intersystem crossing from the acenaphthylene triplet state to the ground state [reaction (10.27i)] can become important for experiments with ethyl iodide.[43b] They found that the quantum yield of dimerization of acenaphthylene rapidly increases (0.006 to 0.17) as the concentration of ethyl iodide increases from 0 to 10 mole % and then slowly decreases as the concentration of ethyl iodide is increased above 10% (30%, 0.13; 60%, 0.088; 100%, 0.056). This is consistent with a large initial increase in dimerization due to an increase in reaction (10.27f) until intersystem crossing from the first singlet to the lowest triplet occurs with unit efficiency. Then a decrease in dimerization results from the decrease in triplet concentration due to the increasing importance of reaction (10.27i). It is obvious, then, that while the use of heavy-atom solvents to control preparative organic photoreactions holds great promise, the applications must be carefully formulated.

It has been possible to employ the heavy-atom solvent effect in determining the rate constants for the various intercombinational nonradiative transitions in acenaphthylene and 5,6-dichloroacenaphthylene.[43b,c,d] These rate constants, which are not accessible in light-atom solvents due to the complexity of the mechanism and the low efficiency of intersystem crossing from the first excited singlet to the first excited triplet, can be readily evaluated under the influence of heavy-atom perturbation.

The following simplified mechanism is consistent with all of the kinetic data, where A = 5,6-dichloroacenaphthylene or acenaphthylene:

$$A + h\nu \longrightarrow A^s \qquad I_a$$

$$A^s \xrightarrow{k_{1c}} A \qquad k_{1c}[A^s]$$

$$A^s \xrightarrow{k_{isc}} A^t \qquad k_{isc}[A^s]$$

$$A^s + HA \xrightarrow{k'_{isc}} A^t + HA \qquad k'_{isc}[A^s][HA]$$

$$A^t + A \xrightarrow{k_2} A_2\ (\textit{cis}\ \text{and}\ \textit{trans}) \qquad k_2[A^t][A]$$

$$A^t + A \xrightarrow{k_{cq}} 2A \qquad k_{cq}[A^t][A]$$

$$A^t + HA \xrightarrow{k''_{isc}} A + HA \qquad k''_{isc}[A^t][HA]$$

$$A^t \xrightarrow{k_d} A \qquad k_d[A^t]$$

$$A^t + Q \xrightarrow{k_q} A + Q \qquad k_q[A^t][Q]$$

A_2 is the dimer, HA is ethyl iodide, and Q is ferrocene. This mechanism is essentially the same as that proposed for the dimerization of acenaphthylene. The steps involving possible singlet excimer formation and singlet dimerization have been excluded since they are at best relatively unimportant processes in heavy-atom solvents. Using the steady-state approximation, we can derive the following expression for Φ_D, the quantum yield of dimerization:

$$\Phi_D = \Phi_{isc} \frac{k_2[A]}{k_2[A] + k_{cq}[A] + k''_{isc}[HA] + k_q[Q] + k_d} \quad (10.28a)$$

By setting Φ_{isc} equal to unity, we obtain the following relationships:

$$\frac{1}{\Phi_D} = \frac{(k_2 + k_{cq})[A] + k_d + k_q[Q]}{k_2[A]} + \frac{k''_{isc}}{k_2[A]} \cdot [HA] \quad (10.28b)$$

$$\frac{1}{\Phi_D} = \frac{k_2 + k_{cq}}{k_2} + \frac{k''_{isc}[HA] + k_d + k_q[Q]}{k_2} \cdot \frac{1}{[A]} \quad (10.28c)$$

From these equations it is obvious that $1/\Phi_D$ should be and, in fact, is a linear function of both [HA] and 1/[A].

The Stern–Volmer plot is represented by the following equation:

$$\frac{\Phi_D{}^0}{\Phi_D} = 1 + k_q\tau[Q] \quad (10.28d)$$

where k_q is the rate constant for the quenching of the triplet, τ is the triplet lifetime, and [Q] is the concentration of ferrocene. Assuming that k_q is diffusion controlled, the slope of the Stern–Volmer plot yields an estimate of the triplet lifetime.

The value of k_2, the rate constant for dimerization, can then be calculated from the following equation:

$$\frac{\Phi_D{}^0}{\tau} \cdot \frac{1}{[A]} = k_2 \quad (10.28e)$$

By substituting this value of k_2 into the analytical expression for the intercept given for the $1/\Phi_D$ versus 1/[A] equation and setting this expression equal to the calculated value of the intercept, an estimate of k_{cq} may be determined. The rate constant for unimolecular decay, k_d, was obtained by setting the intercept given in equation (10.28b) equal to the calculated value of this intercept, and by substituting the values of k_2, k_{cq}, and [A] into the expression for the intercept. Since no quencher was present, k_q[Q] was set equal to zero. The rate constant for heavy-atom quenching, k''_{isc} was determined by substituting the values for k_2 and [A] into the expression for the slope in equation (10.28b), and this was then set equal to its calculated value. The results of these calculations are presented in Table 10.7.

It is not unreasonable that the substitution of chlorines onto the acenaphthylene ring should have such a significant influence on the intersystem crossing processes, k_d and k_{isc}. It is somewhat surprising, however, that the other processes involving spin-inversion were affected to such a small extent.

TABLE 10.7. *Triplet Lifetime and Rate Constants*

Quantity	5,6-Dichloroace-naphthylene	Acenaphthylene
τ, sec	6.6×10^{-7}	2.15×10^{-6}
k_2, M^{-1} sec^{-1}	9.1×10^5	6.58×10^5
k_{cq}, M^{-1} sec^{-1}	4.0×10^6	1.58×10^6
k_d, sec^{-1}	6.8×10^5	9.20×10^4
k''_{isc}, M^{-1} sec^{-1}	1.5×10^5	6.71×10^4

The rate constants for dimerization, concentration quenching, and heavy-atom quenching undergo an approximately two-fold increase as a result of heavy-atom substitution, as compared to the seven-fold increase observed for k_d.

The photodimerization of various acenaphthylene derivatives has also been reported:

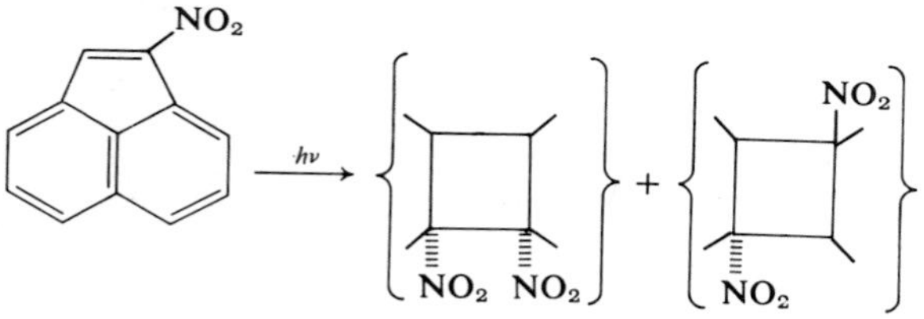

(10.29a)

The dimerization of 1-nitroacenaphthylene has been proposed to arise from an excited singlet state although little experimental evidence has been revealed.

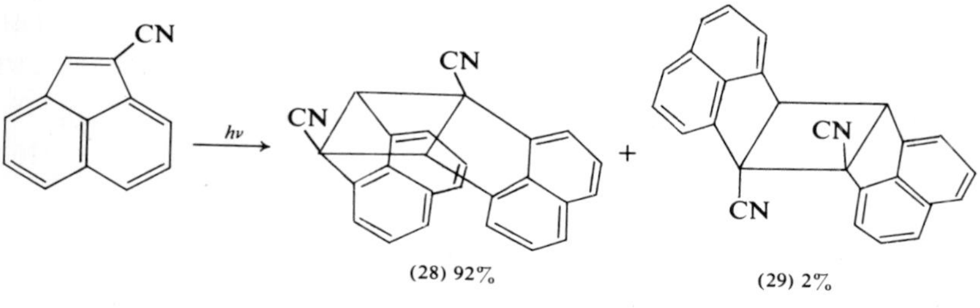

(28) 92%

(29) 2%

(10.29b)

35%

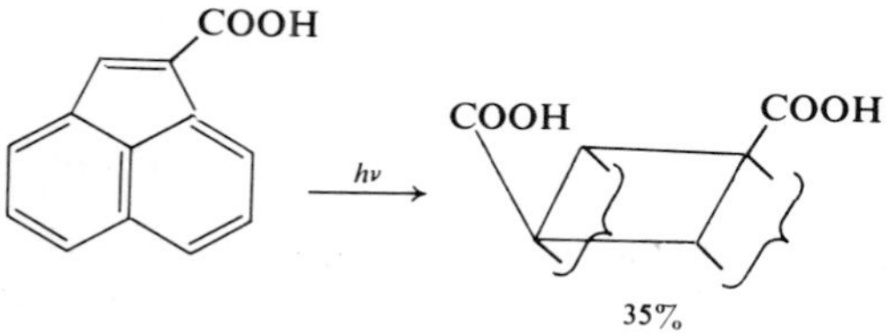

(10.29c)

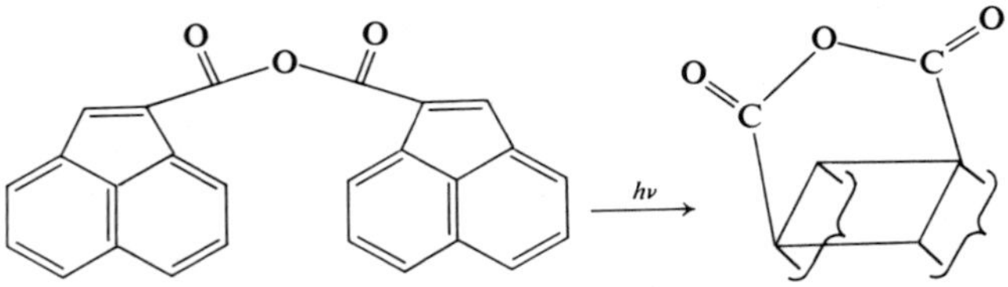

(10.29d)

In the presence of a heavy-atom solvent, cyanoacenaphthylene yielded 80.5% dimer (29) and 4.5% dimer (28), in contrast to the dimer distribution produced in the absence of a heavy atom.[50]

Photodimerizations of other olefins conjugated with aromatic rings are shown below[51–53]:

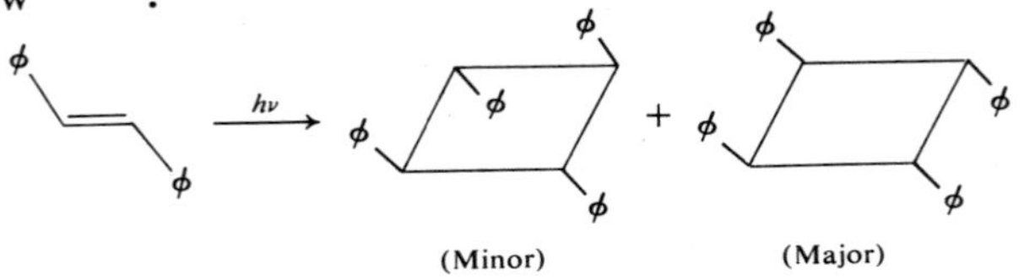

(Minor) (Major)

(10.30)

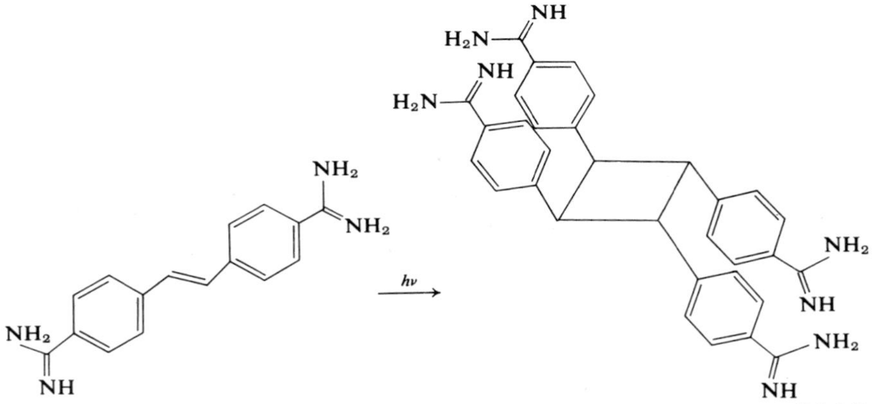

(10.31)

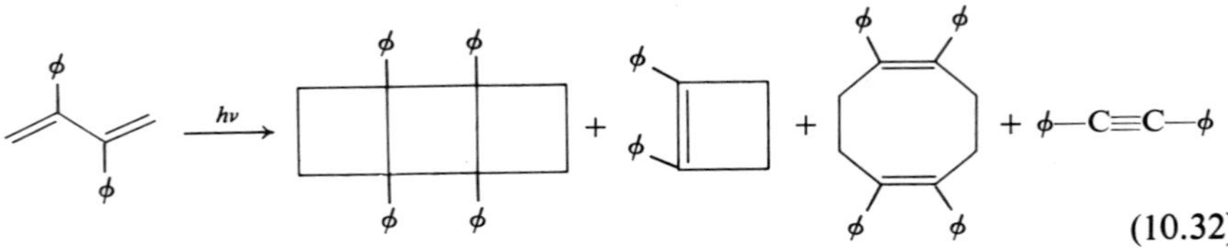

(10.32)

10.1b. *Photocycloaddition Reactions of Olefins and Polyenes*

Many examples of photocycloaddition reactions of olefins and polyenes yielding cyclobutane derivatives have been reported. Irradiation of mixtures of butadiene and 1,1-dichloroethylene in the presence of a sensitizer yields cyclobutane (30) as well as the butadiene dimers discussed earlier[27]:

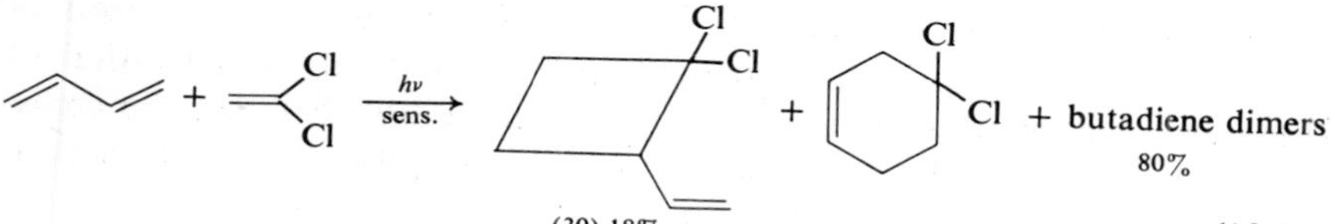

(30) 18%

80%

(10.33)

With 1,1-difluoro-2,2-dichloroethylene, butadiene yielded cyclobutane (31) in the presence of benzophenone:

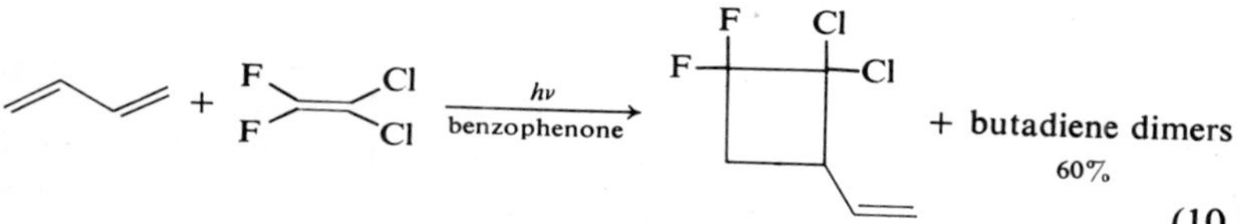

(31) 40%

60%

(10.34)

Both cyclobutane products (30) and (31) are thought to arise from the most stable diradical intermediate[27]:

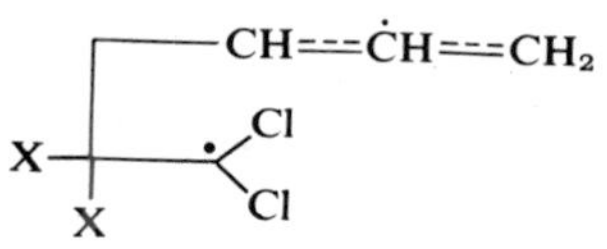

Irradiation of equimolar mixtures of butadiene and 2-acetoxyacrylonitrile in the presence of sensitizers is reported to yield cross-adducts (32)–(34) in addition to butadiene dimers[54]:

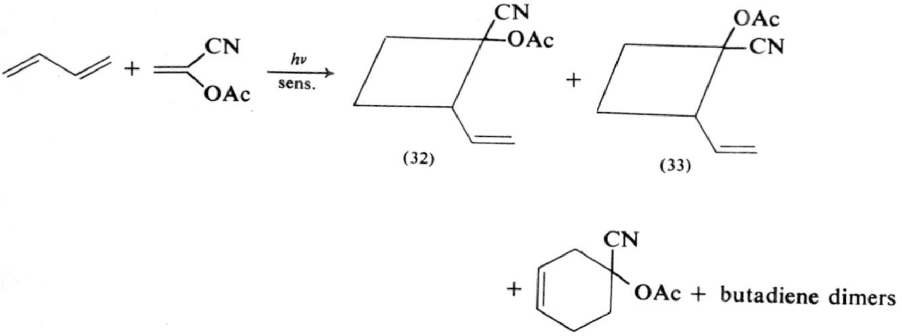

(32)

(33)

(34)

(10.35)

The product distribution in this reaction was found to be dependent upon the triplet energy of the sensitizer. Variation of the relative amounts of cyclobutanes (32) and (33) to cyclohexene (34) with sensitizer triplet energy was found to parallel quite closely that for the cyclobutanes and vinlycyclohexene produced from the dimerization of butadiene, although relatively more cyclobutanes were produced in the cross-addition with sensitizers of triplet energies above 50 kcal/mole than observed in the dimerization of butadiene. This was attributed to a faster ring closure to form the cross-adduct than to form the dimer due to polar contributions in the transition state for coupling of two dissimilar radicals[55]:

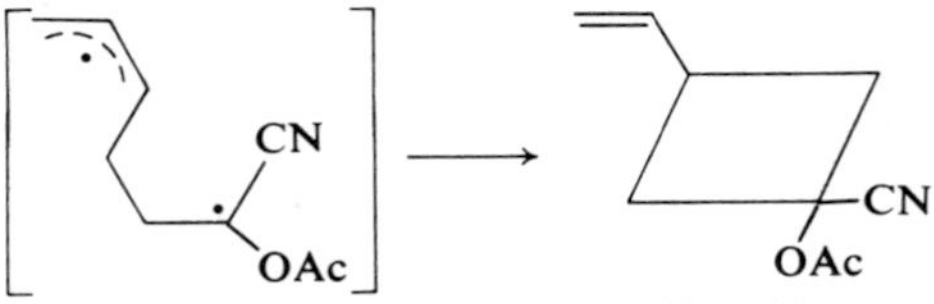

CN OAc → CN OAc (10.36)

(32) + (33)

Similar effects were observed in the photocycloaddition of isoprene and α-acetoxyacrylonitrile[55b,56] but not in the cross-addition of the acrylonitrile and cyclopentadiene.[55b,57]. Other photocycloadditions involving butadiene are shown below[58,59]:

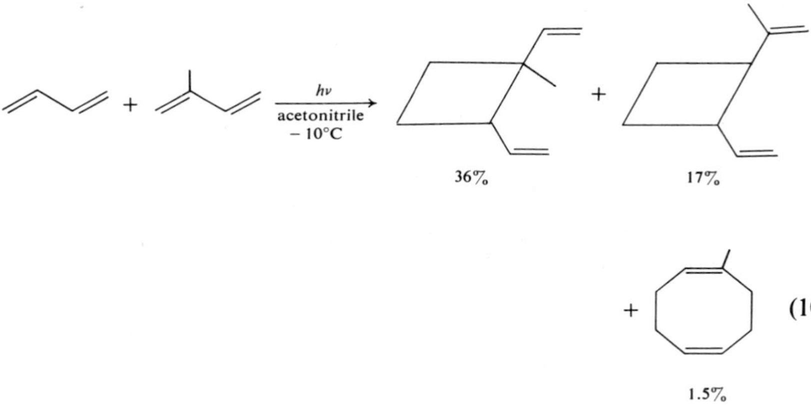

$h\nu$, acetonitrile, − 10°C

36% + 17%

+ 1.5% (10.37)

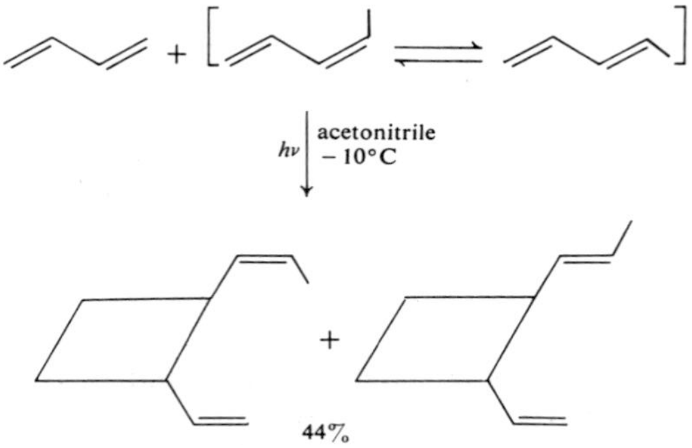

$h\nu$, acetonitrile, − 10°C (10.38)

44%

hν acetophenone −10%C

58% 15%

(10.39)

hν sens.

dimers of butadiene and cyclopentadiene (10.40)

Isoprene and piperylene undergo photochemical cross-addition to olefins to yield products similar to those observed for butadiene[27]:

F Cl F Cl hν benzophenone F Cl F Cl F Cl isoprene dimers

26%

(10.41)

F Cl F Cl hν sens. F Cl F Cl F Cl F Cl

(10.42)

Irradiation of a mixture of hexadiene isomers in 1,1-dichloroethylene with β-acetonaphthone gave four cyclobutane derivatives[54]:

2,4-hexadiene + $CH_2{=}CCl_2$ $\xrightarrow[\beta\text{-acetonaphthone},\ -18°C]{h\nu}$

15.5% + 15.1% + 3.2% + 2.2%

(10.43)

Cyclopentadiene has been reported to undergo photocycloaddition with 1,1-dichloroethylene,[27] difluorodichloroethylene,[27] and 1,2-dichloroethylene[60] to yield the following products, equations (10.44)–(10.46).

The relative reactivity of cyclopentadiene and *cis*-dichloroethylene toward triplet cyclopentadiene was found to be greater than 20:1 while that for cyclopentadiene and *trans*-dichloroethylene is less than 5:1. Thus the *trans* isomer is about four times more reactive toward the triplet cyclopentadiene than the *cis* isomer. An interesting temperature dependence of the product distribution of this reaction has been reported (Table 10.8). The data in Table 10.8 indicate that the relative amount of 1,4 addition [products (39) and (40)] is much more sensitive to temperature than 1,2 addition [products (35)–(38)], especially for the *trans*-olefin. The data also indicate that some rotation about the CHCl-CHCl bond occurs in intermediate radicals derived from both *cis*- and *trans*-dichloroethylene. However, rotational equilibrium is not established at ring closure since the ratios of *cis*-dichlorocyclobutanes

cyclopentadiene + $CH_2{=}CCl_2$ $\xrightarrow[\text{benzophenone},\ 0°]{h\nu}$ 48% + 32% + pentadiene dimers 20%

(10.44)

cyclopentadiene + $F_2C{=}CCl_2$ $\xrightarrow[\text{benzophenone}]{h\nu}$ 60% + 30%

+ pentadiene dimers (10.45)

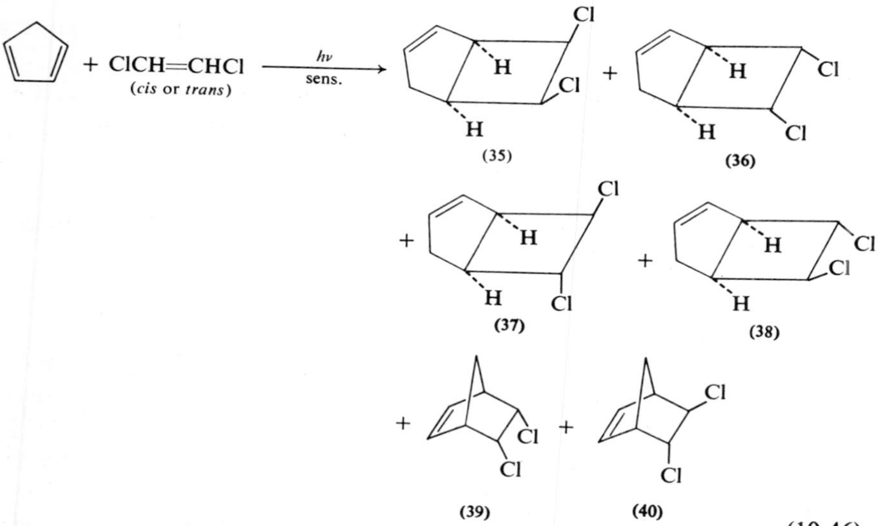

+ ClCH═CHCl (cis or trans) $\xrightarrow[\text{sens.}]{h\nu}$ (35) + (36) + (37) + (38) + (39) + (40) (10.46)

[(35) and (36)] to *trans*-dichlorcyclobutanes [(37) and (38)] are not the same for both stereoisomeric forms of the olefin. These ratios are more nearly equal at 25°C, however.

1,1-Diphenylethylene has been reported to add to several olefins.[61] The cross-addition of this olefin with isobutene yields (41) in 63% yield:

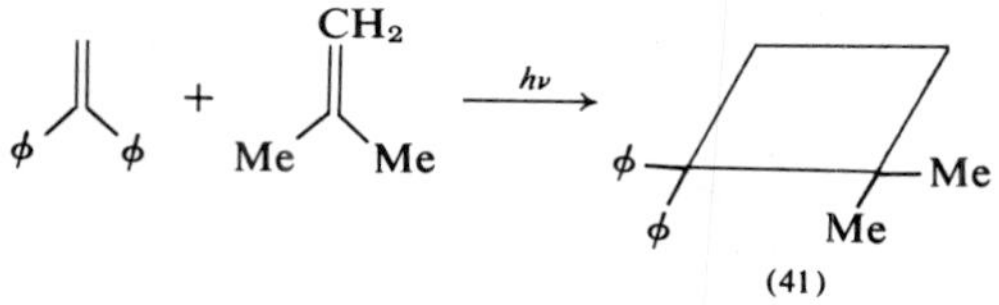

$$\phi_2C{=}CH_2 + CH_2{=}CMe_2 \xrightarrow{h\nu} (41) \qquad (10.47)$$

TABLE 10.8. *Effect of Temperature on the Product Distribution from the α-Acetonaphthone-Sensitized Photoaddition of Cyclopentadiene to cis- and trans-Dichloroethylene*[60]

Olefin	Temp., °C	% Cross-adducts (35)	(36)	(37 + 38)	(39)	(40)
cis-Dichloroethylene	−24.8	2.7	22.3	62.0	1.2	11.8
	1	2.8	21.6	61.9	1.0	12.6
	25.0	3.1	20.3	63.0	0.9	12.7
trans-Dichloroethylene	−24.8	4.1	7.6	24.1	2.2	62.0
	1.7	4.2	10.0	29.3	2.5	54.0
	25.0	3.8	14.1	38.3	2.9	40.8

Product (41) was also obtained in the presence of xanthone as sensitizer. This product could be totally eliminated by the triplet quencher piperylene, indicating (41) to be derived from the diphenylethylene triplet state.

Photoaddition of *trans*-stilbene to tetramethylethylene yields cyclobutane (42) in high yield. *Cis*-stilbene also results from the photolysis:

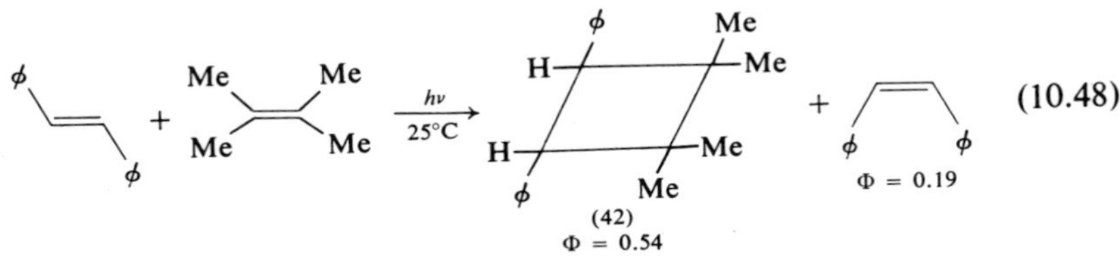

(10.48)

The quantum yield for the formation of the cycloaddition product has been found to be temperature dependent, increasing by a factor of approximately three as the temperature is lowered from 65 ($\Phi = 0.24$) to 5°C ($\Phi = 0.69$). Photolysis of mixtures of the olefin and *trans*-stilbene in the presence of sensitizers yielded no cycloaddition product (42) but rather only *cis*-stilbene. This suggests that the cycloadduct is produced via a singlet reaction. This conclusion is supported by the fact that tetramethylethylene quenches fluorescence from the *trans*-stilbene singlet. A plot of $1/\Phi_{(42)}$ vs. 1/[TME] (TME = tetramethylethylene) is linear. The slope of this plot yields rate constants for cycloadduct formation which show a negative temperature dependence. To account for this fact, a reversibly formed exciplex leading to (42) was proposed in the following mechanism[62]:

$$S + h\nu \longrightarrow S^s$$

$$S^s \longrightarrow S$$

$$S^s \longrightarrow S + h\nu'$$

$$S^s \longrightarrow S^t$$

$$S^s(\textit{trans}) \longrightarrow S(\textit{cis})$$

$$S^s + \text{TME} \rightleftharpoons [S\cdots\text{TME}]^s \longrightarrow \text{adduct}$$

A similar temperature dependence has been observed in the photoaddition of *trans*-stilbene to 1-methylcyclohexene.[63]

The photoaddition of β-nitrostyrene to olefins has been studied by Chapman and co-workers[64]:

NO_2 + hν → (43) NO_2 (10.49)

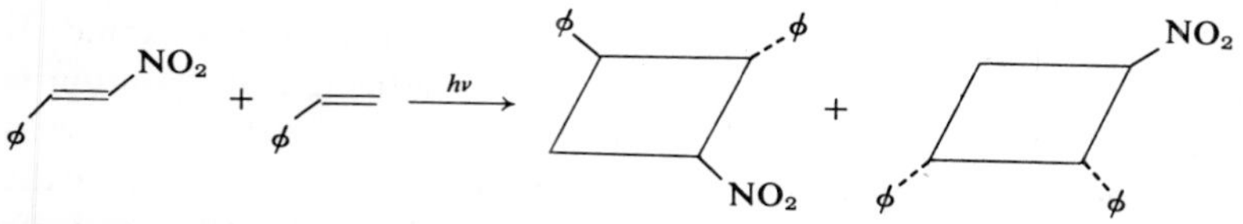

ϕ NO$_2$ + ϕ —hν→ (10.50)

These reactions are stereospecific in that the phenyl and nitrogroups derived from the β-nitrostyrene are always *trans*. In the addition of 1,1-diphenylethylene, the phenyl groups are situated on adjacent carbon atoms in the product (43). Only one isomer is thus formed. Stereospecificity in this reaction is postulated to arise as follows:

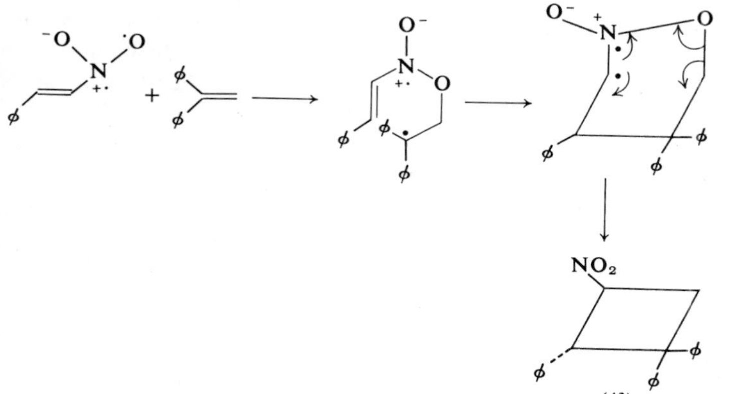

(10.51)

(43)

β-Nitrostyrene adds to 2,3-dimethylbutadiene to produce similar products, probably by a mechanism analogous to that above[65]:

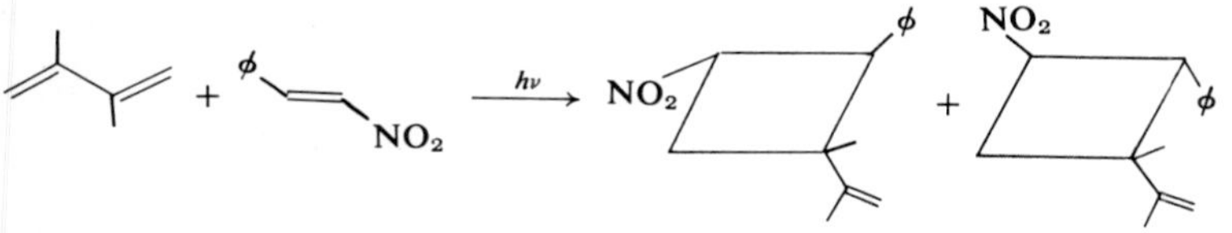

(10.52)

Cis- and *trans*-stilbenes have been reported to add to 2,3-dihydropyran upon photolysis to yield adducts (44) and (45) as well as two stilbene dimers[66]:

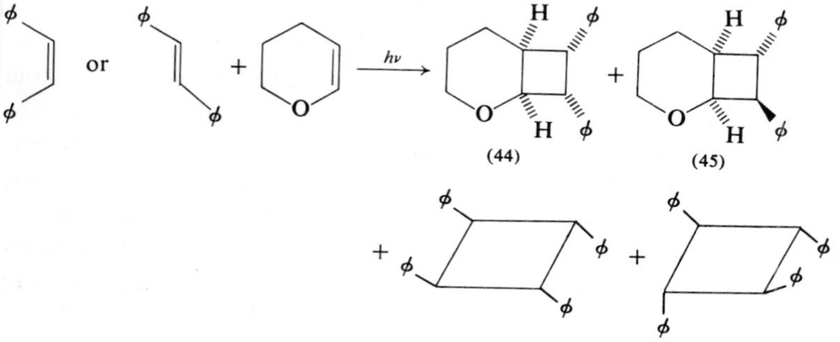

(44) (45)

(10.53)

The same product ratio is obtained regardless of which stilbene isomer is photolyzed. In the presence of oxygen the rate of stilbene isomerization is retarded but no effect on the production of the four products above was observed. No dimers or cycloadducts were produced by photolysis in the presence of the triplet sensitizer triphenylene when the latter absorbed the light. These results effectively eliminate products (44) and (45) from being triplet-derived. The invariance of the product ratio with respect to the ratio of *cis*- and *trans*-stilbenes would also appear to eliminate a singlet reaction, assuming a concerted reaction from this state. The authors conclude, therefore, that formation of (44) and (45) results from vibrationally excited ground states populated through internal conversion from the first excited singlet states of the stilbenes. The effect of 2,3-dihydropyran on stilbene fluorescence was not reported. Another explanation we favor is that the phantom stilbene singlet state is being trapped by 2,3-dihydropyran (see Chapter 9).

Acenaphthylene undergoes photocycloaddition to cyclopentadiene[67] and acrylonitrile[68] to yield the following products:

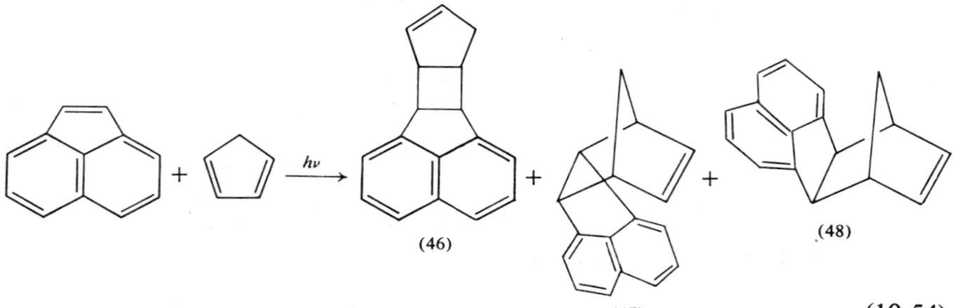

hν (46) (47) (48) (10.54)

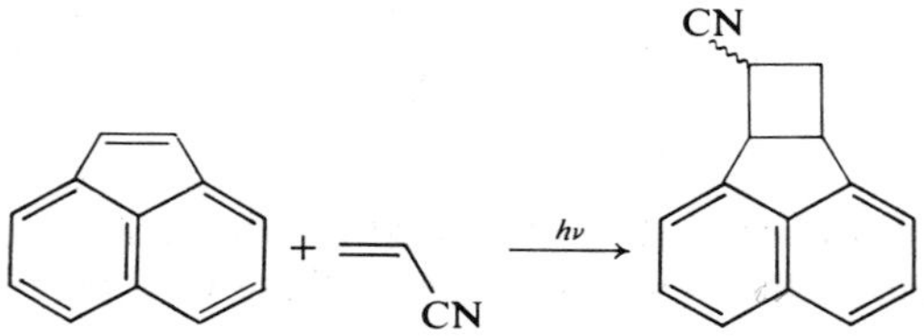

CN hν CN syn + anti (10.55)

The photoaddition of acenaphthylene to cyclopentadiene was shown to be sensitive to the presence of heavy atoms in the solvent, as shown in Table 10.9. The data in Table 10.9 show that product (46) increases from 18% of the total product in cyclohexane to 38% of the total product in 1,2-dibromoethane. This suggests that the [4 + 2] cycloaddition products (47) and (48) and the [2 + 2] product (46) are produced from different excited states. Accordingly, some of the [4 + 2] product has been postulated to arise from either (a) a singlet excited state or (b) a vibrationally excited ground state

TABLE 10.9. *Heavy-Atom Effect on the Photocycloaddition of Acenaphthylene to Cyclopentadiene*[67]

Solvent	(46)[a]	(47)[a]	(48)[a]	Φ
Cyclohexane	1.00	2.6	2.1	0.008
Acetonitrile	1.00	2.9	2.8	0.013
Bromoethane	1.00	0.54	1.5	0.12
Bromobenzene	1.00	0.44	1.3	0.15
1,2-Dibromoethane	1.00	0.44	1.2	0.20

[a] Product ratios normalized to (46) for each run.

(Diels–Alder reactions are thought to be forbidden as a concerted process from electronic excited states).

A heavy-atom effect on the photocycloaddition of acenaphthylene to acrylonitrile has also been observed.[68] The effect of heavy atoms in this case is seen as an apparent increase in the quantum yield of product formation in heavy-atom solvents as opposed to cyclohexane (the time to achieve about 42% reaction in cyclohexane is greater than that required to produce the same conversion in dibromoethane by a factor of ten). An increase in the rate of acenaphthylene intersystem crossing due to heavy-atom perturbation was proposed to explain this increase in reaction rate.

Plummer and Ferree[69] have utilized the photoaddition of 5-bromoacenaphthylene to cyclopentadiene to compare external and internal heavy-atom effects in the acenaphthylene system:

$h\nu$, $\lambda > 366$ nm

Br, Br, H, H, Br, Br, H, H

(49) (50) (51)

(10.56)

The data obtained in this study are shown in Table 10.10. Comparison of the data in Table 10.9 with those in Table 10.10 shows the latter system to be considerably less sensitive to the presence of a heavy atom than the acenaphthylene–cyclopentadiene system. This can also be seen in a plot of Φ for these two systems vs. the concentration of bromine in the solvent (Figure 10.4). The insensitivity of the ratio of products derived from 5-bromoacenaphthylene to the heavy-atom solvents suggests that these products are triplet-derived. Figure 10.4 shows that the quantum yields for product formation in the acenaphthylene system are higher in heavily brominated solvents (e.g., ethylene bromide) than those for product formation in the 5-bromoacenaphthylene system. These effects are explained by the relative magnitudes of the rate constants for intersystem crossing k_{isc} and triplet deactivation k_d:

$$A + h\nu \longrightarrow A^s$$

$$A^s \xrightarrow{k_{isc}} A^t$$

$$A^t \xrightarrow{k_d} A$$

$$A^t + \text{cyclopentadiene} \xrightarrow{k_r} \text{products}$$

If k_{isc} and k_d are much larger in 5-bromoacenaphthylene than in acenaphthylene and both rate constants are relatively large in the former, the quantum yield of reaction would be expected to be less sensitive to small increases in k_{isc}, due to external heavy-atom perturbation, and more sensitive to increases in k_d resulting from external heavy atoms.

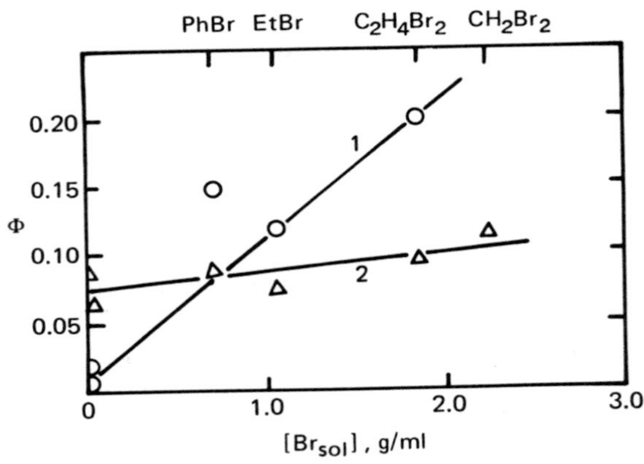

FIGURE 10.4. Plot of Φ for (1) acenaphthylene and (2) 5-bromoacenaphthylene versus concentration of bromine in the solvent. Reproduced with permission from Ref. 69.

TABLE 10.10. *The Photocycloaddition of 5-Bromoacenaphthylene to Cyclopentadiene in Various Solvents*(69)

Solvent	(49)[a]	(50)[a]	(51)[a]	Φ
Cyclohexane	1.00	0.53	1.93	0.067
Acetonitrile	1.00	0.66	3.00	0.086
Bromobenzene	1.00	0.55	1.68	0.087
Ethyl bromide	1.00	0.60	2.00	0.077
1,2-Dibromoethane	1.00	0.64	1.74	0.093
Methylene bromide	1.00	0.54	1.66	0.112

[a] Product ratios normalized to (49) for each run.

10.2. PHOTODIMERIZATION AND PHOTOCYCLOADDITION REACTIONS OF AROMATIC COMPOUNDS

Benzene undergoes photocycloaddition with simple olefins to produce 1,3, 1,2 and 1,4 adducts as shown below for tetramethylethylene:

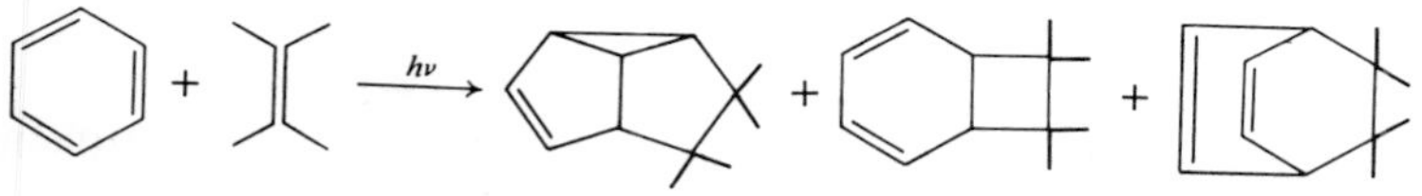

$h\nu$ (10.57)

1,3-Cycloaddition 1,2-Cycloaddition 1,4-Cycloaddition

The products consistently exhibit retention of the original olefin stereochemistry and are probably formed in a concerted manner. An exciplex formed from singlet excited benzene and the ground state olefin (allowing relaxation of the orbital symmetry requirements for concerted 1,3- and 1,4-cycloaddition) has been proposed to account for these products.(126)

Srinivasan and Hill reported an unusual photochemical addition to benzene to form cycloadduct (52)(74):

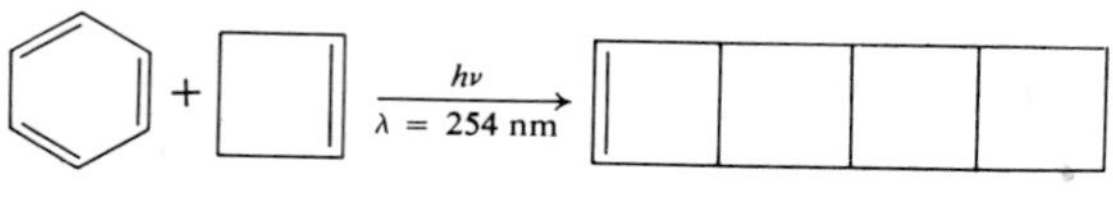

$h\nu$, λ = 254 nm (10.58)

(52)

More recent investigations have revealed the product to be a mixture of (52a) and (52b)(125):

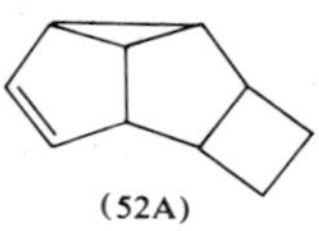

(52A) 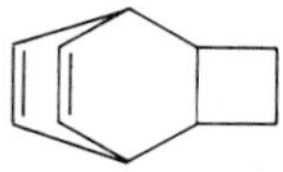 (52B)

Irradiation of a 1:1 mixture of benzonitrile and 2-methyl-2-butene for an extended period yields the 1:1 adduct (53) in a 63% yield:

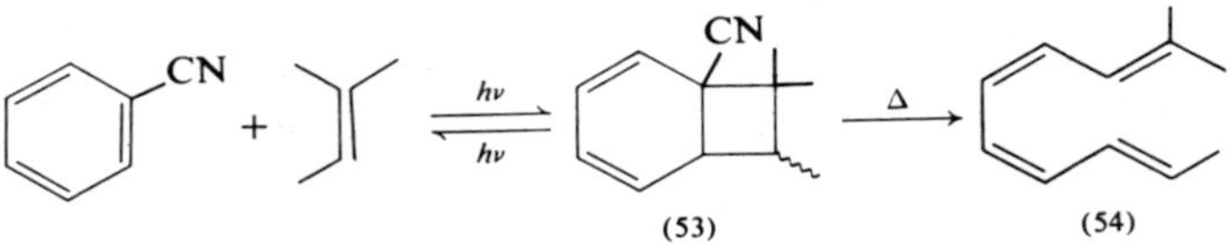

(10.59)

(53) (54)

Irradiation of (53) causes decomposition to starting materials, whereas pyrolysis yields the tetraene (54).[75]

Irradiation of benzene in the presence of butadiene is reported to result in a mixture of photoproducts including (55)–(59)[70–72]:

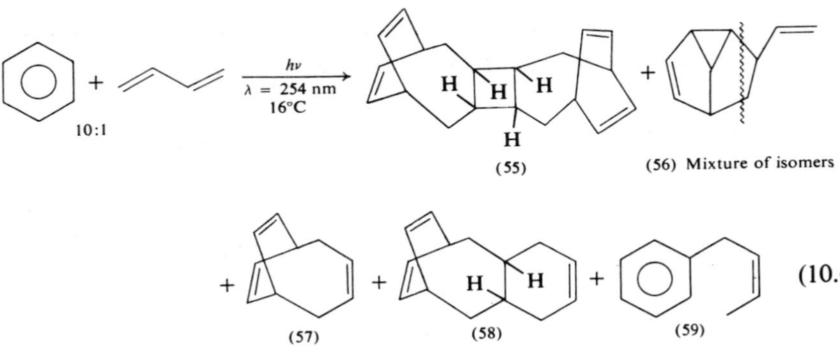

10:1 $h\nu$, λ = 254 nm, 16°C

(55) (56) Mixture of isomers

(57) (58) (59) (10.60)

The mechanism for the formation of the 2:1 adducts and 2:2 adducts is believed to involve [4 + 4] addition of excited benzene to the *trans* diene to yield (60), which either dimerizes or reacts with ground state butadiene[72]:

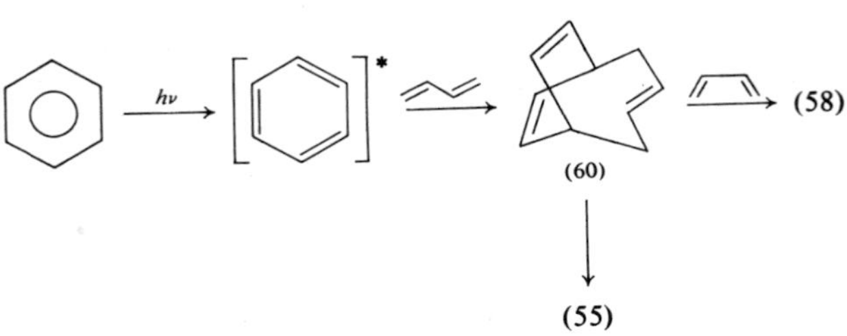

$h\nu$ → (58) (10.61)

(60)

(55)

Irradiation of a 1:1 molar mixture of benzene and butadiene (−80°C in pentane) yielded only (55). Alkylbenzenes are reported to yield similar products with butadiene although structures were not identified.[72]

Mixtures of benzene and isoprene yield products (61) and (62) plus isoprene dimers[70,71]:

(61)
(two isomers)
23%

(10.62)

(62)
46%

Attempts to sensitize this reaction resulted in no product formation, suggesting participation of a benzene singlet in the direct photolysis. Irradiation of isoprene in excess benzene resulted in products similar to those observed with butadiene.[73] Irradiation of toluene and *o*- and *p*-xylene in the presence of isoprene yielded products similar to (62).[72]

Acrylonitrile adds to naphthalene upon irradiation to produce the products shown[76]:

(10.63)

A singlet naphthalene or a singlet exciplex is thought to be the reactive species in this reaction since the quantum yield of cycloaddition parallels the quenching of naphthalene fluorescence by acrylonitrile.

Phenanthrene is reported to add to *cis*- and *trans*-dichloroethylene and acrylonitrile to yield a variety of products (Table 10.11)[77]:

$$+ \text{X—CH=CH—Y} \xrightarrow{h\nu} \qquad (10.64)$$

X = Y = Cl (*cis* and *trans*)
X = H, Y = CN

TABLE 10.11. *Photocycloaddition of Phenanthrene with Dichloroethylenes and Acrylonitrile*[77]

Olefin	Product	Stereochemistry	Yield, %
cis-Dichloroethylene	I	*Cis, trans, cis*	21.9
	II	*Cis, cis, cis*	7.4
	III	*Cis, trans, cis*	3.0
trans-Dichloroethylene	I	—	24.9
	II	—	32.4
	III	—	0.6
Acrylonitrile	V	*Syn*	78.4
	VI	*Anti*	19.6

Naphthalene undergoes photocycloaddition with diphenylacetylene to produce (63) and (64), with the latter thought to result through light absorption by product (63)[78]:

$$+ \phi C{\equiv}C\phi \xrightarrow{h\nu} (63) + (64) \qquad (10.65)$$

Attempts to produce dimers of naphthalene similar to those observed for anthracene (Chapter 2) have been unsuccessful, although three naphthalene derivatives have been reported to produce dimer (65) upon photolysis[79] (the structure of these dimers have been the object of some debate, however):

$$\underset{h\nu}{\overset{\Delta}{\rightleftharpoons}} \text{ or } (65) \qquad (10.66)$$

R = OCH_3, OCH_2CH_3, CN

Why these naphthalene derivatives should dimerize while most others are stable to ultraviolet irradiation is unclear.

Another aromatic derivative reported to dimerize while the parent aromatic hydrocarbon does not is 9,10-dicyano-phenanthrene[80]:

NC CN ... $\xrightarrow{h\nu}$ ... CN CN / CN CN (10.67)

10.3. PHOTODIMERIZATION AND PHOTOCYCLOADDITION REACTIONS OF α,β-UNSATURATED CARBONYLS AND ACID DERIVATIVES

Many interesting examples of photodimerizations and photocycloadditions of α,β-unsaturated carbonyls and acid derivatives yielding cyclobutanes have been reported. In this section, as in Section 10.1, we will first discuss photodimerizations and then photocycloadditions.

10.3a. *Photodimerization of α,β-Unsaturated Carbonyls and Acid Derivatives*

One of the first examples of photodimerization of an α,β-unsaturated carbonyl was that of dibenzylideneacetone reported by Ciamician and Silber. Direct irradiation of this compound in ethanol or in isopropanol/benzene solution yields cyclobutane (66), although irradiation in the presence of uranyl chloride as sensitizer results in (67)[81]:

ϕ—CH=CH—CO—CH=CH—ϕ $\xrightarrow[\text{ethanol or isopropanol/benzene}]{h\nu}$ (66) [ϕ, ϕ, COCH=CHϕ, COCH=CHϕ]

uranyl chloride, $h\nu$ → (67) [ϕ, COCH=CHϕ, ϕCH=CHCO, ϕ] (10.68)

This suggests that (66) is formed from the α,β-unsaturated ketone excited singlet and (67) is formed from the lowest excited triplet state.

The photodimerization of 2-cyclopentenone has been studied by Eaton and Hurt[82] and Ruhlen and Leermakers[83]:

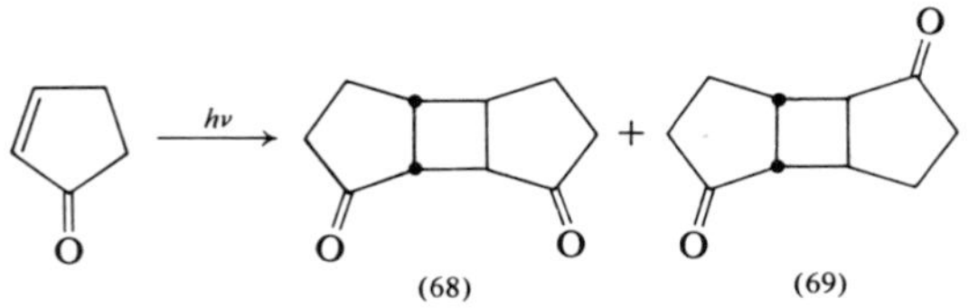

$$\xrightarrow{h\nu} \text{(68)} + \text{(69)} \tag{10.69}$$

This reaction is quenched by piperylene and is sensitized by xanthone (E_t = 74.2 kcal/mole) at a rate directly comparable to that observed in the direct irradiation, implicating a reactive triplet species. A linear Stern–Volmer plot was obtained for the piperylene quenching with the quantum yield for dimerization approaching zero at high quencher concentrations. The product ratio (68):(69) has been found to depend upon both the concentration of the enone and the solvent polarity, the head-to-head dimer (68) becoming increasingly important at higher concentrations of enone and increasing solvent polarity. This behavior is thought to result from increased stabilization of the intermediate leading to (68) over that leading to (69) with high concentrations or high polarity.[82] Experiments in which the rate of dimerization for direct photolysis is compared to the rate observed with sensitizers where the latter absorb varying amounts of the incident light indicate that the quantum yield for intersystem crossing in cyclopentenone must be very near unity.[83a] To account for the low Φ_r compared to Φ_{isc}, reversible formation of a triplet excimer has been proposed.[83b]

The photodimerization of 2-cyclohexenone to yield cyclobutanes (70) and (71) has also been reported[84]:

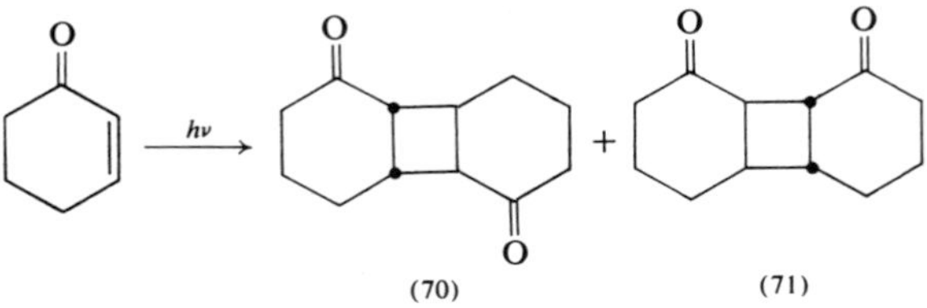

$$\xrightarrow{h\nu} \text{(70)} + \text{(71)} \tag{10.70}$$

As with 2-cyclopentenone, the ratio (70):(71) varies with the molar concentration of the enone, the head-to-head dimer (71) becoming increasingly important at higher concentrations.[133] This reaction is efficiently sensitized by acetophenone, benzophenone, thioxanthone, and naphthalene. The same enone concentration effect was observed in the sensitized photo-dimerization as in the direct photolysis. Similarly, quenching of the dimerization by piperylene was not accompanied by a change in dimer ratio. Systematic examination of the ultraviolet spectrum of 2-cyclohexenone at varying concentrations showed a definite deviation from Beer's law in the absorption band (evidence for aggregated species) at about 366 nm. However, there was no significant difference between the quantum yield for photodimerization at 313 nm (where Beer's law is obeyed) and that at 366 nm. It has been concluded that monomeric and aggregated excited species must equilibrate at a rate rapid in comparison to that required for dimerization.

More recent investigations of the photodimerizations of cyclopentenone and cyclohexenone[134] have revealed that both reactions result from the lowest triplet states of the enones ($E_t \geq 70$ kcal/mole), which is probably $\pi \rightarrow \pi^*$ in character.[133] Both photodimerizations occur via reversibly formed metastable triplet species which can decay to ground state enones or produce dimers:

$$E^t + E \longrightarrow (E\text{–}E)^*$$

$$(E\text{–}E)^* \longrightarrow 2E$$

$$(E\text{–}E)^* \longrightarrow \text{dimers}$$

The metastable species involved in these reactions may be either excited charge-transfer complexes or biradicals, although a charge-transfer complex would be expected to result in a predominance of head-to-head dimers. Dipole effects, on the other hand, would favor formation of head-to-tail dimers, especially in relatively nonpolar solvents:

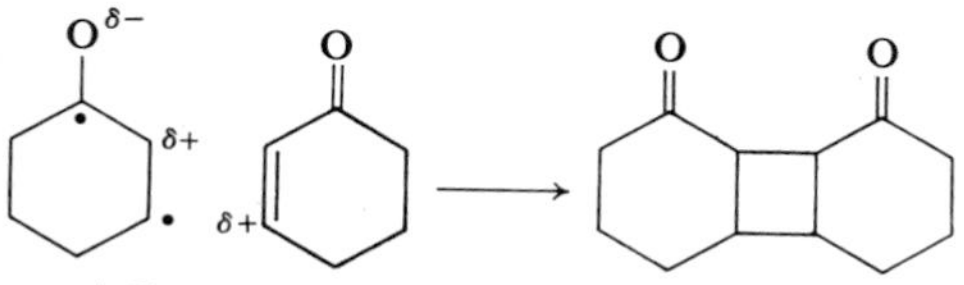

(10.71)

π, π^* Charge-transfer

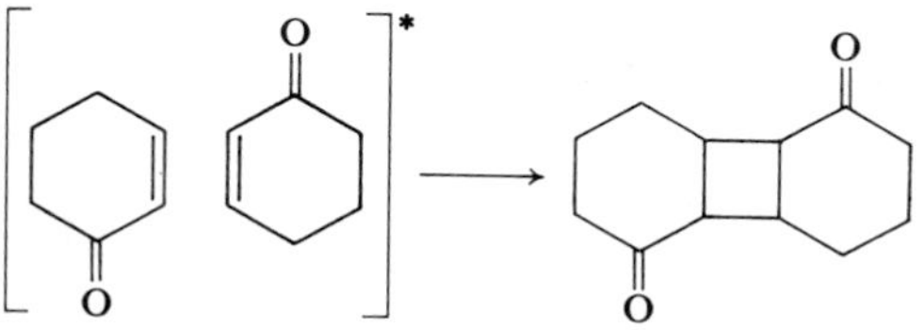

(10.72)

Dipole effects

Comparison of the triplet state rate constants for these two enones indicates that the ratio of triplet deactivation to dimerization for cyclohexenone is 45 times larger than that for cyclopentenone. This is thought to arise from deactivation of the cyclohexenone triplet by twisting about the double bond.[134] Significantly, the larger ring enones cycloheptenone and cyclooctenone do not dimerize at all.[135]

Isophorone yields dimers (72)–(74) upon irradiation [85]:

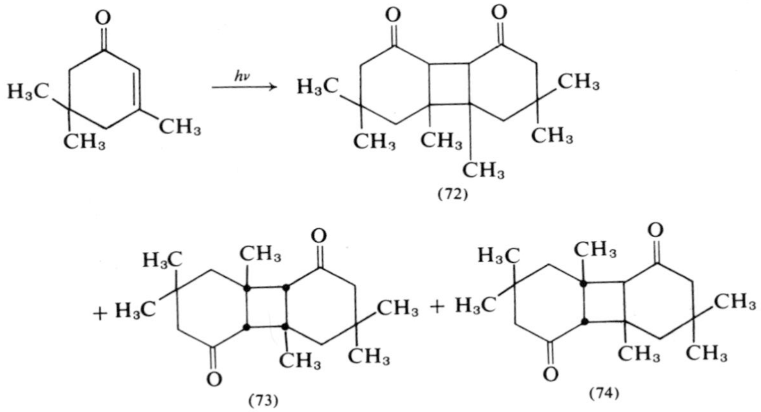

(72)

(10.73)

(73) (74)

Dimers (73) and (74) were formed in approximately equal amounts in all cases, although, as in the cases of 2-cyclopentenone and 2-cyclohexenone, the relative amount of (72) (either *cis-syn-cis* or *cis-anti-cis*) was found to vary substantially with solvent polarity. As in 2-cyclopentenone, this increase in the rate of head-to-head dimerization was attributed to stabilization of the increase in dipole moment in going to the transition state leading to (72) in polar solvents. It is thought that the solvent effect in this case is not associated with the state of aggregation since a plot of Φ_r vs. 1/[isophorone] is linear. A linear Stern–Volmer plot and complete quenching with 0.2 *M* piperylene indicate that the reaction proceeds mainly from the triplet manifold. However, the rates of formation of head-to-head and head-to-tail dimers do not show the same relationship when sensitized by benzophenone as in the direct photolysis. This effect, when combined with different intercepts for head-to-head and head-to-tail dimerizations quenched by piperylene in the Stern–Volmer plot, indicates that two distinct excited triplet states are involved with differing efficiencies of population. The nature of these two triplets has not been disclosed.

Rubin and co-workers have investigated the photodimerization of the 4,6-diene-3-ketosteroid (75)[86]:

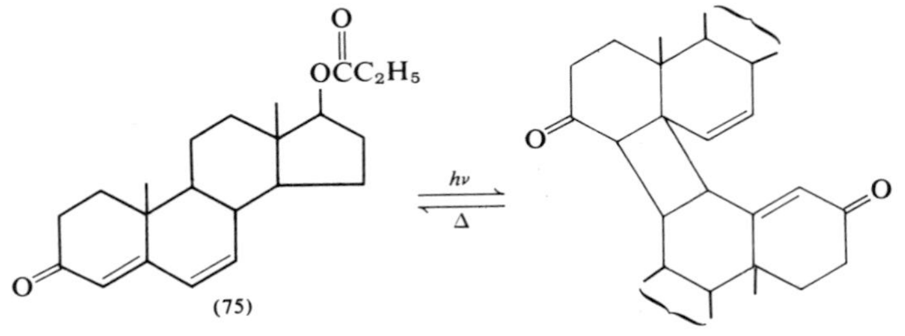

(75)

(10.74)

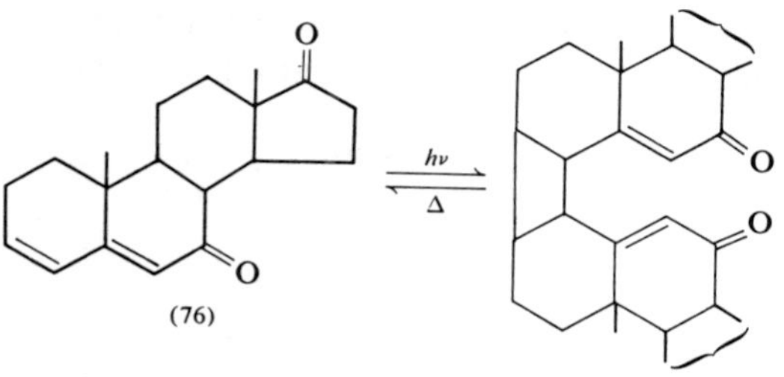

(76)

(10.75)

With two γ,δ double bonds, two α,β double bonds, and the possibilities of *cis* and *trans* ring fusions with *syn* and *anti* configurations, 20 isomeric dimers are possible. Surprisingly, only one product is formed in a head-to-tail fashion. The sole product of the irradiation of the 3,5-diene-7-ketosteroid (76), however, is the head-to-head dimer. The specificity and mode of addition arise presumably through the effect of the specific environment of the chromaphore. The dimerization of (75) is believed to involve the addition of the α,β double bond of a photoexcited molecule to the less hindered γ,δ double bond of a ground state molecule. The photocondensation of (76) with cyclopentene, in which steric hindrance should not be a controlling factor, was found to yield a cyclobutane product involving the α,β bond of the steroid in contrast to dimerization across the γ,δ bond.

Photodimerizations of quinone derivatives are quite common. Representative examples are as follows[87,88]:

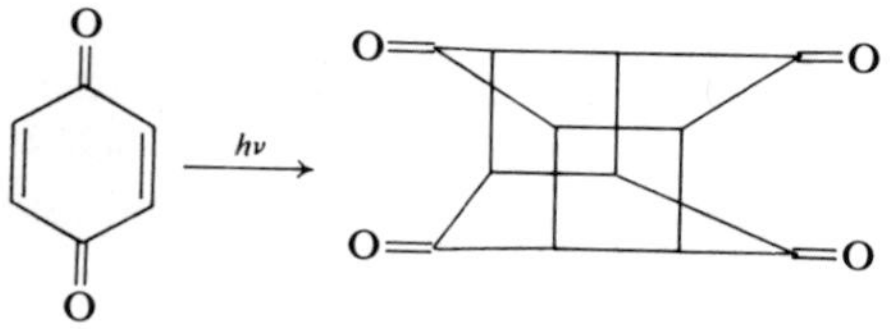

(10.76)

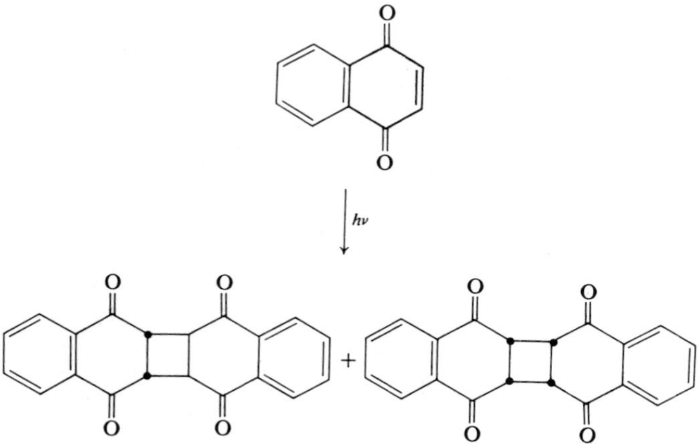

(10.77)

Many derivatives of maleic and fumaric acid are known to undergo photosensitized dimerization.[121] Some derivatives of maleic acid are as follows:

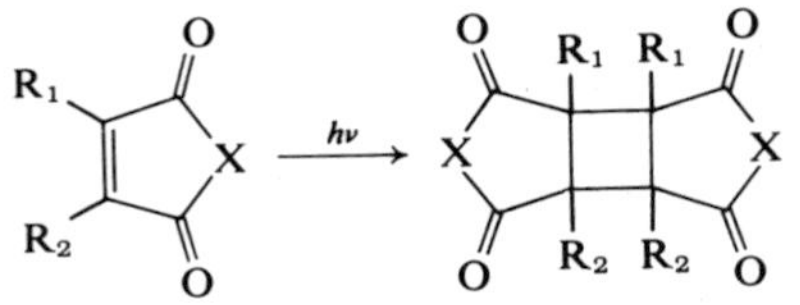

(10.78)

(a) $R_1 = R_2 = CH_3$, $X = O$
(b) $R_1 = CH_3$, $R_2 = H$, $X = O$
(c) $R_1 = R_2 = H$, $X = O$
(d) $R_1 = R_2 = CH_3$, $X = NH$
(e) $R_1 = R_2 = H$, $X = N{-}\phi$
(f) $R_1 = R_2 = H$, $X = N{-}C_6H_{11}$

Other α,β-unsaturated carbonyl and acid derivatives, such as dimethylfumarate, 2,5-dimethylquinone, and *trans*-cinnamic acid, are known to dimerize in the crystalline state and will be discussed in Section 10.4.

2,6-Dimethyl-4-pyrone forms the cage dimer (77) upon irradiation in concentrated solutions in ethanol, benzene, or acetic acid[89]:

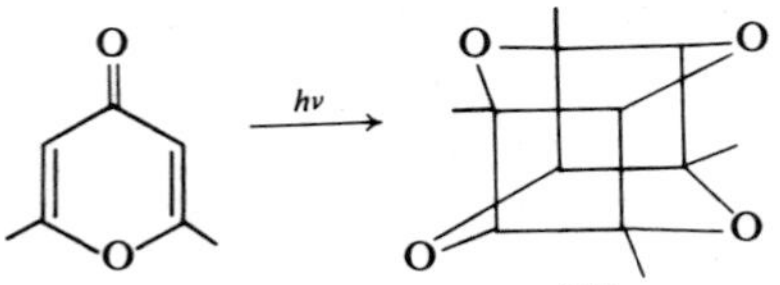

(10.79)

(77)

The photodimerization of coumarin has been rather extensively investigated.[90–92] Irradiation of coumarin in solution yields a mixture of four stereoisomeric dimers (78)–(81) with dimers (78) and (79) strongly predominating:

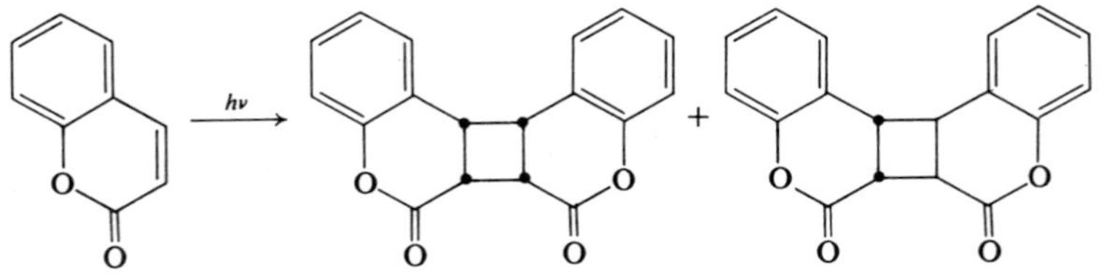

(78) (79)

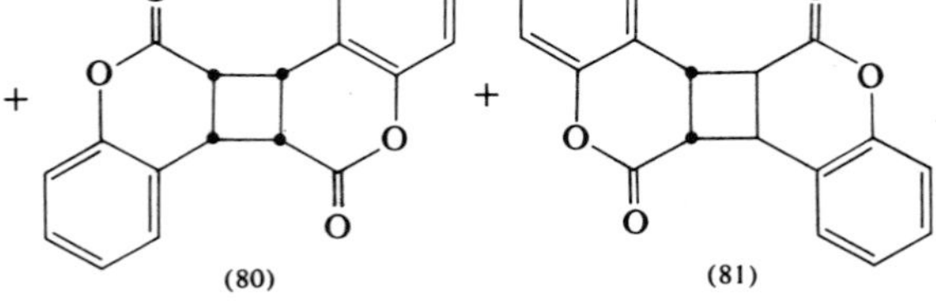

(10.80)

(80) (81)

Irradiation in polar solvents (acetonitrile, dimethylformamide, or methanol) leads to mixtures of (78) and (79) with the former predominating.[90,91] Nonpolar solvents (dioxane, benzene, or ethyl acetate) or the presence of benzophenone as sensitizer lead to a predominance of the *anti* isomer (79).[90,92] Irradiation of coumarin in the presence of the triplet quencher piperylene leads to a significant reduction in the formation of (79). In addition, a dependence of the product distribution on coumarin concentration is observed for direct photolysis in methanol. Thus it is concluded that (79) is formed by a triplet state reaction while (78) is formed predominately by a singlet pathway. The concentration effect on the production of (78) suggests that this isomer may be formed through an intermediate singlet excimer in polar solvents. In nonpolar solvents the formation of (78) is small due to bimolecular self-quenching of the coumarin singlets. The minor products (80) and (81) have been isolated from the direct and sensitized reactions, respectively. Product (80) has been found to be formed similarly to (78), and (81) similarly to (79).[90]

In order to determine the sensitivity of coumarin dimerization to heavy-atom perturbation, the direct photolysis was carried out in butyl chloride, propyl bromide, and ethyl iodide. Comparison of these results with those obtained in nonpolar solvents indicated that there was no significant change in the amount of the triplet-derived product (80) formed in the heavy-atom solvents as would be expected if heavy-atom perturbation were important.[90] More recent investigations,[93b] however, of the formation of (79) in carbon tetrachloride indicate that halocarbon enhancement of the formation of this isomer exists. A direct measure of the intersystem quantum

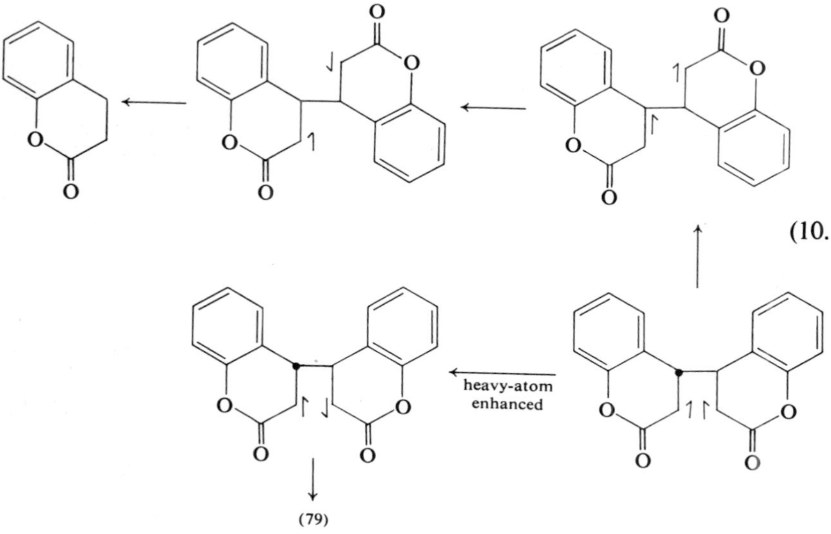

heavy-atom enhanced

(79)

(10.81)

yield (by triplet counting with *cis*-piperylene) indicated this process to be no more efficient for coumarin in carbon tetrachloride than in toluene or methanol. Thus it is proposed that the heavy-atom effect on the formation of (79) results from an increased efficiency of triplet dimerization (see below). Support for this postulate was obtained by the observation that the formation of (79) in the presence of a sensitizer is also enhanced in CCl_4 relative to other solvents [see (10.81) p. 465].

Evidence for the role of a singlet excimer in the formation of (78) was provided by the observation of excimer fluorescence from dilute solutions of coumarin in methylcyclohexane and isopentane glasses at 77°K.[93b]

The photodimerization of some other α,β-unsaturated carbonyls and acid derivatives is shown below[127–130]:

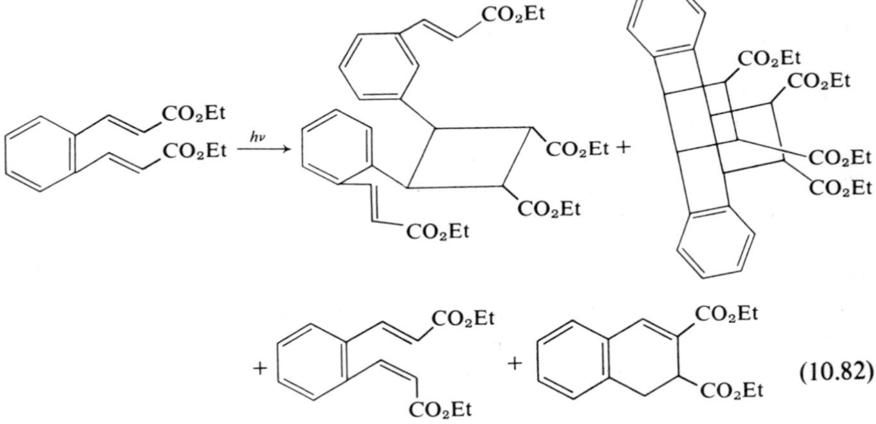

CO_2Et CO_2Et $\xrightarrow{h\nu}$ CO_2Et CO_2Et + CO_2Et CO_2Et CO_2Et CO_2Et CO_2Et CO_2Et

+ CO_2Et CO_2Et + CO_2Et CO_2Et (10.82)

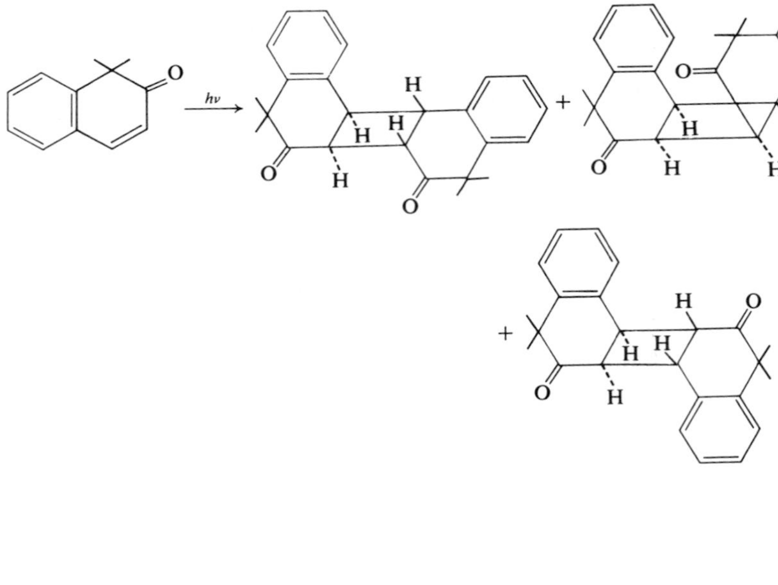

O $\xrightarrow{h\nu}$ H H H H O O + O O H H H H

+ H O H H H O (10.83)

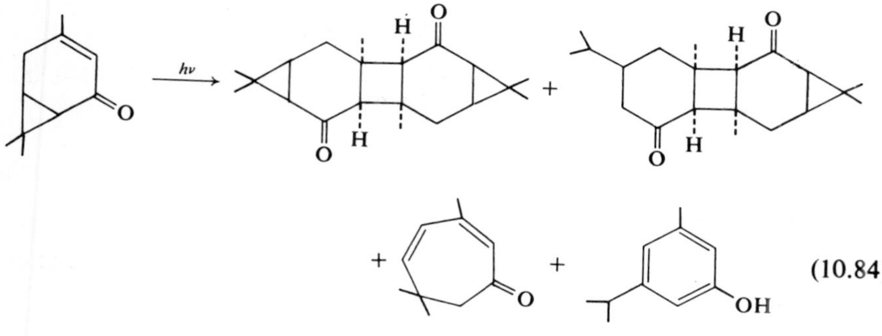

(10.84)

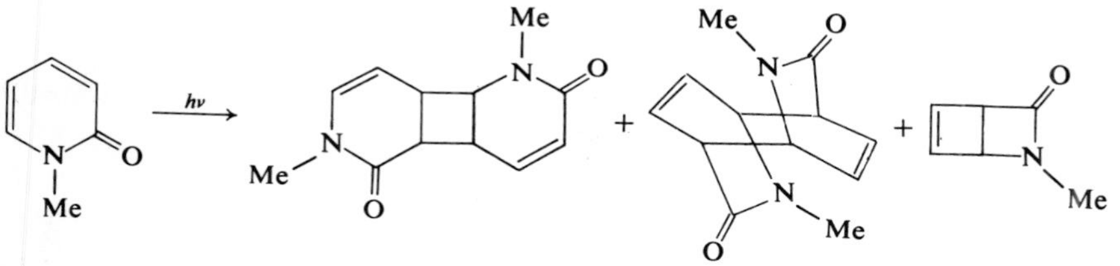

(10.85)

10.3b. *Photocycloaddition Reactions of α,β-Unsaturated Carbonyls and Acid Derivatives*

Most investigations of the photocycloaddition of simple α,β-unsaturated carbonyls have involved the reactions of the ring compounds cyclopentenone and cyclohexenone. The photoaddition of 2-cyclohexenone to a number of different olefins has been reported by Corey and co-workers[94]:

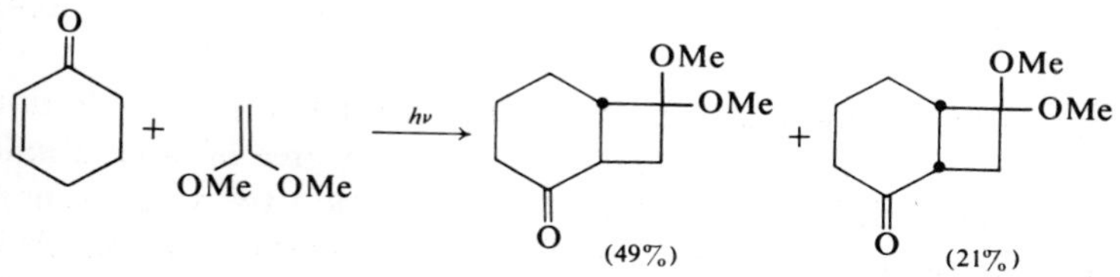

(49%) (21%) (10.86)

(65%) (10.87)

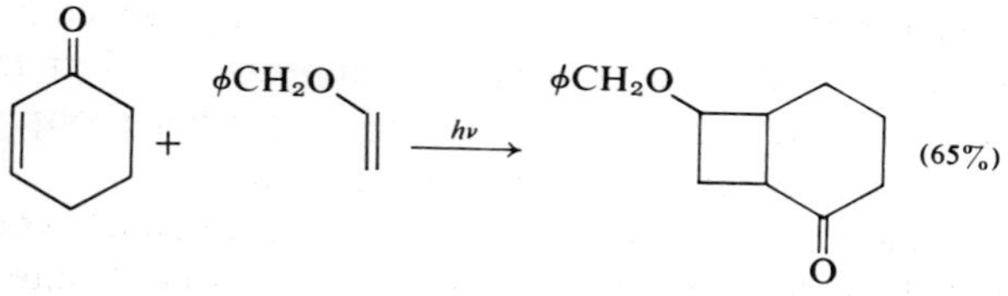

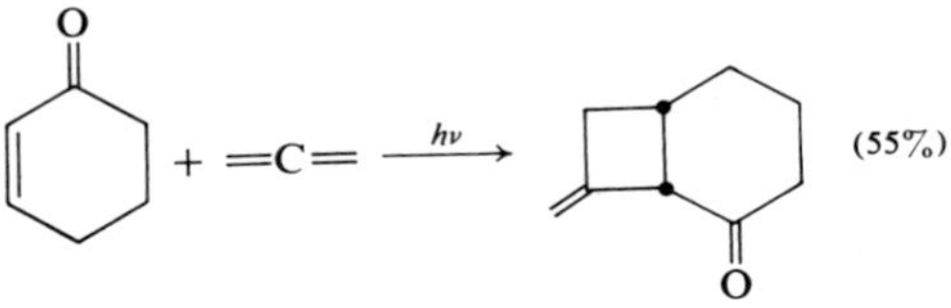

(55%) (10.88)

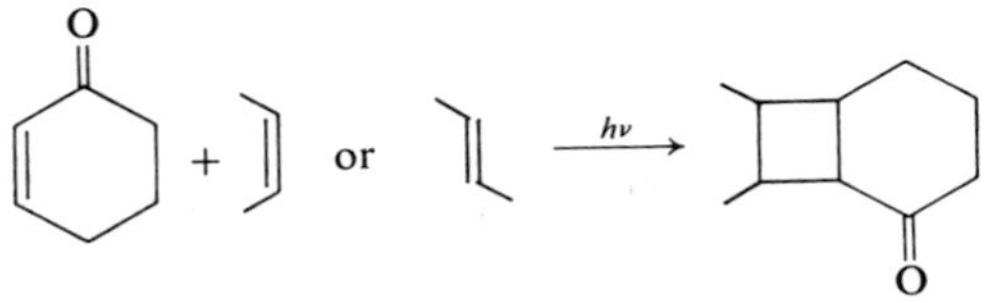

(10.89)

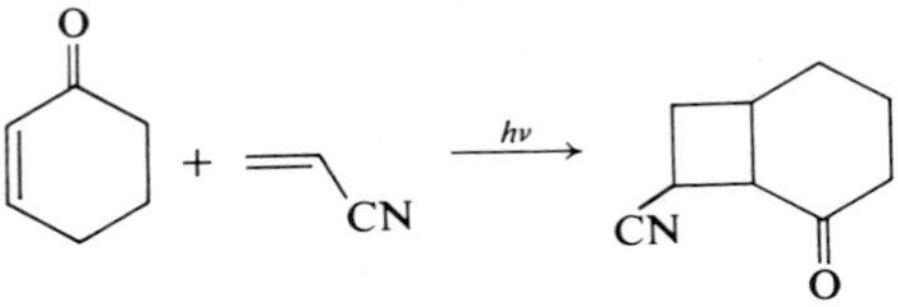

or

(10.90)

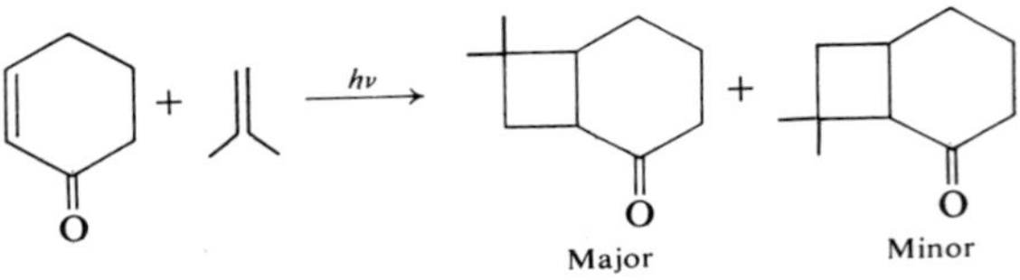

(10.91)

Major Minor

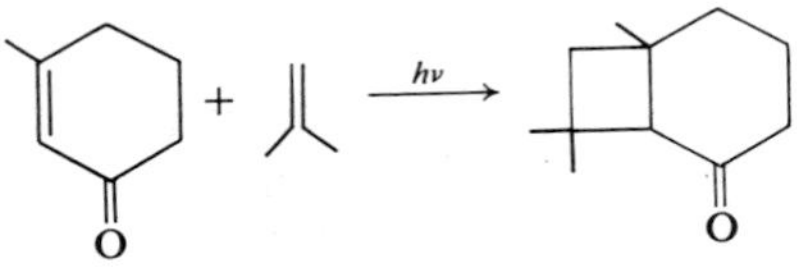

(10.92)

The relative rates of the olefins studied decreases in the order

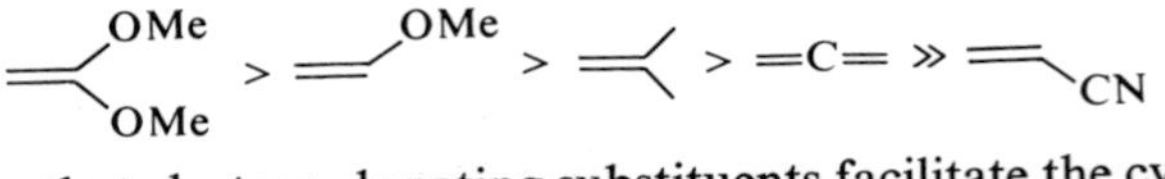

(10.93)

OMe > OMe > > =C= ≫ CN

Thus it is seen that electron-donating substituents facilitate the cycloaddition. Examination of the products formed indicates a high degree of orientational specificity, with the major product of each cycloaddition to cyclohexenone corresponding to addition of the more nucleophilic carbon of the olefin double bond being attached to the α carbons of the enone. The presence of a methyl group in the 3 position of the cyclohexenone appears to affect the mode of addition without altering the rate of addition. A methyl group in the 2 position was found to retard the cycloaddition, however.

The fact that nearly identical product distributions are obtained when cyclohexenone is irradiated in the presence of either *cis*- or *trans*-2-butene effectively eliminates the possibility of a concerted addition of the excited enone to the ground state olefin and suggests a two-step addition involving an intermediate diradical (82) or (83). To explain the orientational selectivity

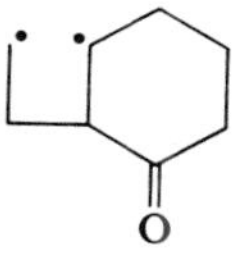

(82) 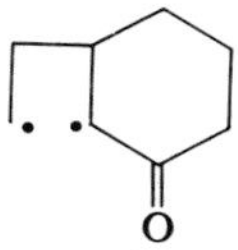 (83)

observed in these reactions, an excited π-complex (exciplex) leading to diradicals (82) or (83) has been proposed:

$$\text{enone}^* + \text{olefin} \rightleftharpoons \underset{[\text{enone}^{*\delta-}\cdot\ \text{olefin}^{\delta+}]}{\text{oriented } \pi\text{-complex}} \rightleftharpoons (82) \text{ or } (83) \longrightarrow \text{cycloaddition product}$$

Assuming that the enone component of the oriented π-complex is in its $n \rightarrow \pi^*$ excited state and the olefin is situated above the α,β double bond of the enone, one can explain the mode of addition by the calculated charge distribution of the $n \rightarrow \pi^*$ excited state:

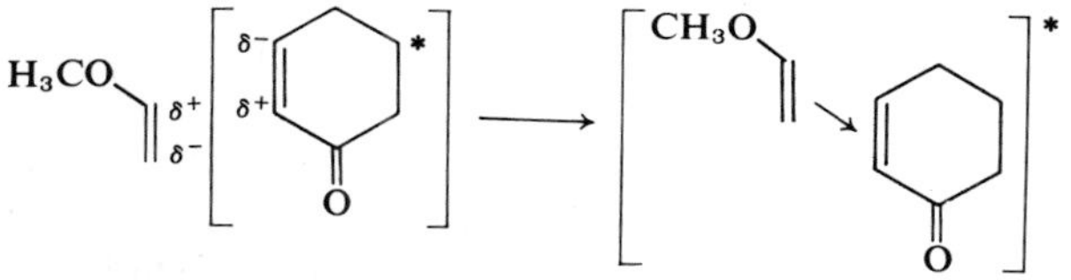

(10.94)

Absorption studies of 2-cyclohexenone–ethoxylethylene solutions failed to reveal evidence of donor–acceptor complex formation. It should be noted, however, that photocycloaddition from ground state π-complexes (such as would be observed from absorption studies) does not correctly predict the observed orientational effects.

The photochemical cycloaddition reactions of 3-methyl, 3-phenyl, and 3-acetoxycyclohexanones with a variety of olefins have been studied by Cantrell, Haller, and Williams.[136] These enones add to the various olefins bearing electron-donating substituents to give bicyclo[4.2.0]octanones in which the orientation of the substituents was that predicted by the oriented π-complex proposed by Corey. Photosensitization studies indicated the products to be triplet-derived although cycloaddition was somewhat less efficient in the presence of the sensitizer. The large amounts of *trans*-fused products observed in this and the earlier study by Corey are thought to arise from highly strained ground state *trans*-cyclohexenone or, alternatively, from two different triplet states of different geometry.[137]

One of the most extensively documented photoadditions is that of maleic anhydride and its derivatives to olefins. Dichloromaleic anhydride adds to butadiene and 2,3-dimethylbutadiene under both direct and sensitized photolysis conditions although product yields are considerably higher in the presence of a sensitizer,[95,96]

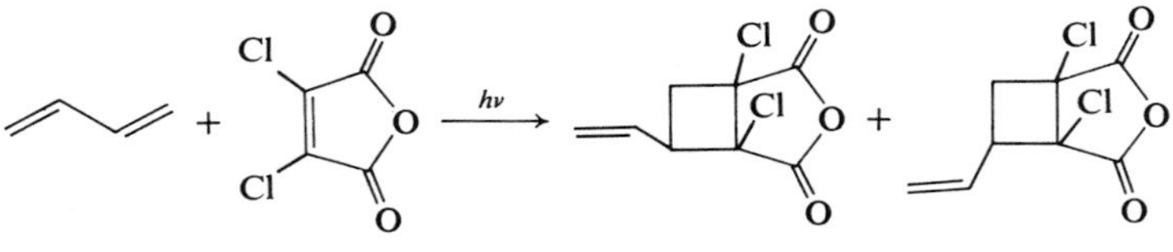

(10.95)

39–42%

Similar products have been isolated from the photocycloaddition of dimethylmaleic anhydride to isoprene.[97]

Cyclopentadiene adds to dimethylmaleic anhydride to produce cyclobutane dimer (84) and two products from 1,4-cycloaddition[97]:

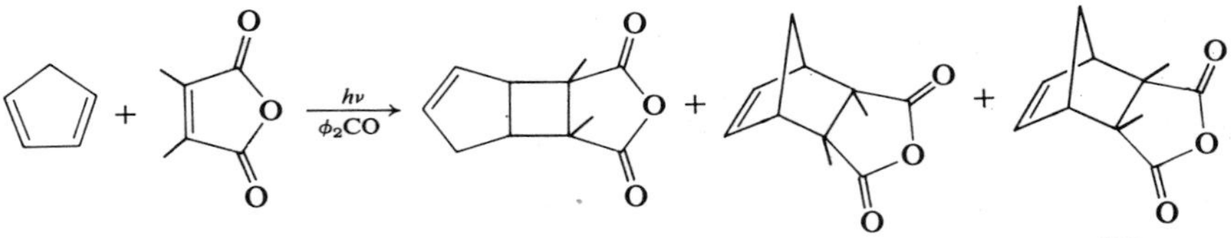

(84) (85) (86)

(10.96)

Endo product (86) is thought to result from thermal addition and is probably not a photoproduct. Cyclohexadiene yields cyclobutanes (87)–(89) and 1,4-cycloaddition product (90) with dimethylmaleic anhydride[97]:

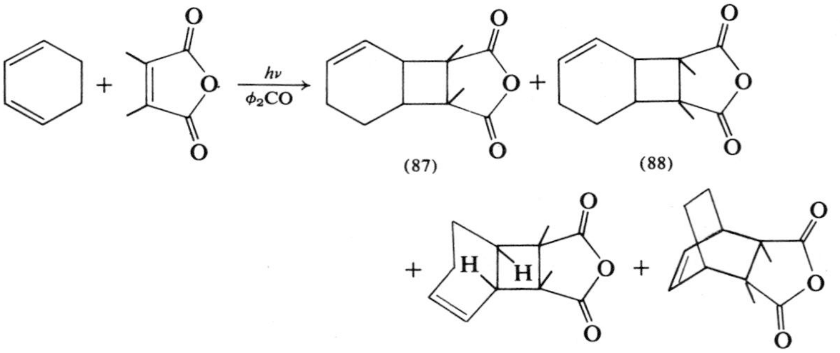

(87) (88)

(89) (90)

(10.97)

Photocycloadditions of maleic anhydride and dimethylmaleic anhydride to various other substituted cyclohexadienes and to cycloheptatriene have also been reported.[58,97,98]

The photocycloaddition of maleic anhydride to cyclohexene has been found to occur through ground state complexes of charge-transfer type[99]:

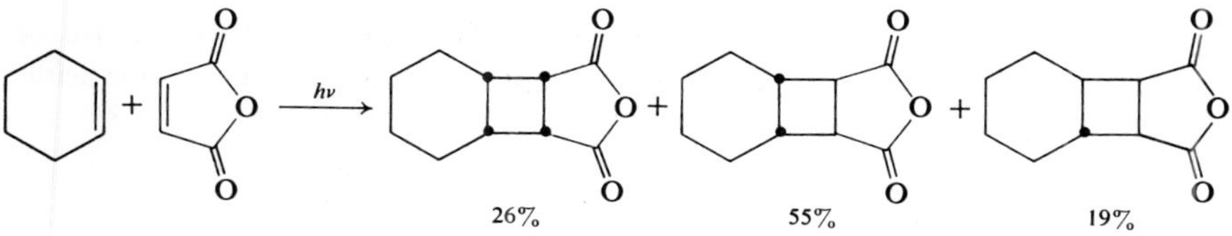

+ $\xrightarrow{h\nu}$ + +

26% 55% 19%

(10.98)

This follows from irradiating mixtures of cyclohexene and maleic anhydride at wavelengths at which neither component of the solution absorbs appreciably individually. The absorption band which appears in these mixtures (λ_{max} = 270 nm) has been assigned as a charge-transfer state. The products of this irradiation (as in those with maleionitrile and fumaronitrile, which also show charge-transfer interactions with cyclohexene) demonstrate a significant degree of specificity when compared to those derived from dimethyl maleate and cyclohexene, for which no charge-transfer complex has been observed:

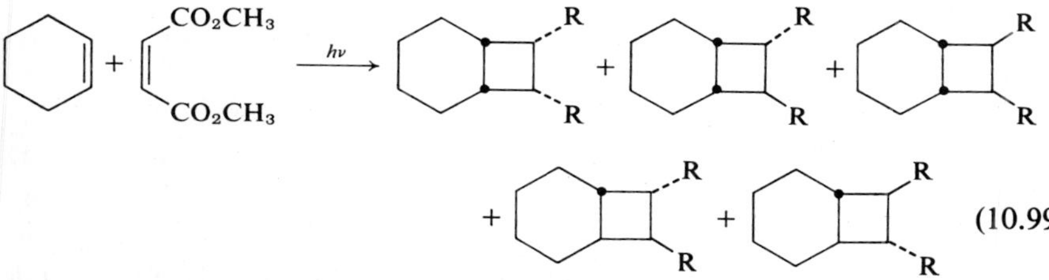

CO_2CH_3 CO_2CH_3 $\xrightarrow{h\nu}$ R R R R R R R R R R

(10.99)

The stereospecificity of the photoaddition of maleic anhydride to cyclohexene can be conveniently rationalized as follows:

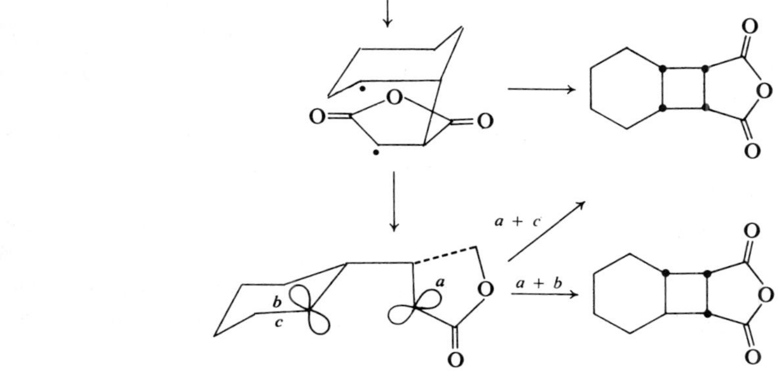

CT $h\nu$ O O O O O O O O O O O O O O O

$a + c$ $a + b$ a b c O O O O O

(10.100)

The photocycloaddition of dimethyl maleate to norbornene yields cyclobutanes (91) and (92), while the photocycloaddition of maleic anhydride to this olefin yields (92) and (93) (after hydrolysis)[100]:

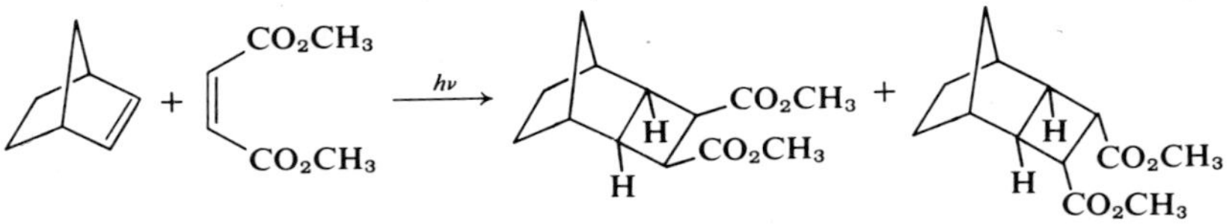

(91) Major (92) (10.101)

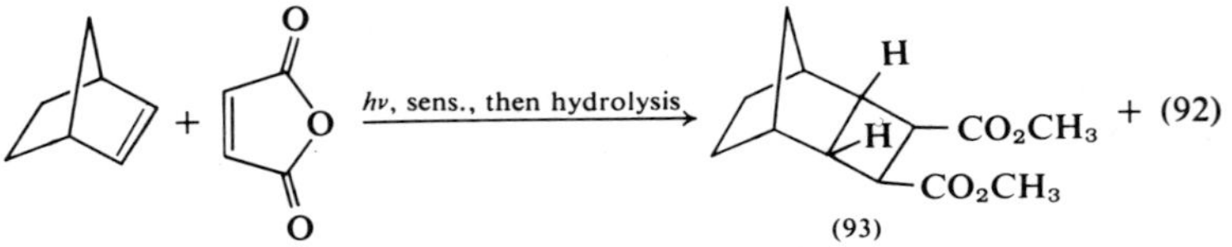

(93) (10.102)

As in the photoaddition of maleic anhydride to norbornene, the *endo* product (93) may arise from a charge-transfer complex.

The photocycloaddition of maleic anhydride to acenaphthylene has been studied by Hartmann and Heine.[107a] Irradiation of acenaphthylene in the presence of maleic anhydride in light-atom solvents (dioxane, acetone, or acetonitrile) yields only dimers or copolymers of acenaphthylene. In heavy-atom solvents (dichloromethane, dibromomethane, or iodomethane), however, dimerization is suppressed and cycloaddition with maleic anhydride predominates:

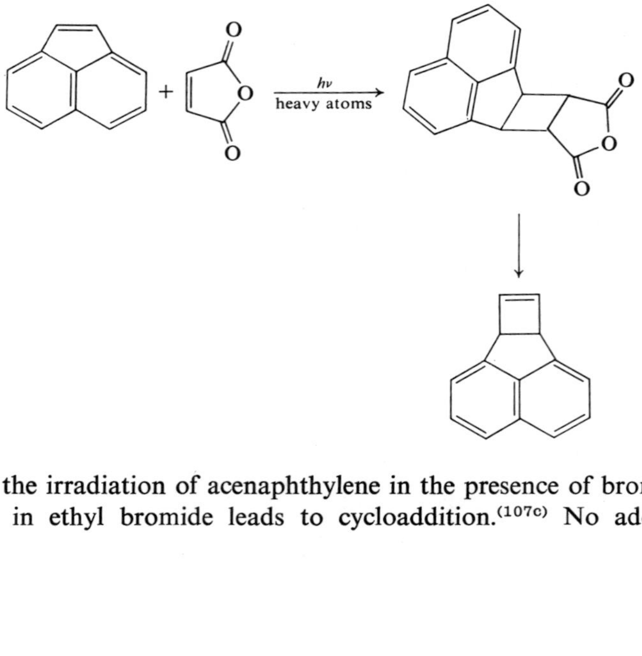

(10.103)

Similarly, the irradiation of acenaphthylene in the presence of bromomaleic anhydride in ethyl bromide leads to cycloaddition.[107c] No adduct was formed from irradiation in the absence of a heavy-atom solvent:

$$\text{acenaphthylene} + \text{bromomaleic anhydride} \xrightarrow[\text{heavy atoms}]{h\nu} \text{adduct} \qquad (10.104)$$

Maleic anhydride adds to benzene upon photolysis to yield a 2:1 adduct (94)[101–104] which has been proposed to arise through stepwise 1,2 and 1,4 additions[101]:

$$\text{benzene} + \text{maleic anhydride} \xrightarrow{h\nu} \text{(94)} \qquad (10.105)$$

(94)

Bryce-Smith and Lodge[103] have found that excitation ($\lambda > 280$ nm) in the region of charge-transfer absorption for complexed benzene maleic-anhydride ($\lambda_{max}^{CT} = 279$ nm) produced (94) although photoactivation of individual benzene or maleic anhydride molecules was virtually prohibited in this wavelength range. In addition, the photoaddition was not significantly inhibited by oxygen, indicating a singlet (or fast triplet) pathway. The reaction, when sensitized by benzophenone, is, however, sensitive to oxygen and in this case all reaction is quenched although the benzophenone absorbs only 30% of the total light. The benzophenone-sensitized reaction is significantly less efficient than the direct irradiation. Hammond and Hardham[104] have asserted that this inefficiency arises by competition between addition of maleic anhydride triplets to benzene and their nonradiative decay.

The photoadditions of maleic anhydride to some other compounds are as follows[38a,105,106]:

$$\text{phenanthrene} + \text{maleic anhydride} \xrightarrow[\text{sens.}]{h\nu} \text{adduct} \qquad (10.106)$$

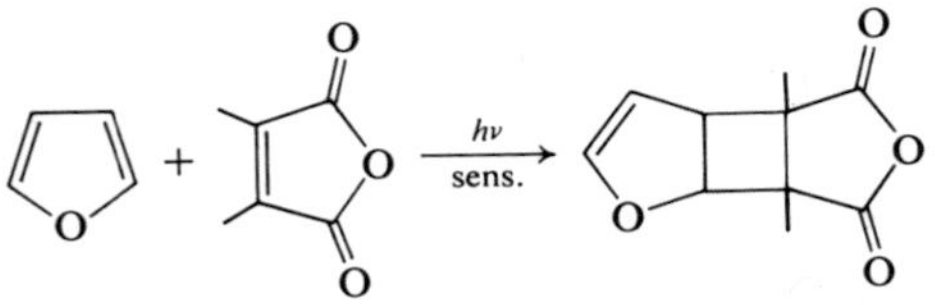

(10.107)

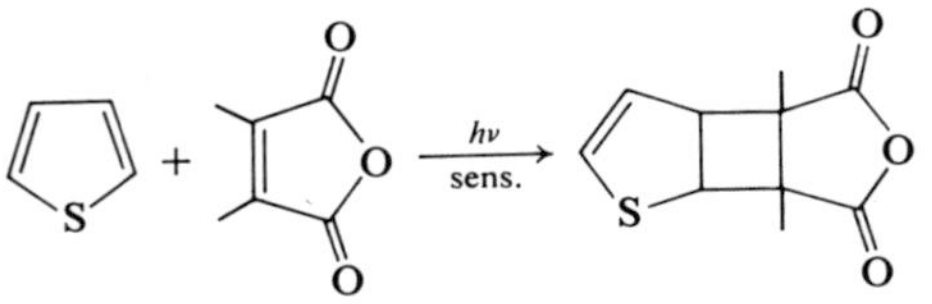

(10.108)

Tetramethylbenzoquinone undergoes photocycloaddition with butadiene to yield a 2:1 adduct (95)[108]:

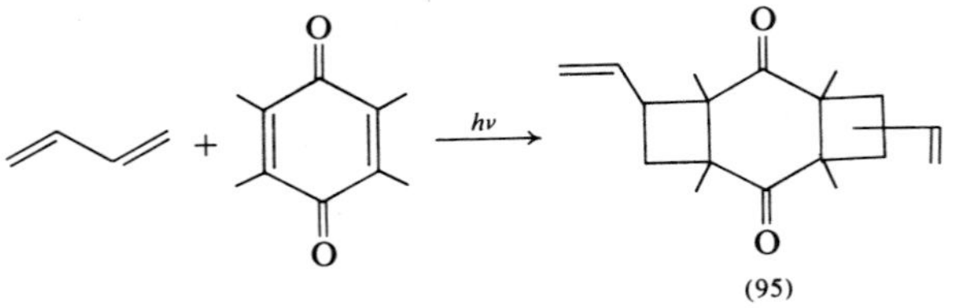

(95)

(10.109)

Chloranil yields 1,2-cycloadducts (96) and (97) with butadiene[109]:

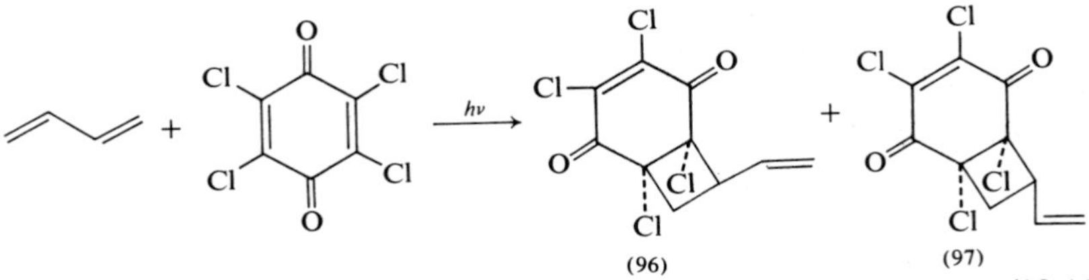

(96) (97)

(10.110)

p-Benzoquinone adds to dimethylbutadiene to yield spiropyran (98),[110] while 2,5-dimethylquinone yields cyclobutane (99)[109] and tetramethylquinone yields (100),[108]

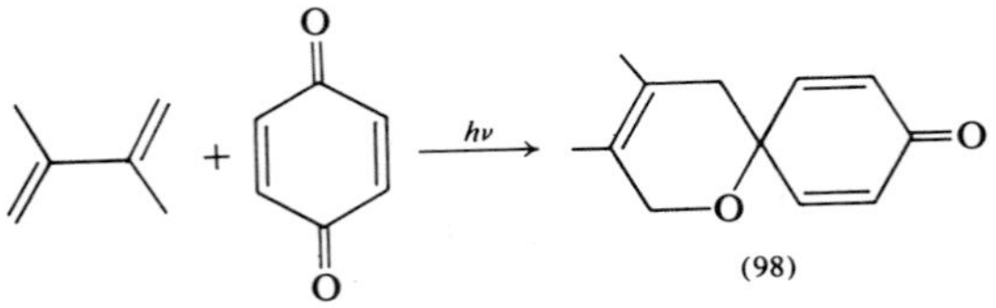

(98)

(10.111)

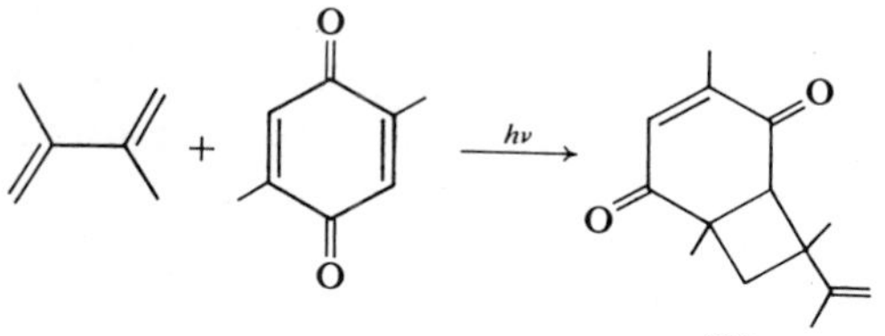

(10.112)

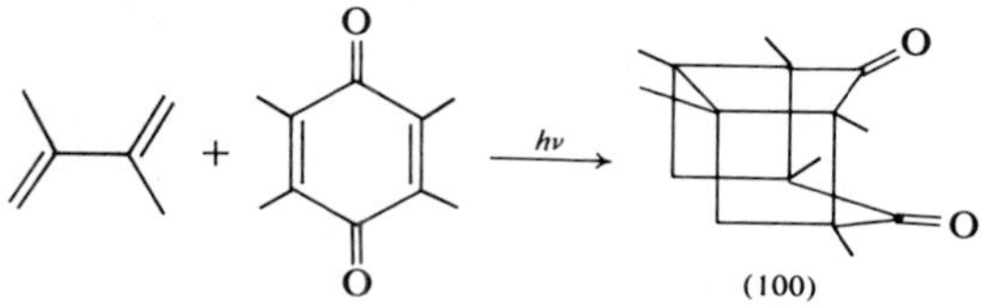

(99)

(10.113)

(100)

Many other photocycloadditions with quinones have been reported.[14]

An elegant example of the use of photochemistry in complex organic synthesis is the preparation the bollweevil phenomone (sex attractant). The key step in both the Zoecon Corporation synthesis and the USDA synthesis involves the formation of a cyclobutane ring by a photoaddition reaction:

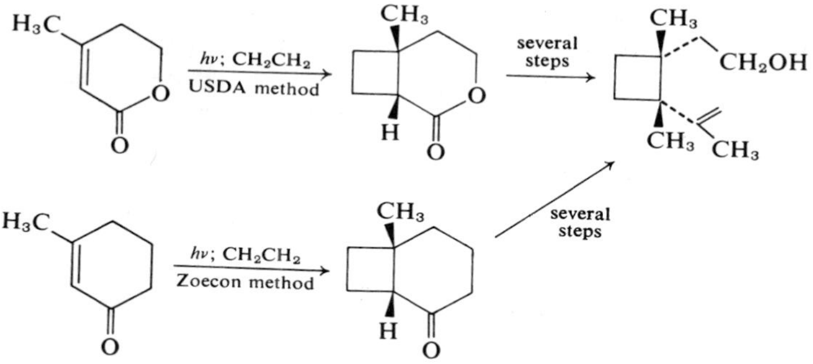

H_3C ... $h\nu$; CH_2CH_2 USDA method ... CH_3 ... H ... O ... several steps ... CH_3 ... CH_2OH ... CH_3 CH_3

H_3C ... $h\nu$; CH_2CH_2 Zoecon method ... CH_3 ... H ... O ... several steps

10.4. DIMERIZATION IN THE SOLID PHASE

Let us now consider a molecule A in the solid phase of a given crystalline modification. Irradiation of this crystalline phase leads to excitation of A to its first excited singlet state, which can (a) react to give product, (b) decay to the ground state by radiative or nonradiative processes, (c) undergo intersystem crossing to the triplet, which can (1) react to give product, or (2) decay to the ground state by radiative or nonradiative processes. In the solid state *topo-chemical* effects dominate steps (a) and (1), and (b) and (2). That is, the fate of the excited molecule, the course of the transformation, and the structure of the product are determined by the geometry of the crystal lattice.

The role of the lattice geometry and its effect on the reaction mechanism, rate, and products has been investigated by Cohen and Schmidt.[111] The following postulate has been developed: *Reaction in the solid state occurs with a minimum amount of atomic or molecular movement*. Thus solid state reactions are controlled by the relatively fixed distances and orientations between potentially reactive centers, as determined by the crystal structure. There should be an upper limit for these distances beyond which reaction will not occur, and bimolecular reactions are expected to involve nearest neighbors. Topochemical influences are expected to be of minor importance in reactions involving long-range migration of electrons or excitation energies but of major importance in such reactions as photodimerization and *cis–trans* isomerization.

Topochemical control of solid state dimerizations is well illustrated by the example of the *trans*-cinnamic acids.[112] The α form of *trans*-cinnamic acid is known to have a molecular separation of 3.6 Å between double bonds and the molecules are arranged in a head-to-tail fashion. β-Cinnamic acid has approximately the same intermolecular distance in the crystal but the molecules are arranged in a parallel head-to-head manner. α-Truxillic (101) and β-truxinic (102) acids are the products expected and observed:

(101) (10.114)

β form (102) (10.115)

The third crystalline form, γ-cinnamic acid, is photochemically stable since the intermolecular distance (4.7–5.1 Å) is apparently too large for bond formation to occur.

Topochemical control is also revealed by closely related compounds showing significant differences in chemical behavior in the solid state. For example, cinnamylidenemalonic acid (103) dimerizes in the solid state to cyclobutane (104), while cinnamylideneacetic acid (105) dimerizes to cyclobutane (106)[113]:

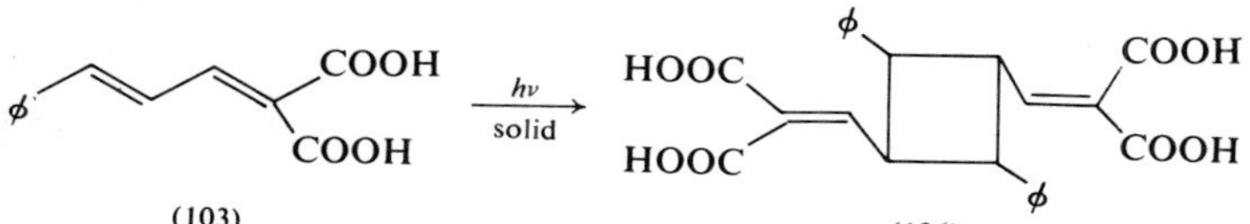

ϕ COOH COOH (103) $\xrightarrow[\text{solid}]{h\nu}$ HOOC HOOC ϕ COOH COOH ϕ (104) (10.116)

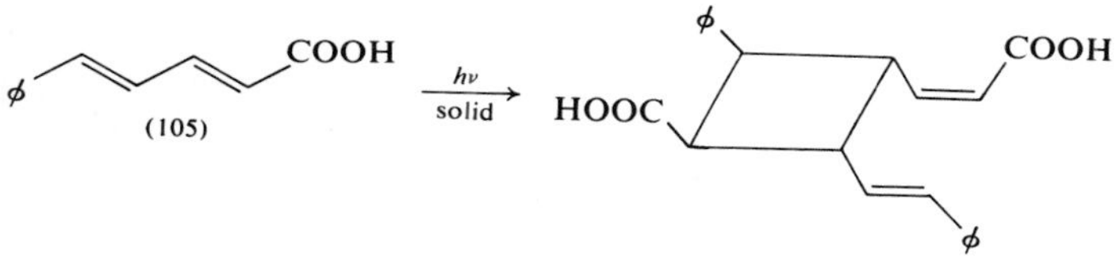

ϕ COOH (105) $\xrightarrow[\text{solid}]{h\nu}$ HOOC ϕ COOH ϕ (106) (10.117)

Crystalline methyl-α-cyanocinnamylideneacetate (107) forms dimer (108) upon irradiation, while the ethyl ester (109) yields an open-chain dimer (110)[114]:

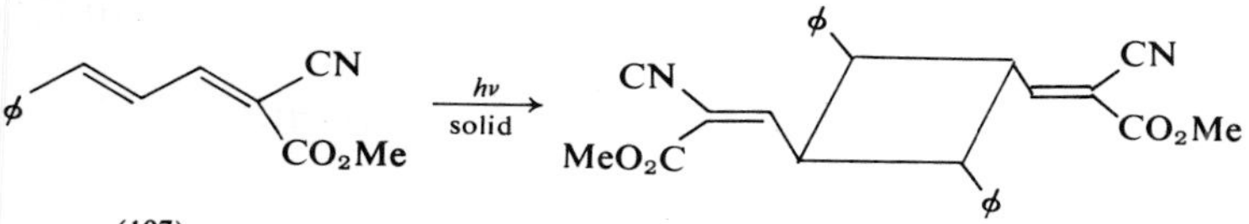

ϕ CN CO_2Me (107) $\xrightarrow[\text{solid}]{h\nu}$ CN MeO_2C ϕ CN CO_2Me ϕ (108) (10.118)

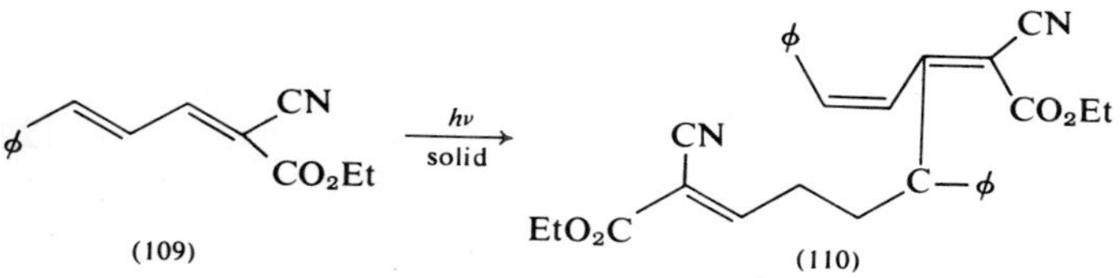

ϕ CN CO_2Et (109) $\xrightarrow[\text{solid}]{h\nu}$ ϕ CN CO_2Et CN C—ϕ EtO_2C (110) (10.119)

Topochemical control will also be revealed in causing a given compound to react differently in the solid phase than in solution. Crystalline 2,5-dimethylbenzoquinone yields dimers (111) and (112) upon irradiation; 2,5-dimethylbenzoquinone in ethyl acetate solution yields primarily the oxetane (113)[115–118]:

hν / solid → (111) 40% + (112) 2%

ethylacetate → (113)

(10.120)

Irradiation of crystalline dimethylfumarate leads selectively to cyclobutane (114), one of four possible isomers[119]:

CH_3O_2C–CH=CH–CO_2CH_3 —hν, solid→ (114)

(10.121)

Similarly, fumaronitrile yields cyclobutane (115) when irradiated in the solid state[120]:

NC–CH=CH–CN —hν, solid→ (115)

(10.122)

Irradiation of crystalline acenaphthylene[121] at temperatures below 25°C results in formation of the *trans*-cyclobutane dimer only. At temperatures above 40°C both *cis* and *trans* dimers are formed, with the *cis* dimer predominating at 60°C (Table 10.12). The formation of the *trans* dimer, which occurs with low quantum yield at all temperatures studied, is thought to

TABLE 10.12. *Photodimerization of Crystalline Acenaphthylene*

Temp., °C	ϕ_{cis}	ϕ_{trans}
0	0.0000	0.0029
20	0.0000	0.0041
40	0.0002	0.0052
50	0.0006	0.0061
60	0.0078	0.0055
65	0.0231	0.0025
70	0.0405	0.0041

arise from migration of the excitation energy to crystal imperfections, allowing dimerization to occur.

Head-to-tail dimers result from the photolysis of dihydroacetic acid, a food additive, in the solid phase[131]:

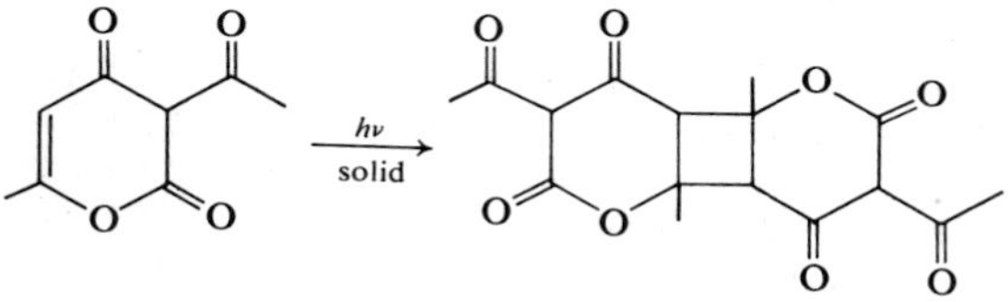

$h\nu$, solid (10.123)

The solid phase dimerization of 2-benzyl-5-benzylidene-cyclopentanone also yields only one product, the antiparallel dimer[132]

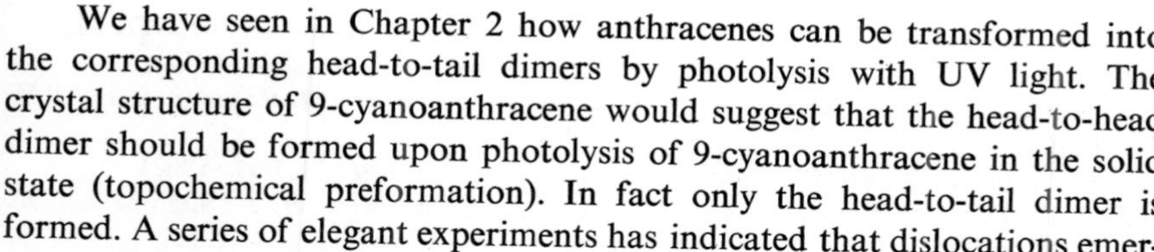

H(Br); $h\nu$, solid; Ar; H; H; Ar; ϕ (10.124)

We have seen in Chapter 2 how anthracenes can be transformed into the corresponding head-to-tail dimers by photolysis with UV light. The crystal structure of 9-cyanoanthracene would suggest that the head-to-head dimer should be formed upon photolysis of 9-cyanoanthracene in the solid state (topochemical preformation). In fact only the head-to-tail dimer is formed. A series of elegant experiments has indicated that dislocations emergent at the *bc* and *ac* faces of the crystal are the preferred sites of photodimerization; in particular, partial dislocations in the (211) planes give rise to a head-to-tail registry of the molecules across contiguous planes, and thereby facilitate the formation of the head-to-tail dimer. This conclusion is based on the study of melt-grown crystals. One half of each crystal was etched to reveal the distribution of emergent dislocations, the other half was irradiated with UV light. Then the distribution of product nuclei was observed using conventional optical, interference contrast (Nonarski), and fluorescence microscopy. The above conclusions were based on a comparison of the dislocation patterns (from etching) and product patterns on corresponding faces of the crystal.[138–140]

REFERENCES

1. A. Mustafa, *Chem. Rev.* **51**, 1 (1951).
2. H. Yamazaki and R. J. Cvetanovic, *J. Amer. Chem. Soc.* **91**, 520 (1969).
3. D. Scharf and F. Korte, *Tetrahedron Lett.*, 821 (1963).
4. R. Srinivasan and F. I. Sonntag, *J. Amer. Chem. Soc.* **87**, 3778 (1965).
5. L. Salem, *J. Amer. Chem. Soc.* **90**, 553 (1968).
6. G. S. Hammond, *Reactivity of the Photoexcited Organic Molecule*, Interscience, New York (1967).
7. N. J. Turro, *Molecular Photochemistry*, Benjamin, New York (1965).
8. G. S. Hammond, N. J. Turro, and A. Fischer, *J. Amer. Chem. Soc.* **83**, 4674 (1961).
9. G. S. Hammond, N. J. Turro, and R. S. H. Liu, *J. Org. Chem.* **28**, 3297 (1963).
10. G. S. Hammond and C. D. Deboer, *J. Amer. Chem. Soc.* **86**, 899 (1964).
11. R. S. H. Liu, N. J. Turro, Jr., and G. S. Hammond, *J. Amer. Chem. Soc.* **87**, 3406 (1965).
12. C. D. Deboer, N. J. Turro, and G. S. Hammond, *Org. Syn.* **47**, 64 (1967).
13. W. G. Herkstroeter, A. A. Lamola, and G. S. Hammond, *J. Amer. Chem. Soc.* **86**, 4537 (1964).
14. W. L. Dilling, *Chem. Rev.*, 845 (1969).
15. W. B. Smith and J. L. Massingill, *J. Amer. Chem. Soc.* **83**, 4301 (1961).
16. J. D. Roberts, *Notes on Molecular Orbital Calculations*, Benjamin, New York (1961).
17. R. S. H. Liu and D. M. Gale, *J. Amer. Chem. Soc.* **90**, 1897 (1968).
18. G. S. Hammond and R. S. H. Liu, *J. Amer. Chem. Soc.* **85**, 477 (1963).
19. M. Heberhold and G. S. Hammond, *Ber. Bunsenges. Phys. Chem.* **72**, 309 (1968).
20. J. J. Dannenberg and J. H. Richards, *J. Amer. Chem. Soc.* **87**, 1626 (1965).
21. J. P. Guillory, C. F. Cook, and D. R. Scott, *J. Amer. Chem. Soc.* **89**, 6776 (1967).
22. N. J. Turro and G. S. Hammond, *J. Amer. Chem. Soc.* **84**, 2841 (1962).
23. N. C. Yang, M. Nussim, M. J. Jorgenson, and S. Morov, *Tetrahedron Lett.*, 3657 (1964).
24. G. O. Schenck and R. Steinmetz, *Bull. Soc. Chim. Belges* **71**, 781 (1962).
25. G. O. Schenck, W. Hartmann, S. P. Mannsfeld, W. Metzner, R. Steinmetz, I. V. Wilucki, R. Wolgast, and C. H. Krauch, *Angew. Chem.* **73**, 764 (1961).
26. D. Valentine, N. J. Turro, Jr., and G. S. Hammond, *J. Amer. Chem. Soc.* **86**, 5202 (1964).
27. N. J. Turro and P. D. Bartlett, *J. Org. Chem.* **30**, 1849, 4396 (1965).
28. G. O. Schenck, S. P. Mannsfeld, G. Schomburg, and C. H. Krauch, *Z. Naturforsch.* **19B**, 18 (1964).
29. H. H. Stechl, *Angew. Chem.* **75**, 1176 (1963); *Ber.* **97**, 2681 (1964).

30. R. Srinivasan and K. A. Hill, *J. Amer. Chem. Soc.* **88**, 3765 (1966).
31. D. M. Lemal and J. P. Lokengard, *J. Amer. Chem. Soc.* **88**, 5934 (1966); B. W. Schafer, R. Criegee, R. Askani, and H. Gruner, *Angew. Chem.* (*Int. Ed. Engl.*) **6**, 78 (1967); K. E. Wilzbach and L. Kaplan, *J. Amer. Chem. Soc.* **87**, 4004 (1965); R. Criegee and R. Askani, *Angew. Chem.* (*Int. Ed. Engl.*) **5**, 519 (1966).
32. W. G. Dauben and D. L. Whalen, *Tetrahedron Lett.*, 3743 (1966).
33. E. E. Van Tamelen and D. Carty, *J. Amer. Chem. Soc.* **89**, 3922 (1967).
34. G. O. Schenck and R. Steinmetz, *Chem. Ber.* **96**, 520 (1963); R. C. Cookson and E. Crandwell, *Chem. Ind.*, 1004 (1958).
35. H. Prinzbach, *Pure Applied Chem.* **16**, 17 (1968).
36. J. E. Baldwin and J. P. Nelson, *J. Org. Chem.* **31**, 336 (1966).
37. H. Notaki, T. Otani, R. Noyori, and M. Kawanisi, *Tetrahedron* **24**, 2183 (1968).
38. (a) G. O. Schenck *et al.*, *Chem. Ber.* **95**, 1642 (1962); (b) J. Bowyer and Q. N. Porter, *Austral. J. Chem.* **19**, 1455 (1965).
39. C. DeBoer, *J. Amer. Chem. Soc.* **91**, 1855 (1969).
40. D. O. Cowan and R. L. Drisko, *Tetrahedron Lett.*, 1255 (1967); D. O. Cowan and R. L. Drisko, *J. Amer. Chem. Soc.* **89**, 3069 (1967).
41. D. O. Cowan and R. L. Drisko, *J. Amer. Chem. Soc.* **92**, 6286 (1970).
42. D. O. Cowan and R. L. Drisko, *J. Amer. Chem. Soc.* **92**, 6281 (1970).
43. (a) R. L. Drisko, Ph.D. Dissertation, The Johns Hopkins University, Baltimore, Maryland (1968); (b) D. O. Cowan and J. Koziar, *J. Amer. Chem. Soc.* **96**, 1229 (1974); (c) D. O. Cowan and J. Koziar, *J. Amer. Chem. Soc.* **97**, 249 (1975); (d) J. Koziar and D. O. Cowan, *J. Amer. Chem. Soc.* **98**, 1001 (1976).
44. E. J. Bowen and J. D. F. Marsh, *J. Chem. Soc.*, 109 (1947).
45. R. Livingston and K. S. Wei, *J. Phys. Chem.* **71**, 541 (1967).
46. I. Hartmann, W. Hartmann, and G. O. Schenck, *Ber.* **100**, 3146 (1967).
47. N. Y. C. Chu and D. R. Kearns, *J. Phys. Chem.* **74**, 1255 (1970).
48. J. A. Berson, Z. Hamlet, and W. A. Mueller, *J. Amer. Chem. Soc.* **84**, 297 (1962).
49. T. S. Cantrell and H. Shechter, *J. Org. Chem.* **33**, 114 (1968).
50. H. Bouas-Laurent, A. Castellan, J. P. Desvergne, G. Dumartin, C. Courseille, J. Gaultier, and C. Hauw, *Chem. Comm.*, 1267 (1974).
51. W. Baker, J. W. Hilpern, and J. W. F. McOmie, *J. Chem. Soc.*, 479 (1961); H. Shechter, W. J. Link, and G. V. D. Tiers, *J. Amer. Chem. Soc.* **85**, 1601 (1963).
52. J. D. Fulton and J. D. Dunitz, *Nature* **160**, 161 (1947); J. D. Fulton, *Brit. J. Pharmacol.* **3**, 75 (1948); A. J. Henry, *J. Chem. Soc.*, 1156 (1946).
53. E. H. White and J. P. Anhalt, *Tetrahedron Lett.*, 3937 (1965).
54. W. L. Dilling and J. C. Little, *J. Amer. Chem. Soc.* **89**, 2741 (1967).
55. (a) W. L. Dilling, *J. Amer. Chem. Soc.* **89**, 2742 (1967); (b) W. L. Dilling, R. D. Kroening, and J. C. Little, *J. Amer. Chem. Soc.* **92**, 928 (1970).
56. W. L. Dilling and R. D. Kroening, *Tetrahedron Lett.*, 5101 (1968).
57. W. L. Dilling and R. D. Kroening, *Tetrahedron Lett.*, 5601 (1968).
58. G. Sartori, V. Turba, A. Valvassori, and M. Riva, *Tetrahedron Lett.*, 211 (1966).
59. G. Sartori, V. Turba, A. Valvassori, and M. Riva, *Tetrahedron Lett.*, 4777 (1966).
60. P. D. Bartlett, R. Helgeson, and O. A. Wersel, *Pure Appl. Chem.* **16**, 187 (1968); P. D. Bartlett, *Science* **159**, 833 (1968).
61. T. S. Cantrell, *Chem. Comm.*, 1633 (1970).
62. O. L. Chapman and W. R. Adams, *J. Amer. Chem. Soc.* **90**, 2333 (1968); O. L. Chapman and R. D. Lura, *J. Amer. Chem. Soc.* **92**, 6352 (1970).
63. J. Saltiel, J. D'Agostino, O. L. Chapman, and R. D. Lura, *J. Amer. Chem. Soc.* **93**, 2804 (1971).

64. O. L. Chapman, A. A. Griswold, E. Hoganson, G. Lenz, and J. Reasoner, *Pure Appl. Chem.* **9**, 585 (1964).
65. O. L. Chapman and G. Lenz, *Org. Photochem.* **1**, 283 (1967).
66. H. M. Rosenberg, R. Rondeau, and P. Servé, *J. Org. Chem.* **34**, 471 (1969).
67. B. F. Plummer and D. M. Chihal, *J. Amer. Chem. Soc.* **93**, 2071 (1971).
68. B. F. Plummer and R. A. Hall, *Chem. Comm.*, 44 (1970).
69. B. F. Plummer and W. I. Ferree, Jr., *Chem. Comm.*, 306 (1972).
70. G. Koltzenburg and K. Kraft, *Angew. Chem.* **77**, 1029 (1965).
71. G. Koltzenburg and K. Kraft, *Tetrahedron Lett.*, 389 (1966).
72. K. Kraft and G. Koltzenburg, *Tetrahedron Lett.*, 4357 (1967).
73. K. Kraft and G. Koltzenburg, *Tetrahedron Lett.*, 4723 (1967).
74. R. Srinivasan and K. A. Hill, *J. Amer. Chem. Soc.* **87**, 4653 (1965).
75. J. G. Atkinson, D. E. Ayer, G. Büchi, and E. W. Robb, *J. Amer. Chem. Soc.* **85**, 2257 (1963).
76. J. J. McCullough, C. Calvo, and C. W. Huang, *Chem. Comm.*, 1176, 1968; J. J. McCullough and C. W. Huang, *Can. J. Chem.* **47**, 757 (1969); R. M. Bowan and J. J. McCullough, *Chem. Comm.*, 948 (1970).
77. T. Miyamoto, T. Mori, and Y. Odaira, *Chem. Comm.*, 1598 (1970).
78. W. H. F. Sasse, P. J. Collin, and G. Sugowdz, *Tetrahedron Lett.*, 3373 (1967); 1689 (1968); R. J. McDonald and B. K. Selinger, *Tetrahedron Lett.*, 4791 (1968).
79. J. S. Bradshaw and G. S. Hammond, *J. Amer. Chem. Soc.* **85**, 3953 (1963); *Austral. J. Chem.* **21**, 733 (1968); M. Sterns and B. K. Selinger, *Austral. J. Chem.* **21**, 2131 (1968); *J. Amer. Chem. Soc.* **91**, 621 (1969).
80. M. V. Sargent and C. J. Timmons, *J. Chem. Soc.*, 5544 (1964).
81. G. Ciamician and P. Silber, *Ber. Dtsch. Chem. Ges.* **42**, 1386 (1909); P. Praetorius and F. Korn, *Ber. Dtsch. Chem. Ges.* **43**, 2744 (1910); G. W. Recktenwald, J. N. Pitts, Jr., and R. L. Letsinger, *J. Amer. Chem. Soc.* **75**, 3028 (1953).
82. P. E. Eaton and W. S. Hurt, *J. Amer. Chem. Soc.* **88**, 5038 (1966); P. E. Eaton, *J. Amer. Chem. Soc.* **84**, 2344 (1962).
83. (a) J. L. Ruhlen and P. A. Leermakers, *J. Amer. Chem. Soc.* **88**, 5671 (1966); **89**, 4944 (1967); (b) P. Wagner and D. J. Bucheck, *Can. J. Chem.* **47**, 713 (1969).
84. E. G. Lam, D. Valentine, and G. S. Hammond, *J. Amer. Chem. Soc.* **89**, 3482 (1967).
85. O. L. Chapman, P. J. Nelson, R. W. King, D. J. Trecker, and A. A. Griswold, *Rec. Chem. Prog.* **28**, 167 (1967).
86. M. B. Rubin, G. E. Hipps, and D. Glover, *J. Org. Chem.* **29**, 68 (1964); M. B. Rubin, D. Glover, and R. G. Parker, *Tetrahedron Lett.*, 1075 (1964).
87. D. Bryce-Smith and A. Gilbert, *J. Chem. Soc.*, 2428 (1964).
88. J. Dekker, P. J. Van Vuurren, and D. P. Venter, *J. Org. Chem.* **33**, 464 (1968).
89. P. Yates and M. J. Jorgenson, *J. Amer. Chem. Soc.* **80**, 6150 (1958); **85**, 2956 (1963).
90. H. Morrison, H. Curtis, and T. McDowell, *J. Amer. Chem. Soc.* **88**, 5415 (1966).
91. C. H. Krauch, S. Farid, and G. O. Schenck, *Ber.* **99**, 625 (1966).
92. G. S. Hammond, C. A. Stout, and A. A. Lamola, *J. Amer. Chem. Soc.* **86**, 3103 (1964).
93. (a) H. Morrison and R. Hoffman, *Chem. Comm.*, 453 (1968); (b) R. Hoffman, P. Wells, and H. Morrison, *J. Org. Chem.* **36**, 102 (1971).
94. E. J. Corey, J. D. Bass, R. LeMehieu, and R. B. Mitra, *J. Org. Chem.* **29**, 68 (1964).
95. H.-D. Scharf and F. Korte, *Chem. Ber.* **99**, 1299 (1966).
96. H.-D. Scharf and F. Korte, *Angew. Chem.* **77**, 1037 (1965).
97. G. O. Schenck, J. Kuhls, and C. H. Krauch, *Ann.* **693**, 20 (1966); *Z. Naturforsch.* **20B**, 635 (1965).

98. L. M. Stevenson and G. S. Hammond, *Pure Appl. Chem.* **16**, 125 (1968).
99. J. A. Barltrop and R. Robson, *Tetrahedron Lett.*, 597 (1963); R. Robson, P. W. Grubb, and J. A. Barltrop, *J. Chem. Soc.*, 2153 (1964).
100. R. L. Cargill and M. R. Willcott, III, *J. Org. Chem.* **31**, 3938 (1966).
101. H. J. F. Angus and D. Bryce-Smith, *Proc. Chem. Soc.*, 326 (1959); *J. Chem. Soc.*, 4791 (1968).
102. G. Grovenstein, Jr., D. V. Rao, and J. W. Taylor, *J. Amer. Chem. Soc.* **83**, 1705 (1961).
103. D. Bryce-Smith, A. Gilbert, and B. Vickery, *Chem. Ind.* (*London*), 2060 (1962); D. Bryce-Smith and J. E. Lodge, *J. Chem. Soc.*, 2675 (1962); D. Bryce-Smith, *Pure Appl. Chem.* **16**, 47 (1968).
104. G. S. Hammond and W. M. Hardham, *Proc. Chem. Soc.*, 63 (1963).
105. D. Bryce-Smith and B. Vickery, *Chem. Ind.*, 429 (1961).
106. G. O. Schenk, W. Hartmann, and R. Steimetz, *Chem. Ber.* **96**, 495 (1963).
107. (a) W. Hartmann and H. G. Heine, *Angew. Chem. Int. Ed.* **10**, 272 (1971); (b) J. Meinwald, G. E. Samuelson, and M. Ikeda, *J. Amer. Chem. Soc.* **92**, 7604 (1970); (c) J. E. Shields, D. Grarilovic, and J. Kopecky, *Tetrahedron Lett.*, 271 (1971).
108. G. Koltzenburg, K. Kraft, and G. O. Schenck, *Tetrahedron Lett.*, 353 (1965).
109. J. A. Barltrop and B. Hesp, *J. Chem. Soc. C*, 1625 (1967).
110. (a) J. A. Barltrop and P. Hesp, *J. Chem. Soc.*, 5182 (1965). (b) R. Zurfluh, C. L. Dunham, V. L. Spain, and J. B. Siddall, *J. Amer. Chem. Soc.* **92**, 425 (1970); *C & EN* **1970** (January 26), 40.
111. M. D. Cohen and G. M. J. Schmidt, *J. Chem. Soc.*, 1996 (1964).
112. M. D. Cohen, G. M. J. Schmidt, and F. I. Sonntag, *J. Chem. Soc.*, 2000 (1964); H. Stobbe and A. Lehfeldt, *Ber.* **58**, 2415 (1925); G. M. J. Schmidt, *Acta Cryst.* **10**, 793 (1957); H. I. Bernstein and W. C. Quimby, *J. Amer. Chem. Soc.* **63**, 1845 (1943).
113. C. N. Ruber, *Ber.* **35**, 2411 (1902); *Ber.* **46**, 335 (1913).
114. M. Reimer and E. Keller, *Am. Chem. J.* **50**, 157 (1913).
115. R. C. Cookson and J. Hudec, *Proc. Chem. Soc.*, 11 (1959).
116. E. H. Gold and D. Ginsburg, *J. Chem. Soc. C*, 15 (1967).
117. R. C. Cookson, D. A. Cox, and J. Hudec, *J. Chem. Soc.*, 4499 (1961).
118. D. Rabinovich and G. M. J. Schmidt, *J. Chem. Soc.*, 2030 (1964).
119. G. W. Griffin, A. F. Vellturo, and K. Furukawa, *J. Amer. Chem. Soc.* **83**, 2725 (1961).
120. G. W. Griffin, J. E. Basinski, and L. I. Peterson, *J. Amer. Chem. Soc.* **84**, 1012 (1962).
121. (a) G. O. Schenck, W. Hartmann, S.-P. Mannsfeld, W. Metzner, and C. H. Krauch, *Chem. Ber.* **95**, 1642 (1962); (b) K. S. Wei and R. Livingston, *J. Phys. Chem.* **71**, 548 (1967).
122. W. Hartmann, *Ber.* **101**, 1643 (1968).
123. R. A. Crallin, M. C. Lambert, and A. Ledwith, *Chem. Comm.*, 682 (1970).
124. G. W. Griffin and U. Heep, *J. Org. Chem.* **35**, 4223 (1970).
125. R. Srinivasan, *IBM J. Res. Develop.* **15**, 34 (1971).
126. K. E. Wilzbach and L. Kaplan, *J. Amer. Chem. Soc.* **93**, 2073 (1971).
127. D. F. Tavares and W. H. Ploder, *Tetrahedron Lett.*, 1567 (1970).
128. J. Carnduff, J. Iball, D. G. Leppard, and J. N. Low, *Chem. Comm.*, 1218 (1969); T. Mukai, T. Oine, and H. Sukawa, *Chem. Comm.*, 271 (1970).
129. P. H. Boyle, W. Cocker, D. H. Grayson, and P. V. R. Shannon, *J. Chem. Soc. C*, 1073 (1971).
130. L. J. Sharp and G. S. Hammond, *Mol. Photochem.* **2**, 225 (1970).

131. N. Sagiyama, T. Sato, H. Kataoka, and C. Kashima, *Bull. Chem. Soc. Japan* **44**, 555 (1971).
132. G. C. Forward and D. A. Whitind, *J. Chem. Soc. C*, 1868 (1969).
133. P. DeMayo, *Acc. Chem. Res.* **4**, 41 (1971).
134. P. J. Wagner and D. J. Bucheck, *J. Amer. Chem. Soc.* **91**, 5090 (1969).
135. P. E. Eaton, *Acc. Chem. Res.* **1**, 50 (1968).
136. T. S. Cantrell, W. S. Haller, and J. C. Williams, *J. Org. Chem.* **34**, 509 (1969).
137. O. L. Chapman, T. H. Koch, F. Klein, P. J. Nelson, and E. L. Brown, *J. Amer. Chem. Soc.* **90**, 1657 (1968).
138. J. M. Thomas and J. O. Williams, *Chem. Comm.*, 432 (1967).
139. D. P. Craig and P. Sarti-Fantoni, *Chem. Comm.*, 742 (1966).
140. M. D. Cohen, Z. Ludmer, J. M. Thomas, and J. O. Williams, *Chem. Comm.*, 1173 (1969).

11

Photoelimination, Photoaddition, and Photosubstitution

This chapter contains discussions of photoelimination, photoaddition, and photosubstitution. Although there may appear to be some degree of overlapping between the first two topics in that the species produced by photoelimination may undergo addition to another substrate, our approach will be to concentrate on the reactions brought about by light absorption rather than subsequent dark reactions.

11.1. PHOTOELIMINATION REACTIONS

Photochemical elimination reactions include all those photoinduced reactions resulting in the loss of one or more fragments from the excited molecule. Loss of carbon monoxide from type I or α-cleavage of carbonyl compounds has been previously considered in Chapter 3. Other types of photoeliminations, to be discussed here, include loss of molecular nitrogen from azo, diazo, and azido compounds, loss of nitric oxide from organic nitrites, and loss of sulfur dioxide and other miscellaneous species.

11.1a. *Photoelimination of Nitrogen*

Compounds containing nitrogen–nitrogen double bonds exhibit extreme photoreactivity due to the stability of the photoelimination product, nitrogen gas. In fact, the photodecomposition of azo compounds occurs so readily that several of these materials, such as azo-bisisobutyronitrile (AIBN), are used commercially as polymerization initiators.

The photochemistry of the simplest member of the azo family, azomethane, has been extensively investigated in the vapor phase[1] and in solution[2]:

$$\text{trans-}H_3C{-}N{=}N{-}CH_3 \xrightarrow[\lambda = 366\text{ nm; vapor phase; } \Phi = 1]{h\nu} 2CH_3\cdot + N_2 \longrightarrow CH_3CH_3 \tag{11.1}$$

$$\text{trans-}H_3C{-}N{=}N{-}CH_3 \xrightarrow[\text{isooctane; } \Phi = 0.17]{h\nu\text{; solution phase;}} \text{cis-}H_3C{-}N{=}N{-}CH_3$$

Photolysis in the gas phase leads to the quantitative production of nitrogen and methyl radicals. Photolysis in solution, however, results in a shift in the absorption spectrum to longer wavelengths due to the production of a new species, which is identified as the *cis*-azomethane (the *trans* configuration is the normal isomer). Similarly, irradiation of *trans*-azoisopropane[3] results in *trans–cis* isomerization to the *cis* isomer:

$$\text{trans-}(CH_3)_2CH{-}N{=}N{-}CH(CH_3)_2 \underset{\Phi = 0.56}{\overset{h\nu}{\underset{\longleftarrow}{\overset{\longrightarrow}{\scriptstyle \Phi = 0.38}}}} \text{cis-}(CH_3)_2CH{-}N{=}N{-}CH(CH_3)_2 \tag{11.2}$$

Irradiation of the *cis* isomer results in the isomerization with quantum yield, equal to 0.56. The fact that the quantum yields of *cis* and *trans* formation are similar and add to unity suggests the presence of a common intermediate, thought to be a twisted $\pi \rightarrow \pi^*$ triplet state. Gas-phase photolysis of *trans*-azoisopropane yields no *cis* isomer, but rather elimination products are produced. As the pressure is increased by the addition of inert gas, however, photoisomerization occurs (for *trans*-azoisopropane pressures of 0.248 Torr and CO_2 pressure of 600 Torr, $\Phi_{\text{dec}} = 0.18$, $\Phi_{cis} = 0.31$). This suggests that in the vapor phase dissociation may occur from an excited singlet species $n \rightarrow \pi^*$ or perhaps from a vibrationally excited state. Intersystem crossing or collisional deactivation of the vibrationally excited triplet by the presence of the inert gas or solvent molecules in solution could lead to the $\pi \rightarrow \pi^*$ triplet, from which isomerization results. Attempts to photosensitize the isomerization using benzene or benzophenone as sensitizer were unsuccessful, however. Photoisomerization was achieved in the presence of naphthalene, although this was shown to occur by singlet–singlet energy transfer.

Photolysis of azoisobutane at room temperature in solution yields products indicative of photoelimination[4]:

$$\mathrm{CH_3-\underset{CH_3}{\overset{CH_3}{\underset{|}{\overset{|}{C}}}}-N{=}N-\underset{CH_3}{\overset{CH_3}{\underset{|}{\overset{|}{C}}}}-CH_3} \xrightarrow{h\nu} \mathrm{N_2 + CH_3-\underset{CH_3}{\overset{CH_3}{\underset{|}{\overset{|}{C}}}}-\underset{CH_3}{\overset{CH_3}{\underset{|}{\overset{|}{C}}}}-CH_3}$$

$$+ \mathrm{CH_3-\overset{CH_3}{\overset{|}{C}}{=}CH_2 + CH_3-\underset{CH_3}{\overset{CH_3}{\underset{|}{\overset{|}{C}}}}-H} \tag{11.3}$$

Irradiation of acetone or methanol solutions of this compound at $-50°C$, however, yields the *cis* photoisomer, which upon warming (0°C), decomposes to yield photoelimination products. This is taken as evidence that the major path at room temperature involves photoisomerization of the *trans* to *cis* isomer and rapid thermolysis of the latter. Similar behavior was observed for azobis(isobutyronitrile), azobis(cyanocyclohexane), and azobis(2-methylpropyl acetate).[4]

Over the past few years considerable effort has been given to attempts to observe spin correlation effects from the radicals produced from the decomposition of azo compounds. It has generally been assumed that direct photolysis of the azo compound leads to singlet state decomposition to produce singlet radical pairs, while triplet-sensitized decomposition results in triplet radical pairs. The subsequent reactions of the radical pairs should depend upon their multiplicity. However, failure to observe spin correlation effects led to the conclusion that spin inversion must occur faster than diffusion of the radicals from the solvent cage[5] or that the presence of the nitrogen molecule produced a damping effect on the interaction between the electron spins in the radical pair.[6] It has now been shown by Engel and Bartlett[7] that the sensitized photolysis of azomethane and azo-2-methyl-2-propane with polynuclear aromatic hydrocarbons occurs via *singlet–singlet energy transfer* to produce singlet azo compounds and thus *singlet* radical pairs. This conclusion is supported by the quenching of the aromatic hydrocarbon fluorescence by azo-2-methyl-2-propane and by the observation that the triphenylene-sensitized decomposition of the azo compound is unaffected by piperylene.

The mode of fission of some azo compounds into alkyl radicals and nitrogen has been studied by Pryor and Smith[8] using the following postulates: (1) A molecule that decomposes by a concerted scission of both C—N bonds will not undergo cage return and will have a rate constant independent of viscosity; (2) a molecule that decomposes by a stepwise scission of the C—N bonds can undergo cage recombination and the rate constant for decomposition will decrease with solvent viscosity increase provided that the lifetime of the radicals produced by the initial homolysis is of the same order

TABLE 11.1 *Effect of Solvent Viscosity on the Rate of Decomposition of p-Nitrophenylazotriphenylmethane*

Solvent	Relative Viscosity, 60°C	Relative k_{obs}, NAT, 60°
Hexane	1.0	1.0
Nonane	2.04	0.83
Hexadecane	7.10	0.68

of magnitude as the time required for diffusion and separation of the radical pair ($\sim 10^{-10}$ sec). Data obtained for the observed rate of decomposition of *p*-nitrophenylazotriphenylmethane (NAT) as a function of solvent viscosity are shown in Table 11.1. The variation of k_{obs} with solvent viscosity would suggest the decomposition of NAT to occur by a stepwise pathway:

$$O_2N\text{—}C_6H_4\text{—}N{=}N\text{—}C\phi_3 \xrightarrow{\Delta} (ArN_2\cdot \ \cdot C\phi_3) \longrightarrow ArN_2\cdot + \cdot C\phi_3 \tag{11.4}$$

Similar results were obtained for the decomposition of phenylazotriphenylmethane; however, the lack of viscosity effect on the decomposition of azocumene suggests this process to be concerted. Other reports[9] also indicate symmetrically substituted azo compounds to decompose thermally by a concerted pathway.

Application of this technique to a study of the photoelimination of azo compounds has been reported by Porter, Landis, and Marrett.[10] Photolysis of the unsymmetrically substituted azo compound (1) in solvents of varying viscosity revealed a dependence of Φ_{dis} on solvent viscosity as shown in Table 11.2. Photolysis of optically active (1) (40% completion) and examination of the remaining azo compound indicated that 26% of the original optical activity had been lost. This is explained by the following mechanism involving stepwise homolysis:

$$\underset{(1)}{(H_3C)(\phi)(C_2H_5)C\text{—}N{=}N\text{—}\phi} \xrightarrow{h\nu} \left[\phi\text{—}N{=}N\cdot \quad \cdot C(H_3C)(C_2H_5)\phi\right] \longrightarrow \phi\text{—}N_2\cdot + \cdot C(H_3C)(C_2H_5)\phi \tag{11.5}$$

recombination → racemized (1)

$$\left[\phi\cdot \quad N_2 \quad \cdot C(H_3C)(C_2H_5)\phi\right] \longrightarrow \phi\cdot + \cdot C(H_3C)(C_2H_5)\phi$$

TABLE 11.2 *Quantum Yields for Disappearance of* (1) *at* 25°*C*

Solvent	Φ_{dis}
Octane	0.044
Decane	0.039
Dodecane	0.035
Hexadecane	0.029

Aromatic azo compounds such as azobenzene undergo *cis–trans* photoisomerization but do not decompose, due to the inherent instability of phenyl radicals.

Many reports concerning the direct and sensitized photolysis of cyclic azo compounds have appeared.[11] Photolysis of *meso* and *d,l* forms of azo compound (2) results in varying mixtures of products (3), (4), and (5):

(2) $\xrightarrow{h\nu}$ (3) + (4) + (5) (11.6)

However, in contrast to linear azo compounds, the cyclic compound (2) yielded product mixtures upon sensitized photolysis distinctly different from those resulting from direct photolysis, as seen in Table 11.3. These differences are thought to result from the greater lifetime of the triplet biradical resulting from triplet photoelimination as opposed to that of the singlet biradical produced in the direct photolysis. The cyclic azo triplet is unable to undergo the *cis–trans* isomerization observed in acyclic azo compounds and hence undergoes concerted homolysis. The parallel spins in the triplet biradical delay ring closure, allowing single bond rotation (thus separating the radical centers) and lead to a greater amount of cleavage product (3).

TABLE 11.3 *Product Distributions for Decomposition of* (2)

Isomer	Decomposition Mode	%(3)	%(4)	%(5)	%Retention
meso	Direct	61	35	3.5	95
d,l	Direct	60	4	33	97
meso	Thioxanthone-sensitized	77	11.5	8	61
d,l	Thioxanthone-sensitized	75	8	12	65

Similar effects have been observed with the azo compound (6)[13]:

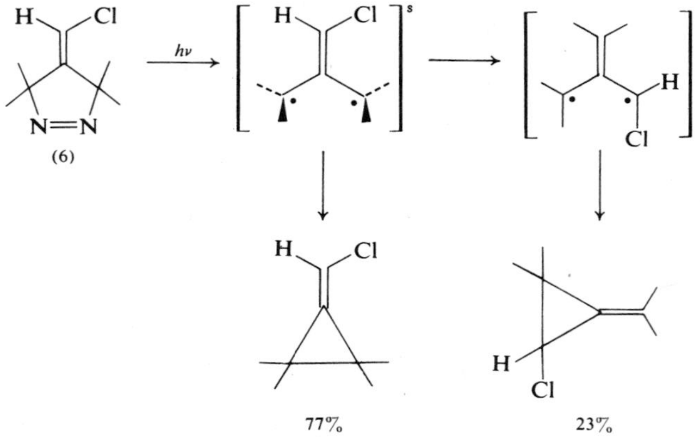

(11.7)

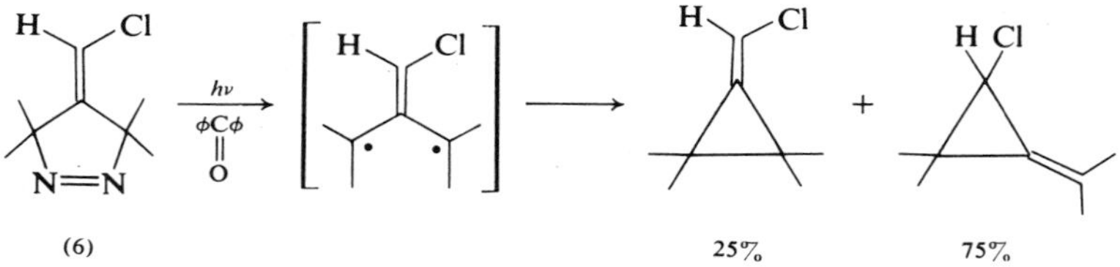

(11.8)

The direct photolysis of (6) was found to involve no triplet intermediates, since the same product distribution was obtained in the presence of the triplet quencher piperylene as in its absence.

In contrast to the strained ring systems discussed above, photolysis of the larger ring system in (7) results in olefin (8) as the major product[14]:

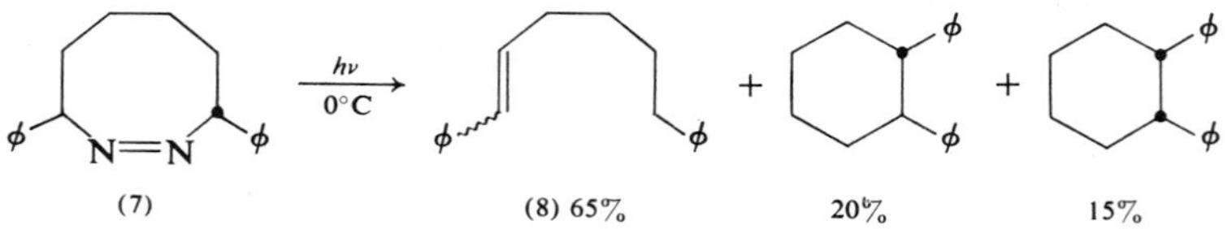

(11.9)

In the presence of benzophenone, (8) was again the major product (>95%) and only trace amounts of the cyclohexane products were produced. These results suggest the intermediacy of a singlet 1,6-hexylene biradical in the direct photolysis and a longer lived triplet 1,6-diradical in the sensitized photolysis. In the triplet biradical more time is available for 1,6-hydrogen transfer to occur prior to spin inversion and hence more olefin (8) is produced. Similar results were reported for the direct and photosensitized photolysis of the 3,8-dimethyl derivative of (7).

The photolysis reactions of some other cyclic azo compounds are shown below[15–19]:

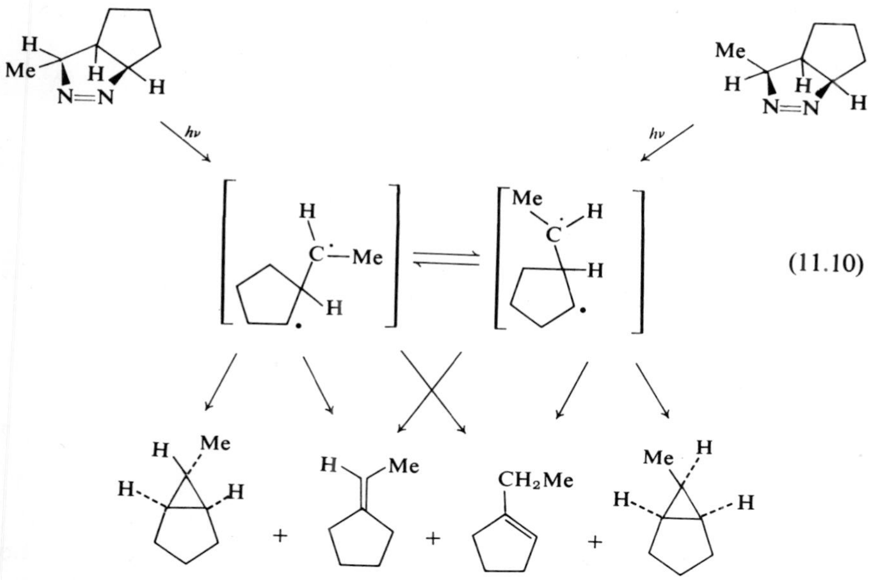

hν hν

(11.10)

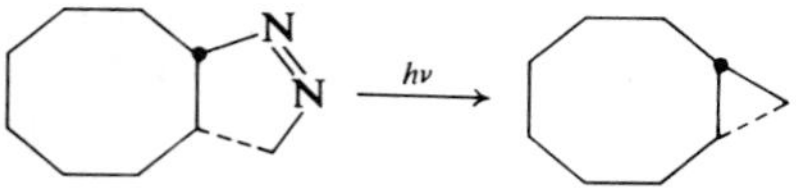

hν

(11.11)

96%

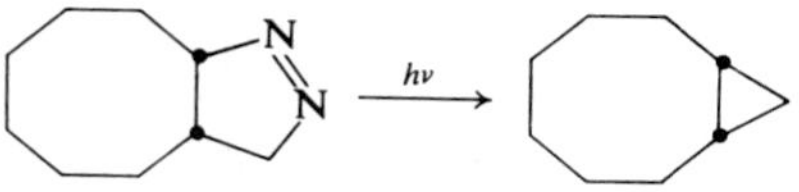

hν

(11.12)

100%

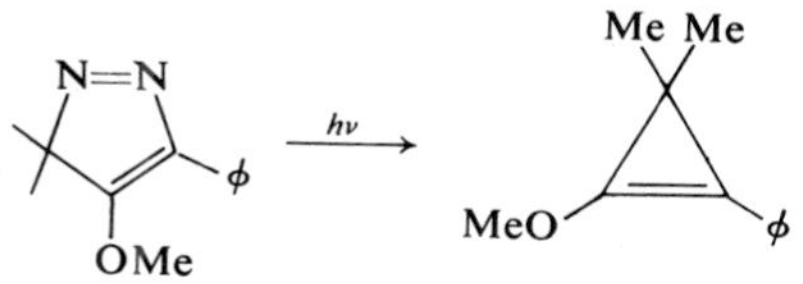

hν

(11.13)

hν

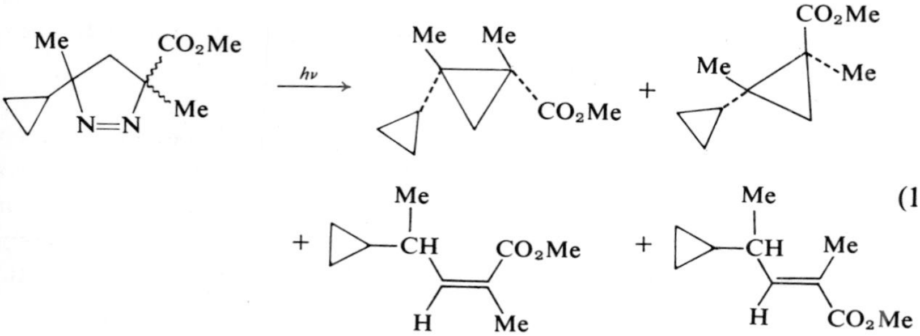

(11.14)

(9) (11.15)

(10) (11.16)

(11) (11.17)

That photodecomposition of some cyclic azo compounds need not proceed via elimination to form singlet or triplet biradicals has been shown by studies of the photoelimination of nitrogen from dihydrophthalazine in rigid glass matrices at $-196°C$.[156] Irradiation of this compound results in rapid formation of a new species identified as *o*-xylylene. No triplet EPR signal was detected under conditions which permitted triplet signals of ground state triplets such as the triphenylene dianion. Melting of the glassy solution containing this photoproduct results in formation of a spiro dimer:

$$\xrightarrow[\text{EPA; } -196°C]{h\nu} \qquad \longrightarrow \tag{11.18}$$

Dihydrophthalazine *o*-Xylylene

PPP calculations indicate the lowest triplet state of *o*-xylylene to be 8000 cm^{-1} above the ground state singlet.

The simplest diazo compound, diazomethane, undergoes loss of nitrogen in the gas or solution phase to yield a divalent carbon compound, methylene:

$$CH_2N_2 \xrightarrow{h\nu} :CH_2 + N_2 \tag{11.19}$$

The nature of the photochemically produced methylene has been the subject of considerable study. Evidence tends to indicate that this species is produced in its singlet state photochemically. Photolysis of diazomethane in the gas phase or in solution in the presence of excess *cis*- or *trans*-2-butene produces cyclopropane products due to methylene insertion in which the original *cis* or *trans* stereochemistry of the olefin is retained. Similarly, the products corresponding to insertion of methylene into a C—H bond in the olefin retain their stereochemistry. It is argued that this stereospecificity can result only via a concerted addition of singlet methylene since a triplet process would be expected to be stepwise and result in isomerization about the double bond:

$$CH_2N_2 + \underset{H\ \ \ \ \ \ \ CH_3}{\overset{H_3C\ \ \ \ \ \ \ H}{C{=}C}} \xrightarrow{h\nu} \text{(cyclopropane: } H_3C,\ H \text{ on one carbon; } H,\ CH_3 \text{ on the other)} + \underset{H\ \ \ \ \ \ \ CH_3}{\overset{CH_3CH_2\ \ \ \ \ \ \ H}{C{=}C}} + N_2 \tag{11.20}$$

$$CH_2N_2 + \underset{H_3C\ \ \ \ \ \ \ CH_3}{\overset{H\ \ \ \ \ \ \ H}{C{=}C}} \xrightarrow{h\nu} \text{(cyclopropane: } H_3C,\ H \text{ on one carbon; } CH_3,\ H \text{ on the other)} + \underset{CH_3CH_2\ \ \ \ \ \ \ CH_3}{\overset{H\ \ \ \ \ \ \ H}{C{=}C}} + N_2 \tag{11.21}$$

Methylene insertion into C—H bonds is believed to be concerted for the singlet species and stepwise for the triplet.(154,155) The C—H insertion of methylene into the ^{14}C-labeled isobutylene shown below results in 92% unrearranged isopentenes and 8% rearranged isopentene [Eq. (11.22)]. Assuming that an additional 8% of the unrearranged isopentene arises from the stepwise addition, it is clear that 84% of the insertion products result from insertion by singlet methylene and 16% by triplet methylene:

$$^{14}CH_2{=}\overset{CH_3}{\overset{|}{C}}{-}CH_3 + {:}CH_2 \xrightarrow{84\%} {}^{14}CH_2{=}\overset{CH_3}{\overset{|}{C}}{-}CH_2CH_3$$

$$\downarrow 16\%$$

$$\left[\ {}^{14}CH_2{=}\overset{CH_3}{\overset{|}{C}}{-}CH_2\cdot \ \longleftrightarrow\ {}^{14}\dot{C}H_2{-}\overset{CH_3}{\overset{|}{C}}{=}CH_2\ \right] + \dot{C}H_3 \xrightarrow{8\%} {}^{14}CH_2{=}\overset{CH_3}{\overset{|}{C}}{-}CH_2CH_3 \quad ; \quad \xrightarrow{8\%} {}^{14}CH_3{-}CH_2{-}\overset{CH_3}{\overset{|}{C}}{=}CH_2 \tag{11.22}$$

Additional evidence that photolysis of diazomethane leads to singlet methylene was obtained by Kopecky, Hammond, and Leermakers.(21) These workers observed that triplet methylene produced by energy transfer decomposition with benzophenone reacts nonstereospecifically with the 2-butenes:

$$(\!:\!CH_2)^t + \underset{cis}{CH_3CH{=}CHCH_3} \xrightarrow{h\nu} \underset{66\%}{\text{cis-1,2-dimethylcyclopropane}} + \underset{34\%}{\text{trans-1,2-dimethylcyclopropane}} \qquad (11.23)$$

Similarly, the direct [Eq. (11.24)] and photosensitized [Eq. (11.25)] decomposition of diazomethane in the presence of cyclohexene yielded product distributions indicative of greater selectivity in the triplet methylene addition:

$$\text{cyclohexene} + CH_2N_2 \xrightarrow{h\nu} \underset{39\%}{\text{norcarane}} + \underset{9\%}{\text{1-methylcyclohexene}} + \underbrace{\text{3-methylcyclohexene} + \text{4-methylcyclohexene}}_{52\%} \qquad (11.24)$$

$$\text{cyclohexene} + CH_2N_2 \xrightarrow[\phi C(=O)\phi]{h\nu} \underset{70\%}{\text{norcarane}} + \underset{\text{trace}}{\text{1-methylcyclohexene}} + \underbrace{\text{3-methylcyclohexene} + \text{4-methylcyclohexene}}_{30\%} \qquad (11.25)$$

Flash photolysis studies[22] have indicated singlet methylene to be produced from the diazomethane-excited singlet upon loss of nitrogen followed by collisional deactivation to the triplet, the ground state multiplicity for this molecule.

Gas-phase photolysis of diazoethane results in mixtures of ethylene, acetylene, and *cis*- and *trans*-2-butene. A mechanism involving the initial formation of ethylidene followed by formation of activated ethylene [which is collisionally deactivated or decomposes to produce acetylene and hydrogen—Eqs. (11.26(b,c,d)] or alternate attack on diazoethane to produce 2-butene [Eq. 11.26(e)] is proposed:

$$CH_3CHN_2 \xrightarrow{h\nu} CH_3CH\!: + N_2 \qquad (11.26a)$$

$$CH_3CH\!: \longrightarrow C_2H_4^* \qquad (11.26b)$$

$$C_2H_4^* \longrightarrow C_2H_2 + H_2 \qquad (11.26c)$$

$$C_2H_4^* + M \longrightarrow C_2H_4 \qquad (11.26d)$$

$$CH_3CHN_2 + CH_3CH\!: \longrightarrow CH_3CH{=}CHCH_3 \ (\textit{cis} \text{ and } \textit{trans}) \qquad (11.26e)$$

In the presence of propene, photolysis of diazoethane yields *cis* and *trans* isomers of 1,2-dimethylcyclopropane:

$$CH_3CHN_2 \xrightarrow{h\nu} CH_3CH: \xrightarrow{\text{cis-2-butene}} \text{cis-1,2-dimethyl-3-methylcyclopropane} + \text{trans isomer} \quad (11.27)$$

No products resulting from C—H insertion of the carbene were found. Also, ethylidene failed to undergo addition to 2-butene, indicating this species to be considerably more selective than methylene.[23]

2-Diazobutane yields 1-butene, *cis*- and *trans*-2-butene, and methylcyclopropane on gas-phase photolysis. Product distributions are independent of pressure (50–200 mm Hg) and the presence of nitrogen up to 2 atm[24]:

$$CH_3CH_2C(N_2)CH_3 \xrightarrow{h\nu} CH_3CH_2\ddot{C}CH_3 \longrightarrow \underset{23.4\%}{CH_3CH_2CH{=}CH_2} + \underset{\substack{cis\,+\,trans \\ 74.2\%}}{CH_3CH{=}CHCH_3} + \underset{2.4\%}{\text{methylcyclopropane}} \quad (11.28)$$

Photolysis of diazocyclopentadiene in *trans*- and *cis*-methyl-2-pentene results in products (12) and (13), respectively, in an essentially stereospecific manner (1–2% of the *trans*-cyclopropane was observed to result from addition to the *cis* olefin)[25]:

$$\text{diazocyclopentadiene} \xrightarrow{h\nu} [\text{cyclopentadienylidene}] \xrightarrow{H_3C(H)C{=}C(H)C_3H_7\ (trans)} \text{(12)}$$

$$[\text{cyclopentadienylidene}] + H_3C(H)C{=}C(H)C_3H_7\ (cis) \longrightarrow \text{(13)} \quad (11.29)$$

(12): spiro cyclopropane with CH_3, H, H, $CH(CH_3)_2$ (trans)

(13): spiro cyclopropane with CH_3, H, $CH(CH_3)_2$, H (cis)

The stereospecificity of addition suggests a singlet[10] carbene although the ground state of cyclopentadienylidene is known to be a triplet. Attempts to produce a triplet species, which would be expected to react nonstereospecifically, in a 4-methyl-*cis*-2-pentene matrix at 77°K or by dilution of mixtures of the azo compound and olefin with hexafluorobenzene or octafluorocyclobutane (inert diluents) were unsuccessful. It was concluded that the singlet carbene produced upon photolysis reacts more rapidly with the olefinic substrate than it crosses to its triplet state. Interestingly, the singlet state of this carbene can assume an aromatic hydrid structure (14):

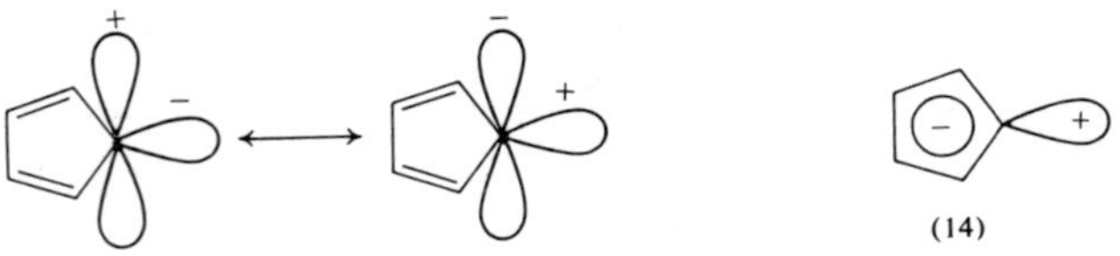

(14)

It might be expected that the aromatic character of (14) would lend unique characteristics to its chemistry. Studies of the olefin addition and C—H insertion reactions of the carbene, however, indicate that it is poorly stabilized and highly reactive.[26]

Two other carbenes, cyclopropenylidene[27] and cycloheptatrienylidene,[28] which have aromatic resonance hydrids (15) and (16), have been studied:

(15)

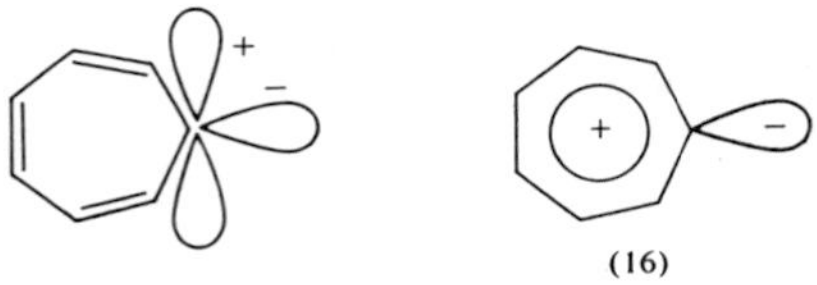

(16)

Resonance hydrids (15) and (16) would suggest that these carbenes should show nucleophilic behavior toward olefins. As predicted, only olefins with electron-poor π bonds have proven to be suitable substrates for these carbenes.

Photolysis of phenyldiazomethane in *cis*- or *trans*-butene leads to nearly stereospecific cyclopropane formation, although some C—H insertion occurs[29]:

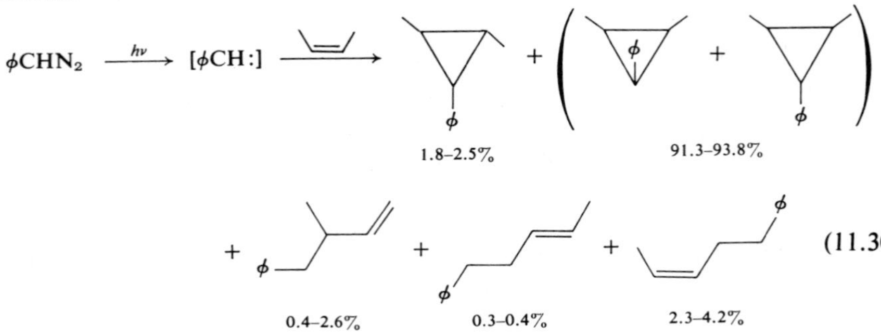

$$\phi CHN_2 \xrightarrow{h\nu} [\phi CH:] \longrightarrow$$

1.8–2.5%

91.3–93.8%

(11.30)

0.4–2.6% 0.3–0.4% 2.3–4.2%

A singlet carbene was proposed to account for this stereoselectivity. Attempts to produce triplet carbene by collisional deactivation with octafluorocyclobutane were unsuccessful and stereospecific addition to olefin still occurred. However, nonstereospecific addition to olefins and larger amounts of olefinic (insertion) products result from irradiation of the phenyldiazomethane in a frozen *cis*-butene matrix at −196°C:

ϕCHN_2 + [cis-2-butene] $\xrightarrow[-196°C]{h\nu}$ [cyclopropanes] 3.2% + ([cyclopropanes]) 43.9% + [olefins] 18.5% + 4.9% + 29% (11.31)

The olefinic products which formally correspond to C—H insertion reactions are thought to arise by stepwise abstraction of hydrogen by triplet carbene and subsequent recombination:

$[\phi CH:]^t$ + [cis-2-butene] ⟶ $\phi\dot{C}H_2$ [allyl radical] ⟶ $\phi\dot{C}H_2$ [allyl radical] → ϕCH_2–[alkene], [alkene]–$CH_2\phi$, [alkene]–$CH_2\phi$ (11.32)

EPR studies of diphenylmethylene and a number of other arylmethylenes have indicated that these carbenes have triplet ground states.[30] Photolysis of diphenyldiazomethane in olefin matrices results in the formation of triplet diphenylmethylene, which undergoes primarily abstraction reactions with the olefins. Cyclopropanes are produced as minor products.

Photolysis of diazoacetonitrile in *cis*-butene results in cyclopropanes (17) and (18)[31]:

$CHN_2CN \xrightarrow{h\nu} [:CHCN] \xrightarrow{\text{cis-butene}}$ (17) 94% + (18) 6% (11.33)

The yield of *trans* product (18) is decreased by the presence of a radical scavenger such as 1,1-diphenylethylene and increased by dilution of the reactants with methylene chloride or butane, indicating this product to result from the triplet carbene. A heavy-atom effect on the carbene intermediate was observed by photolysis of α-methylmercuridiazoacetonitrile. With *cis*-2-butene as the trapping agent either direct photolysis or triplet benzophenone-sensitized decomposition results in formation of cyclopropanes (19) and (20) in a 1:1 ratio:

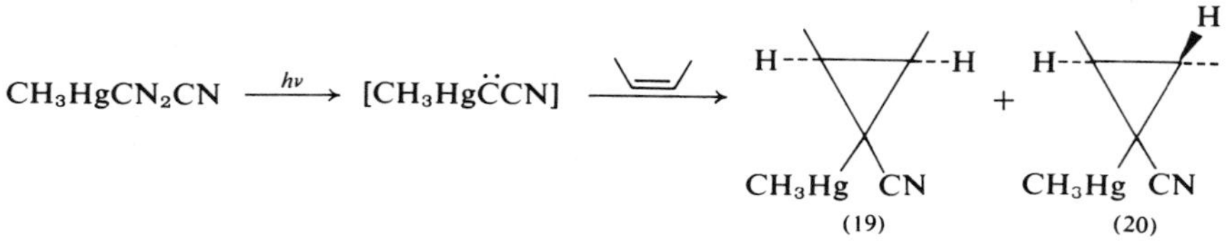

$$CH_3HgCN_2CN \xrightarrow{h\nu} [CH_3Hg\ddot{C}CN] \longrightarrow (19) + (20) \tag{11.34}$$

CH₃Hg CN (19) CH₃Hg CN (20)

Obviously the presence of the mercury substituent greatly accelerates intersystem crossing to the triplet ground state in the singlet cyanomethylene. In light of this report, the failure to observe nonstereospecificity in the additions of CBr_2, CI_2, and CHI to olefins can be understood as arising from the *singlet* nature of the ground states of these carbenes.

Photoelimination of nitrogen from diazoketones is complicated by Wolff rearrangement of the intermediate carbene, as shown below for diazoacetophenone[35]:

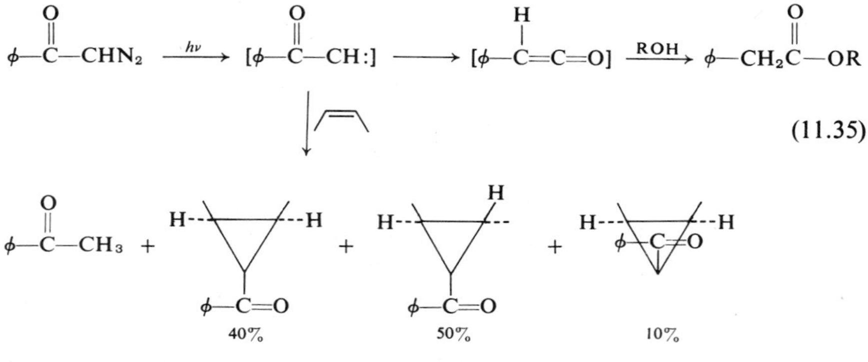

$$\phi{-}\overset{O}{\overset{\|}{C}}{-}CHN_2 \xrightarrow{h\nu} [\phi{-}\overset{O}{\overset{\|}{C}}{-}CH{:}] \longrightarrow [\phi{-}\overset{H}{\overset{|}{C}}{=}C{=}O] \xrightarrow{ROH} \phi{-}CH_2\overset{O}{\overset{\|}{C}}{-}OR \tag{11.35}$$

$$\phi{-}\overset{O}{\overset{\|}{C}}{-}CH_3 + \text{cis-cyclopropyl ketone } (40\%) + \text{trans-cyclopropyl ketone } (50\%) + \text{oxetane-type product } (10\%)$$

40% 50% 10%

The yields of cyclopropanes in this case are low in relation to the amount of acetophenone formed. However, similar cyclopropane product ratios are obtained when photolysis is carried out in the presence of Michler's ketone as sensitizer. Thus the carbene intermediate produced in the direct irradiation is thought to be a triplet, as suggested by the nonstereospecificity of its addition. Whether this intermediate arose from singlet diazoacetophenone (via singlet decomposition and intersystem crossing of the singlet carbene) or by decomposition of the triplet molecule was not determined.

Photolysis of diazoacetone in methanol results in products corresponding to addition to a ketene:

$$\underset{\underset{N_2}{\|}}{H\overset{\overset{O}{\|}}{C}CCH_3} \xrightarrow[\text{MeOH}]{h\nu} \left[H\ddot{C}\overset{\overset{O}{\|}}{C}CH_3\right] \longrightarrow \left[CH_3\overset{\overset{H}{|}}{C}{=}C{=}O\right] \longrightarrow \underset{70\%}{CH_3CH_2COOCH_3} + \underset{2\%}{CH_3OCH_2\underset{\underset{O}{\|}}{C}CH_3} \qquad (11.36)$$

Photolysis of this azoketone in olefinic solutions yields no cyclopropane product. However, a recent report by Skell and Velenty(36) indicates that methylmercuridiazoacetone produces high yields of cyclopropanes upon photolysis in olefinic solution:

$$\underset{\underset{N_2}{\|}}{MeHgC}\overset{\overset{O}{\|}}{C}Me \xrightarrow{h\nu} \left[MeHg\ddot{C}\overset{\overset{O}{\|}}{C}Me\right] \xrightarrow{\text{olefin}} \underset{45\text{–}91\%}{\text{1-(MeHg)-1-acetyl-2,3-dimethylcyclopropane}} \qquad (11.37)$$

The addition of the methylmercuriacetylcarbene to *cis*- and *trans*-butene was found to be completely stereospecific, suggesting that this carbene has a *singlet* ground state (with the heavy mercury atom, relaxation to the ground state should be rapid).

Carboalkoxymethylenes, like acylmethylenes, undergo rearrangement to ketenes as well as the olefin addition and C—H insertion reactions characteristic of methylenes.(37) Thus the photolysis of ethyl diazoacetate in olefinic solvents leads to substantial yields of products, which can be rationalized in terms of a Wolff rearrangement of the carboethoxymethylene followed by cycloaddition of the resulting ethoxyketene to the olefin:

$$EtO\overset{\overset{O}{\|}}{C}CHN_2 \longrightarrow \left[EtO\overset{\overset{O}{\|}}{C}CH{:}\right] \longrightarrow \left[EtO\overset{\overset{H}{|}}{C}{=}C{=}O\right] \qquad (11.38)$$

$$\left[EtO\overset{\overset{O}{\|}}{C}CH{:}\right] \xrightarrow{CH_2CH_2} \text{cyclopropyl-}\overset{\overset{O}{\|}}{C}OEt + CH_2{=}CHCH_2CO_2Et$$

$$\left[EtO\overset{\overset{H}{|}}{C}{=}C{=}O\right] \xrightarrow{CH_2CH_2} \text{2-ethoxycyclobutanone (OEt)}$$

However, the photolysis of methylmercuridiazoacetate in olefins proceeds cleanly to olefin addition products from the intermediate carbene in 70–90% yields[36]:

$$\mathrm{MeHgC(=N_2)COOMe} \xrightarrow{h\nu} \mathrm{MeHg\ddot{C}COOMe} \xrightarrow{\mathrm{CH_2CH_2}} \text{(cyclopropane with MeHg, COOMe)}\ (70\text{–}90\%) + \mathrm{Hg}\ (5\text{–}10\%) \quad (11.39)$$

The mechanism through which α-methylmercury substituents eliminate Wolff rearrangements of acyl- and carboalkoxymethylenes is not clearly evident.

The photolysis of diazo compounds has been applied to some interesting synthetic problems, including the following[38–41]:

(11.40)

$h\nu$; $\mathrm{MeO_2C{\equiv}CO_2Me}$; $h\nu$

ϕ ; $=N_2$; CO_2Me ; CO_2Me ; H

Photolysis of α-diazo-amides to produce *cis*-β-lactams has provided a route to the synthesis of new types of β-lactam antibacterial agents[39]:

(11.41)

CO_2Et ; CN_2 ; N ; O ; $h\nu$; EtO_2C ; H H ; N ; O ; + ; EtO_2C ; H H ; N ; O

The synthesis of morphinandienone (22) and aporphine (23) derivatives has been achieved photochemically through irradiation of diazo compound (21)[40]:

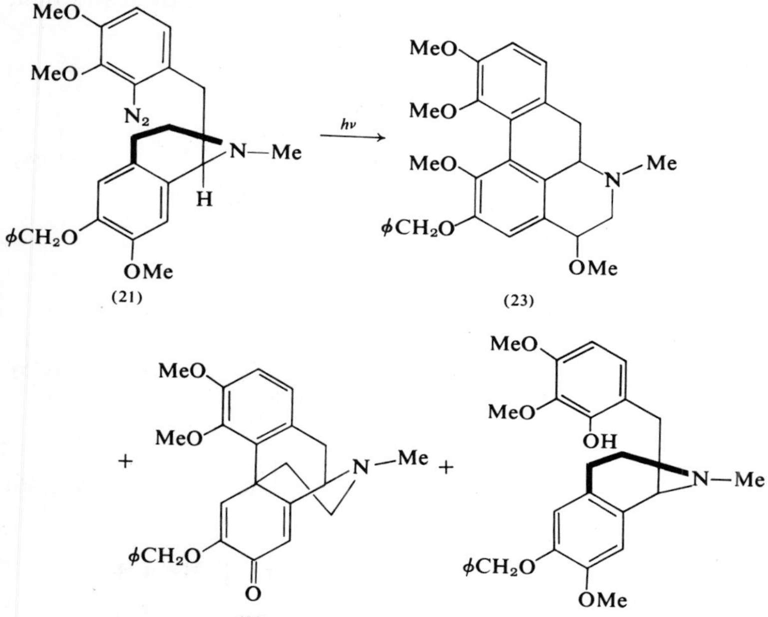

(21) → (23) + (22) + … (11.42)

A synthetic route to the previously difficult to obtain isotryptamine system has been achieved by photolysis of (24)[41]:

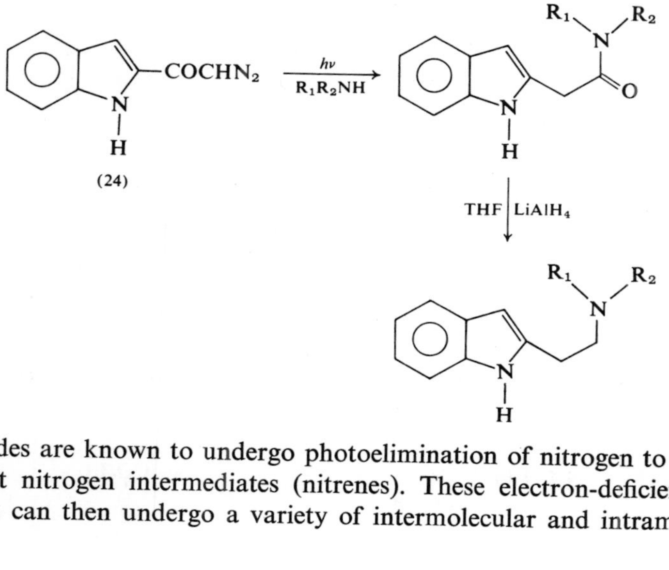

(24) → (11.43)

Azides are known to undergo photoelimination of nitrogen to produce univalent nitrogen intermediates (nitrenes). These electron-deficient intermediates can then undergo a variety of intermolecular and intramolecular reactions, such as insertion into an ethylenic bond, hydrogen abstraction, coupling to form azo compounds, and C—N cleavage:

$$RN_3 \xrightarrow{h\nu} N_2 + RN: \xrightarrow{>C=C<} \text{aziridine } (\text{N–R}) \qquad \text{(11.44)}$$

$$RN: \xrightarrow{R'H} RNH_2 + 2R'\cdot \qquad RN: \longrightarrow RN_2R$$

Some examples follow[42–44]:

$$CH_3CH_2CH_2CH_2N_3 \xrightarrow{h\nu} [CH_3CH_2CH_2CH_2N:] \xrightarrow[\text{benzene}]{} \underset{22\%}{C_4H_9NH\phi} \qquad \text{(11.45)}$$

$$CH_3CH_2CH_2N_3 \xrightarrow{h\nu} \underset{59\%}{CH_3CH_2CH{=}NH} \qquad \text{(11.46)}$$

$$CH_3CH_2CH_2CH_2N_3 \xrightarrow[\text{EtOH}]{h\nu} \left[\begin{matrix} H_2C—CH_2 \\ H_3C \quad CH_2 \\ :N \end{matrix}\right] \longrightarrow \left[\begin{matrix} CH_2—CH_2 \\ \dot{C}H_2 \quad CH_2 \\ \cdot NH \end{matrix}\right] \qquad \text{(11.47)}$$

$$\left[H_3C \ldots :N\right] \longrightarrow \underset{12\%}{CH_3CH_2CH_2CH{=}NH} \qquad \left[\cdot NH \ldots\right] \longrightarrow \underset{22\%}{\text{pyrrolidine (N–H)}}$$

$$CH_3CH_2CH_2CH(N_3)COOH \xrightarrow{h\nu} [CH_3CH_2CH_2CH(N:)COOH] \qquad \text{(11.48)}$$

$$[CH_3CH_2CH_2CH(N:)COOH] \xrightarrow{\text{solvent}} CH_3CH_2CH_2CH(NH_2)COOH \qquad [CH_3CH_2CH_2CH(N:)COOH] \xrightarrow{-CO_2} CH_3CH_2CH_2CH{=}NH$$

$$N_3CO_2Et + \text{cyclohexene} \xrightarrow{h\nu} \text{7-azabicyclo[4.1.0]heptane}\text{N—}CO_2Et \qquad \text{(11.49)}$$

Secondary and tertiary azides commonly undergo rearrangement by substituent migration to imines. The observation that in tertiary azides there is no preference for methyl vs. phenyl vs. substituted phenyl migration upon photolysis supports the intermediacy of a discrete nitrene intermediate (no participation by the migrating group in the original nitrogen loss)[45]:

$$\phi_3C\text{—}N_3 \xrightarrow{h\nu} \left[\phi_3C\text{—}\ddot{N}:\right] \longrightarrow \phi_2C{=}N\phi \qquad (11.50)$$

Table 11.4. *Migratory Aptitudes in the Photolysis of t-Alkyl Azides*[49]

Compound	Migrating groups	Ratio[a]
ϕ—C(CH_3)$_2$—N_3	ϕ, CH_3	$\phi/CH_3 = 0.75$
4-biphenylyl—C(CH_3)$_2$—N_3	Ar, Me	$Ar/CH_3 = 0.69$
4-biphenylyl—C(ϕ)$_2$—N_3	Ar, ϕ	$Ar/\phi = 0.44$
CH_3—C(N_3)(CH_3)—$CH_2CH_2\phi$	ϕCH_2CH_2, Me	$\phi CH_2CH_2/Me = 0.89$

[a] Corrected for statistical preference.

Similar products are obtained from the photosensitized decomposition of the tertiary azides, suggesting that decomposition may result from the triplet azides under both direct and sensitized photolysis.[46] Additional evidence for a discrete nitrene intermediate comes from the observation that this intermediate can be trapped by decomposition of the azides in the presence of good hydrogen donors such as tri-*n*-butyltin hydride and *sec*-butyl mercaptan. Triarylamines result:

$$\phi_3C{-}N_3 \xrightarrow{h\nu} \left[\phi_3C{-}\ddot{N}\!:\right] \xrightarrow{2[H]} \phi_3C{-}NH_2 \qquad (11.51)$$

Further study of the direct decomposition revealed that the reaction could not be quenched by the presence of piperylene; thus, decomposition must result either from singlet-excited azide or from triplet azide at a rate on the order of 10^{11} sec^{-1}.[47]

More recent investigations on the migratory aptitudes of various substituents suggest that a discrete nitrene intermediate may not be formed in azide decompositions (see Table 11.4).[48,49] The data in Table 11.4 indicate that the smaller group migrates preferentially in each case and that the electron-deficient intermediate is not free to attack the more nucleophilic group. This is explained by assuming that the preferred ground state conformation of the alkyl azide determines which group is suitably oriented to migrate with respect to the electron-deficient p_y orbital on the α-nitrogen ($\pi_y \rightarrow \pi_x{}^*$ excited state). If the most important steric interaction is between the large group L and the azide N_2, conformation (25) would be more important than conformation (26) and migration of group M would occur preferentially:

(11.52)

$$\text{(26)} \xrightarrow{h\nu} \text{M}_2\text{C}{=}\text{NL} \tag{11.53}$$

Thus the earlier failure to observe migratory preference with *p*-substituted phenyl groups in triarylmethylazides results from the fact that the presence of the *p*-substituent would have no influence on the ground state conformation of the azides. In support of this hypothesis is the observation that the decomposition and rearrangement of (27) becomes increasingly statistical as the temperature is raised (higher energy conformations become more populated):

$$\text{(27)} \xrightarrow{h\nu} \text{Me}(\phi)\text{C}{=}\text{NH} + \phi\text{N}{=}\text{CH(Me)} + \text{H}(\phi)\text{C}{=}\text{NMe} \tag{11.54}$$

ESR spectra for a number of nitrene intermediates produced by photolysis at 77°K have been reported.[50] Analysis of these spectra has resulted in their assignment to triplet ground state nitrenes.

Photolysis of diazonium salts in alcohols produces aromatic ethers and aromatic hydrocarbons[51,52]:

$$O_2N\text{-}C_6H_4\text{-}N{\equiv}N^{\oplus} \xrightarrow[\text{EtOH, } 0^\circ]{h\nu} \underset{77\%}{C_6H_5NO_2} + \underset{5\%}{p\text{-}EtO\text{-}C_6H_4\text{-}NO_2} + \underset{4\%}{p\text{-}CH_3CH(OH)\text{-}C_6H_4\text{-}NO_2} + p\text{-}HOCH_2CH_2\text{-}C_6H_4\text{-}NO_2 \tag{11.55}$$

These products are thought to result from attack of an intermediate *p*-nitrophenyl radical on the solvent. Evidence for this intermediate was obtained by scavenging the radical with diphenylpicrylhydrazyl, halogens, and nitric oxide.[52] However, the presence of the *p*-nitrophenetole in the products formed in the presence of iodine, which effectively eliminates most other radical products, suggests that another mechanism involving a phenyl carbonium ion may be also operative:

$$O_2N-C_6H_4-N_2^{\oplus} \xrightarrow{h\nu} \left[NO_2-C_6H_4^{\oplus} \right] \xrightarrow{EtOH} \left[O_2N-C_6H_4-\overset{H}{\underset{|}{O}}C_2H_5 \right]^+ \longrightarrow EtO-C_6H_4-NO_2 + H^+ \quad (11.56)$$

Indeed, the acidity of the reaction mixture was found to increase upon continued photolysis, in accord with the carbonium ion mechanism.

In aqueous solutions diazonium salts produce products corresponding to anion addition to an intermediate carbonium ion:

$$O_2N-C_6H_4-N_2^{\oplus}X^- \xrightarrow[H_2O]{h\nu} \left[NO_2-C_6H_4^{\oplus} \right] \xrightarrow{X^- \text{ or } OH^-} \underset{74\%}{NO_2-C_6H_4-OH} + \underset{4\%,\ X = Cl}{NO_2-C_6H_4-X} \quad (11.57)$$

Irradiation of a diazonium compound in EPA glass at 77°K and monitoring the reaction by UV revealed the presence of new absorption bands, which disappeared upon warming to room temperature. Little or no evolution of nitrogen occurred in these experiments. Hence the low-temperature intermediate has been proposed to be a rather stable triplet state of the diazonium compound.

The photolysis of crystalline diazonium tetrafluoroborates and hexafluorophosphates affords a convenient route to aromatic fluorides[53]:

$$CH_3O-C_6H_4-N_2^+BF_4^- \xrightarrow{h\nu} \underset{69\%}{CH_3O-C_6H_4-F} \quad (11.58)$$

$$p\text{-}(CH_3)_2N\text{-}C_6H_4\text{-}N_2^+PF_6^- \xrightarrow{h\nu} p\text{-}(CH_3)_2N\text{-}C_6H_4\text{-}F \quad (72\%) \tag{11.59}$$

Ionic intermediates are proposed in the photolysis of these salts.

11.1b. *Photoelimination of Nitric Oxide from Organic Nitrites*

Organic nitrites display a weak long-wavelength absorption ($n \rightarrow \pi^*$) involving excitation of a nonbonding electron on nitrogen. Photolysis has been found to result in photodissociation of the RO—NO bond to produce alkoxy radicals and nitric oxide:

$$RO—NO \xrightarrow{h\nu} RO\cdot + NO \tag{11.60}$$

The alkoxy radicals produced in this photoelimination can react in several ways:

1. *Fragmentation to produce carbonyl compounds and alkyl radicals*, as shown below for the gas-phase photolysis of *t*-butylnitrite[54]:

$$(CH_3)_3C—ONO \xrightarrow{h\nu} (CH_3)_3C—O\cdot + NO$$

$$(CH_3)_3C—O\cdot \longrightarrow (CH_3)_2C{=}O + CH_3\cdot \xrightarrow{NO} CH_3NO \tag{11.61}$$

$$CH_3NO \xrightarrow{CH_3NO} CH_3(O\leftarrow)N{=}N(\rightarrow O)CH_3$$

2. *Intermolecular hydrogen abstraction:*

$$RO\cdot + \text{solvent} \longrightarrow ROH + \text{solvent}\cdot \tag{11.62}$$

Photolysis of *n*-octylnitrite in heptane solution results in formation of the dimers of 2-, 3-, and 4-nitrosoheptane as well as the dimer of 4-nitroso-1-octanol (intramolecular hydrogen abstraction)[55]:

ONO

$h\nu$

O·

H

+ NO $\xrightarrow[2.1\%]{\text{heptane}}$

NO

+

NO

+

NO

OH

OH

NO

(11.63)

Photolysis of *n*-octylnitrite in solvents that are poor hydrogen donors (e.g., benzene) results primarily in the formation of the dimer of 4-nitroso-1-octanol.

3. *Intramolecular hydrogen abstraction.* As stated above, photolysis of *n*-octylnitrite results in a predominance of the product corresponding to intramolecular hydrogen abstraction. Photolysis of 3-heptyl, 2-hexyl, and 2-pentylnitrites yields intramolecular hydrogen abstraction products in 28%, 31%, and 6% yields, respectively.[56] In each case the product corresponded to hydrogen abstraction via a six-membered transition state:

H R ONO R′ $\xrightarrow{h\nu}$ [R′ O· H R]

[R′ O H · R] ⟶ R′ OH NO R (11.64)

Photolysis of 2-propylnitrite, in which intramolecular hydrogen abstraction cannot occur via a six-membered transition state, results in a 30% yield of 2-propanone. As we shall see, this intramolecular hydrogen abstraction has been extensively applied to steroid syntheses by D. H. R. Barton and hence is commonly referred to as the Barton reaction.

Photolysis reactions of some cyclic nitrites are shown below[57]:

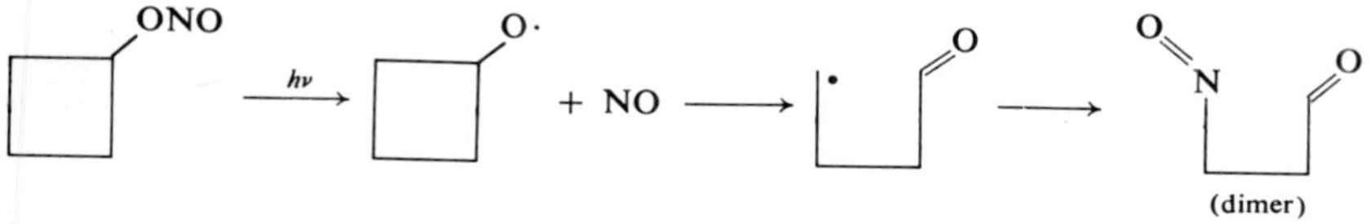

(11.65)

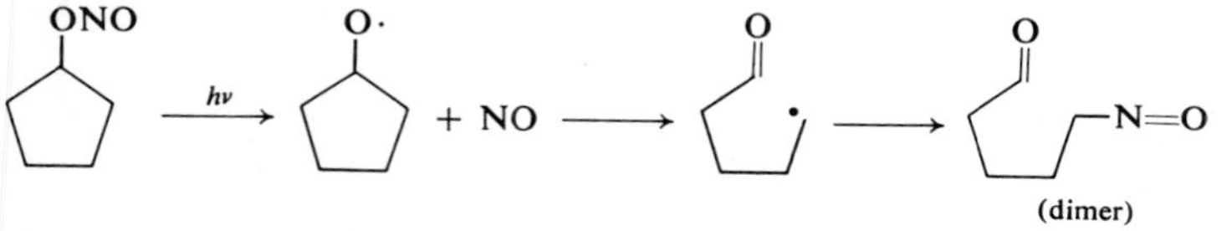

(11.66)

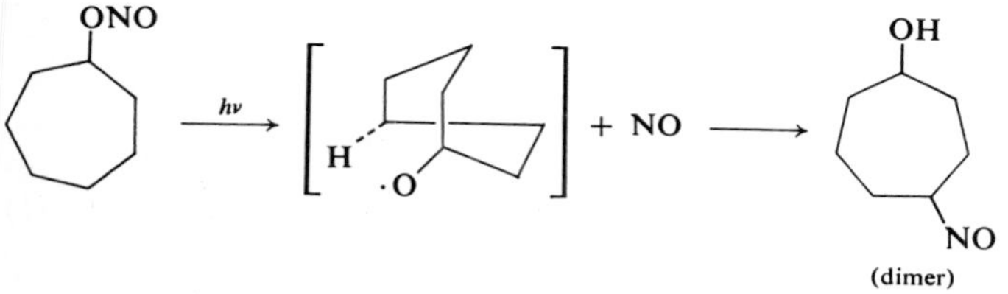

(11.67)

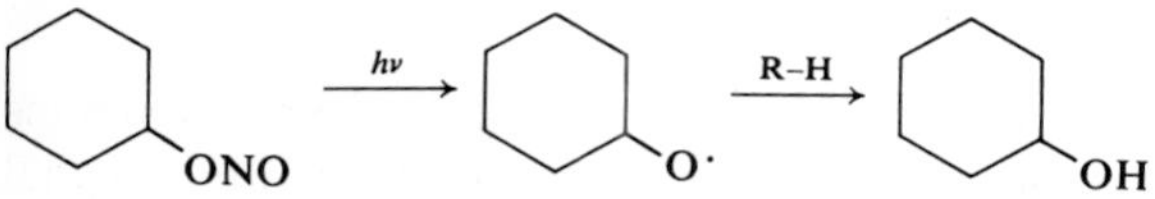

(11.68)

A mechanistic study of this reaction has been carried out by Akhtar and Pechet.[58] Three distinct mechanisms were considered:

(a) Fragmentation of the nitrite into an alkoxy radical and nitric oxide, followed by radical recombination or hydrogen abstraction within the primary solvent cage:

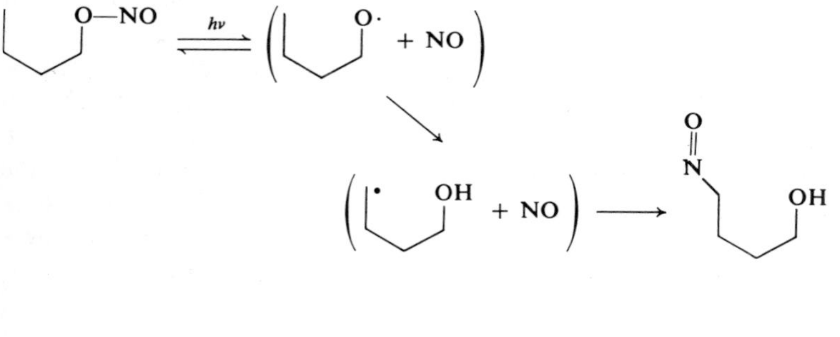

(11.69)

(b) Fragmentation of the nitrite, followed by diffusion from the primary radical cage, followed by combination of "free" radical fragments:

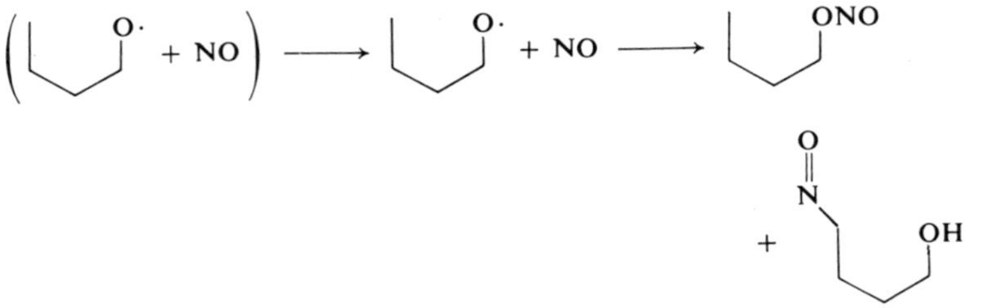

(11.70)

(c) Fragmentation of the nitrite by a partitioning between cage return and diffusion:

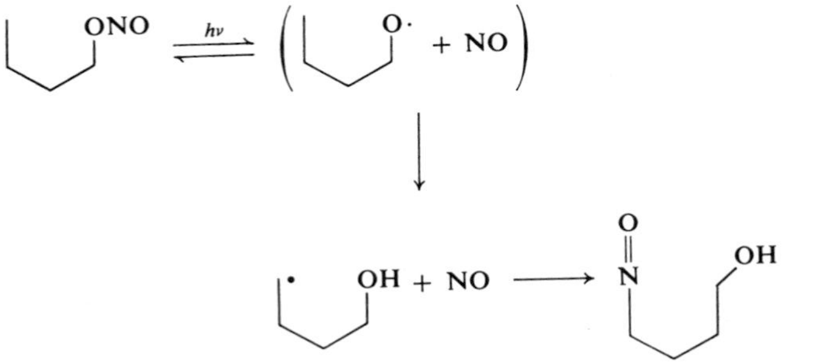

(11.71)

Thus, upon photolyzing a mixture of two nitrites, one of which contains an isotopically labeled nitrogen, the products should show the original isotopic distribution if mechanism (a) is operative, the original isotopic distribution in unreacted nitrite but scrambling in the nitroso product if (c) is operative, and complete isotopic scrambling if (b) is operative.

Irradiation of a mixture of androstane (28) and cholestane (29) followed by oxidation of the product oximes to nitriles and mass spectrometric analysis revealed a N^{15}/N^{14} ratio of 1:1.32 for the product derived from (28) and a N^{15}/N^{14} ratio of 1:1.25 for the product derived from (29). Thus mechanism (a) is effectively eliminated, see top of facing page.

A choice between mechanisms (b) and (c) was achieved by photolyzing a mixture of a ^{14}N-nitrite in the pregnane series and the ^{15}N-nitrite (29) to half-completion. Rearrangement of the nitroso dimers thus obtained to a mixture of oximes followed by oxidation of the pure oxime from (29) to the ketonitrile and mass spectrometric analysis indicated the N^{15}/N^{14} ratio to be 1.15:1.00. Analysis of pure unreacted (29) gave a ratio of 1.00:0.00. These results indicate that mechanism (c) is operative in nitrite photolyses.

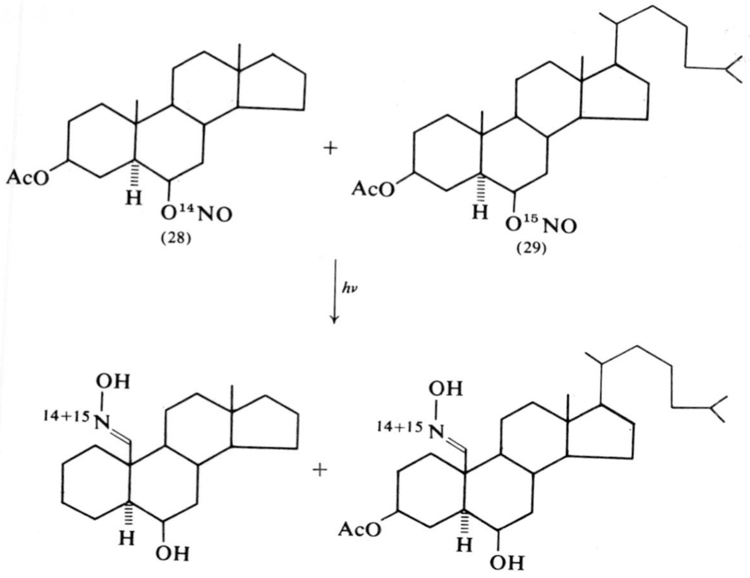

AcO H $O^{14}NO$ (28) + AcO H $O^{15}NO$ (29)

$h\nu$

OH 14+15 N H OH + OH 14+15 N AcO H OH

(11.72)

The photolysis of nitrite esters has been found to be of considerable synthetic utility, particularly in functionalizing steroid methyl groups. The most spectacular application of this reaction has been a three-step synthesis of aldosterone.[59] Irradiation of corticosterone-11-nitrite (30) followed by nitric acid treatment gave aldosterone (32) in a 15% overall yield:

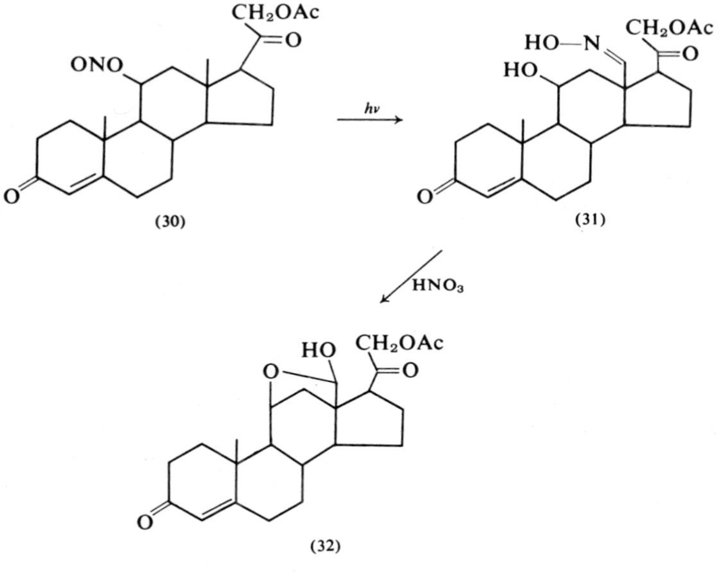

CH_2OAc ONO O (30) $h\nu$ HO—N CH_2OAc HO O (31)

HNO_3

HO CH_2OAc O O (32)

(11.73)

Other synthetic applications of the Barton reaction in the steroid field follow[(59–62)]:

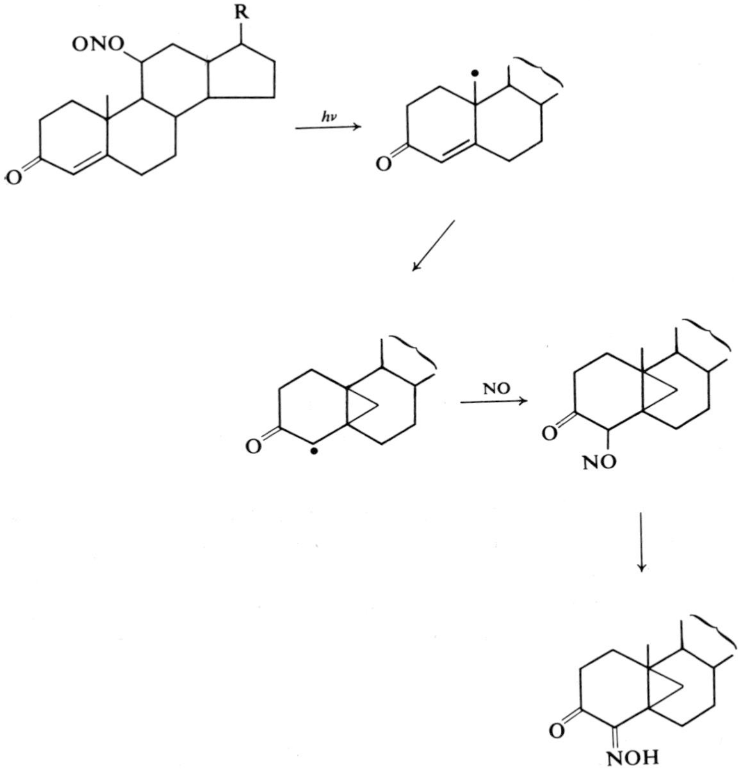

ONO R hν NO NOH (11.74)

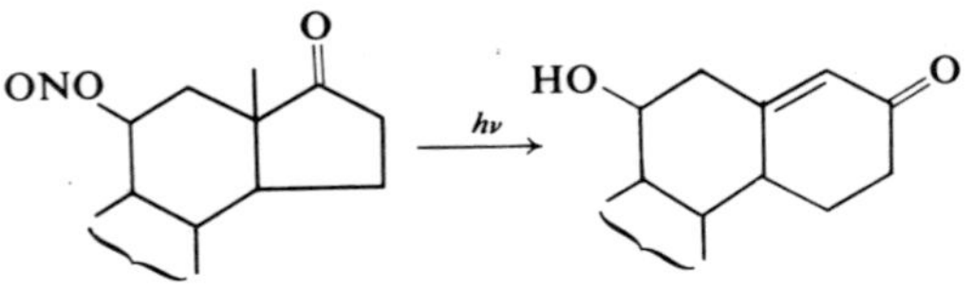

ONO O hν HO O (11.75)

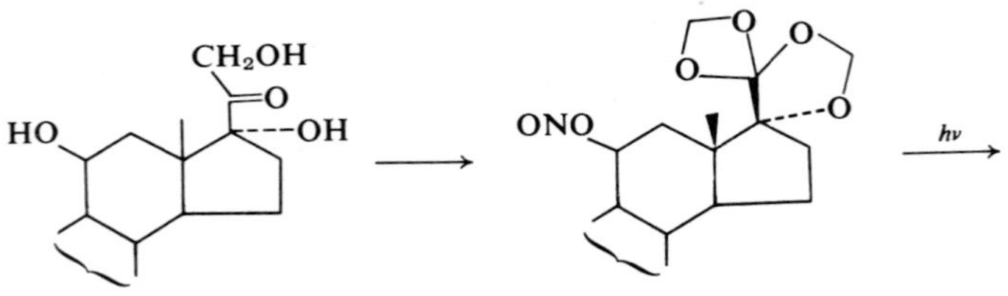

CH_2OH O HO OH ONO O O O O hν (11.76)

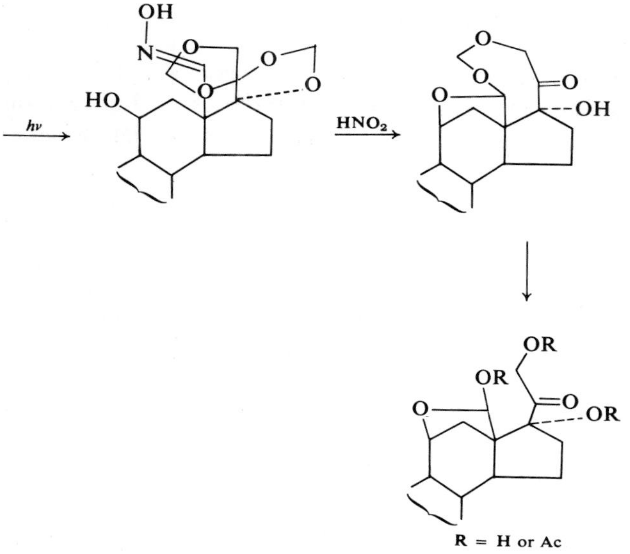

$h\nu$ → HNO$_2$ →

(11.76)

R = H or Ac

$h\nu$, I$_2$ →

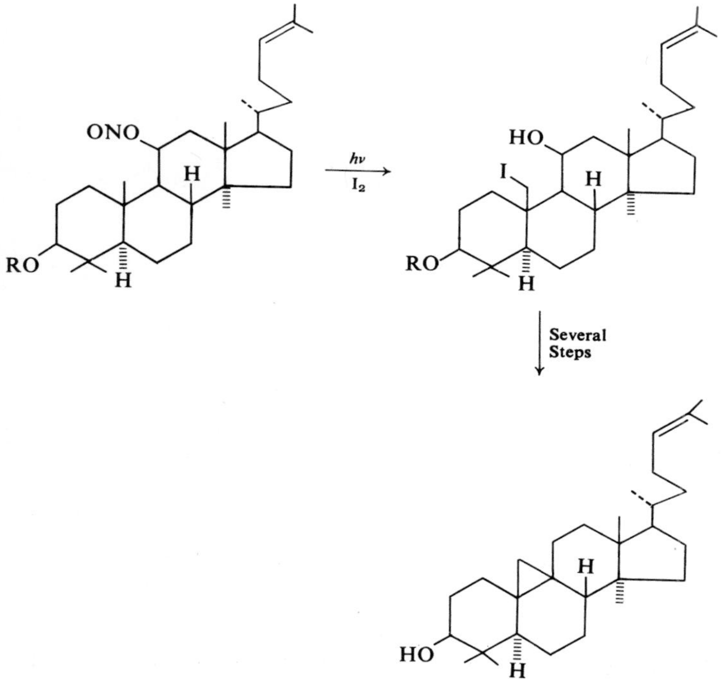

(11.77)

Several Steps

11.1c. *Miscellaneous Photoeliminations*

The photochemistry of unsaturated nitro compounds has been investigated by Chapman and co-workers.[63] Photolysis of 9-nitro-anthracene in degassed solutions leads to anthraquinone and 10,10′-bianthrone:

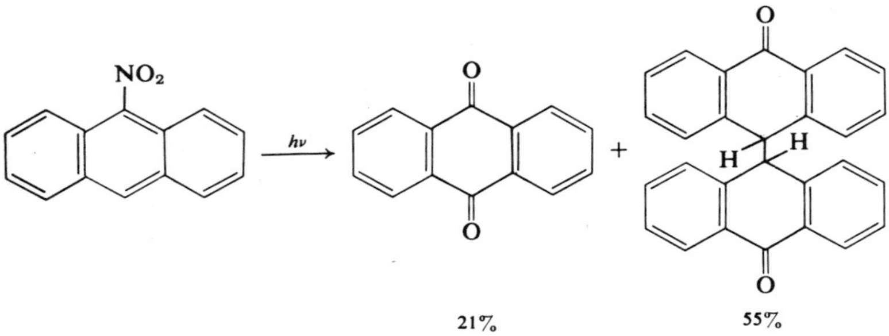

NO_2 $\xrightarrow{h\nu}$ O / O + O / H H / O (11.78)

21% 55%

Irradiation of 9-nitroanthracene under conditions in which gaseous products are continuously removed by nitrogen flushing results in the formation of anthraquinone monooxime in addition to the above products:

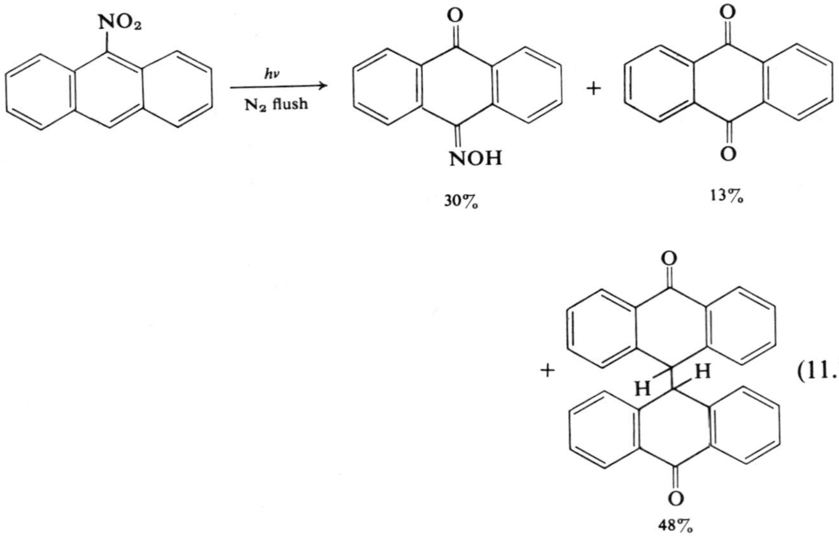

NO_2 $\xrightarrow[N_2 \text{ flush}]{h\nu}$ O / NOH + O / O

30% 13%

+ O / H H / O (11.79)

48%

These results can be understood in terms of the following mechanism:

NO_2 ONO *hν* O· NO O NOH *hν* NO bianthrone O O (11.80)

The nitro compound is isomerized to the nitrite, which can decompose either photochemically or thermally to the 9-anthroxyl radical. This radical can then dimerize or add nitric oxide to the 10 position to form the nitrosoketone. Isomerization of the nitrosoketone yields the monooxime, which then yields anthraquinone photochemically in the presence of NO.

Irradiation of other nitro compounds yields the products shown below, probably by similar mechanistic pathways:

NO_2 CH_3 CH_3 *hν* O CH_3 CH_3 O (11.81)

$$\phi\text{—CH=C}(CH_3)(NO_2) \xrightarrow{h\nu} \phi\text{—}\overset{\displaystyle NOH}{\overset{\|}{C}}\text{—}\overset{\displaystyle O}{\overset{\|}{C}}\text{—}CH_3 \quad 79\% \tag{11.82}$$

Sulfones are known to undergo elimination of sulfur dioxide upon photolysis.[64] Thus sulfone (33) yields hydrocarbon (34) in the presence of a photosensitizer ($\lambda > 320$ nm) or with light of $\lambda < 280$ nm in the absence of sensitizer:

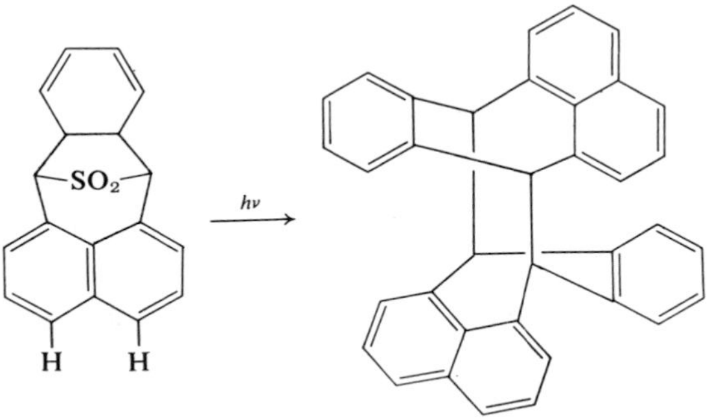

(33) $\xrightarrow{h\nu}$ (34) (11.83)

By studying the photolysis of sulfone (33) in the presence of sensitizers of varying triplet energy, it was found that the triplet state of (33) lies approximately in the range of 53.0–59.5 kcal/mole.

Similarly, (35) yields (36) upon photolysis ($\lambda < 280$ nm) but not with $\lambda > 320$ nm in the presence of a sensitizer:

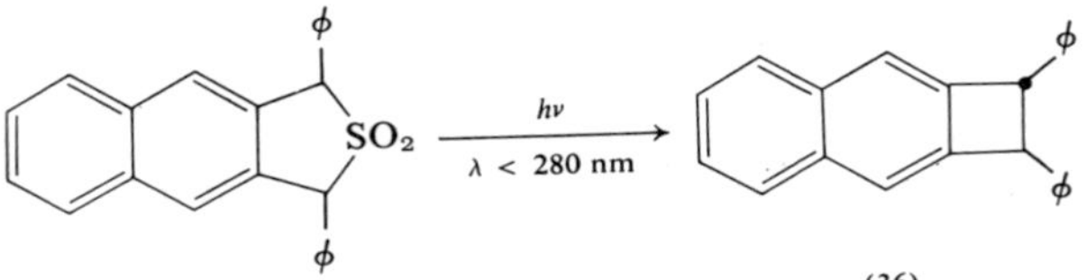

(35) $\xrightarrow[\lambda\, <\, 280\ \text{nm}]{h\nu}$ (36) (11.84)

Sulfone (37) failed to undergo decomposition under any conditions:

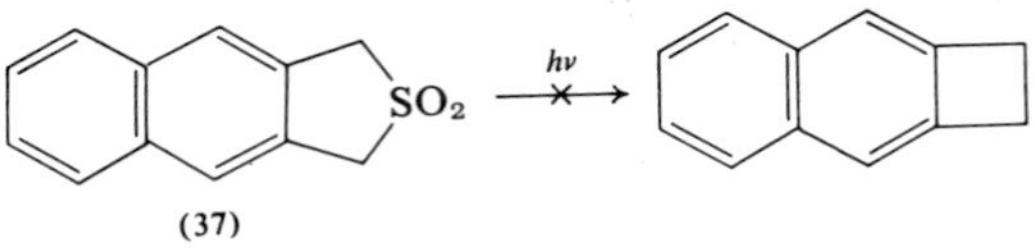

(37) $\xrightarrow{h\nu}\!\!\!\!\!\!\times$ (11.85)

The ease with which these compounds undergo photolytic decomposition is thought to parallel the relative stability of the quinonoid hydrocarbons formed as initial decomposition products.

Other sulfur compounds are known to undergo photoelimination and fragmentation reactions[65–69]:

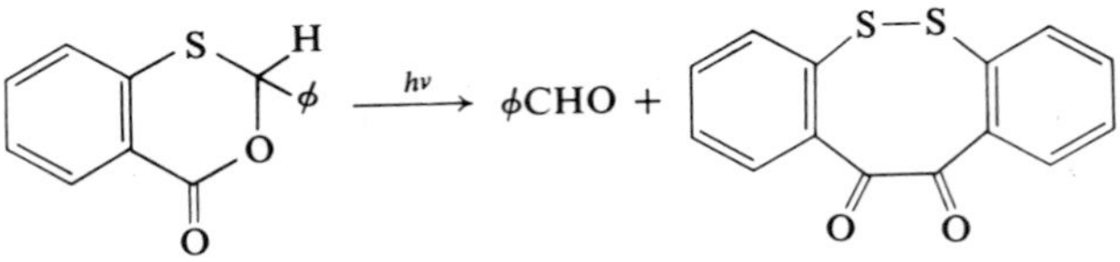

$\xrightarrow{h\nu}$ ϕCHO + (11.86)

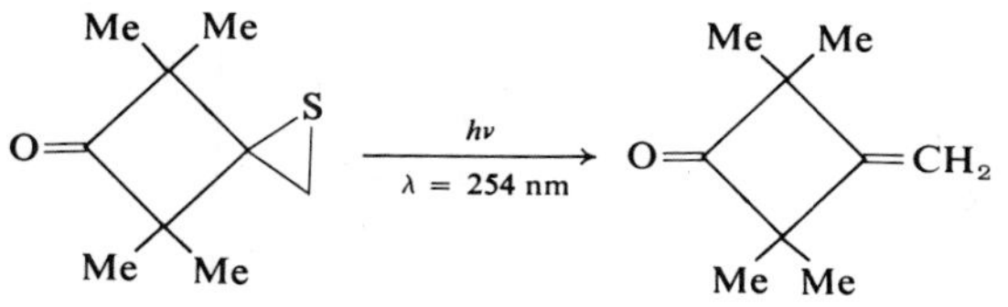

(11.87)

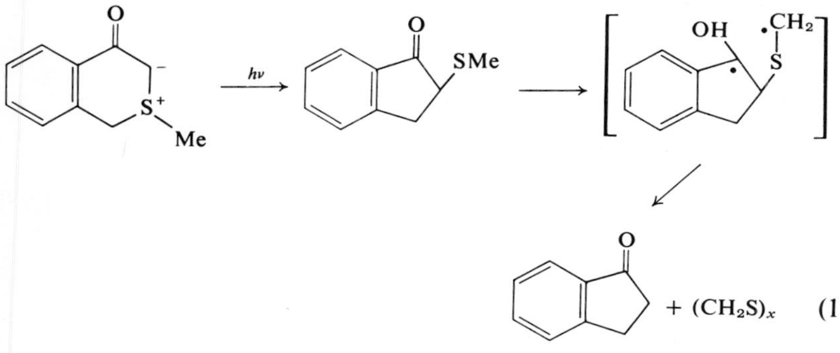

(11.88)

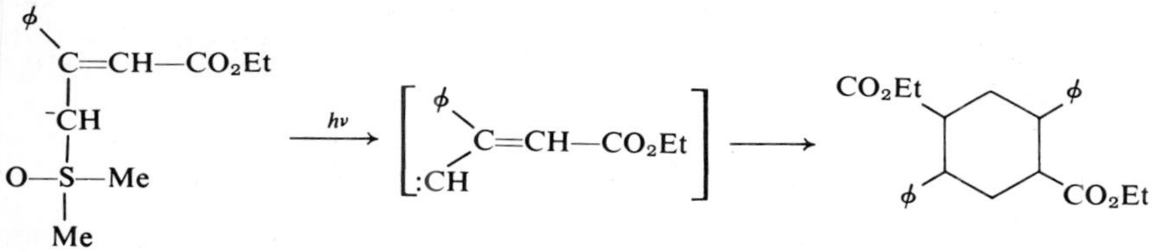

(11.89)

Numerous examples of the photoelimination of CO_2 have been reported.[11] Some examples are shown below[69–72]:

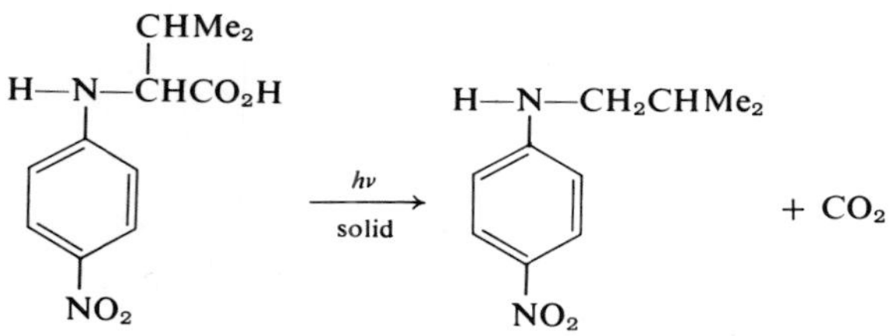

(11.90)

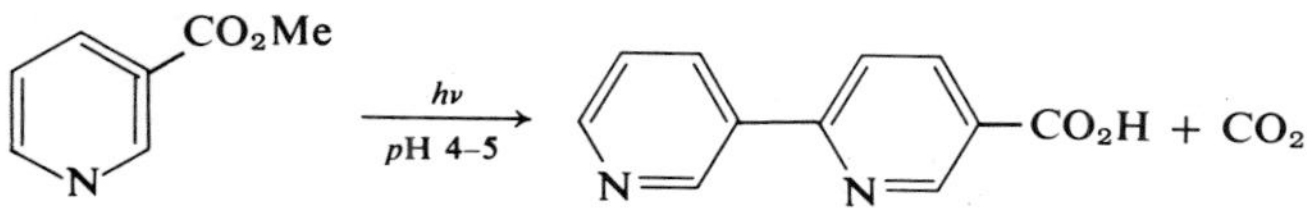

(11.91)

$$\underset{\phi}{\overset{H}{}}\!\!>\!\overset{O}{C\!-\!-\!CH}\!-\!CO_2^- \xrightarrow{h\nu} \phi CH_2CHO + \phi CHO + CO_2 \qquad \phi CH_2CHO \xrightarrow{h\nu} \phi CH_2CH_2\phi \tag{11.92}$$

$$(C_4H_9)_2C\text{(cyclic peroxide)} \xrightarrow[-CO_2]{h\nu} \left[(C_4H_9)_2\overset{O}{C\!-\!C}\!=\!O \longleftrightarrow (C_4H_9)_2\overset{+}{C}\!-\!C(=O)O^- \right] \xrightarrow{MeOH} (C_4H_9)_2\underset{OMe}{C}\!-\!CO_2H \tag{11.93}$$

11.2. PHOTOADDITION REACTIONS

In this section the photoaddition of water, alcohols, carboxylic acids, and miscellaneous other small molecules to various substrates will be considered.

11.2a. *Photoaddition of Water, Alcohols, and Carboxylic Acids*

Numerous examples of the photoaddition of water, alcohols, and carboxylic acids to multiple carbon–carbon bonds have been reported. The photoaddition of water to pyrimidine derivatives is of probable biological significance (see Chapter 12):

$$\text{1,3-dimethyluracil } (H_3C\!-\!N,\ N\!-\!CH_3,\ C\!=\!O) \xrightarrow[H_2O]{h\nu} \text{6-hydroxy-5,6-dihydro-1,3-dimethyluracil } (\ldots\!-\!OH) \tag{11.94}$$

Alcohols add to olefins upon irradiation in two modes, to produce higher alcohols and to produce ethers. In most photochemical additions of alcohols the first mode occurs:

$$RCH{=}CHR + R_2'CHOH \xrightarrow{h\nu} RCH_2\underset{}{\overset{R}{\overset{|}{C}}}H\overset{OH}{\overset{|}{C}}R_2' \tag{11.95}$$

$$RCH{=}CHR + R_2'CHOH \xrightarrow{h\nu} RCH_2\!-\!CHROCHR_2' \tag{11.96}$$

Irradiation of alcohols (ethanol, isopropanol, etc.) with maleic acid in the presence of benzophenone or anthraquinone as sensitizer leads to 1:1 addition products[75]:

$$CH_3CH_2OH + HC(COOH){=}CH(COOH) \xrightarrow[\text{sens.}]{h\nu} \text{product} \quad (11.97)$$

Similarly, isopropanol adds to cyclopentenone upon photolysis to yield ketoalcohol (38)[76]:

$$\text{cyclopentenone} + (H_3C)_2CHOH \xrightarrow{h\nu} \textbf{(38)} \quad (11.98)$$

This product probably arises from initial abstraction of a hydrogen radical from the isopropanol, followed by radical coupling:

$$(11.99)$$

Isopropanol forms 2:1 adducts with acetylenedicarboxylic acid upon photolysis in the presence of benzophenone.[77]

The second mode of addition of alcohols to olefins, to produce ethers, has been found to occur only with cyclic olefins. Thus in 1966 Kropp reported that cyclohexenes (39)–(42) produce tertiary ethers upon photolysis in methanol in the presence of high-energy sensitizers such as benzene, toluene, or xylene[79]:

$$\textbf{(39)} \xrightarrow[\text{MeOH, sens.}]{h\nu} \text{(OCH}_3\text{ ether)} + \text{(exocyclic methylene isomer)} \quad (11.100)$$

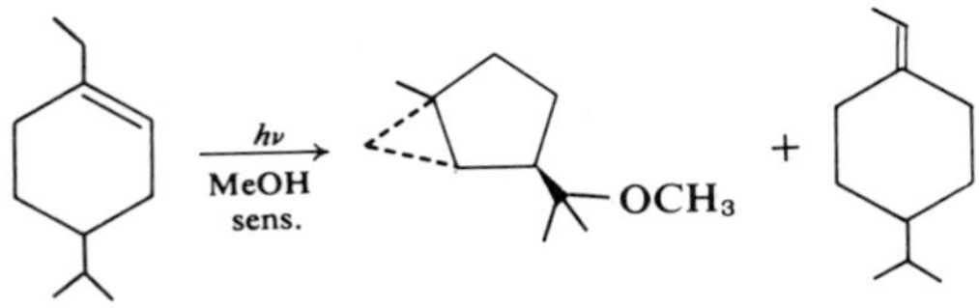

hν, MeOH, sens.

(40)

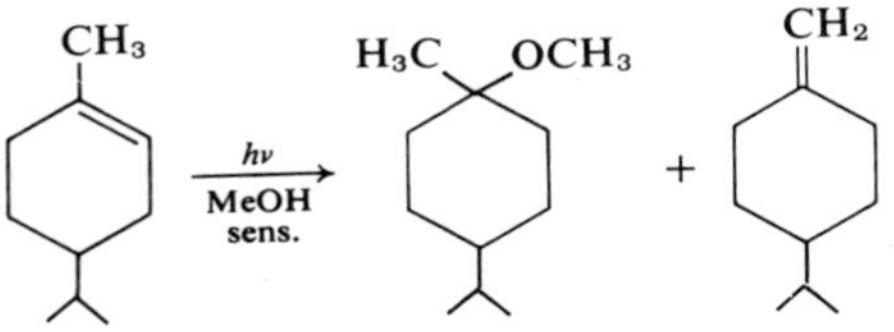

CH_3 / H_3C OCH_3 / CH_2

hν, MeOH, sens.

(11.101)

(41)

(11.102)

OCH_3

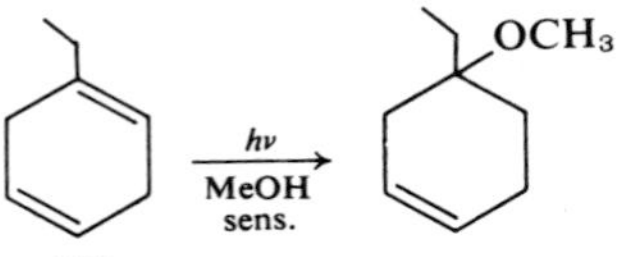

hν, MeOH, sens.

(42)

(11.103)

Labeling studies indicate that protons rather than hydrogen radicals are transferred in these reactions. The incorporation of protons, the formation of C—O bonds, and the Markovnikov direction of addition all indicate an ionic rather than a free radical mechanism. A concurrent report by Marshall and Carroll[78] confirmed this reaction in the photoaddition of a variety of alcohols to 1-menthene (41). Marked retardation of ether formation resulted from photolysis in the presence of oxygen, indicating a triplet precursor in the photoaddition. A reaction pathway is proposed involving an excessively strained olefin triplet, which may give way to an intermediate possessing carbonium ion character[78] or undergo intersystem crossing to a highly energetic ground state *trans* olefin, which then undergoes addition.[79]

Attempts to extend this reaction to the five-membered ring olefins 1-methylcyclopentene and norbornene resulted in 1-methylcyclopentane and methylenecyclopentane for the former and products (43)–(48) for the latter[80]:

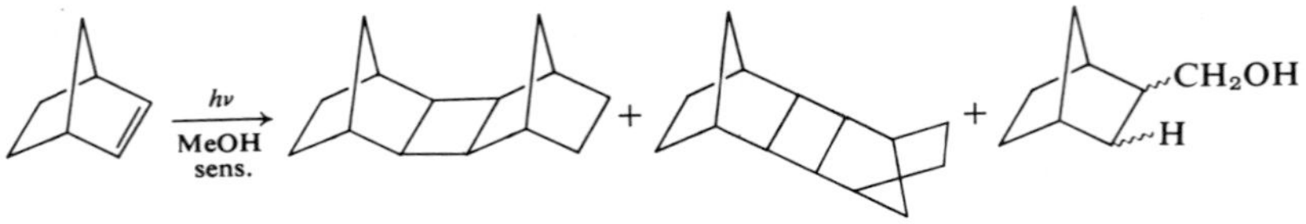

hν, MeOH, sens.

CH_2OH H

(43) (44) (45)

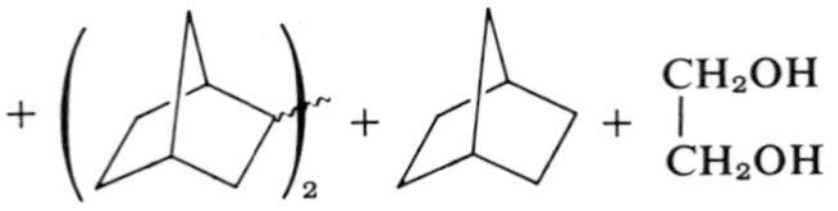

CH_2OH | CH_2OH

(46) (47) (48)

(11.104)

Product (45) is believed to be formed by hydrogen abstraction from methanol by 2-norbornene. In these highly strained olefins the triplet cannot attain its preferred orthogonal orientation or decay to a *trans* olefin. Hence the triplet is sufficiently long-lived and/or energetic enough to abstract a proton. Attempts to extend the reaction to larger ring systems[81] resulted in a 62% yield of ether (49) from 1-methylcycloheptene:

CH_3 (1-methylcycloheptene) $\xrightarrow[\text{MeOH, xylene}]{h\nu}$ H_3C, OCH_3 (49) + CH_2 (trace) (11.105)

Interestingly, the less strained 1-methylcyclooctene afforded neither ether formation, isomerization, nor reduction. Further studies have shown that water[82] and carboxylic acids[81] undergo additions analogous to alcohols:

CH_3 (1-methylcyclohexene) + CH_3COOH $\xrightarrow[\text{sens.}]{h\nu}$ CH_3, $OCCH_3$ (O) 30% + CH_2 20% (11.106)

CH_3, H $\xrightarrow[\text{H}_2\text{O, sens.}]{h\nu}$ CH_3, OH, H 22.4% + OH, CH_3, H 21.2% + CH_3, H 63% (11.107)

CH_3, CH_3, H $\xrightarrow[\text{H}_2\text{O, sens.}]{h\nu}$ CH_3, H, H, CH_2, H 57% + CH_3, H, H, CH_3, OH, H 17.5% + CH_3, H, H, OH, CH_3, H 19.5% (11.108)

That the high degree of torsional and other types of strain inherent in the triplet states or *trans* conformers of cyclohexene and cycloheptene may be responsible for their photochemical behavior is suggested by the reactions of compound (50), a moderately twisted olefin according to molecular models. Compound (50) quantitatively yields bicyclo[3.3.1]non-1-yl acetate (51) within 15 sec after being dissolved in glacial acetic acid[83]:

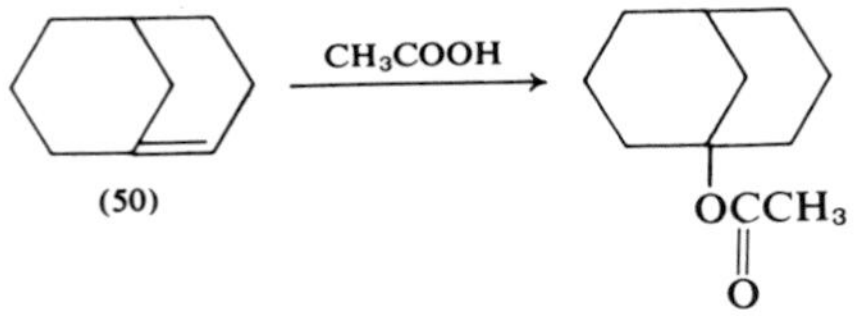

CH_3COOH (50) → (51) $OCCH_3$ (=O) (11.109)

The stereochemistry of addition of these substrates to olefins can be rationalized by three limiting cases of addition to *trans* olefins which yield intermediates (52), (53), and (54):

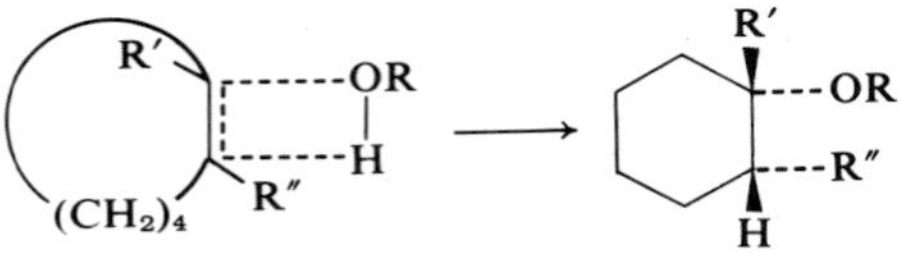

(52) R′, OR, H, R″, $(CH_2)_4$ → R′, OR, R″, H (11.110)

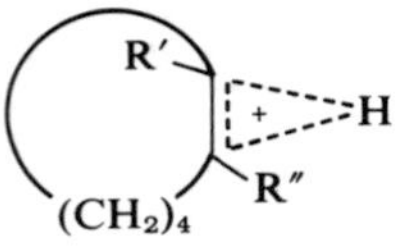

(53) R′, +, H, R″, $(CH_2)_4$

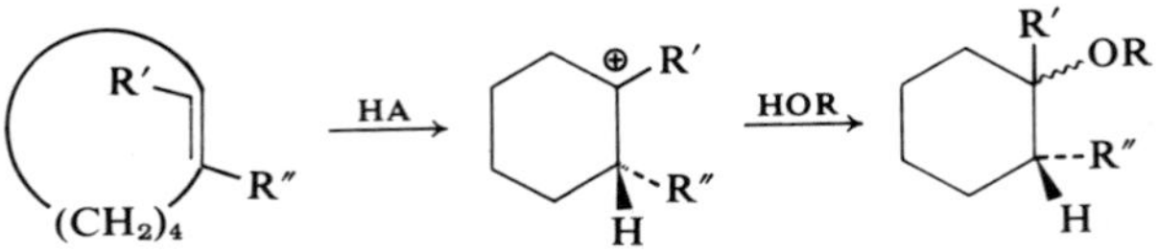

R′, R″, $(CH_2)_4$ —HA→ (54) ⊕ R′, R″, H —HOR→ R′, OR, R″, H (11.111)

A concerted four-center *cis* addition leads to (52) and a *trans* adduct; a *trans* addition, possibly via protonium species, leads to (53) and a *cis* adduct; a stepwise cationic addition leads to (54) and a mixture of *cis* and *trans* adducts. Recent studies by Marshall and Wurth strongly indicate that intermediate (54) is correct. Irradiation of octalin (55) in aqueous *t*-butyl alcohol (D_2O)–xylene results in formation of the equatorially deuterated alcohols (56) and (57) and the equatorially deuterated exocyclic olefin (58):

(11.112)

With octalin (55), rotation to the *trans* chair form (59) followed by protonation from the outside face of the double bond at the less substituted carbon atom leads to the tertiary cation (60) with net equatorial introduction of the proton. Rotation to a *trans* boat form would lead to net axial protonation[84]:

(56) + (57) + (58) (11.113)

A recent report by Kropp[85] indicates that whereas 2-norbornene undergoes radical reactions when irradiated in alcohol solution in the presence of a sensitizer, 2-phenyl-2-norbornene and related analogs undergo polar addition of methanol when irradiated directly (photosensitized irradiation led to no products):

(11.114)

Photolysis in MeOD indicates that all three of the above products arise via initial protonation of the double bond to form the 2-phenyl-2-norbornyl cation, which then undergoes either nucleophilic capture by solvent, hydride abstraction to form hydrocarbon, or deprotonation to regenerate starting olefin. The failure to obtain this reaction under photosensitized conditions (acetophenone, fluorenone), coupled with the observation that methanol substantially quenches the fluorescence of the olefin, suggests that an excited singlet species displaying charge-transfer character is the reactive intermediate in this reaction. Interestingly, 1-phenylcyclohexene and 1-phenylcycloheptene have been observed to undergo radical addition upon direct photolysis in alcohols but ionic addition upon sensitized photolysis.(86) Why norbornene should undergo radical reactions through its triplet, 2-phenylnorbornene should undergo polar addition through its singlet, and 1-phenylcyclohexene should undergo both radical and ionic additions through its singlet and triplet states, respectively, is unclear at this time.

The photoaddition of water, alcohols, and acids to benzene and its derivatives has also been extensively investigated(87–91):

$h\nu$, MeOH; H; CH_3O; 90% (11.115)

$h\nu$, CF_3CH_2OH; H; H; CF_3CH_2O; H; + ; H; CF_3CH_2O; H; H; 2:1 (11.116)

$h\nu$, CH_3COOH; H; H; H; $OCCH_3$; O; + polymer (11.117)

$h\nu$, H_2O; H; H; H; HO; + aldehydes (11.118)

Benzvalene (61)(87) and the benzenonium ion (62) leading to cation (63)(89) have been proposed as intermediates in these reactions:

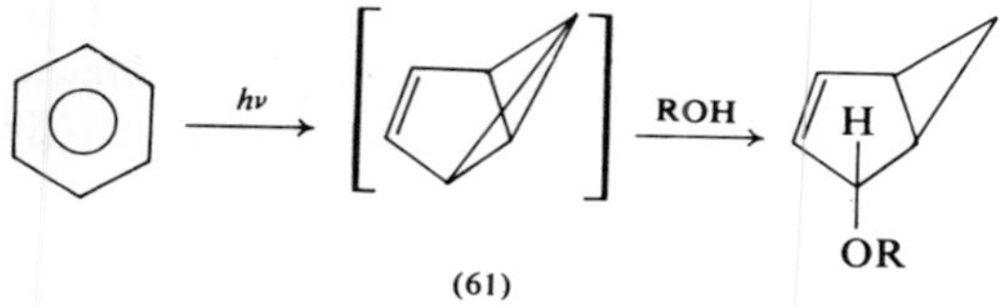

hν, ROH (61) OR H (11.119)

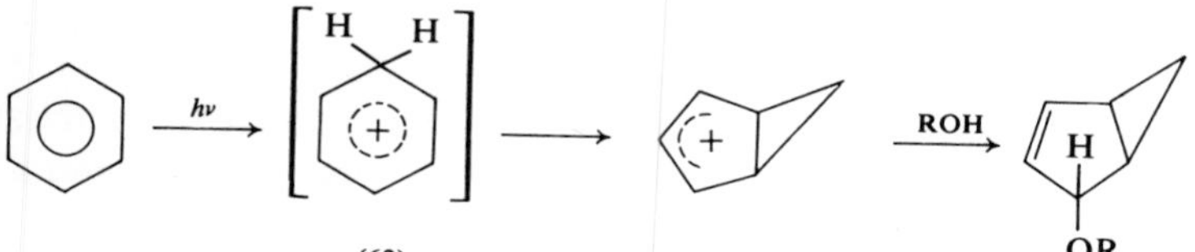

hν, ROH (62) (63) H H + OR H (11.120)

The isolation of benzvalene (61) from the irradiation of benzene at 254 nm and the observation that this compound produces the expected bicyclic ethers when treated with acidified methanol lend credence to the intermediacy of (61).[90] Photolysis of benzene in acetic acid was found to result in formation of acetates (64)–(67), with the product composition changing with time:

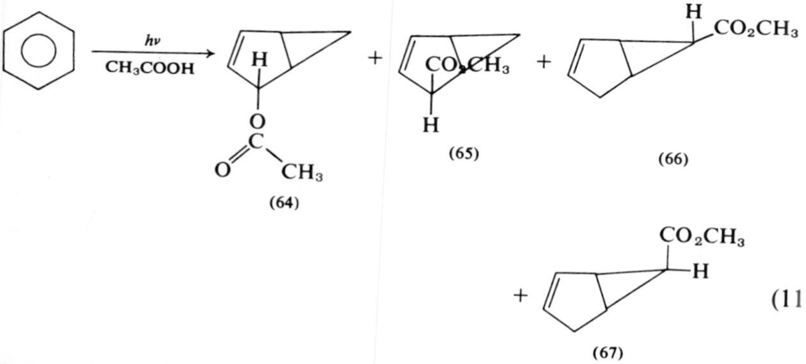

hν, CH_3COOH H O C O CH_3 (64) + CO_2CH_3 H (65) + H CO_2CH_3 (66)

+ CO_2CH_3 H (67) (11.121)

Extrapolation of the data to zero time suggests that the *endo* acetates (65) and (67) are produced in amounts as great as or greater than the *exo* isomer. Solvolysis studies of the bicyclo[3.1.0]hex-2-en-6-yl cation reveal that nucleophilic capture occurs preferentially from the *exo* side to give (66) rather than (67). Similarly, solvolysis of cation (63) leads to *exo* product (64) in at least a 90% yield. Photolysis of benzene in deuteriophosphoric acid results in (68), in which all the deuterium is incorporated into the 6-*endo* position:

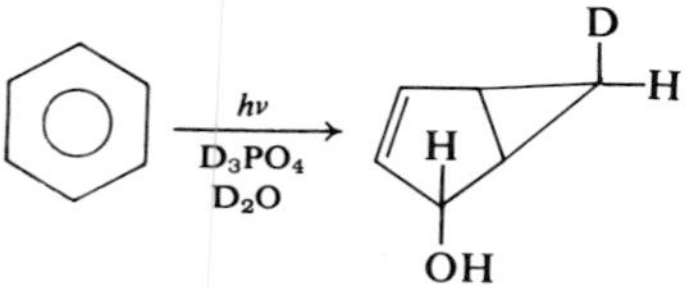

hν, D_3PO_4, D_2O D H H OH (68) (11.122)

The benzenonium ion (62), the proposed intermediate for the formation of (63), has chemically equivalent hydrogens in the methylene group. Hence essentially equal amounts of 6-*exo* and 6-*endo* deuterium should appear in (68) if this mechanism is operative.

11.2b. *Miscellaneous Photoadditions*

Several examples of the photoaddition of amines and amine derivatives to olefins have been reported.[(92)] Butylamine adds to 1-octene upon ultraviolet irradiation to produce 4-aminoalkane:

$$CH_3CH_2CH_2CH_2NH_2 + C_6H_{13}CH{=}CH_2 \xrightarrow{h\nu} C_3H_9\underset{}{\overset{NH_2}{\overset{|}{C}}}HCH_2C_7H_{15} \quad (11.123)$$

Photolysis of piperidine and 1-octene mixtures similarly leads to 2-octylpiperidine through the intermediacy of an amine radical $R\dot{C}HNHR'$.[(93)]

Monochloroamine adds to cyclohexene to produce *trans*-chlorocyclohexylamine and other products from radical recombination[(94)]:

$$NH_2Cl + \text{cyclohexene} \xrightarrow{h\nu} \text{trans-2-chlorocyclohexylamine} + \text{1,2-dichlorocyclohexane} + \text{other products} \quad (11.124)$$

N-Nitrosoamines undergo addition to olefins upon photolysis in the presence of acid to produce α-*t*-aminonitrosoalkanes in which the amine moiety adds to the less substituted carbon of the double bond[(95)]:

$$\text{N-nitrosopiperidine} \xrightarrow[\text{cyclopentene}]{h\nu} \text{2-piperidino-1-nitrosocyclopentane (ON)} \longrightarrow \text{2-piperidinocyclopentanone oxime (HON)} \rightleftharpoons \text{dimer} \quad (11.125)$$

$$(CH_3)_2N{-}N{=}O \xrightarrow[\text{styrene}]{h\nu} C_6H_5CH(NO)CH_2N(CH_3)_2 \longrightarrow C_6H_5C({=}NOH)CH_2N(CH_3)_2 \quad (11.126)$$

N-Substituted amides and lactams possess potentially reactive C—H bonds on carbon atoms alpha to the nitrogen and carbonyl group. These hydrogen atoms are easily abstracted by excited carbonyl compounds (e.g., acetone or benzophenone) to produce radicals which undergo olefin addition[92,96,97]:

$$\phi\overset{O}{\overset{\|}{C}}\phi \xrightarrow{h\nu} \phi\overset{O}{\overset{\|}{C}}\phi^{*} + CH_3CONHCH_3 \longrightarrow \dot{C}H_2CONHCH_3 + CH_3CONH\dot{C}H_2$$

$$\dot{C}H_2CONHCH_3 \xrightarrow{RCH=CH_2} \underset{\text{Minor}}{R(CH_2)_2CH_2CONHCH_3} \qquad CH_3CONH\dot{C}H_2 \xrightarrow{RCH=CH_2} \underset{\text{Major}}{CH_3CONHCH_2(CH_2)_2R} \tag{11.127}$$

$$\text{2-pyrrolidone} + RCH{=}CH_2 \xrightarrow[\text{sens.}]{h\nu} \underset{(69)}{\text{5-}(CH_2CH_2R)\text{-2-pyrrolidone}} + \underset{(70)}{\text{3-}(CH_2CH_2R)\text{-2-pyrrolidone}} \tag{11.128}$$

In the photoaddition of 2-pyrrolidone the 5-alkyl isomer (69) always predominates, usually in a ratio of 2:1. The formation of anti-Markovnikov 1:1 adducts, telomers, and dehydrodimers of structure (71) supports a free radical mechanism. Similarly, formamide undergoes olefin addition under

(71)

direct or photosensitized conditions to yield higher homologous amides[92,98–101]:

$$RCH{=}CH_2 + HCONH_2 \xrightarrow[\text{ketone}]{h\nu} RCH_2CH_2CONH_2 \tag{11.129}$$

With terminal olefins the principal product corresponds to anti-Markovnikov addition. With nonterminal olefins, mixtures of the two possible amides result:

$$\text{4-vinylcyclohexene} + HCONH_2 \xrightarrow[\text{acetone}]{h\nu} \text{4-}(CH_2CH_2CONH_2)\text{cyclohexene} \tag{11.130}$$

$$\text{norbornadiene} + HCONH_2 \xrightarrow[\text{acetone}]{h\nu} \text{(nortricyclyl)}-CONH_2 \tag{11.131}$$

A free radical chain mechanism is proposed for these reactions:

$$HCONH_2 \xrightarrow{h\nu} \cdot CONH_2 \tag{11.132a}$$

$$HCONH_2 \xrightarrow[\text{acetone}]{h\nu} \cdot CONH_2 \tag{11.132b}$$

$$RCH_2{=}CH_2 + \cdot CONH \longrightarrow R\dot{C}HCH_2CONH_2 \tag{11.132c}$$

$$R\dot{C}HCH_2CONH_2 + H{-}CONH_2 \longrightarrow RCH_2CH_2CONH_2 + \dot{C}ONH_2 \tag{11.132d}$$

$$R\dot{C}HCH_2CONH_2 + RCH{=}CH_2 \longrightarrow \underset{\displaystyle CH_2\dot{C}HR}{RCHCH_2CONH_2}, \text{ etc.} \tag{11.132e}$$

The formation of telomers (2:1 to 4:1) supports this mechanism. With α,β-unsaturated esters, β-addition occurs, except for the cinnamates, where α-addition results:

$$C_5H_{11}CH{=}CHCOOR + H{-}CONH_2 \xrightarrow[\text{benzophenone}]{h\nu} \underset{\displaystyle CONH_2}{RCH}{-}CH_2CO_2R \tag{11.133}$$

$$\phi CH{=}CHCO_2C_2H_5 + H{-}CONH_2 \xrightarrow[\text{benzophenone}]{h\nu} \left[\phi{-}\dot{C}H{-}CH\begin{matrix} CONH_2 \\ CO_2C_2H_5 \end{matrix}\right] \xrightarrow{\phi_2\dot{C}OH} \left[\begin{matrix} & CONH_2 & \\ \phi CH & {-}CH{-} & CO_2C_2H_5 \\ \phi_2C(OH) & & \end{matrix}\right] \longrightarrow \begin{matrix} \phi{-}CH & {-}CHCONH_2 \\ \phi{-}C(\phi) & {-}O{-}C{=}O \end{matrix} \tag{11.134}$$

Formamide adds to terminal acetylenes to produce 2:2 adducts, but with nonterminal acetylenes, 2:1 adducts result:

$$RC{\equiv}CH + H{-}CONH_2 \xrightarrow[\text{ketone}]{h\nu} \underset{\displaystyle CONH_2}{RCH}{-}\overset{\displaystyle CH{=}CHR}{CHCONH_2} \tag{11.135}$$

$$RC{\equiv}CR + H{-}CONH_2 \xrightarrow{h\nu} \begin{matrix} RCH & {=} & CHR \\ NH_2{-}C{=}O & & CONH_2 \end{matrix} \tag{11.136}$$

Aldehydes add to olefins on photolysis to yield ketones[102–103]

$$RCHO + R'CH{=}CH_2 \xrightarrow{h\nu} RCOCH_2CH_2R' \qquad (11.137)$$

$$RCHO + \begin{matrix} CHCO_2C_2H_5 \\ \| \\ CHCO_2C_2H_5 \end{matrix} \xrightarrow{h\nu} \begin{matrix} RCOCHCO_2C_2H_5 \\ | \\ CH_2CO_2C_2H_5 \end{matrix} \qquad (11.138)$$

These reactions involve the addition of acyl free radicals $R\dot{C}O$, which may result either from the ejection of a hydrogen atom from the photoexcited aldehyde or through hydrogen abstraction from a ground state aldehyde by a photoexcited molecule.

Ketones add to olefins to produce 2-alkyl derivatives upon photolysis[104–106]:

$$\text{cyclohexanone} + C_6H_{13}CH{=}CH_2 \xrightarrow{h\nu} \text{2-}(CH_2CH_2C_6H_{13})\text{cyclohexanone} \qquad (11.139)$$

$$CH_3\overset{O}{\overset{\|}{C}}CH_3 + \text{norbornene} \xrightarrow{h\nu} \text{2-}(CH_2COCH_3)\text{norbornane} \qquad (11.140)$$

Elimination or abstraction of a hydrogen atom leads to the free radical $R\dot{C}H\overset{O}{\overset{\|}{C}}R$, which undergoes olefin addition.

The photoadditions of halogens, hydrogen halides, and alkylhalides to olefins have been extensively documented.[107] Photohalogenation reactions occur by absorption of light by the halogen, leading to excitation of a nonbonded p electron to an antibonding σ^* excited level, followed by decomposition of the molecule into free radicals:

$$X_2 \xrightarrow{h\nu} [X_2]^{n\to\sigma^*} \longrightarrow 2X\cdot \qquad (11.141a)$$

$$R_2C{=}CR_2 + X\cdot \longrightarrow R_2CX\dot{C}R_2 \qquad (11.141b)$$

$$R_2CX\dot{C}R_2 + X_2 \longrightarrow R_2CXCXR_2 + X\cdot \qquad (11.141c)$$

For example, vinyl chloride undergoes photochlorination to produce 1,1,2-trichloroethane[108]:

$$H_2C{=}CHCl + Cl_2 \xrightarrow{h\nu} ClCH_2CHCl_2 \qquad (11.142)$$

Butadiene derivatives add chlorine under sunlight to produce tetrachlorocompounds[105]:

$$\phi{-}CH{=}CH{-}CH{=}CH{-}COOH + Cl_2 \xrightarrow{h\nu} \phi{-}CHClCHClCHClCHClCOOH \qquad (11.143)$$

1,1,1-Trichloropropene yields dibromides (72) and (73), with the latter resulting from partial isomerization of the intermediate free radical[110]:

$$CCl_3CH{=}CH_2 \xrightarrow[Br_2]{h\nu} [CCl_3\dot{C}H{-}CH_2Br] \longrightarrow [\dot{C}Cl_2CHClCH_2Br]$$

$$[CCl_3\dot{C}H{-}CH_2Br] \xrightarrow{Br_2} \underset{(72)}{CCl_3CHBr{-}CH_2Br} \qquad [\dot{C}Cl_2CHClCH_2Br] \xrightarrow{Br_2} \underset{(73)}{CBrCl_2CHClCH_2Br} \quad (11.144)$$

Photobromination of anhydride (74) results in 56% *trans* (75) and 36% *cis* (76). Compound (76) is thought to result from intermediate radical (77):

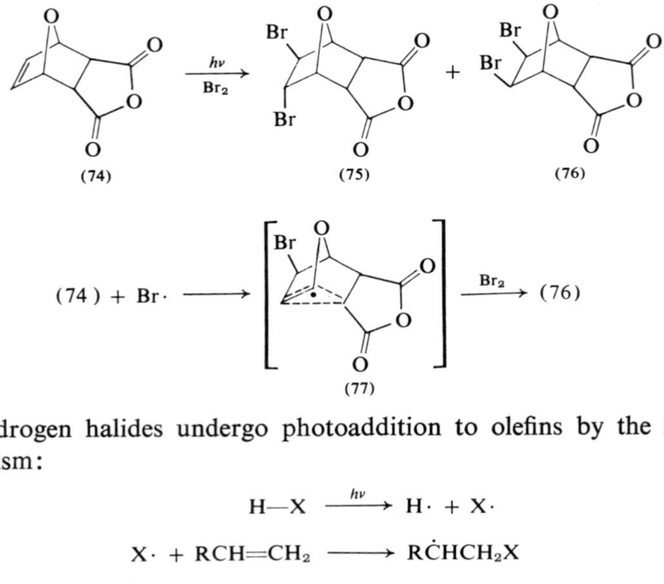

(74) $\xrightarrow[Br_2]{h\nu}$ (75) + (76)

(74) + Br· ⟶ [(77)] $\xrightarrow{Br_2}$ (76) (11.145)

Hydrogen halides undergo photoaddition to olefins by the following mechanism:

$$H{-}X \xrightarrow{h\nu} H\cdot + X\cdot \quad (11.146a)$$

$$X\cdot + RCH{=}CH_2 \longrightarrow R\dot{C}HCH_2X \quad (11.146b)$$

$$R\dot{C}HCH_2X + H{-}X \longrightarrow RCH_2CH_2X + X\cdot \quad (11.146c)$$

Since HCl and HBr absorb light at wavelengths shorter than 290 nm, 254-nm light or photosensitizers (acetone, acetaldehyde, tetraethyl lead, etc.) are commonly employed.[112] Anti-Markovnikov products result[113]:

$$CF_3CH{=}CH_2 + HBr \xrightarrow{h\nu} \underset{90\%}{CF_3CH_2CH_2Br} \quad (11.147)$$

Hydrogen bromide undergoes photoaddition to 1-bromocyclohexene to produce *cis*-1,2-dibromocyclohexane as the sole product.[114] Studies of the photoaddition of HBr to other olefins indicate that essentially stereospecific *trans* addition occurs[115–117]:

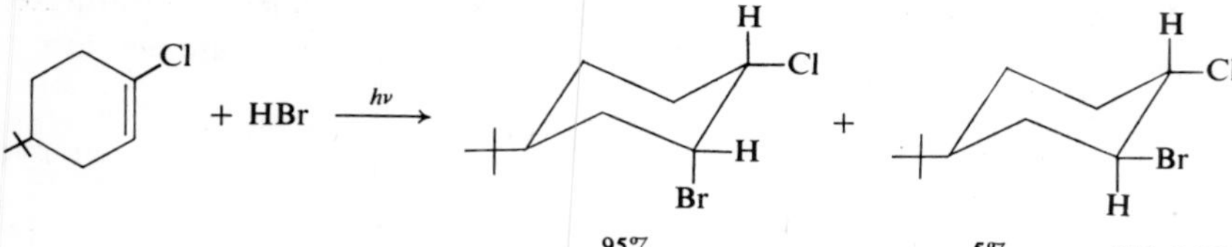

95% 5% (11.148)

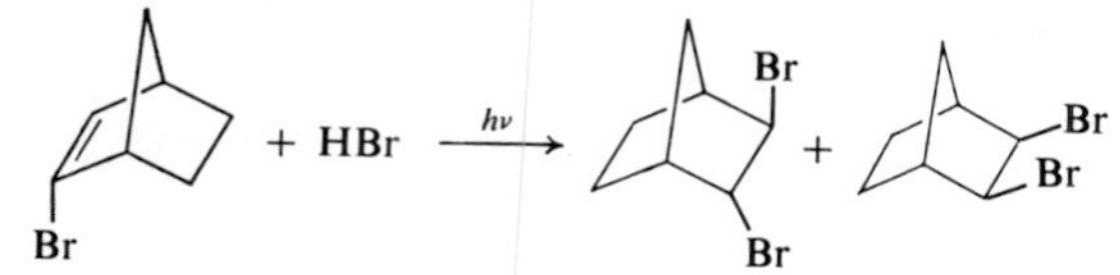

(11.149)

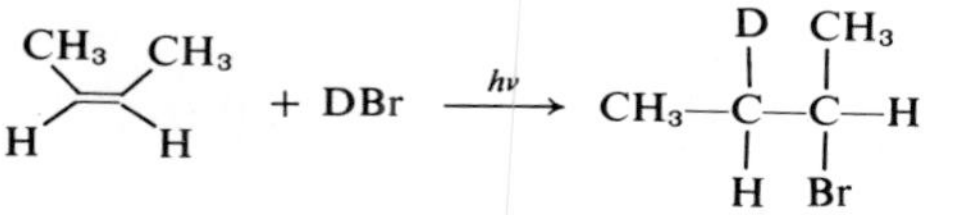

threo (11.150)

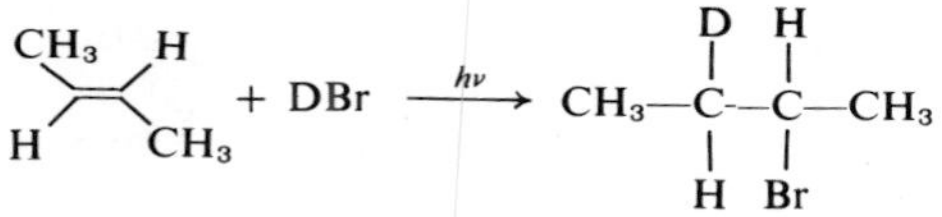

erythro (11.151)

The stereospecificity of these reactions is explained by the intermediacy of bridge structures (78) and (79)[115]:

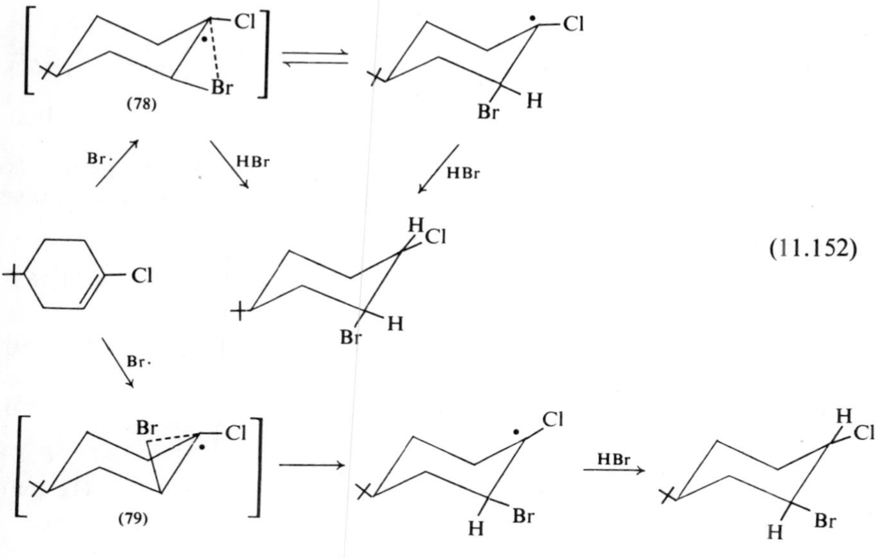

(78) (79) (11.152)

Many examples of the photoaddition of alkyl polyhalides to unsaturated compounds have been reported.[107] Some representative examples are as follows[118–121]:

$$CH_3CH{=}CH_2 + CCl_4 \xrightarrow{h\nu} \underset{55\%}{CH_3CHClCH_2CCl_3} \quad (11.153)$$

$$C_6H_5CH{=}CH_2 + CBr_4 \xrightarrow{h\nu} \underset{96\%}{C_6H_5CHBrCH_2CBr_3} \quad (11.154)$$

$$\text{cyclopentene} + BrCCl_3 \xrightarrow{h\nu} \underset{62\%}{\text{1-bromo-2-(trichloromethyl)cyclopentane}} \quad (11.155)$$

$$CF_3CH{=}CF_2 + ICF_3 \xrightarrow{h\nu} (CF_3)_2CHCF_2I \quad (11.156)$$

$$CF_3CF{=}CF_2 + ICF_3 \xrightarrow{h\nu} CF_3CFICF_2CF_3 \quad (11.157)$$

$$CH_2{=}C{=}CH_2 + ICF_3 \xrightarrow{h\nu} CF_3CH_2CI{=}CH_2 \quad (11.158)$$

Irradiation of chloroform yields a $\cdot CCl_3$ radical, whereas irradiation of bromoform yields $\cdot CHBr_2$ radicals[122]:

$$RCH{=}CH_2 + HCCl_3 \xrightarrow{h\nu} RCH_2CH_2CCl_3 \quad (11.159)$$

$$RCH{=}CH_2 + BrCHBr_2 \xrightarrow{h\nu} RCHBrCH_2CHBr_2 \quad (11.160)$$

Polyhalomethanes[123] and pentafluoroiodoethane[121] add to acetylenes to yield olefins:

$$CH_3(CH_2)_4C{\equiv}CH + BrCCl_3 \xrightarrow{h\nu} CH_3(CH_2)_4CBr{=}CHCCl_3 \quad (11.161)$$

$$CH{\equiv}CH + C_2F_5I \xrightarrow{h\nu} C_2F_5CH{=}CHI \quad (11.162)$$

As with the halogens, the thiols, phosphines, silanes, and germanes undergo photoaddition to olefins via homolytic cleavage. Representative examples follow[124–130]:

$$CH_3CH{=}CH_2 + n\text{-}C_3H_7SH \xrightarrow{h\nu} \underset{96\%}{(n\text{-}C_3H_7)_2S} \quad (11.163)$$

$$CH_2{=}CClCOOCH_3 + CH_3COSH \xrightarrow{h\nu} \underset{84\%}{CH_3COSCH_2CHClCO_2CH_3} \quad (11.164)$$

$$F_2C{=}CF_2 + PH_3 \xrightarrow{h\nu} \underset{86\%}{F_2CHCF_2PH_2} \quad (11.165)$$

$$FClC{=}CF_2 + PH_3 \xrightarrow{h\nu} FClHC{-}CF_2PH_2 \quad (11.166)$$

$$\text{cyclohexene} + HPS(OC_2H_5)_2 \xrightarrow{h\nu} \text{cyclohexane(H)}(PS(OC_2H_5)_2) \tag{11.167}$$

$$CF_2{=}CHF + HSiCl_3 \xrightarrow{h\nu} CHF_2CHFSiCl_3 \tag{11.168}$$

$$CF_2{=}CF_2 + (CH_3)_2SiH_2 \xrightarrow{h\nu} CHF_2CF_2Si(CH_3)_2H \tag{11.169}$$

$$\text{cyclohexene} + \phi_3GeH \xrightarrow{h\nu} \text{cyclohexane(H)}(Ge\phi_3) \tag{11.170}$$

11.3. PHOTOSUBSTITUTION REACTIONS

Photochemical solvolysis reactions are directly analogous to ordinary solvolysis reactions except that the absorption of light energy is necessary for reaction to occur:

$$RX + Y \xrightarrow{h\nu} RY + X \tag{11.171}$$

For example, substituted phenacyl chloride reacts with ethanol upon photolysis to yield compounds (80) and (81)[131]:

$$X\text{-}C_6H_4\text{-}\overset{O}{\overset{\|}{C}}CH_2Cl + EtOH \xrightarrow{h\nu} X\text{-}C_6H_4\text{-}\overset{O}{\overset{\|}{C}}CH_2OEt\ (80) + X\text{-}C_6H_4\text{-}CH_2COOEt\ (81) \tag{11.172}$$

Other examples follow[132–134]:

$$O_2N\text{-}C_6H_4\text{-}OPO_3Na_2 + \text{pyridine} \xrightarrow[H_2O]{h\nu} C_5H_5\overset{\oplus}{N}\text{-}C_6H_4\text{-}OPO_3^{2-}Na^+ + NaNO_2 \tag{11.173}$$

$$C_6H_6 + BX_3 \xrightarrow{h\nu} C_6H_5BX_2 + HX \tag{11.174}$$

X = Br, I

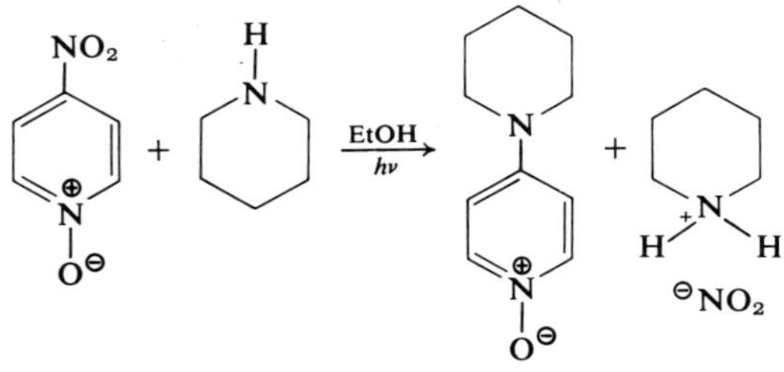

(11.175)

Zimmerman and co-workers[135] have reported that photolysis of *p*-methoxybenzyl acetate in aqueous dioxane results in *p,p'*-dimethoxybibenzyl and bidioxane, probably through the intermediacy of *p*-methoxybenzyl radicals:

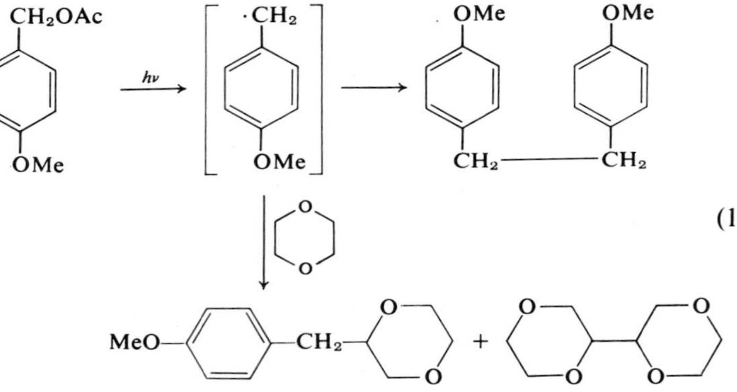

(11.176)

When 3-methoxybenzyl acetate is photolyzed, however, the products expected from free radical coupling are decreased and a substantial amount of 3-methoxybenzylalcohol results. A carbonium ion intermediate has been proposed in the photolysis of this *meta* ester:

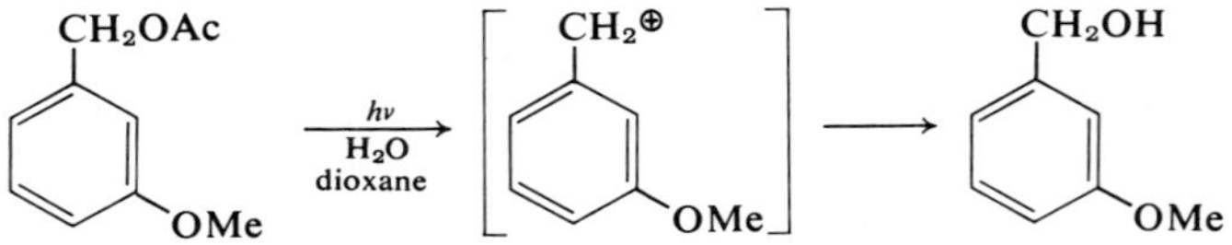

(11.177)

The studies of Havinga and co-workers have resulted in the discovery of many solvolysis reactions which appear to occur via polar or heterolytic pathways[136]:

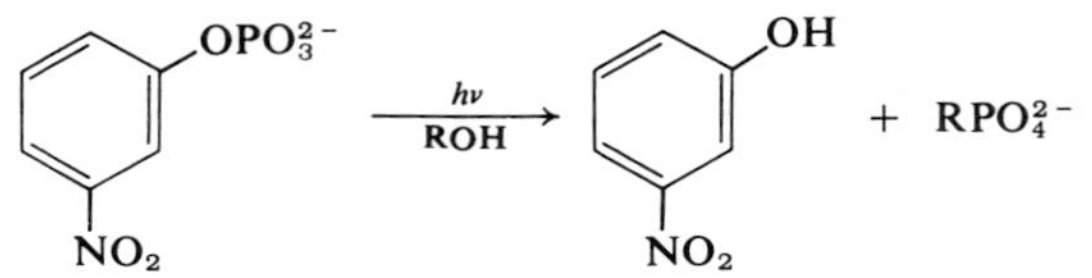

(11.178)

$$m\text{-}O_2NC_6H_4OPO_3^{2-} + {}^{18}OH^- \xrightarrow{h\nu} m\text{-}O_2NC_6H_4{}^{18}O^{\ominus} + HPO_4^{2-} \quad (11.179)$$

$$m\text{-}O_2NC_6H_4OCH_3 + CH_3NH_2 \longrightarrow m\text{-}O_2NC_6H_4NHCH_3 \quad (11.180)$$

$$O_2N\text{-}C_6H_4\text{-}OCH_3 \xrightarrow[CH_3NH_2]{h\nu} O_2N\text{-}C_6H_4\text{-}NHCH_3 \quad (11.181)$$

$$O_2N\text{-}C_6H_4\text{-}NO_2 \xrightarrow[NH_3]{h\nu} O_2N\text{-}C_6H_4\text{-}NH_2 \quad (11.182)$$

The startling observation made in this work was that *meta* derivatives were more reactive toward photosubstitution than *ortho* and *para* isomers, as shown below for the photoreaction of dimethoxynitrobenzene and methyl amine, in contradistinction to the rules of classical ground state nucleophilic substitution:

$$(CH_3O)_2C_6H_3NO_2 \xrightarrow[CH_3NH_2]{h\nu} CH_3O(CH_3NH)C_6H_3NO_2 \quad (11.183)$$

Other examples follow[136,137]:

$$O_2N\text{-}C_6H_4\text{-}OCH_3 \xrightarrow[O_2]{h\nu,\ CN^-} O_2N\text{-}C_6H_3(CN)\text{-}OCH_3 \quad (11.184)$$

$$O_2N\text{-}C_6H_3(OCH_3)\text{-}OCH_3 \xrightarrow[CN^-]{h\nu} O_2N\text{-}C_6H_3(CN)\text{-}OCH_3 \quad (11.185)$$

Electrophilic substitutions reveal *ortho*/*meta* activations by electron donors such as alkyl and methoxy groups:

$$CH_3\text{-}C_6H_5 \xrightarrow[CF_3COOD]{h\nu} CH_3\text{-}C_6H_4D \quad (11.186)$$

$$CH_3O\text{-}C_6H_5 \xrightarrow[CH_3COOD]{h\nu} CH_3O\text{-}C_6H_4\text{-}D\ (ortho) + CH_3O\text{-}C_6H_4\text{-}D\ (meta) \tag{11.187}$$

$$O_2N\text{-}C_6H_5 \xrightarrow[CF_3COOD]{h\nu} O_2N\text{-}C_6H_4\text{-}D\ (para) + O_2N\text{-}C_6H_4\text{-}D\ (meta) \tag{11.188}$$

The positional specificity of these nucleophilic and electrophilic photosubstitutions can be readily understood by examination of the calculated charge densities of some of these molecules in their ground and first excited singlet states. For example, 4-nitrocatechol has the charge densities shown in (82) and (83).[139] Thus on this basis one would predict that 4-nitroveratrole

(82) S_0: N +0.771; ring carbons −0.777, −0.028, −0.012, +0.023, −0.066, +0.020 (with OH groups)

(83) S_1: N +0.614; ring carbons −0.010, −0.067, +0.112, +0.088, −0.044, +0.023 (with OH groups)

(84) should undergo nucleophilic solvolysis at the position *para* to the nitro group in a thermal reaction and *meta* to the nitro group in a photochemical process involving the lowest $\pi \rightarrow \pi^*$ excited singlet. This prediction is borne out by experiment:

$$\text{(84) 4-nitroveratrole} \xrightarrow[(\sim 20°C)]{h\nu,\ OH^-} \text{2-methoxy-5-nitrophenol (OH meta to } NO_2,\ OCH_3 \text{ para)}$$

$$\text{(84)} \xrightarrow[(\sim 80°C)]{\Delta,\ OH^-} \text{2-methoxy-4-nitrophenol (OH para to } NO_2,\ OCH_3 \text{ meta)} \tag{11.189}$$

Anisole has charge densities as shown in (85) and (86). This indicates that ground state anisole should undergo electrophilic substitution at the *ortho*

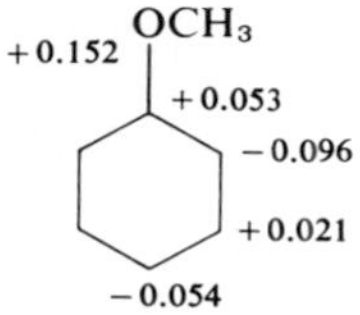

OCH_3 +0.152, +0.053, −0.096, +0.021, −0.054

(85)

S_0

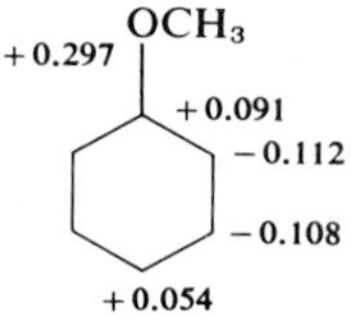

OCH_3 +0.297, +0.091, −0.112, −0.108, +0.054

(86)

S_1

and *para* positions but photochemical $\pi \rightarrow \pi^*$ substitution at the *ortho* and *meta* positions. Equation (11.187) indicates that this is valid. Valence bond diagrams of the first excited singlet states of these molecules look like (87) and (88).

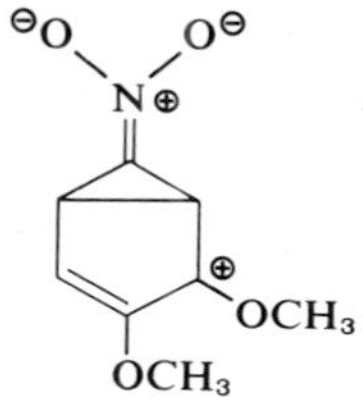

$^{\ominus}O$ $O^{\ominus}$ $N^{\oplus}$ $^{\oplus}OCH_3$ OCH_3

(87)

$^{\oplus}OCH_3$ $:^{\ominus}$

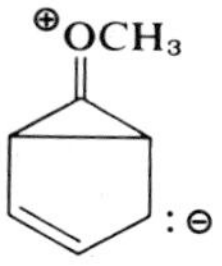

(88)

The *ortho/para* orientation rule of ground state chemistry appears to be followed in the photosubstitution reactions of nitrobenzene derivatives in liquid ammonia[140]:

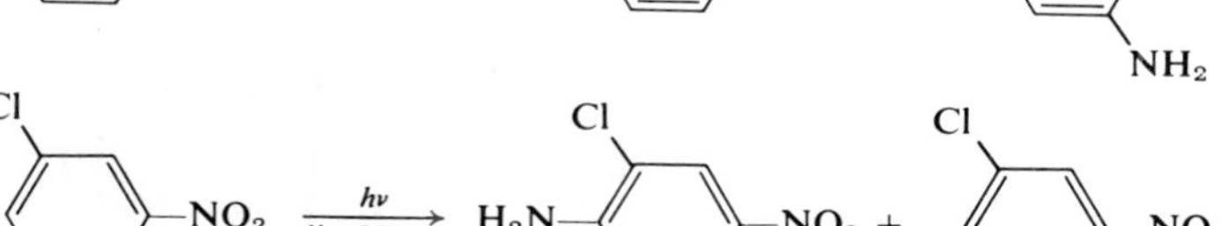

$$C_6H_5\text{—}NO_2 \xrightarrow[\text{liq. } NH_3]{h\nu} H_2N\text{—}C_6H_4\text{—}NO_2 + C_6H_4(NH_2)\text{—}NO_2 \qquad (11.190)$$

$$Cl\text{—}C_6H_4\text{—}NO_2 \xrightarrow[\text{liq. } NH_3]{h\nu} H_2N\text{—}C_6H_4\text{—}NO_2 + Cl\text{—}C_6H_3(NH_2)\text{—}NO_2 \qquad (11.191)$$

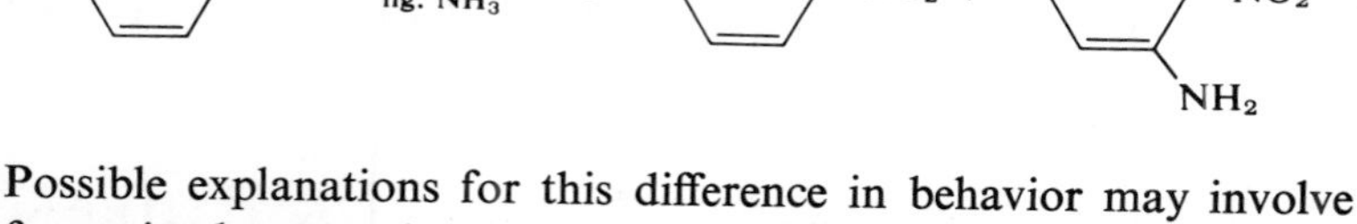

$$Cl\text{—}C_6H_4\text{—}NO_2 \xrightarrow[\text{liq. } NH_3]{h\nu} H_2N\text{—}C_6H_3(Cl)\text{—}NO_2 + C_6H_3(Cl)(NH_2)\text{—}NO_2 \qquad (11.192)$$

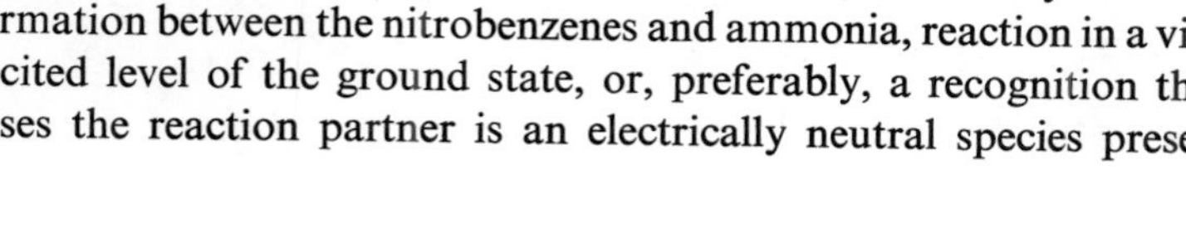

Possible explanations for this difference in behavior may involve complex formation between the nitrobenzenes and ammonia, reaction in a vibrationally excited level of the ground state, or, preferably, a recognition that in these cases the reaction partner is an electrically neutral species present in high concentration. It may well be that the initial attack at the various positions is largely nondiscriminatory and the products reflect the amount of resonance stabilization in the course of reaction (formation of *m*-nitroaniline would be least favored in this case).[139]

The failure to observe photosubstitution in the presence of a sensitizer in which the latter is the principal absorber, the invariance of product quantum yield with wavelengths shorter than 350 nm (onset of $n \rightarrow \pi^*$ absorption), and the observation that chloride and bromide ions (known to catalyze $S \rightarrow T$ intersystem crossing) strongly diminish the quantum yields of these reactions, strongly points to the lowest excited $\pi \rightarrow \pi^*$ singlet state as the reactive species in these transformations. Excitation into the $n \rightarrow \pi^*$ absorption band results in little product formation. A triplet state may, however, be involved in the photoamination of nitrobenzene.[141]

These studies have recently been extended to the naphthalene system.[142] In alkaline media 2-methoxy-5-nitro-, 1-methoxy-6-nitro-, and 2-methoxy-7-nitro-naphthalenes all undergo photosubstitutions to produce naphthols:

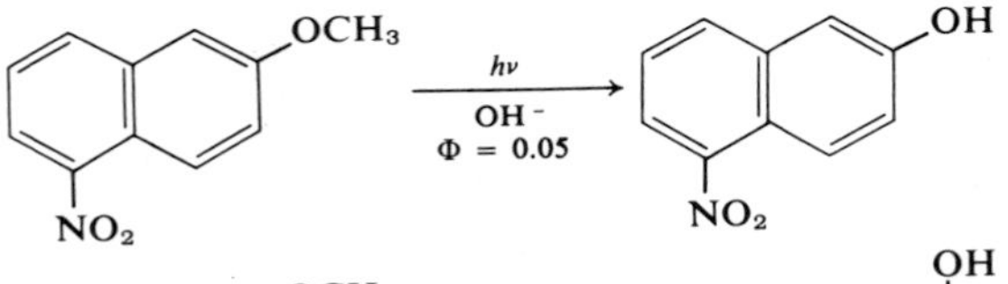

OCH$_3$, NO$_2$ $\xrightarrow[\text{OH}^-,\ \Phi = 0.05]{h\nu}$ OH, NO$_2$ (11.193)

OCH$_3$, NO$_2$ $\xrightarrow[\text{OH}^-,\ \Phi = 0.15]{h\nu}$ OH, NO$_2$ (11.194)

O$_2$N, OCH$_3$ $\xrightarrow[\text{OH}^-,\ \Phi = 0.02]{h\nu}$ O$_2$N, OH (11.195)

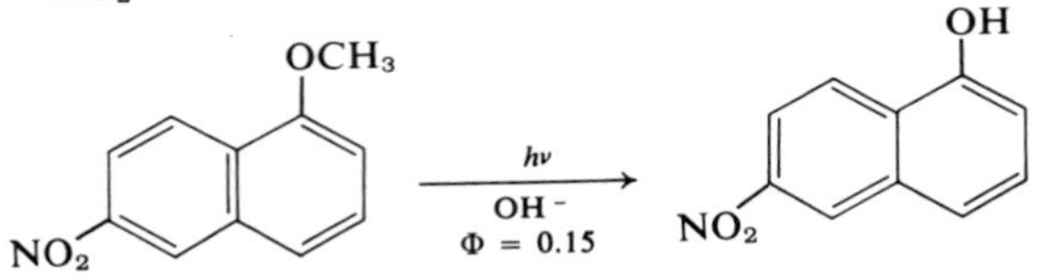

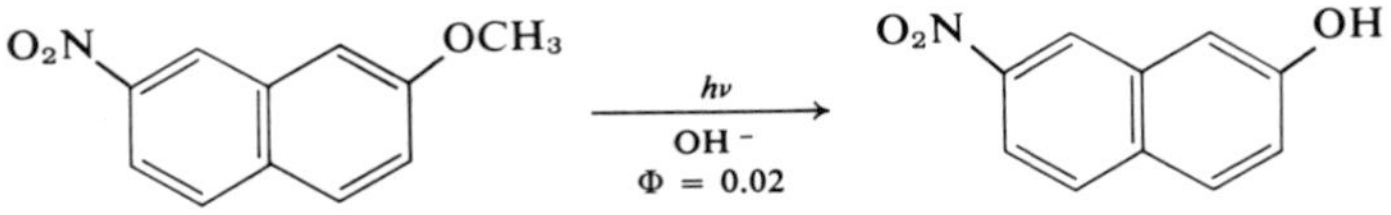

To demonstrate that a position *meta* to the nitro group has a greater rate of reaction in the excited state than a *para* position, two compounds having one methoxy group *meta* and another *para* to the nitro group were irradiated. Both yielded products having an —OH group *meta* to the —NO$_2$:

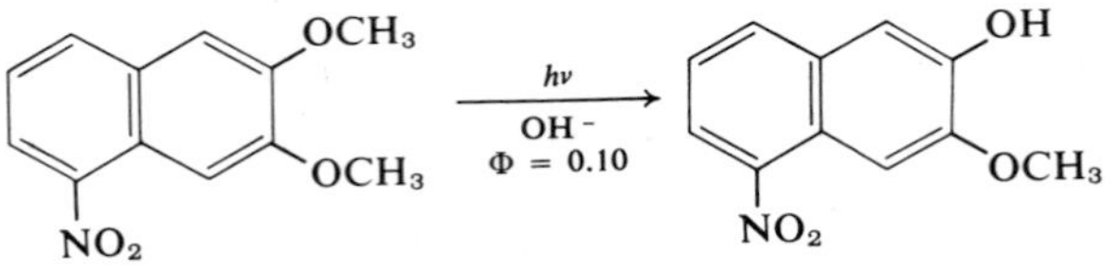

OCH$_3$, OCH$_3$, NO$_2$ $\xrightarrow[\text{OH}^-,\ \Phi = 0.10]{h\nu}$ OH, OCH$_3$, NO$_2$ (11.196a)

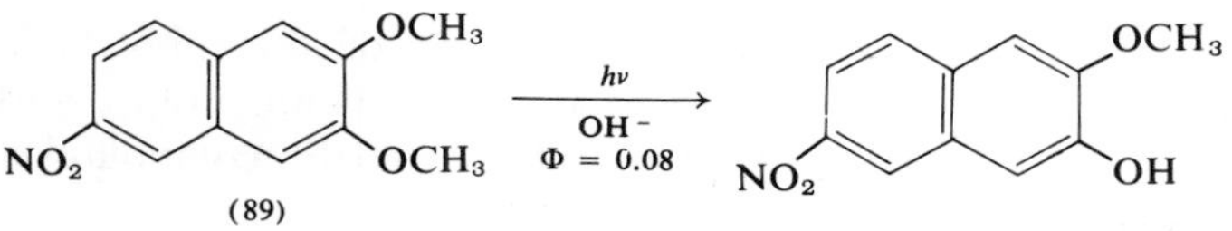

(89) $\xrightarrow[\text{OH}^-,\ \Phi = 0.08]{h\nu}$ (11.196b)

Comparison of these experimental results with the calculated charge densities (S_0 and S_1) at the 2 and 3 positions (Table 11.5) shows that this is the expected result. Except for those compounds discussed below, the failure to observe quenching with triplet quenchers and reaction in the presence of a photosensitizer indicated singlet reactions. Compound (89) was found to also undergo benzophenone-photosensitized substitution, indicating that the triplet state of this compound is also reactive. The reaction, however, was less clean than that observed in the direct photolysis. Similarly, 1,6-dinitronaphthalene was found to undergo both direct and benzophenone-photosensitized substitution:

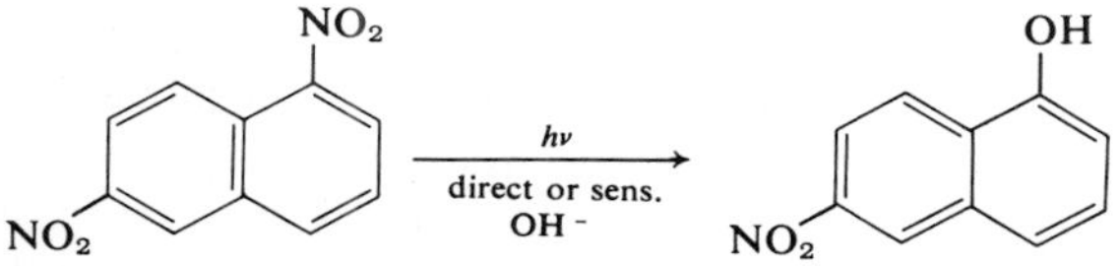

$\xrightarrow[\text{direct or sens.},\ \text{OH}^-]{h\nu}$ (11.197)

Fluorenone and sodium sorbate served to quench the unsensitized reaction, indicating a triplet excited state. Similarly, recent investigations of the photosubstitution of 3,5-dinitroanisole indicate a reactive triplet species[143]:

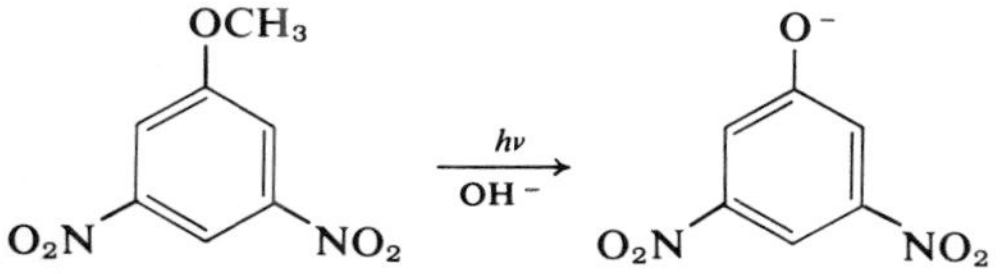

$\xrightarrow[\text{OH}^-]{h\nu}$ (11.198)

TABLE 11.5. *Calculated Charge Distributions for Positions 2 and 3 of* 2,3-*Dimethoxy*-6- (*and* 5-) *Nitronaphthalenes*

Compound	S_0	S_1
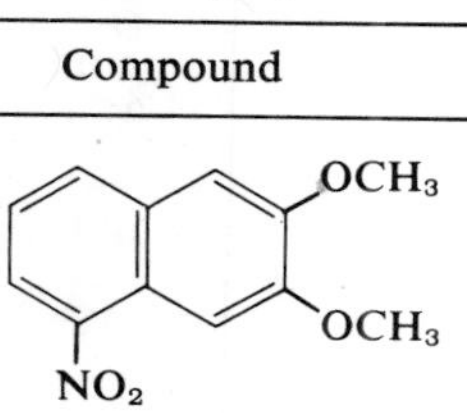 2,3-dimethoxy-5-nitronaphthalene (OCH_3, upper)	+0.001	+0.053
(OCH_3, lower)	+0.011	+0.036
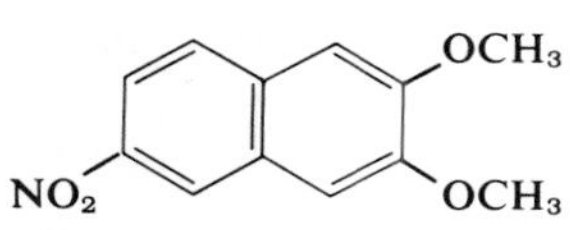 2,3-dimethoxy-6-nitronaphthalene (OCH_3, upper)	+0.011	+0.030
(OCH_3, lower)	+0.000	+0.126

Flash photolysis, ESR, and absorption studies of this photolysis have identified a transient species (leading to product) as a complex ($\lambda_{max} \sim 410$ nm, $\tau \sim 5 \times 10^{-7}$ sec) between the triplet excited aromatic compound and its nucleophilic reaction partner.

Seiler and Wirz have reported a study of the photohydrolysis of eight isomeric trifluoromethylnaphthols.[(144)] Six of the eight isomers studied underwent hydrolysis cleanly to produce quantitative yields of the corresponding hydroxynaphthoic acids:

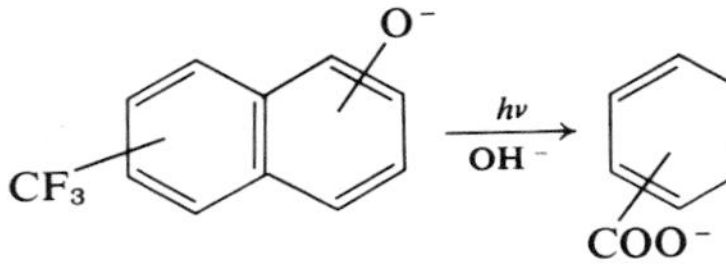

$$CF_3\text{-}C_{10}H_6\text{-}O^- \xrightarrow[OH^-]{h\nu} {}^-OOC\text{-}C_{10}H_6\text{-}O^- \qquad (11.199)$$

Stern–Volmer plots of the quenching of photohydrolysis and of naphthol fluorescence by sodium sorbate and copper(II) and iron(II) sulfate yielded straight lines with slopes equal within experimental error, indicating the quenched state to be the singlet. In addition, the photoreaction could not be sensitized by acetone or triphenylene-1-sulfonic acid. Thus it was concluded that the photoreactive state in these reactions is the first excited singlet. A linear correlation between the log of the rate constant for photohydrolysis and the calculated charge density $q_\mu{}^*$ at the ring carbon bearing the trifluoromethyl substituent was discovered, as seen in Figure 11.1.

Photochemical nucleophilic substitution of some simple anthraquinone derivatives has been reported by Griffiths and Hawkins.[(145)] Irradiation of 1-methoxyanthraquinone (90) in ammonia solution resulted primarily in 1-aminoanthraquinone (91):

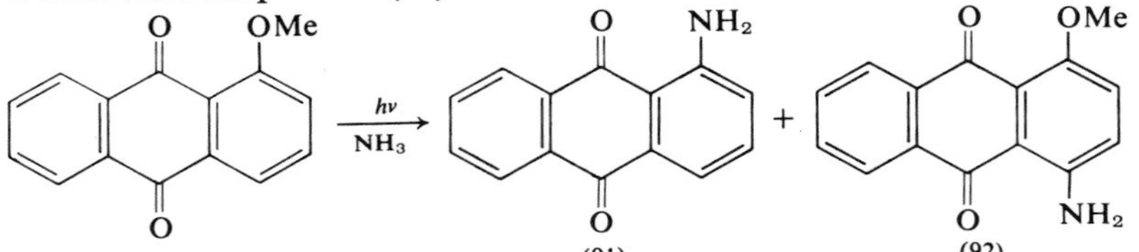

$$\text{(90)} \xrightarrow[NH_3]{h\nu} \text{(91) } 96\% + \text{(92) } 0.5\% \qquad (11.200)$$

Similarly, 2-methoxyanthraquinone (93) resulted in compounds (94) and (95):

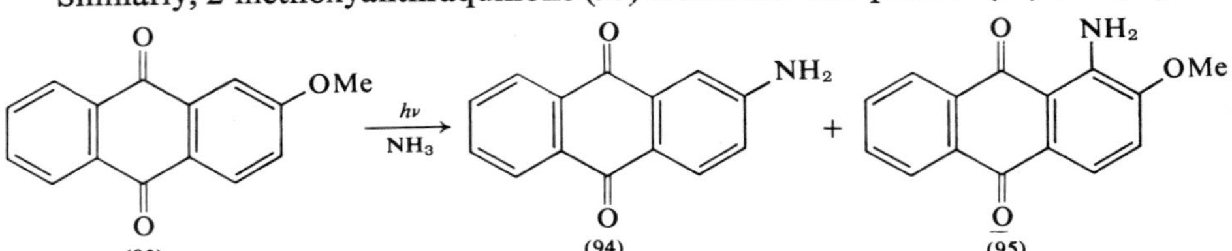

$$\text{(93)} \xrightarrow[NH_3]{h\nu} \text{(94) } 25\% + \text{(95) } 70\% \qquad (11.201)$$

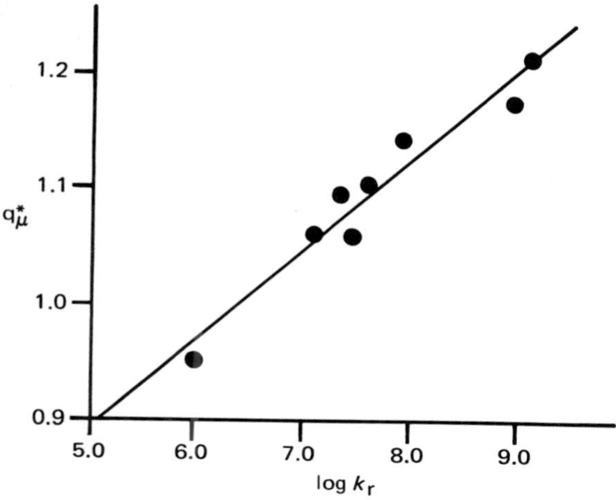

FIGURE 11.1 Correlation between calculated charge density and reactivity for trifluoromethylnaphthalates. (Reproduced with permission from Ref. 144.)

Photosensitization and piperylene quenching studies indicated that these reactions arise from the singlet manifold. To evaluate the possible role of radical intermediates, these reactions were irradiated in the presence of the radical scavenger 2,6-di-*t*-butylphenol. The rate of reaction of (90) was unaffected by the presence of the radical scavenger, whereas the rate of photoamination of (93) was actually *increased* to produce significantly more (95). The product ratio resulting from photolysis of (93) indicates that attack by ammonia at position 1 is favored over attack at position 2 even though the intermediate σ-complex (96) cannot easily eliminate hydride ion to yield (95). Formation of (95) could occur much more readily by homolytic loss of hydrogen from (96), yielding the semiquinone radical (97). Oxidation of (97) to yield (95) should occur readily in the presence of oxygen:

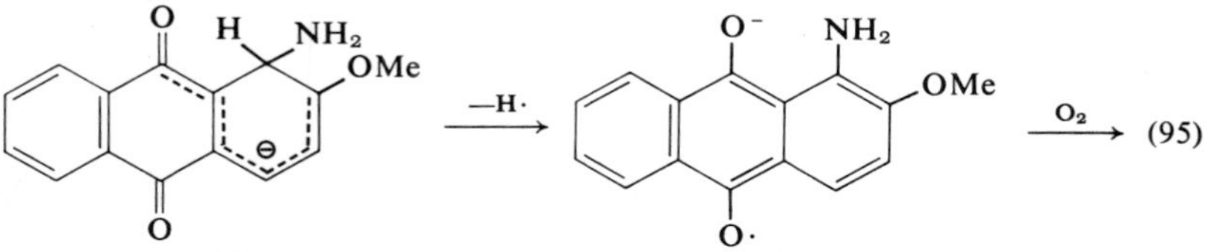

(96) $\xrightarrow{-H\cdot}$ (97) $\xrightarrow{O_2}$ (95)

(11.202)

Predictably, the rate of formation of (95) was drastically reduced in the absence of O_2.

Irradiation of 3-methoxyacetophenone in a solution of sodium cyanide in 5:1 acetonitrile–water is reported to result in (98). When the solvent is predominantly water, (99) and (100) are formed[(146)]:

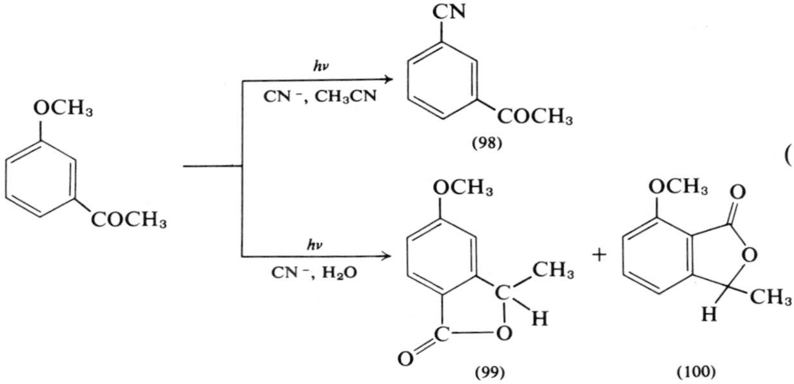

(11.203)

Compounds (99) and (100) are thought to be formed by addition of cyanide ion to the ring *ortho* to the carbonyl, followed by protonation at oxygen, aromatization by tautomerization, hydrolysis of the nitrile, and lactonization upon acidification. The photolysis of 2-methoxyacetophenone, on the other hand, results in rearrangement to 3-methoxyacetophenone:

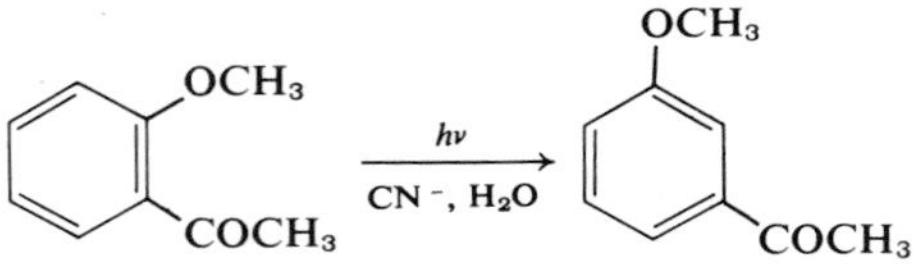

(11.204)

To obtain information concerning the pathway of this rearrangement, 2-methoxyacetophenone was photolyzed in the presence of cyanide ion in deuterium oxide to yield (101):

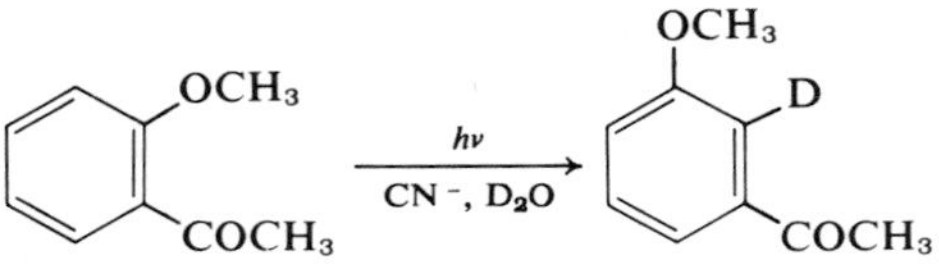

(11.205)

Irradiation of 3,5-dimethylacetophenone yields rearrangement product (102), indicating migration of the acetyl group:

Me, Me, $COCH_3$ $\xrightarrow[CN^-, H_2O]{h\nu}$ Me, Me, $COCH_3$ (102)

$\xrightarrow[CN^-, D_2O]{h\nu}$ Me, Me, $COCH_3$, D (11.206)

No mechanism for this rearrangement, which requires the presence of cyanide ion, has yet been elucidated.

Another type of photochemical substitution reaction is the halogen interchange reaction involving the displacement of a heavier halogen atom from an aromatic ring by a lighter halogen atom.[147] Halogen atoms are thought to be intermediates in this reaction since radical reaction inhibitors, such as azobenzene, azoxybenzene, and nitrobenzene, retard the reaction rate. In addition, the reaction is retarded by electron-withdrawing groups located on the benzene ring[148]:

$$X_2' \longrightarrow 2X\cdot'$$

$$C_6H_5{-}X + \cdot X' \longrightarrow \text{X, X'} \quad (11.207a)$$

Atomic weight of X greater than that of X′

$$\text{X, X'} \longrightarrow C_6H_5{-}X' + X\cdot \quad (11.207b)$$

This reaction is general for chlorine and bromine, although competing reactions make the displacement of iodine less useful[149]:

$$C_6H_5{-}I + Cl_2 \xrightarrow{h\nu} C_6H_5{-}ICl_2 \quad (11.208)$$

Iodine can, however, be displaced cleanly with bromine[150] and with chlorine if iodine monochloride or sulfuryl chloride and benzoyl peroxide are used as chlorinating agents.[151] By competitive experiments[150] the reactivity ratios for the displacement of bromine by chlorine from substituted bromo-benzenes have been found to be in the order *p*-phenyl > *o*-methoxy > *p*-chloro > unsubstituted > p-carbomethoxy > p-cyano > p-nitro. Some typical examples of this photosubstitution are[147,150,152,153]:

$$C_6H_5Br + Cl_2 \xrightarrow{h\nu} C_6H_5Cl \qquad (11.209)$$

$$p\text{-}CH_3C_6H_4Br \xrightarrow[SO_2Cl_2,\ Bz_2O_2]{h\nu} p\text{-}CH_3C_6H_4Cl \qquad (11.210)$$

$$C_6H_5I + Br_2 \xrightarrow{h\nu} C_6H_5Br \qquad (11.211)$$

$$C_6H_5I \xrightarrow[ICl\ \text{or}\ SO_2Cl_2\ +\ \text{peroxide}]{h\nu} C_6H_5Cl \qquad (11.212)$$

REFERENCES

1. E. W. R. Steacie, *Atomic and Free Radical Reactions*, Vol. I, Reinhold, New York (1954).
2. R. F. Hutton and C. Steel, *J. Amer. Chem. Soc.* **86**, 745 (1964).
3. I. I. Abram, G. S. Milne, B. S. Soloman, and C. Steel, *J. Amer. Chem. Soc.* **91**, 1220 (1969).
4. T. Mill and R. S. Stringham, *Tetrahedron Lett.*, 1853 (1969).
5. S. F. Nelsen and P. D. Bartlett, *J. Amer. Chem. Soc.* **88**, 143 (1966).
6. J. R. Fox and G. S. Hammond, *J. Amer. Chem. Soc.* **86**, 4031 (1964).
7. P. S. Engel and P. D. Bartlett, *J. Amer. Chem. Soc.* **92**, 5883 (1970).
8. W. A. Pryor and K. Smith, *J. Amer. Chem. Soc.* **92**, 5403 (1970).
9. S. Seltzer, *J. Amer. Chem. Soc.* **83**, 2625 (1961); **85**, 14 (1963).
10. N. A. Porter, M. E. Landis, and L. J. Marrett, *J. Amer. Chem. Soc.* **93**, 795 (1971).
11. A. Gilbert, in *Photochemistry*, D. Bryce-Smith, ed., The Chemical Society, London, Vol. I (1970), Vol. II (1971), Vol. III (1972).
12. P. D. Bartlett and N. A. Porter, *J. Amer. Chem. Soc.* **90**, 5317 (1968).
13. S. D. Andrews and A. C. Day, *Chem. Comm.*, 667 (1966).

14. C. G. Overberger and J. W. Stoddard, *J. Amer. Chem. Soc.* **92**, 4922 (1970).
15. R. G. Bergman and W. L. Carter, *J. Amer. Chem. Soc.* **91**, 7411 (1969).
16. T. Aratani, Y. Nakanisi, and H. Nozaki, *Tetrahedron* **26**, 4339 (1970).
17. D. P. G. Hamon and L. J. Holding, *Chem. Comm.*, 1330 (1970).
18. T. Sanjiki, M. Ohta, and H. Kato, *Chem. Comm.*, 638 (1969).
19. J. A. Berson, and S. S. Olin, *J. Amer. Chem. Soc.* **92**, 1086 (1970).
20. P. S. Skell and R. C. Woodworth, *J. Amer. Chem. Soc.* **78**, 4496 (1956); R. C. Woodworth and P. S. Skell, *J. Amer. Chem. Soc.* **81**, 3383 (1959).
21. K. R. Kopecky, G. S. Hammond, and P. A. Leermakers, *J. Amer. Chem. Soc.* **84**, 1015 (1962).
22. G. Herzberg and J. Shoosmith, *Nature* **183**, 1801 (1959); W. Braun, M. Bass, and M. Pilling, *J. Chem. Phys.* **52**, 5131 (1970).
23. H. M. Frey, *J. Chem. Soc.*, 2293 (1962).
24. H. M. Frey and I. D. R. Stevens, *J. Amer. Chem. Soc.* **84**, 2647 (1962).
25. R. A. Moss and J. R. Przybyla, *J. Org. Chem.* **33**, 3817 (1968).
26. R. A. Moss, *J. Org. Chem.* **31**, 3296 (1966).
27. W. M. Jones, M. E. Stowe, E. E. Wells, Jr., and E. W. Lester, *J. Amer. Chem. Soc.* **90**, 1849 (1968).
28. W. M. Jones and C. L. Ennis, *J. Amer. Chem. Soc.* **89**, 3069 (1967); T. Nazakawa and K. Isobe, *Tetrahedron Lett.*, 565 (1968).
29. R. A. Moss and U.-H. Dolling, *J. Amer. Chem. Soc.* **93**, 954 (1971).
30. A. M. Trozzolo, *Acc. Chem. Res.* **1**, 329 (1968).
31. P. S. Skell, S. J. Valenty, and P. W. Humer, *J. Amer. Chem. Soc.* **95**, 5041 (1973).
32. P. S. Skell and A. Y. Garner, *J. Amer. Chem. Soc.* **78**, 3409 (1956).
33. J. P. Oliver and U. V. Rao, *J. Org. Chem.* **31**, 2696 (1966).
34. N. C. Yang and T. A. Marolewski, *J. Amer. Chem. Soc.* **90**, 5644 (1968).
35. D. O. Cowan, M. A. Couch, K. R. Kopecky, and G. S. Hammond, *J. Org. Chem.* **29**, 1922 (1964).
36. P. S. Skell and S. J. Valenty, *J. Amer. Chem. Soc.* **95**, 5042 (1973).
37. T. DoMinh and O. P. Strausz, *J. Amer. Chem. Soc.* **92**, 1766 (1970).
38. H. Duerr and G. Scheppes, *Chem. Ber.* **103**, 380 (1970); H. Duerr, L. Schrader, and H. Seidl, *Chem. Ber.* **104**, 391 (1971).
39. G. Lowe and J. Parker, *Chem. Comm.*, 577 (1971).
40. T. Kametani, K. Fukumoto, and K. Shishido, *Chem. Ind.* (London), 1566 (1970).
41. V. Snieckus and K. S. Bhandari, *Tetrahedron Lett.*, 3375 (1969).
42. D. H. R. Barton and L. R. Morgan, Jr., *J. Chem. Soc.*, 622 (1962).
43. R. M. Moriarty and M. Rahman, *Tetrahedron* **21**, 2877 (1965).
44. W. Lwowski, and T. Mattingly, *Tetrahedron Lett.*, 277 (1962); W. Lwowski, T. J. Maricich, and T. W. Mattingly, *J. Amer. Chem. Soc.* **85**, 1200 (1963).
45. W. H. Saunders, Jr. and E. A. Caress, *J. Amer. Chem. Soc.* **86**, 861 (1964).
46. F. D. Lewis and W. H. Saunders, Jr., *J. Amer. Chem. Soc.* **89**, 645 (1967).
47. F. D. Lewis and W. H. Saunders, Jr., *J. Amer. Chem. Soc.* **90**, 7031 (1968).
48. R. M. Moriarty and R. C. Reardon, *Tetrahedron* **26**, 1379 (1970).
49. R. A. Abramovitch and E. P. Kyba, *J. Amer. Chem. Soc.* **93**, 1537 (1971).
50. G. Smolinsky, E. Wasserman, and W. A. Yager, *J. Amer. Chem. Soc.* **84**, 3220 (1962); A. M. Trozzolo, R. W. Murray, G. Smolinsky, W. A. Yager, and E. Wasserman, *J. Amer. Chem. Soc.*, **85**, 2526 (1963); B. Singh and J. S. Brinen, *J. Amer. Chem. Soc.* **93**, 540 (1971).
51. J. G. Calvert and J. N. Pitts, *Photochemistry*, Wiley, New York (1966).

52. W. E. Lee, J. G. Calvert, and E. W. Malmberg, *J. Amer. Chem. Soc.* **83**, 1928 (1961).
53. R. C. Petterson, A. DiMaggio, A. L. Hebert, T. J. Haley, J. P. Mykytka, and I. M. Sarkar, *J. Org. Chem.* **36**, 631 (1971).
54. P. Tarte, *Bull. Soc. Roy. Liege* **22**, 226 (1953); B. G. Gowenlock and J. Trotman, *J. Chem. Soc.*, 4190 (1955); 1670 (1956).
55. P. Kabasakalian and E. R. Townley, *J. Amer. Chem. Soc.* **84**, 2711 (1962).
56. P. Kabasakalian, E. R. Townley, and M. D. Yudis, *J. Amer. Chem. Soc.* **84**, 2718 (1962).
57. P. Kabasakalian and E. R. Townley, *J. Org. Chem.* **27**, 2918 (1962).
58. M. Akhtar and M. M. Pechet, *J. Amer. Chem. Soc.* **86**, 265 (1964).
59. D. H. R. Barton and J. M. Beaton, *J. Amer. Chem. Soc.* **82**, 2641 (1960); **83**, 4083 (1961).
60. H. A. Reimann, A. S. Capomaggi, T. Strauss, E. P. Oliveto, and D. H. R. Barton, *J. Amer. Chem. Soc.* **83**, 4481 (1961).
61. M. Akhtar, D. H. R. Barton, J. M. Beaton, and A. G. Hortmann, *J. Amer. Chem. Soc.* **85**, 1512 (1963).
62. D. H. R. Barton, D. Kumari, P. Welzel, L. J. Danks, and J. F. McGhie, *J. Amer. Chem. Soc. C*, 332 (1969).
63. O. L. Chapman, A. A. Griswold, E. Hoganson, G. Lenz, and J. Reasoner, *Pure Appl. Chem.* **9**, 585 (1964).
64. M. P. Cava, R. H. Schlessinger, and J. P. Van Meter, *J. Amer. Chem. Soc.* **86**, 3173 (1964).
65. A. O. Pederson, S. O. Lawesson, P. D. Klemmensen, and J. Kolc, *Tetrahedron* **26**, 1157 (1970).
66. J. G. Pacifici and C. Diebert, *J. Amer. Chem. Soc.* **91**, 4595 (1969).
67. R. H. Fish, L. C. Chow, and M. C. Caserio, *Tetrahedron Lett.*, 1259, (1969).
68. Y. Kishida, T. Hiraoka, and J. Ide, *Chem. Pharm. Bull. Tokyo*, 1591, (1969).
69. P. H. MacFarlane and D. W. Russell, *Tetrahedron Lett.*, 725 (1971).
70. F. Takeuchi, T. Fujiimori, and A. Sugimori, *Bull. Chem. Soc. Japan* **43**, 3637 (1970).
71. S. P. Singh and J. Kagan, *J. Org. Chem.* **35**, 2203, 3839 (1970).
72. W. Adam and R. Rucktaeschel, *J. Amer. Chem. Soc.* **93**, 557 (1971).
73. S. Y. Wang, *Fed. Proc.* (Suppl. 15) **24**, 71 (1965).
74. A. Stoll and W. Schlientz, *Helv. Chim. Acta* **38**, 585 (1955).
75. G. O. Schenck, G. Koltzenburg, and H. Grossmann, *Angew. Chem.* **69**, 177 (1957); R. Dulon, M. Vilkas, and M. Pfau, *Compt. Rend.* **249**, 429 (1959).
76. M. Pfau, R. Dulon, and M. Vilkas, *Compt. Rend.* **254**, 1817 (1962).
77. G. O. Schenck and R. Steinmetz, *Naturwiss.* **47**, 514 (1960).
78. J. A. Marshall and R. D. Carroll, *J. Amer. Chem. Soc.* **88**, 4092 (1966).
79. P. J. Kropp, *J. Amer. Chem. Soc.* **88**, 4091 (1966).
80. P. J. Kropp, *J. Amer. Chem. Soc.* **89**, 3650 (1967).
81. P. J. Kropp and H. J. Krauss, *J. Amer. Chem. Soc.* **89**, 5199 (1967).
82. J. A. Marshall and M. J. Wurth, *J. Amer. Chem. Soc.* **89**, 6788 (1967).
83. H. Faubl, quoted in Ref. 84.
84. J. A. Marshall, *Acc. Chem. Res.* **2**, 33 (1969).
85. P. J. Kropp, *J. Amer. Chem. Soc.* **95**, 4611 (1973).
86. P. J. Kropp, *J. Amer. Chem. Soc.* **91**, 5783 (1969).
87. L. Kaplan, J. S. Ritscher, and K. E. Wilzbach, *J. Amer. Chem. Soc.* **88**, 2881 (1966).
88. E. Farenhorst and A. F. Bickel, *Tetrahedron Lett.*, 5911 (1966).
89. D. Bryce-Smith, A. Gilbert, and H. C. Longuet-Higgens, *Chem. Comm.*, 240 (1967).

90. K. E. Wilzbach, J. S. Ritscher, and L. Kaplan, *J. Amer. Chem. Soc.* **89**, 1031 (1967).
91. J. A. Berson and H. M. Hasty, Jr., *J. Amer. Chem. Soc.* **93**, 1549 (1971).
92. D. Elad, in *Organic Photochemistry*, Vol. 2., O. L. Chapman, ed., Marcel Dekker, New York (1969).
93. W. H. Urry, O. O. Juveland, and F. W. Stacey, *J. Amer. Chem. Soc.* **74**, 6155 (1952); W. H. Urry and O. O. Juveland, *J. Amer. Chem. Soc.* **80**, 3322 (1958).
94. Y. Ogata, Y. Izana, and H. Tomioka, *Tetrahedron* **23**, 1590 (1967).
95. Y. L. Chow, C. Colon, and S. C. Chen, *J. Org. Chem.* **32**, 2109 (1967).
96. D. Elad and J. Sinnreich, *Chem. Ind.* (*London*), 768 (1965).
97. J. Sinnreich and D. Elad, *Tetrahedron* **24**, 4509 (1968).
98. J. Rokach and D. Elad, *J. Org. Chem.* **31**, 4210 (1966).
99. G. Friedman and A. Komen, *Tetrahedron Lett.*, 3357 (1968).
100. J. Rokach, C. H. Krauch, and D. Elad, *Tetrahedron Lett.*, 3253 (1966).
101. C. H. Krauch, J. Rokach, and D. Elad, *Tetrahedron Lett.*, 3209 (1965).
102. M. S. Kharasch, H. W. Urrey, and B. M. Kuderna, *J. Org. Chem.* **14**, 248 (1949).
103. T. M. Patrick, *J. Org. Chem.* **17**, 1009, 1269 (1952).
104. M. S. Kharasch, J. Kuderna, and W. Nudenberg, *J. Org. Chem.* **18**, 1225 (1953).
105. P. DeMayo, J. B. Stothers, and W. Templeton, *Can. J. Chem.* **39**, 488 (1961).
106. W. Reusch, *J. Org. Chem.* **27**, 1882 (1963).
107. G. Sosnovsky, *Free Radical Reactions in Preparative Organic Chemistry*, Macmillan, New York (1964).
108. R. Schmitz and H. J. Schumacher, *Z. Physik. Chem.* (*Leipzig*) **B52**, 72 (1942).
109. S. A. Fasech, *J. Chem. Soc.*, 3708 (1953).
110. A. N. Nesmeyanov, P. Kh. Freidlina, and V. N. Kost, *Izv. Akad. Nauk SSSR, Ser. Khim.*, 1205 (1958).
111. J. A. Berson and R. Swidler, *J. Amer. Chem. Soc.* **76**, 4060 (1954).
112. W. E. Vaughan, F. F. Rust, and T. W. Evans, *J. Org. Chem.* **7**, 477 (1942).
113. A. L. Henne and M. Nager, *J. Amer. Chem. Soc.* **73**, 5527 (1951).
114. H. L. Goering, P. I. Abell, and B. F. Aycock, *J. Amer. Chem. Soc.* **74**, 3588 (1952).
115. P. D. Readio and P. S. Skell, *J. Org. Chem.* **31**, 753 (1966).
116. N. A. Lebel, *J. Amer. Chem. Soc.* **82**, 623 (1960).
117. P. S. Skell and R. G. Allen, *J. Amer. Chem. Soc.* **81**, 5383 (1959).
118. E. C. Kooyman, *Rec. Trav. Chim.* **70**, 684, 867 (1951).
119. M. S. Karasch, E. V. Jensen, and W. H. Urrey, *J. Amer. Chem. Soc.* **68**, 154 (1946); **69**, 1100 (1947).
120. E. A. I. Heiba and L. C. Anderson, *J. Amer. Chem. Soc.* **79**, 4940 (1957); M. S. Kharasch and H. N. Friedlander, *J. Org. Chem.* **14**, 239 (1949).
121. (a) R. N. Haszeldine and B. R. Steele, *J. Chem. Soc.*, 3005 (1955).; (b) R. N. Haszeldine and K. Leedham, *J. Chem. Soc.*, 3483 (1952).
122. C. Walling, *Free Radicals in Solution*, Wiley, New York (1957).
123. E. I. Heiba and R. M. Dessau, *J. Amer. Chem. Soc.* **89**, 3772 (1967).
124. F. F. Rust and W. E. Vaughan, U.S. Pat. 2,392,294 (1946).
125. W. J. Peppel, U.S. Pat. 2,408,095 (1947); W. A. Lazier, A. A. Pavlie, and W. J. Peppel, U.S. Pat. 2,422,246 (1947).
126. G. M. Burch, H. Goldwhite, and R. N. Haszeldine, *J. Chem. Soc.*, 1083 (1963).
127. A. N. Pudovik and I. V. Konvalova, *Zh. Okshch. Khim.* **29**, 3342 (1959).
128. R. N. Haszeldine and J. C. Young, *J. Chem. Soc.*, 4553 (1960).
129. A. M. Geyer and R. N. Haszeldine, *J. Chem. Soc.*, 1038 (1957).
130. R. Fuchs and H. Gilman, *J. Org. Chem.* **22**, 1009 (1957).
131. J. C. Anderson and C. B. Reese, *Tetrahedron Lett.*, 1 (1962).

132. R. L. Letsinger and O. B. Ramsey, *J. Amer. Chem. Soc.* **86**, 1447 (1964); R. L. Letsinger, O. B. Ramsey, and J. H. McCain, *J. Amer. Chem. Soc.* **87**, 2945 (1965).
133. R. A. Bowie and O. C. Musgrave, *Proc. Chem. Soc.*, 15 (1964).
134. R. M. Johnson and C. W. Rees, *Proc. Chem. Soc.*, 213 (1964).
135. H. E. Zimmerman and S. Somasekhara, *J. Amer. Chem. Soc.* **85**, 922 (1963); H. E. Zimmerman and V. E. Sandel, *J. Amer. Chem. Soc.* **85**, 915 (1963).
136. E. Havinga, R. O. De Jongh, and W. Dorst, *Rec. Trav. Chim.* **75**, 378 (1956); E. Havinga, *Kon Ned. Akad. Wetenschap, Verslag afdeling Nat.* **70**, 52 (1961); E. Havinga and R. O. de Jongh, *Bull. Soc. Chim. Belg.* **71**, 803 (1962); D. F. Nijhoff and E. Havinga, *Tetrahedron Lett.*, 4199 (1965); R. O. De Jongh and E. Havinga, *Rec. Trav. Chim.* **85**, 275 (1966); M. E. Kronenberg, A. van der Heyden, and E. Havinga, *Rec. Trav. Chim.* **86**, 254 (1967); G. H. D. van der Stegen, E. J. Poziomek, M. E. Kronenberg, and E. Havinga, *Tetrahedron Lett.*, 6371 (1966).
137. R. L. Letsinger and J. H. McCain, *J. Amer. Chem. Soc.* **88**, 2884 (1966).
138. D. A. de Bie and E. Havinga, *Tetrahedron* **21**, 2359 (1965).
139. E. Havinga and M. E. Kronenberg, *Pure Appl. Chem.* **16**, 137 (1968).
140. A. Van Vliet, M. E. Kronenberg, and E. Havinga, *Tetrahedron Lett.*, 5957, (1966).
141. A. van Vliet, J. Cornelisse, and E. Havinga, *Rec. Trav. Chim.* **88**, 1339 (1969).
142. G. M. J. Biejersbergen van Henegouven and E. Havinga, *Rec. Trav. Chim.* **89**, 907 (1970).
143. G. P. de Gunst and E. Havinga, *Tetrahedron* **29**, 2167 (1973).
144. P. Seiler and J. Wirz, *Tetrahedron Lett.*, 1683 (1971).
145. J. Griffiths and C. Hawkins, *Chem. Comm.*, 111 (1973).
146. R. L. Letsinger and A. L. Colb, *J. Amer. Chem. Soc.* **94**, 3665 (1972).
147. A. Eibner, *Chem. Ber.* **36**, 1229 (1903).
148. B. Miller and C. Walling, *J. Amer. Chem. Soc.* **74**, 4187 (1957).
149. C. Willegerodt, *J. Prakt. Chem.* **2**, 33 (1886); **2**, 154 (1886).
150. B. Milligan, R. L. Bradow, J. E. Rose, H. E. Hubbert, and A. Roe, *J. Amer. Chem. Soc.* **84**, 158 (1962).
151. L. J. Andrews and R. M. Keefer, *J. Amer. Chem. Soc.* **80**, 1723 (1958); R. M. Keefer and L. J. Andrews, *J. Amer. Chem. Soc.* **80**, 5350 (1958).
152. G. L. Goener and R. C. Nametz, *J. Amer. Chem. Soc.* **73**, 2940 (1951).
153. R. C. R. Bacon and H. A. O. Hill, *J. Chem. Soc.*, 1097 (1964).
154. W. V. E. Doering and H. Prinzbach, *Tetrahedron* **6**, 24 (1959).
155. H. M. Frey and G. B. Kistiakowsky, *J. Amer. Chem. Soc.* **79**, 6373 (1957); H. M. Frey, *Proc. Chem. Soc.*, 318 (1959).
156. C. R. Flynn and J. Michl, *J. Amer. Chem. Soc.* **95**, 5802 (1973).

12

An Introduction to Photobiology

Thus far in this book we have discussed one- or two-component photochemical systems which because of their relative simplicity lend themselves quite well to laboratory study. Consequently the mechanisms of many of the photoreactions we have discussed have been elucidated in exquisite detail. As we turn our attention in this chapter to some photochemical aspects of living systems, we shall find much more complex situations in which mechanistic details are just now beginning to be obtained. In some systems, such as those which exhibit phototaxis or phototropism, so little is known that our treatment must as a consequence be limited to only a brief discussion of these phenomena. The topics we will consider here are photosynthesis, vision, phototaxis and phototropism, and damage and subsequent repair of damage by light. Due to space limitations, a discussion of the very fascinating area of bioluminescence must be omitted.

12.1. PHOTOSYNTHESIS

The process occurring in plants and algae by which water is oxidized to molecular oxygen and carbon dioxide is converted to carbohydrates in the presence of light is called photosynthesis. In addition to the products oxygen and carbohydrate, light energy is stored chemically in adenosine triphosphate (ATP) for later use for a variety of purposes. The production of oxygen and carbohydrate can be simply represented by the following equation:

$$6CO_2 + 6H_2O \xrightarrow[\text{chlorophyll}]{h\nu} C_6H_{12}O_6 + 6O_2 \qquad (12.1a)$$

$$CO_2 + H_2O \xrightarrow[\text{chlorophyll}]{h\nu} (CH_2O) + O_2 \qquad (12.1b)$$

where (CH_2O) represents one-sixth of a carbohydrate molecule. In the absence of light the process represented by these reactions is energetically unfavorable, requiring a minimum of about 112 kcal/mole of carbon dioxide converted.(1) From a knowledge of the wavelengths of light effective in photosynthesis, one can calculate that at least three to four quanta of light must be supplied to provide sufficient energy to effect the reduction of a single CO_2 molecule. In actuality the number of quanta necessary may be as high as 8 ± 1, due to known energy losses.(2)

A study of photosynthetic organisms other than green plants has revealed that certain bacteria, such as the purple sulfur bacteria, utilize H_2S instead of H_2O as a reductant in photosynthesis. The product obtained is elemental sulfur instead of oxygen:

$$CO_2 + 2H_2S \xrightarrow{h\nu} (CH_2O) + 2S + H_2O \qquad (12.2)$$

Other substrates which can be utilized as reducing agents by various photosynthetic bacteria are hydrogen, isopropanol, thiosulfate, and selenium.(3)

12.1a. *The Photosynthetic Apparatus*

The photosynthetic apparatus in green plants and algae is located in the chloroplast, which is a flattened, double-membraned structure about 150–200 Å thick.(4,5) The two flat membranes lie one above the other and are united at their peripheries. These double-membraned structures have been termed "thylakoids" (from the Greek "sacklike").(6) Each membrane of the thylakoid consists of a water-insoluble lipoprotein complex which contains the light-absorbing chlorophyll and other pigments utilized in photosynthesis.

By using light flashes of an intensity such that the rate of light absorption approached the rate at which the energy could be consumed by photochemical reactions and by spacing the flashes more than 0.04 sec apart to allow full recovery after each flash (a dark recovery time of about 0.02 sec was required in the system studied), Emerson and Arnold(7) were able to show that the product of one flash was never more than one molecule of oxygen for about 2500 molecules of chlorophyll in the organism. Assuming that eight light quanta are necessary to produce one oxygen atom, the data indicate that a set of 300–400 chlorophyll molecules is responsible for just one oxidation–reduction event. This implies the existence of photosynthetic units or reaction centers which serve as light-gathering antennas. This conclusion has been confirmed by Kok, using short light flashes from a high-pressure mercury arc source and monitoring O_2 production. The results of these experiments indicate that 2000–5000 molecules of chlorophyll are present per molecule of oxygen produced.[8]

In this concept of the photosynthesis reaction center, it is postulated that only those chlorophyll molecules situated in close proximity to substrates involved in the photosynthesis process are actually capable of initiating a photochemical event upon absorption of light energy. The function of other chlorophyll molecules as well as other pigments found in photosynthetic organisms is to absorb light and to transfer the energy to photochemically active sites. This is diagrammed in Figure 12.1, in which the photochemically active chlorophyll molecule is designated as *P*.

Studies of wavelength regions utilized by green plants, algae, and photosynthetic bacteria indicate that nearly all of the usable electromagnetic radiation is indeed absorbed by the various pigment systems of these organisms. Wavelengths longer than about 950 nm (infrared) are so strongly absorbed by atmospheric water vapor as to be ineffective, as are wavelengths less than 300 nm, which are sufficiently energetic to damage living systems. Green plants utilize light from 400 to 700 nm. Certain bacteria are capable of the successful utilization of wavelengths as long as 950 nm. In each organism the pigment responsible for the longest wavelength absorption is the

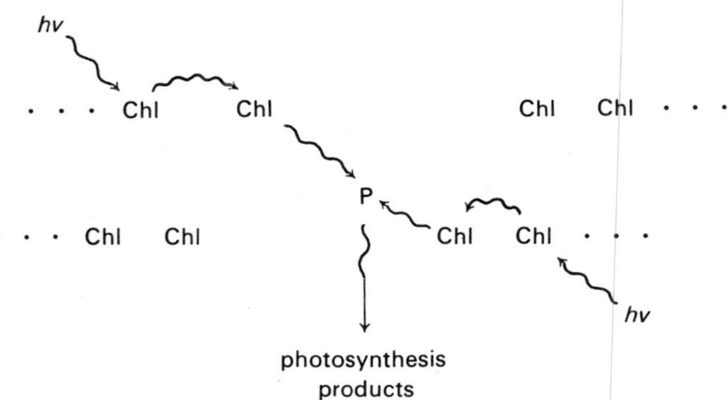

FIGURE 12.1. Schematic diagram of a photosynthesis reaction center. Light is absorbed by pigments in the light-gathering antenna and absorbed energy is transferred to a photochemically active site *P*, where it is utilized to initiate photosynthetic reactions.

photochemically active pigment and essentially acts as an energy sink, accepting energy from pigments absorbing at higher wavelengths and utilizing the transferred energy to initiate photochemical events leading to photosynthesis products. In green plants the photochemically active pigment is of course chlorophyll, whereas in bacteria it is a similar molecule called bacteriochlorophyll.

Thus far we have used the word chlorophyll as if this term related to a unique chemical species. In actuality there are a number of structurally related molecules present in photosynthetic organisms which are collectively referred to as chlorophyll. The general chlorophyll structure is:

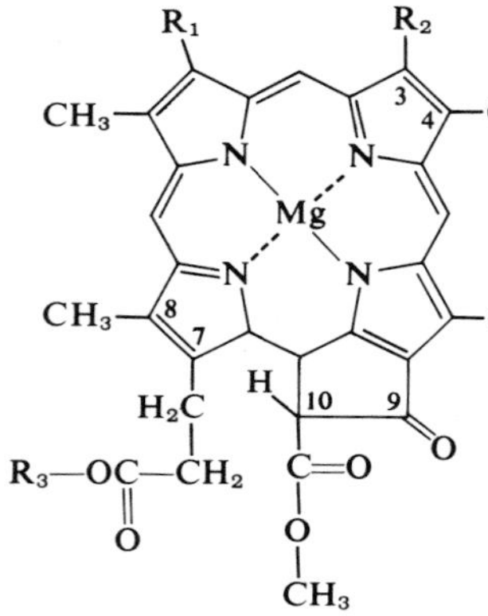

The various chlorophyll types are listed in Table 12.1. An alternate resonance form is available for all the chlorophylls listed in Table 12.1 except for bacteriochlorophyll. Chlorophyll *a* is found in all green plants. In most land plants and algae one also finds chlorophyll *b* or *d*. Thus a more correct statement than the one made previously would be that *chlorophyll a* is the photochemically active pigment in all green plants, as shown by experimentation.[9]

TABLE 12.1 *Various Chlorophyll Derivatives*

Chlorophyll	R_1	R_2	R_3	$\Delta^{3,4\,a}$	$\Delta^{7,8\,a}$
Chlorophyll *a*	$CH_2{=}CH-$	CH_3-	$C_{20}H_{39}$	u	s
Chlorophyll *b*	$CH_2{=}CH-$	$\overset{O \atop \|}{C}H-$	$C_{20}H_{39}$	u	s
Chlorophyll *d*	$\overset{O \atop \|}{C}H-$	CH_3-	$C_{20}H_{39}$	u	s
Bacteriochlorophyll	CH_3CO-	CH_3	$C_{20}H_{39}$	s	s

[a] u = unsaturated, s = saturated (dihydro derivative).

A second major group of compounds associated with photosynthetic organisms are the carotenoids. These compounds are practically all derivatives of the same linear skeleton composed of eight isoprenoid (C_5) units combined such that the two methyl groups at the molecular center are 1:6 to each other and other methyl groups are in 1:5 positions:

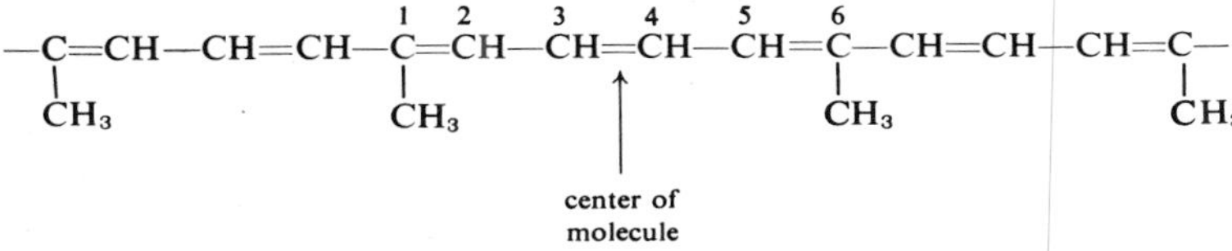

```
                              1  2     3     4     5    6
—C=CH—CH=CH—C=CH—CH=CH—CH=C—CH=CH—CH=C—
 |                            |                         |                            |
CH3                        CH3          ↑          CH3                        CH3
                                         |
                                    center of
                                    molecule
```

The carotenoids are subdivided into two main groups, the carotenes (hydrocarbon carotenoids) and the xanthophylls (oxygenated carotenoids). The structure of β-carotene is as follows:

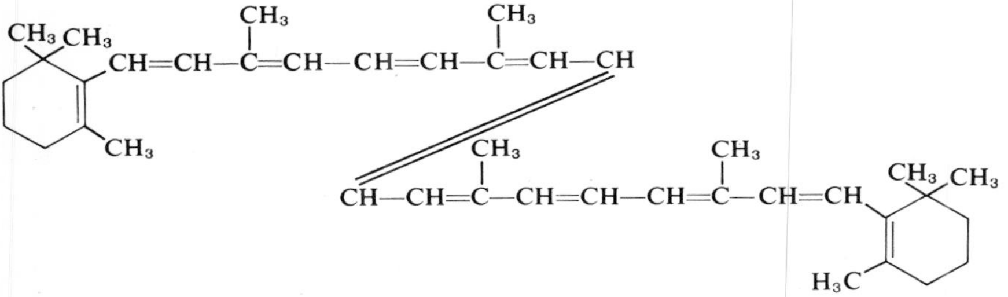

```
                        CH3              CH3
CH3 CH3                  |                |
     CH=CH—C=CH—CH=CH—C=CH—CH
  (ring)                                      //
     CH3                       CH3              CH3
                                |                |                CH3 CH3
                       CH—CH=C—CH=CH—CH=C—CH=CH
                                                          (ring)
                                                       H3C
```

In green plants β-carotene is the major carotenoid pigment.

A third group of supplementary pigments, which is confined to blue-green algae, red algae, and cryptomonads, is the phycobilins, which are large protein aggregates containing firmly bound bile pigments.[10]

It should be noted that even though a number of distinct chlorophylls have been identified in photosynthetic organisms and chlorophyll *a* has been found to be the photochemically active member of the family, even chlorophyll *a* appears to assume several different forms within an individual organism. This may perhaps result from shifts of the absorption spectra of molecules located in differing electrostatic environments (as, for example, a chlorophyll *a* molecule at an active site as opposed to a chlorophyll *a* molecule acting as a light harvester.) Indeed the absorption spectrum of chlorophyll *a* in ether is significantly shifted from the *in vivo* spectrum. Spectral information concerning the photosynthetic pigments present in various organisms is collected in Table 12.2. Sager has reported a ratio of chlorophyll to total carotenoid of 14.8 and a ratio of chlorophyll to β-carotene of 23.6 in normal strains of the species *Chlamydomonas*.[12]

Since the function of most of the chlorophyll and the accessory pigments is thought to be the harvesting of light, one would expect the action spectrum

TABLE 12.2. *In vivo Spectral Data for Various Photosynthetic Pigments*[11]

Pigment	λ_{max},[a] nm	λ,[b] nm
Chlorophyll *a* (extracted)	665	705
Chlorophyll *a*, form 673	673	—
Chlorophyll *a*, form 684	684	—
Chlorophyll *a*, form 695	695	—
Chlorophyll *a*, form 705	705	—
Chlorophyll *b*	650	685
Chlorophyll *c*	640	680
β-Carotene	482	520
Fucoxanthin	470	530
Phycoerythrin	566	600
Phycocyanin	615	650–670

[a] Position of the longest wavelength maximum.
[b] Approximate position of the end of long-wavelength absorption.

for photosynthesis to closely resemble the absorption spectrum of the tissue, allowing, of course, for small differences due to varying energy transfer efficiencies of the several components. It should be possible to correct for these differences at individual wavelength regions by increasing the intensity of light absorbed by components from which transfer may be inefficient. Thus one would expect to be able to obtain equal photosynthesis rates using red or green light by adjusting the intensities of the radiation absorbed at each wavelength. However, Blinks has reported that although the intensities of red and green light could be adjusted to yield equal photosynthesis rates (rate of oxygen evolution and carbon dioxide assimilation), the change from one wavelength to another results in a transitory disturbance in the rate. Furthermore, plots of transient rate versus time in proceeding from green to red light reveal different graphs for oxygen evolution and carbon dioxide assimilation.[13]

Other evidence indicating that the light-gathering process in photosynthesis is not as simple as originally believed has been reported by Emerson and Lewis. These workers observed that the quantum efficiency of photosynthesis declined at a steeper rate than the total absorption on the long-wavelength side of the red absorption band of chlorophyll *a*.[14] Later studies by Emerson indicated that this "red drop effect" could be counteracted by simultaneous irradiation with light of wavelength below 680 nm in addition to the far-red light. This enhancement phenomenon was observed to occur even if the light of the two wavelength regions was presented as alternating flashes separated by several seconds.[15] This observation has also been noted by several other workers.[16] In fact Myers and French[17] have shown that light at 650 and 700 nm taken together yields *more* photosynthesis than the sum of the rates at each individual wavelength (total light absorbed being the same in all cases). Again it was observed that the two wavelengths need not be applied simultaneously but could be given as alternate flashes. This indicates that the photochemical products of the reactions initiated at one wavelength can persist long enough to interact with the products produced by irradiation with the other wavelength.

Studies of photosynthesis in red and blue-green algae have revealed the phycobilins to be unusually effective in promoting photosynthesis. Action spectra for oxygen evolution in these organisms showed that light absorbed by the phycobilins was more efficiently used than light absorbed by chlorophyll *a* itself although the action spectrum for chlorophyll *a* fluorescence paralleled the one for photosynthesis, indicating that the light energy was indeed delivered to chlorophyll *a*. However, it was also established that the phycobilins are able to sensitize chlorophyll *a* fluorescence more efficiently than other chlorophyll *a* molecules. To resolve this dilemma, it was proposed that the algae must contain chlorophyll *a* in two states, one which is fluorescent (accessible to the phycobilins) and photochemically active and one which is nonfluorescent (inaccessible to the phycobilins) and photochemically inactive. Direct absorption of light can occur by both types of pigment but the excitation energy absorbed by the phycobilins is transferred directly to the active form of the chlorophyll. The inactive form was then identified as responsible for the red drop effect by assuming its absorption to occur along the far-red side of the chlorophyll *a* absorption band. Enhancement of the rate of photosynthesis by simultaneous illumination with far-red and shorter wavelength light could arise by activation of the inactive form of the chlorophyll through the synergistic action of the higher energy radiation.[18]

The findings discussed above have led workers to propose that photosynthesis involves the operation of two pigment systems, which are called S_I and S_{II}. Each system contains a large number of pigment molecules, most of which act as light gatherers and are photochemically inert. Each pigment system is responsible for one of two distinct photochemical reactions, neither of which is capable of sustaining photosynthesis alone, since the products of one reaction ultimately result in substrates for the other. Thus we have a two-quantum process whereby intermediate chemical (dark) steps link the two primary photochemical events. The Blinks effect and the Emerson effect can now be seen to arise from the relative degree of excitation of each of the pigment systems, S_I having its absorption primarily in the far-red and S_{II} absorbing primarily at wavelengths less than 680 nm. In green plants and green algae S_{II} pigments include chlorophyll *b* and a special form of chlorophyll *a* absorbing at 670 nm. In red and blue-green algae S_{II} pigments include the phycobilins. The S_I pigments consist of special forms of chlorophyll *a* absorbing to the red of those of S_{II}.

12.1b. *A Mechanistic Model for Photosynthesis*

To begin our discussion of a mechanistic model for photosynthesis, let us return to Eq. (12.1b) for the overall process of photosynthesis in green plants:

$$CO_2 + H_2O \xrightarrow[\text{chlorophyll}]{h\nu} (CH_2O) + O_2$$

It should be recalled that for purple sulfur bacteria the equation is [Eq. (12.2)]

$$CO_2 + 2H_2S \xrightarrow[\text{bacterio-chlorophyll}]{h\nu} (CH_2O) + 2S + 2H_2O$$

To make the two systems directly analogous, Van Niel[19] suggested that the reaction for photosynthesis in green plants should be written as

$$CO_2 + 2H_2O \xrightarrow[\text{chlorophyll}]{h\nu} (CH_2O) + H_2O + O_2 \qquad (12.3)$$

This reaction can now be seen as consisting of two parallel reactions, the oxidation of H_2O to produce molecular oxygen, and the reduction of CO_2 to produce a carbohydrate precursor. Since organisms capable of the reduction of CO_2 *in the dark* in the presence of O_2 are known, Van Niel concluded that the unique feature of photosynthesis is not the reduction of CO_2 nor the oxidation of substrate, but rather the photochemical reactions sensitized by chlorophyll which produce the oxidizing and reducing species capable of driving the organism's metabolic processes. In other words, the excited chlorophyll acts to produce an oxidizing species, represented below as $\oplus$, and a reducing species, represented as $\ominus$:

$$h\nu \rightsquigarrow \text{chlorophyll} \qquad (12.4a)$$

$$\text{chlorophyll}^* \longrightarrow \oplus \text{ and } \ominus \qquad (12.4b)$$

$$\oplus + CO_2 \longrightarrow \longrightarrow (CH_2O) \qquad (12.4c)$$

$$\ominus + H_2A \text{ (oxidizable substrate)} \longrightarrow \longrightarrow A \qquad (12.4d)$$

This of course is a grossly oversimplified picture, as we shall see later, but it serves as a starting point for our discussion. In actuality, four reducing equivalents and four oxidizing equivalents are required to convert one molecule of CO_2 to carbohydrate and to liberate one molecule of O_2 from water.

The oxidation–reduction nature of photosynthesis has been experimentally confirmed. Early confirmation was obtained from the experiments of Hill.[4] Hill was able to show that green leaf extracts were able to effect the evolution of oxygen under light when artificial electron acceptors other than CO_2 were present. Suitable electron acceptors (Hill reagents) include ferricyanide, benzoquinone, phenol indophenol, and methylene blue. Thus the assimilation of CO_2 was shown to be separable from the evolution of oxygen by replacement with a suitable electron acceptor. Gaffron has shown that by incubating green algae under hydrogen, the algae could be adapted to assimilate CO_2 in light using hydrogen as an oxidizable substrate resulting in the evolution of little or no oxygen. This shows that oxygen evolution can be separated from the photosynthesis pathway, as can the assimilation of carbon dioxide.[18] Other experiments which confirm the redox nature of photosynthesis will be discussed later in this section.

The fact that CO_2 assimilation can be separated from O_2 evolution (and vice versa) in photosynthetic organisms, coupled with the development of the dual pigment concept, suggests that perhaps one light reaction is responsible for the production of a strong reductant capable of initiating the sequence of reactions leading to CO_2 reduction, while the other light reaction produces an oxidant capable of initiating the reaction sequence leading to the production of molecular oxygen from water. This concept is shown schematically as follows:

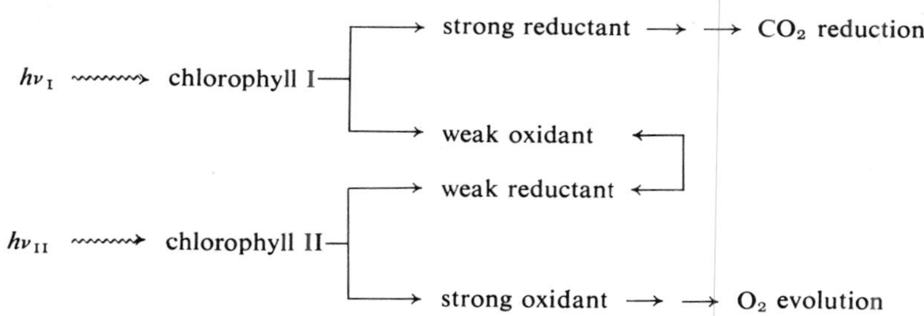

This scheme can account for the observed photosynthesis products if combination of the weak oxidant with the weak reductant is coupled to energy storage by ATP production (photophosphorylation). Since four reducing and four oxidizing equivalents are necessary for each process, eight quanta of light would be required for the evolution of one molecule of O_2, in good agreement with the generally accepted figure discussed earlier.

Recent experimental results indicate that photosynthesis probably does indeed result from a scheme similar to the one presented above, although a large number of the individual participants in the scheme remain unknown to date. Let us now briefly review these experimental findings.

What is believed to be the energy trap for the S_I pigment system has been reported by Kok and Hoch.[20] These workers observed that a minor pigment with maximum absorbtion near 700 nm is reversibly bleached by light. Chemical titration of the pigment showed it to be a one-electron redox agent with a potential of about 400 mV. Removal of the bulk chlorophyll *a* from the chloroplasts by extraction with a mixture of acetone and water resulted in a sevenfold increase in the ratio of this pigment to chlorophyll *a*. Spectra of the extracted chloroplasts during illumination indicated that a bleaching at 430 nm accompanies that at 705 nm, suggesting that the pigment is a specialized form of chlorophyll *a*. Assuming that the extinction coefficient of this pigment (termed P_{700}) and that of chlorophyll *a* are the same, one can calculate that there is one molecule of P_{700} for approximately 300 molecules of chlorophyll *a* in the unextracted chloroplasts. The photochemical oxidation of P_{700} appears to be coupled with the oxidation of cytochrome *f in vivo*, although the primary electron acceptor associated with these events was not determined.

Further investigations of P_{700} and cytochrome *f* (Cyt *f*) in chloroplasts have been carried out by Witt and co-workers.[21] It was found that both P_{700} and Cyt *f* are oxidized by illumination (reversibly at room temperature and irreversibly at 120°K). A brief flash of light of about 10^{-5} sec duration results in a bleaching at 705 nm that decays in about 10^{-3} sec. Appearance of absorption due to oxidized Cyt *f* does not occur immediately upon flashing but instead parallels the decay of the P_{700} reaction. Thus it appears that the excited P_{700} transfers an electron to an acceptor (χ) and an electron is subsequently returned to P_{700} from Cyt *f*:

$$h\nu \rightsquigarrow P_{700}\cdot\chi\cdot\text{Cyt } f \longrightarrow P^{*}_{700}\cdot\chi\cdot\text{Cyt } f \longrightarrow P^{\oplus}_{700}\cdot\chi^{\ominus}\cdot\text{Cyt } f \longrightarrow P_{700}\cdot\chi^{\ominus}\cdot\text{Cyt } f^{\oplus} \tag{12.5}$$

It has been found that adding the herbicide dichlorophenyldimethylurea (DCMU) to chloroplast preparations results in a blockage of electron flow from S_I and S_{II}, thus uncoupling the two systems.[22,23] Other methods of separating the two systems in order to study their separate functions include use of mutant strains which lack certain components involved in electron transport, the washing out of soluble components, and physical separation of subparticles obtained from the treatment of chloroplasts with detergents.[18] From studies of this type it has been found that S_I of green plants is responsible for the reduction of CO_2. Since $NADPH_2$ (reduced nicotinamide adenine dinucleotide phosphate) is the reductant involved in enzymatic reactions leading to CO_2 reduction, it is considered to be the terminal product of electron transfer in S_I. Studies have linked the reduction of NADP to a class of nonheme iron compounds called ferredoxin, which occurs in plants and bacteria and is believed to act as a one-electron redox agent.[(24,25)] That ferredoxin may be closely related to the primary electron acceptor χ has been shown by Rumberg and Witt.[(23)] These workers observed that when chloroplasts are illuminated with far-red light (S_{II} therefore not activated), P_{700} remains oxidized. Upon subsequent illumination with red light or upon the addition of NADP plus ferredoxin, P_{700} returns to its reduced state. An additional role of the ferredoxin in this experiment was thought to be to serve as an electron trap for the primary acceptor χ. Accordingly, χ is commonly referred to as ferredoxin-reducing substance (FRS).

Illumination of detergent-treated chloroplasts, in which only S_I is active, has been observed to result in the oxidation of plastocyanin, a blue copper protein containing two copper atoms per molecule present in a ratio of one plastocyanin to 30 chlorophyll molecules.[(26)] Removal of plastocyanin from the chloroplasts results in the loss or significant depression of electron transfer reactions, including loss of NADP photoreduction activity.[(27)] In addition, it has been observed that in algae plastocyanin is affected by both red and far-red light, being reduced by the former and oxidized by the latter.[(28)] This information suggests that plastocyanin may occupy a position linking the electron transport pathways of S_I and S_{II}. The fact that there is no net change in the oxidation state of S_I when isolated from S_{II} indicates that the electron flow pathway in this system is probably cyclic.

Recent investigations of the primary photoprocesses occurring in S_{II} have led to the discovery of what is thought to be the energy trap in the S_{II} pigment system with the observation by Döring and co-workers of absorbance changes at 435 and 682–690 nm due to chlorophyll *a* upon irradiation.[(29)] The observation of these changes in the presence of DCMU (thought to act directly at the reaction center by inhibiting electron transport) suggested that the function of the energy trap in S_{II} was merely to transfer energy to a primary electron donor rather than actual electron transfer by chlorophyll *a* as postulated for P_{700} in S_I.[(29,30)] Other workers[(31)] have suggested the inhibition by DCMU to occur further along the S_{II} electron transport pathway and consequently, in analogy with P_{700} and the postulated energy trap in photosynthetic bacteria (P_{870}),[(18)] have attributed electron transfer characteristics to the primary photocenter in S_{II} (designated as P_{680}).

The discovery by Knaff and Arnon[(32)] of a light-induced photooxidation at $-189°C$ requiring short-wavelength light has provided information as to a possible primary electron donor for S_{II}. The photooxidized substance has been identified as a form of cytochrome *b* absorbing at 559 nm (cytochrome b_{559}). Pretreatment of the spinach chloroplasts with ferricyanide to oxidize all cytochromes resulted in no absorption changes upon illumination at 557 nm at −189°C. In addition, irradiation by light absorbed exclusively by S_I yielded no evidence of the photooxidation as did irradiation at room temperature. However, irradiation of Tris-washed chloroplasts (Tris-washing blocks electron flow from water and hence oxygen evolution[33]) yielded absorbance changes indicative of oxidation of cytochrome b_{559}.

Information pertaining to the primary electron donor in S_{II} has also been recently reported.[31,34–37] Irradiation of chloroplasts at liquid nitrogen temperature is reported to result in the reduction of a component absorbing at 550 nm (C_{550}) and the oxidation of cytochrome b_{559}.[34] The light-induced changes (which involved an absorbance increase at 543 nm as well as bleaching at 546 nm) could also be achieved by the addition of strong reductants, such as dithionite.[35] Both the reduction of C_{550} and the oxidation of Cyt b_{559} are thought to arise from S_{II} due to their requirement of red light and their lack of observation in mutants not having S_{II} activity.[34] That C_{550} is the component of S_{II} responsible for receiving electrons from P_{680} is still not certain, although correlations among the absorbance changes of C_{550}, the oxidation of Cyt b_{559}, and fluorescence yield as a function of the redox potential of the chloroplast samples would indicate a very close relationship.[31] The studies of Butler[31] indicate C_{550} to be related to a carotenoprotein.

Figure 12.2a summarizes the foregoing discussion by placing the photo-

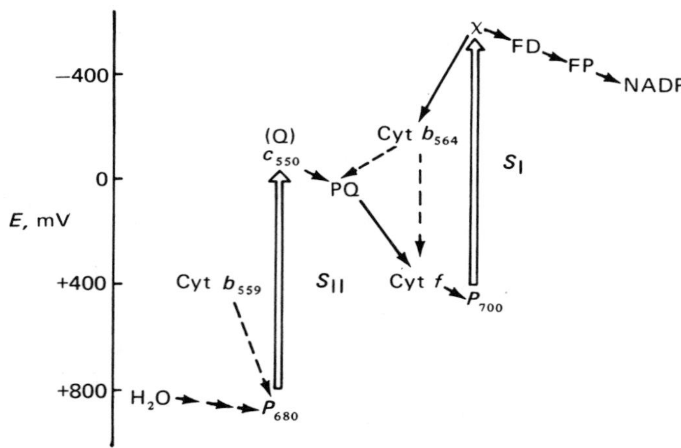

FIGURE 12.2a. Photosynthetic Z-scheme for green plants. Abbreviations not included in the text are PQ, plastiquinone; Cyt b_{564}, a form of cytochrome b absorbing at 564 nm; FD, ferredoxin; FP a flavoprotein. Long vertical arrows indicate steps arising from photoactivation of pigment reaction centers; dashed arrows indicate uncertain pathways.[18]

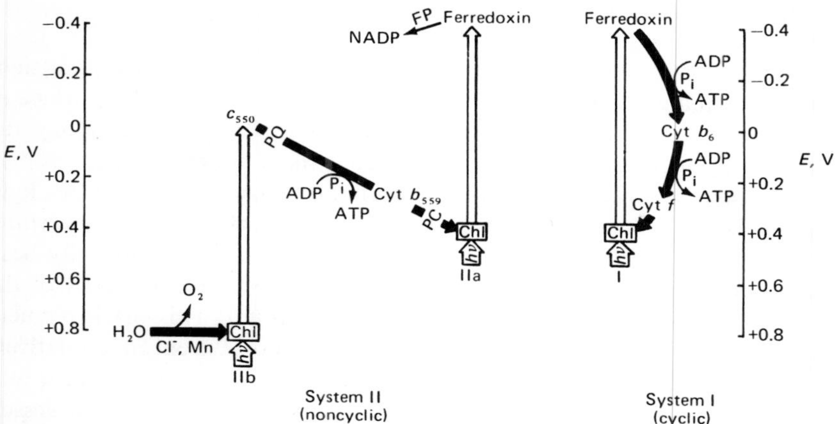

FIGURE 12.2b. Scheme for three light reactions in plant photosynthesis proposed by Knaff and Arnon.[39]

synthetic compounds implicated to date in a traditional Z-scheme.[18,31] The vertical direction in this scheme is used to designate relative redox potentials. Space limitations prohibit discussion of some of the members of this scheme; the interested reader is refered to Ref. 38. It should be noted that other photosynthesis schemes have been proposed.[38] One of these, proposed by Knaff and Arnon[39,40] (Figure 12.2b), develops a concept of three separate light reactions being involved in photosynthesis. The series formulation or Z-scheme of the type presented in Figure 12.2a appears to have received the greatest support to date, however.[11,18]

12.2. THE PHOTOCHEMISTRY OF VISION

The most versatile light detector known is the human eye and its accompanying nervous system. Capable of detecting and measuring light intensities varying over 13 orders of magnitude, the dark-adapted eye is capable of being stimulated by a single photon.[11b,41] For seeing, the minimum number of photons that must be absorbed within a short time period is only about ten.[42] Absolute light intensities are not measured by the eye because of its large range of light adaptation; however, at normal light intensities the eye can differentiate between sources which differ in intensity by as little as 2%.[11b] The architecture and photobiology of the human eye will be the subjects of this section.

12.2a. *Anatomy of the Human Eye*

Similar in some respects to a camera, the eye consists of a light-tight cavity containing a light-sensitive detector located along the inner surface of the posterior chamber. The visual field is mapped as an image along the surface of this detector (the retina), which consists of a layer of light-sensitive cells. As in a camera, a lens is used to form an image. The amount of light entering the eye is controlled by the iris, a circular muscle of variable aperture. In strong light contraction of the iris ensures that only the optically best, center portion of the lens is used. The light-sensitive receptor cells of the retina are set in a bed of black-pigmented epithelium. This black pigment serves to reduce the effects of scattered light in the eye. Except for a relatively small region in the center part of the retina, called the fovea, the receptor cells are covered by a layer of nerve cells, optic nerve fibers, and blood vessels through which light must pass to be detected by the receptors. Being relatively free of obstructing tissues, the fovea provides the clearest vision. The optic nerve fibers pass from the eye in a bundle containing no receptor cells. Hence this area acts as a "blind spot" where light can cause no visual stimulation. A diagram of the human eye is shown in Figure 12.3.[18]

The vertebrate retina contains two classes of light-sensitive receptor cells called rods and cones. The rod is an elongated cylindrical cell containing several hundred thylakoids which support the visual pigment. The pigment system in the rod is confined to internal membranes situated close to the outer membrane of the cell. In the other type of visual receptor, the cone, the pigment is situated in the external membrane itself. In the cone the external

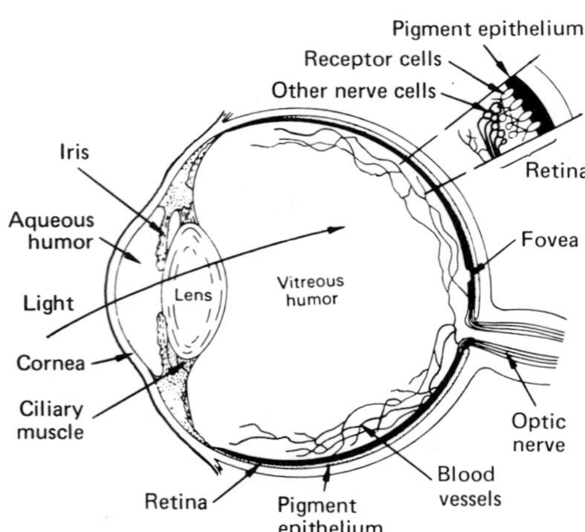

FIGURE 12.3. A diagrammatic view of the eye. (Reproduced with permission from Ref. 18.)

membrane containing the pigment system is folded to resemble the stack of thylakoids in a rod. The human fovea (about 0.3 mm in diameter) contains about 10,000 cones and no rods.[(18)] The pigment epithelium as a whole contains approximately 125 million rods and 6.5 million cones.[(11b)] Each cone in the fovea appears to be connected to a separate nerve fiber, accounting for the high visual acuity of the foveal region. On the other hand, as many as 10,000 rods converge to a single nerve fiber, resulting in reduced visual acuity of the rods relative to the cones. However, since the rods pool light-activated signals to fewer points than the cones, the rod system imparts greater light sensitivity than the latter and is superior for vision in very weak light. The cones appear to be specialized for color vision as well as for high visual acuity.[(11b)]

12.2b. *The Visual Pigments and the Chemistry of Vision*

The visual pigment present in rods has been termed rhodopsin and consists of 11-*cis*-retinal, a derivative of vitamin A_1, and a lipoprotein called opsin. Recent evidence[(43)] suggests that in native rhodopsin the retinal chromophore is covalently bonded to a phosphatidylethanolamine residue of the lipid portion of opsin. The structure of 11-*cis*-retinal is as follows:

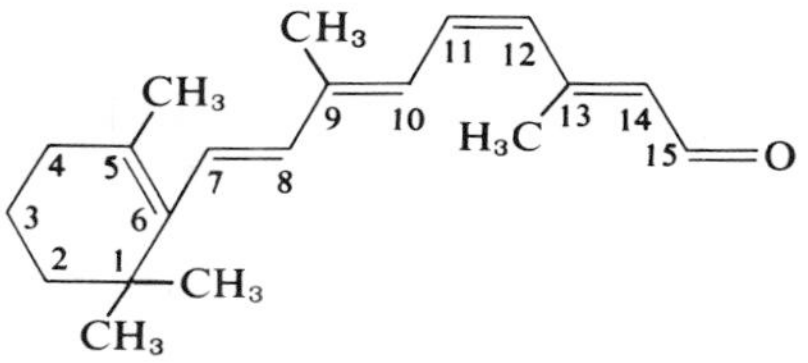

Rhodopsin extracted from rod segments is a red-purple color absorbing maximally about 500 nm. It was first observed by Boll in 1877 as a reddish-purple pigmentation of excised frog retinas which bleached on exposure to white light.[(44)] That rhodopsin is indeed the pigment responsible for vision is evidenced by the correspondence of the absorption spectra of rhodopsin with the action spectrum for sight in dim light.[(18)]

The bleaching of rhodopsin has been found to lead to the all-*trans* form of retinal through several intermediate steps.[(11b,45)] These steps are temperature dependent; consequently low temperatures must be used to observe the intermediate products. A solution of rhodopsin is subjected to a flash of light at liquid nitrogen temperature and intermediates are detected by changes in the absorption spectrum. The first intermediate, with an absorption maximum at 543 nm, believed to be an all-*trans*-retinal bound to opsin, has been termed prelumirhodopsin. Warming the sample to a temperature greater than

$$\text{rhodopsin} \xrightarrow[-196^\circ C]{h\nu} \text{prelumirhodopsin} \xrightarrow[> -140^\circ C]{} \text{lumirhodopsin}$$

$$\text{lumirhodopsin} \xrightarrow{> -40^\circ C} \text{metarhodopsin I}$$

$$\textit{trans}\text{-retinal} \xleftarrow[> 0^\circ C]{} \text{metarhodopsin II} \underset{> -15^\circ C}{\overset{H^+}{\rightleftharpoons}} \text{metarhodopsin I}$$

FIGURE 12.4. Intermediates in the conversion of rhodopsin to *trans*-retinal.

−140°C results in the appearance of an absorption centered at 497 nm due to an intermediate called lumirhodopsin. Above −40°C this changes to metarhodopsin I (λ_{max} = 487 nm) and above −15°C metarhodopsin I establishes an equilibrium with another form, absorbing at 380 nm, called metarhodopsin II. Finally, above 0°C the absorbance due to all-*trans*-retinal appears (λ_{max} = 387 nm). These results are summarized in Figure 12.4 and Table 12.3.

Each step of the process of conversion of rhodopsin to all-*trans*-retinal and opsin is thought to be photochemically reversible, regenerating the 11-*cis* form of retinal and thus yielding rhodopsin again directly. Since the intermediates involved in this conversion absorb maximally in different regions of the spectrum, irradiation at a given wavelength region should result in a photostationary state in which the relative populations of each of the intermediates depend upon their individual absorbances in the wavelength range.[11b,18]

Bleaching is reversed in the dark and the red-purple color of rhodopsin returns. This is thought to occur by the reduction of all-*trans*-retinal to vitamin A_1 (retinal), which diffuses from the rod into the pigment epithelium, where it is converted enzymatically to the 11-*cis* isomer of vitamin A_1. The enzymatic isomerization is followed by diffusion back into the rod, oxidation to 11-*cis*-retinal, and combination with opsin to form rhodopsin. This process is shown schematically in Figure 12.5.[11b]

TABLE 12.3. *Spectral Data for Rhodopsin and Its Bleaching Intermediates*[18]

Species	λ_{max}, nm
Rhodopsin	498
Prelumirhodopsin	543
Lumirhodopsin	497
Metarhodopsin I	487
Metarhodopsin II	380
trans-Retinal	387

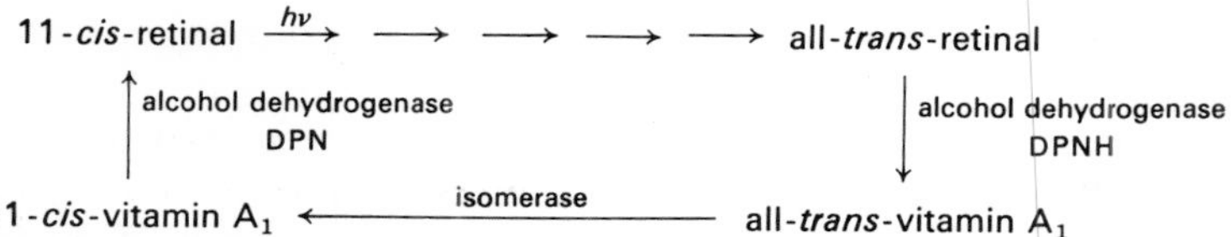

FIGURE 12.5. Cycle of reactions involved in the bleaching of rhodopsin by light and subsequent dark restoration.

Theories of the origin of color vision assume the presence of different visual pigments present in the cones which respond to various regions of the visible spectrum. These lead to different patterns of activity among the receptor cones which are interpreted by the brain in terms of color. The work of Stiles in the analysis of color discrimination has led to the conclusion that color vision is based on three kinds of receptor: two major types with absorption maxima at 540 and 580 nm and a minor type with a maximum at 440 nm.[53] Rushton has studied color vision in subjects with defective color vision by measuring absorbance changes in cones caused by bleaching the pigments with strong light flashes.[52] One type of color-blind subject was found to have cone pigments absorbing at 540 nm. In another type of subject this pigment was absent and a pigment with maximum about 570 or 580 nm was present. Both pigments are present in subjects with normal color vision. These pigments have been called chlorolabe ($\lambda_{max} = 540$ nm) and erythrolabe ($\lambda_{max} = 580$ nm). Direct evidence in humans for the third type of pigment ("cyanolabe") suggested by Stiles has not yet been obtained, although Marks *et al.* have shown evidence of a blue-absorbing pigment in primates.[46] Studies of color vision in the carp by Kaneko have identified three kinds of cone in this species, one responding to blue light ($\lambda_{max} = 450$ nm), one to green ($\lambda_{max} = 520$ nm), and one to yellow light ($\lambda_{max} = 620$ nm). The actual pigments responsible for the three types of color identification have not been isolated and characterized. It is interesting to note that for color vision in the chicken retina there is only one type of pigment, a combination of retinal and cone opsin (iodopsin). Color differentiation appears to arise from filtration of the incident light by three species of cones containing yellow, green, and red droplets of oil.[18]

For a more detailed discussion of vision, including the transduction and processing of visual information, the reader is refered to Ref. 18, Chapter 3 and references therein.

12.3. PHOTOTAXIS AND PHOTOTROPISM

Phototropism is defined as the directed growth or bending of fixed organisms in response to light. Phototaxis is light-induced locomotion of organisms or cellular components. We will begin our discussion with phototropism.

Most experimental work on phototropsin has been carried out on oat seedlings and the mold *Phycomyces*. In *Phycomyces* it is the sporangiophore, the stalk supporting the sphore case, which is the object of study. The sporangiophore grows to a height of several centimeters at a rate of about 3 mm/hr. When illuminated from the side the sporangiophore bends toward the light during growth. Bending arises from a faster growth along the back side as opposed to the front side. Since the sporangiophore is nearly transparent and corresponds optically to a spherical lens, the differential growth observed cannot arise from the front side receiving more light since the intensity of the light is actually higher along the back. Thus it appears that phototropism results from faster growth occurring in regions receiving higher light intensities. This conclusion is supported by the following results. Sporangiophores immersed in a medium having a high index of refraction (such that parallel light rays diverge instead of converge) bend away from the light (negative phototropism). The phototropic response to UV light is also negative. This is explained by the presence of gallic acid, which absorbs UV light strongly about 280 nm. Uniform illumination results in straight growth.[(18)]

The situation is more complex in the oat coleoptile (the growing tip of the seedling). In this case if one side is illuminated while the other is shaded from the light, the seedling grows *toward* the light (opposite to what is observed in *Phycomyces* sporangiophores if the back side is shaded). Thus growth in the coleoptile appears to be faster on the shaded side. If, however, diffusion is prohibited from one side of the seedling to the other by insertion of a glass plate, there is no phototropic response. It appears therefore that light produces a substance which must diffuse away from the light source and effects a stimulation of growth at the opposite side.

These results have stimulated the search for plant growth hormones. One growth hormone which has received much attention in this regard is auxin, indoleacetic acid. Went[(47)] has proposed that phototropic curvature toward light is due to an excess of auxin on the dark side and a deficiency due to destruction of auxin by light on the light side. Briggs *et al.* have shown that although the concentration of auxin upon illumination increases on the dark side, the total amount of auxin remains constant.[(49)] Thus it appears that if an imbalance in auxin concentration is responsible for differential growth in the oat seedlings upon irradiation, the observed phototropic response must result from a phototactic response involving migration of auxin from the illuminated side of the plant to the shaded side. Let us now turn our attention temporarily to a discussion of phototaxis.

Phototaxis in photosynthetic bacteria is well illustrated by the behavior of *Rhodospirillum*. This organism can swim in either direction along its axis by reversing the orientation and sense of direction of its flagella, fibrous protein structures on either end of the organism which can contract and cause motion. If the bacterium enters an unilluminated area it will stop swimming, pause for a short period, and resume swimming in the opposite direction. The *Rhodospirillum* can be fooled by bringing a dark boundary up from behind. In this case the reversal of motion will bring the organism deeper into darkness. Thus it appears that the cell responds simply to a decrease in light intensity reaching some photoreceptive area of the organism without regard to the direction from which the light has come. This type of phototactic response is called *phobic*. Other organisms exhibit what is called a *topic* response, in which the direction of movement becomes oriented toward the light source. Flagellated green algae, such as *Euglena*, exhibit topic phototaxis through a series of phobic responses. This behavior is thought to result from two structures, a light-sensitive photoreceptor and a bright orange spot (the stigma) located nearby. Shading of the photoreceptor region (at the base of the flagellum) causes a phobic response. The cell rotates while swimming, causing the photoreceptor to be shaded by the stigma on every revolution. If the organism is swimming directly toward the light source, the photoreceptor will not be shaded. Thus the intermittent shading of the photoreceptor is thought to cause a series of movements which result in alignment of the direction of swimming toward the source.[18,49,50]

Phototaxis in *Rhodospirillum* has been found to be related to photosynthesis in this organism due to similarities in the action spectra for these two processes. Factors which interfere with respiration and photosynthesis also appear to interfere with the phototactic response, supporting the idea that a sudden decrease in energy supply to the flagellar apparatus results in phototaxis.[50] Action spectra for phototaxis in *Euglena* indicate that blue light is effective while red light is not.[51] Thus in this organism the phototactic mechanism does not appear to be directly related to photosynthesis, although studies show that the phototactic mechanism results in placing the organism in light containing wavelengths (800–900 nm) utilized by bacteriochlorophyll for photosynthesis. Search for the identity of the photoreceptor pigments in *Euglena* and other phototactic organisms as well as those responsible for phototropic responses are complicated for several reasons: Concentration of the pigments is so small as to make direct absorption measurements impractical, action spectra are distorted by screening pigments and do not reflect the absorption spectra of the photoreceptor pigments, and phototactic behavior in certain green algae appears to be also governed by the concentrations of alkali and alkaline earths ions in the medium.[50] Some details of action spectra for phototaxis and phototropism in various organisms are shown in Table 12.4.[18,51]

TABLE 12.4. *Action Spectra Maxima for Phototaxis or Phototropism of Various Organisms*[18,51]

Response	λ_{max}, nm		
	Principal	Secondary	Other
Phototaxis			
Platymonas	493	435	340, 270
Ulva	485	435	—
Gonyaulax	475	—	—
Prorocentrum	570	—	—
Euglena[a]	495	425	—
Euglena[b]	410	—	—
Phototropism			
Avena	475, 445	425	370
Phycomyces	485, 450	420	380

[a] Normal.
[b] Mutant lacking stigma or chloroplasts.

Although fragmentary evidence points to a linkage to metabolic processes such as photosynthesis and respiration, the mechanisms of phototaxis and phototropism remain largely unknown and must await the results of further research in this very interesting area.

12.4. DAMAGE AND SUBSEQUENT REPAIR BY LIGHT (PHOTOREACTIVATION)

Visible and ultraviolet radiations are known to cause a wide range of damage to living systems, as exemplified by the inactivation of enzymes and viruses, the killing of bacteria, genetic mutations and chromosome defects, and carcinogenesis. Most research in this area has concentrated on the deleterious effects due to UV absorption, especially in the region below 300 nm. Of all cellular components the most vulnerable to the effects of UV radiation in this range are thought to be the polynucleotides. The vulnerability of these components arises from two different areas. First, the polynucleotides are known to contain genetic information for synthesis of all cellular components, including the nucleic acids themselves, without which the cell will be unable to survive. Second, and more applicable to our discussion, is the fact that absorption studies of cellular components indicate the nucleic acids to be the primary absorbers of radiation in this wavelength range. Indeed, action spectra for mutation production and for killing of cells follow the polynucleotide absorption spectrum very closely. Our discussion in this section will center first on the photochemistry of the nucleic acids and second on methods by which the resulting damage is repaired.

12.4a. *The Photochemistry of the Nucleic Acids*

The photochemistry of the polynucleotides has been elucidated primarily by studies of the photochemical behavior of the individual pyrimidine and purine bases (the ribose and phosphate groups would not be expected to undergo photochemical reactions in this wavelength range). These studies have shown the pyrimidines (cytosine and thymine) to be roughly ten times more sensitive to UV than the purines (adenine and guanine.) Thus we would expect most of the photochemistry of the nucleic acids to result from the action of light on the pyrimidines.

In 1949 Sinsheimer and Hastings[54] reported the almost complete loss of the 260-nm absorption of the pyrimidines upon irradiation in aqueous solution. This effect is now known to be caused by the photochemical addition of water across the 5,6 double bond, as shown below for cytosine:

$$\xrightarrow[H_2O]{h\nu} \qquad (12.6)$$

This reaction has also been shown to occur in cytidine, cytidylic acid, uracil, uridine, and uridylic acid (found in RNA) but reportedly not in thymine, thymidine, or thymidylic acid.[55] The photohydration has been found to be partially reversible, dehydration being nearly complete at extremes of temperature and *p*H.

Very important information concerning the photochemistry of the nucleic acids was furnished by the report of Beukers and Berends that irradiation of frozen solutions of thymine produces a stable photoproduct corresponding to dimerization of the thymine.[56] This photoproduct has been subsequently identified as a *cis-syn-cis* dimer:

$$\xrightarrow[\lambda\,=\,254\text{ nm}]{h\nu} \qquad (12.7)$$

Formation of the photodimer is reversible, the dimer being split with high quantum yield with light of wavelength about 235 nm.[62] Similar dimerizations have been observed with several other pyrimidine derivatives, as

TABLE 12.5. *Photochemistry of Some Polynucleotides and Their Pyridine Components*[55,57–61,64]

Substrate	Percent hydration[a]	Percent dimerization[b]
Uracil	50	70
Uridine	95	20
Uridylic acid	95	40
Cytosine	60–90	0
Cytidine	100	0
Cytidylic acid	100	0
Thymine	0	85
Thymidine	0	50
Thymidylic acid	0	40
Poly-*r*U	65	15
Poly-*d*T	0	35
DNA (double stranded)	0	6.5

[a] In aqueous solution.
[b] In frozen aqueous solution.

indicated in Table 12.5. Studies of model polynucleotides as well as DNA indicate that the production of photodimers constitutes the major light reaction in the nucleic acids. In DNA primarily thymine dimers are produced while in RNA both uracil and thymine dimers may result (in addition to photohydration).[63]

12.4b. *Photoreactivation*

Photoreactivation was first discovered in 1948 by Kelner,[65] who reported significantly higher survival rates for microbial colonies which had been exposed to UV radiation and then exposed to diffuse daylight during storage as opposed to those colonies that had been irradiated and kept in the dark. This phenomenon was also noted by Dulbecco in studies of the bacterial virus *T*2 infecting *E. coli B* cells.[66] Further experiments indicated that post-irradiation treatment with near-UV light was responsible for the remarkable recovery of the irradiated cells. This is shown in Figure 12.6, in which the fraction of colony formers is plotted against initial UV dose for cells kept dark after this dose and before plating (curve I) and cells subjected to this dose and then to illumination from a 500-W tungsten lamp (filtered through $CuCl_2$ solution) for 60 min before plating (curve II). The data indicate that approximately 60% of the effect of the far-UV dose can be erased by the post-irradiation illumination.[67]

The wavelengths of light responsible for producing the recovery of colony-forming ability lie in the blue-violet and near-UV regions of the spectrum. Action spectra for photoreactivation appear to be specific to the species under study, the action spectrum for photoreactivation of *E. coli B/r* differing considerably from that of *S. griseus*, as indicated in Figure 12.7.

The development of the method of bacterial transformation (the exposing of transformable cells to dilute DNA from a mutant strain possessing some distinguishable character, such as streptomycin resistance, and the resulting appearance of the mutant trait in the cellular population)(70) has proved to be a valuable asset to the study of photoreactivation. When *extracts* of photoreactivable cells are mixed with far-UV-irradiated DNA *in vitro*, illumination of the mixture restores the transforming activity of the irradiated DNA, as shown in Figure 12.8.(72,73) An increase in transforming activity occurs only when the DNA has first been inactivated by UV radiation. As can be seen in Figure 12.8, recovery reaches a final level considerably less than 100%. DNA which has been maximally reactivated by one system, *E. coli* extracts, for example, is not further reactivated by extracts from another photoreactivable organism, indicating that the damage repaired by these systems is probably identical.

Studies on the active agent of the yeast extract have shown it to be nondialyzable, heat labile, and inactivated by trypsin and chymotrypsin as well as silver ions and *p*-hydroxymercuribenzoate.(71) These and other properties suggest that the active agent is an enzyme. Sedimentation studies have shown that the yeast photoreactivating enzyme combines with UV-irradiated DNA, in which condition it is more resistant to heat inactivation and inactivation due to silver ions and *p*-hydroxymercuribenzoate.(75) The

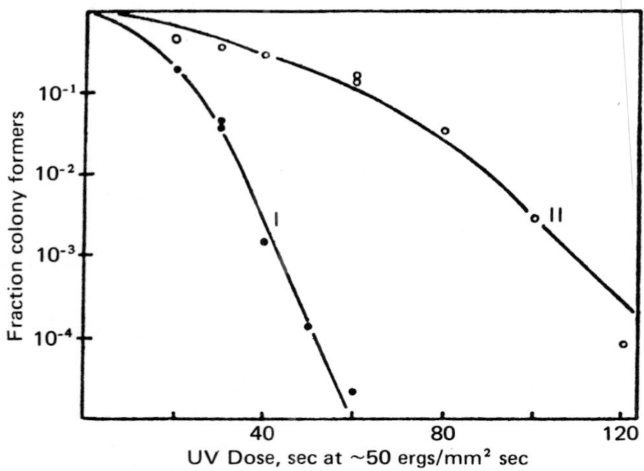

FIGURE 12.6. Photoreactivation of recovery from the effects of far-UV irradiation in *E. coli B/r* (Reproduced with permission from Ref. 67.)

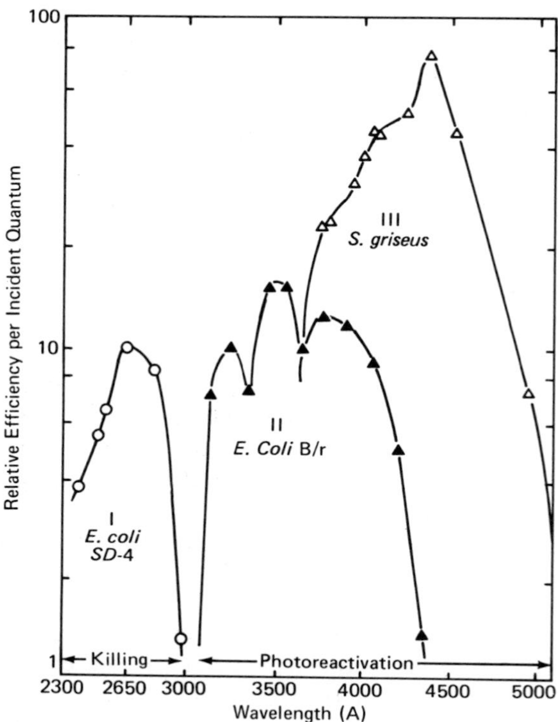

FIGURE 12.7. Action spectra for killing (*E. coli* SD-4; curve I) and for photoreactivation (*E. coli B/r*, curve II; *S. griseus*, curve III). (Reproduced with permission from Ref. 64.)

enzyme–DNA complex is dissociated by photoreactivating illumination. Illumination of mixtures of irradiated DNA and partially purified extracts of yeast photoreactivating enzyme followed by hydrolysis of the DNA have revealed that thymine dimers are split.[63,76] No decrease is produced by dark incubation with photoreactivating enzyme or by illumination with heated enzyme. Further work with a variety of model polynucleotides has shown that uracil and cytosine dimers are also split by the photoreactivating enzyme from yeast in the presence of light. Thus it appears that recovery mitigated by photoreactivating enzyme results from splitting of pyrimidine dimers. There is no evidence for repair of other UV-induced damage, such as pyrimidine hydration or strand-breaks, by the photoreactivating enzyme, although dark repair systems capable of working on these types of damage in DNA by excision of the damaged region and subsequent resynthesis are known for certain organisms.[76] These enzymes are also capable of repairing photoreactivable damage (thymine dimers). Of course in this case repair is also by excision and resynthesis rather than by direct splitting as in photoreactivation.

The chromophore responsible for absorption of the photoreactivating illumination has been subject to much speculation and study. Some workers believe the chromophore to result from the complex formed between the photoreacting enzyme and the irradiated DNA. In this view differences in photoreactivation action spectra for different organisms arise from shifts in the absorption maxima due to electrostatic effects. Others believe the enzyme to contain a chromophoric substance bound to the enzyme.[71] Recent evidence tends to favor the latter theory. Minato and Werbin[77] have reported absorption (λ_{max} = 380 nm) and fluorescence spectra (λ_{max} = 485–490 nm, excitation λ_{max} = 385 nm) of the chromophore in a photoreactivating enzyme which had been purified 70,000-fold from Baker's yeast. The molecular weight of the enzyme, which was shown to be active in photoreactivation, was determined to be 53,000 by gel filtration. Partial dissociation of the enzyme into a chromophore fraction and a protein fraction by chromatography on Sephadex G-100 with salt-free buffer was reported. A further report by these workers described the excitation and fluorescence spectra of the chromophore associated with the photoreactivating enzyme from the blue-green algae *Anacystis Nidulans*.[78] The purified photolyase from this species exhibited an absorption maximum at 418 nm, in good agreement with the fluorescence excitation maximum (420 nm) and the maximum in the action

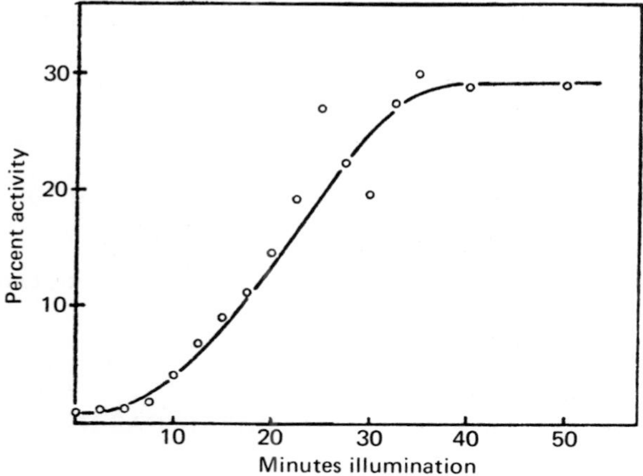

FIGURE 12.8. Transformation activity (streptomycin resistance) of a DNA–yeast extract mixture as a function of illumination time. (Reproduced with permission from Ref. 74.)

spectrum for photoreactivation (438 nm) in this species. The identities of the chromophores in these two enzymes have not been established.

Repair of UV damage to such higher life forms as mammals and humans does not appear to involve photoreactivation but rather enzymes of the dark-repair type. A rare skin disease, zeroderma pigmentosum, results from the absence of dark-repair enzymes in human epidermis. Such a lack of repair leads to a high incidence of malignancy due to the effects of the sun. Damage to all life forms by the sun would be far worse were it not for the blanket of ozone about the earth, which acts as a filter for the most harmful UV rays. The presence of photoreactivating enzyme systems in bacteria, algae, and other lower life forms probably arose by evolution prior to the development of our atmosphere as it now exists.

REFERENCES

1. E. I. Rabinowitch, *Photosynthesis and Related Processes*, Vol. II, Part I, Interscience, New York, (1951).
2. R. Emerson and R. Chalmers, *Plant Physiol.* **30**, 504 (1955); R. Emerson, *Ann. Rev. Plant. Physiol.* **9**, 1 (1958); E. L. Yuan, R. W. Evans, and F. Daniels, *Biochim. Biophys. Acta* **17**, 185 (1955).
3. C. B. Van Niel, in *Advances in Enzymology*, Vol. I, F. F. Ford and C. H. Werkman, eds., Interscience, New York (1941).
4. R. Hill, *Proc. Roy. Soc.* (*London*) **B 127**, 192 (1939); D. I. Arnon, F. R. Whatley, and M. B. Allen, *J. Amer. Chem. Soc.* **76**, 6324 (1954).
5. S. Granick, in *The Cell*, Vol. II, J. Brachet and A. E. Mirsky, eds., Academic Press, New York (1961).
6. W. Menke, *Ann. Rev. Plant Physiol.* **13**, 27 (1962).
7. R. Emerson and W. Arnold, *J. Gen. Physiol.* **15**, 391 (1932); **16**, 191 (1932).
8. B. Kok, *Biochim. Biophys. Acta* **21**, 245 (1956).
9. C. S. French, in *Handbuch der Pflanzenphysiologie*, Vol. 5, W. Ruhland *et al.*, eds., Springer-Verlag, Berlin (1960).
10. M. D. Kamen, *Primary Processes in Photosynthesis*, Academic Press, New York (1963).
11. (a) A. T. Jagendorf, *Surv. Biol. Progr.* **4**, 181 (1962); (b) H. H. Seliger and W. D. McElroy, *Light: Physical and Biological Action*, Academic Press, New York (1965).
12. R. Sager, *Brookhaven Symp. Biol.* **11** (BNL C 28), 101 (1959).
13. L. R. Blinks, in *Research in Photosynthesis*, H. Gaffron *et al.*, eds, Interscience, New York (1957).
14. R. Emerson and C. M. Lewis, *Am. J. Botany* **30**, 165 (1943).
15. R. Emerson, *Science* **125**, 746 (1957); R. Emerson, R. Chalmers, and C. Cederstrand, *Proc. Natl. Acad. Sci. U.S.* **43**, 133 (1957).
16. Govindjee, *Photosynthetic Mechanisms of Green Plants*, NAS–NRC Publication 1145, Washington, D.C. (1963).
17. J. Myers and C. S. French, *J. Gen. Physiol.* **43**, 723 (1960).
18. R. K. Clayton, *Light and Living Matter*, Vol. 2: *The Biological Part*, McGraw-Hill, New York (1971).

19. C. B. Van Niel, *Arch. Microbiol.* **3**, 1 (1931); *Photosynthesis in Plants*, J. Franck and W. E. Loomis, eds., Iowa State Press, Ames, Iowa (1949).
20. B. Kok and G. Hoch, in *Light and Life*, W. D. McElroy and B. Glass, eds., The Johns Hopkins Press, Baltimore, Maryland (1961).
21. A. Müller, B. Rumberg, and H. T. Witt, *Proc. Roy. Soc.* (*London*) *B* **157**, 313 (1963); B. Rumberg, P. Schmidt-Mende, J. Weikard, and H. T. Witt, *Photosynthetic Mechanisms of Green Plants*, NAS–NRC Publication 1145, Washington, D.C. (1963).
22. B. Ke, *Biochim. Biophys. Acta* **88**, 289 (1964).
23. B. Rumberg and H. T. Witt, *Z. Naturforsch.* **19b**, 693 (1964).
24. K. T. Fry, R. A. Lazzarini, and A. San Pietro, *Proc. Natl. Acad. Sci. U.S.* **50**, 652 (1963).
25. F. R. Whatley, K. Tagawa, and D. I. Arnon, *Proc. Natl. Acad. Sci. U.S.* **49**, 266 (1963).
26. S. Katoh and A. Takamiya, *Plant Cell Physiol.* (*Tokyo*) **4**, 335 (1963).
27. S. Katoh and A. Takamiya, *Biochim. Biophys. Acta* **99**, 156 (1965).
28. Y. de Kouchbovsky and D. C. Fork, *Proc. Natl. Acad. Sci. U.S.* **52** 232 (1964).
29. G. Döring, H. H. Stiehl, and H. T. Witt, *Z. Naturforsch.* **22b**, 639 (1967).
30. Govinjee, G. Döring, and R. Govindjee, *Biochim. Biophys. Acta* **205**, 303 (1970).
31. W. L. Butler, *Biophys. J.* **12**, 851 (1972); *Acc. Chem. Res.* **6**, 177 (1973).
32. D. B. Knaff and D. I. Arnon, *Proc. Natl. Acad. Sci. U.S.* **63**, 956 (1969).
33. T. Yamashita and T. Horio, *Plant Cell Physiol.* **9**, 268 (1968); T. Yamashita and W. L. Butler, *Plant Physiol.* **43**, 1978 (1968).
34. D. B. Knaff and D. I. Arnon, *Proc. Nat. Acad. Sci. U.S.* **63**, 963 (1969).
35. K. Erixon and W. L. Butler, *Photochem. Photobiol.* **14**, 427 (1971).
36. N. K. Boardman, J. M. Anderson, and R. G. Hiller, *Biochim. Biophys. Acta* **234**, 126 (1971).
37. D. S. Bindall and D. Sofrova, *Biochim. Biophys. Acta* **234**, 371 (1971).
38. G. Forti, M. Avron, and A. Melandri, eds., *Proceedings of the 2nd International Congress on Photosynthesis Research*, Dr. W. Junk N. V. Publishers, The Hague, Netherlands (1972).
39. D. B. Knaff and D. I. Arnon, *Proc. Natl. Acad. Sci. U.S.* **64**, 715 (1969).
40. D. I. Arnon, D. B. Knaff, B. D. McSwain, R. K. Chain, and H. Y. Tsujimoto, *Photochem. Photobiol.* **14**, 397 (1971).
41. S. Hecht, S. Shlaer, and M. H. Pirenne, *J. Gen. Physiol.* **25**, 819 (1942).
42. M. H. Pirenne, The Eye, in *The Visual Process*, Vol. 2, H. Davison, ed., Academic Press, New York, 1962.
43. R. P. Poincelot, P. Millar, P. G. Kimbel, and E. W. Abrahamson, *Nature*, **221**, 256 (1969).
44. F. Boll, *Arch. Anat. Physiol., Physiol. Abt.* **4** (1877).
45. G. Wald, in *Light and Life*, W. D. McElroy and B. Glass, eds., The Johns Hopkins Press, Baltimore, Maryland (1961).
46. W. B. Marks, W. H. Dobelle, and E. F. MacNichol, Jr., *Science* **143**, 1181 (1964).
47. F. W. Went, *Rec. Trav. Botan. Neerl.* **25**, 1 (1928).
48. W. R. Briggs, R. D. Tocher, and J. F. Wilson, *Science* **126**, 210 (1957).
49. H. S. Jennings, *Behavior of the Lower Organisms*, University of Indiana Press, Bloomington, Indiana (1962).
50. R. K. Clayton, in *Photophysiology*, Vol. II, A. C. Giese, ed., Academic Press, New York (1964).
51. P. Halldal, *Physiol. Plantarum* **14**, 133 (1961).
52. W. A. H. Rushton, in *Photophysiology*, Vol. II, A. C. Giese, ed., Academic Press, New York (1964).

53. W. S. Stiles, *Proc. Natl. Acad. Sci. U.S.* **45**, 100 (1959).
54. R. L. Sinsheimer and R. Hastings, *Science* **110**, 525 (1949).
55. D. Shugar, in *The Nucleic Acids: Chemistry and Biology*, Vol. III, E. Chargaff and J. N. Davidson, eds., Academic Press, New York (1960).
56. R. Beukers and W. Berends, *Biochim. Biophys. Acta* **41**, 550 (1960).
57. K. C. Smith, in *Photophysiology*, Vol. II, A. C. Giese, ed., Academic Press, New York (1964).
58. P. A. Swenson and R. B. Setlow, *Photochem. Photobiol.* **2**, 419 (1963).
59. J. K. Setlow, in *Current Topics in Radiation Research*, Vol. 2, M. Ebert and A. Howard, eds., North-Holland, Amsterdam (1966).
60. D. L. Wulff, *Biophys. J.* **3**, 355 (1963).
61. S. Y. Wang, *Fed. Proc.* **24**, S–71 (1965).
62. R. Beukers, J. Ijlstra, and W. Berends, *Rec. Trav. Chim.* **78**, 833 (1959).
63. R. B. Setlow and W. L. Carrier, *J. Mol. Biol.* **17**, 237 (1966).
64. J. Jagger, *Introduction to Research in Ultraviolet Photobiology*, Prentice-Hall, Englewood Cliffs, New Jersey (1967).
65. A. Kelner, *J. Bacteriol.* **56**, 157 (1948); **51**, 73 (1948).
66. R. Dulbecco, *Nature* **163**, 949 (1949).
67. A. Kelner, *J. Bacteriol.* **58**, 511 (1949).
68. A. Kelner, *J. Gen. Physiol.* **34**, 835 (1951).
69. J. Jagger and R. Latarjet, *Ann. Inst. Pasteur* **91**, 858 (1956).
70. R. O. Hotchkiss, in *The Nucleic Acids*, Vol. II, E. Chargaff and J. N. Davidson, eds., Academic Press, New York (1955); S. Zamenhof, *Progr. Biophys. in Biophys. Chem.* **6**, 86 (1956).
71. C. S. Rupert, in *Photophysiology*, Vol. II, A. C. Giese, ed., Academic Press, New York (1964).
72. S. H. Goodgal, C. S. Rupert, and R. M. Herriott, in *The Chemical Basis of Heredity*, W. D. McElroy and B. Glass, eds., Johns Hopkins Press, Baltimore, Maryland (1957).
73. C. S. Rupert, S. H. Goodgal, and R. M. Herriott, *J. Gen. Physiol.* **41**, 451 (1958).
74. C. S. Rupert, *J. Gen. Physiol.* **43**, 573 (1960).
75. C. S. Rupert, *J. Gen. Physiol.* **45**, 725 (1962).
76. (a) A. Wacker, *J. Chim. Phys.* **58**, 1041 (1961); D. L. Wulff, and C. S. Rupert, *Biochem. Biophys. Res. Comm.* **7**, 237 (1962); (b) R. C. Hanawalt, in *An Introduction to Photobiology*, C. P. Swanson, ed., Prentice-Hall, Englewood Cliffs, New Jersey (1969).
77. S. Minato and H. Werbin, *Biochemistry* **10**, 4503 (1971).
78. S. Minato and H. Werbin, *Photochem. Photobiol.* **15**, 97 (1972).

Index